Michelle Cyglenski

Chemistry for Engineers

An Applied Approach

Mary Jane Shultz

Tufts University

Houghton Mifflin Company   Boston   New York

Publisher: Charles Hartford
Executive Editor: Richard Stratton
Senior Developmental Editor: Rita Lombard
Senior Project Editors: Cathy Labresh Brooks, Carol Merrigan
Manufacturing Manager: Karen Fawcett
Marketing Manager: Laura McGinn

Cover image: © Zakim Bridge Lighting Design: The Mintz Lighting Group, Inc. Photographer:
Ken Douglas

About the cover
The Leonard P. Zakim Bunker Hill Bridge, the widest cable-stayed bridge in the world, has become a symbol of Boston. Chemistry plays an essential role in efforts to improve structural materials such as concrete and metals used in bridge construction.

Printed in the U.S.A.

Library of Congress Catalog Card Number: 2002109670

Instructor's Exam Copy:
ISBN 13: 978-0-618-73064-3
ISBN 10: 0-618-73064-8

For orders, use student text ISBNs:
ISBN 13: 978-0-618-27194-8
ISBN 10: 0-618-27194-5

123456789-VH-09 08 07 06

Brief Contents

Contents

The first question that must be answered by a new chemistry text is, "Why another text?" The short answer is that many students take only one semester of chemistry, and this is increasingly common among engineering students. Virtually all chemistry texts are written for a two-semester sequence and therefore are not designed to meet the needs of students taking a one-semester course.

The long answer is that this text was created in response to my engineering students' lament that nothing in general chemistry is relevant to their profession. As I thought about it from their perspective, I realized they were closer to the mark than I would like to admit; the connections between chemistry and issues in engineering professions were somewhat buried. Why not write a text that brings those connections, those applications, to the forefront? Why not start with the material properties of elemental metals and connect them with the electronic structure of the atom, with the color of diodes and connect it with interatomic interactions, with the use of freons in air-conditioning and connect it with molecular shape and intermolecular interactions, and, at the same time, call upon the experience that students have to make the connections more tangible. For example, a white shirt appears to glow under a black light, but a red shirt appears black. The basis for this observation is the interaction of light with matter. Finally, why not seal the connections by encouraging the student to apply chemical knowledge to solve real-world engineering problems?

The goal is to have the student say, "Ah, that is why this works. What happens if . . . ?" Chemistry then becomes very relevant. A search for a text to support an applied, need-to-know approach made it clear that a gap needed to be filled. This text was created so core materials applications could be coupled with environmental and biochemical applications. It is the result of collaboration with my engineering colleagues, constant feedback from my students, and reviews from chemistry and engineering faculty and their students from across the country.

Approach

This text adopts a conceptual, goal-oriented approach to teaching the fundamental topics of introductory chemistry with an emphasis on problem solving in order to help students apply chemistry in an engineering context. Broadly, the fundamental topics are atomic structure, molecular structure, and intermolecular interactions. As an example of the applied approach, atomic structure and periodicity are used as a basis for examination of mechanical properties of elemental metal wires. The bridge to molecular structure is built by examining properties of semiconductors then polymeric materials. Intermolecular interactions—electrochemistry, solution formation, and kinetics—are approached through materials degradation, battery design, environmental pollution and remediation, and biological connections.

Wherever possible, chemistry is introduced using an engineering application. In addition to early chapters on atoms and molecules, this book also presents topics such as reaction thermodynamics, equilibrium, oxidation-reduction reactions, and kinetics.

Applications tied to these later topics include biochemical digestion (the ATP cycle), the setting of concrete, environmental pollution, the self-cleansing of the Great Lakes, corrosion, batteries, catalytic converters, and enzymes.

Organization

In general, there are two perspectives in introductory chemistry—start with a macroscopic, phenomenological perspective and follow it with the atomic-molecular basis of the macroscopic or begin with atoms and molecules and lead up to the macroscopic. This text follows the latter approach but interweaves the macroscopic with the atomic-molecular. The illustration program is designed to cement the macroscopic-microscopic connection. To aid the student in imaging a reaction, the text is permeated with molecular-level illustrations of both physical and chemical changes. Illustrations feature visualization on many scales, e.g., from the macroscopic with skating on ice, to the hexagonal structure of a snowflake, to the molecular shape of water.

Chapter one sets up the atomic-level approach with reactions involving gases. This Fundamentals chapter can be used either as a succinct review or a quick survey. The heart of the text begins with the Foundations chapter: the periodic table, the electron, and the hydrogen atom. The Periodicity exposition is centered on the connection between macroscopic and microscopic properties.

Chapters one through three cover the core topics upon which bonding interactions and advanced applications are built. Advanced topics include thermodynamics, equilibrium, electrochemistry, coordination chemistry, polymers, and kinetics. The advanced applications are independent of each other, which allows the instructor to choose topics according to their course needs, schedule, and student preparation.

The treatment of thermodynamics emphasizes energy and energy dispersal, topics that are often treated separately due to the sophistication of the latter. They are brought together here to lay the foundation for equilibrium. Applications include hot packs, cold packs, welding, and the ATP cycle. Electron transfer and electrochemistry are presented through the roles they play in batteries and fuel cells. Fuel cells in particular are becoming increasingly important as fossil fuels diminish and the resulting emissions become a greater concern.

Magnetic materials, colorful gems, and transport of ions across membranes form the basis for discussion of coordination chemistry. The rich variety of carbon compounds is explored through polymers. The treatment of kinetics emphasizes the time scale of a reaction as reflected in the deployment of an airbag, the setting of concrete, and the rusting of automobiles. The time scale of each determines its application.

Pedagogical Features

Engineering-relevant applications, problem-solving, and conceptual understanding are three integrated and yet distinct themes woven throughout the text and underscored by several features. These features work together as a **built-in study guide** that helps students develop their analytical and critical evaluation skills.

Real-World Applications

Real-world applications to mechanical, civil, environmental, and biochemical engineering recur in different contexts throughout the text, making the chemistry relevant to the students in this course.

- Each chapter is organized around a real-life application intended to engage students in the material.
- Applications are embedded in the running text and relate to topics introduced in that section.
- *Apply It* interactive exercises require students to become actively involved using materials easily obtained from a source such as a local hardware store. For example, one early activity asks students to bend copper and steel wires to get a tangible sense of the metals' properties and to test their malleability and ductility.
- *Applied Exercises* at the end of each chapter cover topics from the entire chapter and challenge students to apply concepts to real-world contexts.

Problem-Solving

- *Skill Development Objectives* appear at the beginning of the chapter after a list of chapter concepts. They outline key skills students should have mastered by the end of the chapter. These objectives are revisited in worked examples and in end-of-chapter exercises.
- *Worked Examples* involve specific section topics and model a step-by-step approach to problem solving. Within each example, a *Plan* section emphasizes the need for an outline and prompts students to perform certain tasks and/or reminds them of key ideas, equations, or terms they should take note of before solving the problem. The *Implementation* section mirrors the *Plan,* executing the steps to arrive at the solution. Many examples include a comment about the implications of the solution. All *Worked Examples* refer to related exercises at the end of the chapter. These exercises give students an opportunity to practice what they have just learned.
- Over 80 *end-of-chapter exercises* cover basic skills learned in each chapter, concepts, specific section topics, and real-world applications. Solutions to the odd-numbered exercises are included at the end of the text.

Conceptual Understanding

- The *Conceptual Focus* at the beginning of each chapter outlines the concepts to be learned. The Key Idea at the end of the chapter recaps the Conceptual Focus.
- *Concept Questions* challenge students to further consider the ideas underlying the chemistry in the section and act either as a review of the material just learned or as a prompt to build on a concept and apply it in a particular context.
- *Conceptual Exercises* appear at the end of the chapter and test students' understanding of concepts. At least one of the *Conceptual Exercises* challenges students to demonstrate the relationship among key concepts by making a diagram or concept map.
- *Integrative Exercises,* also at the end of each chapter, require students to synthesize a number of concepts and skills and apply what they have learned.

End-of-Chapter Features

The end-of-chapter material offers the following useful and comprehensive review tools:

- A *Checklist for Review* with a list of *Key Terms* and *Key Equations* and where they appear in the chapter
- A *Chapter Summary* that outlines the key themes in the chapter
- The *Key Idea(s)* learned in the chapter
- *Concepts You Should Understand*

- *Operational Skills* students should be able to perform
- *End-of-Chapter Exercises.* Several question types begin with *Skill Building Exercises* keyed to specific sections and covering topics within those sections. *Conceptual Exercises* follow and test students' understanding of concepts throughout the chapter. At least one of the *Conceptual Exercises* challenges students to demonstrate the relationship among key concepts by making a diagram or concept map. *Applied Exercises* cover topics from multiple sections in the chapter and require students to apply concepts in an engineering context. *Integrative Exercises* require students to synthesize a number of concepts and skills and to apply what they've learned. Solutions to all odd-numbered end-of-chapter questions are in the back of the text.

Design, Photo, and Illustration Program

Recognizing the compelling effect of visual elements, we have taken full advantage of color in the text's design and in the artwork, utilizing molecular modeling programs to generate authentic orbitals, surface charge maps, and molecular shapes. Numerous graphical illustrations of pertinent data are included in nearly every chapter. Illustrations emphasize atomic-molecular level interactions relevant to the topic and are designed to fuse macroscopic observations with microscopic action. The extensive original art and the illustrative photos bring chemistry to life in the context of engineering. Students can "imagine the reaction."

Complete Text Support Package

For Students:

SMARTHINKING® live, online tutoring. This online service provides personalized, text-specific tutoring and is available during typical study hours when students need it most: Sunday-Thursday from 2 pm-5 pm and 9 pm-1 am EST.*

With SMARTHINKING, students can submit a question to get a response from a qualified e-structor, usually within 24 hours; use the whiteboard, with full scientific notation and graphics; preschedule time with an e-structor; view past online sessions, questions, or essays in an archive on their personal academic homepage; and view their tutoring schedule. E-structors help students with the process of problem-solving rather than supply answers. Instructors can package SMARTHINKING with the text, and each student will receive a unique passkey to access the SMARTHINKING website: smarthinking.com/houghton.html.

Limits apply. Terms and hours of SMARTHINKING service are subject to change.

Online Study Center (OSC). Access the Online Study Center for this text by visiting college.hmco.com/pic/shultz1e. The OSC enriches, enhances, and facilitates learning by taking the text's main themes—problem-solving, understanding concepts, and real-world applications—to an interactive level. The OSC includes study outlines of key chapter concepts, Houghton Mifflin's interactive ACE self-quizzes, flashcards, and student visualizations— animations and videos with self-test exercises to help mastery of difficult concepts.

Student Solutions Manual. It includes chapter overviews and solutions to even-numbered end-of-chapter problems.

Laboratory Guide. For those courses with a lab component, the manual provides a series of exercises focused on chemistry connected to engineering. For example, one exercise examines the conductivity and density of elemental wires. Another creates conducting polymeric thin films, and a follow-up uses the same procedure to create a photo-imaging film. Principles of solubility and miscibility are used to synthesize nanostructured materials including a highly luminescent solution. Students have an opportunity to exercise a measure of control over macroscopic properties by designing polymers with differing water-absorbing capacity. Guidelines are included for directing student-selected group projects. Engineering students are increasingly working in groups, and this experience provides an opportunity to integrate their fundamental chemical knowledge with group problem solving. For example, one suggested project challenges students to design a coating for seeds that both prevents nutrient leakage and enhances sprouting rates.

For Instructors:

HM ClassPrep CD-ROM with HM Testing. This combined CD includes both the *Computerized Test Bank* and the *HM ClassPrep* instructor resources. The computerized test bank contains questions in multiple-choice formats. Our *HM Testing* program offers delivery of test questions in an easy-to-use interface compatible with both MAC and WIN platforms. Word versions of all *test bank* files are also provided. The *HM ClassPrep Instructor CD-ROM* provides one location for all text-specific preparation materials that instructors might want to have available electronically. It contains Power-Point lecture outlines and art from the textbook as well as electronic versions of the *Instructors Resource Manual.*

HM ClassPresent 2006 is a new tool that helps instructors locate and organize useful animations and videos for easy export into a variety of presentation formats or for presentation directly from the CD-ROM. The CD also features a library of high-quality, scalable lab demonstrations and animations covering core chemistry concepts, arranged by chapter and topic. The resources within it can be browsed by thumbnail and description or searched by chapter, title, or keyword. Full transcripts accompany all audio commentary to reinforce visual presentations and to cater to different learning styles.

Complete Solutions Manual. This provides worked-out solutions to all problems and exercises in the text. It is most appropriately used as an instructor's solutions manual but is available for sale to students at the instructor's discretion.

Blackboard/WebCT Course. This course cartridge contains a wealth of animation, practice, and tutorial tools as well as instructor preparation, presentation, and testing materials. These include interactive tutorials/visualizations, ACE practice tests, PowerPoint presentations of the art from the text, and HM Testing (computerized test bank) content.

Online Teaching Center (OTC). Access the Online Teaching Center for this text by visiting college.hmco.com/pic/shultz1e. Instructors and teaching assistants can access lecture preparation and presentation resources such as PowerPoint presentations, overhead transparency pdfs, suggested demonstrations, suggested syllabi, and animations.

Acknowledgments

I thank my colleagues and my students for the many contributions they have made to this textbook through correspondence, questionnaires, and classroom testing of the material. I especially thank my many students at Tufts University, who both learned from and contributed to early versions of the text. Their "can do," "challenge me" attitude was a constant source of pleasure and inspiration. Former graduate students Steve Baldelli, now a Professor at the University of Houston, and Cheryl Schnitzer, now a Professor at Stonehill College, were genuine colleagues in the early stages of this work. I also thank the following reviewers for their suggestions and comments; they have helped us greatly in shaping this text:

Ludwig Bartels, *University of California, Riverside*
Wolfgang Bertsch, *University of Alabama*
James Carr, *University of Nebraska, Lincoln*
Paul Chirik, *Cornell University*
Thomas Greenbowe, *Iowa State University*
Curt Hare, *University of Miami*
Julie Harmon, *University of South Florida*
Richard Nafshun, *Oregon State University*
Williams Reiff, *Northeastern University*
Joel Russell, *Oakland University*
Karl Sohlberg, *Drexel University*
Joyce Solochek, *Milwaukee School of Engineering*
Pamela Wolff, *Carleton University*

Additionally, my thanks to Paul Chirik, Cornell University, who along with students Kaitlyn Gray, Jacqueline Hacker, Westin Kurkancheek, and Margaret Kuo helped ensure the accuracy of the text and problems.

Partial financial support for development of this course was provided by the National Science Foundation (NSF   ) and the PEW Charitable Trusts. Special thanks go to Susan Hixson of the NSF for her continued encouragement.

I am indebted to the Houghton Mifflin staff and several others for their dedicated and conscientious efforts in the production of *Chemistry for Engineers.* I especially thank Charles Hartford, Publisher; Richard Stratton, Executive Editor; Rita Lombard, Senior Development Editor; Carol Merrigan and Cathy Brooks, Senior Project Editors; Laura McGinn and Katherine Greig, Marketing Managers; Erin Lane, Marketing Coordinator; Jessyca Broekman and Jill Haber, Art Editors; Jean Hammond, Designer; and all those who have been involved in this text's development and production. A special thanks to Ben Roberts who believed in the vision of this project from its very early stages. Finally, I'd like to acknowledge the loving and constant support of my family. My children, Chris, Kim, and Ryan, not only read the text but also endured many hours of discussion about the chemistry of numerous phenomena. And a major thanks to my husband, Fred, whose insight into effective teaching and pedagogical tools is a constant inspiration.

Chemistry for Engineers weaves three distinct themes throughout the text to help students develop their quantitative and qualitative skills:

Problem solving

Conceptual understanding

Engineering – relevant applications and examples

Chapter 2
Foundations

2.1 **The Periodic Table**
Origin of the Periodic Table
Navigating the Periodic Table
Notation and Terminology
2.2 **The Components of the Atom**
⚙ **APPLY IT**
The Electron
Electron Charge and Electron Mass
Mass Distribution within the Atom
The Nucleus and the Electron
Major Components of the Atom
and the Periodic Table
2.3 **Light – Unlocking the Electronic
Structure of Atoms**
Light, Color, and Energy
⚙ **APPLY IT**
Line Spectra and the Energy
of the Atom
2.4 **Model of the Atom**
Ionization Energy
Line Spectra Revisited
The Quantum Mechanical Model
Two-Dimensional Waves
Three-Dimensional Waves
Quantum Numbers
2.5 **Energy Levels of Multielectron
Atoms**
Energy Ordering
⚙ **APPLY IT**
Orbital Electron Capacity: Pauli
Exclusion Principle
Electron Configuration
Core and Valence Electrons
Configuration Notations
Periodic patterns

CONCEPTUAL FOCUS

- Examine patterns in the periodic table.
- Investigate the connection between electricity and binding atoms into molecules.
- Relate light or electromagnetic radiation to energy.
- Build a model for electrons in atoms.

A stadium-shaped corral of iron atoms on a copper surface confines electrons on the surface, creating a rippling wave pattern of charge that can be detected with an instrument known as a scanning tunneling microscope. Electrons play a key role in chemical reactions.

SKILL DEVELOPMENT OBJECTIVES

- Determine the portion of the mass of an object that is due to electrons (Worked Example 2.1).
- Connect energy with light (Worked Example 2.2).
- Calculate the energy of hydrogen's electron (Worked Example 2.3), and relate energy to emitted radiation (Worked Example 2.4).
- Use nodes to classify and label electron waves (Worked Example 2.5); use quantum numbers to label orbitals (Worked Examples 2.6, 2.7).
- Calculate the effective charge holding an electron in the atom (Worked Example 2.8).
- Use transition energy to determine the energy of electronic states in multielectron atoms (Worked Example 2.9).
- Connect quantum numbers and the number of electrons in a shell (Worked Example 2.10).

39

Chapter Outlines give students a useful road map to navigate each chapter.

Apply It Exercises allow students to **apply textbook concepts to real life situations** using commonplace objects. For example, an activity in chapter one asks students to bend copper and steel wire to get a sense of their properties and to test their malleability and ductility.

This text has a built-in study guide to help improve analytical and evaluation skills. Students can methodically work through each chapter and its features to review, apply, and revisit material relevant to their course and professions.

The Conceptual Focus feature underscores the major **concepts** to be learned in the chapter and is reinforced by in-chapter *Concept Questions* and end-of-chapter *Conceptual Exercises*.

Skill Development Objectives outline key **skills** students should have mastered by the end of the chapter. They are revisited in *Worked Examples* and in end-of-chapter *Exercises*.

Integrated Problem-Solving Approach

Chemistry for Engineers offers students many opportunities to work through hands-on practical problems.

Fundamentals chapter This optional chapter at the beginning of the text sets up the atomic-level approach with reactions involving gases. It can be used as a succinct review or a quick survey.

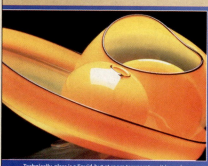

Chapter 1
Fundamentals

Technically, glass is a liquid, but at room temperature it is so viscous it looks and feels like a solid. At high temperature, glass becomes softer and can be worked like taffy. The plastic-like property allows artists like Dale Chihuly to craft stunningly beautiful pieces like this apricot

---

WORKED EXAMPLE 2.1 *Relative Mass of Electron and Atom*

The electron accounts for only a small portion of the mass of an atom, but just how small a portion is it? To put this number in perspective, let us scale it up. The hydrogen atom consists of one proton and one electron. Multiplying the mass of the hydrogen atom by 3×10^{28} — more than a billion billion billion — it would weigh about 110 pounds, or the weight of a small adult. In this magnified atom, what is the mass of the electron?

Plan

■ Use the data: The mass of an electron is 1/1837 of the mass of an atom.

Implementation

■ Multiply the weight of the atom times 1/1837 to give the weight of the electron. (110 pounds) $\times$ (1/1837) = 0.0599 pound.

Convert this to an appropriate unit.
There are 16 ounces in a pound; 0.0599×16 ounces = 0.958 ounces, o ounce—about the mass of a mouse!

Worked Examples titled for easy reference, involve specific topics and with *Plan* and *Implementation* sections, model a step-by-step approach to problem solving. A reference to appropriate follow-up exercises appears after each *Worked Example* giving students an opportunity to practice what they have just learned.

End-of-Chapter Exercises

Over 80 end-of-chapter *Exercises* cover skills learned in the chapter with a range of emphases (applied/conceptual/integrative).

Identify those that are redox reactions. Indicate the oxidation and reduction half-reactions.

91. Write the half-reaction taking place at each electrode when molten KCl is electrolyzed.

92. Suggest a method to obtain Mn from an ore containing Mn_2O_3. Can Mn be reduced with coke? With CO?

■ CONCEPTUAL EXERCISES

93. Draw a diagram indicating the relationship among the following terms: oxidation, reduction, redox, anode, cathode, salt bridge, cell potential. All terms should have at least one connection.

94. Draw a diagram indicating the relationship among the following terms: corrosion prevention, Pilling-Bedworth ratio, sacrificial anode, galvanic protection, tin cans, protective oxide, porous oxide. All terms should have at least one connection.

95. A box of baking soda ($NaHCO_3$) is a handy and effective fire extinguisher in the kitchen. At high temperature, baking soda decomposes, generating CO_2 gas in an endothermic reaction (ΔH = 135.6 kJ/mol). Baking soda is said to be double-acting in extinguishing the fire. Explain this statement.

96. Magnesium is a very active metal that can be oxidized with oxygen to MgO, nitrogen to Mg_3N_2, water to $Mg(OH)_2$ and CO_2 to MgO. Write the redox reaction associated with each of these four oxidations. As a result of these processes, a magnesium fire is particularly hazardous. Explain why water or a CO_2 fire extinguisher should not be used with a magnesium fire. How would you extinguish a magnesium fire?

✱ APPLIED EXERCISES

97. Anyone who has encountered the odor of a skunk remembers the pungent fragrance for a long time—even longer if the odor resulted from an encounter between a pet and the skunk. Two compounds responsible for the skunk odor are the thiols H_7C_4SH and $H_{11}C_5SH$. One remedy for the family pet is to

Emphasis on Conceptual Understanding

Starting with *Conceptual Focus* objectives, students work through various features such as *Concept Questions, Chapter Summary, Key Ideas,* and *Concepts You Should Understand* as well as practice *Conceptual* and *Integrative Exercises* to enhance their understanding of key chapter concepts.

Concept Questions appear throughout each chapter and serve either as a review of the material just learned or as a prompt to build on a concept and apply it to a particular situation.

CONCEPT QUESTION You are told that an atom of an unknown element has an atomic mass of 32 amu. Can you identify the element from this information? What other information do you need? ■

■ CONCEPTUAL EXERCISES

99. Make a diagram showing relationship among the following terms: metals, nonmetals, semimetals, alkali metals, alkaline earths, halogens, noble gases, periodic law. All terms must be interconnected.

100. Draw a diagram indicating the relationships among the following terms: effective nuclear charge, electron configuration, octet, ionization energy, electron affinity, electronegativity. All terms should be interconnected.

Conceptual Exercises appear at the end of each chapter, testing students' understanding of key concepts. At least one of the *Conceptual Exercises* challenges students to create a concept map to demonstrate the relationship(s) between key ideas.

■ INTEGRATIVE EXERCISES

105. Joining metal pipes or metal components in a circuit involves soldering—melting a metal to join metal surfaces. Three common solder materials are silver, tin, and lead. Find these metallic elements on the periodic table. Based on their chemical properties, which would you use to solder an electric circuit? On what properties(s) did you base your choice? Lead was formerly used to solder water pipes. What property made lead a good choice? Why is the use of lead now banned?

106. Viscosity is the resistance to flow. Compare the viscosity of the three phases of water. We do not commonly think of solids as flowing. Give an example illustrating the flow of ice.

107. Diamond and graphite are two forms of elemental carbon. The density of graphite is 2.2 g/cm^3 and that of diamond is 3.513 g/cm^3. Graphite is an excellent lubricant. Diamond is the hardest material known. Use the aforementioned data to discuss the interatomic interactions in diamond compared to those in graphite.

108. Cite evidence about atomic size, ionization energy, and electron affinity that suggests that $n - 1$ shell d-orbital electrons nearly perfectly shield the nuclear charge.

Integrative Exercises at the end of each chapter require students to synthesize their knowledge and understanding of a number of key chapter concepts.

Engineering-Relevant Application and Examples

The text offers an active learning and discovering approach to its content through a variety of real-world application and examples.

7.1

Case Study: Cold Packs, Hot Packs, and Welding

Injuries in sporting accidents are often treated with an "instant" cold pack. Such a pack consists of a tough outer pouch containing water (often with a blue dye to signify cold) and an inner pouch containing a salt. The term *salt* is used in the generalized chemical sense, meaning a substance consisting of oppositely charged ions held together by coulombic forces. Magnifying the cold-pack salt by a hundred million times reveals that it consists of two ions (Figure 7.2), both of which are made up of several atoms. One ion is NH_4^+, called the ammonium ion; the other is NO_3^-, called the nitrate ion. When the inner pouch of the cold pack is broken (Figure 7.3), water from the outer pouch mixes with the salt, separating the oppositely charged ions and surrounding them with water molecules. The cold sensation accompanying this process indicates that heat is drawn into the pack from the area around it—including the injured area.

While sprains and pulled muscles are often treated with cold therapy, heat is usually recommended for muscle aches or spasms. One method for producing the heat is to use a hot pack. Like cold packs, hot packs consist of an outer pouch containing water and an inner pouch containing a salt. In this case, the salt is often $CaCl_2$, a solid consisting of Ca^{2+} and Cl^- ions. For this salt, forming the solution results in heat flowing out, just as heat flows out when wood or other fuel is burned.

The amount of heat produced in a chemical reaction can be quite substantial. For example, in remote areas welding of railroad rails is done using the reaction of metallic aluminum with iron oxide:

$$\text{(7.1)} \qquad Fe_2O_3(s) + 2Al(s) \longrightarrow 2Fe(l) + Al_2O_3(s)$$

The reaction in Equation (7.1) generates so much heat that the elemental iron produced is molten (Figure 7.4)! The molten iron flows around the rails, and the rails conduct the heat away, thereby solidifying the iron and fusing the rails together. Because all of the oxygen needed to form Al_2O_3 comes from the Fe_2O_3, once started the reaction is self-sustaining and can take place in the absence of added oxygen. It can even be used to weld underwater. This reaction is referred to as the thermite reaction.

In the following sections, these three reactions are examined in more detail, connecting the heat flow with differences in interactions among the reactants and those

Figure 7.2 The term "salt" is often used in chemistry to denote a solid consisting of oppositely charged ions. The ions are held together by coulombic attraction between the positively charged ion and the negatively charged ion. The salt illustrated here consists of NH_4^+ ions and NO_3^- ions. Both of these ions contain several elements covalently bonded together. The charge indicates that the group of ions has collectively either lost (+) or gained (−) an electron. The group NH_4^+ is called an ammonium ion, and the group NO_3^- is called a nitrate ion. Both are commonly found in the environment, in the foods we eat, and in our bodies.

Real-life Application

Each chapter includes real-life applications intended to engage students in the material.

APPLY IT Go to a sporting goods store or a ski shop and obtain a product called a "hand warmer." Two varieties of hand warmer exist: One contains a solution and is reusable; the other contains a solid and is designed for one-time use. For the solution variety of hand warmer, activate the warmer. Describe your observations. How is the product restored to its original solution state? What are the components of the solid hand warmer? Propose a mechanism for the solid hand warmer.

Apply It exercises

– designated by a practical "gear-like" icon – require students to apply concepts to real-life situations, using tools that are commonplace or that may be easily obtained from a source such as the local hardware store.

Applied Exercises

Each chapter has a specific end-of-chapter **Applied Exercises** section as well as an icon designating applied-type problems in other sections.

✹ APPLIED EXERCISES

73. Metal ions—particularly copper, iron, and nickel—catalyze the reaction of oxygen with fats in foods, resulting in rancidity. The metal ions come from a variety of sources, including soils and machinery used to process the foods. Comment on the use of $EDTA^{4-}$ to guard against oxidation. Consult the table of complex ion formation constants (Appendix A). Which metal ions are effectively sequestered by $EDTA^{4-}$ in the presence of Ca^{2+} ions in many foodstuffs?

74. Magnetite, Fe_3O_4, is a ferroelectric material. The magnetite structure consists of an FCC lattice of oxygen atoms with iron in both the octahedral and tetrahedral holes. The unit cell consists of eight FCC cubes of oxygen. Eight tetrahedral sites are occupied by Fe^{3+}, eight octahedral sites are occupied by Fe^{3+}, and eight octahedral sites are occupied by Fe^{2+}.
    a. Show that the unit cell is consistent with the formula Fe_3O_4. How many formula units are in a unit cell?
    b. Which is larger, the octahedral hole or the tetrahedral hole?
    c. The magnetization from the octahedral sites is larger than that from the tetrahedral sites. Magnetization from the

Visualization Tutorials

Visualization tutorials are videos and animations available for students and instructors using this text. They are accessible on the *Online Study Center* or *Online Teaching Center* at college.hmco.com/pic/shultz1e. They are also available on the instructor HMClass Present 2006: General Chemistry CD-ROM.

Topics include the following:

- Boyle's Law
- Buffers
- Changes of State
- Charles's Law
- Conservation of Mass and Balancing Equations
- Electrochemical Half-Reactions in a Galvanic Cell
- Formation of Ionic Compounds
- Heterogeneous Catalysis
- Homogeneous Catalysis
- Hybridization: sp
- Le Châtelier's Principle
- Limiting Reactant
- Magnetic Levitation by a Superconductor
- Ostwald Process
- Structure of a Gas
- Structure of a Liquid
- Structure of a Solid
- Synthesis of Nylon
- The Ideal Gas Law
- Thermite Reaction
- Voltaic Cell: Anode Reaction
- Voltaic Cell: Cathode Reaction
- VSEPR

A complete listing is available on the Online Study and Online Teaching Centers.

Chapter 1
Fundamentals

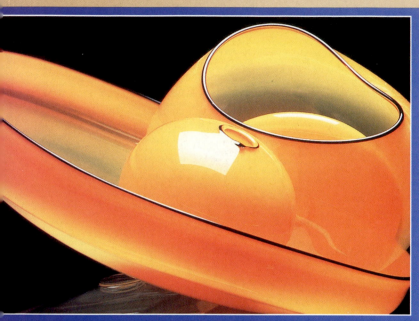

Technically, glass is a liquid, but at room temperature it is so viscous it looks and feels like a solid. At high temperature, glass becomes softer and can be worked like taffy. The plastic-like property allows artists like Dale Chihuly to craft stunningly beautiful pieces like this apricot basket.

CONCEPTUAL FOCUS

■ Develop an atomistic view of substances.
■ Connect macroscopic observations with gases with the atomistic view of reactions.

SKILL DEVELOPMENT OBJECTIVES

■ Use macroscopic observations to fill in reaction formulas (Worked Example 1.1).
■ Relate macroscopic dimensions to atomic dimensions and convert between mass and moles (Worked Examples 1.2–1.4).
■ Use the gas laws (Worked Examples 1.5–1.7, 1.9–1.11).
■ Convert among mass, moles, and (Worked Example 1.8).

■ Decode formulae and balance reactions (Worked Examples 1.12, 1.14–1.15).
■ Calculate energy (Worked Examples 1.16–1.19).
■ Visualize relationships among terms (Worked Example 1.20).

This is an amazingly exciting time to delve into chemistry because solutions to some of the most vexing problems facing mankind are possible. *Possible* because the large-scale principles that determine macroscopic outcomes and properties are known so that solutions are within reach. Yet the application of those principles is sufficiently complex that solutions for many problems are not even close to being in hand. For example, recently the human genome was sequenced. *In principle* cures for disease and repair of injuries are now possible. However, there remain a myriad of nuances, many not yet recognized, that stand between us and the achievement of that objective.

The overall goal of this text is to enable the student to acquire the basic knowledge and skills needed to employ chemical principles in solving a variety of problems, from generating better building materials to designing faster, more reliable computers, repairing damaged tissue, and remediating pollution. Many of these problems require manipulating substances at vastly different size scales, ranging from the atomic level to the macroscopic. Chemists visualize problems on all these levels, so the text contains illustrations that connect these various levels (Figure 1.1).

1.1   Chemistry, Chemistry Everywhere (with apologies to Samuel Taylor Coleridge)

Look around and notice the tremendous variety of materials you encounter every day. Contrast the flexibility of the pages of this book with the hardness of the desktop (Figure 1.2). The characteristics of each material are appropriate to its function,

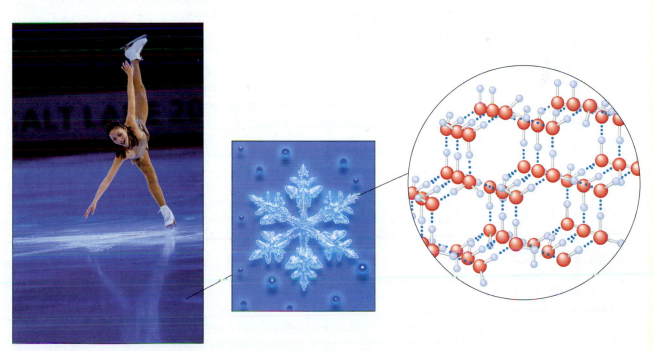

Figure 1.1 Ice is shown at three size scales. The macroscopic surface of ice provides a low friction glide for the skater. The hexagonal shape of the snowflake presents a macromolecular view of ice. At the atomic-molecular level, ice has hexagonal channels into which other molecules can be trapped. The properties of ice at the molecular scale enable the skater's graceful glide.

and—it turns out—these contrasting characteristics result from variations in the atomic-level structure of each material. Materials are often designed for a particular function, so attention is usually focused on the macroscopic character. However, the macroscopic properties depend upon the microscopic properties (Figure 1.3). To design the properties of a substance on the macroscopic level, the material must be manipulated on the atomic-molecular level.

1.2   Atoms as Building Blocks: Atomic Theory

Atom The smallest particle that an element can be subdivided into and retain the chemical properties of that element.

Compound A substance with constant composition that can be broken down into elements by chemical processes.

Molecule A bonded collection of two or more atoms of the same or different elements. The smallest particle of a compound having the properties of a compound.

The word **atom** comes from the Greek *atmos,* meaning "indivisible." Although it is now known that atoms are composed of yet smaller particles, an atom remains the smallest entity that an element can be broken down into and retain the properties of that element. Atoms can be imaged with two types of instruments: a scanning tunneling microscope (STM) and an atomic force microscope (AFM).

A **compound** is a substance that can be broken down into two or more elements. Common table salt is a compound composed of the elements sodium and chlorine. Water is composed of hydrogen and oxygen. Some elements are found as associated atoms, called a **molecule.** The term "molecule" is often used to refer both to compounds and to elemental molecules.

At the heart of the atomistic view of materials is the notion that the fundamental building blocks for matter are atoms. The atomistic view is based on a series of observations with gases. From 1803 to 1808, John Dalton proposed that:

1. All material is made of atoms.
2. An atom is the smallest particle of an element that can take part in a reaction.
3. The atoms of the same element are identical.
4. The atoms of different elements are different.
5. A compound is made up of atoms of different elements in a fixed whole number ratio.

Up to this point, Dalton's theory is correct (with the exception that it neglects isotopes). However, Dalton went further by proposing two more points, which turned out to be wrong, and which took the larger chemistry community 50 years to correct. First, he proposed that all elemental gases are monatomic. Second, Dalton proposed that water consists of one atom of hydrogen and one atom of oxygen. At least part of the reason for this error is that the process of arriving at the correct conclusion is neither straightforward nor simple. (Bear this in mind as you construct your own coherent understanding of chemistry.)

Think about how atomic masses can be determined in the absence of a picture of the atom. Individual atoms are very small—far too small to count or weigh. The small unit weight problem can be solved by weighing a large number of atoms. However, this merely shifts the problem to one of determining the number of atoms in the sample.

A partial solution to this problem is to realize that only relative masses are important. Then arbitrary assignment of the mass of one element determines the masses of other elements in relation to the assigned one.

Figure 1.2 The physical properties of the desktop and the pages of the book are very different, yet the characteristics of each result from properties at the atomic-molecular level.

CONCEPT QUESTIONS It is observed that 1 g of hydrogen reacts completely with 8 g of oxygen to produce 9 g of water. Assigning hydrogen a mass of 1, what other information is needed to assign a mass to oxygen? How would you determine the missing information? ∎

The missing piece of information is the combining ratios of atoms in a molecule or compound. Reactions with gases provide the key to this information. In the early 1800s,

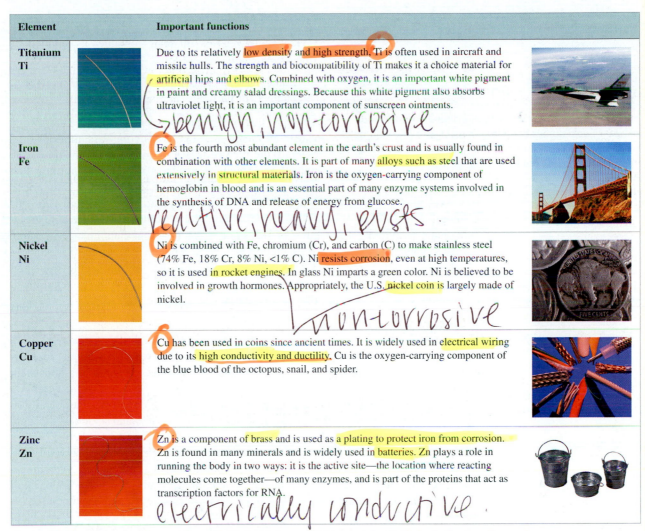

Element		Important functions	
Titanium **Ti**		Due to its relatively low density and high strength, Ti is often used in aircraft and missile hulls. The strength and biocompatibility of Ti makes it a choice material for artificial hips and elbows. Combined with oxygen, it is an important white pigment in paint and creamy salad dressings. Because this white pigment also absorbs ultraviolet light, it is an important component of sunscreen ointments. *benign, non-corrosive*	
Iron **Fe**		Fe is the fourth most abundant element in the earth's crust and is usually found in combination with other elements. It is part of many alloys such as steel that are used extensively in structural materials. Iron is the oxygen-carrying component of hemoglobin in blood and is an essential part of many enzyme systems involved in the synthesis of DNA and release of energy from glucose. *reactive, heavy, rusts*	
Nickel **Ni**		Ni is combined with Fe, chromium (Cr), and carbon (C) to make stainless steel (74% Fe, 18% Cr, 8% Ni, <1% C). Ni resists corrosion, even at high temperatures, so it is used in rocket engines. In glass Ni imparts a green color. Ni is believed to be involved in growth hormones. Appropriately, the U.S. nickel coin is largely made of nickel. *non-corrosive*	
Copper **Cu**		Cu has been used in coins since ancient times. It is widely used in electrical wiring due to its high conductivity and ductility. Cu is the oxygen-carrying component of the blue blood of the octopus, snail, and spider.	
Zinc **Zn**		Zn is a component of brass and is used as a plating to protect iron from corrosion. Zn is found in many minerals and is widely used in batteries. Zn plays a role in running the body in two ways: it is the active site—the location where reacting molecules come together—of many enzymes, and is part of the proteins that act as transcription factors for RNA. *electrically conductive*	

Figure 1.3 Five common elemental metals (Ti, Fe, Ni, Cu, and Zn) have a range of physical properties, and each has many uses in society. In addition, Fe, Cu, and Zn are essential to the proper functioning of the body. Elements are commonly named by those who discover and isolate them. Ti is from the Latin *Titans,* the first sons of the earth in mythology; Fe from the Latin *ferrum;* Cu from the Latin *cuprum;* Zn from German (obscure origin); Ni from the German *kupfernickel* (copper Satan or Old Nick's copper), so-called because miners mistakenly thought it to contain copper.

Joseph Gay-Lussac made two key observations involving gas-phase reactions that would ultimately lead to combining ratios:

1. Two volumes of hydrogen react with one volume of oxygen to yield two volumes of water.
2. One volume of hydrogen reacts with one volume of chlorine to yield two volumes of hydrogen chloride.

To proceed from these observations to the balanced chemical equation, it is necessary to know something about how gases fill a container.

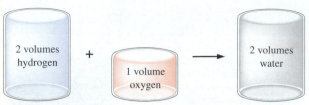

Figure 1.4 Gay-Lussac's observation with hydrogen and oxygen: Two volumes of hydrogen react with one volume of oxygen to yield two volumes of water. This is a macroscopic view of the reaction between hydrogen and oxygen to produce water.

Avogadro's Hypothesis

A view of how gases fill their container is supplied by Amadeo Avogadro's hypothesis:

> Equal volumes of different gases contain equal numbers of particles (in modern language).

Although this hypothesis was put forth in 1811, it took nearly 50 years to sort conflicting assumptions and generate the picture that has stood the test of time. The major impediment to progress was the assumption that elemental gases had to be monatomic—that is hydrogen (H), oxygen (O), nitrogen (N), and chlorine (Cl) are monatomic. The reason for this assumption was a lack of understanding of what holds a homonuclear diatomic molecule—one containing two atoms of the same element—together. Finally in 1860, Stanislao Cannizzaro made a presentation at a conference in Karlsruhe, Germany, in which he did not worry about what held homonuclear diatomic molecules together, but simply asserted that elemental gases had to be diatomic. (Noble gases did not enter into the picture because they were unknown at this time!)

An Atomistic View

Cannizzaro's bold statement, together with Gay-Lussac's observations about reactions of hydrogen, oxygen, and chlorine, plus Avogadro's hypothesis, results in a consistent picture.

Avogadro's hypothesis and the assumption that elements do not change identity in chemical reactions indicate the ratio of hydrogen to oxygen in water as follows (Figure 1.4): The number of atoms of hydrogen in two volumes is twice the number of oxygen atoms in one volume. Because the number of atoms of each element is conserved in any chemical reaction, there are twice as many atoms of hydrogen as atoms of oxygen in water. Using subscripts to indicate the smallest whole-number ratio for the molecular unit means that the formula for water is H_2O, or two atoms of hydrogen and one atom of oxygen (Figure 1.5).

Figure 1.5 A molecule of water consists of two atoms of hydrogen and one atom of oxygen.

CONCEPT QUESTION Does the result of this experiment indicate whether hydrogen and oxygen gases are monatomic or diatomic? ■

The number of volumes of hydrogen consumed and volumes of water produced are the same. Avogadro's hypothesis thus implies that the number of molecules of water produced in this reaction must be the same as the number of molecules of hydrogen consumed. As there are two atoms of hydrogen in water, there must also be two atoms of hydrogen in elemental hydrogen gas. Similar reasoning indicates that oxygen must also have two atoms in its elemental unit. The result is an atomistic picture of the reaction (Figure 1.6).

Using the symbol for each element this is

(1.1) $$2H_2 + O_2 \longrightarrow 2H_2O$$

where the subscript indicates how many atoms of each element are involved in the molecule. In particular, elemental hydrogen is H_2 and elemental oxygen is O_2.

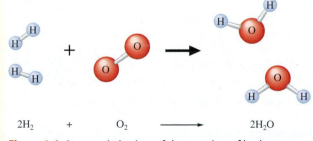

$$2H_2 \quad + \quad O_2 \quad \longrightarrow \quad 2H_2O$$

Figure 1.6 An atomistic view of the reaction of hydrogen and oxygen to produce water. Two molecules of diatomic hydrogen gas react with one molecule of diatomic oxygen gas to yield two molecules of water.

CONCEPT QUESTIONS Draw representations of the reaction of hydrogen and chlorine to produce hydrogen chloride at the macroscopic level and the atomic level. Explain how this reaction supports the conclusion that hydrogen and chlorine are diatomic gases. What is the elemental composition of hydrogen chloride? ■

WORKED EXAMPLE 1.1   *Reaction Formulas Fill in Information*

Avogadro's law provided the essential link between macroscopic observations about volumes of reacting gases and molecular formulas. At constant temperature and pressure, how many milliliters of oxygen gas are required to convert 1000 mL of carbon monoxide (CO) into carbon dioxide (CO_2)? How many milliliters of CO_2 are produced?

Plan

- Write the balanced reaction.
- The molecular density of a gas is independent of the gas.

Implementation

- Reaction:

$$2CO + O_2 \longrightarrow 2CO_2$$

- Carbon monoxide has one atom of oxygen; carbon dioxide has two. This indicates that one oxygen atom is required. Since molecular oxygen comes as O_2, one molecule of elemental oxygen combines with two molecules of CO to produce two molecules of CO_2.

The volume of O_2 required is half the volume of CO or 500 mL. One CO_2 molecule is produced for every molecule of CO consumed, so the volume of CO_2 produced is the same as the original volume of CO: 1000 mL.

See Exercises 16–18, 63.

The Mole

Mole, mol The number equal to the number of carbon atoms in exactly 12 grams of pure ^{12}C: Avogadro's number. One mole represents 6.022×10^{23} units.

Avogadro's number (N_A) The number of atoms in exactly 12 grams of pure ^{12}C, equal to 6.022×10^{23}.

The bridge between a macroscopic sample—a liter of gas or a piece of wire that can be held in hand—and the atomistic view is a quantity called the mole. A **mole,** abbreviated **mol,** contains 6.022137×10^{23} particles, a number referred to as **Avogadro's number** (N_A) in recognition of the importance of Avogadro's hypothesis to the development of chemistry. Perhaps 6.022137×10^{23} particles per mole seems to be a huge number; however, an individual atom is infinitesimally small. To imagine just how small a single atom is, think of an atom as a hard sphere like a ball bearing and look at the end of a piece of wire like the heavy-duty wire used in household circuits. Think about magnifying the wire many times (Figure 1.7). A magnification of nearly 1 million times is required before *any* substructure is apparent. Only after magnification by 10 million times—seven orders of magnitude—are the individual atoms visible. Magnification of 10 million would take 1 cm—the thickness of a finger—to 100 km. About 10 trillion (1×10^{13}) atoms cover the end of a 1-mm diameter wire! (See Worked Example 1.2.)

WORKED EXAMPLE 1.2   *The Number of Atoms on the End of a Wire*

The radius of a Cu atom is 120 pm (1 pm = 1×10^{-12} m and is the standard unit for atomic dimensions). Calculate the approximate number of copper atoms on the surface of the end of an 18-gauge, 1-mm diameter wire.

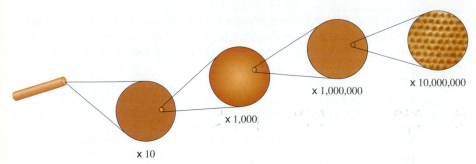

Figure 1.7 Successive magnification of the end of a wire reveals the building blocks that make up the metal. The structure first appears as a vague array, then as individual building blocks—these are atoms. By the second magnification, the wire is 1000 times bigger than the original wire and has a diameter of 1 m, but individual atoms are still not visible. A magnification of a 1,000,000 times barely reveals atoms. At this magnification, the original wire has become 1 km in diameter. An additional power of 10 is required to see the individual atoms that have become 2.4 mm in diameter—just a bit larger than the end of the original wire. At this magnification, 10 million or 10^7, the wire would be 10 km across. There are more than 10 trillion atoms on the end of this wire.

Plan
- How many just-touching copper atoms fit this diameter?
- Area of a circle $= \pi r^2$.

Implementation
- The radius of a copper atom in metallic copper is 120 pm and the diameter is 240 pm (240×10^{-12} m). The wire diameter is 1 mm $= 1 \times 10^{-3}$ m. Assuming that each atom just touches the next, there are

$$\frac{1 \times 10^{-3} \text{ m}}{240 \times 10^{-12} \text{ m/atom}} = 4 \times 10^6 \text{ atoms}$$

or 4 million atoms across the diameter of the wire. In units of atoms, the radius of the wire is 2,000,000 atoms.

- Area $= \pi r^2 = \pi \times (2 \times 10^6)^2 = 1.2 \times 10^{13}$.

Thus more than 10 trillion atoms cover the end. To put this number in context, if each atom is magnified to 1 mm in diameter (the diameter of the original wire), about half a billion would cover a football or soccer field. Ten trillion would cover the field to a depth of 10 m.

See Exercises 7–8.

Atomic mass unit, (amu)
1.660540×10^{-24} g, $\frac{1}{12}$ the mass of carbon-12, approximately the mass of a hydrogen atom.

Atomic mass The weighted average mass in atomic mass units of the atoms in a naturally occurring substance.

Molar mass Mass per mole of an atom or molecule.

In the case of an elemental material, such as copper, a particle is a single atom. Because the mass of a single atom is very small, atomic mass is usually specified in **atomic mass units (amu).** One amu is equal to $1.6605402 \times 10^{-24}$ g (1 g/ 6.022137×10^{23}), or about the mass of a single hydrogen atom. Combining the small mass of a single atom with the large number of atoms in a mole results in a conveniently measurable mass for a mole. The mass of one mole of an element is called the **atomic mass** or the **molar mass.** For example, one mole of copper atoms has a mass of 63.546 g; this is the atomic mass of copper. Table 1.1 lists the names, symbols, and masses of the elements.

CONCEPT QUESTIONS Draw representations of the reaction of hydrogen and chlorine to produce hydrogen chloride at the macroscopic level and the atomic level. Explain how this reaction supports the conclusion that hydrogen and chlorine are diatomic gases. What is the elemental composition of hydrogen chloride? ■

WORKED EXAMPLE 1.1   *Reaction Formulas Fill in Information*

Avogadro's law provided the essential link between macroscopic observations about volumes of reacting gases and molecular formulas. At constant temperature and pressure, how many milliliters of oxygen gas are required to convert 1000 mL of carbon monoxide (CO) into carbon dioxide (CO_2)? How many milliliters of CO_2 are produced?

Plan
- Write the balanced reaction.
- The molecular density of a gas is independent of the gas.

Implementation
- Reaction:

$$2CO + O_2 \longrightarrow 2CO_2$$

- Carbon monoxide has one atom of oxygen; carbon dioxide has two. This indicates that one oxygen atom is required. Since molecular oxygen comes as O_2, one molecule of elemental oxygen combines with two molecules of CO to produce two molecules of CO_2.

The volume of O_2 required is half the volume of CO or 500 mL. One CO_2 molecule is produced for every molecule of CO consumed, so the volume of CO_2 produced is the same as the original volume of CO: 1000 mL.

See Exercises 16–18, 63.

The Mole

Mole, mol The number equal to the number of carbon atoms in exactly 12 grams of pure ^{12}C: Avogadro's number. One mole represents 6.022×10^{23} units.

Avogadro's number (N_A) The number of atoms in exactly 12 grams of pure ^{12}C, equal to 6.022×10^{23}.

The bridge between a macroscopic sample—a liter of gas or a piece of wire that can be held in hand—and the atomistic view is a quantity called the mole. A **mole**, abbreviated **mol**, contains 6.022137×10^{23} particles, a number referred to as **Avogadro's number** (N_A) in recognition of the importance of Avogadro's hypothesis to the development of chemistry. Perhaps 6.022137×10^{23} particles per mole seems to be a huge number; however, an individual atom is infinitesimally small. To imagine just how small a single atom is, think of an atom as a hard sphere like a ball bearing and look at the end of a piece of wire like the heavy-duty wire used in household circuits. Think about magnifying the wire many times (Figure 1.7). A magnification of nearly 1 million times is required before *any* substructure is apparent. Only after magnification by 10 million times—seven orders of magnitude—are the individual atoms visible. Magnification of 10 million would take 1 cm—the thickness of a finger—to 100 km. About 10 trillion (1×10^{13}) atoms cover the end of a 1-mm diameter wire! (See Worked Example 1.2.)

WORKED EXAMPLE 1.2   *The Number of Atoms on the End of a Wire*

The radius of a Cu atom is 120 pm (1 pm = 1×10^{-12} m and is the standard unit for atomic dimensions). Calculate the approximate number of copper atoms on the surface of the end of an 18-gauge, 1-mm diameter wire.

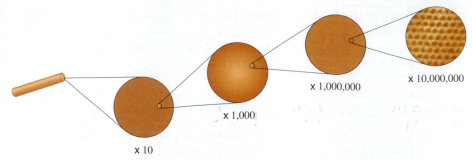

x 10

x 1,000

x 1,000,000

x 10,000,000

Figure 1.7 Successive magnification of the end of a wire reveals the building blocks that make up the metal. The structure first appears as a vague array, then as individual building blocks — these are atoms. By the second magnification, the wire is 1000 times bigger than the original wire and has a diameter of 1 m, but individual atoms are still not visible. A magnification of a 1,000,000 times barely reveals atoms. At this magnification, the original wire has become 1 km in diameter. An additional power of 10 is required to see the individual atoms that have become 2.4 mm in diameter — just a bit larger than the end of the original wire. At this magnification, 10 million or 10^7, the wire would be 10 km across. There are more than 10 trillion atoms on the end of this wire.

Plan

■ How many just-touching copper atoms fit this diameter?
■ Area of a circle = πr^2.

Implementation

■ The radius of a copper atom in metallic copper is 120 pm and the diameter is 240 pm (240×10^{-12} m). The wire diameter is 1 mm = 1×10^{-3} m. Assuming that each atom just touches the next, there are

$$\frac{1 \times 10^{-3} \text{ m}}{240 \times 10^{-12} \text{ m/atom}} = 4 \times 10^6 \text{ atoms}$$

or 4 million atoms across the diameter of the wire. In units of atoms, the radius of the wire is 2,000,000 atoms.

■ Area = $\pi r^2 = \pi \times (2 \times 10^6)^2 = 1.2 \times 10^{13}$.

Thus more than 10 trillion atoms cover the end. To put this number in context, if each atom is magnified to 1 mm in diameter (the diameter of the original wire), about half a billion would cover a football or soccer field. Ten trillion would cover the field to a depth of 10 m.

See Exercises 7–8.

Atomic mass unit, (amu)
1.660540×10^{-24} g, $\frac{1}{12}$ the mass of carbon-12, approximately the mass of a hydrogen atom.

Atomic mass The weighted average mass in atomic mass units of the atoms in a naturally occurring substance.

Molar mass Mass per mole of an atom or molecule.

In the case of an elemental material, such as copper, a particle is a single atom. Because the mass of a single atom is very small, atomic mass is usually specified in **atomic mass units (amu)**. One amu is equal to $1.6605402 \times 10^{-24}$ g (1 g/6.022137×10^{23}), or about the mass of a single hydrogen atom. Combining the small mass of a single atom with the large number of atoms in a mole results in a conveniently measurable mass for a mole. The mass of one mole of an element is called the **atomic mass** or the **molar mass**. For example, one mole of copper atoms has a mass of 63.546 g; this is the atomic mass of copper. Table 1.1 lists the names, symbols, and masses of the elements.

Table 1.1 Elements and Atomic Masses

Element		Mass (g)	Element		Mass (g)
Ac	Actinium	227	Fe	Iron	55.845
Al	Aluminum	26.98154	Kr	Krypton	83.80
Am	Americium	243	La	Lanthanum	138.9055
Sb	Antimony	121.760	Lr	Lawrencium	262
Ar	Argon	39.948	Pb	Lead	207.2
As	Arsenic	74.9215	Li	Lithium	6.941
At	Astatine	210	Lu	Lutetium	174.967
Ba	Barium	137.33	Mg	Magnesium	24.305
Bk	Berkelium	247	Mn	Manganese	54.9380
Be	Beryllium	9.01218	Mt	Meitnerium	266
Bi	Bismuth	208.9804	Md	Mendelevium	258
Bh	Bohrium	262	Hg	Mercury	200.59
B	Boron	10.81	Mo	Molybdenum	95.94
Br	Bromine	79.904	Nd	Neodymium	144.24
Cd	Cadmium	112.41	Ne	Neon	20.1797
Ca	Calcium	40.078	Np	Neptunium	237
Cf	Californium	251	Ni	Nickel	58.6934
C	Carbon	12.011	Nb	Niobium	92.9064
Ce	Cerium	140.115	N	Nitrogen	14.0067
Cs	Cesium	132.9054	No	Nobelium	259
Cl	Chlorine	35.4527	Os	Osmium	190.3
Cr	Chromium	51.996	O	Oxygen	15.9994
Co	Cobalt	58.9332	Pd	Palladium	106.42
Cu	Copper	63.546	P	Phosphorus	30.97376
Cm	Curium	247	Pt	Platinum	195.08
Db	Dubnium	262	Pu	Plutonium	244
Dy	Dysprosium	162.50	Po	Polonium	209
Ds	Darmstadium	271	K	Potassium	39.0983
Es	Einsteinium	252	Pr	Praseodymium	140.9077
Er	Erbium	167.26	Pm	Promethium	145
Eu	Europium	151.965	Pa	Protactinium	231.0359
Fm	Fermium	257	Ra	Radium	226
F	Fluorine	18.99840	Rn	Radon	222
Fr	Francium	223	Re	Rhenium	186.207
Gd	Gadolinium	157.25	Rh	Rhodium	102.9055
Ga	Gallium	69.723	Rb	Rubidium	85.4678
Ge	Germanium	72.61	Ru	Ruthenium	101.07
Au	Gold	196.9665	Rf	Rutherfordium	261
Hf	Hafnium	178.49	Sm	Samarium	150.36
Hs	Hassium	265	Sc	Scandium	44.9559
He	Helium	4.00260	Sg	Seaborgium	263
Ho	Holmium	164.9303	Se	Selenium	78.96
H	Hydrogen	1.00794	Si	Silicon	28.0855
In	Indium	114.818	Ag	Silver	107.8682
I	Iodine	126.9045	Na	Sodium	22.98977
Ir	Iridium	192.2017	Sr	Strontium	87.62

Element		Mass (g)	Element		Mass (g)
S	Sulfur	32.066	W	Tungsten	183.84
Ta	Tantalum	180.9479	Uub	Unununbium	
Tc	Technetium	98	Uuu	Unununium	272
Te	Tellurium	127.60	U	Uranium	238.029
Tb	Terbium	158.9253	V	Vanadium	50.9415
Tl	Thallium	204.383	Xe	Xenon	131.29
Th	Thorium	232.0381	Yb	Ytterbium	173.04
Tm	Thulium	168.9342	Y	Yttrium	88.9059
Sn	Tin	118.710	Zn	Zinc	65.39
Ti	Titanium	47.867	Zr	Zirconium	91.224

WORKED EXAMPLE 1.3   *Plant Nutrients*

Phosphorus (P) is an essential nutrient for many plants. How many grams of phosphorus are in one mole of $Ca_3(PO_4)_2$? How many pounds of $Ca_3(PO_4)_2$ are needed to apply 100 lb of phosphorus to the soil?

Plan

- Determine the number of phosphorus atoms in $Ca_3(PO_4)_2$.
- Use atomic mass to convert to mass.
- Determine the molecular mass of $Ca_3(PO_4)_2$.
- Use ratios.

Implementation

- One mole of $Ca_3(PO_4)_2$ contains two moles of phosphorus.
- The atomic mass of phosphorus is 30.97 g/mol, so $Ca_3(PO_4)_2$ contains

$$\frac{2 \text{ moles P}}{\text{mol Ca}_3(\text{PO}_4)_2} \times 30.97 \text{ g/mol P} = 61.94 \text{ g P/mol Ca}_3(\text{PO}_4)_2$$

- Molecular mass of $Ca_3(PO_4)_2 = [(3 \times 40.08) + (2 \times 30.97) + (8 \times 16.00)] = 310.2$ g/mol

- $$\frac{310.2 \text{ g Ca}_3(\text{PO}_4)_2}{61.94 \text{ g P}} = \frac{x \text{ lb Ca}_3(\text{PO}_4)_2}{100 \text{ lb P}} \Rightarrow x = 501 \text{ lb}$$

The weight of $Ca_3(PO_4)_2$ required is about five times the amount of phosphorus that is supplied. The primary reason for this inefficiency is the large mass due to the three calcium atoms. Replacing calcium with sodium or lithium reduces the inefficiency considerably.

See Exercises 64, 65.

WORKED EXAMPLE 1.4   *Oxygen Generation*

Oxygen can be generated by heating a solid if the oxygen is loosely bound in the solid. One solid that yields oxygen when heated is $KClO_3$. How many grams of O_2 can be produced from 10 g of $KClO_3$?

Plan

- Calculate the molar mass of $KClO_3$.
- Determine the mass of three oxygen atoms.
- Determine the fractional mass due to oxygen and multiply by the total mass.

Implementation

- The molar mass is the sum of the masses due to each of the elements. In this case, one mole of K per mole of $KClO_3$ means 39.098 g/mol is due to K. One mole of Cl per mole of $KClO_3$ means 35.453 g/mol is due to Cl. Three moles of O per mole of $KClO_3$ means $3 \times 15.999 = 47.997$ g/mol is due to O. The total mass is $(39.098 + 35.453 + 47.997)$ g/mol = 122.548 g/mol.
- The mass of three moles of oxygen atoms is 47.997 g.
- In $KClO_3$, the fraction of the molar mass that is due to oxygen is $47.997/122.548 = 0.39166$. Thus, of the 10 g of $KClO_3$, 3.9 g is due to oxygen. This is the mass of oxygen that can be produced by the decomposition.

See Exercises 53 – 54.

1.3   Gas Laws

As indicated above, gases played a key role in development of the atomistic view of chemical reactions. Here, the gas laws are reviewed in the context of understanding pumping water uphill. In many areas of the world, water is pumped to fields by the same method used at least as far back as Aristotle's time. A piston is fitted tightly inside a vertical tube. Lifting the piston creates a rough vacuum. When placed in a river or lake, water rises in the tube like a beverage in a drinking straw. The limit to how high the water can be lifted in this fashion is about 34 ft. Getting water out of a deep gorge requires a series of such pumps. The purpose of this section is to understand the 34-ft limit.

Pressure-Volume Relationship: Boyle's Law

As early as 1660, Robert Boyle observed that as the pressure on a gas goes up, the volume goes down if the temperature remains constant.

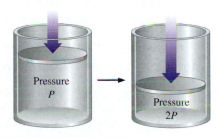

More specifically, the volume decrease is inversely proportional to the pressure increase:

(1.2a)

$$\frac{\text{new volume}}{\text{old volume}} = \frac{\text{old pressure}}{\text{new pressure}}$$

River

or in symbolic form

Boyle's law The volume of a given sample of gas at constant temperature varies inversely with the pressure.

(1.2b)

$$\frac{V_2}{V_1} = \frac{P_1}{P_2}$$

Equation (1.2b) is known as **Boyle's law.**

WORKED EXAMPLE 1.5 *Pump It Up*

A bicycle pump is 50 cm long with a cross-sectional area of 5 cm². How much pressure must the cyclist exert to reduce the volume of air in the pump to $^1/_{10}$ of the original volume?

Plan

■ Use the rearranged Boyle's law:

(1.3)

$$P_2 = P_1 \times \frac{V_1}{V_2}$$

Implementation

■ There are four variables in Boyle's law: V_1, V_2, P_1, and P_2. Volume information is given explicitly, but pressure information is not. The problem is to determine the final pressure; however, what is the initial pressure? Since the pump starts in the atmosphere, the initial pressure must be 1 atm. Equation (1.3) indicates that the ratio of the initial volume to the final volume is needed, so volume units must cancel. The information about the pump dimensions is extraneous. One atmosphere times the volume ratio is the required pressure.

The final volume is $^1/_{10}$ of the original volume. Symbolically, $V_2 = 0.1 \times V_1$. Hence,

$$P_2 = P_1 \times \frac{V_1}{V_2} = P_1 \times \frac{V_1}{0.1 \times V_1} = P_1 \times 10 = 10 \text{ atm}$$

Of course, the pump is open (to the bicycle tube) so the pressure is not quite this high. The answer is sensible because to reduce the volume the pressure must be increased.

See Exercises 19–22.

CONCEPT QUESTION A 1-ft-diameter cylinder is closed on one end and fitted with a moveable piston on the other. With 1 atm applied pressure, the piston is located 20 ft from the closed end. Where is the piston located if the pressure is reduced to 0.15 atm? ■

Temperature-Volume Relationship: Charles's Law

Boyle's law is now recognized as part of the more general description of the equation for ideal gases. Yet it was more than 100 years later, in 1779, that Guillaume Amontons observed that the ratio of the pressure to the temperature is a constant if the volume is constant. A year later, the French balloonist Jacques Charles arrived at the conclusion that the ratio of volume to temperature is constant if pressure is constant. These two results are stated symbolically in Equations (1.4) and (1.5).

(1.4)

$$\frac{P_1}{T_1} = \frac{P_2}{T_2}$$

Amontons's observation
Constant *V*

(1.5)
$$\frac{V_1}{T_1} = \frac{V_2}{T_2}$$
Charles's law
Constant P

Charles's law The volume of a given sample of gas at constant pressure is directly proportional to the temperature in absolute (K).

Equation (1.5) is known as **Charles's law.**

It should be noted that these relationships resulted from observations. The fundamental basis for these observations was not formulated for another 70 years!

In these relationships, the temperature needs to be positive. The normal scale used is an absolute temperature scale called the Kelvin scale, in which units are designated as K. The relationship between °C and K is

(1.6)   temperature in K = temperature in °C + 273.15

Notice that the notation for the Kelvin scale has no degree (°) symbol.

WORKED EXAMPLE 1.6   *Absolute Zero*

Charles's law indicates that at fixed pressure the volume decreases as the temperature is lowered. Thus, at some temperature, a given volume will decrease to zero. Further lowering of the temperature will result in a negative volume. Because a negative volume is not physically reasonable, there must be a lowest temperature; this point is referred to as absolute zero. Use the data given in the table below to determine the value of absolute zero. All data are taken at $P = 1$ atm.

Temperature (°C)	Gas Volume (cm³)
0	1000
−10	963
−30	890
−50	817
−150	451
−240	121
−270	11.0

Plan

- Use Equation (1.5).
- Extrapolate to zero volume.
- Determine the slope and calculate the intercept.

Implementation

- Charles's law specifies the relationship between temperature and volume at fixed pressure. Plotting T versus V and extrapolating to zero volume yields absolute zero.

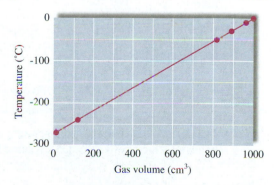

- Intercept: $-273\ °C$. Within the significant digits given, this is the same as the actual value of $-273.15\ °C$.
- Alternatively, the data indicate that the volume decreases by $989\ cm^3$ for a $270\ °C$ change in temperature, an average of $(989\ cm^3/270\ °C) = 3.66\ cm^3/°C$. So determining how many more units the temperature needs to drop to send the volume to zero determines absolute zero.

Know	Need
cm^3 and $cm^3/°C$	$°C$

Dividing the remaining volume by the average change yields the remaining temperature drop needed.

Unit analysis:

$$\frac{cm^3}{cm^3/°C} = °C \qquad \frac{11.0\ cm^3}{3.66\ cm^3/°C} = 3.00\ °C$$

Absolute zero is $-270\ °C - 3.00\ °C = -273\ °C$.

Make sure to convert to Kelvin for gas-law calculations involving temperature!

See Exercises 23–25.

CONCEPT QUESTION   A 1-ft-diameter cylinder is closed on one end and fitted with a moveable piston on the other. At room temperature (70 °F or 21 °C), the piston is located 20 ft from the closed end. Where is the piston located when the cylinder is immersed in a cold mountain stream at 50 °F? ■

Combined Gas Law

Combining Charles's and Boyle's laws produces a very useful result:

(1.7)
$$\frac{V_1}{V_2} = \frac{T_1 P_2}{T_2 P_1} \quad \text{or} \quad \frac{V_1 P_1}{T_1} = \frac{P_2 V_2}{T_2}$$

No gas obeys this relationship exactly. However, for moderate pressures and temperatures, many gases nearly obey this equation.

WORKED EXAMPLE 1.7   *Combined Gas Laws*

A balloon is filled with 11.0 L of gas at 23 °C and a pressure of 760 torr. When the balloon reaches an altitude of 20.0 km, the pressure is 63.0 torr and the temperature is 223 K. What is the volume of the balloon?

Plan
- Convert T from °C to K.
- Given: V_1, P_1, T_1, P_2, and T_2; need V_2: Rearrange Equation (1.7) to $V_2 = \dfrac{V_1 P_1 T_2}{T_1 P_2}$.

Implementation
- °C to K: 23 + 273 = 296 K

- $V_2 = \dfrac{V_1 P_1 T_2}{T_1 P_2} = \dfrac{(11.0\ L)(760\ torr)(223\ K)}{(296\ K)(63.0\ torr)} = 100\ L$

Is the answer sensible? The temperature decreases by 25% and the pressure decreases by about a factor of 12. Hence a volume increase of about a factor of 10 is reasonable.

See Exercises 29–30.

The Universal Gas Constant

To complete the gas law, the equation describing the behavior of gases, we return to Avogadro's hypothesis that equal volumes of different gases contain equal numbers of particles if the temperature and pressure are the same. Another way of stating this hypothesis is that, for a given pressure and temperature, the atomic or molecular density of all gases is the same. This leads directly to **Avogadro's law:**

Avogadro's law Equal volumes of gases at the same temperature and pressure contain the same number of particles.

> The volume of a gas maintained at constant temperature and pressure is directly proportional to the number of moles of the gas.

Summarizing, the volume, V, of a gas is directly proportional to the number of moles, n, (Avogadro's law); directly proportional to the temperature, T, (Charles's law); and inversely proportional to the pressure, P, (Boyle's law).

(1.8)
$$V \propto \frac{nT}{P}$$

Or in Avogadro's terms,

(1.9)
$$\frac{n}{V} \propto \frac{P}{T}$$

Molar density moles per unit volume

The **molar density**, n/V, of all gases is the same if the temperature and pressure are the same. This proportionality can be turned into an equality if we can determine the proportionality constant. In principle, the constant can be determined by measuring the density. However, an issue arises with this approach, that is, molar density is not what is commonly referred to as density. Instead, density generally refers to **mass density**—mass per unit volume—given in units of grams per liter for gases. The mass density of every gas is different, so every gas has a characteristic proportionality constant.

Mass density Mass per unit volume. Typical Units: (Solids and liquids) g/cm³ or (gases) g/L.

(1.10)
$$\frac{P}{T} = (\text{gas-dependent constant}) \times \text{density}$$

CONCEPT QUESTION How is mass density converted to molar density? ■

The mole provides the link between mass density and molar density.

(1.11)
$$\text{Molar density} = \frac{\text{mass density}}{\text{atomic mass}} \qquad \text{Units: mol/cm}^3 = \frac{\text{g/cm}^3}{\text{g/mol}}$$

Converting the *mass* density to *molar* density results in

(1.12)
$$\frac{P}{T} = (\text{universal constant}) \times \frac{n}{V}$$

Gas constant, R The proportionality constant in the ideal gas law; 0.08206 L·atm/K·mol or 8.3145 J/K·mol.

A universal constant applicable to all gases at moderate pressures and temperatures, it is given the symbol R and called the universal gas law constant or, more simply, the **gas constant.**

(1.13)
$$R = 0.08206 \text{ L} \cdot \text{atm/K} \cdot \text{mol} = 8.3145 \text{ J/K} \cdot \text{mol}$$

The beauty of R is that it is applicable to all gases: monatomic, diatomic, or polyatomic. Putting R in Equation (1.12) yields

(1.14)
$$\frac{P}{T} = R\frac{n}{V} \qquad \text{or} \qquad PV = nRT$$

Ideal gas law An equation of state for a gas, where the state of the gas is its condition at a given time; expressed by $PV = nRT$, where P = pressure, V = volume, n = moles of the gas, R = the universal gas constant, and T = absolute temperature. This equation expresses behavior approached by real gases at moderate T and low P.

Equation (1.14) is called the **ideal gas law** or the ideal gas equation. At 0 °C and 1 atm pressure, the molar density of typical gases is 0.0446 mol per liter.

At temperatures near room temperature and pressures around one atmosphere, many gases are ideal gases. An **ideal gas** is one with no interactions between the particles. For moderate temperatures and pressures, real gases very closely follow the ideal gas law.

CONCEPT QUESTION Describe in your own words what is meant by mass density and by atomic density. Under what circumstances is each the appropriate unit to use? ∎

Avogadro's number provides the final link to the atomistic view of gases—the link between molar density and atomic density. Multiplication of the molar density by Avogadro's number gives the atomic density:

Ideal gas A gas with no interactions between the particles.

$$(1.15) \qquad \frac{0.0446 \text{ mol}}{\text{L}} \times \frac{6.022137 \times 10^{23} \text{ particles}}{\text{mol}} = 2.69 \times 10^{22} \text{ particles/L}$$

Standard temperature and pressure (STP) The condition 0 °C and 1 atm of pressure. At STP one mole of gas occupies 22.4 L.

For the monatomic gases, the particles are atoms, so the atomic density of He through Rn is constant (Figure 1.8).

In many practical problems, the pressure is near one atmosphere and the temperature near 0 °C. One atmosphere and 0 °C is called **standard temperature and pressure (STP).** Inserting STP into the gas law, Equation (1.14) indicates that at STP, one mole occupies 22.4 L—a handy benchmark to know when doing calculations involving gases.

WORKED EXAMPLE 1.8   *The Relationship Between Mass and Moles*

A pharmaceutical company is synthesizing a zinc supplement. The current RDA (recommended daily allowance) for zinc is 15 mg per day for an adult.

(a) How many moles of zinc are required for each adult?
(b) How many atoms of zinc should an adult consume each day?

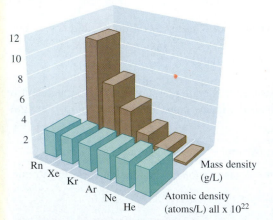

	Helium He	Neon Ne	Argon Ar	Krypton Kr	Xenon Xe	Radon Rn
Mass density (g/L)	0.1787	0.9009	1.783	3.741	5.8612	9.91
Atomic mass (amu)	4.003	20.18	39.948	83.80	131.29	222
Atomic density (atoms/L) all x 10^{22}	2.69	2.69	2.69	2.69	2.69	2.69

Figure 1.8 Mass density and atomic density of several gases at 1 atm pressure and 0 °C. The mass density decreases for the gases Rn through He. In contrast, the atomic density is constant.

Plan

■ Unit analysis:

$$\text{Moles} = \text{mass/atomic mass} = \frac{g}{g/mol}$$

$$\text{Atoms} = \text{moles} \times \text{atoms/mole}$$

Implementation

■ (a) How many moles of Zn is 15 mg?

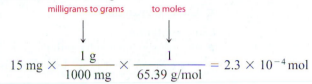

$$15 \text{ mg} \times \frac{1 \text{ g}}{1000 \text{ mg}} \times \frac{1}{65.39 \text{ g/mol}} = 2.3 \times 10^{-4} \text{ mol}$$

(b) How many atoms is 15 mg?

$$2.3 \times 10^{-4} \text{ mol} \times \frac{6.022045 \times 10^{23} \text{ atoms}}{mol} = 1.4 \times 10^{20} \text{ atoms}$$

Atoms are so small that many are required to create several milligrams of material. It is easier to think in terms of 0.00023 mole than 1.4×10^{20} atoms.

See Exercise 53.

Applications

Origin of Atmospheric Pressure So far, all pressures have been expressed in atmosphere units. This pressure is the force exerted by a column of air above a unit area at sea level. One method for measuring this pressure is to immerse one end of an evacuated tube in an open container of a liquid. The liquid rises in the tube until the gravitational force exerted by it equals the force exerted by the atmosphere. An apparatus based on this principle is called a Torricelli barometer (Figure 1.9). The liquid in this apparatus is often mercury, which rises to a height of 760 millimeters at sea level. Hence pressure is sometimes expressed in units of millimeters of mercury. One millimeter of mercury is called one **torr** in recognition of Torricelli's invention of this apparatus.

Torr Another name for millimeter of mercury (mm Hg).

The force exerted by a column of liquid in the earth's gravitational field is given by the mass of the column times the gravitational acceleration, g (9.80665 m/s^2).

(1.16)
$$F = mass \times g$$

The mass of the liquid column is the volume (or height, h, times the cross-sectional area, A) times the density, ρ,

(1.17)
$$mass = volume \times \rho = (h \times A) \times \rho$$

The pressure is the force exerted per unit area

(1.18)
$$P = \frac{F}{A} = \frac{(h \times A \times \rho) \times g}{A}$$

(1.19)
$$P_{atm} = \rho g h$$

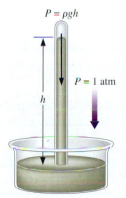

$P = \rho g h$

$P = 1$ atm

h

Figure 1.9 Illustration of evacuated tube in dish of liquid — Torricelli's barometer.

WORKED EXAMPLE 1.9   *The 34-ft Limit*

A Torricelli barometer is made with water as the operating liquid and the apparatus is at sea level. How high does the column of water rise?

Data: Density of water $= 1.000$ g/cm^3

Plan

- Rearrange Equation (1.19) to $h = \dfrac{P_{atm}}{\rho g}$.

- Use unit analysis.

Implementation

- Water rises in the tube until the force exerted by the column of water equals that of a column of air at sea level. The height is given by

$$h = \frac{P_{atm}}{\rho g}$$

- Unit analysis:

Know	Need
P (atm)	h (m)
g (m/s^2)	
ρ (g/cm^3)	

Converting length units to meters results in denominator units of kg·m/s^2. Conversion of pressure to newtons per square meter (1 N $=$ 1 kg·m/s^2) yields height in meters. One atmosphere is 1.01325×10^5 N/m^2.

Unit analysis:

$$h = \frac{P_{atm}}{\rho g} = \frac{atm}{(g/cm^3)(m/s^2)} \times \frac{N/m^2}{atm} \times \frac{kg \cdot m/s^2}{N} \times \frac{g}{kg} \times \left(\frac{m}{cm}\right)^3 = m$$

Insert numerical data:

$$h = \frac{P_{atm}}{dg} = \frac{1 \text{ atm}}{(1.000 \text{ g/cm}^3)(9.80665 \text{ m/s}^2)} \times \frac{1.01325 \times 10^5 \text{ N/m}^2}{1 \text{ atm}}$$

$$\times \frac{1 \text{ kg} \cdot \text{m/s}^2}{1 \text{ N}} \times \frac{1000 \text{ g}}{kg} \times \left(\frac{m}{100 \text{ cm}}\right)^3$$

$$= 10.33 \text{ m}$$

10.33 m $=$ 33.90 ft or about 34 ft. This is the origin of the 34 ft limit for the piston pump: Air pressure pushes the water up the tube until the mass of the column of water equals that of the column of air.

Note: The number of significant digits in the answer is limited by the number of significant digits in the density of water because the conversion factors (such as centimeters per meter) are exact numbers.

See Exercise 41.

CONCEPT QUESTION   The 34-ft limit is based on the piston pump being placed in a stream at sea level, or close to it. How high can the water be pumped if it is placed in a mountain stream where the local pressure is 630 torr? (This corresponds to an elevation similar to Denver, CO.) ■

The Weight of the Earth's Atmosphere   Although it may not seem apparent while standing on the earth, the atmosphere overhead has a considerable mass. Can this mass be determined? For simplicity, let us assume that a person occupies an area of 1 ft². If the volume and density of the atmosphere were known, the mass of the atmosphere overhead could be determined. However, as anyone who has visited the mountains knows, the density of the atmosphere decreases with height. It seems that the answer to this simple question is not so simple. However, let us take stock of what is known about the atmosphere. The atmosphere is composed of molecules, and each molecule has some mass. Indeed, the force due to this mass in the gravitational field of the earth is what supports the mercury column in Torricelli's barometer! So the atmosphere supports a column of mercury 76 cm high. Assume that the cross-sectional area of the column is 1 in².

Know	Need
volume = 1 in² × 76 cm	cm³

Conversion: 2.54 cm per inch

$$\text{volume} = (1 \text{ in.})^2 \times \left(\frac{2.54 \text{ cm}}{\text{in.}}\right)^2 \times 76 \text{ cm} = 490 \text{ cm}^3$$

The density of mercury is 13.56 g/cm³, so the mass of the column of mercury is 6.64 kg, or (1 kg = 2.2 lb) 14.6 lb/in². (This number may be familiar!) Since the assumption is that a person occupies an area of 1 ft², multiply this value by 144 in². The result is 2100 lb, just over a ton.

Given this result, here is a question to ponder: Why do people not collapse under this weight? Try the following experiment for a clue. Take an empty aluminum soda can and put a small amount of water in it. Put the can on a heat source until the water boils. Water will then displace most of the air in the can. Using tongs or an insulated glove, invert the can and quickly put the top into a container of water. See the solution section for an explanation.

At 1 ton/ft², what is the total mass of the earth's atmosphere? To answer this question, the surface area of the earth is needed: 5.1×10^8 km² or 5.5×10^{15} ft². The mass of the atmosphere is 5.5×10^{15} tons, or about 6 billion megatons!

CONCEPT QUESTION   Aristotle's breath: a question to ponder[†] How likely is it that you are breathing in a molecule of nitrogen that was in the last breath of Aristotle? Molecular nitrogen is rather nonreactive, so assume that the nitrogen in Aristotle's last breath is still in the atmosphere. Given the long time since his last breath, assume that all the nitrogen molecules contained in it are distributed uniformly throughout the atmosphere. ■

This question is chunked into smaller questions in Exercises 42–44.

WORKED EXAMPLE 1.10   ***Buoyancy — The Hindenburg***

Did the makers of the Hindenburg take an unnecessary risk in filling the dirigible with hydrogen rather than helium? Hint: Why does an iceberg float? Because ice is less dense than water. Keeping this result in mind, consider what happens if some of the air overhead is replaced with a container filled with something less dense than air. Assuming the container is not too heavy, it will float as anyone who has held a helium-filled balloon knows. This is the principle behind a hydrogen- or helium-filled blimp. Compare the lifting power of hydrogen versus helium in this application.

[†] The source of this problem is Professor Herschbach's Chem. Zen class at Harvard University.

Data

> Average molecular weight of air = 28.8 g/mol
> Molecular weight of hydrogen = 2.02 g/mol
> Molecular weight of helium = 4.00 g/mol

Plan

- Estimate the answer.
- Force = Δmg, Δm = mass difference, g = gravitational acceleration.

Implementation

- At first glance, it seems that hydrogen should have twice the lifting power of helium because the molecular mass of hydrogen is half that of helium. However, in a blimp, the gas filling the blimp displaces air. One mole of hydrogen displaces one mole of air, replacing 28.8 g with 2.02 g. The difference, 26.8 g, in the earth's gravitational field is the lifting power of hydrogen. For helium, the difference drops to 24.8 g. The ratio of the lifting power of hydrogen to that of helium is about 27/25 or about 8% greater lifting power for hydrogen.
- For hydrogen:

$$26.8 \text{ g} \times 9.81 \text{ m/s}^2 \times \frac{1 \text{ kg}}{1000 \text{ g}} \times \frac{1 \text{ N}}{\text{kg} \cdot \text{m/s}^2} = 0.263 \text{ N}$$

For helium:

$$24.8 \text{ g} \times 9.81 \text{ m/s}^2 \times \frac{1 \text{ kg}}{1000 \text{ g}} \times \frac{1 \text{ N}}{\text{kg} \cdot \text{m/s}^2} = 0.243 \text{ N}$$

The lifting power of hydrogen versus helium is 0.263 N/0.243 N = 1.06. The lifting power of hydrogen is about 6% more than the lifting power of helium.

Given the difference in lifting power, which would you use in a blimp—hydrogen or helium?

See Exercises 26, 45.

WORKED EXAMPLE 1.11   *Flying*

The Wright brothers would probably be impressed but not surprised at today's airplanes. They understood that slicing a curved wing through the air results in a difference in air pressure above and below the wing. A jumbo jet weighs about 350 tons. Determine the pressure differential required to lift this weight given that the wing area is 3500 ft^2.

Plan

- Determine weight per unit area.
- Determine the fraction of an atmosphere.

Implementation

- With a wing area of 3500 ft^2, the lift per square foot needs to be 0.1 ton.
- At one atmosphere pressure, the atmospheric mass is 15 lb/in^2, or 1 t/ft^2. Thus, a pressure differential of only 0.1 atm lifts the plane!

The area of an airplane's wing is proportional to the loaded weight of the airplane.

See Exercise 46.

1.4   Molecules and Representations

Empirical formula Simplest whole number ratio that represents the composition of the substance.

Ion An atom or group of atoms with a charge.

Ball-and-stick model A molecular model that represents atoms as balls (ball size not significant) and bonds as sticks showing bond relationships clearly.

Structural formula Indicates how atoms are linked together, but does not contain the three-dimensional information of the ball-and-stick model.

Condensed structural formula A compact version of a structural formula showing how atoms are grouped together.

Figure 1.10 The compound water is depicted by two blue balls representing hydrogen atoms and a single red ball representing oxygen.

H
|
H   O   H
| | |
H — C — C — C — H
| | |
H   H   H

Figure 1.11 A structural formula shows how atoms are linked together. The molecule shown here is isopropanol, the most volatile component of rubbing alcohol.

Since atoms and molecules are so small, it is useful to have a method for representing them. The most compact method of representing molecules is by a chemical or molecular formula. A chemical formula indicates the elements involved in the compound and their number—elements by their atomic symbol and the number of atoms by a subscript. For example, hydrogen sulfide (H_2S), the substance that gives rotten eggs their noxious odor, contains two atoms of hydrogen and one atom of sulfur. Solids are often represented by their **empirical formula:** the simplest whole number ratio that represents the composition of the solid. For example, common table salt is represented as NaCl, which indicates that for every atom of sodium there is one atom of chlorine. Polyatomic **ions,** which are groups of bonded atoms with a charge, are enclosed in parentheses. For example, the formula for calcium phosphate, a component of bone, is $Ca_3(PO_4)_2$. The polyatomic ion PO_4^{3-} is indicated as a unit by enclosing it in parentheses. There are three calcium ions, Ca^{2+}, for every two phosphate units, PO_4^{3-}, in solid $Ca_3(PO_4)_2$.

A number of different representations have been developed to aid in communicating and visualizing chemistry. In particular, atoms are often represented as balls. Atoms in compounds are linked together by bonds, often depicted as lines or sticks connecting the balls that represent the atoms. For example, water (Figure 1.10) consists of two atoms of hydrogen and one atom of oxygen. The colors chosen for atoms are not agreed upon, although chemists most often represent oxygen with a red ball and hydrogen with a white or blue ball. In reference to the balls and lines connecting them, this representation is called a **ball-and-stick model.**

A **structural formula** indicates how the atoms are linked together, but does not contain the three-dimensional information of the ball-and-stick model. For example, the most volatile component of rubbing alcohol is isopropanol with the structural formula shown in Figure 1.11.

Molecules with many atoms become quite cumbersome to draw with either a ball-and-stick or structural model, yet the molecular formula does not convey sufficient information. As a consequence, these molecules are often shown as a **condensed structural formula.** For example, a component of wine and other alcoholic beverages is ethanol which has the molecular formula C_2H_6O. There are a number of ways of arranging these nine atoms, and each arrangement has different properties. The condensed structural formula CH_3CH_2OH is much more informative, indicating three hydrogen atoms around a carbon attached to a second carbon with two hydrogen atoms and one OH group. The same molecular formula belongs to dimethyl ether, a relative of the ether used as an anesthetic. The structural formula of dimethyl ether is CH_3OCH_3.

The element carbon is fairly unique in forming quite large molecules in which carbon is bonded to carbon. A more abstract representation of such molecules is the **line formula.** A line formula represents bonds as lines; carbon atoms occupy positions at the junctions between lines, and hydrogen atoms are not shown explicitly (Figure 1.12). Every carbon atom has four connections or bonds, and those not shown explicitly are occupied by hydrogen atoms. Atoms of elements other than carbon and hydrogen are shown by their symbols (Figure 1.13).

In addition to forming long chains, carbon can form a rich variety of ring structures and carbon atoms can be connected by multiple bonds (Figure 1.14). Multiple bonds are shown by multiple lines.

Line formula Representation of an organic compound where bonds are shown as lines, carbon atoms (not shown explicitly) occur at the junction of lines, hydrogen atoms fill required valence, and other atoms are represented by their chemical symbol.

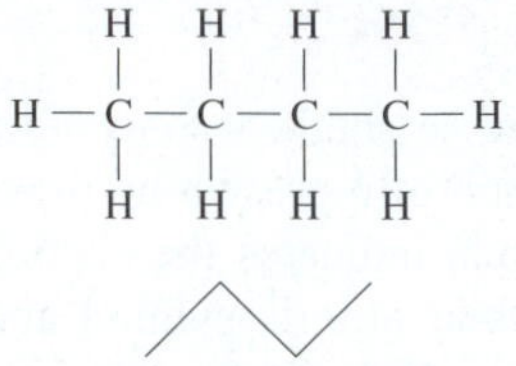

Figure 1.12 A line formula is a very compact representation of a molecule. Butane, often used in cigarette lighters, is shown as both a structural formula and a line formula.

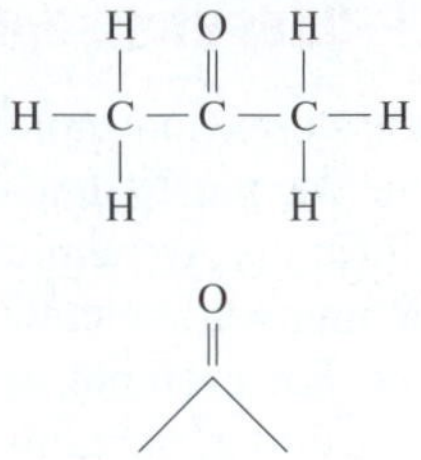

Figure 1.13 Acetone, the volatile component of nail polish remover, contains one oxygen atom. Atoms other than carbon and hydrogen are shown explicitly in a line formula.

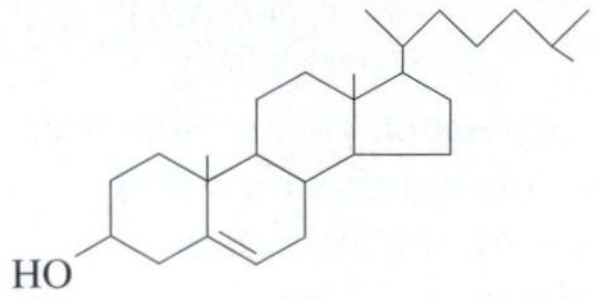

Figure 1.14 Carbon forms a large variety of ring and multiple-bond structures. Shown here is cholesterol, the precursor for sex hormones and other steroids.

WORKED EXAMPLE 1.12 *Decoding a Line Formula*

A line drawing of the molecule cholesterol is shown in Figure 1.14. How many carbon atoms are in cholesterol? How many hydrogen atoms?

Plan
- Carbon atoms are located at the junctions of lines.
- Carbon has four connections; connections not shown are to hydrogen atoms.

Implementation
- The ring on the left has six vertices representing six carbon atoms. The next ring also has six vertices, including two in common with the first ring; this gives an additional four carbon atoms. The next ring adds four more carbon atoms. The next ring is a five-membered ring, so it adds three carbon atoms. The chain at the top adds eight carbon atoms, and two more are located at the ends of the two lines off the rings. The total number of carbon atoms: $6 + 4 + 4 + 3 + 8 + 2 = 27$.
- One hydrogen is shown explicitly. The first ring has seven more, the next ring has five, the next ring has five, the five-membered ring has five, the chain has 17, and the two carbon atoms attached to the rings add six. The total number of hydrogen atoms: $1 + 7 + 5 + 5 + 5 + 17 + 6 = 46$.

See Exercises 49–51.

Shape

Space-filling model A model of a molecule showing the relative sizes of the atoms and their relative orientations.

The three-dimensional shape of a molecule is very important in determining its interactions with other molecules. The ball-and-stick model is one representation of this shape. It is particularly effective for showing the angle formed by two bonds and for showing bond lengths (Figure 1.15a). It is often helpful to picture how the atoms fit together and

positive

neutral

negative

(a)   (b)   (c)   (d)

Figure 1.15 (a) The ball-and-stick model is very effective for showing the angle formed by two bonds or the distance between two atoms. (b) A space-filling model represents the shape of the molecule slightly more accurately and indicates how the atoms fit together with one another. (c) A density isosurface illustrates the distribution of electrons giving a sense of how electrons spread throughout the molecule. (d) A density potential plot depicts the net charge at each point on the isosurface. Regions with negative charge are shown in red and positive regions are shown in blue. Following the colors of the rainbow, green regions are neutral.

Density isosurface Represents the molecular structure with a surface of constant electron density.

Density potential Charge distribution showing charge with colors from red for the most negatively charged regions to green for neutral regions to blue for the most positively charged regions.

fill space. The **space-filling model** (Figure 1.15b) illustrates this property. The space-filling model also gives a somewhat more accurate picture of the shape of a molecule, although it may be more difficult to picture where all the atoms are because some are in the back of the structure.

A representation that gives a sense of how the electrons are spread over the molecule is a **density isosurface** (Figure 1.15c). Chemists often need to know the charge distribution on the surface of the molecule (Figure 1.15d); this is shown in a **density potential** plot. Regions with a negative potential are shown in red and positive regions are shown in blue. The scale follows the colors of the rainbow, with green representing neutral regions.

1.5   Stoichiometry

Molecular stoichiometry Quantitative relationship among the atoms that constitute a molecule.

Reaction Transformation of one or more molecules into a different set of molecules. A physical reaction involves a change in physical state with no change in molecular identity.

Reaction stoichiometry Quantitative relationship among the molecules involved in a reaction.

Law of definite proportions A given compound always contains exactly the same proportion of elements by mass.

There are two quantitative relationships that are essential to chemistry. The first is the quantitative relationship among the atoms that constitute in a chemical substance. This relationship is called the **molecular stoichiometry.** Along with the shape, the molecular stoichiometry determines the identity of a substance. Fundamentally, chemistry involves processes, called **reactions,** that change one or more chemical substances, one or more molecules, into different molecules. The quantitative relationship between the substances involved in a chemical reaction is called the **reaction stoichiometry.**

Molecular Identity

Stoichiometry reflects an atomistic view—that is, the view that molecules are composed of units called atoms. The atomistic view is based on the law of definite proportions. Articulated by Joseph Proust in the early 1800s, the **law of definite proportions** states:

Different samples of a pure chemical substance always contain the same proportion of elements by mass.

For example, water always consists of eight times the mass of oxygen compared with the mass of hydrogen present. This mass ratio holds whether the water is created on earth or on the other side of the universe, and whether the water is created by combining elemental hydrogen and oxygen or by rearranging other compounds to release one atom of oxygen and two atoms of hydrogen.

Although any one substance always has the same mass ratio, elements can combine in different mass ratios to form distinct substances. For example, carbon monoxide has a carbon-to-oxygen mass ratio of $12:16$ or $3:4$. This substance is a poisonous gas. Carbon and oxygen can also combine with a mass ratio of carbon to oxygen of $12:32$, producing carbon dioxide. This gas gives soda its fizz. Different mass ratios generate distinct substances. In the early 1800s, it was observed that the ratio of the mass ratio of elements in compounds that have different combining ratios is a small whole number. The mass of oxygen relative to the mass of carbon in carbon dioxide is twice the mass ratio of oxygen to carbon in carbon monoxide. Observing these small whole-number ratios provided strong evidence that matter is constituted from indivisible particles—atoms. In 1808, John Dalton, a Quaker schoolteacher, published his observations on combining ratios in what has become known as the **law of multiple proportions:**

> Elements can combine in different ways to form different substances. The ratio of the mass ratio of an element in the different substances is a small whole number.

For example, in NO the mass ratio of oxygen to nitrogen is $16:14$, and in NO_2 the mass ratio is $32:14$. The latter compound has twice as much oxygen as the former compound, and the ratio of $32:14$ to $16:14$ is 2.

Law of multiple proportions A law stating that when two elements form a series of compounds, the ratios of the masses of the second element that combine with 1 g of the first element can always be reduced to small whole numbers.

Reaction Proportions

Compounds are broken down, rearranged, or assembled in chemical reactions. A reaction equation specifies the starting substances, called the **reactants,** and the ending substances, called the **products,** of a reaction. Reactants and products are separated by an arrow pointing from reactants to products. In a chemical reaction, atoms are rearranged but are neither created nor destroyed. In a balanced reaction, there are always the same numbers of atoms of each element among the reactants as there are among the products. A balanced reaction is a direct result of conservation of mass. In contrast, compounds can be created or destroyed as a result of the atomic rearrangement; *neither moles nor molecules are conserved but atoms are.*

For example, hydrogen and oxygen combine to form water. Hydrogen and oxygen exist as diatomic molecules—H_2 and O_2, respectively. So the skeletal reaction is

Reactant Starting substance in a chemical reaction.

Product Final substance in a chemical reaction.

(1.20)
$$H_2 + O_2 \longrightarrow H_2O$$

Equation (1.20) is not balanced: There are two atoms of oxygen among the reactants, but only one among the products. Multiplication of water, H_2O, by 2 balances the oxygen but leaves hydrogen unbalanced, with four atoms among the products and only two among the reactants. Multiplication of both hydrogen and water by 2 balances all elements:

(1.21)
$$2H_2 + O_2 \longrightarrow 2H_2O$$

Stoichiometry Quantitative relationship between the quantities of reactants consumed and products formed in a chemical reaction.

The factors of 2 are called stoichiometric coefficients as is the factor of 1, which is not explicitly written. Reaction **stoichiometry** refers to the coefficients in the balanced reaction equation.

Balancing Reactions

For practical purposes of balancing reaction equations, it is helpful to divide reactions into those in which electrons are not transferred and those involving electron exchange. If electrons are not exchanged in a reaction, then the reaction consists of shuffling elements. For example, calcium carbonate, $CaCO_3$, is a component of sea shells, limestone, marble, calcite, pearls, and chalk. In the environment, many ions are found in water. Liquid water mixed with other substances is called an **aqueous solution.** Calcium ions (Ca^{2+}) and carbonate ions (CO_3^{2-}) ions are among those commonly found in aqueous solution and $CaCO_3$ results when these come together

> **Aqueous solution** A homogeneous solution with water as the major component.

$$(1.22) \qquad Ca^{2+}(aq) + CO_3^{2-}(aq) \longrightarrow CaCO_3(s)$$

The notation (aq) refers to an ion in aqueous solution and (s) refers to a solid. An example of this reaction in action is the formation of the spectacular stalactites and stalagmites in limestone caverns. Formation of a solid from its ions is called a **precipitation reaction.**

> **Precipitation reaction** A reaction in which ions in a solution come together to form a solid.

Equation (1.22) is a balanced reaction: There is one mole of calcium and one mole of carbonate ions on each side of the reaction. The total charge on both sides of the reaction is zero. Equation (1.22) is a **net ionic equation:** It indicates only those species explicitly taking part in the reaction. In the laboratory you might form $CaCO_3$ by starting with a solution of $Ca(NO_3)_2$ and Na_2CO_3. Then the **molecular equation** describing the reaction is

> **Net ionic equation** An equation for a reaction in aqueous solution, where strong electrolytes are written as ions, showing only those components that are directly involved in the chemical change.

$$(1.23) \qquad Ca(NO_3)_2(aq) + Na_2CO_3(aq) \longrightarrow CaCO_3(s) + 2NaNO_3(aq)$$

Writing out all the ions involved in reaction (1.23) results in the **total ionic equation**

> **Molecular equation** An equation representing a reaction in aqueous solution showing the reactants and products in undissociated form, whether they are strong or weak electrolytes.

$$(1.24) \qquad Ca^{2+}(aq) + 2NO_3^{-}(aq) + 2Na^{+}(aq) + CO_3^{2-}(aq) \longrightarrow$$
$$CaCO_3(s) + 2Na^{+}(aq) + 2NO_3^{-}(aq)$$

NO_3^{-} and Na^{+} are termed **spectator ions** in reference to their nonparticipation. Their role in the reaction is to make the solutions electrically neutral.

> **Total ionic equation** An equation representing a reaction in aqueous solution showing all ions present in the solution.

Calcium carbonate can also be formed from the reaction between solid CaO and gaseous CO_2

$$(1.25) \qquad CaO(s) + CO_2(g) \longrightarrow CaCO_3(s)$$

> **Spectator ion** An ion that does not participate in a reaction. It is present to maintain overall electrical neutrality.

The calcium carbonate formed in reactions (1.22) through (1.24) is indistinguishable from that formed in reaction (1.25). This is an example of the law of definite proportions in action.

In all the reactions indicated, the number of atoms of each *element* on both sides of the reaction is the same, and the charge of any ion on both sides is the same, so the number of electrons associated with each element remains the same. These nonelectron-transfer reactions are usually balanced by inspection. Once balanced, the stoichiometry indicates how many molecules or empirical formula units of each type are involved. For example, Equation (1.23) indicates that one formula unit of $Ca(NO)_3$ reacts with one formula unit of Na_2CO_3 to yield one formula unit of $CaCO_3$ and two formula units of $NaNO_3$.

WORKED EXAMPLE 1.13   *Multiple Combining Ratios*

Carbon and oxygen combine in two different ratios, as CO and as CO_2. How many grams of oxygen gas are required to convert 1.00 mg of carbon monoxide (CO) into carbon dioxide (CO_2)?

Plan

- Write the balanced reaction.
- Use mass ratios.

Implementation

- Reaction:

$$2CO + O_2 \longrightarrow 2CO_2$$

Check that the reaction is balanced:

C: two atoms of C in 2CO $\rightarrow$ two atoms of C in $2CO_2$.
O: two atoms of O in 2CO and two in O_2 for a total of four $\rightarrow$ four atoms of O in $2CO_2$.

- Mass ratio C:O in CO is 12:16. Mass ratio C:O in CO_2 is 12:32. One milligram CO contains

$$\frac{16 \text{ g O}}{(16 + 12) \text{ g CO}} \times 1.00 \text{ mg CO} = 0.57 \text{ mg O}$$

To convert CO to CO_2 requires a mass of oxygen equal to that already present, or 0.57 mg.

See Exercises 55–59.

Reactions involving transfer of electrons require a little extra attention. For example, one of the first batteries was made by Alexander Volta and is based on the reaction between copper ions, $Cu^{2+}(aq)$, and metallic zinc (Zn). Electrical energy results from transfer of electrons from zinc metal to copper ions. Even more energy per unit weight can be obtained from Al and Cu^{2+}. The skeletal reactions for each of these are

(1.26) $Cu^{2+}(aq) + Zn(s) \longrightarrow Cu(s) + Zn^{2+}(aq)$

and

(1.27) $Cu^{2+}(aq) + Al(s) \longrightarrow Al^{3+}(aq) + Cu(s)$

Of these, only the first, reaction (1.26), is balanced. Reaction (1.27) might appear to be balanced: One copper and one aluminum appear on both sides of the reaction. However, the charge is not balanced. An atom becomes positively charged due to the loss of an electron. With two positive charges, the reactants have lost two electrons. The products have three positive charges, so they have lost three electrons. Electrons are not balanced in Equation (1.27). Multiplying Cu^{2+} on the reactant side by three increases the number of electrons lost among the reactants to six. Multiplying Al^{3+} by two on the product side indicates a loss of six electrons among the products. Multiplying Al on the reactant side by two and Cu on the product side by three balances each of these. Thus, the balanced reaction is

(1.28) $$3Cu^{2+}(aq) + 2Al(s) \longrightarrow 2Al^{3+}(aq) + 3Cu(s)$$

Charge balancing is the same as balancing electrons—another example of conservation of mass.

WORKED EXAMPLE 1.14   *Writing Equations, Identifying Spectators*

The insoluble material calcium phosphate, $Ca_3(PO_4)_2$, is a component of bones. The NO_3^- ion forms soluble combinations with both Na^+ and Ca^{2+}. PO_4^{3-} with Ca^{2+} forms the insoluble mineral portion of bones. Write the balanced molecular, total ionic, and net ionic equations for the reaction that occurs when a $Ca(NO_3)_2$ solution is added to a Na_3PO_4 solution.

Plan
- Identify reactants and products.
- Use conservation of mass.
- Identify ions.
- Eliminate spectator ions to give the net ionic equation.

Implementation
- The reactants are $Ca(NO_3)_2$ and Na_3PO_4. The products are the solid $Ca_3(PO_4)_2$ and the remaining species are Na^+ and NO_3^-.
- $Ca_3(PO_4)_2$ requires three Ca^{2+} ions and two PO_4^{3-} ions. Two molecules of Na_3PO_4 are required to produce two PO_4^{3-} ions. Similarly, three molecules of $Ca(NO_3)_2$ are required to produce three Ca^{2+} ions.

$$3Ca(NO_3)_2 + 2Na_3PO_4 \longrightarrow Ca_3(PO_4)_2 + 6NaNO_3 \qquad \text{Balanced}$$

This is the balanced molecular reaction.

- All species except $Ca_3(PO_4)_2$ are soluble, so among the reactants, the ions are Ca^{2+}, NO_3^-, Na^+, and PO_4^{3-}. Among the products, the ions are Na^+ and NO_3^-. The total ionic equation is

$$3Ca^{2+}(aq) + 6NO_3^-(aq) + 6Na^+(aq) + 2PO_4^{3-}(aq) \longrightarrow$$
$$Ca_3(PO_4)_2(s) + 6Na^+(aq) + 6NO_3^-(aq)$$

- Ions common to reactants and products are the Na^+ and NO_3^- ions. Eliminate these:

$$3Ca^{2+}(aq) + 2PO_4^{3-}(aq) \longrightarrow Ca_3(PO_4)_2(s)$$

See Exercises 72–73.

Limiting Reactant

Often when molecules come together, one or the other molecule is in short supply. An analogy is making salad dressing. The usual ratio is one part vinegar to three parts oil by volume. Suppose the pantry has a pint of vinegar and a pint of oil. The maximum that can be used in making salad dressing is one-third of the vinegar. Oil is in short supply. In chemistry, the proportions are determined by stoichiometry rather than taste, and the substance in short supply is called the **limiting reactant.**

Limiting reactant The reactant that is present in sufficiently low concentration so that it limits the extent of the reaction.

WORKED EXAMPLE 1.15 *In Short Supply*

Ammonia is important in the manufacture of fertilizer. The molecular formula of ammonia is NH_3 and the mass ratio of nitrogen to hydrogen is $14:3$. Production of ammonia requires breaking the strong bond in N_2, which takes considerable energy. Hydrogen gas is easier to break apart. What is the maximum amount of ammonia that can be produced from 1 kg of nitrogen? At the start of an ammonia production run, a plant has 1 kg nitrogen and 100 g hydrogen. Should the operator produce more hydrogen before beginning the run?

Plan

- Use the mass ratio.
- Determine the limiting reactant.

Implementation

- The mass ratio of nitrogen to ammonia is $14:17$. One kilogram of nitrogen can produce

$$(17/14) \times 1 \text{ kg} = 1.2 \text{ kg ammonia.}$$

- Using 1 kg nitrogen to produce 1.2 kg ammonia requires 0.2 kg (200 g) hydrogen. The plant operator needs to double the amount of hydrogen on hand.

In practice, ammonia plants use a large excess of hydrogen to ensure that all nitrogen activated is made into ammonia.

See Exercises 72–74.

1.6 Naming Compounds

Nomenclature The system of naming compounds.

Communicating in chemistry is facilitated by a chemical vocabulary to name compounds. Naming compounds is called **nomenclature** from the Latin *nomen* (name) and *calare* (to call). There are millions of known chemical substances, so naming them all would require hundreds of pages. Only a few basic rules are given here.

Substances that have been known for a long time often have common names. Water (H_2O) and ammonia (NH_3) are two examples. The rules for other substances divide compounds into two broad categories: organic—those compounds containing carbon, hydrogen and often other elements such as oxygen, nitrogen, or sulfur. All other compounds are classified as inorganic. In this section, rules for naming inorganic compounds are given. It is convenient to divide inorganic compounds into three types: ionic compounds, acids, and binary molecular compounds.

Ionic compounds most often consist of a combination of two different ions. We have encountered a few ionic substances in the examples given earlier. The overall compound is neutral and consists of a positive ion, called the **cation,** and a negative ion, called the **anion.**

Cation An atom or group of atoms with a positive charge.

Anion An atom or group of atoms with a negative charge.

Rule	Examples
Cations:	Na^+ sodium ion
a. Cations formed from metal ions have the same name as the metal.	Zn^{2+} zinc ion
	Al^{3+} aluminum ion
b. Some metal ions can form cations with different charges. The newer method for naming these encloses the charge as a	Fe^{2+} iron (II) ion
	Fe^{3+} iron (III) ion

Rule	Examples
Roman numeral in parentheses. An older nomenclature takes the Latin name for the metal and adds *-ous* for the lower charge and *-ic* for the higher charge.	Cu^{2+} copper (II) ion Co^{3+} copper (III) ion Fe^{2+} ferr*ous* ion Fe^{3+} ferr*ic* ion Cu^{2+} cupr*ous* (II) ion Co^{3+} cuper*ic* (III) ion
c. Cations formed from nonmetal atoms have names that end in *-ium*	NH_4^+ ammon*ium* H_3O^+ hydron*ium*
Anions a. Monatomic anions are named by replacing the end of the element name with *-ide*. A few simple polyatomic anions have common names that end in *-ide*.	Cl^- chlor*ide* O^{2-} ox*ide* ion N^{3-} nitr*ide* ion OH^- hydrox*ide* CN^- cyan*ide*
b. Polyatomic anions often contain oxygen. The ending *-ate* is used for the most common ion and *-ite* for the ion with the same charge but a different number of oxygen atoms. Some elements form four different ions with oxygen. These use prefixes: The ion with one more oxygen than the *-ate* ion adds *per* to the ion ending in *-ate* and the ion with one fewer oxygen atom than the *-ite* ion adds the prefix *hypo-*.	NO_3^- nitr*ate* NO_2^- nitr*ite* ClO_3^- chlor*ate* ClO_2^- chlor*ite* ClO_4^- *per*chlor*ate* ClO^- *hypo*chlor*ite*
c. Anions formed from oxygen-containing polyatomic ions and H^+ are named by adding the word hydrogen or dihydrogen to the name of the ion. The older nomenclature adds the prefix *bi-* to the name of the anion. Note that addition of H^+ reduces the ion charge by one.	HCO_3^- hydrogen carbon*ate* $H_2PO_4^-$ dihydrogen phosph*ate* HCO_3^- *bi*carbonate HPO_4^{2-} *bi*phosphate
Ionic Compounds Compounds consisting of two ions are named by naming the cation followed by the anion.	$CaCO_3$ calcium carbonate $MgHPO_4$ magnesium biphosphate
Acids For naming purposes an acid is a substance that dissolves in water to yield hydrogen ions (H^+). Acids containing anions ending in *-ide* have a prefix hydro- and the *-ide* ending is changed to *-ic*.	HCl *hydro*chlor*ic* acid HF *hydro*fluor*ic* acid
Acids containing anions with names ending in *-ate* or *-ite* have associated acids with an *-ic* or *-ous* respectively	HNO_3 nitr*ic* acid HNO_2 nitr*ous* acid $HClO$ *hypo*chlor*ous* acid.
Binary Compounds These are named by giving the name of the element to the lower left of the periodic table first, and the name of the second element is given an *-ide* ending. Greek prefixes are used to indicate the number of atoms of each element (except that mono is not used). When the prefix ends with an *a* or *o* and the second element begins with a vowel, the *a* or *o* is often dropped.	CO carbon monox*ide* CO_2 carbon diox*ide* N_2O_4 dinitrogen tetrox*ide* N_2O_5 dinitrogen pentox*ide*

1.7   Energy

Energy plays an important role in chemistry. Energy is involved in doing work, heating an object, lighting a space, or powering a laptop. Four types of energy are important in chemical interactions: kinetic, potential, electromagnetic, and electrical.

Kinetic energy ($^1/_2 mv^2$) Energy due to the motion of an object; dependent on the mass of the object and the square of its velocity.

Kinetic energy is associated with the motion of an object. An object of mass m traveling with a velocity v has kinetic energy:

(1.29)
$$\text{kinetic energy} = 1/2 \, mv^2$$

A ball rolling down a hill has kinetic energy. Atoms and molecules also have kinetic energy due to their motion.

WORKED EXAMPLE 1.16   *Fastball Energy*

The first baseball pitcher to pitch at more than 100 mph was Nolan Ryan on September 7, 1974, when his pitch was officially clocked at 100.8 mph. (This record is now approaching 105 mph.) What is the kinetic energy (in joules) of a fastball?

Data
Mass of a baseball = 142.5 g.

Plan
■ 1 J = 1 kg · m²/s². Convert mph to m/s; convert mass to kg.
■ Use Equation (1.29), kinetic energy = 1/2 mv^2.

Implementation
■ Unit analysis: distance: (mi) × (ft/mi) × (in/ft) × (cm/in) × (m/cm) × (km/m) = km
       time: h × (min/h) × (s/min) = s
       1000 g = 1 kg

$$100.8 \text{ mi} \times \frac{5280 \text{ ft}}{\text{mi}} \times \frac{12 \text{ in}}{\text{ft}} \times \frac{2.54 \text{ cm}}{\text{in}} \times \frac{1 \text{ m}}{100 \text{ cm}} = 162{,}200 \text{ m}$$

$$1 \text{ h} \times \frac{60 \text{ min}}{\text{h}} \times \frac{60 \text{ s}}{\text{min}} = 3600 \text{ s}$$

■ Kinetic energy = 1/2(0.1425 kg)(162,200 m/3600 s)² = 144.6 J

Being hit with 144 J of energy would really hurt. That is why batters wear helmets!

See Exercises 77 – 78.

Potential energy Energy due to position or composition.

Coulombic energy $E = (9.00 \times 10^{18} \text{ J} \cdot \text{m/C}^2) \times \frac{q_1 q_2}{r}$ where E is the energy of interaction between a pair of ions, expressed in joules; r is the distance between the ion centers in nm; and q_1 and q_2 are the numerical ion charges.

Potential energy is associated with position. A boulder on the top of a mountain has potential energy in the gravitation field of the earth. An object of mass m located at a height h has potential energy:

(1.30)
$$\text{potential energy} = mgh$$

where g is the gravitational constant, 9.81 m/s². Gravitational potential energy plays little direct role in chemistry. In contrast, another form of potential energy, the attraction of oppositely charged objects and repulsion of like charged objects, called the **coulombic energy**, plays a central role. A particle of charge q_1 separated from a second particle with charge q_2 by a distance r has a potential energy of

(1.31)
$$\text{Coulomb potential} = -k \frac{q_1 q_2}{r}$$

where k (a constant) = $1/4 \, \pi\varepsilon_0$; ε_0 (8.854×10^{-12} C²/J · m) is a measure of how effectively one charge penetrates to another and is called the vacuum permittivity. Coulombic interactions are arguably *the* dominant interactions in chemistry.

WORKED EXAMPLE 1.17   *The Potential of Books*

A typical textbook weighs 1.5 kg. A typical desktop is located 30. in. off the floor. What is the potential energy (in joules) of the book on the desktop?

Plan

- Convert in. to m.
- Use Equation (1.30), potential energy = *mgh*.

Implementation

- Unit analysis: in $\times$ (cm/in.) $\times$ (m/cm) = m
  30. in. $\times$ (2.54 cm/in.) $\times$ (1 m/100 cm) = 0.76 m
- Potential energy = (1.5 kg) $\times$ (9.81 m/s^2)(0.76 m) = 11 J

The fastball has more than 10 times the energy of a textbook sitting on the desktop.

See Exercises 79–80.

WORKED EXAMPLE 1.18   *Charged Energy*

The charge on an electron is very small, 1.60218×10^{-19} C. Nonetheless, the energy in the electron's field can be considerable if the distance is very small. Determine the repulsive energy for two electrons held at arm's length (about 2 m) and two electrons at a typical bonding distance of 150 pm.

Plan

- Use Equation (1.31), potential energy $= -k\dfrac{q_1 q_2}{r}$; $k = \frac{1}{4}\pi\varepsilon_0$.

Implementation

- Energy at arm's length $= -\dfrac{1}{4\pi \cdot 8.854 \times 10^{-12}\ \text{C}^2/\text{J}\cdot\text{m}}$

  $\times \dfrac{(1.60218 \times 10^{-19}\ \text{C})^2}{2\ \text{m}} = -1.2 \times 10^{-28}\ \text{J}$

  Energy at bonding distance $= -1.54 \times 10^{-18}\ \text{J}$

The negative sign indicates that the energy is repulsive. At bonding distance, the interaction energy is 10 orders of magnitude greater than at arm's length.

See Exercises 81–84.

Electromagnetic energy A form of the energy carried by oscillating, mutually perpendicular electric and magnetic fields.

Electromagnetic energy consists of oscillating, mutually perpendicular electric and magnetic fields. Electromagnetic radiation includes visible light, infrared radiation, ultraviolet radiation, X rays, radio waves, and other forms. Electromagnetic waves carry energy through space and interact both with charged particles and with the magnetic fields created by charged particles in atoms and molecules. Much of what is known about atoms and molecules today is a result of observing interactions with electromagnetic radiation.

WORKED EXAMPLE 1.19   *Solar Power*

The average intensity of sunlight at mid-latitude on a clear day is about 1 kW/m^2 at noon. What is the maximum energy (in joules) that a 12 m^2 solar collector can collect in 1 h?

Plan

■ Convert W to J.

■ Multiply by area and time.

Implementation

■ 1 W = 1 J/s; 1 kW = 1000 W

■ $\dfrac{100\ \text{W}}{\text{m}^2} \times \dfrac{1\ \text{J/s}}{\text{W}} \times 12\ \text{m}^2 \times 1\ \text{h} \times \dfrac{60\ \text{min}}{\text{h}} \times \dfrac{60\ \text{s}}{\text{min}} = 4.3 \times 10^7\ \text{J}$

The sun supplies a fantastic amount of energy, which explains the attractiveness of solar power.

See Exercises 85 – 86.

Principle: Energy Is Conserved

A substance can have a combination of kinetic and potential energy, and these two forms of energy can be interconverted. For example, a boulder at the top of a mountain has a large potential energy. As it rolls down the slope, it acquires a large kinetic energy. When it comes to rest in the valley, the large kinetic energy of the boulder is transformed into kinetic energy of the particles that the boulder ran into. This kinetic energy is then dissipated as heat. Similarly, a pendulum at the top of its swing has a large potential energy. This potential is converted to kinetic energy as the pendulum swings to the bottom of its arc, then reverts to potential energy at the top of the opposite side. In the absence of friction, the pendulum goes forever.

One form of energy can be converted into another form, but the total energy remains constant. This principle is known as the **conservation of energy:**

> Energy can be interconverted among its various forms, but the total energy remains constant.

Conservation of energy Energy can be converted from one form to another but can be neither created nor destroyed so the total energy remains constant.

Conservation of energy is an important principle and is used to probe interactions between atoms and molecules.

1.8   Concept Connections

The myriad concepts in chemistry are interrelated in numerous ways. A useful tool for visualizing these relationships is a **concept map.** Worked Example 1.20 illustrates the construction of a concept map for seven concepts or ideas presented in this chapter.

Concept map A visual representation of the relationships or connections between concepts.

WORKED EXAMPLE 1.20   *Relationships, Relationships . . .*

A concept map is a visual representation of the connections between concepts. It consists of terms placed in boxes or ovals (called nodes) connected by arrows. The arrows are labeled with a short phrase or word providing the link between the concepts. Shown on page 31 is a concept map containing the following nodes: Atom, Atomic theory, Atomic/molecular density, Avogadro's law, Gas law $PV = nRT$, and Mole. A complete map has every concept connected to at least one other concept and all terms interrelated. That is, the map does not separate into two or more unconnected parts. The map shown is not complete. Add links to the map to make it more complete. Note that a concept map is not unique, as many maps are possible with the same set of nodes.

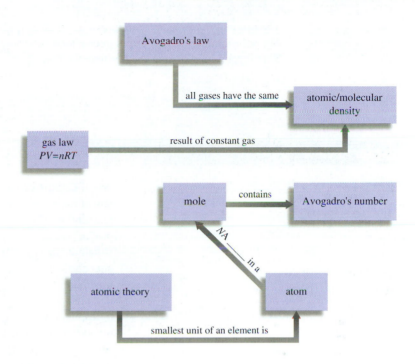

Plan
- Identify and label relationships.

Implementation

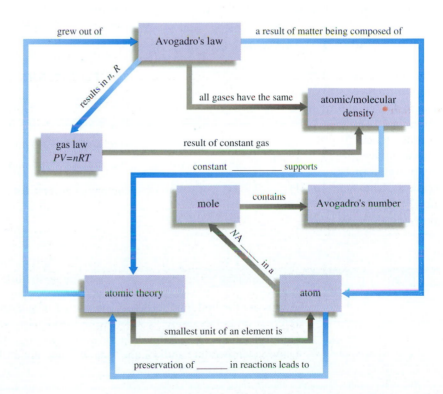

The original map separated into two disconnected sets of nodes, making it incomplete. The solution map has all terms interrelated; several terms are related in both directions. This second map is much richer than the first.

See Exercises 87–88.

Checklist for Review

KEY TERMS

atom (p. 2)
compound (p. 2)
molecule (p. 2)
mole, mol (p. 5)
Avogadro's number, N_A (p. 5)
atomic mass unit, amu (p. 6)
atomic mass (p. 6)
molar mass (p. 6)
Boyle's law (p. 10)
Charles's law (p. 11)
Avogadro's law (p. 13)
molar density (p. 13)
mass density (p. 13)
gas constant, R (p. 13)
ideal gas law (p. 14)
ideal gas (p. 14)
standard temperature and pressure (STP) (p. 14)
torr (p. 15)
empirical formula (p. 19)
ion (p. 19)
ball-and-stick model (p. 19)
structural formula (p. 19)
condensed structural formula (p. 19)
line formula (p. 19)
space-filling model (p. 20)
density isosurface (p. 21)
density potential (p. 21)
molecular stoichiometry (p. 21)
reaction (p. 21)
reaction stoichiometry (p. 21)

law of definite proportions (p. 21)
law of multiple proportions (p. 22)
reactant (p. 22)
product (p. 22)
stoichiometry (p. 22)
aqueous solution (p. 23)
precipitation reaction (p. 23)
net ionic equation (p. 23)
molecular equation (p. 23)
total ionic equation (p. 23)
spectator ion (p. 23)
limiting reactant (p. 25)
nomenclature (p. 26)
cation (p. 26)
anion (p. 26)
kinetic energy (p. 28)
potential energy (p. 28)
coulombic energy (p. 28)
electromagnetic energy (p. 29)
conservation of energy (p. 30)
concept map (p. 30)

KEY EQUATIONS

$PV = nRT$

$P_{atm} = \rho g h$

kinetic energy $= \frac{1}{2} m v^2$

potential energy $= mgh$

coulombic energy $= -k \dfrac{q_1 q_2}{r}$

Chapter Summary

Atoms are the building blocks for the myriad of substances found all around us. This is referred to as the atomistic view of the world. The atomistic view has its roots in the study of gases. The relationship between pressure and volume, Boyle's law, and that between temperature and volume, Charles's law, are unified in the universal gas law.

There are several ways of representing molecules, compounds, and solids. The empirical formula or molecular formula uses the symbol for the element and denotes the number of atoms of each element as subscripts. This very compact notation sometimes leads to ambiguity about the arrangement of atoms in the substance. This arrangement can be shown with a ball-and stick model, a structural formula, or a simple line formula. Any of these can be used in a reaction formula. A balanced reaction formula is based on the notion that matter is conserved. Hence, the number of atoms of each substance involved in a reaction must be the same among the reactants as among the products.

KEY IDEA

The building blocks for matter are atoms.

CONCEPTS YOU SHOULD UNDERSTAND

- Avogadro's hypothesis
- atomistic composition of matter
- Boyle's law
- Charles's law
- ideal gas law
- law of definite proportions
- law of multiple proportions
- limiting reactant
- conservation of energy

OPERATIONAL SKILLS

- Generate a macroscopic picture of gas-phase reactions (Worked Example 1.1)
- Relate atomic dimensions to macroscopic dimensions (Worked Example 1.2)
- Convert among mass, moles, and atoms. Convert density between atomic and molecular units (Worked Examples 1.3 and 1.4)
- Use the gas laws (Worked Example 1.5 through 1.7)
- Calculate and use mass ratios and moles (Worked Example 1.8)
- Relate density and pressure (Worked Example 1.9)
- Calculate buoyancy (Worked Example 1.10 and 1.11)
- Decode formulas (Worked Example 1.12)
- Balance reactions (Worked Example 1.13)
- Identify spectator ions (Worked Example 1.14 and 1.15)
- Identify limiting reactant (Worked Example 1.15)
- Calculate energy: kinetic (Worked Example 1.16), potential (Worked Example 1.17), coulombic (Worked Example 1.18), and electromagnetic (Worked Example 1.19)
- Generate concept maps of terms (Worked Example 1.20)

Exercises

A blue exercise number indicates that the answer to that exercise appears at the back of the book.

■ SKILL BUILDING EXERCISES

1.2 The Mole

1. What is Avogadro's number? What is the relationship between Avogadro's number and the mole?

2. Complete the following sentence: One mole of copper (Cu) contains _____ atoms and its mass is _____ grams.

3. The molecular mass of water is 18 g/mol.
   a. How many molecules are in a cup (250 mL) of water? (The density of water is 1 g/mL at 0 °C.)
   b. Estimate the number of cups of water in the world's oceans. The following data may be useful: The earth's radius is 4000 mi, three-fourths of the earth's surface is covered with oceans, the average ocean depth is 5 km, 1 mi is 8/5 of 1 km. (Assume a spherical earth.)

c. Comment on the relationship between the number of cups of water in the earth's oceans and the number of molecules in a cup of water. Which is larger? By what factor?

4. The aim of this exercise is to illustrate the enormity of Avogadro's number. Estimate how deeply one mole of M&M candies would cover the earth. The following data may be useful: The volume of an M&M candy is 1 mL and the radius of the earth is 4000 miles. (Assume a spherical earth.)

5. Atomic scale dimensions are often given in ångstroms (Å), pm, or nm. 1 Å = 10^{-8} cm, 1 pm = 10^{-12} m, and 1 nm = 10^{-9} m. The atomic radius of nickel (Ni) is 1.246 Å. What is the radius in nm? In pm?

6. Chromium (Cr) is used to chrome plate plumbing fixtures. The atomic radius of chromium is 185 pm. By what factor must a Cr atom be magnified to be the size of a baseball (2.9 in. diameter)?

7. The radius of a silver (Ag) atom is 134 pm. If a mole of silver atoms are laid in a line, just touching, how far does the line stretch?

8. The radius of an aluminum (Al) atom is 118 pm. How many Al atoms, just touching, make a line 1 in. long? (2.54 cm = 1 in.) (a) How many are required to cover a square 1 cm on a side if the atoms pack in a square array? (b) How many cover a square if they pack in an hexagonal array?

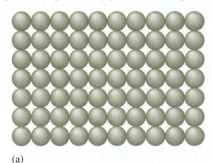

(a)

(b)

9. Give the units of the following:
a. mass density
b. molecular density
c. atomic density

10. What information do you need to convert from
a. mass density to molecular density?
b. molecular density to atomic density?
c. atomic density to mass density?

11. A tablet of a certain pain reliever has a mass of 0.41 g and contains 200 mg of an active ingredient. A bottle contains 80 tablets. What is the mass of the tablets in the container? What is the mass of a mole of tablets? Compare this to the mass of the earth: 6×10^{24} kg.

12. Tungsten is commonly used as the filament in light bulbs. The density of W is 19.3 g/cm³. The mass of a piece of W is 1.26 kg. What is the volume of the piece? The filament in a light bulb has a 1-mm diameter. How many feet of filament can be made from this piece?

13. The potassium content of an apple is 159 mg. How many moles of potassium are in the apple?

14. To protect Fe from corrosion, iron sheets are often coated with zinc. The coating requires 2.0×10^{15} atoms/cm². How many grams of zinc are required to coat two sides of a 10 cm × 10 cm sheet?

1.2 Atoms as Building Blocks: Atomic Theory and Avogadro's Hypothesis

15. Using Gay-Lussac's observation that the reaction of one volume of hydrogen with one volume of chlorine gives two volumes of hydrogen chloride, and Avogadro's hypothesis
a. deduce the ratio of hydrogen to chlorine in hydrogen chloride.
b. determine the molecular formula for elemental hydrogen and chlorine.

16. What volume of hydrogen gas is required to produce 2.5 mL of hydrogen bromide gas? (Molecular formula: HBr.)

17. It is observed that 1000 mL of hydrogen gas reacts with 1000 mL of fluorine gas to produce 2000 mL of hydrogen fluoride gas. What is the molecular formula of hydrogen fluoride? What are the molecular formulae of hydrogen and fluorine gases?

18. Hydrogen chloride (HCl) and ammonia (NH_3) combine to form a solid. It is observed that 1000 mL of gaseous HCl reacts completely with 1000 mL of gaseous NH_3. What is the molecular formula of the solid produced in this reaction?

1.3 Pressure-Volume Relationship: Boyle's Law

19. A helium tank is to be used to fill balloons for a party. The tank is 1 m high, it has a cross-sectional area of 450 cm², and the pressure is 40 atm. How many 30-cm diameter balloons will this tank fill? Assume temperature is constant, and the stretchiness of the balloon adds no pressure.

20. A spaceship's maneuvering engine burns hydrogen and oxygen. Hydrogen is to be delivered to this engine at 1.1 atm pressure and 1500 L is required for a planned maneuver. What pressure must the holding tank be charged with if its volume is 40 L? Assume constant temperature.

21. A natural-gas-burning car fills its piston with gas at 0.85 atm. Prior to ignition, the piston volume is reduced to 20% of its original volume. What is the pressure prior to ignition? Assume the temperature does not change until after ignition.

22. A racing bicycle requires a pressure of about 5 atm in its tire. ☀ If a bicycle pump has the same volume as the bicycle tire, how many times must the cyclist stroke the pump to properly pressurize the tire? Assume temperature is constant and the tire starts empty.

1.3 Temperature-Volume Relationship: Charles's Law

23. A car tire is inflated to 22 lb/in^2 on a winter day when the ☀ temperature is 0 °C. After driving at high speed, the temperature in the tire increases to 50 °C. What is the tire pressure? Assume that the tire does not expand.

24. To demonstrate the gas laws, a balloon is filled with helium at room temperature, 23 °C. This balloon is then immersed in liquid nitrogen at 77 K. What is the relative volume of the balloon at this lower temperature? (Relative volume is the ratio of the final volume to the initial volume. The pressure is 1 atm throughout.)

25. A spherical balloon is filled to a 5-cm radius at 23 °C. The balloon is then taken outside on a winter day when the temperature is 0 °C. What is the diameter of the balloon at the colder outdoor temperature?

26. You want to take a hot-air balloon ride on a summer day ☀ when the temperature is 25 °C. To what temperature must you heat the air inside a 40-ft diameter balloon to lift two people (average weight 150 lb) and the basket and balloon, which together weigh 100 lb?

1.3 Combined Gas Law

27. What portion of a mole of hydrogen gas is contained in a 2-L vessel at STP?

28. How many moles of helium are needed to fill a balloon of volume 90 L at 22 °C and 1 atm pressure?

29. Predict whether the volume of a gas decreases, increases, or stays the same in each of the following scenarios:
 a. Pressure increases from 200 torr to 300 torr, and temperature increases from 200 °C to 300 °C
 b. Pressure decreases from 760 torr to 350 torr, and temperature decreases from 250 °C to −23 °C
 c. Pressure decreases from 2 atm to 1 atm, and temperature increases from 300 K to 600 K

30. A car tire is inflated at 22 °C to a pressure of 1.8 atm. After ☀ the car has been driven for several hours, the volume of the tire expands from 7.2 L to 7.8 L and the pressure increases to 1.9 atm. What is the tire's temperature?

31. Sulfur dioxide (SO_2) is generated from combustion of coal. At 20 °C and 1 atm pressure, the density of SO_2 is 2.66 g/L.
 a. How many molecules of SO_2 are in 1 L at 20 °C and 1 atm?
 b. How many atoms of oxygen are in 1 L of SO_2 at 20 °C and 1 atm pressure?

32. One milliliter of carbon monoxide gas contains 8.92×10^{-5} moles of carbon monoxide. One milliliter of water contains 5.55×10^{-2} moles of water. Which contains more atoms?

33. The volume of a gas is 200 L at 12 °C and 750 mm Hg. What volume will it occupy at 40 °C and 720 mm Hg? Assume ideal behavior.

34. Does the volume of a gas increase, decrease, or stay the same when the pressure decreases from 760 to 350 mm Hg and the temperature decreases from 250 °C to −50 °C?

35. A weather balloon is filled with gas at 30 °C at sea level ☀ (1 atm pressure) and rises to an altitude at which the temperature is −20 °C and the pressure is 0.60 atm. Does the volume increase, decrease, or stay the same?

36. A motorist inflates a car tire to 29 psig (pounds per square ☀ inch gauge) on a cold winter day when the temperature is −10 °C (psig refers to the pressure greater than 1 atm, which is 14.7 psi, pounds per square inch, at sea level). After driving for some time, the tire warms to 30 °C. Assuming that the tire volume remains constant, what is the pressure after warming? What is the pressure if the tire radius expands by 1%? (Assume the tire is a doughnut of circular cross section.)

37. Ozone, O_3, consists of three atoms of oxygen. Its density is ☀ 2.144 g/L. Calculate the molecular density and the atomic density of ozone at STP.

38. There are two oxides of carbon: carbon dioxide (CO_2; ☀ molecular mass 44.01 g/mol) and carbon monoxide (CO; molecular mass 28.01 g/mol). The densities of these gases at 1 atm pressure and 0 °C are 1.977 g/L and 1.250 g/L, respectively. Determine the molecular density of each of these gases. Compare these densities with the density predicted by the ideal gas law.

39. Calculate the volume of one mole of a gas at 1 atm pressure and 0 °C.

40. Nitrogen forms several oxides. At STP, what are the densities of NO_2, N_2O_4, and N_2O_5? How many volumes of oxygen are required to convert one volume of NO_2 into N_2O_5?

1.3 Gas Laws: Applications

41. Suppose that the liquid in the open container in a barometer is ☀ water. To what height, in feet, does the water rise at sea level?

42. If the average molecular weight of air is 29 g/mol, how many ☀ molecules are in the earth's atmosphere?

43. Each breath of air is about $^3/_4$ pint of air. How many ☀ molecules are in each breath? (Use $P = 1$ atm, $T = 273$ K.)

44. How likely is it that you are breathing a molecule of nitrogen ☀ contained in Aristotle's last breath? (Air is about 80% nitrogen.)

45. Balloons are often used to carry instruments into the strato- ☀ sphere to monitor conditions there. One such balloon has a volume of 45,000 m^3. The balloon is to carry its payload of 50 kg (including the mass of the balloon) to an altitude of 30 KM where the temperature is −46 °C and the pressure 8.36 torr. By how many degrees must the sun heat the air inside the balloon for it to attain this altitude? (Air is 80% N_2 and 20% O_2).

46. A hawk has a wing area of about 1.5 ft^2 and weighs about ☀ 1 lb. What pressure differential is required to support the hawk? Explain how a hawk rides a thermal.

47. Which is denser at 0 °C, ice or water? How do you know?

48. The density of liquid nitrogen is 809 g/L at 77 K (the boiling point of nitrogen). The density of nitrogen gas at 77 K is 4.566 g/L at 1 atm pressure. A student decides to conduct an experiment and puts 75 mL of liquid nitrogen into a 2-L soda bottle and caps it tightly. The burst pressure of the soda bottle is 5 atm. Will the bottle explode? (Do not try this at home!)

1.4 Molecules and Representations

49. The molecule benzene, a carcinogen, has the following structure:

How many carbon atoms and how many hydrogen atoms are in one benzene molecule?

50. The molecule cyclohexane has the structure shown below. How many carbon atoms and how many hydrogen atoms are contained in one cyclohexane molecule?

51. The molecule benzoyl peroxide is used as an initiator for polymer formation. It has the structure shown below. How many carbon atoms and how many hydrogen atoms are in one benzoyl peroxide molecule? What other elements are found in benzoyl peroxide?

52. The molecule caffeine is found in many beverages. Caffeine has the structure shown below. How many carbon atoms and how many hydrogen atoms are in one caffeine molecule? What other elements are found in caffeine?

1.5 Stoichiometry: Balancing Reactions

53. Calculate the molar mass of CCl_4. How many grams of Cl_2 are required to produce 1 kg of CCl_4?

54. Propane, C_3H_8, is a component of natural gas. What is the molar mass of propane? How many grams of carbon are in 10 kg of propane?

55. What mass of hydrogen gas is required to produce 2.5 mg of hydrogen bromide gas (molecular formula, HBr)?

56. Balance the following reaction:

$$K + Cl_2 + O_2 \longrightarrow KClO_3$$

57. Balance the following reaction:

$$Na + O_2 \longrightarrow Na_2O$$

58. Identify which of the following reactions are not balanced. Balance them.
    a. $Ag_2O + Cu \rightarrow CuO + Ag$
    b. $H_2 + Br_2 \rightarrow HBr$

c. $S + O_2 \rightarrow SO_2$
d. $KClO_3 \rightarrow KCl + O_2$

59. Ammonia, NH_3, is produced by combining hydrogen, H_2, with nitrogen, N_2. Balance the reaction for the production of ammonia.

60. Marble consists of $CaCO_3$ and results from reaction of CaO with CO_2. This reaction is responsible for storing large amounts of atmospheric CO_2.
    a. Write the balanced reaction for formation of $CaCO_3$.
    b. How many pounds of CO_2 are stored in a ton of marble?

1.5 Stoichiometry: Reaction Proportions

61. Ozone, O_3, is an unstable molecule that decomposes to form O_2. How many volumes of O_2 are produced from one volume of O_3? Make macroscopic and molecular-level pictures for this reaction. Write the balanced chemical reaction.

62. Propane, C_3H_8, is a constituent of natural gas. Combustion of propane produces carbon dioxide, CO_2, and water, H_2O. What mass of oxygen is required to burn 1 lb of propane? How many grams of CO_2 are produced?

63. Nitrogen and oxygen can combine to form several oxides of nitrogen. Use the following data to determine the molecular formula of each.

	Volume of Nitrogen	Volume of Oxygen	Volume of Product	Name of Product
a.	5.60 L	2.80 L	5.60 L	nitrous oxide
b.	6.30 L	6.30 L	12.60 L	nitric oxide
c.	4.52 L	9.04 L	9.04 L	nitrogen dioxide
d.	5.30 L	10.60 L	5.30 L	dinitrogen tetroxide
e.	3.60 L	9.00 L	3.60 L	dinitrogen pentoxide

64. Aluminum sulfate is a component of fertilizer used on acid-loving plants. Calculate the molar mass of $Al_2(SO_4)_3$. How many grams of sulfur are in 50 kg of $Al_2(SO_4)_3$?

65. One of the factors that contributes to the stability of aluminum objects is the stability of the oxide coat, formula Al_2O_3, on aluminum. How many grams of oxygen are required to completely oxidize 1 kg of aluminum? An aluminum object is found to gain 4.7 g per kg. What fraction of the aluminum has oxidized?

66. Sulfuric acid, H_2SO_4, is one of the top chemicals produced each year. One method for production of sulfuric acid is to burn sulfur (S) in air producing SO_2. SO_2 further oxidizes to SO_3, and SO_3 dissolves in water to produce H_2SO_4. Balance the reactions involved in the production of sulfuric acid.

67. Octane, C_8H_{18}, is a constituent of gasoline. Combustion of octane, C_8H_{18}, in gasoline produces carbon dioxide, CO_2, and water, H_2O. What mass of oxygen is required to burn 1 lb of octane? How many grams of CO_2 are produced?

68. Marble is composed of $CaCO_3$. $CaCO_3$ can be made from the reaction of CaO with CO_2. How many pounds of CO_2 are locked up in 50 lb of marble?

69. One step in the production of iron is roasting the oxide, Fe_2O_3, with coke, essentially elemental carbon. In the process, carbon is first oxidized to CO by oxygen in the air.

The CO subsequently removes oxygen from Fe_2O_3 and is converted to CO_2. How many pounds of carbon are required to produce 1 ton of iron by this process? How many pounds of carbon dioxide are produced?

70. Label the following reactions as molecular, ionic, or net ionic reactions. Write net ionic and molecular reactions for any total ionic reactions listed.
    a. $2HCl + Zn \rightarrow H_2 + ZnCl_2$
    b. $Pb^{2+} + 2I^- \rightarrow PbI_2$
    c. $2Ag^+ + 2NO_3^- + 2Na^+ + CO_3^{2-} \rightarrow Ag_2CO_3 + 2Na^+ + 2NO_3^-$
    d. $H_2SO_4 + H_2O \rightarrow HSO_4^- + H_3O^+$

71. Label the following reactions as molecular, ionic, or net ionic reactions. Write net ionic and molecular reactions for any total ionic reactions listed.
    a. $NH_3 + H_2O \rightarrow NH_4^+ + OH^-$
    b. $2H^+ + 2Br^- + Ba^{2+} + 2OH^- \rightarrow 2H_2O + 2Br^- + Ba^{2+}$
    c. $2Na + 2H_2O \rightarrow H_2 + 2OH^- + 2Na^+$
    d. $H_2SO_4 + Ni^{2+} + 2Cl^- \rightarrow NiSO_4 + 2H^+ + 2Cl^-$

72. One component of wet rust is $Fe(OH)_2$. How many grams of oxygen, O_2, are in 1 kg of $Fe(OH)_2$?

73. Ammonium nitrate, NH_4NO_3, can be made from the acid–base reaction between ammonia, NH_3, and nitric acid, HNO_3. How many pounds of nitric acid are required to produce 50 lb of ammonium nitrate?

74. One step in the production of lead from its ore is production of lead oxide, PbO. Elemental lead is produced from PbO by reduction with coke, essentially elemental carbon. In addition to metallic lead, this reaction produces carbon dioxide. How many pounds of carbon are required to produce 1 ton of lead by this process? How many pounds of carbon dioxide are produced?

75. In an experiment, Gay-Lussac showed that one volume of nitrogen plus one volume of oxygen produces two volumes of nitric oxide. Determine the ratio of nitrogen to oxygen in nitric oxide and determine the formula for elemental nitrogen and oxygen. Draw a molecular picture of this reaction and write the chemical formula of nitric oxide.

76. One volume of nitrogen gas combines with three volumes of hydrogen to produce two volumes of ammonia. Determine the ratio of nitrogen to hydrogen in ammonia and the formula for elemental hydrogen and nitrogen. Draw a molecular-level picture of this reaction and write the chemical formula for ammonia.

1.7 Energy

77. A mosquito is a very small and annoying insect that cannot fly very fast. Determine the kinetic energy (in joules) of a mosquito flying at 3 mph (a typical mosquito weighs 1 mg). Although a mosquito does not fly very fast, its wings beat very rapidly giving the hum heard at its approach. A typical mosquito beats its wings 1000 times per second. A typical wing span of a mosquito is 6 mm. The wing tip has a mass of 1 μg. Determine the kinetic energy of the wing tip. What portion of the expended energy propels the mosquito forward and what portion keeps it aloft?

78. An electron has a resting mass of 9.1094×10^{-28} g. What is the kinetic energy (in joules) of an electron traveling at half the speed of light?

79. Stair walking is considered to be a good form of exercise. Determine the potential energy gain (in joules) from walking up a typical flight of stairs for an adult weighing 140 lb. Where does the energy come from? How many flights of stairs must the typical adult climb to expend 100 Cal? A typical flight of steps is 3 m high. 1 Cal (food calorie) = 1000 cal; 1 cal = 4.184 J.

80. New York is also known as the "Big Apple." What is the potential energy of an apple sitting on the top of the Empire State Building in New York? Height of the Empire State Building = 1250 ft; mass of an apple = 250 g.

81. The electron and the proton in a hydrogen atom are 52.9 pm apart. What is the energy of interaction between the electron and the proton?

82. How far apart do an electron and a proton have to be so that the energy of the coulombic attraction is the same as the energy of the proton in the earth's gravitational field when the proton is 1 m above the earth's surface? Is gravity an important source of energy on an atomic scale?

83. The earth carries a net charge of -4.3×10^5 C. The force due to this charge is the same as if all the charge were concentrated at the center of the earth. How much charge would you have to place on a 10-g mass 1 m above the earth's surface for the coulombic and gravitational energies to be equal?

84. Two balloons are rubbed against a wool object, charging them with 0.01 μC. The balloons are tied together with a 10-in. string and are held 6 in. apart due to the like charges. What is the mass of each balloon?

85. A laser pointer delivers 0.1 mW in a 0.9-mm diameter beam. How much energy falls on the screen that the pointer is aimed at in 1 min? Assume that the energy is equally divided between the electric and magnetic fields. How much energy is in the electric field?

86. Microwave ovens are used to cook food. A typical microwave oven is rated at 625 W. If the energy is spread uniformly throughout the 1 ft³ cavity, how much is incident on a 50-in³ potato that you want to bake?

■ CONCEPTUAL EXERCISES

87. Make a diagram showing relationships among the following terms: Avogadro's law, mole, Avogadro's number, amu, and atomic mass. All terms must be interconnected.

88. Make a diagram showing relationships among the following terms: density, atomic mass, mole, molar density, and atomic density. All terms must be interconnected.

✷ APPLIED EXERCISES

89. Chocolate napoleons, beef Wellington, and baked brie are examples of foods based on puff pastry. The puff is a result of a phase transformation of the water in the pastry, a process known as steam leavening. The vapor pressure of water at room temperature is about 20 torr. Puff pastries are usually baked at about 400 °F. What volume is occupied by one mole of liquid water at room temperature? What volume is occupied by steam at 400 °F? Give an explanation for why puff pastries are baked at 400 °F rather than 212 °F.

90. One of Dalton's proposals is that the atoms of an element are identical. We now know this statement is not quite true. While the chemical characteristics of all atoms of an element are identical, some elements have atoms that differ in atomic mass. For example, most hydrogen atoms have an atomic mass of 1 amu, but a few have a mass of 2 amu. Those with atomic mass 2 are commonly referred to as deuterium and given the symbol D. Substitution of D for H in water preserves the properties of water, including the crystal structure of ice. Will a D_2O ice cube float in a glass of H_2O water?

91. The buoyant force is defined as the difference between the weight of an object and the weight of fluid it displaces. Use the following data to determine the buoyant force per liter of a hydrogen-filled balloon and the same balloon filled with helium. Air is approximately 78% nitrogen and 22% oxygen by volume. The molecular mass of nitrogen is 28 g/mol and that of oxygen is 32 g/mol. The molecular mass of hydrogen is 2 g/mol and that of helium is 4 g/mol.

92. A floating object has zero net buoyant force. (See Exercise 91 for the definition of buoyant force.)
   a. The average density of an iceberg is 0.86 times that of sea water. What fraction of the iceberg volume is submerged?
   b. A glass of water with an ice cube floating in it is filled exactly to the top rim. Will the glass overflow when the ice cube melts? Why or why not?

■ **INTEGRATIVE EXERCISES**

93. Aluminum foil is sold in supermarkets in rolls $66^2/_3$-yd long, 12-in. wide, and 6.5×10^{-4}-in. thick. The density of Al is 2.6989 g/cm³. What is the mass of a roll of Al foil? How

many atoms are in a 1-in. square of foil? The radius of Al is 118 pm. Estimate the thickness of the foil in atoms.

94. Gold leaf consists of an extremely thin sheet of gold, approximately 0.1-μm thick.
   a. The radius of a gold atom is 134 pm. Estimate the number of layers of gold atoms in gold leaf.
   b. A ceremonial plate with a diameter of 14 in. is to be covered with gold leaf. How many moles of gold are required?
   c. If the plate, made of stainless steel (density 7.8 g/cm⁻³), is $^1/_8$-in. thick, what fraction of the mass of the finished plate is due to the gold leaf covering? (Density of gold is 19.3 g/cm³.)

95. Lava lamps were quite popular in the 1960s and enjoyed a resurgence in the 1990s. A lava lamp contains two liquids, one of which is usually water. The other liquid is usually an oil that is denser than water so that it settles to the bottom of the lamp. When the lamp is switched on, heat from the bulb warms the oil and it rises.
   a. What happens to the density of the oil when it is heated?
   b. What happens to the density of water as it is heated?
   c. Expansion of a liquid with temperature is called the coefficient of thermal expansion. What is the relationship of the coefficient of thermal expansion of water and that of the oil?

96. The mass of a 250-mL beaker is 190 g when empty. Determine its mass when filled with
   a. water (density 0.9970 g/cm³ at 20 °C)
   b. alcohol (density 0.7893 g/cm³ at 20 °C)
   c. mercury (density 13.546 g/cm³ at 20 °C)
   d. Explain why mercury-containing pollutants are usually found in the mud at the bottom of a lake or stream.

Chapter 2
Foundations

CONCEPTUAL FOCUS

- Examine patterns in the periodic table.
- Investigate the connection between electricity and binding atoms into molecules.
- Relate light or electromagnetic radiation to energy.
- Build a model for electrons in atoms.

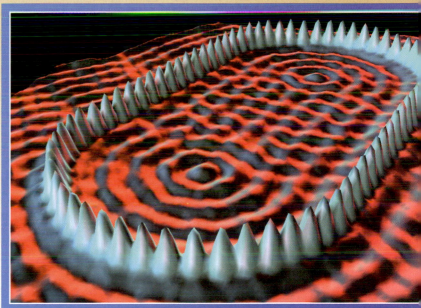

A stadium-shaped corral of iron atoms on a copper surface confines electrons on the surface, creating a rippling wave pattern of charge that can be detected with an instrument known as a scanning tunneling microscope. Electrons play a key role in chemical reactions.

SKILL DEVELOPMENT OBJECTIVES

- Determine the portion of the mass of an object that is due to electrons (Worked Example 2.1).
- Connect energy with light (Worked Example 2.2).
- Calculate the energy of hydrogen's electron (Worked Example 2.3), and relate energy to emitted radiation (Worked Example 2.4).
- Use nodes to classify and label electron waves (Worked Example 2.5); use quantum numbers to label orbitals (Worked Examples 2.6, 2.7).
- Calculate the effective charge holding an electron in the atom (Worked Example 2.8).
- Use transition energy to determine the energy of electronic states in multielectron atoms (Worked Example 2.9).
- Connect quantum numbers and the number of electrons in a shell (Worked Example 2.10).

2.1 The Periodic Table

The table you see in Figure 2.1 is a cornerstone of chemistry. It is the single most powerful organizer of chemical information ever to be devised. The periodic table enjoys this lofty status because it can be used to predict the physical and chemical properties of the elements. To use the periodic table effectively its underlying organizational structure needs to be unlocked. Start with the outline of the table: There are two elements in the first row, eight in the second and third rows, 18 in the fourth and fifth rows, and 32 in the sixth and seventh rows. These numbers—2, 8, 8, 18, 18, 32, and 32—are referred to as "magic numbers" and result from the fundamental structure of the atom.

The fundamental structure of the atom is explored by starting with the major components. The major components, however, do not explain the magic numbers. Their explanation requires a deeper probe into the structure of the atom, specifically into the electronic structure of the atom. The tool used to probe the electronic structure is light and the connection between light and energy. These are all topics explored in this chapter.

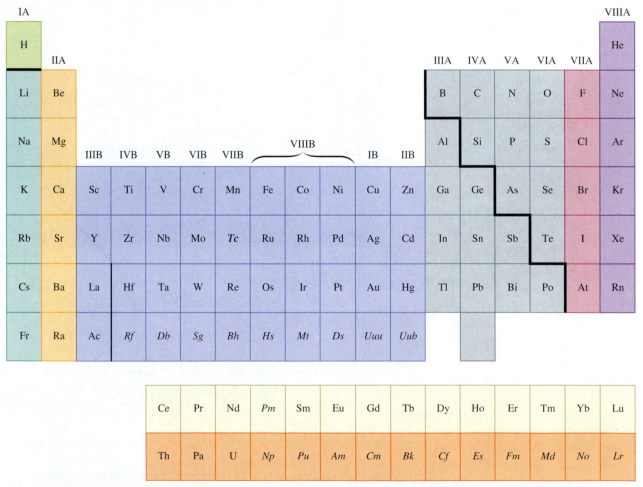

Figure 2.1 The periodic table of the elements lists all known elements in rows and columns. Elements that are in italic type are not found naturally on earth but have been made in nuclear reactions.

Origin of the Periodic Table

Imagine being presented with this challenge: Assemble a 120-piece puzzle with half the pieces missing and without edges or corners as guideposts. You might start by grouping pieces of similar color and lining up pieces with a linear motif like a pole or fence. In similar manner scientists of the 1800s, most notably Dimitri Mendeléeff[†] (Figure 2.2) and Lothar Meyer (Figure 2.3) grouped elements with similar properties and ordered them by atomic mass. Mendeléeff's table was more successful than Meyer's primarily because he considered chemical characteristics as well as physical properties. For example, copper, silver, and gold are easily isolated in pure form since they share the chemical characteristic of not readily combining with other elements. Due to this low reactivity, these elements were used for coins as early as the Roman Empire. Other examples include bromine (Br), chlorine (Cl), and iodine (I). Bromine is extracted from the sea, as are iodine and chlorine, due to the chemical property that solids containing these elements tend to be soluble in water. Grouping elements in a column according to their properties with mass increasing down the column (Figure 2.4), Cu, Ag, and Au constitute most of one column and Cl, Br, and I make up most of another.

Recognizing several such patterns, in 1869 Mendeléeff published a table with all 61 known elements in a grand array. The brilliance of Mendeléeff's table was that it left spaces for elements not yet discovered—it had *predictive* capability. Importantly, the periodic pattern of chemical properties recognized by Mendeléeff meant that properties

Figure 2.2 Dimitri Mendeléeff was a Russian chemist who orgnized the elements in an array that included spaces for elements yet to be identified.

Figure 2.3 Lothar Meyer was a German chemist who organized the known elements in an array similar to that of Mendeléeff, but based primarily on physical properties.

---

† Numerous spellings of Mendeléeff's name exist because transliteration from the Cyrillic alphabet is so difficult. This is the spelling used by Mendeléeff when he visited England in 1887.

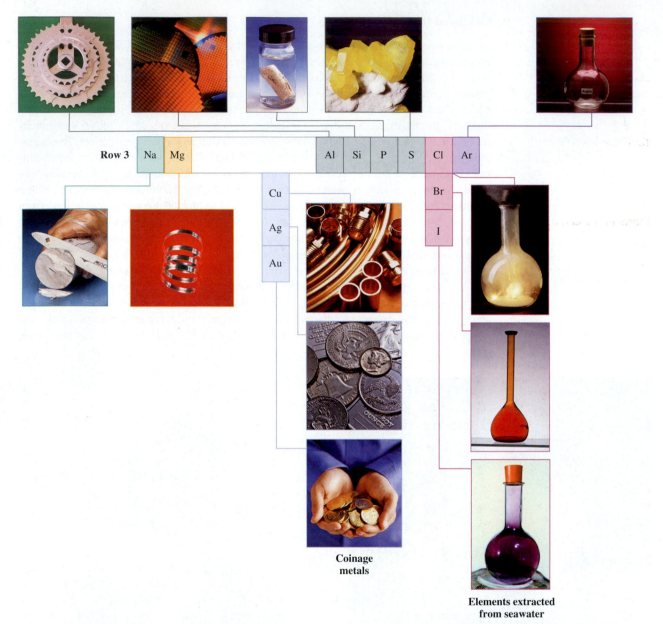

Figure 2.4 Elements with similar physical and chemical characteristics are grouped and ordered by increasing atomic mass. Across a row, properties vary systematically.

of known elements could be used to determine where in nature these undiscovered elements were likely to be found and how to separate them from the other substances found with them. Mendeléeff's table was widely embraced primarily due to this predictive power; a power that underlies the continued utility of the periodic table.

Navigating the Periodic Table

Notation and Terminology Just as in Mendeléeff's table, the columns of elements in today's periodic table have similar chemical and physical properties, and are called **groups** or **families.** Within a group, elements are listed in order of increasing atomic

Group A vertical column of elements in the periodic table showing similar properties. Also called a *family* of elements.

Period Row in the periodic table.

Main group Elements in the groups labeled IA, IIA, IIIA, IVA, VA, VIA, VIIA, and VIIIA in the periodic table.

Transition elements Denotes those elements from Sc to Zn, Y to Cd, La, Hf to Hg, and Ac, Rf to the not-yet-discovered element #112.

Lanthanides (actinides) A group of 14 elements following lanthanum (actinium) in the periodic table.

mass. In this two-dimensional array, mass generally increases across a row, called a **period,** although there are a few exceptions. The periods are simply numbered from top to bottom, from Period 1 to Period 7. Labeling the groups is a little more complex, and there are several conventions. A common convention, and the one used in this book (Figure 2.1), labels the first two and last six columns, known as **main group** elements, with the letter "A" and Roman numerals I through VIII. Between the two sets of main-group elements are the **transition elements,** labeled with the letter "B" and numbered I through VIII (with three columns numbered VIII). An alternative convention for naming the groups is to label the first two main groups and the first eight transition element groups with "A" and number them I through VIII (with three columns numbered VIII) and to label the last two transition groups plus the last six main groups with "B". A third convention avoids the "A"/"B" conflict by labeling the columns 1–18. The elements that have been pulled out of their position in Periods 6 and 7 and shown at the bottom of the table are called **lanthanides** (or rare earth elements) and **actinides** in reference to lanthanum (La) and actinium (Ac) that precede them in the sixth and seventh rows, respectively. These elements are pulled out of the main table to keep its width manageable.

Periodic Table Terminology

Groups or families	Column
Period	Row
Main-group elements	Columns 1 and 2 and 13–18, labeled "A"
Transition elements	Columns 3–12, labeled "B"

CONCEPT QUESTIONS   Which group is headed by carbon (C)? Which element is at the top of group IIIA? Sodium (Na) is in which period? ∎

2.2   The Components of the Atom

The structure of the atom and its variation among the elements are keys to understanding the periodic variation in elemental properties and decoding the wealth of information contained in the periodic table. In the early 1800s as Mendeléeff and others were organizing the then-known elements into the periodic table, other scientists were conducting studies destined to reveal the internal structure of the atom. Among them was an English chemist named Humphrey Davy (Figure 2.5). Davy's experiments revealed the importance of electrical forces in binding atoms in liquids and solids. Davy found that passing an electric current through numerous materials caused them to break down into their constituent elements.

APPLY IT   Obtain a 9-volt battery and a snap lead. Expose a bit of wire on the free end of each lead. Connect the snaps to the battery, taking care not to short the leads across your hand. Place the exposed ends in a container of water (a small amount of salt added to the water speeds up the reaction by aiding current flow). What do you observe?

Figure 2.5 Humphrey Davy was a pioneer in the field of electrochemistry.

The Electron

One substance that Davy experimented with was water. Early experiments with gases had shown that each water molecule contains one oxygen atom and two hydrogen atoms (Figure 2.7). Davy found that passing an electric current through water produces a stream of bubbles at both leads.

With leads from a battery well separated in air (Figure 2.6), no current flows. When the leads are immersed in water, however, the circuit is completed, current flows, and bubbles are produced (Figure 2.8). The volume of gas produced at one lead is twice the volume produced at the other—a proportion consistent with a water molecule consisting of two atoms of hydrogen and one atom of oxygen. A simple test, such as thrusting a glowing splint into a container of one gas and hearing a loud pop, indicates that the gas produced is hydrogen. (Don't try this at home!) Similarly, glowing hot steel wool bursts into flame when thrust into the other gas, indicating that the gas is oxygen. Twice as much hydrogen as oxygen is produced:

$$\textbf{(2.1)} \qquad \text{water} + \text{electricity} \longrightarrow 2 \text{ volumes of hydrogen gas} \\ + 1 \text{ volume of oxygen gas}$$

Water molecules hold no net electrical charge; they are *neutral*. Water molecules must be neutral because all water molecules are alike, and if they held charge, the like charges would repel one another.

9 V battery

Figure 2.6 No current flows through air between leads that are connected to the battery.

No current flow

Magnified 100,000,000 times

H₂O

Figure 2.7 Magnifying the familiar material, water, a hundred million times reveals that a molecule of water is a compound consisting of two atoms of hydrogen and one atom of oxygen: H_2O.

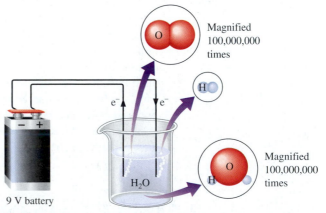

Magnified 100,000,000 times

Magnified 100,000,000 times

9 V battery

Figure 2.8 Current flow (symbolized by e^-, indicating a flow of negative charge) breaks water into its constituent elements: hydrogen at one lead and oxygen at the other.

Electrode A solid electrical conductor through which an electric current enters or leaves an electrolytic cell or other medium.

Electrolysis The splitting of a compound into its components, using electricity.

Immersing the leads, technically referred to as **electrodes,** into the water splits water into its constituent elements (H and O) using electricity. This process, called **electrolysis,** is an important industrial method for extracting a metal such as aluminum from a mineral and for purifying a metal such as copper, which has to be very pure for use as electrical wiring. Electrolysis is also used to isolate chlorine from brine or concentrated sea water for subsequent use in making plastic plumbing fixtures, in the manufacture of pharmaceuticals, and to decontaminate water supplies.

In the late 1800s Michael Faraday made a general observation about electrolysis: The amount of material broken into its constituent elements is proportional to the amount of electricity passing through it. Further, the mass of material generated at a given lead is proportional to the amount of electricity and to the atomic mass of the element divided by a small integer: 1, 2, or 3.

The idea that electricity is the flow of *packets* of negative charge is consistent with Faraday's observation that the mass of material produced is related to the amount of electricity and the atomic mass. In 1891 George Stoney suggested the name *electron* for these packets of electrical charge. The observation that the flow of electrons breaks water and other compounds into the constituent elements suggests that electrons bind atoms into molecules. George Stoney extended this suggestion by postulating that electrons are, in fact, part of the atom.

CONCEPT QUESTION How is the idea that electricity is the flow of *packets* of negative charge consistent with Faraday's observation that the mass of material produced is related to the amount of electricity and the atomic mass? ■

Electron Charge and Electron Mass In a series of experiments, Joseph John Thomson (Figure 2.9), usually referred to as J. J. Thomson, showed that the electron is part of the atom. Thomson also measured the charge-to-mass ratio. This work was so significant that Thomson received the Nobel Prize in 1906 and is generally credited with the discovery of the electron.

Once the charge-to-mass ratio was determined, additional experiments were needed to determine either the charge or the mass. In 1909, Robert Millikan (Figure 2.10) carried out a set of exacting experiments and determined the charge of an electron, thus making it possible to calculate the mass. The modern value is 9.1094×10^{-28} grams, which is 1/1837 the mass of the hydrogen atom, the lightest of all atoms. This result indicates that although the electron is part of the atom, it accounts for only a very small part of the mass.

To summarize:

- The electron is a part of all atoms.
- The electron is very light compared with the rest of the atom.
- The electron has a negative charge associated with it.
- An electric current is a flow of electrons that can result in molecules being broken into their constituent elements.

Figure 2.9 J. J. Thomson determined many properties of the electron, including the charge to mass ratio. He received the Nobel Prize in physics in 1906 for this work.

Figure 2.10 Robert Millikan determined the mass of the electron. This work contributed to Millikan winning the Nobel Prize in physics in 1923.

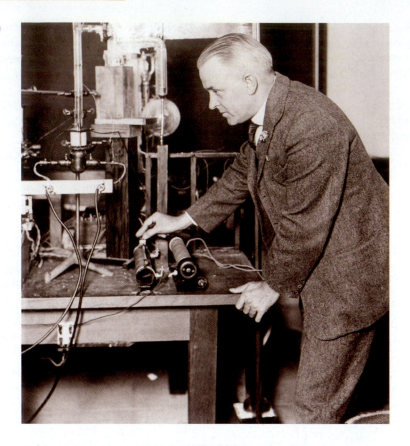

WORKED EXAMPLE 2.1   *Relative Mass of Electron and Atom*

The electron accounts for only a small portion of the mass of an atom, but just how small a portion is it? To put this number in perspective, let us scale it up. The hydrogen atom consists of one proton and one electron. Multiplying the mass of the hydrogen atom by 3×10^{28}—more than a billion billion billion—it would weigh about 110 pounds, or the weight of a small adult. In this magnified atom, what is the mass of the electron?

Plan

■ Use the data: The mass of an electron is 1/1837 of the mass of an atom.

Implementation

■ Multiply the weight of the atom times 1/1837 to give the weight of the scaled-up electron. (110 pounds) $\times$ (1/1837) = 0.0599 pound.

Convert this to an appropriate unit.
There are 16 ounces in a pound; 0.0599×16 ounces = 0.958 ounces, or about an ounce—about the mass of a mouse!

See Exercises 7–11.

Mass Distribution within the Atom

It was intuitive to expect that if the electron were responsible for only a very small portion of the mass of the atom, it should also be responsible for only a small part of the volume. However, imagine doing the following experiment, one actually done by Ernest Rutherford and coworkers Marsden and Geiger between 1906 and 1909. Take some fast-moving particles that are each about 7000 times the mass of an electron, and direct them at a very thin sheet of some material. Rutherford used α particles and gold foil. The α particle was known to be much more massive than an electron and to have a charge of $+2$. Gold foil is easily obtained in pure form and can be made into a very thin sheet just a few thousand atoms thick.

CONCEPT QUESTION If the radius of an α particle is about one-third that of a gold atom and the mass is uniformly distributed, what is the expected outcome of shooting a stream of α particles at a gold sheet? ∎

The α particles are much more massive than an electron. Our everyday experience of objects colliding suggests that the far-lighter electrons in the atoms should not deflect or scatter the particles. This situation is comparable to launching a bowling ball at several cotton balls: The bowling ball will roll along hardly perturbed by the smaller, lighter objects. At the time of Rutherford's experiment, the model for the atom indicated that the mass of the remainder of the atom is spread uniformly throughout the entire volume of the atom. Like shooting a cannon ball at a wall of cannon balls only a few balls thick, Rutherford expected the gold atoms to be jostled aside and the α particles to pass through the gold foil relatively undisturbed, though perhaps with some slight deflection (Figure 2.11a).

Instead, Rutherford saw almost all the α particles pass through the foil with *no* deflection (Figure 2.11b). The fatal blow to the model of a uniform mass distribution, however, was the observation that a few α particles (about one in 10^5) were scattered with a large deflection. Some even bounced directly back! These results were as surprising as if a bowling ball bounced back upon colliding with a piece of tissue. A uniform distribution of mass simply cannot account for observations of mostly undeflected α particles with one part in 10^5 being deflected at a large angle.

CONCEPT QUESTION If Rutherford had used a different metal that could not be made as thin as Au, such as an Al sheet 1 μm (10^{-6} m) thick, what portion of the α particles would be scattered? ∎

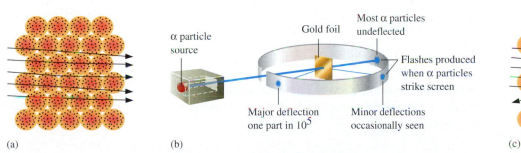

(a)   (b)   (c)

Figure 2.11 (a) The plum-pudding model predicts that nearly all particles will be transmitted with only slight deflection. (b) Schematic of Rutherford's scattering apparatus and scattering results: The large-angle, scattered portion is approximately one part in 10^5. The gold foil is about 0.5 μm (10^{-6} m) thick. This corresponds to about 2000 atomic layers. Hence in a foil one atom-layer thick, about one part in 10^8 is deflected. (c) Rutherford concluded that some part of the atom must be dense and have a cross-sectional area of only 0.00000001 times as great as the cross-sectional area of the atom. Hence, the diameter of the nucleus is only 1/10,000 as great as the diameter of the atom (the square root of 0.00000001 is 1/10,000). Here the nucleus is shown greatly enlarged relative to the size of the atom.

← Approximately 10^{-10} m →

Approximately 10^{-14}m

● Proton
● Neutron

Figure 2.12 Schematic magnified cross section of an atom. The nucleus contains virtually all the mass of the atom, but is only about one part in 10^8 of the cross-sectional area. The remainder of the atom consists of a very diffuse cloud of electrons.

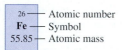

26 —— Atomic number
Fe —— Symbol
55.85 —— Atomic mass

Nucleus The small, dense center of positive charge in an atom that is responsible for most of the mass.

Proton A positively charged particle in an atomic nucleus.

Atomic number Z The number of protons in the nucleus of the atom; determines the identity of the element.

Neutron An uncharged particle of mass slightly greater than the proton.

Periodic law The properties of the elements are periodic functions of their atomic numbers.

The Nucleus and the Electron Given the small number of α particles deflected, the dense parts of the gold foil must be extremely small and account for about one part in 10^5 of the area of the material. Rutherford named the dense lump the atom's **nucleus.** The atom's electrons (Figure 2.12) fill the space around the nucleus. To put the relative size in context, the nucleus can be compared to a ball bearing a millimeter in diameter in a 30-foot-wide lecture hall or to a pea on a football field. The α particles hitting the dense part—the nucleus—scatter over large angles, while those encountering only the more diffuse part—the electrons—go straight through.

Although the nucleus is often illustrated as a small sphere, it actually contains several particles. The first nuclear particle to be recognized was the proton. The atom is neutral, so there must be a positive charge for every electron and its negative charge. This positive charge is carried by the proton. The **atomic number,** represented by the symbol **Z**, is equal to the number of protons in the nucleus. In the periodic table, the atomic number or number of protons increases by one for each successive element. The other major component of the nucleus, the **neutron,** eluded discovery until 1932, when James Chadwick detected neutrons emitted from bombardment of beryllium with high-energy α particles. The neutron is an uncharged particle of slightly greater mass than the proton. These three particles—the electron, the proton, and the neutron—are the major constituents of an atom (Table 2.1).

CONCEPT QUESTION You are told that an atom of an unknown element has an atomic mass of 32 amu. Can you identify the element from this information? What other information do you need? ■

Major Components of the Atom and the Periodic Table

In the periodic table, the number above the element is the atomic number, the number of protons in the nucleus of the element. Elements are arranged in the periodic table in order of increasing atomic number, which leads to the **periodic law:**

> The properties of the elements are periodic functions of their atomic numbers.

The atomic number is more than just a number denoting the ordinal position of the element; it is a count of the protons in the nucleus of each and every atom of a given element—a count unique to each element. Atomic mass is due primarily to the number of particles in the nucleus and is a measure of the number of protons *and neutrons.* The relatively light electrons, whose numbers increase from element to element in parallel with the protons, fill most of the volume of the atom. Despite this information about the components of the atom, nothing thus far explains why the negatively charged electron does not simply collapse into the nucleus, or how the structure of the atom results in periodic variation in properties of the elements. A more complete model of the electrons in atoms is required to explain both the stability of atoms and the periodic variation of properties.

Table 2.1 Atomic Particle Mass and Charge

Particle	Mass (g)	Mass (amu)	Charge in Coulombs (C)	Relative Charge
electron (e^-)	9.1094×10^{-28}	0.00054858	-1.60218×10^{-19}	-1 (negative)
proton (p or p^+)	1.6726×10^{-24}	1.0073	1.60218×10^{-19}	$+1$ (positive)
neutron (n or n^o)	1.6749×10^{-24}	1.0087	0	none (neutral)

2.3 Light – Unlocking the Electronic Structure of Atoms

The primary probe used to unlock information about electrons in atoms is light. Picture New Year's Eve at midnight. Fireworks burst across the sky; bright yellows, blues, crimsons, and whites cascade earthward as the fireworks crack open with an explosion of color (Figure 2.13). These vibrant colors are produced by hot atoms of metallic elements such as sodium and lithium, which are heated as the firework explodes. The yellow color characteristic of sodium (Na) can be produced by sprinkling some common table salt (NaCl) into a gas flame — for example, that of a gas stove. Yellow street lamps also owe their distinctive color to sodium. In contrast, neon (Ne) lights have an orange hue, and mercury (Hg) lights, most often seen as fluorescent lights, feature a bluer color. Electrons in the atoms are responsible for these colors. Analyzing the colors of light produced by hot atoms generates a model for how the electrons fill the vast majority of the volume of an atom.

Visible light — including sodium's yellow light — is but a small slice of a much larger spectrum of radiant energy known as **electromagnetic radiation.** Electromagnetic radiation shuttles energy in and out of matter. The relationship between electromagnetic radiation and energy is the key to understanding atomic structure and atomic behavior.

Electromagnetic radiation Radiant energy consisting of oscillating and magnetic fields; includes visible light, x rays, and radiowaves.

Continuous spectrum Continuous bands of color produced, e.g., by a prism being inserted into a shaft of white light.

Disperse To separate.

Wavelength, λ The distance between two consecutive peaks or troughs in a wave.

Figure 2.13 The brilliant colors of fireworks are a display of energy emitted from atoms.

Light, Color, and Energy

A prism inserted into a shaft of white light (Figure 2.14) produces beautiful, continuous bands of color. Just as in a rainbow, the colors start from red and range through yellow, green, and blue to violet (magenta). That is, they form a **continuous spectrum.** The spectrum results from white light, consisting of all colors, being separated or **dispersed** into its constituents by some medium. In the case of a rainbow in the sky, the dispersing medium is water in the form of droplets, and the light source is the sun. Visible light, like all electromagnetic radiation, can be described in terms of waves of energy. These waves repeat in space and can be characterized by the distance over which the wave pattern repeats, known as the **wavelength** (Figure 2.15) and denoted by the Greek letter lambda, **λ**. Our eye receives visible light waves and responds with a perception of color that depends on the wavelength. For example, what is perceived as red is light with a wavelength of approximately 650 nm (nm stands for nanometer, one-billionth of a meter); in such light, a distance of 650 nm separates one wave crest from the next. When waves of all colors of light combine, it is seen as white light.

CONCEPT QUESTIONS The warming, infrared radiation from the sun has a wavelength of about 5 μm (μm stands for micrometer, one-millionth of a meter). How many crests occur per centimeter? How does this compare with visible light? ∎

Figure 2.14 A shaft of sunlight is broken into the rainbow of colors by a glass prism. As light passes through the glass, red light is bent the least and violet (magenta) light the most.

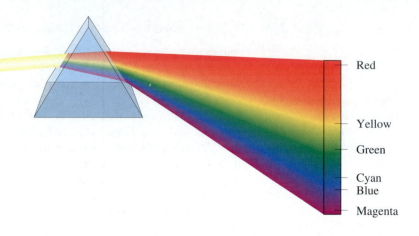

Red

Yellow

Green

Cyan
Blue

Magenta

APPLY IT Obtain a long flexible cord such as a phone cord or a window shade cord. Attach one end securely to a stationary object. Grasping the other end, attempt to produce the patterns seen in Figure 2.15. Note the effort required to produce each pattern.

Light such as that emitted by the various metallic elements in fireworks can tell a great deal about the structure of atoms, but first the wavelength that characterizes the light and the energy of the light need to be connected. As an analogy for this connection, picture a jump rope. The arc between the ends forms a half-wave. If the rope is rotated rapidly and viewed from a distance, it forms a sustained wave and appears not to move. A sustained wave is called a **standing wave.** The half-wave of a jump rope is the longest wavelength for a standing wave that can occur between the two fixed ends. Just as in a string instrument, it is called the **fundamental wave** (Figure 2.15a). Over the same distance, the wave in Figure 2.15b consists of one whole wavelength, which is called the first overtone. The wavelength thus *decreases* from panels (a) through (c). The effort required to sustain these waves *increases* from panels (a) through (c). This is a general result: The shorter the wavelength, the higher the energy. For light, the relationship between wavelength and energy is

Standing wave A stationary sustained wave such as on a string of a musical instrument.

Fundamental wave The longest wavelength for a standing wave that can occur between two fixed ends.

(2.2)
$$E = \frac{hc}{\lambda}$$

where h is Planck's constant, 6.63×10^{-34} J·s (1 J = 1 kg·m²/s²), and c is the speed of light, 2.9979×10^8 m/s. The wavelength of light in the visible region of the electromagnetic spectrum ranges from 750 nm for extreme red to 400 nm for extreme violet (magenta) (Table 2.2); red light has lower energy than violet (magenta) light.

The common, constant speed, c, of all electromagnetic radiation results in an alternate description of light in terms of a variable called the **frequency** of the wave and denoted by the Greek letter nu, ν. Frequency is the number of wavelengths moving past a stationary observer per second (Figure 2.16). Given that the speed of light is fixed, an inverse relationship exists between the wavelength and the frequency of the wave,

Frequency, ν The number of waves (cycles) per second that pass a given point in space.

(2.3)
$$\nu = c/\lambda$$

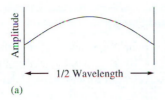

(a)

1/2 Wavelength

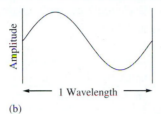

(b)

1 Wavelength

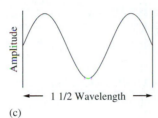

(c)

1 1/2 Wavelength

Figure 2.15 It takes little effort to form a wave like that in panel (a) with a flexible cord. The pattern in panel (b) is more difficult to form, and that in (c) is quite difficult.

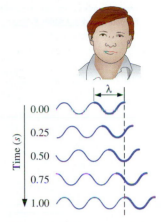

Figure 2.16 The sinusoidal oscillation of a light wave as it moves past an observer. The number of wavelengths that pass the observer per second define the frequency. In the example shown above, one second elapses between crests or oscillations, and the frequency is 1.0 Hertz. A frequency of 1.0 Hertz is a very low frequency. Visible light oscillates at 10^{14} to 10^{15} Hertz.

The unit of frequency is the hertz, defined as the number of wavelengths passing a point per second and abbreviated Hz or s^{-1}. A direct relationship exists between energy and frequency as shown by combining Equations (2.2) and (2.3):

(2.4) $$E = h\nu$$

For visible light, the energy ranges between 3×10^{-19} J and 5×10^{-19} J.

Table 2.2 Color and Energy of Visible Light Form a Continuous Spectrum
The wavelengths listed are near the center of the range that most people would perceive as that color.

Wavelength (nm)	650	560	490	450	430	400
Frequency (Hz)	4.62×10^{14}	5.36×10^{14}	6.12×10^{14}	6.67×10^{14}	6.98×10^{14}	7.50×10^{14}
Energy (J)	3.06×10^{-19}	3.56×10^{-19}	4.05×10^{-19}	4.42×10^{-19}	4.63×10^{-19}	4.97×10^{-19}
Energy (eV)	1.91	2.22	2.53	2.76	2.89	3.10
Color	red	yellow	green	cyan	blue	violet (magenta)

WORKED EXAMPLE 2.2 *Frequency, Wavelength, and Energy*

a. AM radio stations often give their frequency as part of their name. The unit for this frequency is kilohertz (kHz = 10^3 s^{-1}). A radio station broadcasts at a frequency of 680 kHz. Calculate the wavelength for 680 kHz. Calculate the energy of this wave in joules (J).
b. The wavelength of the yellow light from sodium is 590 nm. Calculate its energy in joules.
c. X rays have a much shorter wavelength than visible light. Calculate the energy of a 10-nm wavelength X ray in joules.

Plan

■ (a) Invert Equation (2.3), $\nu = c/\lambda$, as $\lambda = c/\nu$ to calculate the wavelength.
■ (a) Use Equation (2.4), $E = h\nu$, to calculate the energy from the frequency.
■ (b) and (c): Use Equation (2.2), $E = hc/\lambda$.

Implementation

(a)

■ $\lambda = c/\nu = (2.9979 \times 10^8$ m/s$)/(680 \times 10^3$ $s^{-1}) = 441$ m
■ $E = h\nu = (6.63 \times 10^{-34}$ J·s$)(680 \times 10^6$ $s^{-1}) = 4.51 \times 10^{-28}$ J
■ $E = \dfrac{hc}{\lambda}$

(b) $E = \dfrac{(6.63 \times 10^{-34}\,\text{J·s})(2.9979 \times 10^8\,\text{m/s})}{(590\ \text{nm})} \times \dfrac{10^9\ \text{nm}}{\text{m}} = 3.37 \times 10^{-19}$ J

Yellow light has a considerably shorter wavelength and is significantly more energetic than radio waves.

(c) $E = \dfrac{(6.63 \times 10^{-34}\,\text{J·s})(2.9979 \times 10^8\,\text{m/s})}{(10\ \text{nm})} \times \dfrac{10^9\ \text{nm}}{\text{m}} = 1.99 \times 10^{-17}$ J

X rays are much more energetic than visible light waves.

See Exercises 19, 21, 22.

Electromagnetic radiation provides a window into the characteristics of electrons in atoms (Figure 2.17). For example, consider the origin of the yellow color observed from heating table salt (NaCl). The flame deposits energy into the NaCl. Some of that energy is absorbed by the sodium atoms, then released (emitted) in the form of yellow light.

Figure 2.17 Color, light, and the connection to energy provide a window into the inner workings of the atom.

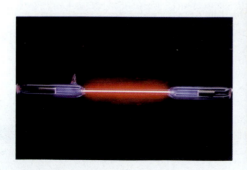

A sodium atom contains 11 electrons, each with a unit of negative charge. Due to the complex mutual repulsion among these 11 electrons, unraveling the connection between the yellow color of sodium and the structure of sodium's 11 electrons is an involved process. To start this process, consider the color emitted by hydrogen, a simpler atom consisting of only one electron and one proton.

Line Spectra and the Energy of the Atom

Line spectrum Emitted light consisting of discrete wavelengths

In contrast to the yellow emission characteristic of hot sodium atoms, the color emitted by hot hydrogen atoms is hot pink. Dispersing the hot-pink light from hydrogen with a prism (Figure 2.18) produces a very different result from the continuous colors of the rainbow. Only four wavelengths, or four energies, of visible light are emitted by hydrogen. The four distinct well-separated colors emitted by hydrogen are referred to as the **line spectrum** of hydrogen. In 1885 a Swiss school teacher named Johann Balmer noticed that the frequencies of the four visible lines from hydrogen fit a very simple formula:

(2.5)
$$\nu = C\left(\frac{1}{2^2} - \frac{1}{n^2}\right)$$

where $n = 3, 4, 5, 6$, and C is a constant equal to $3.289 \times 10^{15}\ \text{s}^{-1}$. Extending beyond the visible spectrum are many more similarly distinct lines. Balmer noted that these lines fit a more general form of Equation (2.5):

(2.6)
$$\nu = C\left(\frac{1}{n_1^2} - \frac{1}{n_2^2}\right)$$

where n_1 and n_2 are integers with $n_2 > n_1$.

The line spectra of the elements other than hydrogen are more complex and cannot be described by a simple, Balmer-type equation. Nonetheless, each element has a unique spectral signature that acts as a fingerprint and is used to identify elements in materials. For example, the unique absorption spectrum of lead is used to measure the amounts of lead in blood to diagnose lead poisoning. The composition of stars and other celestial bodies is determined by the spectral signatures of their components. Historically these fingerprint spectra posed a stringent test for any theory proposed to explain the structure of atoms.

Figure 2.18 (a) Light from a hydrogen lamp when dispersed by a prism is split into its components. (b) The emission spectrum of hydrogen consists of just four colors of visible light: red, green, cyan, and violet (magenta).

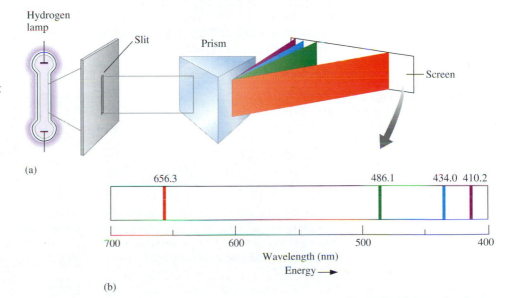

Consider the implications of a *line* spectrum. The photons in the emitted light have only a few distinct frequencies, ν. Frequency is related to the energy of the photon, $E = h\nu$, so the light has restricted energies. Because energy is conserved, the energy of the atom before it emits radiation must equal the energy of the atom after emission plus the photon energy:

(2.7) $$E_{\text{before}}^{\text{atom}} = E_{\text{after}}^{\text{atom}} + h\nu \text{ or } \Delta E^{\text{atom}} = E_{\text{before}}^{\text{atom}} - E_{\text{after}}^{\text{atom}} = h\nu$$

These restricted energy differences imply that the atom itself has only specific, isolated energies. Rutherford's model of the atom contains no clue about these restricted energies.

The restricted energies are usually associated with the electron. But why are the energies restricted, and what happens within an atom when energy is emitted? As a model, consider the restricted energies as being variably spaced steps on a ladder (Figure 2.19a). It is possible to stand on the steps, but not between them. Movement can occur one step at a time or steps can be skipped. Movement can be up or down. Upward movement corresponds to energy input, and downward movement corresponds to energy removal. The steps of the ladder represent the allowed energies, called **energy levels,** of the atom. Energy levels are illustrated in a schematic called an **energy-level diagram** (Figure 2.19). The energy difference, which is the distance between steps, is called a **transition energy**, and a set of transition energies is the line **spectrum.** Figure 2.19b shows a portion of the spectrum for the ladder. The bottom step of the ladder represents the lowest energy possible.

Energy level Allowed energy for an electron in an atom.

Energy-level diagram Illustrates energy levels.

Transition energy Energy difference between two electronic states.

Spectrum The collection of frequencies or energies emitted by an atom, ion, molecule, or solid.

CONCEPT QUESTION It is 3026 miles from Boston, Massachusetts, to Los Angeles, California. In a long day of driving, you have gone 625 miles. What other information is necessary for you to know how close you are to Los Angeles? ■

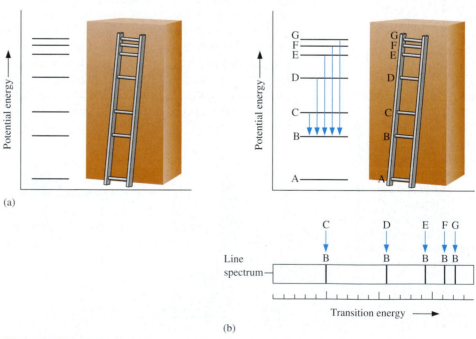

(a)

(b)

Figure 2.19 (a) We can view the restricted energy of an atom as steps on a ladder. To the left of the ladder is an energy-level schematic for the steps. (b) The line spectrum relates to the difference between the positions of the steps. Transition energies for five transitions ending on level B are shown here as a line spectrum below the energy-level schematic.

State Describes the location of the electron relative to the nucleus.

Ground state The lowest possible energy state of an atom or molecule.

Excited states Higher energy orbits.

Each line in the line spectrum of hydrogen (Figure 2.18b) relates to an energy *difference* between two steps on hydrogen's energy ladder, but does not provide any information on the energy value (position) of each step. To deduce the energy of the steps on the ladder from the line spectrum, the energy value of at least one step is needed. This situation is analogous to knowing that you grew six inches in the last year. Of course, this is not enough information to determine your height. To determine your height, you need to know how tall you were last year, in other words, you need a reference point. Similarly, transition energies are differences (analogous to your growth last year). To generate the energy-level diagram, however, the energy of at least one energy state is needed as a reference point.

CONCEPT QUESTION How do the energies of the series of transitions ending on level A in Figure 2.19 compare with those ending on level B? Indicate them on the schematic line spectrum in Figure 2.19. ∎

2.4 Model of the Atom

The first successful attempt to describe the atom in detail and to explain the line spectrum came from Niels Bohr (Figure 2.20) in 1913. Although the Bohr model has since been replaced by a more correct and complex model of the atom, some important features of it are retained in the current model. Specifically, the Bohr model provides a reference point for measuring energies and a model for locating the steps relative to the reference point. (Algebraic details of the model are developed in Exercises 31 to 36.) The Bohr model treats the electron as a particle, a small sphere of negative charge, that orbits around the nucleus much like a planet orbits around the sun. It models the attraction between the electron and the nucleus as the coulombic attraction between unlike charges, and this coulombic attraction is retained in the current model.

The major result of Bohr's work was an explanation of the line spectrum; the essential conclusion from Bohr's work is that the electron's energy is related to the distance of the orbit of the negatively charged electron from the positively charged nucleus. In the more correct model, the role of the orbit radius is replaced by the radius of the electron cloud. In both models, high energy corresponds to the electron being located farther from the nucleus than it is at low energy. In the Bohr model, when the electron moves from a high-energy orbit to a low-energy one, energy equal to the difference in orbit energies is emitted. If the distance of the orbits from the nucleus is limited to specific distances, then the possible energies of the atom are restricted as well. Bohr called these allowed energy orbits **states,** with the lowest-energy orbit being the **ground state** and higher-energy orbits **excited states.** All of the states are designated with integers, n, beginning with $n = 1$ for the ground state. A schematic of the

Figure 2.20 Niels Bohr in the early 1920s. Bohr developed a planetary model for the hydrogen atom that successfully explained its spectrum. This achievement was so significant that he received the Nobel Prize in physics in 1922 and is now immortalized with the naming of element number 107, bohrium.

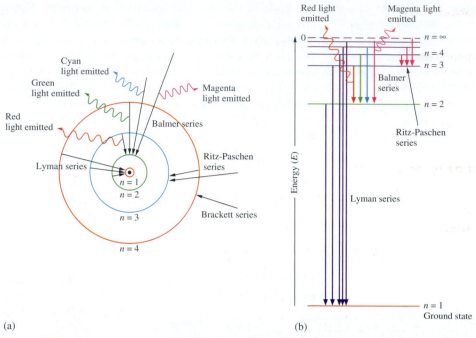

(a)   (b)

Figure 2.21 (a) Bohr's model shows the electron increasingly far from the nucleus in higher-energy orbits. Energy is emitted in the form of electromagnetic radiation when the electron changes orbit and reduces the distance to the nucleus. The nucleus is represented by the black circle in the center, shown greatly magnified relative to the electron orbits. (b) The energy of the electron is shown in this energy-level diagram, with orbits labeled by integers and transitions indicated by arrows. An electron in the $n = 1$ orbit of hydrogen occupies the lowest-energy state and is nearest the nucleus. The visible lines in the hydrogen spectrum are the four lowest-energy transitions in the Balmer series, which ends on the second level of hydrogen. These were the only lines known when Balmer generated his formula for the hydrogen spectrum. The series of lines ending on the lowest-energy level — the ground state — of hydrogen form the Lyman series. Those ending on $n = 3$ are the Ritz-Paschen series. Those ending on $n = 4$ are the Brackett series.

allowed orbits is shown in Figure 2.21a and the corresponding energy states in Figure 2.21b. There is a large energy difference between the lowest energy state, $n = 1$, and the next energy state, $n = 2$. After that point, higher-energy states are found increasingly closer together.

CONCEPT QUESTIONS Red light is emitted when the electron falls from energy level 3 to energy level 2. Violet (magenta) is emitted when the electron falls from energy level 6 to energy level 2. In which energy state, 6 or 2, is the electron closer to the nucleus? In which state is the electron less strongly attracted to the nucleus? ∎

Ionization Energy

If enough energy is put into the atom, the electron can be moved so far away from the nucleus that it is no longer part of the atom. In principle, this point is at infinity. For practical purposes, however, it is considered to be where the attraction between the electron and the nucleus is essentially zero; the electron is free of the atom. The minimum energy

Ionization energy The energy required to remove an electron from an atom, ion, or molecule. Also known as binding energy or ionization potential.

that removes an electron from the atom is the **ionization energy.** Each element has a characteristic ionization energy value. The energy of the electron is measured relative to the free electron. An electron just freed from the attraction to the nucleus is said to have an energy of zero: This is the essential reference energy. All states where the electron is associated with the nucleus therefore have *negative* energy values. Bohr's model provided an equation for the energy of the states of the hydrogen atom and this same equation is produced by the current model

(2.8)
$$\text{Electron energy} = -\frac{hcR}{n^2}$$

Rydberg constant Fundamental constant that relates to the energy states of hydrogen, $R = 1.097 \times 10^7 \text{ m}^{-1}$.

where h is Plank's constant, c is the speed of light, R is a constant called the **Rydberg constant** $(1.097 \times 10^7 \text{ m}^{-1})$, and n is an integer that indicates the orbit number of the electron $(n = 1, 2, 3, \ldots)$. Note the negative sign. When the electron is bound to the nucleus, the energy of the electron, also referred to as the energy of the atom, is *negative.*

To remove the electron from the atom, to elevate it to, a zero-energy state, requires energy input:

(2.9)
$$\text{Ionization energy for level } n = \frac{hcR}{n^2}$$

Electron-volt, eV Energy required to move an electron from a region of a given potential to a region one volt higher in potential.

For example, for the $n = 1$ state of the hydrogen atom, the ionization energy is 2.18×10^{-18} J. To remove the electron from a ground state hydrogen atom requires input of 2.18×10^{-18} J. Since this is a very small number, it is convenient to use a smaller unit, the electron-volt. An **electron-volt,** abbreviated eV, is the energy required to move an electron from a region of a given potential to a region one volt higher in potential. One eV is equal to 1.6022×10^{-19} J. In eV units, the ionization energy of ground-state hydrogen is 13.6 eV. (Note that volts and electron-volts do not describe the same quantity. An eV is an energy unit, while a V is a unit of electrical potential.)

CONCEPT QUESTION How much energy is required to remove the electrons from a mole of hydrogen atoms? ■

WORKED EXAMPLE 2.3   *Energies: The Hydrogen Atom*

When electrical energy is input to a tube of hydrogen, atoms become excited. Suppose some hydrogen atom has its electron in the $n = 4$ state while another's electron is in the $n = 2$ state. Further energy input to each atom breaks these electrons free to become part of the current running through the tube. How much energy is required to break each of these electrons free?

Plan
■ Use the following information: Ionization energy $= \dfrac{hcR}{n^2}$.

In this example, $n = 2$ and $n = 4$.

Implementation
■ $n = 2 \Rightarrow$ Ionization energy $= \frac{1}{4}(hcR)$
$$= (\tfrac{1}{4})(6.63 \times 10^{-34} \text{ J} \cdot \text{s})(2.9979 \times 10^8 \text{ m/s})$$
$$\times (1.09 \times 10^7 \text{ m}^{-1})$$
$$= 5.42 \times 10^{-19} \text{ J}$$
$n = 4 \Rightarrow$ Ionization energy $= \frac{1}{16}(hcR)$
$$= (\tfrac{1}{16})(6.63 \times 10^{-34} \text{ J} \cdot \text{s})(2.9979 \times 10^8 \text{ m/s}^{-1})$$
$$\times (1.09 \times 10^7 \text{ m}^{-1})$$
$$= 1.35 \times 10^{-19} \text{ J}$$

Four times as much energy is required to remove the electron in the $n = 2$ state because it is closer to the nucleus and therefore more strongly attracted.

See Exercises 39, 40, 112.

Line Spectra Revisited

In the ladder analogy, energy is emitted when stepping from a higher energy step to a lower one. Similarly, the hydrogen atom emits energy, specifically a photon, when the electron drops from a higher energy level to a lower one. The energy of the photon, $h\nu$, is equal to the energy difference between the two states, $\Delta E = h\nu$. The corresponding frequency of light can be calculated from the allowed energy levels. Using this information, the equation for emission from hydrogen is the same as Balmer's observational formula for hydrogen:

(2.10)
$$\text{Energy of state } n = -\frac{hcR}{n^2}$$

(2.11)
$$\text{Energy of state } 2 = -\frac{hcR}{2^2}$$

(2.12)
$$\text{Energy difference} = h\nu = hcR\left(\frac{1}{2^2} - \frac{1}{n^2}\right)$$

The values of n in Balmer's formula relate to the orbit number in Bohr's model, and the constant C is equal to cR, the speed of light times the Rydberg constant. The four lines of the visible spectrum for hydrogen are emitted when the electron moves from orbit 3, 4, 5, or 6 to orbit 2. In any one hydrogen atom, only one of these transitions is occurring at any given time. But a hydrogen lamp contains so many hydrogen atoms that all transitions are occurring, and all four colors are observed at once.

WORKED EXAMPLE 2.4   *Transition Energies and Energy Levels*

The second line of the visible series for hydrogen occurs at 486.1 nm. Calculate the energy of the 486.1-nm transition in hydrogen. The transitions in the visible region of the spectrum all end on the $n = 2$ level. From what level does the 486.1-nm transition originate?

Plan

- Use the information: $\Delta E = hc/\lambda$.
- Use Equation (2.5) to determine n.

Implementation

- 486.1 nm = 4.861×10^{-7} m

$$\Delta E = \frac{hc}{\lambda} = \frac{(6.6262 \times 10^{-34}\,\text{J}\cdot\text{s})(2.9979 \times 10^8\,\text{m/s})}{4.861 \times 10^{-7}\,\text{m}} = 4.086 \times 10^{-19}\,\text{J}$$

$$E = 4.086 \times 10^{-19}\,\text{J} \times \frac{1\,\text{eV}}{1.6022 \times 10^{-19}\,\text{J}} = 2.5502\,\text{eV}$$

- Use Equation (2.5) to determine n:

$$\nu = C\left(\frac{1}{2^2} - \frac{1}{n^2}\right) = Rc\left(\frac{1}{2^2} - \frac{1}{n^2}\right)$$

$$\Rightarrow n = \sqrt{\frac{1}{0.25 - (^{\nu}\!/_{Rc})}}$$

$$= \sqrt{\frac{1}{0.25 - \left(\dfrac{(6.167 \times 10^{14}\,\text{s}^{-1})}{(1.097 \times 10^{7}\,\text{m}^{-1})(2.9979 \times 10^{8}\,\text{m/s})}\right)}} = 4$$

See Exercises 44–51.

The Quantum Mechanical Model

The major limitation of the Bohr model is that it treats the electron as a particle and leaves unanswered the question of how the electron fills the vast majority of the volume of the atom. The current model treats electrons as both waves and particles. When should the electron be thought of as a particle, and when should it be envisioned as a wave? In practice, the wave properties are particularly prominent when the electron is confined to a small space such as within an atom or between the atoms of a crystal. Electrons always carry a unit negative charge. When thinking of packets of charge, it is useful to think of an electron as a particle. Both models are useful, however, and both are used. As we are currently concentrating on the structure of electrons within an atom, the focus here is on the wave model.

One of the major features of the wave model is that the electron wave envelops the nucleus and fills the three-dimensional space around the nucleus. The mathematical formulation of the electron as a wave was proposed by an Austrian physicist, Erwin Schrödinger (Figure 2.22), in 1926 and is known as the **quantum mechanical model** of the atom. Schrödinger's equation for the electron involves complex mathematics. The solutions describing the possible electron waves are called **wave functions** and are denoted by the Greek letter psi, ψ. The results of Schrödinger's equation provide a model for the electron that is consistent not only with observed atomic spectra, but also with interactions among atoms.

Schrödinger's model provides an interpretation for the wave function: The wave function, ψ, is the **amplitude** of the electron wave at any point in space. The amplitude of the electron wave is analogous to the height of a physical wave such as a flexible cord; ψ locates the crests and troughs of the electron wave. The square of the amplitude, $|\psi|^2$ (the wave function is complex and $|\psi|^2$ denotes the square of this complex function), is a measure of the probability of finding the electron in a small volume—that is, it is the electron wave **probability density.** The electron's position cannot be located precisely, but $|\psi|^2$ indicates where the electron is most likely to be found.

For electromagnetic waves, a direct correlation exists between the frequency of the wave and its energy. As the frequency of a wave increases, the number of points per unit length where the wave has zero amplitude, called **nodes,** increases (Figure 2.15). Not counting the two endpoints, the fundamental wave has no nodes within the wave (Figure 2.15a). Inputting more energy creates a higher-frequency wave. Along with the higher frequency comes an increase in the number of nodes (Figures 2.15b and c). The number of nodes in a three-dimensional wave is also correlated with the energy of the wave and serves to classify and label the many wave functions of the electron.

To aid in visualizing the three-dimensional waves of the electron, first consider two-dimensional, physical waves. An example occurs on a drum top. The notation used to label the nodes in these two-dimensional waves is analogous to that used for the three-dimensional electron waves.

Figure 2.22 Erwin Schrödinger developed a model for electrons in an atom as waves. He published this model in 1926 and received the Nobel Prize for this work in 1933.

Quantum mechanical model Mathematical formulation of the electron as a wave.

Wave function, ψ A function of the coordinates of an electron's position in three-dimensional space that describes the properties of the electron.

Amplitude Magnitude of the wave function of the electron wave at a point in space.

Probability density, $|\psi|^2$ Measure of the probability of finding the electron in a small volume.

Node Point of an orbital having zero electron probability

Two-Dimensional Waves

A drum creates sound when the drum top is hit and oscillates. Two-dimensional waves are created on the drum top, and the energy input to the drum dictates the form of those waves. A series of fast-frame photos of the drumhead effectively freezes it at various points in its motion. Figure 2.23 shows a rendition of a drumhead oscillating in the fundamental or lowest-energy state. Frames on the left catch the drumhead with an upward, or positive, displacement (equivalent to a crest); frames on the right have a downward, or negative, displacement (equivalent to a trough).

Hitting the drumhead harder puts more energy into the wave. Figure 2.24 shows a "freeze frame" of several higher-energy, higher frequency modes. These modes have characteristically increasing numbers of nodes. The highest amplitude is in the center of the drum. As a two-dimensional analogue of the atom, the nucleus is in the center of the drum and the nodes (shown as white lines in the freeze frame) encircle the nucleus. Rather than appearing like a planet orbiting the sun, these two-dimensional analogues of the electron wave look like the ripples created when an object strikes the surface of a pond.

The lower part of Figure 2.24 represents the same information as the upper part, but this time as contour plots. A contour plot shows the height of the wave (the amplitude) in the same way that an elevation map used by hikers shows the height of the land. When the amplitude of the wave changes slowly, the contour lines are relatively far

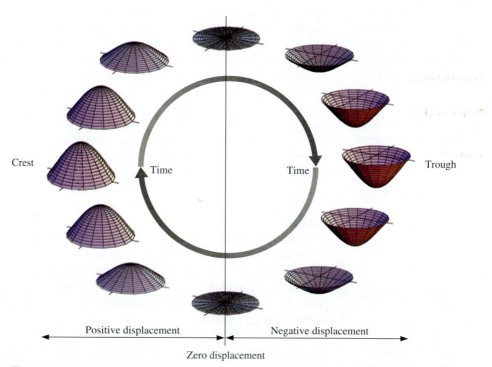

Crest   Time   Time   Trough

Positive displacement   Negative displacement

Zero displacement

Figure 2.23 In the oscillation of a drumhead with no friction, the drumhead cycles indefinitely from a crest to flat to a trough, and back again. The mode shown here is the fundamental (lowest-energy) mode. There are no nodes—no point where the displacement remains at zero and the top is stationary.

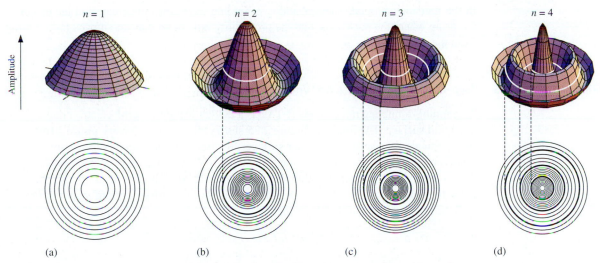

Figure 2.24 Four successively higher energy modes of a drum shown as surface plots and contour plots. Contour plots are similar to maps with elevation lines. On the contour plots, the zero contours are shaded dark. Vertical lines connect the zero-amplitude points (nodes) on the two types of plots.

Positive phase The crest of a wave.

Negative phase The trough of a wave.

Mode Refers to the entire wave.

apart. When the amplitude changes rapidly, the lines are close together. The darkest lines on the contour plots indicate the nodes of the wave where the amplitude is zero. Nodes separate the "mountains" from the "valleys" (analogous to the crests and troughs of one-dimensional waves). Putting an elevation scale on the amplitude, the node corresponds to zero, the mountains to positive numbers, and the valleys to negative numbers. In wave terminology, mountains and valleys are of opposite phase, mountains have **positive phase,** and valleys have **negative phase.** The contours in Figure 2.24 are all circles: The wave is the same height at all equal distances from the center, and the nodes are circular rings.

The geometrical shape of the nodes is used to classify the various waves. The waves are referred to as **modes.** Waves containing circular *nodes* are called *s* waves or *s modes. Mode* and *node* are similar words but they refer to very different objects. *Mode* refers to the entire wave while *node* refers to just the zero-amplitude portion of the wave. The number of nodes is correlated with the energy of the mode, so energy increases from wave (a) through (d) in Figure 2.24. The wave in Figure 2.24a is the fundamental mode shown in Figure 2.23, and corresponds to the $n = 1$ orbit of the Bohr model. The wave in Figure 2.24d corresponds to the $n = 4$ orbit. For a single electron, the energy is entirely determined by the total number of nodes, Bohr's n. Indeed, n is the total number of nodes plus one. The waves in Figure 2.24 have been labeled according to this scheme. In three dimensions, the same relationship holds: The total number of nodes is $n - 1$.

Consider the two-dimensional waves of the drum as being the electron in a sort of flat-land atom. The positively charged nucleus is a miniscule dot at the center of the drum, and the wave surrounding the nucleus is the negatively charged electron. In Schrödinger's wave model, the square of the wave amplitude, $|\psi|^2$, represents the probability of finding the electron in a given volume. This language, including the use of the word "probability," reflects the fuzzier nature of the electron compared with everyday objects.

In the fundamental mode the probability of finding the electron is greatest at the center of the drum where the nucleus is, and the probability falls rapidly as one moves farther away from the nucleus.

WORKED EXAMPLE 2.5   *Two-Dimensional Electron Wave*

A scanning-tunneling microscope (STM) is a device that detects the electron density in a solid surface. The image in Figure 2.25 shows the electron density inside a ring of 48 iron atoms placed on a copper surface. Electrons on the copper surface are relatively loosely bound. Because the iron atoms form a barrier to electron motion, these electrons are corralled within the ring of iron atoms. The ripples on the surface show the variation of the electron density within the ring. What value of n should be assigned to this mode of the electron?

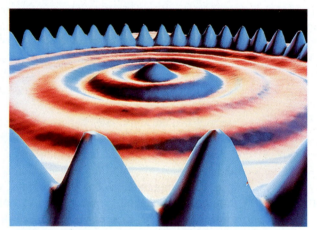

Figure 2.25 Image of a copper surface with a ring of 48 iron atoms, taken with a scanning tunneling microscope (STM). The light-to-dark shading in the STM image corresponds with the high-to-low electron density, respectively. The image is therefore the electron density for an electron confined to a circular region. The mode of an electron confined to a circular region looks like a mode of an oscillating drum.

Plan

- Electron density is the square of the electron amplitude such that crests and troughs both have high values of the density. The nodes remain zero, and are dark regions on the image.
- The total number of nodes is $n - 1$.

Implementation

- Counting the nodes (dark regions) outward from the center of the image, there are four dark rings.
- n is one more than the number of nodes, so $n = 5$.

See Exercises 56, 57.

Three-Dimensional Waves

Three dimensions are required to describe electrons enveloping the nucleus and filling the space surrounding the nucleus. Although waves are more difficult to picture in three dimensions than in two, the same principles hold. The lowest-energy wave has no nodes. Higher-energy waves have more nodes. In three dimensions, nodes can be oriented in different directions. Keep in mind that electron waves are *complex* (in the mathematical sense; i.e., they have imaginary aspects), while a string or drum oscillation is real. The three-dimensional waves that describe electrons are referred to as **orbitals,** to contrast them with Bohr's orbits. Their mathematical description comes from the wave functions of Schrödinger's equation.

An electron in the $n = 1$ ground state, having the lowest energy, has no nodes and is spherically symmetric. This is referred to as a $1s$ orbital (Figure 2.26). Since the electron is a wave, it does not have sharply defined edges, like those of a soccer ball. Instead, the electron wave is more like a cotton ball that is most dense in the center. Imagine trying to measure the diameter of a cotton ball. How large should the sphere representing the cotton ball be drawn? For the electron, the sphere that is conventionally chosen is a bounding surface within which there is a 90% probability of finding the electron. For the ground state of hydrogen, this sphere has a radius of 79 pm.

Figure 2.27a illustrates the $1s$ orbital (the $1s$ state) of Figure 2.26 being sliced by a plane. Cutting through the center with a contour density map shows that the amplitude of the electron wave (ψ) is greatest at the center of the sphere, where the nucleus is found, and falls off in all directions from the center, much like the fundamental drum

Orbital A specific wave function for an electron in an atom. The square of this function gives the probability distribution for the electron.

Figure 2.26 An *s* orbital is spherically symmetric. The 1*s* orbital has no nodes, just like the lowest-energy drum mode.

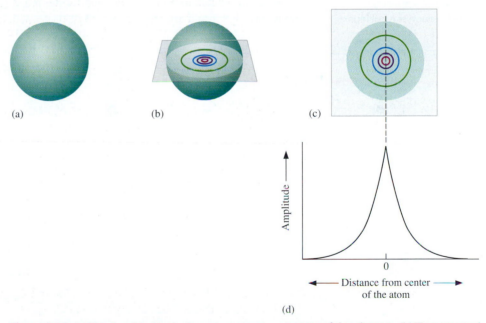

(a)   (b)   (c)

Amplitude

0

⟵ Distance from center ⟶
of the atom

(d)

Figure 2.27 (a) The 1*s* orbital is the lowest-energy, $n = 1$, state of the electron. (b) The spherical surface has been cut by a contour to show the space-filling nature of the 1*s* electron wave. The wave amplitude is illustrated with colors of the rainbow, ranging from green for near-zero contours through magenta for the highest amplitude. (c) Looking down on the electron contour plot clarifies the circular symmetry of the electron density. (d) A plot of the amplitude of the wave as a function of the distance from the center of the sphere illustrates the exponential drop in density with distance away from the nucleus. Note that the nucleus is a minute speck in the center of this sphere, and the electron wave has the greatest amplitude immediately around the nucleus. The wave has no node.

mode. This reduction of amplitude is very rapid; in fact, it is exponential. The square of the wave amplitude, $|\psi|^2$, at each point gives the probability of finding the electron at that point. Thus, for the $1s$ state, the probability of finding the electron is very high at the center of the sphere—at the nucleus—and decreases rapidly moving away from the nucleus. Indeed, the electron density is greatest at the nucleus for all s orbitals, and all s orbitals are spherical. The $1s$ orbital is the lowest-energy orbital for an electron.

One Node The next more energetic electron wave can have either of two types of nodes. One type is a spherical node analogous to the drum s wave. If the node is a sphere, then the orbital is spherically symmetric, and the orbital is called a $2s$ orbital (Figure 2.28). This electron wave envelops the nucleus. A contour plot (Figure 2.28c) slicing through the center of the spherical electron cloud emphasizes the circular symmetry and, as for all s orbitals, the amplitude peaks at the nucleus. Due to the spherical node, the $2s$ orbital contains two regions of large amplitude and looks like two concentric spheres with greater electron amplitude in the inner sphere (Figure 2.28c). Like the mountains and valleys of the two-dimensional drum wave, the inner and outer spheres are separated by the node and are of opposite phase. The $2s$ orbital extends farther from the nucleus than the $1s$ orbital does—four times as far for the hydrogen atom.

The three-dimensional electron wave can also have a node that slices through the nucleus (Figure 2.29), with opposite phases then appearing on the two sides of the nucleus. Nodal planes cutting through the nucleus are called **azimuthal** nodes because the plane forms an arc as it slices through the sphere surrounding the nucleus. The number of azimuthal nodes is denoted by the letter ℓ. Orbitals with a single planar node have $\ell = 1$ and are labeled p modes; those with two azimuthal nodes ($\ell = 2$)

Azimuthal A nodal plane cutting an arc through the bounding sphere. Associated orbital has a nodal plane through the nucleus.

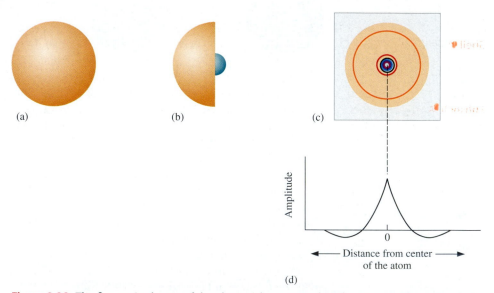

(a)   (b)   (c)

Amplitude

0

← Distance from center →
of the atom

(d)

Figure 2.28 The first excited state of the electron has one node. When the node is a spherical node, the state consists of two large-amplitude spheres and is called a $2s$ orbital. The $2s$ orbital is shown whole (a) and after peeling away half of the outer sphere (b). Slicing through the equator of the wave (c) reveals the spherical symmetry of the wave's interior. (d) The amplitude of the electron wave is shown with colors: green corresponds to the node; colors from green to magenta correspond to a crest; colors from green to red represent a trough. The center of the sphere is separated from the outer edges by a node (zero amplitude). The inner sphere is said to be of opposite phase from the outer sphere.

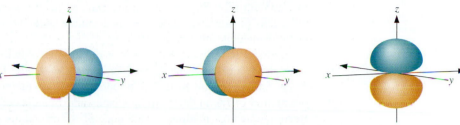

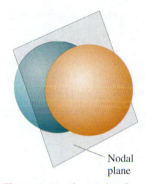

Figure 2.29 The node of an electron orbital can be a plane that cuts through the center of the atom.

Figure 2.30 The three *p* orbitals illustrated by the bounding surface that contains 90% of the electron wave. Orbitals are labeled according to their direction. For example, the orbital with lobes pointing along the *x*-axis is called a p_x orbital. The crest is depicted in turquoise, the trough in yellow. For example, in the p_x orbital the amplitude in the positive *x*-direction is opposite of that in the negative *x*-direction. The set of three *p* orbitals constitutes the *p* subshell.

are labeled *d* modes; and those with three ($\ell = 3$) are called *f* modes. Extending this notation, an *s* mode has ℓ equal to zero—that is, it has no azimuthal nodes.

All *p* orbitals have one planar node. In three dimensions there are three possible perpendicular directions, and a nodal plane can be located in any of these directions. As a result, there are three possible orientations of the *p* orbitals. A *p* orbital can have a nodal plane defined by the *x*- and *y*-axes; the electron density is then located along the *z*-axis, and the orbital is labeled a p_z orbital. Similarly, the planar node can be defined by the *x*- and *z*-axes with density along the *y*-axis; it is then called a p_y orbital. The third combination is a p_x orbital (Figure 2.30).

CONCEPT QUESTION Where is the nodal plane of a p_x orbital? ■

Shell All electron orbitals with a common value of *n*.

Subshell All the orbitals of a given shell, *n*, with a common value of ℓ.

For electron waves, the set of orbitals with a common number of nodes constitutes a **shell.** In the second shell, $n = 2$, so there can be only one node: The number of nodes is $n - 1$. Since all nodes are either spherical or planar, the single *s* orbital and the three *p* orbitals constitute the entire second shell. The *s* orbital of the second shell contains one spherical node. The *p* orbitals in the second shell contain a single planar node.

Orbitals with a common value of *n* and ℓ constitute a **subshell** of the *n*th shell. In the second shell, one 2*s* orbital constitutes the 2*s* subshell. The three 2*p* orbitals constitute the 2*p* subshell. The second shell has two subshells.

A major difference between the *s* orbital and the *p* orbitals of a given shell is that an *s* orbital has a *maximum* amplitude at the origin (the nucleus), whereas each *p* orbital has a node or *zero amplitude* at the nucleus (Figures 2.28). Hence an *s* orbital electron has considerable amplitude, ψ, and density, $|\psi|^2$, at the nucleus. In contrast, a *p* orbital electron can be found close to the nucleus but never at the nucleus. Although it is tempting to think of the electron in a *p* orbital as a particle hopping from one side of the nucleus to the other, near the nucleus it is better to think of the electron as a wave. The *p* wave has a crest on one side of the nucleus and a trough on the other. The nucleus sits on the node between the crest and the trough.

CONCEPT QUESTION The longest wavelength transition in the Lyman series of the hydrogen atom corresponds to a transition from a 2*p* orbital to the 1*s* orbital. Draw a schematic of the two orbitals involved in this transition. Label the orbitals with their energy [Equation (2.8)]. Where is this energy located on an energy-level diagram for hydrogen? ■

The bounding surface indicates the overall geometry of a *p* orbital. The interior structure can be examined by slicing a *p* orbital with a plane and mapping the amplitude

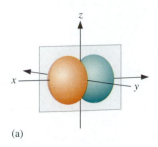

(a)

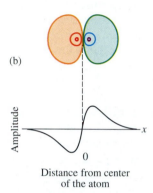

(b)

(c)

Figure 2.31 (a) A p_x orbital shown with a slicing surface in the x-z plane. (b) The contour map shows the node in the y-z plane. The maximum amplitude gathers near the nucleus, but there is a node, zero amplitude, exactly at the nucleus. (c) The p_y and p_z orbitals are similar, with lobes and nodes rotated 90°.

in a contour plot. Figure 2.31 shows such a slice through the p_x orbital. In all of these renditions of orbitals, the nucleus is a minute speck at the origin. All p orbitals have a nodal plane that runs through the origin.

Indeed, all orbitals other than s orbitals have at least one nodal plane that runs through the nucleus. The nodal plane along with the negative charge of the electron results in a magnetic field. The orientation of this magnetic field is determined by the orientation of the nodal plane or planes and is indicated by an integer denoted as m_ℓ. Because three orientations for the node of a p orbital are possible, there are three possible values of m_ℓ ($+1, 0, -1$).

Two or More Nodes

Increasing energy goes hand-in-hand with more nodes. Orbitals with $n = 3$ are next higher in energy in hydrogen, and each of these has two nodes. The nodes can be both spherical, one spherical and one planar, or both planar. If both nodes are spherical, the orbital is called a $3s$ orbital. If one is spherical and one planar, the obital is a $3p$ orbital.

The final option for two nodes is for both nodes to be planar. All planar nodes pass through the nucleus. With two planar nodes, $\ell = 2$. The orbital is called a d orbital—in this case, a $3d$ orbital. Five d orbitals are possible (Figure 2.32), and the d subshell is formed from these five d orbitals.

Figure 2.33 shows the result of slicing through a d orbital with a plane through the nucleus. The contour plot shows that the amplitude of each lobe increases rapidly from the node at the nucleus. After reaching the extreme value, the amplitude approaches zero rapidly as the distance from the nucleus increases.

Quantum Numbers

Energy increases as the number of nodes increases. High-energy electron orbitals thus have many nodes. The quantum mechanical model provides a powerful, compact notation for specifying these orbitals, called **quantum numbers.** Three quantum numbers indicate an orbital: n, ℓ, and m_ℓ. Exacting spectroscopic measurements indicate a fourth quantum number that has a small effect on the energy of the electron; it is called the spin magnetic quantum number. Electron spin discusses briefly below.

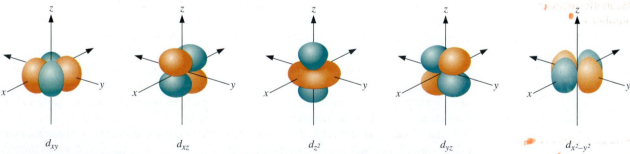

Figure 2.32 The five $3d$ orbitals. The turquoise and yellow lobes are crests and troughs, respectively. The labels for these orbitals reflect the directions in which the quadratic function has a maximum value. For example, the quadratic function xz has a maximum between the x- and z-axes, as has the d_{xz} orbital. Similarly, the function $x^2 - y^2$ has a maximum along the x- and y-axes, as has the $d_{x^2-y^2}$ orbital. The product of x times z has its extreme value when x and z are both positive and when both are negative. The d_{xz} orbital has a crest in both the $+x+z$ quadrant and in the $-x-z$ quadrant. The product is negative in the other two quadrants, and the orbital has a trough there. The set of five d orbitals constitutes the d subshell.

Figure 2.33 (a) A plane slicing through one of the *d* orbitals. (b) A contour plot of amplitude in the plane shown in (a) reveals the two nodal planes running through the nucleus.

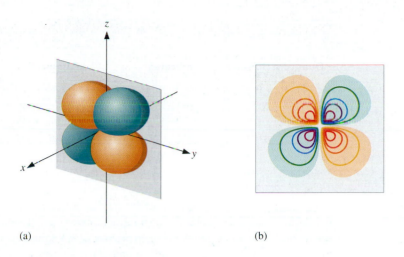

(a) (b)

Quantum numbers Compact notation for specifying orbitals.

Principal quantum number, *n* The quantum number relating to the size and energy of an orbital; it can have any positive integer value.

Azimuthal quantum number, *ℓ* Second quantum number that is equal to the number of planar (azimuthal) nodes. $\ell \leq n - 1$.

Magnetic quantum number, *m*ℓ The quantum number relating to the orientation of an orbital in space relative to the other orbitals with the same *ℓ* quantum number. It can have integral values between *ℓ* and −*ℓ*, including zero.

Spin magnetic quantum number, *m*s Fourth quantum number that contributes a small energy correction: values $\pm^1/_2$.

The first three quantum numbers specify the orbital. The **principal quantum number, *n*,** specifies the shell and indicates the total number of nodes; the number of nodes is equal to $n - 1$. For hydrogen the energy is correlated with *n* as described by Equation (2.8). Higher values of *n* correspond to larger and higher-energy orbitals. The values of *n* are positive integers: 1, 2, 3, ...

The **azimuthal quantum number, *ℓ*,** is equal to the number of planar (azimuthal) nodes. There can be no more *nodal planes* than *total nodes*. Thus *ℓ*, which is an integer, has values from 0 to $n - 1$. The letter code indicates the value of $\ell : \ell = 0$ is designated by *s*, $\ell = 1$ by *p*, $\ell = 2$ by *d*, and $\ell = 3$ by *f*. Higher values of *ℓ* continue alphabetically: *g, h,* . . . . This letter code originated with spectroscopic work done in the 1890s characterizing the line spectra of the elements. The letter *s* corresponds to sharp, *p* to principal, *d* to diffuse, and *f* to fundamental lines. The set of orbitals with common *ℓ* values constitute a subshell, and each shell *n* has *n* subshells.

Planar nodes can be oriented in several directions in three-dimensional space. This orientation does not affect the energy of an electron but it does affect its spatial geometry and the orientation of the associated magnetic field. The orientation of the planar nodes is denoted by the **magnetic quantum number, *m*ℓ.** The magnetic quantum number has integer values between +*ℓ* and −*ℓ* ($\ell, \ell - 1, \ldots, 0, -1, \ldots, -\ell$). Since m_ℓ is restricted to integer values between +*ℓ* and −*ℓ*, there are a total of $2\ell + 1$ values for m_ℓ. This is the basis for the number of orbitals in each subshell. For example, for $\ell = 0$, m_ℓ can be only 0; hence there is one *s* orbital. Each higher value of *ℓ* adds two more orbitals, giving the series 1*s* orbital, 3*p* orbitals, 5*d* orbitals, and so on. With $2\ell + 1$ values of m_ℓ, there are $2\ell + 1$ orbitals in each subshell. With *n* subshells in shell *n*, there are n^2 orbitals in shell *n*.

The fourth quantum number, called the **spin magnetic quantum number, *m*s,** contributes a small energy correction and is consistent with the electron behaving as though it is spinning on its axis. Even though this language makes it seem as if the electron is a particle, it should be thought of as a wave. Nothing in our macroscopic everyday experience can precisely describe spin. It is as though the electron turns on its axis in the way the earth revolves on its axis as it orbits the sun. Spin can be clockwise or counterclockwise and is labeled as $\pm^1/_2$. There is only a very small energy difference between spin $= +^1/_2$ and spin $= -^1/_2$. With two possible values for spin, there are $2n^2$ unique quantum numbers for each shell *n*.

Table 2.3 Four Quantum Numbers Indicate the Electron Orbital and Spin

Quantum Number	Indicates	Values	Number of Possible Values
n	shell	$1, 2, 3, \ldots$	∞
ℓ	subshell $s, p, d, f, \ldots$	$0, 1, 2, \ldots, n-1$	n
m_ℓ	orbital orientation	$-\ell, \ldots, 0, \ldots, \ell$	$2\ell + 1$
m_s	spin	$-\frac{1}{2}, +\frac{1}{2}$	2

CONCEPT QUESTIONS All the transitions involved in the Balmer series terminate in the second shell with either the 2s or 2p orbitals. Green light emission starts with the electron in $n = 4$. In the transition the quantum number ℓ must either increase or decrease by one. What initial values can ℓ have if the electron ends in the 2s orbital? What initial values can ℓ have if the electron ends in the 2p orbital? ∎

A full set of quantum numbers for an electron is n, ℓ, m_ℓ, and m_s (Table 2.3).

In summary, the complete set of quantum numbers is as follows:

- **Principal quantum number:** n
  Specifies the shell, the *total number of nodes* (equal to $n - 1$), and the energy
  $n = 1, 2, 3, \ldots$
- **Azimuthal quantum number:** ℓ
  Specifies the *number of planar nodes* and, therefore defines the orbital shape
  Orbitals are described as $s, p, d, f, g, \ldots$ corresponding to $\ell = 0, 1, 2, 3, 4, \ldots$
  $\ell = n - 1, \ldots, 0$
- **Magnetic quantum number:** m_ℓ
  Specifies the *orientation of the planar nodes*
  $m_\ell = \ell, \ell - 1, \ldots, 0, \ldots, -\ell$
- **Spin quantum number:** m_s
  Specifies the *electron spin* direction
  $m_s = \pm \frac{1}{2}$

WORKED EXAMPLE 2.6   *Connections Among Quantum Numbers*

How many states are possible for the first excited state of hydrogen?

Plan
- Determine the principal quantum number, n.
- Determine the allowed combinations of the remaining quantum numbers.
- Alternatively, determine the number of orbitals (n^2) and the number of states ($2n^2$).

Implementation
- For the ground state, $n = 1$. Therefore, for the first excited state, $n = 2$.
- If $n = 2$, then allowed values of ℓ are 1 and 0. If $\ell = 1$, then $m_\ell = 1, 0, -1$. If $\ell = 0$, $m_\ell = 0$. Every state also has $m_s = \pm \frac{1}{2}$. The possibilities are listed in the following table. The total number of states possible is eight.

n		**2**		
ℓ		1		0
m_ℓ	1	0	-1	0
m_s	$\pm \frac{1}{2}$	$\pm \frac{1}{2}$	$\pm \frac{1}{2}$	$\pm \frac{1}{2}$

- Using the alternative method, for $n = 2$, $n^2 = 4$, and the number of states equals $2n^2 = 8$.

See Exercises 72–79.

WORKED EXAMPLE 2.7   *Connecting Orbitals to Quantum Numbers*

What is the lowest shell (smallest principal quantum number) that contains an f orbital?

Plan
- Determine the value of ℓ for an f orbital.
- Use the information: $n - 1 \geq \ell$.

Implementation
- Orbital letter designations are connected with the value of ℓ: $\ell = 0$ is designated by s, $\ell = 1$ by p, $\ell = 2$ by d, and $\ell = 3$ by f. Thus, for an f orbital, $\ell = 3$.
- For $\ell = 3$, $n - 1 \geq 3 \Rightarrow n \geq 4$. The lowest shell that contains an f orbital is the $n = 4$ shell.

See Exercises 68–71.

2.5   Energy Levels of Multielectron Atoms

Orbitals of multielectron atoms retain the same shape as hydrogen's orbitals. However, the orbital sizes and energies depend on the nuclear charge and the number of electrons, which are quantities unique to each element. For example, the energies of the three electrons in lithium (which has three protons in its nucleus) and the 11 electrons in sodium (which has 11 protons in its nucleus) are quite different, and this difference is clearly seen in the contrast between the yellow color emitted by sodium and the vivid red of lithium. Indeed, each element in the periodic table has its own characteristic emission that can be used to detect the presence of the element. For example, the toxic contaminants cadmium (Cd) and chromium (Cr) in soils or drinking water are detected via an atomic emission spectrograph, an instrument that analyzes the radiation emitted when a sample is heated in a flame or plasma. The arrangement of electrons in the orbitals of the atoms of an element is responsible for the different electron energies, and the electron energies determine the color emitted by Na, Li, and other elements.

Progressing from one element to the next in the periodic table, two important properties change: The number of protons in the nucleus increases by one, and the number of electrons increases by one to maintain a neutral atom. Just as in the hydrogen atom, each electron in a multielectron atom is attracted toward the nucleus, and a larger positive charge from the additional protons tends to create a stronger electron-nuclear attraction. However, unlike in hydrogen, the presence of more than one electron introduces an electron-electron repulsion from the like charges of the electrons. This repulsion opposes the stronger electron–nuclear attraction, affecting the electron energy and size of the orbitals. This interplay of coulombic attraction between each electron and the nucleus and the mutual repulsion between electrons determines the energy state (energy level) of electrons in the atom.

In the ground state, hydrogen's lone electron is found in the most strongly bound state available—the $1s$ orbital. The electrons of a multielectron atom also occupy the lowest energy states available. The electrons do not all crowd into the $1s$ orbital, however. A limited number of them can occupy any given orbital. A multielectron atom in its ground state has electrons in the most strongly bound, lowest-energy state available for each electron—known as the ground-state **electron configuration.** To determine which orbitals are available—the electron configuration—both the energy ordering for electrons in various orbitals and the number of electrons that can occupy each orbital are needed.

Electron configuration
Arrangement of electrons in an atom or ion.

Energy Ordering

Electron energy in multielectron atoms differs from that in one-electron atoms or ions in two important ways. First, electrons are like-charged particles and repel each other. *Repulsion* is greatest for electrons in the same orbital, as they are found in the same region of space around the nucleus. Similarly, electrons in the same subshell repel each other quite strongly due to their close proximity. Electrons in different shells are separated the most and repel least.

Second, the energy of electrons in multielectron atoms differs from the energy of the single electron in hydrogen due to **shielding.** *Shielding* occurs when some of the electron density of the inner electrons—those with a lower shell number, n—comes between the outer electrons (larger n) and the nucleus, effectively screening the nuclear attraction felt by the outer electrons. As an analogy, when you are driving through fog, the beam from the car headlights cannot penetrate as far as on a clear night. Similarly, the charge from the nucleus, shrouded by inner-shell electrons, does not penetrate effectively to the outer shell electrons. Shielding effectiveness depends both on the shell and on the shape of the electron orbital. Due to the wave nature of the electron, electrons interpenetrate. An electron in an orbital with a large probability density at the nucleus (e.g., an s-orbital electron) penetrates through the shield created by inner-shell electrons more effectively than do electrons in orbitals with a node at the nucleus, such as a p-orbital electron.

Shielding. Diminishment of the nuclear charge felt by an electron due to the presence of other electrons.

A variation of Equation (2.9) is used to determine how effectively the inner-shell electrons shield the nuclear charge. Consider lithium. In the absence of shielding, the ionization energy of the second-shell electron in Li is

(2.13)   Ionization energy, shell $n = \dfrac{Z^2}{n^2}$ (13.6 eV) $= \dfrac{3^2}{2^2}$ (13.6 eV) $= 30.6$ eV

However, the measured ionization energy of Li is only 5.39 eV. The difference, more than 25 eV, is due to shielding of the nuclear charge by the electrons of the first shell—a fairly effective shield. To quantify the shielding, imagine that the lithium atom consists of a nucleus with an **effective charge, Z_{eff},** and a single electron in the second shell. This hypothetical lithium atom is like the hydrogen atom because it has only one electron. The question is: What charge, Z_{eff}, is needed for the ionization energy to be 5.39 eV?

Since the atom has only one electron, results of the Bohr model can be applied. Modifying and rearranging Equation (2.13) gives

Effective nuclear charge, Z_{eff}
The effective charge holding an electron in the atom under a Bohr model of the atom, i.e., all charge treated as a point charge at the center of the atom.

(2.14) $Z_{\text{eff}} = n \sqrt{\dfrac{\text{ionization energy}}{13.6 \text{ eV}}}$

The electron is in the 2s orbital ($n = 2$), so this expression becomes

(2.15)
$$Z_{eff} = 2\sqrt{\frac{5.39 \text{ eV}}{13.6 \text{ eV}}} = 1.26$$

More than one unit of positive charge holds the electron on the hypothetical Li atom. Let us put this number in context. An effective charge of 1.00 indicates completely effective shielding: The two 1s electrons balance two of the charges in the Li nucleus. Since Z_{eff} is greater than one, a portion of the nuclear charge gets through the two 1s-orbital electrons, but the effective charge is significantly less than the $+3$ nuclear charge.

WORKED EXAMPLE 2.8   *Shielding in Cesium*

Despite having 55 protons attracting the electrons in cesium (Cs), the ionization energy of cesium is quite low, 3.894 eV, which is typical of the Group IA alkali metals. Assume that the outermost electron in Cs is in the sixth shell. Determine how well the 54 inner electrons shield the outermost electron from the $+55$ charge of the Cs nucleus.

Plan

■ Use Equation (2.14), $Z_{eff} = n\sqrt{\dfrac{\text{ionization energy}}{13.6 \text{ eV}}}$ with $n = 6$.

Implementation

■ $Z_{eff} = 6\sqrt{\dfrac{3.894 \text{ eV}}{13.6 \text{ eV}}} = 3.21$

If screening were perfect, the effective charge would be one, so only 2.21 of the 55 charges leak out. On average, the core electrons are 96% effective at shielding a unit of positive charge.

See Exercises 91–92.

 APPLY IT Observe light from a car with quartz-halogen headlights, a fluorescent light, a neon light, and an incandescent bulb. Describe the color of each of these light sources. How do you distinguish between a car with quartz-halogen headlights and one with tungsten headlights? Describe how energy is input and name the source of the emission. Which of these four light sources can be described as an atomic emission source, that is, which generates light by an electric discharge through a gas?

WORKED EXAMPLE 2.9   *Color, Energy, and Shielding*

The wavelength of the vivid red color from Li is 670.8 nm. This photon results from a transition from the 2p to the 2s level. What is the difference in ionization energy of an electron in the 2p and 2s orbitals of Li? What is the ionization energy of the 2p state in Li? What is Z_{eff} for a 2p electron?

Plan

■ Use the equation $E = \dfrac{hc}{\lambda}$ and the conversion factor 1 eV $= 1.6022 \times 10^{-19}$ J.

■ Energy of the 2p state $= -5.392$ eV $+$ energy difference between 2s and 2p states.

■ Use Equation (2.14): $Z_{\text{eff}} = n \sqrt{\dfrac{\text{ionization energy}}{13.6 \text{ eV}}}$

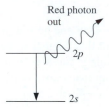

Red photon out

2p

2s

Implementation

■ The red-colored, 670.8-nm photon is emitted when the electron drops from the 2p level to the 2s level. The energy of this photon corresponds to the difference in energy between the 2p and 2s states.

$$E = \frac{hc}{\lambda} = \frac{(6.6262 \times 10^{-34}\, \text{J} \cdot \text{s})(2.9979 \times 10^{8}\, \text{m/s})}{(670.8 \text{ nm})(10^{-9}\, \text{m/nm})}$$

$$\times \frac{1 \text{ eV}}{1.6022 \times 10^{-19}\, \text{J}} = 1.848 \text{ eV}$$

■ 2p state energy = -5.392 eV + 1.85 eV = -3.54 eV; ionization energy = 3.54 eV.

■ $Z_{\text{eff}} = 2 \sqrt{\dfrac{3.54 \text{ eV}}{13.6 \text{ eV}}} = 1.02$

A 2p orbital electron in Li is less stable than a 2s orbital electron. The 2p electron in Li is nearly totally shielded from the added charge in the Li nucleus, thus feeling essentially only one charge.

See Exercises 93, 94.

This greater penetration to the core by s-orbital electrons compared with p-orbital electrons is a general feature that repeats in every period. In every shell (Figure 2.34), the s subshell is the lowest-energy subshell and fills before other subshells. In general, electrons in orbitals with fewer nodes at the nucleus penetrate through core electrons more effectively and thus experience a larger Z_{eff} than those with more nodes.

These two effects—electron-electron repulsion and shielding of the nuclear charge—account for the energy ordering of electronic states for multielectron atoms. The details of determining this order are quite complex because the presence of multiple electrons complicates the picture sufficiently that Schrödinger's equation cannot be solved exactly for any element other than hydrogen. There are, however, general patterns to the results. Due to the shielding effect, in multielectron atoms the electrons in different subshells (s, p, d, . . .) within a shell have somewhat different energies. In contrast, energy for hydrogen depends only on n. The energy for an electron in different subshells depends on the orbital shape—the number of planar nodes—as the number of nodes determines how effectively the electron penetrates through inner-shell electrons to the positive charge of the nucleus.

Despite the complex balance that determines energy ordering, the pattern for main-group elements—the first two and last six elements in each period—is fairly simple. Figure 2.34 illustrates the typical energy ordering for single-electron and multielectron

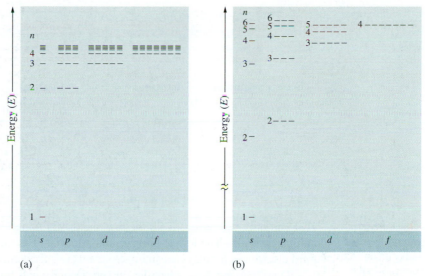

Figure 2.34 (a) Energy levels in hydrogen depend only on the shell or principal quantum number, n. (b) In multielectron atoms, as electrons occupy orbitals nearer to the nucleus, outer electrons are shielded from the nuclear charge. The $n = 2$ and higher levels are drawn approximately to scale. The $1s$ orbital is significantly more negative than shown. The relative energy of different orbitals depends on the number of electrons in other orbitals. Some elements have a different order, and we will discuss these later in the chapter.

atoms. Beyond hydrogen and helium, the $n = 1$ shell electrons are strongly attracted to the nucleus, and there is a large energy gap between them and the $n = 2$ electrons. As n increases, the energy levels get closer together, and the energies of some subshells overlap subshells with a lower value of n. For example, the d orbitals are very close in energy to the next lower shell's s orbital. Similarly, the $4f$, $5d$, and $6s$ orbitals are close in energy for sixth-period metals.

Orbital Election Capacity: Pauli Exclusion Principle

> **Pauli exclusion principle**
> No more than two electrons can occupy one orbital: no two electrons can have the same set of quantum numbers.

In 1925 the Austrian physicist Wolfgang Pauli formulated a principle based on the property that electrons avoid each other. The **Pauli exclusion principle** states that two electrons can occupy the same orbital only if their spins are opposite. This principle has significant implications for the number of electrons that can occupy a subshell and a shell. An orbital is specified by three quantum numbers: n, ℓ, and m_ℓ (Table 2.4). Spin is specified by the fourth quantum number, m_s. These four quantum numbers constitute a full set. Thus the Pauli exclusion principle says that two electrons can occupy the same orbital—have the same first three quantum numbers—only if the last quantum number—the spin—is different. But spin can have only two values: $m_s = \frac{1}{2}$ and $m_s = -\frac{1}{2}$. *As a consequence, only two electrons can occupy any given orbital.* No two electrons can have the same four quantum numbers, so each electron has a unique identity—its own address.

Returning to the lithium atom, the Pauli principle indicates why two of its three electrons occupy the $1s$ orbital, and the third electron must go into the second shell. The two $1s$ electrons have identical values for n, ℓ, and m_ℓ, but one has $m_s = +\frac{1}{2}$ and the other $m_s = -\frac{1}{2}$.

$$n = 1, \ell = 0, m_\ell = 0, m_s = +\tfrac{1}{2} \quad \text{and} \quad n = 1, \ell = 0, m_\ell = 0, m_s = -\tfrac{1}{2}$$

Table 2.4 Quantum Numbers
The quantum numbers that specify an orbital are related:
m_ℓ values are limited by ℓ, and ℓ values are limited by n.

Quantum Number	Indicates	Values	Number of Possible Values
n	shell	$1, 2, 3, \ldots$	∞
ℓ	subshell: $s, p, d, f, \ldots$	$0, 1, 2, \ldots, n-1$	n
m_ℓ	orbital orientation	$-\ell, \ldots, 0, \ldots, \ell$	$2\ell + 1$
m_s	spin	$-\frac{1}{2}, +\frac{1}{2}$	2

Core electrons Inner electrons in an atom; electrons of the previous rare gas plus any filled d or f subshell.

Valence electrons Those electrons beyond the previous rare gas core minus any filled d or f orbitals.

Two and only two electrons fill the first shell. Additional electrons must go into higher-numbered shells. The lowest-energy shell after $n = 1$ is the second shell. Therefore, the third electron in lithium goes into the second ($n = 2$) shell.

There can be eight electrons in the second shell and 18 in the third shell (see Worked Example 2.10). Now look at the periodic table (Figure 2.1 and inside cover). The first row contains just two elements, H and He. The second row contains eight elements. For the first time, we see a connection between the number of orbitals in each shell and the number of elements in the rows of the periodic table.

CONCEPT QUESTION Write the quantum numbers for each element in the first two rows of the periodic table. What can you conclude about the last element in each row? ∎

WORKED EXAMPLE 2.10 *Allowed Quantum Numbers*

The third shell can hold 18 electrons. What are the quantum numbers for these electrons?

Plan
■ The third shell implies $n = 3$, so $\ell = 0, 1$, or 2 (s, p, d).
■ There are one orbital in an s subshell, three orbitals in a p subshell, and five orbitals in a d subshell.
■ Each orbital can hold two electrons.

Implementation
■ With $n = 3$, ℓ can be 2, 1, or 0.
■ With $\ell = 0$, m_ℓ can be only 0. With $\ell = 1$, m_ℓ can be 1, 0, or -1. With $\ell = 2$, m_ℓ can be 2, 1, 0, -1, or -2.
■ Each m_ℓ value can have $m_s = \pm\frac{1}{2}$. Listing these systematically:

n					3				
ℓ	$0(s)$		$1(p)$				$2(d)$		
m_ℓ	0	1	0	-1	2	1	0	-1	-2
m_s	$\pm\frac{1}{2}$	$\pm\frac{1}{2}$	$\pm\frac{1}{2}$	$\pm\frac{1}{2}$	$\pm\frac{1}{2}$	$\pm\frac{1}{2}$	$\pm\frac{1}{2}$	$\pm\frac{1}{2}$	$\pm\frac{1}{2}$

See Exercises 95–96.

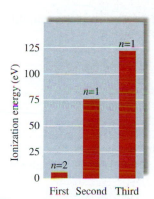

Figure 2.35 The ionization energies of the three electrons in the lithium atom demonstrate the shielding of the outer-shell electron by the two inner-shell electrons. There is a large difference from the first ionization energy corresponding to removing the $n = 2$ shell electron and the energy needed to remove the two $n = 1$ shell electrons. The difference in ionization energy between the two $n = 1$ shell electrons is primarily due to electron–electron repulsion.

Electron Configuration

Core and Valence Electrons The limited number of electrons in a shell forces electrons into higher shells even in the ground state. This is quite different from the situation for hydrogen, where the electron occupies higher shells only if it is excited. Forcing electrons into higher shells changes the energy required to remove electrons from the atom—that is, the ionization energy. Consider the data for lithium shown in Figure 2.35. Removing the first electron takes relatively little energy, 5.39 eV (520 kJ/mol). It takes much more energy to remove the second electron, 75.64 eV. This pattern—a low first ionization energy followed by a jump in ionization energy for the second electron—repeats for all Group IA elements (Figure 2.36). The second ionization energy is always greater than the first, the third higher than the second, and so on. With a complete shell, however, the jump is much greater. Similarly, Group IIA elements have a large jump after the second ionization energy, Group IIIA elements have a large jump after the third ionization energy, and so on. This pattern—a few electrons with relatively low ionization energies followed by a larger jump to electrons with higher ionization energies—is the basis for division of the electrons in an atom into loosely held valence electrons and tightly held core electrons. The **core electrons** consist of the configuration of the previous noble gas (Group VIIIA element) and are difficult to remove. The **valence electrons** are those beyond the previous noble gas configuration and are relatively easy to remove.

CONCEPT QUESTION How many valence and core electrons do each of the following elements have: Na, K, Ca? Use the data in the Appendix: Ionization Energies of the Elements to support your answer. ∎

Configuration Notations When discussing electrons in an atom, writing out the quantum numbers for all the electrons in an element quickly becomes cumbersome. Chemists have therefore developed a shorthand notation. The principal quantum number, n, indicates the shell number and is shown by the leading digit; the azimuthal quantum number, ℓ, indicates the subshell and is denoted by $s, p, d, \ldots$; and the number of electrons in the subshell is indicated by a superscript.

(shell number) (subshell)$^{\text{number of electrons in subshell}}$

Thus the configuration for hydrogen with one electron in the first shell's s subshell is $1s^1$, "one s one." For He, $1s^2$ indicates two electrons in an $\ell = 0$ orbital in the first shell.

For elements with many electrons, this notation is often shortened to indicate core electrons—those of the previous noble gas configuration—by giving the noble gas symbol in square brackets and the valence electrons in full. In this notation, the configuration for Li is [He]$2s^1$. Sometimes it is useful to show these valence electrons graphically so that spin is also represented. The graphic is called a **box diagram** and indicates each orbital as a box. Electrons in the orbital are shown as arrows: An upward-pointing arrow (↑) denotes a spin of

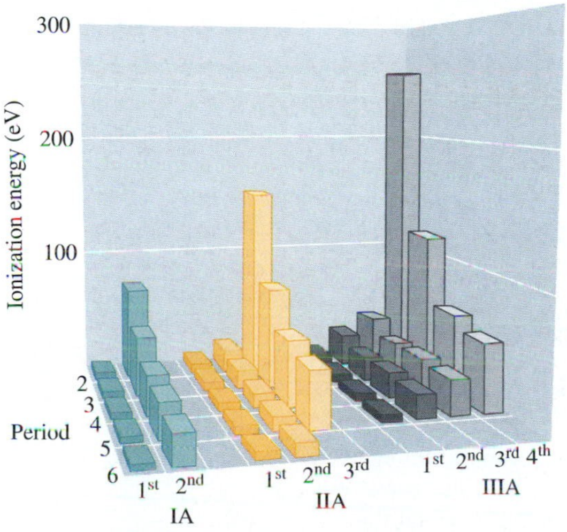

Figure 2.36 Ionization energy jumps after the first ionization in Group IA, after the second ionization in Group IIA, and after the third ionization in Group IIIA.

Box diagram Graphic representation of the valence electrons in which each orbital is represented as a box, and electrons in the orbital are shown as arrows.

$+\frac{1}{2}$, and a downward-pointing arrow ($\downarrow$) denotes a spin of $-\frac{1}{2}$. In this notation, the full configurations of hydrogen and lithium are

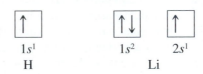

$1s^1$ $1s^2$ $2s^1$

H   Li

Periodic patterns Examining the electron configurations of the elements across the periodic table (Figure 2.37) reveals that the arrangement of the electrons is correlated with the position of the element in the periodic table.

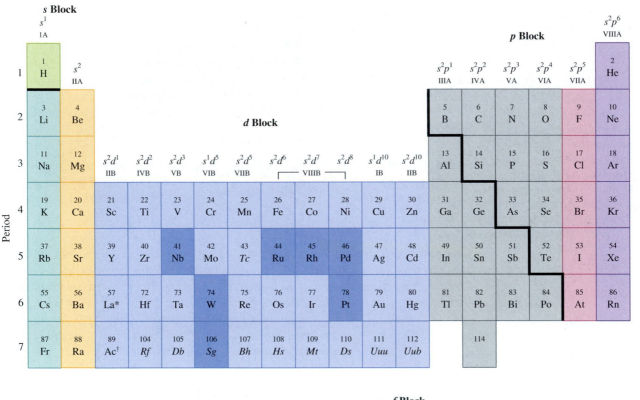

Figure 2.37 Electron filling occurs in blocks in the periodic table. The electron configuration for the valence electrons is shown at the head of each group (the filled *d* or *f* subshells have been omitted to save space). The first two elements of each period fill the valence *s* orbital, the ten transition elements fill the *d* subshell, and the last six elements fill the *p* subshell. (Dark blue shading of elements in the transition series indicate exceptions to the configuration noted at the top of the group. The configuration of all elements is shown in the table on the inside cover.) The 14 elements from periods 6 and 7 displayed at the bottom of the table are the *f*-block elements. Filling of *d* and *f* orbitals in *f*-block elements is irregular.

Checklist for Review

KEY TERMS

Group or family (p. 42)
period (p. 43)
main group (p. 43)
transition elements (p. 43)
lanthanides (p. 43)
actinides (p. 43)
electrode (p. 45)
electrolysis (p. 45)
nucleus (p. 48)
proton (p. 48)
atomic number, Z (p. 48)
neutron (p. 48)
periodic law (p. 48)
electromagnetic radiation (p. 49)
continuous spectrum (p. 49)
disperse (p. 49)
wavelength, λ (p. 49)
standing wave (p. 50)
fundamental wave (p. 50)
frequency, ν (p. 50)
line spectrum (p. 53)
energy level (p. 54)
energy-level diagram (p. 54)
transition energy (p. 54)
spectrum (p. 54)
state (p. 55)
ground state (p. 55)
excited state (p. 55)
ionization energy (p. 57)
Rydberg constant (p. 57)
electron-volt, eV (p. 57)
quantum mechanical model (p. 59)

wave function, ψ (p. 60)
amplitude (p. 60)
probability density, $|\psi|^2$ (p. 60)
node (p. 60)
positive phase (p. 61)
negative phase (p. 61)
mode (p. 61)
orbital (p. 63)
azimuthal (p. 64)
shell (p. 65)
subshell (p. 65)
quantum numbers (p. 67)
principal quantum number, n (p. 67)
azimuthal quantum number, ℓ (p. 67)
magnetic quantum number, m_ℓ (p. 67)
spin quantum number, m_s (p. 67)
electron configuration (p. 70)
shielding (p. 70)
effective nuclear charge, Z_{eff} (p. 70)
Pauli exclusion principle (p. 73)
core electrons (p. 74)
valence electrons (p. 74)
box diagram (p. 76)

KEY EQUATIONS

$$E = \frac{hc}{\lambda} \qquad \nu = c/\lambda \qquad E = h\nu$$

$$\text{Electron energy} = -\frac{hcR}{n^2} \quad \text{Ionization energy} = \frac{hcR}{n^2}$$

$$Z_{eff} = n\sqrt{\frac{\text{ionization energy}}{13.6 \text{ eV}}}$$

Chapter Summary

The periodic table is the most powerful organizer of chemical information ever devised. Its development was based on correlation-of-property observations, both physical and chemical. The regular progression of physical and chemical properties, however, does more than merely organize the elements. The regularity as well as the "magic numbers" (2, 8, 8, 18, 18, 32, 32), which relate to the number of elements between repeating patterns, suggest that a systematic variation in the structure of atoms affects interaction between atoms.

Examining interactions between atoms reveals that atoms are held together in larger structures, such as molecules and solids, by a force that is electrical in nature. From the mid-1800s to the turn of the twentieth century, scientists examined these electrical forces and made several discoveries that led to the current model of the atom. Michael Faraday concluded that the electron is a packet of negative charge, George Stoney conjectured that the electron is part of the atom, and J. J. Thomson proved that the electron is indeed part of the atom. Ernest Rutherford explored the substructure of the atom and obtained the surprising result that the electrons, despite their small mass, account for the vast majority of the volume of the atom. Rutherford's scattering experiment led to the current picture of the atom as consisting of a minute but massive nucleus surrounded by a diffuse electron cloud.

Understanding the electron arrangement in an atom and its variation from element to element requires a probe of the structure of these very light particles. The primary probe is electromagnetic radiation and the connection between electromagnetic radiation and energy.

The modern view of atomic structure provides insight into the similar properties of families of elements—namely, members of families have the same valence electron configuration. Coulombic factors that determine the valence configuration energy ordering include the attraction of the nucleus for the electrons, the repulsion of other electrons, and the shielding of the nuclear charge by core electrons. In addition to the coulombic factors, the electron configuration is determined by the limited number of electrons that can fit into each subshell. Guidelines for determining the electron configuration of elements are as follows:

- The most stable orbitals, that is those with the most negative energy, fill first.
- No two electrons have the same set of quantum numbers (Pauli Exclusion Principle).
- Among orbitals with the same energy, electrons fill each orbital singly before pairing in any one orbital (coulombic repulsion).

KEY IDEA

Atoms consist of negatively charged electrons that account for most of their volume but little of the mass of the atom, positively charged protons that balance the negative charge of the electron, and uncharged neutrons. Electrons play a key role in binding atoms together.

CONCEPTS YOU SHOULD UNDERSTAND

- Periodic law.
- Binding between atoms is electrical in nature.
- The wavelength or frequency of electromagnetic radiation is connected with the energy of the photons.
- The electron waves—orbitals—have geometrical shapes specified by a set of four quantum numbers.

OPERATIONAL SKILLS

- Connect mass with particles (Worked Example 2.1).
- Connect wavelength, frequency and energy (Worked Example 2.2).

- Relate energy levels to ionization energy (Worked Example 2.3).
- Relate transition energies to emitted radiation (Worked Example 2.4).
- Determine wave mode from characteristics of nodes (Worked Example 2.5).
- Determine number of allowed orbitals and states in each shell, and label waves with quantum numbers (Worked Examples 2.6, 2.7).
- Determine the effective charge from the ionization energy of an electron (Worked Example 2.8).
- Determine the ionization energy of an excited state from that of the ground state and the excitation energy or wavelength; convert between energy and wavelength (Worked Examples 2.9, 2.10).

Exercises

A blue exercise number indicates that the answer to that exercise appears at the back of the book.

■ SKILL BUILDING EXERCISES

2.1 Navigating the Periodic Table

1. Without referring to the periodic table, how many elements are in the third period? How many in the fifth period? What principle of organization did you use to answer this question?

2. Without referring to the periodic table, give the atomic number of an element that falls in the same group as an element with atomic number (a) 3, (b) 32, (c) 15, (d) 25, and (e) 44.

2.2 The Components of the Atom

3. Discuss the evidence that supports the conclusion that atoms are held together in molecules by a force that is electrical in nature.

4. In the plum-pudding model for the atom the mass is uniformly distributed. If the plum-pudding model were correct, what scattering pattern would one observe from shooting a beam of α particles at a collection of atoms?

5. What experimental evidence supports a model of the atom consisting of a nonuniform distribution for matter?

6. Describe Rutherford's scattering experiment. Do the results of the scattering experiment provide evidence for the relative size or the relative mass of different components of the atom?

7. The mass of an automobile is approximately one ton (2000 lb). If the automobile represents the nucleus of the atom, what is the mass of the electron? Assume that the automobile is approximately spherical and 2 m in diameter. What volume does this electron fill? How much denser is the proton than the electron?

8. If you were to build a scale model for the atom using a baseball as the nucleus (6 cm diameter), how big would the atom be? If the nucleus is as big as the Empire State Building (443.2 m), how big is the atom? Compare this to the diameter of the earth (6000 km).

9. A glass of water contains about 200 g of water. What is the mass of the electrons in the glass of water? The volume of the glass of water is 200 cm^3. What volume is occupied by the nuclei?

10. The density of copper is 8.96 g/cm^3. Calculate the volume of one mole of copper. Give dimensions of a cube containing one mole of copper. What fraction of the cube volume is due to the nuclei? What volume is occupied by the nuclei?

11. The density of Al is 2.702 g/cm^3. How many moles of aluminum atoms are in a cube of aluminum 1 cm on a side? What volume is occupied by the nuclei?

12. The atomic mass of Ar, atomic number 18, is greater than that of K, atomic number 19.

a. Is K lighter than expected or Ar heavier than expected? Explain your answer.

b. Atoms of a given element that differ in mass are denoted as follows. The number of protons plus neutrons is denoted as a leading superscript, the atomic number as a leading subscript, and the element is indicated by its symbol. There are five isotopes of Ar: $^{36}_{18}$Ar, $^{37}_{18}$Ar, $^{38}_{18}$Ar, $^{39}_{18}$Ar, and $^{40}_{18}$Ar. Explain how these differ. Which isotope is primarily responsible for the atomic mass of Ar (39.948 amu)?

13. There are four isotopes of iron (Fe) with atomic masses of 53.939 amu, 55.935 amu, 56.935 amu, and 57.933 amu. Indicate the notation for each of these isotopes.

14. Ar, K, and Ca all have isotopes of mass close to 40 amu. How many protons, electrons, and neutrons are in each?

2.3 Light, Color, Energy

15. Sketched below are several waves. Which has the highest energy? Which has the greatest intensity? What is the wavelength of each of these waves? How many nodes does each have in 1 m?

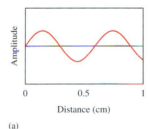

(a)

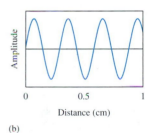

(b)

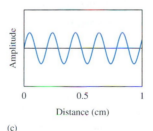

(c)

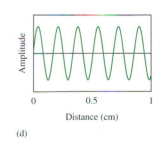

(d)

16. Which is of higher energy, blue or red light? Which has a longer wavelength? Which has a higher frequency?

17. One end of a 2-m rope is tied to a wall. You shake the other end so that you create a standing wave of one-half a wavelength. If this rope were an electromagnetic wave, what would be the frequency? In what part of the electromagnetic spectrum would you find this wave?

18. The laser light used in compact disc players has a wavelength of 780 nm. In what region of the electromagnetic spectrum

is this light? What is the energy of this light? To "read" a feature, the wavelength of light must be less than half the feature size. How many features per millimeter of track can 780-nm light read?

19. Excited lithium atoms emit radiation at 670.8 nm.
   a. Calculate the frequency of this radiation.
   b. Calculate the energy (in eV) of this radiation.
   c. What color is this radiation?

20. Each metal in a fireworks display produces a characteristic color. What are the frequency and color associated with each of the following:
   a. Ba^{2+}, $\lambda = 551$ nm   d. Li^+, $\lambda = 671$ nm
   b. Cs^+, $\lambda = 456$ nm   e. Ca^{2+}, $\lambda = 649$ nm
   c. Na^+, $\lambda = 590$ nm   f. Sr^{2+}, $\lambda = 661$ nm

21. Sodium street lamps give off light at a frequency of 5.08×10^{14} Hz. What is the wavelength, in nanometers, of this light? What is the energy of one photon from the sodium emission (eV is the appropriate unit for the energy of a photon)? What is the energy of a mole of such photons?

22. We encounter a large number of waves every day without realizing it. One such source is radio waves. A typical frequency for radio transmission is 680 kHz. How much energy does one mole of radio waves carry? X rays used by doctors to view bones are much more energetic than radio waves. Determine the energy of one mole of X ray photons (3.8×10^{17} Hz).

2.3 Light — Unlocking the Electronic Structure of Atoms

23. The Balmer equation is often referred to as an empirical equation. Why?

24. Indicate the experimental evidence that supports a model of electrons occupying discrete energy levels in an atom.

25. A series of lines in the spectrum of hydrogen occurs at 656.3 nm, 486.1 nm, 434.0 nm, and 410.2 nm. What is the next line in the series?

26. What is the shortest-wavelength radiation that hydrogen can emit? Are any lines of the series terminating on $n = 1$ in the visible region of the spectrum?

27. A tube containing hydrogen atoms in the $n = 3$ shell does not absorb visible light. The longest wavelength absorbed is 1875 nm. What state is the hydrogen atom in after absorbing a 1875-nm photon?

28. What are the values of the quantum numbers for the initial and final states for hydrogen in the Balmer series? Draw an energy-level diagram for hydrogen, and indicate the transitions of the Balmer series on the diagram.

29. A sample of hydrogen atoms is prepared in the $n = 4$ state. These hydrogen atoms relax to the ground state by emission of one or more photons.
   a. How many spectral lines are expected?
   b. Give the frequency of at least three of these lines.
   c. Are any of the emissions in the visible region of the spectrum?

   Draw an energy-level diagram indicating transitions to support your conclusions.

30. Hydrogen is prepared with all the atoms in the sample in the $n = 4$ state. What is the minimum energy required to ionize a single hydrogen atom, in electron-volts (eV)? How much energy, in joules, is required to ionize a mole of $n = 4$ hydrogen atoms?

2.4 Model of the Atom: Bohr's Model

Note: Exercises 31 – 36 contain the details of the Bohr model of the atom. These are interrelated and should be assigned as a set.

31. The ad hoc assumption made by Bohr to limit electron orbits is that the angular momentum is limited to integer multiples, n, of $\hbar$ ($=h/2\pi$). The angular momentum for a particle of mass m traveling at velocity v on a circle of radius r is mvr. Solve for the allowed velocities when the angular momentum is equal to an integer multiple of $\hbar$.

32. A particle on a circular path is continuously accelerated toward the center to maintain the circular motion. The force required to produce this acceleration for a particle of mass m and velocity v on a circle of radius r is mv^2/r. In the Bohr model for the electron in hydrogen, this force is supplied by the force of coulombic attraction. The coulombic force between two particle of charge $\pm e$ separated by a distance r is $(1/4\pi\varepsilon_0)(e^2/r^2)$, where ε_0 is a measure of how effectively one charge penetrates to the other, referred to as the vacuum permittivity. Equating these two forces leaves two unknowns: the velocity and the radius. Use the result of Exercise 32 to eliminate the velocity and solve for the allowed radii.

33. The kinetic energy of a particle of mass m traveling at velocity v is $\frac{1}{2}mv^2$. Use the results of Exercise 32 for the velocity and radius to express the kinetic energy in terms of the integer n and fundamental constants.

34. The potential energy of a particle of charge $-e$ a distance r from another particle of charge $+e$ is $-e^2/4\pi\varepsilon_0 r$. Use the results of Exercise 32 to express this potential energy in terms of the integer n and fundamental constants.

35. The total energy of a particle is the sum of the potential and kinetic energies. Use the results of Exercises 33 and 34 to determine the energy of hydrogen's electron in terms of n and fundamental constants. Use this result to express the Rydberg constant, R, in terms of other fundamental constants.

36. Show that the result of the Bohr model—the solution to Exercise 35—is consistent with the Balmer equation, Equation (2.6).

2.4 Ionization Energy

37. Describe how (a) electron energy and (b) distance from the nucleus vary with increasing values of n.

38. Consider two hydrogen atoms, one with the electron in the $n = 1$ shell and the other with the electron in the $n = 5$ shell.
   a. Which atom is in the ground-state configuration?
   b. Which orbit has a larger radius?
   c. Which atom has the larger ionization energy?

39. The ionization energy of a ground-state hydrogen atom is 13.6 eV. Can a ground-state hydrogen atom absorb less than 13.6 eV? Explain, showing any energy-level diagram you use.

40. What is the ionization energy of an electron in the fifth Bohr orbit in hydrogen?

41. One of the emission lines from hydrogen has a wavelength of 97.25 nm.
   a. In what region of the electromagnetic spectrum can this photon be found?
   b. After emission of the 97.25-nm photon, the hydrogen atom is in its ground state. What shell does the electron start out in?
   c. What is the ionization energy of the hydrogen atom in the excited state?

42. The difference between the wavelengths of adjacent spectral lines emitted by hydrogen becomes smaller as n gets larger.
   a. Compare the wavelength difference for the emission between the $n = 3$ and $n = 2$ levels to that between the $n = 4$ and $n = 2$ levels.
   b. Continue the series in part (a) for emission from $n = 8$, 7, 6, and 5 to $n = 2$.
   c. What is the convergence limit (large n to $n = 2$) of the series?
   d. Compare your result in part (c) with the ionization limit for $n = 2$ [Equation (2.11)].

2.4 Line Spectra Revisited

43.
   a. Discuss the mechanism for how an atom absorbs a photon.
   b. Discuss the mechanism for how an atom emits a photon.
   c. Draw a schematic of a ground-state atom absorbing a photon.
   d. How does the schematic for emitting a photon differ from the absorption schematic in part (c)?

44. Calculate the energy (in eV) of the three lowest-energy transitions in the Lyman series (the series that ends on $n = 1$) of hydrogen. The wavelengths of these transitions are 121.6 nm, 102.6 nm, and 97.3 nm. Draw an energy-level diagram of these transitions.

45. The lines in the visible spectrum of hydrogen result from electron transitions ending on the second level or $n = 2$ shell. The energy of the $n = 3$ to $n = 2$ transition was calculated in the text. Use Equation (2.6), or (2.12) to calculate the energy of the three remaining hydrogen transitions in the visible region. Calculate the wavelengths of these transitions.

46. Suppose a sample of hydrogen gas is excited to the $n = 5$ level. As the atoms return to the ground state, they emit light.
   a. How many lines are there in the spectrum?
   b. Draw an energy-level diagram indicating these transitions.
   c. Calculate the three lowest-energy transitions, and indicate them on your drawing.

47. Use the general equation for the energy levels for hydrogen to calculate the frequency of light emitted from a hydrogen atom when n_1 is the ground state, and the atom's excited state corresponds to $n_2 = 5$. What is the frequency of electromagnetic radiation emitted? In what region of the electromagnetic spectrum is this emission located?

48. Refer to the transition in Exercise 47.
   a. Calculate the energy of the atom before and after the emission (to three significant figures).
   b. Draw an energy-level diagram and indicate the transition.

49. A line in the hydrogen spectrum appears at 656.28 nm. Determine the value of n_1 and n_2 for this line. To which spectral series does it belong? After emission of the 656.28-nm photon, is the hydrogen atom in its ground state? If not, what energy photon must it emit to return to the ground state?

50. A tube of hydrogen atoms with their electrons in the $n = 1$ shell does not absorb visible light. The absorption line of longest wavelength appears at 121.6 nm. What is the energy of this transition (in eV), and in what state does the electron reside after absorption?

51. For hydrogen, calculate the energy of the transition from the $n = 3$ to $n = 1$ level.

52. Hydrogen-like ions are those with only one electron such as He^+ and Li^{2+}. The Balmer formula for the transitions in these ions can be written as $\nu = Z^2 cR \left(\dfrac{1}{n_1^2} - \dfrac{1}{n_2^2} \right)$. Use this formula to calculate the lowest-energy transition in the Lyman and Balmer series for He^+ and Li^{2+}.

2.4 The Quantum Mechanical Model

53. What are the limitations of the Bohr model? Cite two important results obtained from quantum mechanics that cannot be derived from the Bohr model.

54. What is the difference between an orbit and an orbital? Sketch the $n = 1$ orbit and the $n = 1$ orbital. What is the difference between electron amplitude and electron density?

55. Explain in your own words the following: (a) ψ and (b) $|\psi|^2$.

2.4 Two-Dimensional Waves

56. Figure 2.38 illustrates the maximum displacement of a mode of the drumhead. Label this mode with n and ℓ.

57. Figure 2.39 illustrates a mode of a drumhead. Label this mode with n and ℓ. Is this mode more or less energetic than that shown in Figure 2.38?

58. Describe how the wave motion on a drumhead is analogous to an electron wave. Refer to Figures 2.23 and Figure 2.24. Describe at least one important difference between the drumhead and an electron.

59. Sketch 2s, 3p, and 3d orbitals, giving a sense of the relative size and nodes of each.

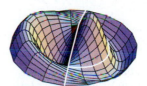

Figure 2.38 A two-dimensional wave as might occur on a drumhead.

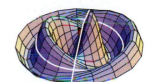

Figure 2.39 A two-dimensional wave as might occur on a drumhead.

60. The $3p$ to $2s$ transition for hydrogen results in a red photon in the Balmer series. Draw a schematic of this transition that includes the relative size and number of spherical and planar nodes in each of the two orbitals. Calculate the difference in energy for an electron in each of these two orbitals.

61. How many nodal surfaces does a $4s$ orbital have? A $5s$ orbital? How many nodes are spherical? How many are planar? Draw a cross section of a $4s$ orbital and a $5s$ orbital showing the nodes, crests, and troughs.

62. How many nodal surfaces does a $4s$ orbital have? A $4p$ orbital? A $4d$ orbital? How many nodes are spherical? How many are planar? Draw a cross section of a $4s$ orbital, $4p$ orbital, and a $4d$ orbital showing the nodes, crests, and troughs.

63. How many nodal surfaces does a $4s$ orbital have? A $3d$ orbital? How many nodes are spherical? How many are planar? Draw a cross section of a $4s$ orbital and a $3d$ orbital showing the nodes, crests, and troughs.

64. How many nodes can an orbital in the fourth shell have? How many subshells can there be in the fourth shell?

2.4 Quantum Numbers

65. How many quantum numbers are required to specify an orbital? What are they?

66. What do each of the four quantum numbers specify?

67. What feature of an orbital is determined by each of the following:
a. the principal quantum number
b. the azimuthal quantum number
c. the magnetic quantum number

68. For each of the following specify the value of the quantum numbers n and ℓ:
a. $3d$ b. $2p$ c. $4f$ d. $3s$ e. $4p$

69. Indicate which of the following cannot occur and explain why:
a. a $2d$ orbital
b. a $6g$ orbital
c. a $6p$ orbital

70. For each of the following sets of quantum numbers, name the orbital:
a. $n = 4, \ell = 2$
b. $n = 3, \ell = 1$
c. $n = 6, \ell = 5$
d. $n = 1, \ell = 0$
e. $n = 3, \ell = 2$

71. Select the following combinations of quantum numbers that are impossible, and state why that combination cannot occur. For each impossible set, indicate a modification that makes it possible.
a. $n = 4, \ell = 4, m_\ell = -2, m_s = \frac{1}{2}$
b. $n = 1, \ell = 0, m_\ell = -1, m_s = -\frac{1}{2}$

c. $n = 4, \ell = 4, m_\ell = 2, m_s = 0$
d. $n = 9, \ell = 5, m_\ell = 0, m_s = \frac{1}{2}$
e. $n = 3, \ell = 2, m_\ell = 2, m_s = \frac{1}{2}$

72. Give the possible quantum numbers for an electron in the following:
a. a $2p$ orbital
b. a $5d$ orbital

73. There can be a maximum of eight unique sets of four quantum numbers in the second shell. What are they?

74. The third shell can have $2 + 6 + 10 = 18$ unique sets of quantum numbers. What are they?

75. What are the possible quantum numbers for all (a) $2s$, (b) $2p$, and (c) $3d$ orbitals? If no two electrons can have the same set of four quantum numbers, what is the maximum number of electrons that can be accommodated in each of these subshells?

76. What are the possible quantum numbers for orbitals in the $2p$ subshell? For orbitals in the $3p$ subshell? How do the number of orbitals in the $2p$ and the $3p$ subshell compare? How does the principal quantum number affect the number of orbitals in a subshell?

77. If no two electrons can have the same set of four quantum numbers, what is the maximum number of electrons (a) in a d subshell, (b) in an f subshell and (c) in an h subshell? Explain your answer.

78. a. How many unique sets of four quantum numbers can there be in the fifth shell?
b. How many sets can there be in the $n = 6$ shell? Justify your answers by grouping the sets into subshells.

79. What are the possible quantum numbers for an electron in a $3d$ orbital?

80. How many nodes are in a $6f$ orbital? How many of those nodes are spherical, and how many are planar?

81. Indicate the total number of nodes, number of planar nodes, and number of spherical nodes in the following orbitals: $4s$, $3p$, $5f$, $10g$.

82. Prove the following statements: There are n subshells in shell n. There are n^2 orbitals in shell n.

Questions 83–86 utilize the geometric description of the orbitals. The electron orbitals are waves, and the shapes of these waves are given by the geometric functions listed below. For example, an s orbital is the same in all angular directions, so no angle, θ or ϕ, appears in the description of the s orbital wave function. Electron density is the square of the amplitude. When more than one orbital on the same atom is occupied, electron densities are additive.

Amplitude of Orbitals:
s orbitals

(2.16)
$$s(\theta, \phi) = \frac{1}{2}\sqrt{\frac{1}{\pi}}$$

p orbitals

(2.17) $$p_z(\theta, \phi) = \frac{1}{2}\sqrt{\frac{3}{\pi}} \cos\theta;$$

$$p_x(\theta, \phi) = \frac{1}{2}\sqrt{\frac{3}{\pi}} \sin\theta \cos\phi;$$

$$p_y(\theta, \phi) = \frac{1}{2}\sqrt{\frac{3}{\pi}} \sin\theta \sin\phi$$

d orbitals

(2.18) $$d_{z^2}(\theta, \phi) = \frac{1}{4}\sqrt{\frac{5}{\pi}} (3\cos^2\theta - 1);$$

$$d_{xz}(\theta, \phi) = \frac{1}{2}\sqrt{\frac{15}{\pi}} \cos\theta \sin\theta \cos\phi;$$

$$d_{yz}(\theta, \phi) = \frac{1}{2}\sqrt{\frac{15}{\pi}} \cos\theta \sin\theta \sin\phi;$$

$$d_{x^2-y^2}(\theta, \phi) = \frac{1}{4}\sqrt{\frac{15}{\pi}} \sin^2\theta \cos2\phi;$$

$$d_{xy}(\theta, \phi) = \frac{1}{4}\sqrt{\frac{15}{\pi}} \sin^2\theta \sin2\phi$$

Spherical Coordinates:

r = distance out from center.
θ = rotation down from *z*-axis.
ϕ = rotation from *x*-axis.

(2.19) $$x = r \sin\theta \cos\phi$$

(2.20) $$y = r \sin\theta \sin\phi$$

(2.21) $$z = r \cos\theta$$

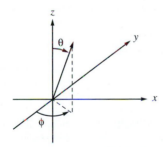

83. Plot the amplitude along the *x*-axis for an electron in a p_x orbital. What is the smallest value of *n* for which a *p* orbital can exist?

84. Plot the amplitude in the *x*-*y* plane for an electron in a $3d_{xy}$ orbital.

85. Show that the sum of the electron densities in a set of three *p* orbitals is a sphere.

86. Show that the sum of the electron densities in a set of five *d* orbitals is a sphere.

2.5 Energy Levels of Multielectron Atoms

87. List the factors that determine the ionization energy for the ground state of a multielectron atom.

88. Explain how the number of planar nodes determines the energy ordering of 4*s*, 4*p*, 4*d*, and 4*f* orbitals.

89. In the hydrogen atom there is no difference in energy between a 2*s*- and a 2*p*-orbital electron, but in the carbon atom a 2*s*-orbital electron is lower in energy than a 2*p*-orbital electron. Explain why.

90. Explain why the outer electron (or electrons) in Na and Mg is (are) a 3*s* electron(s) rather than a 3*p* electron(s).

91. Arrange Na, Al, P, and K in order of highest to lowest effective charge on the outermost electron.

92. Calculate the effective nuclear charges for these five elements: Ti, Fe, Ni, Cu, and Zn.

93. The wavelength of the yellow light emitted by sodium is 589 nm and is due to a transition in the valence shell of sodium from the 3*p* orbital to the 3*s* orbital. The 3*p* to 3*s* transition in Mg occurs at 383 nm. Is the transition in Mg higher or lower in energy than the 3*p* to 3*s* transition in Na? Explain why the energy is higher or lower. Calculate the ionization energy of an electron excited to the 3*p* state in Mg. Determine Z_{eff} for a 3*p* electron in Mg.

94. a. The characteristic emission from Ca occurs at 422.7 nm and corresponds to an electron going from a 4*p* orbital to a 4*s* orbital. Calculate the energy of the electron in the 4*p* orbital. Calculate Z_{eff} for the electron in the 4*p* orbital. Identify the data that you need to answer this question, and look it up in the text.

b. The same 4*p* to 4*s* transition in potassium occurs at 769 nm. Calculate the energy of an electron in the 4*p* orbital of potassium and Z_{eff} for this electron.

c. Explain the differences between your answers to parts (a) and (b).

2.5 Pauli Exclusion Principle

95. State the Pauli exclusion principle. How does it help determine the configuration of the elements? Why does the number of elements in successive periods of the periodic table increase by the progression 2, 8, 8, 18, 18, 32, 32?

96. How many electrons can have the following in common?
a. $n = 2, \ell = 0$.
b. $n = 5, m_\ell = 3$.
c. 4*f*
d. 3*p*

2.5 Electron Configuration

97. Core electrons are not involved in chemical reactions. Explain why. What is the relationship between the core electrons and the valence electrons of an atom?

98. How do the number of core and valence electrons change from left to right across a period? How do they change top to bottom down a group?

99. How many valence electrons are in each of the following:
a. Fe   c. W   e. Ba
b. Sn   d. Ni

100. Use the periodic table to decide how many core electrons are in each of the following: P, Br, Rb, and Ti.

101. How many core electrons does F have? How many does Ge have? Write the valence configuration of each element.

102. Compare the first and second ionization energies of helium with the ionization energy of hydrogen. Why is the first ionization energy of helium less than expected from Coulomb's law?

2.5 Periodic Patterns

103. In your own words, explain the relationship between position in the periodic table and electron configuration. Use this relationship to explain the periodic law.

104. Without referring to the periodic table, in which group and which period would you look to find an element with 25 electrons? An element with 56 electrons? In which group and in which period do you expect element 120 (not yet synthesized or discovered)?

105. Graph the effective charges of the Group IA elements. Rationalize the trend.

106. Explain why the second ionization energy is always greater than the first. What is the significance of the jump in ionization energy from the first to the second ionization for the members of Group IA?

■ CONCEPTUAL EXERCISES

107. Make a diagram showing the interconnections among the following terms: mode, node, wave, orbital, shell, line spectrum. Every term must have at least one connection.

108. Make a diagram showing relationships among the following terms: wavelength, frequency, energy, photon, color, fundamental wave, overtone. All terms must be interconnected.

109. Draw a diagram indicating the relationships among the following terms: electron configuration, shielding, Pauli exclusion principle, valence electrons, core electrons, ionization energy. All terms should be interconnected.

☀ APPLIED EXERCISES

110. Light bulbs emit a broad spectrum of light; that is, they emit photons with a broad range of wavelengths. Suppose that 10% of the photons from a 75-watt (one watt is one joule per second) bulb are emitted with an average wavelength of 500 nm (middle of the green range, where our eyes are most responsive). How many photons per second are emitted by the bulb?

111. Microwaves, as used in microwave ovens, have a wavelength of 0.125 m.
  a. What is the energy of a single microwave photon?
  b. What is the energy of a mole of microwave photons?
  c. If 28.8 kJ is required to heat a 100-g cup of tea from 23 °C to 95 °C, how many moles of microwave photons are required?

112. A current operating for one-half hour deposits 0.800 g of silver. How many moles of silver are deposited? How long

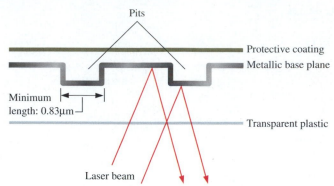

Figure 2.40 The structure of a CD shows the pits that contain the information. Information is read by reflecting a laser beam off the pits and the metallic base plane. Pits are exactly $\frac{1}{4}\lambda$ deep, where λ is the wavelength of the laser used to read the features.

would the current have to flow to deposit one mole of silver? What is the current in amperes?

113. An electric light bulb is operated under conditions such that one ampere of current (1 C/s) is passing through the filament. How many electrons pass through the filament each second? (Remember that the charge of an electron is -1.60×10^{-19} C.)

114. Figure 2.40 illustrates the structure of a CD. Show that if the pit depth is $\frac{1}{4}\lambda$, where λ is the wavelength of the laser used to read the information, the portion of the beam reflected off the pit will interfere with that reflected off the metal base plane. The $0.83\,\mu m$ minimum pit length is related to the wavelength of the laser used to read the information, which is 780 nm for standard CDs. Estimate the minimum pit length for a DVD that is read with a laser of wavelength 635 nm. The minimum pit length limits the amount of information that can be stored on disk. How much more information can be stored on a DVD track compared with a CD track?

115. The tendency of atoms to be in the lowest possible state results in most hydrogen atoms on earth beingin their lowest-energy state. The diameter of a hydrogen atom in this state is 79 pm. In interstellar space, hydrogen atoms can be found in highly excited states, and transitions between adjacent energy levels can be studied using radio astronomy. One of the longest wavelength transitions observed is labeled the 272 α transition. The 272 α transition corresponds to a transition from $n = 273$ to $n = 272$. The diameter of the hydrogen atom is proportional to n^2. What is the diameter of the hydrogen atom in the $n = 272$ state? A certain region in space is being scanned for these very highly excited hydrogen atoms. At what frequency and wavelength should a telescope be set to observe this transition?

116. If a pot boils over while being heated over a gas flame, the flame often appears yellow. What element or elements are responsible for the color? What does the color tell us about electrons in the atoms of this element?

Figure 2.41 A drum in one mode of oscillation. The white material collects at the nodes.

117. The northern lights or aurora borealis gets some of its fantastic colors from atomic emission of oxygen atoms. Two emissions occur in the visible region, one at 557.7 nm and another at 630.0 nm. What colors do these emissions contribute to the aurora borealis? What is the energy source for the aurora borealis? Look up the ionization energy of oxygen, and suggest which electronic levels might contribute to these emissions.

■ **INTEGRATIVE EXERCISES**

118. Hydrogen is most often placed in Group IA. What properties of hydrogen support this placement? In many ways hydrogen is like the Group VIIA elements. Discuss the rationale for placing hydrogen in Group VIIA. Some periodic tables show hydrogen in the top center of the table in a group by itself. Cite properties that justify this placement.

119. For each of the following statements, cite the supporting experimental evidence.
a. The atom is made up of electrons, protons, and neutrons.
b. The atomic number is numerically equal to the number of electrons in the neutral atom.
c. The mass and positive charge of the atom are concentrated in a small, dense nucleus.

120. Have you ever observed the shimmering display of the aurora borealis? The impressive colors originate when solar flares send energetic particles, mainly electrons, toward the earth at about 400 km/s. Light is emitted after these energetic electrons collide with oxygen and nitrogen in the earth's atmosphere. Nitrogen is responsible for light at wavelengths of 391.2 nm and 470.0 nm. What colors does nitrogen contribute to the aurora? Oxygen is responsible for light at wavelengths of 557.7 nm, 630 nm, and 636.4 nm. What colors does oxygen add to the aurora?

121. Figure 2.41 shows the mode of a drum oscillating in one of its higher-energy states. The top of the drum is covered with a thin layer of dark particles that collect at the nodes. Treating the drum as a flat-land atom, what mode is the drum oscillating in?

Chapter 3
Physical and Chemical Periodicity

Patterns are ubiquitous in natural systems such as this nautilus shell. Like the nautilus shell, the periodic table spirals in expanding rows.

CONCEPTUAL FOCUS

- Connect periodic physical and chemical properties of the elements with their electron configurations.

SKILL DEVELOPMENT OBJECTIVES

- Use Lewis dot representations to predict combining ratios (Worked Examples 3.1, 3.2).
- Calculate ion effective charge from electron affinity (Worked Example 3.3).
- Explain reaction products based on electronegativity (Worked Example 3.4).

- Determine the atomic radius in an elemental solid (Worked Example 3.5).
- Predict properties from periodic patterns (Worked Examples 3.6 and 3.7).

3.1   Trends in the Periodic Table

Periodicity in chemical properties formed the basis of the arrangement of the periodic table developed in Mendeléeff's time. With the modern view of the atom and the electron, the periodic table can be viewed as an arrangement based on the periodicity of electron configurations (Figure 3.1). The periodicity in electron configuration profoundly affects both the attraction of an atom for its valence electrons and the attraction for an additional electron. These two attractions form the basis not only for periodic chemical properties—after all, chemical interactions are electrical in nature—but also for periodic physical properties.

Noble gas A Group VIIIA element.

Alkali metal The IA metals. Members of the group: Li, Na, K, Rb, Cs, Fr

Groups

Exploration of several groups illustrates periodicity (Figure 3.2). Consider the members of group VIIIA, called the **noble gases,** and the members of the succeeding group IA, called the **alkali metals.** Each noble gas is normally found as an isolated atom in the gas phase. Immediately following each noble gas element, the first element in the next

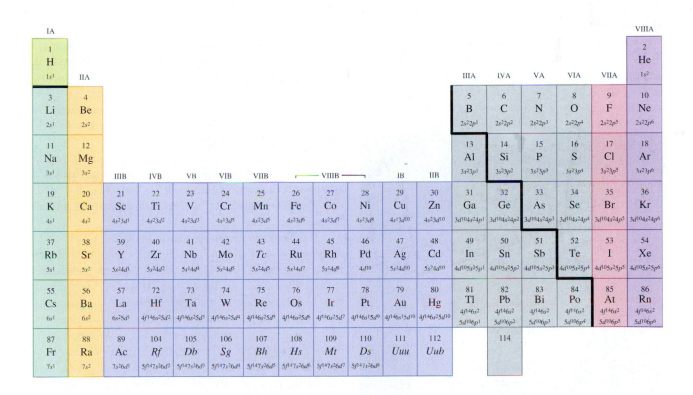

Figure 3.1 The electron configuration of all the elements in the periodic table. Only the electrons beyond the previous rare gas core are shown.

Figure 3.2 The periodic table with the following groups highlighted: noble gases (Group VIIIA), alkali metals (Group IA), alkaline earths (Group IIA), and halogens (Group VIIA).

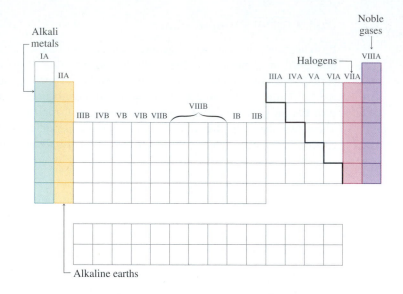

Alkaline earth A IIA metal. Member of the group: Be, Mg, Ca, Sr, Ba, Ra

Halogen Members of the family: F, Cl, Br, I, At.

period is a member of the alkali metal group: elements that are so reactive they are never found in isolation in nature. For example, when pieces of elemental lithium (Li), sodium (Na), or potassium (K) are dropped onto the surface of water, they skitter across the surface, generating bubbles of hydrogen gas and starting a fire (Figure 3.3). The noble gases have a valence shell configuration of s^2p^6—filled s and p subshells. In contrast, the alkali metals have a single electron in their valence shell—an s^1 configuration.

Each alkali metal is followed by a member of Group IIA, which is known as the **alkaline earths.** Each alkaline earth is less reactive with water, is harder, melts at a higher temperature, and is denser than its predecessor in the alkali metal family.

The elements found together in the sea—Cl, Br, and I along with fluorine (F) and astatine (At)—form the **halogen** family, or Group VIIA. The Group VIIA elements all have a valence shell configuration of s^2p^5. These elements are all toxic to breathe, but the halogens also have some properties that differ. Fluorine and chlorine are gaseous elements, while bromine is a liquid and I a solid at room temperature. Moving from period to period, the elements' valence electrons are in a shell with a value of n that is one greater than that for the previous element in the group. For example, Cl has valence electrons in the $n = 3$ shell, whereas those of Br are in the $n = 4$ shell, and those of I are in the $n = 5$ shell. The valence electrons are therefore generally farther away from the nucleus as you proceed down a group, a trend that accounts for the differences in properties.

Periods

The physical and chemical properties of the elements vary across a period (Figure 3.4). For example, in the third period, sodium (Na) is a metallic solid that is putty soft. A piece of magnesium (Mg) is easily bent, but is much

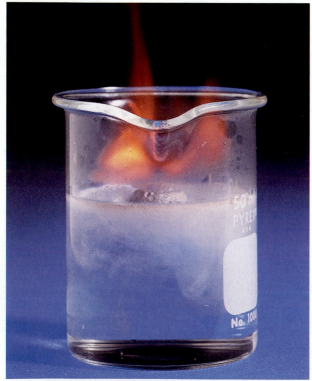

Figure 3.3 Dropping a small piece of Li, Na, or K onto water results in a vigorous reaction evolving gaseous hydrogen and a large quantity of heat—enough to ignite the hydrogen.

Figure 3.4 Physical form of the elements at room temperature and one atmosphere pressure. Most elements are solid at room temperature, a few are gases, and a small number are liquids.

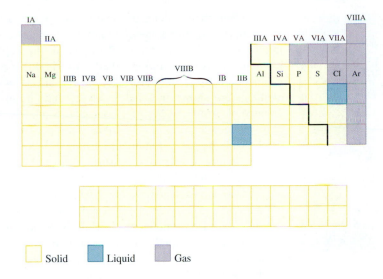

Solid   Liquid   Gas

harder than a piece of Na. Al is typical of many metals: It is easily machined into various shapes and conducts electricity well. Aluminum's neighbor, Si, is the foundation element of the semiconductor industry; it is a brittle solid. Next to Si are phosphorus (P) and sulfur (S), both insulating solids. Finally, chlorine (Cl) and argon (Ar) are gases. This pattern—soft metal to increasingly hard metal to brittle semimetal to insulating solid and finally to gas—is repeated for the main-group elements in each period. The periodic variation becomes more subtle through the transition metals from Period 4 and beyond. Nonetheless, a regular variation, repeated in subsequent periods, occurs across each and every period.

Progressing across a period, the electronic configuration goes from a single valence electron through a filled s subshell to increasing numbers of electrons in the p subshell. The period ends with filled s and p subshells. Transition elements have one or two electrons in the valence s subshell. The additional valence electrons in the transition metals are found in the $(n-1)d$ subshell. The d subshell electrons are located nearer to the nucleus than are the valence s electrons and nearly completely shield the outer-shell s electrons from the additional nuclear charge from one element to the next. The subtle variation of properties among the transition elements is due to this near-charge shielding.

3.2   Electrons and Bonding

The Octet Rule

Focusing specifically on the members of Group VIIIA, the Massachusetts-born chemist Gilbert N. Lewis (Figure 3.5) made the following observation. Helium does not combine with other elements; it is unreactive. With eight more electrons, neon is unreactive, and with another eight electrons, argon is also unreactive. Lewis conjectured that with eight electrons a core is filled, and additional electrons start a new layer. Examining compounds of the elements, Lewis noted that the IA metals tend to combine in a 1:1 ratio with the halogens. IIA metals combine in a 1:2 ratio with the halogens, but a 1:1 ratio with members of Group VIA. These and similar observations led to the **octet rule:**

Octet rule In a compound, atoms tend to achieve a noble gas configuration by acquiring or sharing electrons.

The representative elements (main group) achieve a noble-gas configuration (i.e., eight valence electrons) in *most* of their compounds, except for hydrogen, which has only two valence electrons in its corresponding noble gas structure.

Figure 3.5 Gilbert Newton Lewis developed the octet rule while working as a teaching assistant at Harvard University.

Further, the elements achieve this noble-gas configuration, an octet, via electron(s) from atoms bonded to them by gaining, donating, or sharing electrons with the bonded atoms. To assist in the application of the octet rule, Lewis developed a notation called a **Lewis dot structure** that represents the valence electrons as dots. The Lewis dot structures of the main-group elements are shown in Table 3.1. Note that Group IA elements all have one valence electron and therefore one dot. Similarly, Group IIA elements have two valence electrons and two dots, Group IIIA elements have three, and so on. Indeed, the logic in numbering the Groups IA, IIA, and so forth is based on the number of valence electrons.

In a compound, the atoms acquire, donate, or share electrons with the neighboring atoms to achieve the illusion of having eight electrons. In the Lewis dot notation, the transferred or shared electrons are shown as dots between the symbols for the two atoms involved in the bond. For example, salt is shown as Na∶Cl∶, where the single valence electron from Na fills in the remaining spot for eight electrons around chlorine. Since sodium has no other valence electrons, it can be viewed as having lost its single valence electron to chlorine, resulting in an electron configuration that is the same as neon—a noble-gas configuration.

Octet Eight valence electrons in an $s^2 p^6$ configuration.

Lewis dot structure A diagram of an atom or molecule depicting valence electrons as dots.

Table 3.1 Lewis Dot Structures of the Elements

IA								VIIIA
H·	IIA		IIIA	IVA	VA	VIA	VIIA	He∶
Li·	Be·		·B·	·C·	·N∶	∶O∶	∶F∶	∶Ne∶
Na·	Mg·		·Al·	·Si·	·P∶	∶S∶	∶Cl∶	∶Ar∶
K·	Ca·		·Ga·	·Ge·	·As∶	∶Se∶	∶Br∶	∶Kr∶
Rb·	Sr·		·In·	·Sn·	·Sb∶	∶Te∶	∶I∶	∶Xe∶
Cs·	Ba·		·Tl·	·Pb·	·Bi∶	∶Po∶	∶At∶	∶Rn∶
Fr·	Ra·							

WORKED EXAMPLE 3.1   *Salts*

Many people must limit their intake of sodium due to high blood pressure. For this purpose, a salt substitute was developed that has potassium rather than sodium in combination with chlorine. Write the Lewis dot structure of KCl. In this structure, potassium has the same electron configuration as what element? The chloride salt of calcium, $CaCl_2$, is often used to melt ice. In the Lewis dot structure for $CaCl_2$, calcium has the same electron configuration as what element?

Plan

■ Potassium and calcium have few valence electrons, so tend to lose electrons to the previous rare gas configuration.

Implementation

■ Losing one valence electron produces the potassium cation, which has the same electron configuration as argon. Losing two valence electrons from calcium produces a calcium cation that has the same electron configuration as argon.

KCl is very similar to NaCl except that the cations are different sizes. The different size affects how these ions cross a cell membrane, and this affects blood pressure. $CaCl_2$ produces three ions per unit formula, which makes it more effective at melting ice.

See Exercises 26 – 28.

For Cl_2, the representation is $:\!\overset{..}{Cl}\!:\!\overset{..}{Cl}\!:$, showing two electrons between the two chlorine atoms. Each chlorine atom has six valence electrons in addition to the electrons between the two atoms. The electrons between the atoms are counted among the electrons for each of the atoms; they are a shared pair.

CONCEPT QUESTIONS   Write the Lewis dot structure of O_2. How many electrons are shared between the two oxygen atoms for each to have an octet? Write the Lewis dot structure for N_2. How many electrons are shared in N_2? ■

Lewis dot structures can also provide insight into combining ratios. Consider nitrogen and oxygen and the compounds they form with hydrogen. Nitrogen has five valence electrons, oxygen six. Hydrogen has a single valence electron. Hydrogen tends to have two electrons in its compounds while nitrogen and oxygen each have eight. For nitrogen to have eight electrons in combination with hydrogen, three hydrogen atoms are required. Hence the compound between nitrogen and hydrogen is expected to have the formula NH_3. NH_3 is known as ammonia, which plays an important role in the path between nitrogen in the air (N_2) and the proteins of living organisms. Similar reasoning suggests that the formula expected for the compound formed by hydrogen and oxygen is H_2O, also known as water. Worked Example 3.2 below indicates how the octet rule and Lewis dot structure provide insight into combining ratios.

WORKED EXAMPLE 3.2   *Combining Ratios*

The Lewis dot structure and the octet rule are powerful aids for understanding the combining ratio for elements in a compound. Use the octet rule to explain why potassium forms a binary compound with oxygen that has the formula K_2O, and magnesium forms a compound with oxygen having the formula MgO. Draw the Lewis dot structures of these compounds.

Plan
- Draw the atomic Lewis dot structures.
- Distribute electrons so that each atom's structure is isoelectronic with the corresponding noble gas.

Implementation
- The atomic Lewis dot structures are $K\,\overset{.}{}$, $\overset{.}{Mg}\overset{.}{}$, and $:\!\overset{.}{O}\!:$. For K_2O, oxygen gets two electrons, one from each potassium. For MgO, both electrons come from Mg.
- The Lewis dot structures are $K:\!\overset{..}{O}\!:\!K$ and $Mg:\!\overset{..}{O}\!:$.

See Exercises 26 – 28.

The Lewis dot structure and octet rule for NaCl, $Na:\!\overset{..}{Cl}\!:$, suggest a picture in which the lone valence electron from sodium in fact transfers to chlorine simultaneously, giving sodium an electron configuration like that of neon and chlorine a configuration like that of argon. If this picture is real, then chlorine must have an attraction for an additional electron, and it must not be too difficult to remove the single valence electron from sodium. The

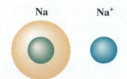

Na   **Na⁺**

Figure 3.6 A sodium atom ([Ne]$3s^1$) has a single electron in the valence shell. Removal of that electron leaves a Na⁺ ion that is isoelectronic to Ne; Na⁺ has a filled first shell and a filled second shell.

attraction for an additional electron and the ease of removal are discussed and quantified in the next section. Since this balance is very important for chemistry, they are combined in a concept called electronegativity, which is also discussed in the next section.

Ionization Energy

Removal of an electron from an atom leaves the atom with a positive charge; it produces a cation (a positive ion). This transformation is expressed as

(3.1) $$\text{Atom}(g) + \text{energy} \longrightarrow \text{Ion}^+(g) + e^-(g)$$

Energy input is required to remove the electron, an energy referred to as the ionization energy. To explore periodic patterns in ionization energy, it is helpful to keep the picture uncomplicated by other factors. So, ionization energy is defined for an isolated atom—an atom in the gas phase. The (g) in Equation (3.1) references this. Since the electrons account for most of the volume of an atom, the cation is smaller than the original atom (as shown schematically for Na in Figure 3.6).

The periodic pattern of ionization energies is shown in Figure 3.7. Ionization energy generally increases across a period from left to right and up a group from the bottom to the top of the table. (Values of ionization energies are given in the Appendix Table 9.) These trends are a direct result of two effects:

■ At the top of the table, the valence electrons are in a shell with a lower value of n; they are closer to the nucleus. With the negatively charged electron closer to the positively charged nucleus, the electron is more difficult to remove. Ionization energy generally *increases up a group*.

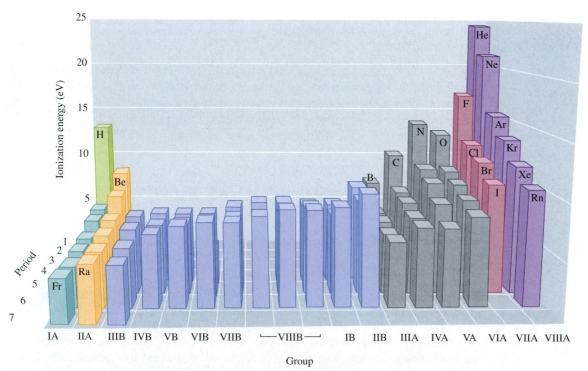

Figure 3.7 The highest ionization energies are found in the upper-right part of the periodic table; the lowest appear in the lower-left part of the periodic table. There is little variation among the transition elements due to very efficient screening of the core charge by the added *d*-orbital electron, which belongs to an inner shell.

Cl

Cl⁻

Figure 3.8 The Cl atom has an electron configuration of $[Ne]3s^23p^5$. Capturing an electron completes the valence shell, forming Cl⁻. The electron configuration of Cl⁻ is $[Ne]3s^23p^6$, which consists of filled s and p subshells around filled first and second shells. Cl⁻ is isoelectronic with Ar.

Electron affinity Measure of the energy of an atom's attraction for an additional electron. Energy is released in the reaction, $Atom(g) + e^-(g) \rightarrow Ion^-(g) + energy$.

■ Across a period, ineffective shielding by same-shell electrons increases the effective nuclear charge, which draws the valence electrons closer to the nucleus. The ionization energy generally *increases from left to right across a period.*

The highest ionization energies are therefore found in the upper-right part of the periodic table, and the lowest are in the lower-left part. A significant drop in ionization energy is observed from the end of one period to the start of the next. This variation is a powerful validation of the notion that electrons occupy successive shells, and the filling of the next shell begins at the start of a period.

CONCEPT QUESTION In each of the following pairs, which element has the larger ionization energy: Cs or Ba; Si or Ge; Ca or Fe; and Fe or Zn? ■

Electron Affinity

In a chemical reaction, if an electron is removed from one atom, it must end up as part of another atom. Chemical reactions do not produce unassociated electrons. Adding an electron to a neutral atom unbalances the electron-proton count and forms a negatively charged ion, an anion. Although atoms are electrically neutral, most can accommodate an additional electron.

CONCEPT QUESTION Write the electron configuration of Cl and the configuration of a Cl⁻ anion. Which neutral atom is isoelectronic with Cl⁻? ■

The process of capturing an electron is represented by the following reaction:

(3.2) $Atom(g) + e^- \longrightarrow Ion^-(g) + energy$

This reaction is shown schematically for chlorine in Figure 3.8. Notice that writing Equation (3.2) in reverse

(3.3) $Ion^-(g) + energy \longrightarrow Atom(g) + e^-$

looks like the ionization energy for the negative ion. Indeed, Equation (3.3) is used to determine how strongly an electron is held by the ion. The energy is called the **electron affinity** of the atom.

Electron affinity for main-group elements is illustrated in Figure 3.9, and values are listed in Table 3.2. As with ionization energy, electron affinity generally increases when

Figure 3.9 Electron affinity of the main-group elements. The electron affinity generally increases from the lower left to the upper right, but has several dips to near-zero values due to filled shells and subshells. Electron affinity varies little down a group due to the common group electron configuration.

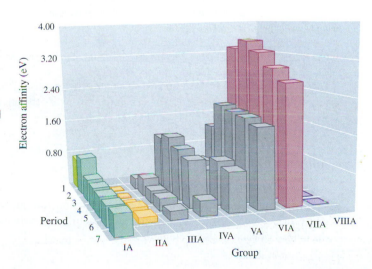

Table 3.2 Electron Affinity of the Main-Group Elements in eV (top) and kJ/mol (bottom).

H 0.754 72.77								He ~0
Li 0.918 88.57	Be ~0		B 0.277 26.7	C 1.263 121.8	N ~0	O 1.461 141.0	F 3.401 328.2	Ne ~0
Na 0.548 52.90	Mg ~0		Al 0.440 42.5	Si 1.385 133.6	P 0.746 72.02	S 2.077 200.4	Cl 3.613 348.6	Ar ~0
K 0.501 48.38	Ca 0.025 2.369	...	Ga 0.301 29	Ge 1.233 119	As 0.808 78	Se 2.021 195.0	Br 3.363 324.5	Kr ~0
Rb 0.486 46.88	Sr 0.148 14.3	...	In 0.301 29	Sn 1.112 107.3	Sb 1.046 100.9	Te 1.971 190.2	I 3.059 295.1	Xe ~0
Cs 0.472 45.50	Ba 0.145 14	...	Tl 0.207 20	Pb 0.364 35.1	Bi 0.946 91.3	Po 1.866 180	At 2.798 270	Rn ~0

moving up and to the right in the periodic table. The values of the electron affinity for most elements are lower than the values for typical ionization energies, and several electron affinity values are near zero. These trends are a direct result of two factors:

- ■ The outermost electron in the negative ion is repelled by the electrons of the neutral atom, and electrons of the neutral atom shield most of the nuclear charge.
- ■ Valence electron configurations with filled shells, or filled or half-filled subshells, effectively shield the nuclear charge. When the outer subshell is filled, the added electron is forced into a higher-energy, less stable subshell, so it is held very loosely (or not at all) by the ion.

CONCEPT QUESTIONS Write the electron configuration of Mg and Ne. Which orbital is occupied next in each of these elements? Do you expect these elements to strongly attract an additional electron? ■

CONCEPT QUESTION Why do the Group IIA elements have low electron affinity values? ■

CONCEPT QUESTION If nuclear shielding is perfect, the effective nuclear charge attracting the additional electron in an anion is zero, while the effective nuclear charge for the outer electron in an atom is 1. Why? ■

WORKED EXAMPLE 3.3   *Anion Effective Charge*

An anion forms because the valence electrons of a neutral atom do not completely shield the nuclear charge. Hence the electrons of an approaching atom are attracted to the atom. What is the effective charge attracting the extra electron to F or Cl to form the anions of each of these elements?

Plan

- Effective charge is determined from the ionization energy—in this case, the ionization energy of the anion:

- Use $Z_{eff} = n\sqrt{\dfrac{\text{ionization energy}}{13.6 \text{ eV}}}$

Implementation

- The ionization energy of the anion is the electron affinity of the atom. For F this is 3.401 eV, for Cl this is 3.613 eV. For F^-, $n = 2$. For Cl^-, $n = 3$.

- For F^-, $Z_{eff} = 2\sqrt{\dfrac{3.401 \text{ eV}}{13.6 \text{ eV}}} = 1.000$; for Cl^-, $Z_{eff} = 3\sqrt{\dfrac{3.613 \text{ eV}}{13.6 \text{ eV}}} = 1.546$

The energy required to remove the electron from fluoride and chloride is nearly the same. But since the chloride electron is farther out from the nucleus, the effective charge is larger.

See Exercises 37–38.

Electronegativity

Electronegativity The tendency of an atom to attract electrons in a molecule or solid, i.e., in competition with other atoms.

Recognizing the importance to chemistry of gaining and losing electrons, chemists have developed a parameter called **electronegativity.** Electronegativity is defined as the tendency to attract electrons in a compound, that is, in competition with other atoms. Electronegativity values are empirically based and are roughly proportional to the mean value of the first ionization energy and the electron affinity. Several electronegativity scales exist. Chemists most often use the one based on a scale—generated by Linus Pauling—on which values range from a low of 0.7 for francium to a high of 3.98 for fluorine (Table 3.3). The variation of the electronegativity for main-group elements is shown in Figure 3.10. The highest electronegativity values are in the upper-right corner of the periodic table and generally decrease in both directions from the maximum for fluorine to the minimum at the lower-left corner of the periodic table.

Table 3.3 Electronegativity Values of the Main-Group Elements

H 2.20								He
Li 0.98	Be 1.57		B 2.04	C 2.55	N 3.04	O 3.44	F 3.98	Ne
Na 0.93	Mg 1.31		Al 1.61	Si 1.90	P 2.19	S 2.58	Cl 3.16	Ar
K 0.82	Ca 1.00	...	Ga 1.81	Ge 2.01	As 2.18	Se 2.55	Br 2.96	Kr
Rb 0.82	Sr 0.95	...	In 1.78	Sn 1.96	Sb 2.05	Te 2.1	I 2.66	Xe
Cs 0.79	Ba 0.89	...	Tl 1.8	Pb 1.8	Bi 1.9	Po 2.0	At	Rn
Fr 0.7	Ra 0.9	...						

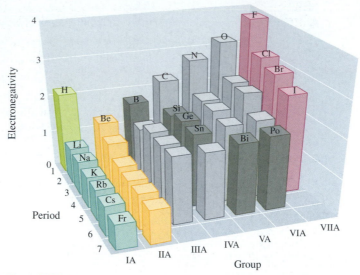

Figure 3.10 Electronegativity of main-group elements shown on the Pauling scale. Elements with electronegativity values in the range 1.90–2.04 are indicated in dark gray. These elements separate the metallic elements on the left of the periodic table from the nonmetallic elements on the right. Hydrogen, a typical nonmetal, has an electronegativity greater than two but is on the left due to its single, s-orbital valence electron.

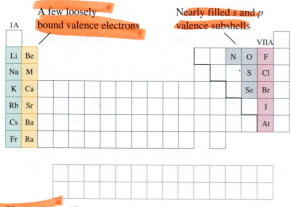

Figure 3.11 Elements on the left side of the periodic table have a few loosely held valence electrons. Those on the right side of the table have nearly filled valence s and p subshells.

Electronegativity values vary more smoothly than either electron affinity or ionization energy (Figure 3.12). An element with a high electron affinity and a high ionization energy both attracts additional electrons strongly and holds its own electrons tightly. When in competition with other elements, these elements strongly attract electrons, so they have high electronegativity values. The highest value among all the elements is that of fluorine, which has an electronegativity value of 3.98. On the opposite end of the scale, an element with a weak attraction for additional electrons and a weak hold on its own valence electrons, has a low electronegativity value. The lowest electronegativity value is 0.7 for francium—a Period 7, Group IA element.

3.3 Chemical Trends

The chemical trends that form the basis for organizing the periodic table can be viewed in terms of a combination of the electron configuration of the elements, the octet rule, and the electronegativity. For example, elements in Group IA have a single valence electron. Removal of that electron leaves a configuration with filled s and p subshells—an octet. In contrast, elements in Group VIIA are one electron short of a filled s^2p^6 configuration. An encounter between a Group IA element and a Group VIIA element results in a 1:1 compound in which the Group VIIA element has gained an electron, resulting in an octet. Since the elements in Group VIIIA have filled s and p valence shells, they have

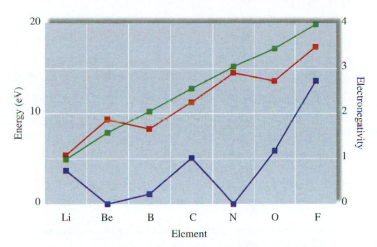

Figure 3.12 Ionization energy, electron affinity (here multiplied by four to show it on the same scale as the ionization energy), and electronegativity values for the second-period elements. Both the ionization energy and the electron affinity have a finer structure due to details of the electron configuration. In contrast, the electronegativity parameter smoothes out these finer variations and rises steadily from Li in Group IA to F in Group VIIA.

an octet when in isolation and are very unreactive; They are found as isolated atoms in the gas phase.

The Group IA elements all have low electronegativity values while the Group VIIA elements have high values. Since electronegativity is the tendency to attract electrons in a compound, the Group VIIA elements tend to gain an electron while the Group IA elements tend to lose one. The resulting substance consists of ions and is referred to as an ionic compound. Ionic compounds are often solids, so the formula for them is given as an empirical formula: the simplest whole number ratio that indicates the composition of the substance.

CONCEPT QUESTION Why do Group IA elements tend to form $2:1$ compounds with Group VIA elements, such as Na_2O? Predict the ratio of aluminum to oxygen in the empirical formula for aluminum oxide. ■

Electronegativity values can also give insight into the fate of electrons among elements with more comparable electronegativity. Consider chlorine and bromine. The electronegativiy of chlorine is 3.16 and that of bromine is 2.96. Combining a clear and colorless solution containing bromide ions, $Br^-(aq)$, with a clear and pale yellow solution containing chlorine molecules, Cl_2 results in a dark brown solution. This color change indicates production of molecular bromine.

CONCEPT QUESTIONS Draw the Lewis dot structures for molecular chlorine, Cl_2, and for the chloride ion, Cl^-. Does chlorine have an octet in both structures? How does the electron configuration of chlorine in Cl_2 differ from that of chlorine in Cl^-? ■

In a halogen ion, the halogen has an octet in its valence shell without the need to share. Production of molecular bromine indicates that bromide ion has transformed into molecular bromine since a chemical reaction does not change the identity of an element. Two bromide ions have sixteen valence electrons (eight each) while molecular bromide has only fourteen electrons. The two electrons lost from two Br^- have been incorporated into another substance, and the only candidate is molecular chlorine, which transformed to two Cl^-. The reaction is

(3.4)
$$2\,Br^-(aq) + Cl_2 \longrightarrow Br_2 + 2\,Cl^-(aq)$$

a reaction consistent with the greater electronegativity of chlorine. This reaction can be thought of as chlorine atoms and bromine atoms attracting an electron. With a greater attraction for an electron, chlorine captures the electron to become chloride.

CONCEPT QUESTION What do you expect will be the outcome of combining a solution containing bromine molecules with a solution containing iodide ions? Write any reaction that occurs. ■

WORKED EXAMPLE 3.4   *Fluorine, Chlorine, and Water*

The reactions of many materials with water are important. More specifically, reactions of halogens with water are of interest because water is often treated with halogens, e.g., chlorine to kill bacteria or fluoride to prevent tooth decay. The halogens are all in the same group, so are expected to react similarly with water. However, there are some important differences. The reaction of fluorine with water produces HF and oxygen

(3.5)
$$2F_2(g) + 2H_2O(l) \longrightarrow 4HF(aq) + O_2(g)$$

In contrast, the reaction of chlorine with water produces HCl and HOCl rather than O_2. HOCl is the active ingredient in laundry bleach and the sanitizing solution for swimming pools.

(3.6)
$$Cl_2(g) + H_2O(l) \longrightarrow HCl(aq) + HOCl(aq)$$

Why do the fluorine and chlorine reactions with water differ? Hint: use electronegativity.

Plan
■ List the electronegativity of oxygen, fluorine, and chlorine.
■ Use the Lewis dot structure to count electrons.
■ Use tendency to attract electrons to explain reaction.

Implementation
■ The electronegativities are: oxygen 3.4, fluorine 3.98, and chlorine 3.16. In order from highest to lowest these are fluorine, oxygen, and chlorine.
■ F_2 and Cl_2 have a total of fourteen valence electrons (seven for each halogen). In HF and HCl, due to the electronegativity difference between the halogen and hydrogen, the halogen has eight electrons in the valence shell. In water, oxygen has eight electrons while molecular oxygen has a total of twelve electrons (six for each oxygen).
■ Fluorine has the greatest electronegativity, so tends to be in a form with eight electrons, i.e., HF. To acquire eight electrons (rather than sharing), another element must lose one or more electrons. Since hydrogen has lost its valence electron to oxygen in water, the only electron source is oxygen. Fluorine is more electronegative than oxygen, so attracts the electron from oxygen. Oxygen must then share and ends up as molecular oxygen. The electronegativity of chlorine is less than that of oxygen, so cannot attract an electron away from oxygen to form O_2. Instead, one chlorine atom from Cl_2 attracts the electron from hydrogen in water to form HCl, and the other replaces hydrogen in water to form HOCl.

See Exercises 51–52.

CONCEPT QUESTIONS   The elements Cu, Ag, and Au are referred to as coinage metals due to their use in coins from ancient times. Compare the electronegativity of these three elements with Fe, another metal commonly found on earth. Of these four elements, which tends to attract electrons most strongly? Which the least? ■

The coinage metals tend to be found in elemental form in the environment. Compared with other metals, the electronegativity values for Cu, Ag, and Au are high hence, tend not to lose their valence electrons. Since metals react by losing electrons, the coinage metals are relatively unreactive and are found in elemental form. In contrast, the electronegativity of the Group IA and IIA elements is among the lowest in the periodic table. As a result, the alkali metal and the alkaline earth elements tend to lose their valence electron(s) to almost any other element, hence are almost never found in elemental form.

Size

The electronegativity, ionization energy, and electron affinity all decrease in progressing from period to period in Group VIIA. With a few exceptions when values are close, the same trend is observed for all the main groups. Progressing from period to period, the valence shell is increasingly far from the nucleus. This increased separation of the negatively charged electrons from the positively charged nucleus is the basis for the trends in which the size of the atoms and ions increases from period to period.

Recall that measuring the size or the radius of an atom is an exercise akin to measuring the diameter of a cotton ball—there is no hard, defined "edge" to the atom. As a result, there are many ways to define the size of an atom. For example, the size could be defined from the volume it occupies in a molecule or a solid. With this definition, the "size" is affected by interaction with other atoms. Hence, the atomic radius is defined as the size of the atom free of all interactions; that is it is the radius of the sphere containing 90% of the electron density for the free atom. With this definition, the periodic variation of atomic size of the elements (Figure 3.13) is directly related to the electronic configuration.

Atoms of the Group IA elements are the largest in each period, and size generally decreases across the period. This trend may appear contradictory—size decreases despite adding electrons. However, electrons in the same shell ineffectively shield the simultaneously added nuclear charge. The outer electrons therefore experience more of the added nuclear charge. All are drawn toward the nucleus, which decreases the atom's size. This increased attraction increases the ionization energy, the electron affinity, and the electronegativity. Size increases down a group due to filling of successive shells. The largest atomic size is therefore found in the lower-left portion of the periodic table. The radius varies only slightly across the transition elements since the *d*-orbital electrons are in an inner shell and effectively shield the outer *s*-orbital electrons from the added nuclear charge.

CONCEPT QUESTIONS   Explain the significance of the change in size between (a) the elements K and Rb, and (b) the elements Zn and Ga. Would you expect the elements Kr and Rb to have similar chemical properties? Why or why not? Would you expect the radioactive element francium (Fr) to be larger or smaller than radium (Ra)? How do you expect the size of the man-made element bohrium (Bh), which has a fleeting existence, to compare with that of rhenium (Re)? ■

Ion Size

When an atom gains or loses an electron, the resulting ion is a different size than the neutral atom. Like charges repel, and losing an electron diminishes the coulombic repulsion among electrons in the atom. Thus a cation—a positively charged ion—resulting from the loss of an electron is smaller than the original atom (Figure 3.14). The reduction in

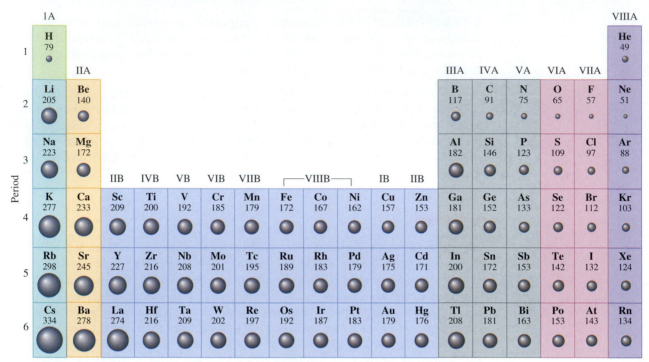

Figure 3.13 The atomic radius of elements (in picometers) increases down a Group and decreases across a period. In contrast to the trend seen with ionization energy, the smallest atoms are found in the upper-right part of the table, and the largest in the lower-left part. The atomic radius is defined as the radius of the sphere containing 90% of the electron density for the free atom.

size can be quite dramatic. Indeed, reduction to one-third to one-half the atom size is seen for the Group IA metals. In Group IA metals, the single outer electron is quite diffuse, with low electron density compared with that of the remainder of the electron cloud.

On the other side of the periodic table, the elements have relatively large electron affinities and tend to form negative ions—that is, anions. In contrast to electron loss, electron addition results in a larger ion than the original atom (Figure 3.14). The added electron contributes to the overall electron-electron repulsion and is held loosely in the ion. Again, the size difference can be quite significant, about a factor of two. Ion size significantly affects the role that an ion plays. For example, the size difference between Na^+ and K^+ controls the electrolyte balance across cell membranes. This balance is disrupted when cell walls are destroyed, such as happens in burn victims. Survival of the patient is critically dependent on maintenance of this balance.

WORKED EXAMPLE 3.5   *Converting Mass Density to Molar and Atomic Density*

Iron (Fe) and titanium (Ti) are comparably stiff materials. Fe is both more abundant and less costly to produce than Ti, so many structural materials are based on Fe. Despite the advantages enjoyed by Fe, Ti is the material of choice in aircraft and rocket hulls due to its lower density. Determine the atomic density of each of these materials. Titanium atoms pack in metallic titanium occupying 74% of the volume. Calculate the radius of a titanium atom in metallic titanium. The atomic radius of titanium is 200 pm. How does the radius

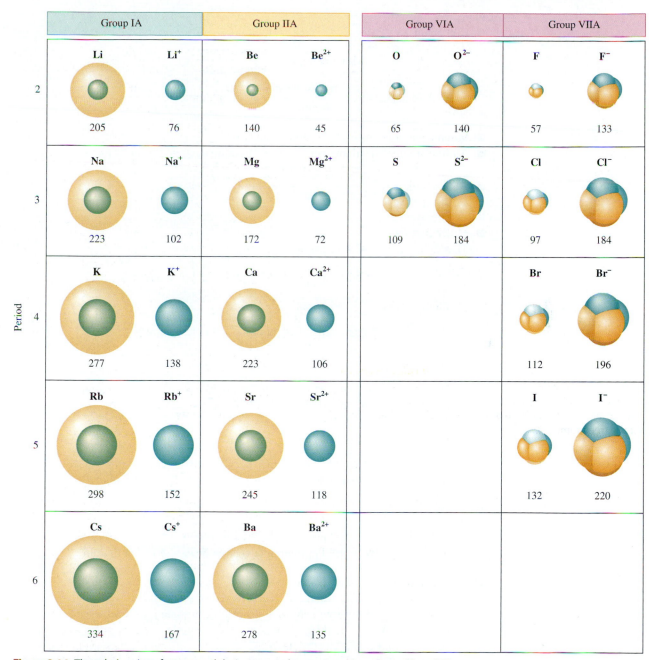

Figure 3.14 The relative size of atoms and their commonly associated ions. Group IA and IIA elements tend to lose one (IA) or two (IIA) electrons, forming cations (positive ions) that are smaller than the atom. Group VIA and VIIA elements tend to gain one (VIIA) or two (VIA) electrons, forming anions (negative ions) that are larger than the atom. Radii (in picometers) are indicated below the atom or ion. Orbitals shown as semitransparent are half-filled; nontransparent orbitals are filled.

in metallic titanium differ from the atomic radius? Why are these radii different? Iron packs slightly less efficiently with only 68% of space filled. Compare the atomic radius of iron (172 pm) with the radius in metallic iron.

Data   Ti density = 5.54 g/cm^3
Fe density = 7.874 g/cm^3

Plan
- Unit analysis

$$\text{Molar density} = \frac{\text{mass density}}{\text{atomic mass}} = \frac{\text{g/cm}^3}{\text{g/mol}}$$

$$\text{Atomic density} = \frac{\text{mass density}}{\text{atomic mass}} \times \text{atoms/mole} = \frac{\text{g/cm}^3}{\text{g/mol}} \times \frac{\text{atoms}}{\text{mol}}$$

- Volume per atom in solid = 1/atomic density
- Atomic volume = (volume per atom) × (fraction occupied by atom)
- Atomic radius = $\sqrt[3]{\dfrac{3 \times \text{atomic volume}}{4\pi}}$

Implementation
- Molar density:
  Ti: atomic mass = 47.867 g/mol

$$5.54 \text{ g/cm}^3 \div 47.867 \text{ g/mol} = 0.116 \text{ mol/cm}^3$$

  Fe: atomic mass = 55.845 g/mol

$$7.874 \text{ g/cm}^3 \div 55.845 \text{ g/mol} = 0.1410 \text{ mol/cm}^3$$

  Atomic density:

$$\text{Ti: } 0.116 \text{ mol/cm}^3 \times 6.022137 \times 10^{23} \text{ atoms/mol} = 6.99 \times 10^{22} \text{ atoms/cm}^3$$
$$\text{Fe: } 0.1410 \text{ mol/cm}^3 \times 6.022137 \times 10^{23} \text{ atoms/mol} = 8.491 \times 10^{22} \text{ atoms/cm}^3$$

- Volume per atom in the solid

$$\text{Ti: } (1/6.99 \times 10^{22} \text{ atom/cm}^3) = 1.43 \times 10^{-23} \text{ cm}^3\text{/atom}$$
$$\text{Fe: } (1/8.491 \times 10^{22} \text{ atom/cm}^3) = 1.178 \times 10^{-23} \text{ cm}^3\text{/atom}$$

- Atomic volume

$$\text{Ti: } 1.43 \times 10^{-23} \text{ cm}^3\text{/atom} \times 0.74 = 1.06 \times 10^{-23} \text{ cm}^3\text{/atom}$$
$$\text{Fe: } 1.178 \times 10^{-23} \text{ cm}^3\text{/atom} \times 0.68 = 8.01 \times 10^{-24} \text{ cm}^3\text{/atom}$$

- Atomic radius

$$\text{Ti: } \sqrt[3]{\frac{3 \times 1.06 \times 10^{-23} \text{ cm}^3}{4\pi}} = 1.36 \times 10^{-8} \text{ cm} = 136 \text{ pm}$$

$$\text{Fe: } \sqrt[3]{\frac{3 \times 8.01 \times 10^{-24} \text{ cm}^3}{4\pi}} = 1.24 \times 10^{-8} \text{ cm} = 124 \text{ pm}$$

The mass density of Fe is greater than that of Ti because the atomic mass of iron is greater than that of Ti, and more atoms are packed into a given volume. The radius of titanium in metallic titanium is greater than the radius of iron in metallic iron. Both radii are considerably smaller than the respective atomic radii, indicating that interatomic interactions have diminished the electron density of the metal atoms in the solids.

See Exercise 54.

3.4   Trends in Physical Properties

Classification of the Elements

Elements in the periodic table are often classified into three broad categories (Figure 3.15): metals, nonmetals, and semimetals. The vast majority of the elements are metals. Those at the right-hand end of each period are nonmetals with fewer nonmetals for each succeeding period. Between the metals and the nonmetals are one or two elements called semimetals with properties that are intermediate between metals and nonmetals.

The metallic elements are those with low electronegativity values, the nonmetals those with high electronegativity values. Semimetallic elements have intermediate values of electronegativity.

Metal An element that is malleable and ductile. Metals are good conductors of heat and electricity.

Malleable Capable of being shaped or formed or pounded into thin sheets, as by hammering or pressure.

Ductile The ability to be drawn into a wire.

Metals The essence of a **metal** is that it is **malleable**—can be pounded into a sheet—and **ductile**—can be drawn into a wire. Approximately 75% of the elements are metals and are found on the left-hand side of the periodic table. Most are solids at room temperature, have a silvery shiny sheen, and conduct heat and electricity well. Metals are used for wiring, structural materials (especially in the transportation industry), and numerous home appliances (Figure 3.16). Metallic character becomes more pronounced from top to bottom and from right to left in the periodic table.

Metals are associated with low electronegativity values. Low electronegativity means that metals have little tendency to attract electrons, and the low ionization energy means that the valence electrons are loosely held. The characteristic properties of a metal are due to the bonding caused by this loose hold resulting in relatively facile movement of metal atoms over each other in the solid. Bonding in metals is explored further in chapter 4.

Nonmetal An element that is neither malleable nor ductile.

Nonmetals Nonmetals are found on the right-hand side of the periodic table and are varied in physical state. Most are gases (11), some are solids (5), and one (bromine) is a liquid at room temperature. In the solid form, they are generally brittle and poor conductors of heat and electricity; as a consequence, they are known as insulating solids.

Nonmetals include oxygen and carbon, both of which are essential for life. The nonmetal carbon exists in three elemental forms (Figure 3.17). In the form of diamond, it is the hardest material known and is widely used for industrial cutting tools and everlasting gems. In the form of graphite, carbon is a conductor, though neither ductile nor malleable, hence is not a metal. Graphite's sheet-like structure makes it an excellent

Figure 3.15 The metallic elements are shaded yellow, the nonmetals are unshaded, and the semimetals are shaded green.

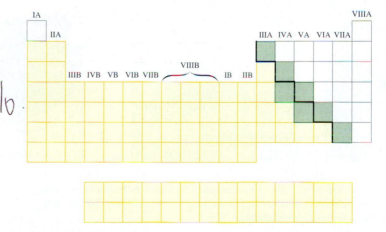

75% # of metals outweigh others

Figure 3.16 Manufacturing of wires, airplane hulls, and sheet metal for appliances all utilize the ductility and malleability of metals.

Allotropes Two or more forms of an element that differ in the ways the atoms are linked, e.g., carbon as graphite and diamond.

lubricant. The final form of carbon, called buckminsterfullerene, has only recently been discovered and is not yet widely used.

Oxygen makes life on earth possible in two ways. As O_2, oxygen makes up 21% of the atmosphere and is essential for respiration. In the form of ozone, (O_3), it filters out harmful radiation from the sun.

Different forms of the same element are called **allotropes.** Ozone, O_3, and oxygen gas, O_2, are allotropes of oxygen just as diamond, graphite, and buckminsterfullerene are allotropes of carbon. As illustrated in the carbon and oxygen examples, allotropes can have profoundly different properties.

Nonmetallic character becomes more pronounced in moving from the bottom to top and from left to right in the periodic table. Due to their high electronegativity values, nonmetals have a strong attraction for electrons. As a result, bonding between nonmetallic elements tends to be localized between the atoms. Nonmetallic bonding is explored in Chapter 5.

Semimetal, metalloid The smallest set of elements falling between the metals and nonmetals in the periodic table.

Semimetals (Metalloids) Semimetals (also called **metalloids**) account for the smallest number of elements and fall between the metals and the nonmetals in the periodic table. Semimetals have some properties that are common to metals and some that are common to nonmetals. All semimetals are solid at room temperature, are brittle, and are poor

Figure 3.17 Three forms of carbon: the gem stone diamond, the lubricant graphite, and the most recently discovered form called Buckminsterfullerene (referred to as "bucky balls"). Bucky balls are so named because the molecular structure resembles domes designed by the architect Buckminster Fuller. Bucky balls occur naturally in soot.

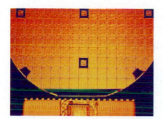

Figure 3.18 Silicon is the most extensively used semi-conducting material. Shown here are (top) a processed silicon wafer and (bottom) a silicon microchip.

conductors of electricity. They include the semiconductors silicon, Si (Figure 3.18), and germanium, Ge. The conductivity of semiconductors is very low at low temperature, but increases with increasing temperature. The increasing conductivity with temperature is in contrast to the properties of metals, for which conductivity decreases with temperature. Semiconductors are distinct from semimetals. In addition to the semiconductors Si and Ge, semimetals include boron, bismuth, and polonium. (The basis for classification of a substance as a semiconductor is explored further in Chapter 5.)

In general, metallic character is associated with few valence s and p electrons and increases as those electrons are located farther and farther from the nucleus. Nonmetallic character is associated with nearly filled valence s and p subshells and becomes more pronounced when the valence electrons are found near the nucleus.

Physical Form of the Elements

Examination (Figure 3.4) of the physical form—the phase—of the elements also reveals a periodic variation. Most elements are solids, some are gases, and a few are liquids at room temperature. With the exception of the first period, every period begins with an element that is a solid and ends with an element that is a gas. This variation is based on differing interactions among the atoms of the element (Figure 3.19).

CONCEPT QUESTION   Although not yet discovered or synthesized, another element probably exists below astatine (At, #85) in the periodic table. Do you expect this element to be a solid, liquid, or gas at room temperature? Explain your reasoning. ■

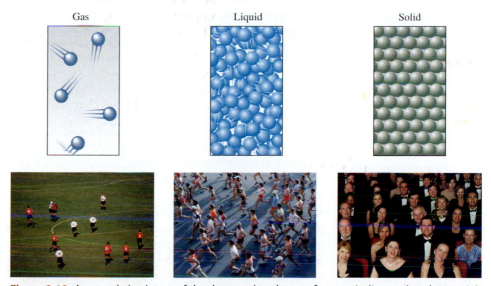

Figure 3.19 An atomistic picture of the three major phases of matter indicates that the spatial relationships among the atoms are very different. In the solid, atoms are arranged in a regular, repeating pattern. The solid holds its shape without the assistance of a container. Atoms in the liquid are about the same distance apart as in the solid, but the regular repeating arrangement is missing and the spatial location of atoms with respect to one another is more flexible. A liquid therefore easily conforms to the shape of its container. The atoms still stick together, and no top is needed on the container. In the gas phase, the distance between atoms is about an order of magnitude larger than in the liquid or solid phase. Atoms in the gas phase are the most free, with little or no cohesion between atoms. A gas needs a tight-fitting lid to keep the atoms from flying off in all directions.

Interactions and the Three Phases of Matter By definition, a solid holds its shape without the support of a container. Hence, the interactions between atoms in a solid must be relatively strong, keeping the atoms fixed with respect to one another. In contrast, atoms in a liquid flow readily over one another to adopt the shape of the container. Atoms or molecules in the gas phase fill whatever container they are in, hence interactions between gas-phase atoms or molecules are very weak.

> **Phase transition** Change from one phase of a substance to another phase.
>
> **Phase change** An alteration in the spatial relationship among atoms or molecules in a substance, e.g. liquid to solid.

With the possible exception of helium, all elements become solid at sufficiently low temperature. Raising the temperature adds energy to a substance, and with sufficient energy all substances become gaseous. The process of transforming from a solid to a liquid and from a liquid to a gas is referred to as a **phase transition** or a **phase change**. Elements that are liquid at room temperature have a melting point—a phase transition from solid to liquid—at a temperature below room temperature. Similarly, elements that are gaseous at room temperature have both a melting point and a boiling point—a phase transition from a liquid to a gas—that are below room temperature. The basis for the regular variation in interactions and the physical state of the elements can be envisioned by considering the melting point of the elements (Figure 3.20).

CONCEPT QUESTION What do you think is happening within a metallic solid as it heats up and melts? ■

Melting Point

For some applications, the melting point is a very important consideration. For example, sealing connections at low temperature require a material that remains resilient when cold—a principle that was missed in designing the O-ring joints for the space shuttle *Challenger*. Indium (In) wire is often used to make a low-temperature seal.

CONCEPT QUESTION What are two applications in which a high melting point is desirable? ■

At the temperature of a phase transition, interactions between the atoms are fundamentally altered. For example, imagine what happens on the atomic level when a solid melts—transforms into a liquid (Figure 3.19).

In a solid, atoms cling tenaciously. As the solid is heated, the heat energy causes the atoms to move. The more heat that is added, the more the atoms move, until finally the cohesion between them is overcome, and the regular arrangement of the atoms is disrupted. This situation is analogous to sitting in a comfortable chair—a certain amount of energy is needed to get up and move around. In the solid the atoms are organized in a regular, repeating array. As more and more energy is added to the atoms, more atoms are able to move greater distances. When enough are moving, the regular array is disrupted. When the solid melts to become a liquid, the atoms still cling to one another, but the heat energy causes them to move so much that they are constantly and randomly jostling one another in a dizzying dance, rather than being frozen in place. Adding yet more heat energy increases the jostling in the liquid still more, until finally the jostling is sufficient that the atoms break free of one another. At that point, the liquid becomes a gas, and there is no connection between the atoms or molecules to keep them together.

An examination of the melting points of the elements (Figure 3.20) reveals a periodic variation. The group IA metals all have quite low melting points. Relatively little energy is required to disrupt the fixed array of atoms in the solid. From these low values at the start of each period, the melting point increases up to a maximum about midperiod and decreases again at the end of the transition elements. In the fourth and fifth periods, this steady progression is interrupted by a dip at manganese (Mn) and technetium (Tc).

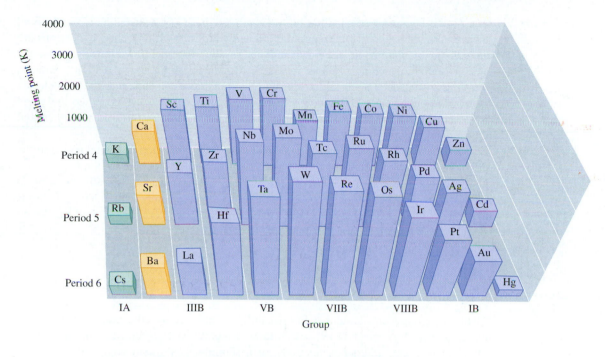

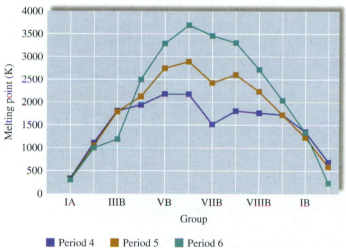

Figure 3.20 The melting point of the metallic elements has a regular variation with position in the periodic table. From a low value at the start of each period, the melting point rises to a maximum about halfway through the transition elements; it then falls back down at the end of the transition elements. Each period starts and ends with a lower melting point than the previous period. However, the peak in temperature increases for each successive period.

The sixth period contains the extremes; both the highest-melting-point element, tungsten (W), and the lowest-melting-point metal, mercury (Hg) are found in this period.

CONCEPT QUESTION   Elements in the seventh period are radioactive and rare, so many of their properties have not been determined. Locate the element with the highest melting point in the fourth, fifth, and sixth periods. Which of the following elements is predicted to have the highest melting point: rutherfordium (Rf), dubnium (Db), seaborgium (Sg), bohrium (Bh), hassium (Hs), or meitnerium (Mt)? ■

At the beginning of each period, the elements have few valence electrons to bind them into a solid, and the melting point is relatively low. As more valence electrons are added, the melting point climbs. About halfway across the transition elements, the steady climb in melting point is reversed. Midway through the transition elements, the $(n - 1)d$ subshell is half filled. Beyond the transition elements, into the p block elements, the outer p subshell contains increasing numbers of electrons. When the s and p subshells are filled, the atoms of the element have little attraction for each other, and the element is a monatomic gas.

The decrease in melting point midway through the transition elements is due to the effective shielding of the core charge by a half-filled d subshell and to bonding in metals. Bonding in metals is explored more fully in the next chapter. Briefly, bonding in metals consists of an attraction of the metal core for the loosely held valence electrons that form a negatively charged sea of charge in which the cores are embedded. When shielding between the core and this charge cloud is more effective, the attraction is weaker, and the melting point is lower.

WORKED EXAMPLE 3.6   *Predicting Properties*

The main-group element number 114 has only recently been synthesized and has a fleeting existence. Is element 114 expected to be a metal that loses electrons easily, a nonmetal that attracts additional electrons, or a semimetal? Explain your reasoning. If a large number of atoms could be synthesized, would element 114 be solid, liquid, or gas at room temperature? Predict the melting point of element 114.

Plan
- Locate element 114.
- Identify the electronegativity of the elements in the same group.
- Examine the melting point of the smaller members of Group IVA.

Implementation
- Element 114 is in Group IVA, period seven. Element 114 is to the left of the continuation of the dividing line between metals and nonmetals, so element 114 is expected to be a metal.
- Lead is the element just above element 114 in Group IVA. The electronegativity of lead is 1.8, just slightly less than the electronegativity of the larger members of Group IVA. Thus the electronegativity of element 114 is expected to be about 1.9, consistent with being a metal.
- The melting point of the immediately preceding members of Group IVA is Sn, 505 K; and Pb, 601 K. The melting point is predicted to be in the 500–600 K range. Element 114 is likely to be solid at room temperature.

See Exercise 86–90.

Mechanical Properties

Most familiar materials are made of a combination of elements. For example, steel contains not only iron, but also carbon, chromium, vanadium, and occasionally other elements in various proportions. A few common materials are nearly pure elements. For example, copper used for electrical wire is at least 99.999% pure elemental copper and is easily obtained in a variety of gauges or thicknesses.

Copper and steel wire exhibit a big difference in flexibility. Connecting the macroscopic properties such as flexibility to the microscopic structures of these and other

Figure 3.21 The flexibility of five common elemental metal wires varies considerably.

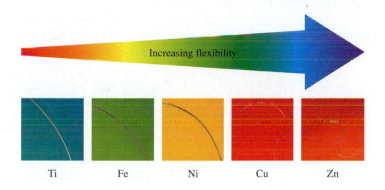

Increasing flexibility

Ti   Fe   Ni   Cu   Zn

materials requires an understanding of how atoms stick together, which in turn is related to the fundamental character of each elemental atom.

Flexibility

The effort required to pound a metal into a sheet, to draw it into a wire, or to bend a wire once drawn varies significantly from metal to metal. Consider five common metals: iron (Fe), copper (Cu), titanium (Ti), nickel (Ni), and zinc (Zn). Figure 3.21 shows typical results of trying to bend these elemental wires. Which metal is the best choice for electrical wiring? Which is preferred for an aircraft hull?

The resistance of these five metals to a bending force is very different (Figure 3.21). Titanium is very stiff, making it useful for airplane hulls and artificial joints. Zinc is easily deformable, almost waxy. Iron is nearly as stiff as titanium. The copper wire is reasonably flexible—a useful property for electrical wiring. If copper were as inflexible as titanium or steel, it would be difficult to wind it around screws to make electrical contact or make into flexible wires. Nickel is intermediate in flexibility between iron and copper.

CONCEPT QUESTIONS What do you think is happening within the wire when you apply stress? Why do you think each wire responds differently to applied stress? ∎

As a metal wire is bent, atoms on the inside of the curve are brought closer together while those on the outside of the curve are stretched further apart (Figure 3.23). Some lines of atoms slide over the neighboring line toward the outside of the curve. A combination of the attraction between atoms and the packing of atoms in a metal explains why this distortion/sliding is more facile in Cu and Zn than in Fe, Ti, or steel.

On the atomic level there are similarities between bending and melting. In both cases, the connections between an atom and its neighbors are disrupted, and this disruption requires energy. To melt a solid, the energy is supplied by heat. Atoms that interact strongly require more energy and therefore must be heated to a higher temperature to break away from their neighbors. In bending, the energy is supplied by the bender, but for a wire to bend, atoms need to break away from and slide over their neighbors. Among the five common metals, an inverse correlation exists between melting point and flexibility (Figure 3.24). That is, increasing flexibility correlates with a decreasing melting point.

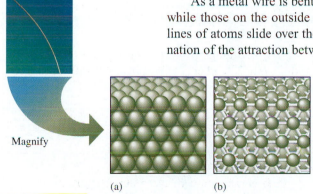

Magnify

(a)   (b)

Figure 3.22 This hexagonal close pack layer is an idealized atomic structure. (a) In a space-filling model, the sphere representing an atom just touches the spheres of its nearest neighbors. (b) In a ball-and-stick model, the balls representing the atoms are somewhat smaller, and sticks connect nearest neighbors.

bending distorts the close packed layers.

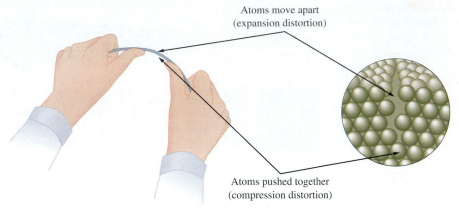

Atoms move apart
(expansion distortion)

Atoms pushed together
(compression distortion)

Figure 3.23 Bending a wire distorts the hexagonal close pack layers. Two types of distortion within the layers of atoms are illustrated — atoms moving apart (expansion) and atoms being pushed together (compression). This distortion is most clearly seen in the area of greatest distortion.

CONCEPT QUESTION   Tungsten is often used as the filament in an electric light bulb. Do you expect tungsten to be a flexible or a stiff solid? Explain your reasoning. ■

The difference in stiffness for various wires suggests that the tendency of layers of atoms in elemental materials to slide, and the ease with which the sliding propagates, varies from one elemental metal to the next. The question then becomes, What is it about the cohesion in some metals that enables easier sliding in them than in other metals? The answer is not simple. Indeed, the concepts needed to answer this question are developed in the next chapter.

Figure 3.24 There is a correlation between stiffness and melting point in wires composed of five common metallic elements.

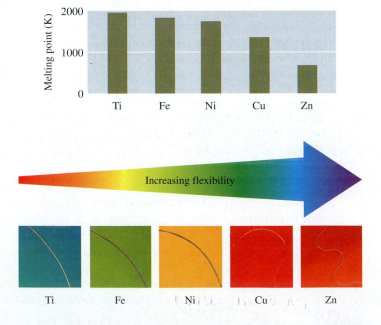

Increasing flexibility

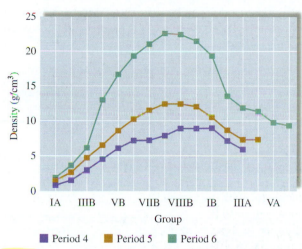

Figure 3.25 Comparison of the density of metals in Periods 4, 5, and 6 of the periodic table shows that the density varies periodically, with a rise and fall repeated in each period.

Density

Density profoundly affects the utility of different elements. For example, Fe and Ti are both members of the fourth period. Due to differences in their atomic structure, Fe is nearly twice as dense as Ti. However, since the melting point of Ti is about the same as that of Fe, the strength of Ti is comparable to that of Fe. Thus, despite the high costs, Ti (and not Fe) is used in rockets, airplanes, and other applications where lightweight strength is a priority. Any material made from elements in the sixth period is very heavy due to the high density of sixth period elements. Hence, applications employing these elements use only small amounts and take advantage of their other characteristics. For example, tungsten is used as a thin filament in light bulbs due its high melting point. Gold is used as a decorative material due to its beauty, low chemical reactivity, and ability to be made extremely thin, which means that it adds little weight to the object it coats.

CONCEPT QUESTION What is an application where a high density is desirable? ■

The regular periodic variation in density (Figure 3.25) was one of the properties used by Mendeléeff to predict the density of yet-to-be-discovered elements. Atomic mass and density increase down a group. Within a period, density increases then decreases despite the mass increase. Similar to the variation in the melting point, this increase then decrease is connected with the electronic structure of the atom.

CONCEPT QUESTION Suppose that none of the lanthanides or actinides had been discovered. What periodic properties would point to their existence? ■

WORKED EXAMPLE 3.7   *Predicting From Patterns*

Mendeléeff could predict the atomic mass of Ga because the masses of the series of elements preceding it increase by one to three mass units per element. There are nearly ten mass units between the preceding element zinc (Zn) and the next known element arsenic (As), so it is reasonable that the mass increase would be on the larger side—three more mass units than zinc is about 68. The density of Zn is 7.133 g/cm³ and that of As is 5.73 g/cm³, so predicting a density in between—about 6 g/cm³—is reasonable. (Density is listed on the periodic table in the back of the book.) How might Mendeléeff have predicted the atomic mass and density of the missing element #75, rhenium (Re)?

Plan
■ Locate element #75 in the periodic table.
■ Use mass and density data.

Implementation
■ Element #75 is between tungsten (W) and osmium (Os), which were both known to Mendeléeff. The difference in atomic mass between W (183.84 amu) and Os (190.3 amu) is about six. So the atomic mass of Re should be about three units greater than that of W. The predicted mass of Re is 187 amu.

■ The density of W is 19.3 g/cm³ and that of Os is 22.57 g/cm³. The difference in density from W to Os is about 3 g/cm³, so the density of Re is predicted to be 1.5 g/cm³ greater than W or about 21 g/cm³.

The actual atomic mass of Re is 186.207 amu, and the actual density is 21.02 g/cm³.

See Exercises 88 – 90.

Checklist for Review

KEY TERMS

noble gases (p. 87)
alkali metals (p. 87)
alkaline earths (p. 88)
halogens (p. 88)
octet rule (p. 89)
octet (p. 90)
Lewis dot structure (p. 90)
electron affinity (p. 93)
electronegativity (p. 95)
atomic radius (p. 99)
metal (p. 103)
malleable (p. 103)

ductile (p. 103)
nonmetal (p. 103)
allotrope (p. 104)
semimetal (p. 104)
phase transition (p. 106)
phase change (p. 106)

KEY EQUATIONS

ionization energy:

$$\text{Atom}(g) + \text{energy} \longrightarrow \text{Ion}^+(g) + \text{e}^-(g)$$

electron affinity:

$$\text{Atom}(g) + \text{e}^- \longrightarrow \text{Ion}^-(g) + \text{energy}$$

Chapter Summary

The chemical and physical properties of the elements vary periodically. This periodic variation has its basis in the periodic variation in the electron configuration of the elements. Each period begins with valence electrons occupying a shell with a principle quantum number that is one greater than the pervious period.

The periodic variation in electron configuration leads to a periodic variation in chemical properties. At the beginning of each period, the elements have few electrons in the valence shell, these are held loosely and are lost to form cations in chemical reactions. The number of loosely held valence electrons is the same as the group number: one in Group IA, two in Group IIA, and three in group IIIA. At the other end of the period, the elements have nearly filled s and p valence subshells. Since the subshells are not filled, the effective nuclear charge is large. Hence the valence electrons are tightly held, and the atom has a strong attraction for additional electron(s).

The common combining ratios for elements on opposite sides of the periodic table can be predicted based on the Lewis dot structure. Main group elements tend to have eight electrons in their compounds except for hydrogen, which has two electrons in the valence shell of its corresponding noble gas.

KEY IDEA

The periodic variation of the electron configuration of the elements is the basis for the periodic variation in chemical and physical properties of the elements.

CONCEPTS YOU SHOULD UNDERSTAND

■ octet rule
■ electronegativity

OPERATIONAL SKILLS

- Use electron configuration to predict combining ratios (Worked Examples 3.1 and 3.2).
- Calculate the effective charge for an anion based on the electron affinity of the element (Worked Example 3.3).
- Predict relative reactivity for gaining or loosing an electron based on electronegativity values (Worked Example 3.4).
- Determine the size of an atom in an elemental solid (Worked Example 3.5).
- Use periodic patterns to predict melting point, physical form, and chemical reactivity of the elements (Worked Examples 3.6 and 3.7).

Exercises

A blue exercise number indicates that the answer to that exercise appears at the back of the book.

■ SKILL BUILDING EXERCISES

3.1 Trends in the Periodic Table

1. What main-group family of elements was missing from Mendeléeff's table?

2. In the periodic table, find four exceptions to the trend that elements are listed in the periodic table in order of increasing atomic mass.

3. What is meant by the term "coinage metals"? Which elements are these? Where are they located in the periodic table (in which group)?

4. Which elements are found along with chlorine in the sea?

5. At the time that Mendeléeff assembled the periodic table the noble gases, Group VIIIA, were not yet discovered. What periodic trends might have indicated that such a group was missing?

6. Mendeléeff was surprised that an entire group of elements, the noble gases, was missing from the periodic table that he published. Based on what we know about atomic structure at this point, discuss the possibility that a group of elements has not yet been discovered.

7. Potassium (K), rubidium (Rb), and cesium (Cs) are three successive members of the alkali metals. The atomic mass of K is 39.0983 and that of Cs is 132.9054. Without looking at a periodic table, predict the atomic mass of Rb.

8. Without referring to the periodic table, predict the mass of the element that lies between each of the following pairs.
   a. S (mass 32.066) and Te (mass 127.60)
   b. Cd (mass 112.41) and Sn (mass 118.710)
   c. Hg (mass 200.59) and Pb (mass 207.2)
   d. Si (mass 28.0855) and Sn (mass 118.710)

9. In which period are the following elements?
   a. sodium (Na)
   b. tungsten (W)
   c. oxygen (O)
   d. sulfur (S)

10. In which group are the following elements?
    a. chlorine (Cl)
    b. carbon (C)
    c. silicon (Si)
    d. cadmium (Cd)
    e. lead (Pb)

11. List five main-group elements. Indicate the name, symbol, period, and group of each.

12. List five transition elements. Indicate the name, symbol, period, and group of each.

13. List five elements that are metals. Indicate the name, symbol, period, and group of each.

14. List five elements that are nonmetals. Indicate the name, symbol, period, and group of each.

15. List five elements that are semimetals. Indicate the name, symbol, period, and group of each.

16. Indicate the group in the periodic table where you would look for the following:
    a. an element with a very low ionization energy
    b. an element with a high ionization energy and a low electron affinity
    c. a very small atom
    d. an atom that forms an ion that is smaller than the parent atom
    e. an atom that forms an ion that is larger than the parent atom

3.2 Electrons and Bonding

17. With perfect shielding, the effective charge for the outer electron of an atom is one, while for the outer electron of a -1 anion it is zero. Explain why. What is the effective charge for the outer electron for perfect shielding on a $+1$ cation? Explain your answers.

18. K^-, Ar, and Cl^- are isoelectronic. Predict the relationship among the ionization energies for this series—that is, which is largest? What data would you use to check your prediction?

3.2 The Octet Rule

19. Name three ions that are isoelectronic with Ar. Give the electron configuration of the parent atom in each case.

20. Identify an ion with each of the following electron configurations: $[Ar]3d^2$, $[Ar]3d^9$, and $[Ar]3d^{10}$.

21. What are the electron configurations of the following ions or elements: Cr, Cu, Mo, and Co^{2+}? To which block of the periodic table do they belong?

22. Give the electron configuration of the following ions: Na^+, S^{2-}, Fe^{2+}, and Fe^{3+}.

23. Write the electron configurations for S and the S^{2-} ion. What atom is isoelectronic with S^{2-}? Al forms a +3 ion. Write the electron configuration for this ion. What atom is isoelectronic with it?

24. In what way are the electron configurations of metallic elements similar? How does the electron configuration of a typical metallic element differ from that of a nonmetal?

25. Write the electron configurations of Cl and the Cl^- anion. Which neutral atom is isoelectronic with Cl^-?

26. What are the expected ion charges for S, Br, Sr, and Al? Which of these ions is the smallest? Which is the largest?

27. Predict the molecular formula for the compound formed from the following pairs of elements:
   a. Al and F   b. Ca and N   c. Ge and H   d. P and H

28. Predict the molecular formula for the compound formed from the following pairs of elements:
   a. Br and Cl   b. Ti and Cl   c. Ga and Cl   d. Al and O

3.2 Ionization Energy

29. Predict whether the ionization energy of S is greater or less than that of P. Explain your prediction, and check the data.

30. Write the reaction that defines the ionization energy.

31. Where on the periodic table would you look for the element with the highest first ionization energy? The lowest?

Answer questions 32 and 34 without reference to Figure 3.7 or Ionization Energies of the Elements in the Appendix, then check your answers with the figure and table.

32. Arrange the following in order of ionization energy from highest to lowest: Li, P, Rb, Ni.

33. Which of the following has the lowest first ionization energy: Li, K, Rb, or Cs?

34. Which of the following has the highest first ionization energy: Na, Si, P, or Cl?

35. Graph the first ionization energies for the third-period elements. Rationalize the variation.

36. Graph the first ionization energies of the fourth-period main-group elements. Rationalize the variations.

37. Determine the effective charge for the outer electron in the Br^- and I^- ions. Compare these with F^- and Cl^- calculated in Worked Example 3.3.

38. Determine the effective charge for the outer electron in the hypothetical N^- and O^- ions.

3.2 Electron Affinity

39. Write the reaction that defines the electron affinity. Which element has the highest electron affinity?

40. Refer to Table 3.2. Explain why the electron affinity of Group IIA and VA elements is low. Why is the electron affinity of Group IIIA higher than that of Group IIA?

41. Why do the halogens (Group VIIA) have the highest electron affinities?

42. Write the electron configurations of Mg and Ne. Which orbital is occupied next in each of these elements? Do you expect these elements to strongly attract an additional electron?

43. Predict whether the electron affinity of P is greater than or less than that of S.

44. Arrange the following in order of electron affinity from largest to smallest: Al, P, S.

45. List four elements (aside from the noble gases) that are likely to have a near-zero electron affinity.

46. Which period 5 and 6 transition elements are likely to have a near-zero electron affinity?

3.2 Electronegativity

47. Give the definition of electronegativity. How does electronegativity differ from electron affinity?

48. The noble gases received their name due to their tendency to not combine with other elements to form compounds. In the 1960s a series of compounds were formed that included the noble gases Xe and Kr. Based on electronegativity values, which element is most likely to form a compound with Xe or Kr? Predict at least two elements that are good candidates for forming compounds with Xe or Kr.

49. Arrange the elements C, H, O, and F in order of increasing electronegativity. If two elements are nearly equal, indicate this with a $\cong$ symbol.

50. Arrange the elements C, Si, and Ge in order of increasing electronegativity. If two elements are nearly equal, indicate this with a $\cong$ symbol.

3.3 Chemical Trends

51. Write the reaction that is expected if molecular chlorine is bubbled through a solution containing I^- ions.

52. A solution contains Cl^-, Br^-, and I^- ions. What can be added to the solution to generate molecular iodine without affecting the other two ions? What can be added to the solution to generate molecular bromine and iodine without affecting the chloride ions?

3.3 Size

53. The density of aluminum (Al) is 2.6989 g/cm^3. How many moles are in a cube of aluminum 1 cm on a side? How many atoms?

54. The density of vanadium (V) is 6.11 g/cm^3. Assume that the atoms fill 68% of the volume. What is the volume of a vanadium atom? Assume that the atom is spherical. What is the radius of the vanadium atom?

55. The atomic density of gallium is 5.264×10^{22} atoms/cm^3. Calculate the mass density of gallium.

56. The mass density of Na is 0.971 g/cm^3 and that of Mg is 1.738 g/cm^3. Calculate the atomic density of each.

57. The radius of Cu is 117 pm. Approximately how many copper atoms are on the top surface of a 1-cm copper cube? The density of copper is 8.96 g/cm^3. How many atoms are in the cube?

58. Explain why Group IIA elements are smaller than their Group IA counterparts in the same period.

59. Explain the significant change in size between (a) the elements K and Rb, and (b) the elements Zn and Ga. Would you expect the elements Kr and Rb to have similar chemical properties? Why or why not?

60. Explain the significant change in size between the elements Ar and K. Do you expect Kr to have chemical properties similar to those of Ar? Do you expect Rb to be chemically similar to K? Why or why not?

61. Explain the correlation between ionization energy and atomic size.

Answer Exercises 62–64 without referring to Figure 3.14, then check your answers with the figure.

62. Arrange the following in order of size from largest to smallest: As, Ca, Cs, S.

63. Which of the following is the largest: Mg, Ca, Sr, or Ba? Explain why.

64. Which of the following is the largest: Te, Sr, In, or Sb? Explain why.

3.3 Ion Size

65. S^{2-} and Ar are isoelectronic. What do you expect for the relative sizes of S^{2-} and Ar? Justify your answer.

66. Two definitions of the size of an atom, the atomic radius and the covalent radius, are given in this chapter. Which choice is appropriate for determining the effect of nuclear charge on size for an isoelectronic series?

67. In each of the following pairs, select the larger atom or ion: K or Zn; Br$^-$ or Rb$^+$; Cr^{6+} or Al^{3+}; Si or Ge.

68. O^{2-}, F$^-$, and Na$^+$ are isoelectronic. Which ion is the largest? Which is the smallest? Which ions are larger than their parent atoms? Which are smaller?

3.4 Trends in Physical Properties

69. The element technetium (#43) does not occur naturally on earth. Explain how the atomic mass, density, and melting point of Tc can be predicted.

70. Discuss the possibility that the recently synthesized and as yet unnamed element Uub is a liquid at room temperature. Predict the density of Uub.

3.4 Classification of the Elements

71. Classify the following as metals, nonmetals, or semimetals.
a. fluorine (F)
b. germanium (Ge)
c. silver (Ag)
d. mercury (Hg)
e. xenon (Xe)
f. boron (B)

72. Classify the following as metals, nonmetals, or semimetals.
a. cobalt (Co)   c. barium (Ba)   e. krypton (Kr)
b. tellurium (Te) d. bromine (Br)   f. vanadium (V)

3.4 Melting Point

73. Draw an atomic-level cartoon of a metallic solid and this same material in the liquid state. Put a scale on your drawing indicating the distance between atoms in the solid compared to the liquid, and indicate the relative motion of atoms in each of these phases.

74. The melting point of potassium (K) is 336 K, that of rubidium (Rb) is 312 K, and that of cesium (Cs) is 302 K. Discuss the possibility that francium (Fr) is a liquid at room temperature.

75. Without referring to the periodic table, in the following pairs choose the one with the lower melting point.
a. in the same period, an element in Group IA or IIA
b. in the same period, an element in Group IA or VIB
c. Cr or Zn
d. Mg or Ba
e. Y or W
f. Sr or Mo

76. Which element has the highest melting point? Which metal has the lowest identified melting point?

3.4 Mechanical Properties

77. Draw an atomic-level cartoon sequence of bending and breaking a metal wire.

78. Among the metallic elements, which do you expect to be the least flexible? Justify your answer.

79. The melting point of zirconium (Zr) is 2125 K. Predict where the flexibility of Zr falls among the five metals discussed in this chapter.

80. The melting point of cadmium (Cd) is 594 K. Predict where the flexibility of Cd falls among the five metals in Figure 2-3.

81. Sodium (Na) metal is so soft it can be cut with a knife. Do you predict that the melting point is higher or lower than that of Zn? Justify your answer.

82. The melting point of gold is 1064 °C and that of silver is 961 °C. Which do you predict to be more flexible, a silver or a gold wire of the same diameter? Justify your answer.

83. The melting point of magnesium (Mg) is 649 °C. Do you expect Mg to be harder or softer than Zn? than Ni? Justify your answers.

84. Among the metallic elements, which do you expect to be the most flexible? Justify your answer.

3.4 Density

85. The mass density of aluminum (Al) is 2.6989 g/cm^3. Calculate the atomic density.

86. Mendeléeff predicted an element between aluminum (Al) and indium (In). This missing element was forecasted to have a density of about 6.0 g/cm^3 and an atomic mass of approximately 68. Explain how Mendeléeff might have predicted these numbers. What is the missing element, and what is its actual density and mass?

87. In Mendeléeff's time there was a missing element between silicon (Si: mass 28.1 amu, density 2.33 g/cm³) and tin (Sn: mass 118.7 amu, density 7.31 g/cm³). Predict the density and mass of the missing element. What element is it?

88. Explain how Mendeléeff might have predicted the density and atomic mass of element #31, Ga.

89. Explain how Mendeléeff might have predicted the density and atomic mass of element #21, Sc.

90. Explain how Mendeléeff might have predicted the density and atomic mass of element #84, Po.

91. From the mass density of Na, Mg, and Al, calculate the atomic density of each. What is the variation for these three elements from Period 3?

92. From the mass density of K, Ca, and Ga, calculate the atomic density of each. Do these three elements follow the variation determined in Exercise 91?

93. Without consulting density data, for each of the following pairs choose the one with the greater density.
   a. elements in the same period in Group IA or IIA
   b. elements in the same period in Group IA or IB
   c. Na or Cs
   d. Ti or Hf
   e. Ti or Fe

94. In Period 4, in which group do you expect to find the metallic element with the greatest density? The element with the lowest density?

95. Make a plot of the atomic density of the metals in the first transition series (Sc to Zn) versus atomic number. Describe the general trend and any exceptions to it.

96. Plot the atomic density of the metals in the second (Y to Cd) and third (La to Hg) transition series against atomic number. Compare these with the first series, and comment on any trends that you observe.

Challenge Problems

97. A half-filled subshell is a particularly effective shield of the nuclear charge. In this worked example/exercise combination, we explore the basis for this effective shield. The quantum mechanical description of atoms indicates that the shape of a combination of occupied orbitals is the shape of the *sum of the electron densities*. The electron density for a given electron wave is the square of the amplitude. (The amplitude of the electron waves is given at the end of Chapter 2 before Exercise 83).

Example

What is the shape of a combination of two *p* orbitals (e.g., p_x and p_y)?

Plan

The density for a single orbital is amplitude squared: $|\psi|^2$. Density for a combination of orbitals is the sum of the densities for each orbital.

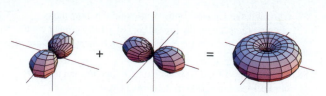

Figure 3.26 A p^2 combination is a torus.

$$\psi(p_x) = \frac{1}{2}\sqrt{\frac{3}{\pi}}\ \sin\theta\ \cos\phi \quad \text{and}$$

$$\psi(p_y) = \frac{1}{2}\sqrt{\frac{3}{\pi}}\ \sin\theta\ \sin\phi$$

Implementation

$$|\psi(p_x(\theta, \phi))|^2 + |\psi(p_y(\theta, \phi))|^2 = \frac{3}{4\pi}\sin^2\theta\cos^2\phi$$

$$+ \frac{3}{4\pi}\sin^2\theta\sin^2\phi = \frac{3}{4\pi}\sin^2\theta$$

Notice that there is no ϕ dependence, indicating that the resulting orbital has constant density about the *z*-axis (Figure 3.26). This shape is called a torus—sort of like a doughnut with a very small "hole." In the *x-y* plane, as one density drops, the other picks up, resulting in a constant density on any circle about the origin as shown in Figure 3.26.

Exercise

Show that addition of the third *p* orbital, the p_z orbital, for a p^3 combination fills the space above and below the plane of the torus to make a sphere.

98. Show that the sum of the densities for five *d* orbitals is a sphere.

■ CONCEPTUAL EXERCISES

99. Make a diagram showing relationship among the following terms: metals, nonmetals, semimetals, alkali metals, alkaline earths, halogens, noble gases, periodic law. All terms must be interconnected.

100. Draw a diagram indicating the relationships among the following terms: effective nuclear charge, electron configuration, octet, ionization energy, electron affinity, electronegativity. All terms should be interconnected.

☀ APPLIED EXERCISES

101. It is now illegal to use aluminum wire in household circuits. Based on physical properties included in this chapter, discuss the advantages and disadvantages of aluminum versus copper wire for household applications.

102. The purity of a gold object is specified in carats: the number of parts by mass of gold in twenty-four parts of an alloy. A twenty-four-carat gold object is thus pure gold. The density of twenty-four-carat gold is 19.3 g · cm³. Eighteen-carat gold is commonly alloyed with silver and has a density of 17.2 g · cm³. A ring has a 2-cm outer diameter, and is 3-mm long and 2-mm thick. How accurately must the ring be weighed to determine if it is pure gold or only eighteen-carat?

103. The mass of a 250-mL beaker is 190 g when empty. Determine its mass when filled with
(a) water (density 0.9970 g · cm⁻³ at 20 °C)
(b) alcohol (density 0.7893 g · cm⁻³ at 20 °C)
(c) mercury (density 13.546 g · cm⁻³ at 20 °C)
Explain why mercury-containing pollutants are usually found in the mud at the bottom of a lake or stream.

104. Filaments in modern light bulbs are made of tungsten. In developing the light bulb, Thomas Edison experimented with more than 6000 types of materials prior to his first successful bulb. What property of metals focused his attention on the metallic elements? Edison narrowed his search to platinum wire. Platinum was not successful. Use the fact that when a light bulb "burns out" the dark spot on the glass envelope is tungsten metal that has sublimed (vaporized from the solid) to suggest a reason for platinum's failure as a bulb filament.

■ INTEGRATIVE EXERCISES

105. Joining metal pipes or metal components in a circuit involves soldering—melting a metal to join metal surfaces. Three common solder materials are silver, tin, and lead. Find these metallic elements on the periodic table. Based on their chemical properties, which would you use to solder an electric circuit? On what properties(s) did you base your choice? Lead was formerly used to solder water pipes. What property made lead a good choice? Why is the use of lead now banned?

106. Viscosity is the resistance to flow. Compare the viscosity of the three phases of water. We do not commonly think of solids as flowing. Give an example illustrating the flow of ice.

107. Diamond and graphite are two forms of elemental carbon. The density of graphite is 2.2 g/cm³ and that of diamond is 3.513 g/cm³. Graphite is an excellent lubricant. Diamond is the hardest material known. Use the aforementioned data to discuss the interatomic interactions in diamond compared to those in graphite.

108. Cite evidence about atomic size, ionization energy, and electron affinity that suggests that $n - 1$ shell d-orbital electrons nearly perfectly shield the nuclear charge.

Chapter 4
Metallic Bonding and Alloys

The Guggenheim Museum in Bilbao, Spain, was designed by Frank Gehry and is a work of art in itself. Its skin consists of thin titanium plates — a unique application of a metal.

CONCEPTUAL FOCUS

- Connect deformation of metals with atomic-level interactions. Link conductivity and density with atomic-level interactions.

SKILL DEVELOPMENT OBJECTIVES

- Determine atomic size from density (Worked Example 4.1).
- Visualize the spatial arrangement of atoms in the solid, calculate packing efficiency, and locate slipping planes in solid structures (Worked Examples 4.2 and 4.3).
- Calculate elongation (strain) from load (stress) (Worked Example 4.4).
- Select metals for applications by mechanical criterion (Worked Examples 4.4 and 4.5).
- Evaluate the potential for alloy formation (Worked Examples 4.6 to 4.8).

etals are an important class of materials used in construction, transportation, and the infrastructure of our modern society. Look around you and note the variety of uses of metals: from fasteners—screws, bolts, nails—to case work—refrigerators, stoves, cars—to conductive elements—circuits, light bulb filaments, fuses. It is difficult to imagine life without metals. The flexibility, density, ability to conduct, and melting point of metals are all linked to the atomic-level structure—the electron configuration—of the constituent atoms.

The hallmark distinguishing physical characteristics of metals are that they are ductile—they can be drawn into wires; malleable—they can be flattened into sheets; and flexible—most can be bent extensively without breaking. The first stage of bending a metal is called the **elastic deformation** stage. Macroscopically, the elastic deformation stage is characterized by restoration of the original shape upon removal of the force (Figure 4.1). Steel I-beams in buildings undergo only elastic deformation under normal conditions. Indeed, building design is constrained by the requirement that expected loads do not exceed the elastic limit; otherwise, the building would sag and fail. As a greater force is applied, the metal further deforms. This deformation remains after the force is removed in what is called the **plastic deformation** stage. Bending a copper wire around a screw to make a contact involves plastic deformation, for example. Malleability and ductility both require plastic deformation.

The macroscopic elastic and plastic deformation stages are linked with the microscopic, atomic-level structure by the connections or cohesion between the metal atoms. These connections are referred to as metallic bonding. Visualizing metallic bonding begins with the idealized close packed structure (Figure 4.2). In the ball-and-stick representation, the lines between the atoms denote connections to the nearest neighbors. In a space-filling model, atoms are represented as spheres and nearest-neighbor atoms just touch one another. In the elastic stage, the connection—the bonding—to the nearest neighbors rules. Nearest-neighbor bonding depends on the electron configuration of the metal. In the second stage of metal deformation, the plastic deformation stage, planes of atoms begin sliding. At this point, the crystal structure takes center stage. Alloy formation alters the ease of sliding and strongly affects plastic deformation.

Additional characteristics influence the selection of a particular metal in a specific application. Two often-used criteria are conductivity and density. Conductivity and density depend on both the electronic structure of the atom and the crystal structure of the solid.

Elastic deformation Distortion of a solid such that it returns to the original shape when the distorting force is removed.

Plastic deformation A distortion of a solid past the elastic deformation stage. The solid remains in an altered shape after the distorting force is removed.

(a)   (b)

Figure 4.1 (a) A flag pole is designed to bend and flex in a strong wind. When calm returns, the flag pole returns to its upright shape in an example of an elastic deformation. (b) A hurricane produces winds sufficiently strong that the flag pole can suffer a permanent bend. The flag pole is plastically deformed.

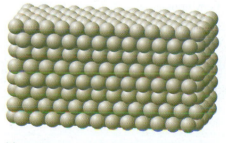

(a)

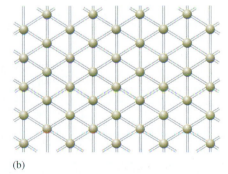

(b)

Figure 4.2 (a) A chunk of metal in an idealized close packed structure in a space-filling model. (b) A single layer of a hexagonally packed metal in a ball-and-stick model. The balls represent the atoms. Note the hexagonal arrangement of atoms around every atom.

4.1   Bonding in Metals

Metals are a class of elements that have relatively low first ionization energies and for which the cation formed from loss of one or more electrons is generally much smaller than the neutral atom. Imagine forming a metallic solid by removing the valence electrons and packing the resulting cations tightly in a solid array. Because the mass of the atom is nearly all contained in the nucleus (Figure 4.3), one of the first results of packing the cations (as opposed to atoms) together is to increase the density of the metal. Worked Example 4.1 explores this increase in density for titanium.

WORKED EXAMPLE 4.1   *Density*

When designing an aircraft, metal density is an important parameter to consider, and the density is partly determined by the effective size of the metal atoms. Treat solid Ti as a packed array of spherical atoms. There are always spaces between packed spheres, so assume that the atoms fill 74% of the volume (this 74% filling factor is due to the crystal structure, which will be explored a bit later). What is the density of this hypothetical solid Ti? (The radius of an isolated Ti atom is 200 pm.) Compare the hypothetical density with that for Ti, 4.54 g/cm³.

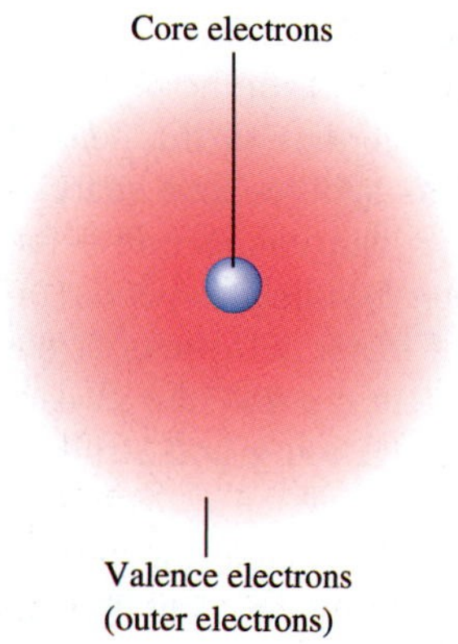

Plan

- Use the atomic radius to determine the atomic volume: $V = \dfrac{4}{3}\pi r^3$.
- Account for filling factor: Multiply by 1/0.74.
- Determine the volume of one mole: Multiply by Avogadro's number.
- Determine density: $\dfrac{\text{molar mass}}{\text{molar volume}}$.

Implementation

- Atomic volume $= \dfrac{4}{3}\pi\left(200 \text{ pm} \times \dfrac{1 \text{ cm}}{1 \times 10^{10} \text{ pm}}\right)^3 = 3.35 \times 10^{-23} \text{ cm}^3$

- Filling factor 74%; $\dfrac{3.35 \times 10^{-23} \text{ cm}^3}{0.74} = 4.53 \times 10^{-23} \text{ cm}^3$

- Volume of one mole $= (6.022 \times 10^{23} \text{ particles/mol}) \times (4.53 \times 10^{-23} \text{ cm}^3/\text{particle})$
  $= 27.3 \text{ cm}^3/\text{mol}$

- Density $= \dfrac{47.8\ \text{g/mol}}{27.3 \text{ cm}^3/\text{mol}} = 1.75 \text{ g/cm}^3$

The actual density of Ti is nearly three times that of the hypothetical solid formed from densely packed atoms. This ratio indicates that the effective volume of Ti in the solid is one-third that of an isolated Ti atom; the radius is about 70% of the radius of the isolated atom. Metal atoms lose part of their volume (part of their valence electrons), when they form a solid.

See Exercises 1–3.

Core electrons

Valence electrons
(outer electrons)

- Nucleus accounts for 99.99% of the mass of the atom.
- Electron cloud accounts for 99.99% of the volume.
- Outer electrons are valence electrons.
- Inner electrons are core electrons.

Figure 4.3 The major components of an atom. The nucleus accounts for 99.99% of the mass of the atom. The electron cloud accounts for 99.99% of the volume. The outer electrons are valence electrons. The inner electrons are core electrons.

Because all the atoms are the same in an elemental metallic solid, the electrons do not transfer from one atom to another to become positive and negative ions. Instead, the electron waves become less localized—like the spreading of a ripple on a pond—and

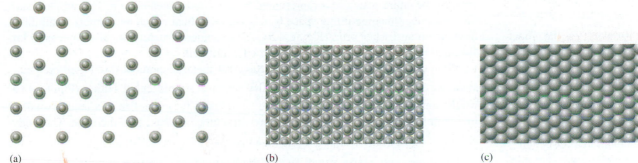

(a)   (b)   (c)

Figure 4.4 Imagine forming a metallic solid starting with the isolated atoms in a regular arrangement. (a) The atoms are far apart and the electron clouds are unaffected by the other atoms. (b) The atoms are brought closer together and their electron clouds just touch. (c) At the separation found in the metallic solid, the electron cloud extends over the entire solid and is no longer associated with a single atom. These itinerate electrons leave the cores with a positive charge. The positively charged cores are held together by coulombic attraction to the negatively charged electrons in the extended wave.

extend over many atoms (Figure 4.4c). Imagine how these extended electron waves arise from a collection of isolated atoms (Figure 4.4a) in which the electrons are confined to the region around each atom. Bringing the atoms closer together (Figure 4.4b) causes the valence electron wave of each atom to overlap that of neighboring atoms. When the atoms are tightly packed (Figure 4.4c), the valence electron waves combine, forming extended waves that encompass all the atoms. The valence electrons occupy these extended waves, and each atom becomes positively charged. The pool of negative charge in the extended wave surrounds the positive ions and is termed an **electron sea.** The electron sea acts like the sticky caramel holding a popcorn ball together. In this case, the caramel—the electron cloud—is negatively charged and the remainder of the atom is positively charged. This nondirectional coulombic bonding is termed **metallic bonding.**

Electron sea Pool of negative charges in an extended wave surrounding positive ions.

Electrons in the extended electron waves that constitute the electron sea are not bound to a particular atom within the solid, so these electrons are more mobile than those bound to an atom. Some characterstic properties of a metal result from the mobility of these electrons. For example, a voltage across an elemental metal wire easily sets the mobile electrons in motion. Moving electrons constitute an electric current, so metals have a high *conductivity.* The nondirectional bonds between atoms mean that it is relatively easy for atoms to shift positions and move over one another, accounting for the characteristic *ductility* and *malleability.*

Metallic bonding Nondirectional bonding consisting of a collection of cations floating in an electron sea held together by the coulombic attraction between the cations and the electron sea.

Although the electron-sea model describes major trends in the physical properties of metals, the electron configuration is needed to understand details. The electron orbitals have an associated direction. When these orbitals overlap, there is a contribution to the bonding between two neighboring atoms that is directed between the two atoms. The more localized *d*-orbital electrons supplement the delocalized electron sea, producing exceptions to general trends.

Crystal Structure of Metals

Metallic elements form extended, three-dimensional structures. The arrangement of the atoms within this extended structure forms a repeating spatial pattern. Since the pattern

Crystal lattice The repeating pattern of many solid materials, such as metals, when viewed at the atomic level.

Unit cell The smallest unit that when stacked together infinitely in three dimensions reproduces the solid.

Body-centered cubic, BCC An arrangement of atoms with atoms on each corner of a cube and on in the body center of the cube.

Face-centered cubic, FCC Arrangement of atoms in a solid in which the unit cube consists of atoms on every corner plus one in each face of the cube. Atoms in one layer are packed in equilateral triangles such that any one atom is surrounded by six other atoms in that layer. Layered in an ABCABC pattern.

Hexagonal close pack, HCP Arrangement of a solid in which the atoms within a layer are packed together in equilateral triangles such that each atom is surrounded by six other atoms in that layer. Layer above and below are directly over each other for an ABAB repeat pattern.

repeats, the entire structure is represented by the small, repeated unit. The simplest such unit is a cube. Imagine filling space by stacking identical cubes indefinitely in all directions, as in an Escher print (Figure 4.5). The extended structure is called a **crystal lattice,** and the repeated unit structure is called a **unit cell.**

Two of the most common crystal lattices for metallic elements are based on a cube symbolized as ▱. A solid structure based on a simple cubic lattice (Figure 4.6a) would consist of atoms at each of the eight corners of the cube. Few elements crystallize in a simple cubic lattice, but two common structures are based on the cube. The **body-centered cubic (BCC)** lattice, symbolized as ▱ (Figure 4.6b), adds one atom in the center of the cube. Starting with a simple cube, the **face-centered cubic (FCC),** lattice symbolized as ▱ (Figure 4.6c), adds an additional atom in the center of each of the six faces of the cube.

The remaining common lattice is called a **hexagonal close pack (HCP)** lattice, symbolized by ⬡ (Figure 4.6d). The name of the HCP lattice is based on the arrangement

Figure 4.5 An Escher print, "Fish," demonstrates a two-dimensional unit cell (indicated with the dark lines).

of atoms in a plane, where every atom is surrounded by six other atoms at the vertices of a hexagon. These planes play a major role in the plastic stage of deformation. The FCC and HCP lattices are closely related with respect to packing atoms in a plane, a topic that is explored later in this chapter.

An important, but less common solid structure is the diamond  structure. In addition to the structure of the gem diamond, it is the structure of tin in one of its solid forms and is the fundamental structure formed by semiconductors. The tetragonal structure is derived from the cubic structure by compressing one dimension analogous to the relationship between a square and a rectangle. The room-temperature form of tin is tetragonal. The very different mechanical and electrical properties of tin in its two crystal structures are directly related to these two different arrangements of the tin atoms.

The lattice structures of metallic solid elements are indicated by symbols in Figure 4.7.

Figure 4.5 *continued*   The three-dimensional print, "Kubische ruimteverdeling" translated as "cubic space division," has a cube as the unit cell. Can you identify it?

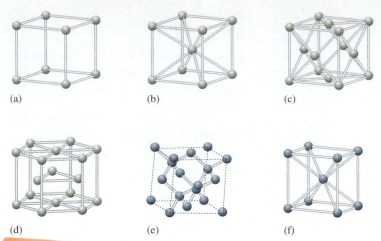

(a)   (b)   (c)

(d)   (e)   (f)

Figure 4.6 The unit cells for the crystal lattices of metallic elements.
The top row shows cubic lattice unit cells. (a) The simple cubic lattice features atoms at each of the eight corners of the cube. (b) A body-centered cubic lattice adds one atom in the center of the cube; hence the name "body centered." (c) Starting with the simple cube, a face-centered cubic unit cell adds an atom in the center of each of the six faces of the cube.
Lattices in the lower row are not based on a cube. (d) A hexagonal close pack lattice consists of staggered planes of atoms in a hexagonal pattern. (e) The diamond structure is named for the atom arrangement in the gem diamond, one form of elemental carbon. (f) The tetragonal structure shown is a body-centered tetragonal and is derived from the body-centered cubic lattice by compressing front to back.

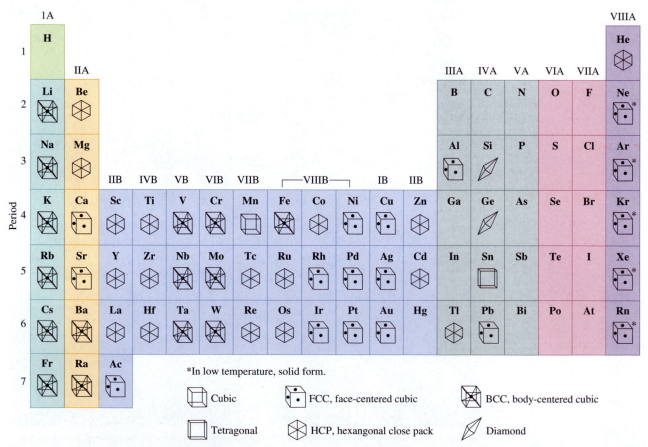

Figure 4.7 The crystal structure of the metallic elements; the semimetals Si and Ge; and the noble gases in their low-temperature, solid form.

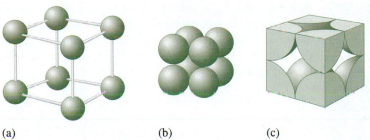

(a)   (b)   (c)

Figure 4.8 Three representations of the unit cell of a simple cubic solid. Each corner of the cube is occupied by an atom, with the center of the atom at the corner point. Space is filled with these cubes stacked in all three directions. (a) A ball-and-stick model emphasizes the cube structure. (b) A space-filling model indicates the space filled by atoms modeled as touching spheres. (c) Truncating the atoms at the surface of the unit cell emphasizes the repeating unit. In this representation, only one-eighth of each corner atom is within the unit cell. With eight corner atoms that are each one-eighth in the cell, there is *one atom in the simple cubic unit cell.*

Several forms are used to depict the unit cell. A ball-and-stick model is shown in Figures 4.6 and 4.8a. Figure 4.8 also shows a simple cubic unit cell in two other representations. In a space-filling representation (b), atoms are represented as spheres that touch, filling as much volume as possible for the given structure. To completely fill space by repeating the same unit with no overlap, atoms are truncated at the cube surface (Figure 4.8c). The truncated structure shows that in a simple cubic structure, only one-eighth of each corner atom is contained within the cube. As a result, each unit cell contains only *one* atom (one-eighth of eight equals one). All three pictures represent the same unit cell. The choice of picture is determined by the question to be answered. For example, the truncated, space-filling model more clearly shows the space between the atoms in the hard-sphere model, while the basic geometry may be more easily seen in the ball-and-stick model.

Hard-sphere model Picture of atoms as rigid and having definite edges.

Although atoms are softer than ball bearings or marbles, it is often helpful to picture them as having definite edges like these more familiar objects. Picturing atoms as having definite edges is referred to as a **hard-sphere model.** In metallic solids, the hard sphere represents the cation. The electron sea interpenetrates this cation array, providing the cohesion needed to keep the entire structure intact. An important difference among the various crystal structures is the number of nearest neighbors—atoms just touching—for any given cation; another difference is the packing efficiency—the fraction of the volume of the unit cube that is filled with atoms. The number of nearest neighbors is called the **coordination number.** The coordination number varies from a high of 12 for cubic close pack and face-centered cubic to as few as 4 for the diamond structure. Packing efficiency is greatest for the face-centered and hexagonal close pack lattices.

Coordination number Number of nearest neighbors or the number of atoms bonded to the central atom.

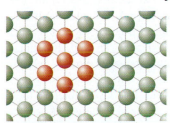

Figure 4.9 One layer of a hexagonal close pack structure in which one hexagon is colored in maroon to emphasize the hexagonal array.

Close Pack Lattices

Most metals crystallize in one of the two closely related structures: hexagonal close pack (HCP) or face-centered cubic (FCC). Although the hexagonal layers are not apparent from the unit cell of the FCC lattice, both FCC and HCP lattices contain hexagonal layers of atoms (Figure 4.9). This hexagonal structure is very common among packed, spherical objects. For example, oranges or grapefruit at the grocery

Figure 4.10 A scanning tunneling microscope (STM) image of oxygen atoms on the surface of a rhodium crystal.

store pack in a hexagonal structure because this structure holds the most objects in the least volume. In reference to the dense packing, hexagonal packed structures are called **close pack** structures. A scanning tunneling microscope (STM) image of a rhodium surface truncated with oxygen atoms (Figure 4.10) reveals the hexagonal packing of the top layer of atoms.

Close pack Crystal structure in which the atoms are densely packed occupying the least total volume in a *crystal lattice*.

 APPLY IT Obtain a large number of spherical objects, such as a bag of marbles or ping-pong balls. Put these in a box and try to duplicate the pattern shown in Figure 4.11a. Compare that challenge to duplicating the pattern in Figure 4.11b. Outline the unit cell in each case. In these two-dimensional structures, how many nearest neighbors does an atom in the center of each array have?

(a)   (b)

Figure 4.11 A collection of spherical objects is packed in (a) a simple square array and (b) a close pack layer.

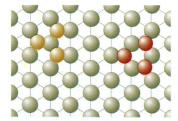

Figure 4.12 In hexagonal packing, cations in the second layer sit in triangular hollows of the first layer. For the next layer, two triangular sites are created: one over atoms in layer 1 (shown in gold to highlight these sites) and one over triangular hollows of layer 1 (shown in maroon to highlight these sites).

To form a three-dimensional structure, the hexagonal first layer is covered with more hexagonal layers. In the second layer (Figure 4.12), each atom rests in a triangular space, called a threefold-hollow site, formed by three atoms of the first layer.

The difference between the HCP and FCC structures is contained in the relationship of the third layer to the first two layers. Look closely at the first two layers. Two types of triangular spaces appear in the second layer (Figure 4.12): one that has an atom in the first layer under it and one that has no atom in the first layer. The third layer atoms can sit in either of these triangular spaces. In the hexagonal close pack structure, the third layer of atoms sits in triangular spaces directly over the first layer of atoms (referred to as ABA packing). In a face-centered cubic structure, the third layer of atoms sits in hollows with no atom in the first layer (referred to as ABC packing). In both structures, each atom has three nearest neighbors in the layers above and below it, and six nearest neighbors in the same layer. Thus the coordination number is 12 in both structures. Twelve is the maximum coordination number for packing identical, spherical objects, so both HCP and FCC are close pack structures.

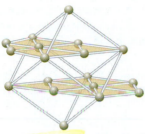

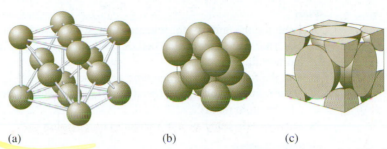

(a)   (b)   (c)

Figure 4.14 The relation-ship of the close pack planes with the unit cell in an FCC lattice is shown by standing the unit cell on one of its corners with the diagonally opposite corner directly above it. The plane contain-ing the lowest corner is the A plane of the ABC stacking, the B and C planes lie above the corner, and the upper corner lies directly over the bottom corner.

Figure 4.13 Three representations of a face-centered cubic unit cell. (a) The ball-and-stick model shows atoms as balls connected to their nearest neighbors by sticks. (In this depiction, the cube is also outlined with sticks.) (b) A space-filling model emphasizes touching of face diagonal atoms. Note that cube edge atoms do not touch. (c) Atoms truncated at the cube surface illustrate the unit cube.

Face-Centered Cubic Figure 4.13 shows the FCC unit cell with ball-and-stick, space-filling, and truncated models.

CONCEPT QUESTION How many atoms are in an FCC unit cell? ∎

The relationship between the FCC unit cell and the packing planes is rather intricate (Figure 4.14). The structure gets its name from the unit cell, but plastic deformation is dependent on these close pack planes.

WORKED EXAMPLE 4.2   *Visualizing Solid Structures*

The unit cell contains a wealth of information about the geometric arrangements of cations in the solid metal. For many practical applications, it is necessary to know the density. To unravel the relationship between the unit cell dimensions and the atomic arrangement, it is necessary to determine the number of atoms in each unit cell. How many atoms are in a face-centered cubic unit cell?

Plan
■ Group the atoms by the portion of each that is within the unit cell.
■ Determine the number of atoms in each group.
■ Multiply and add the fractions.

Implementation
■ There are two types of atoms in an FCC unit cell (Figure 4.13): corner atoms and face center atoms. The corner atoms are common to eight unit cells, so one-eighth of each is within each unit cell. Face atoms are common to two unit cells, so one-half of each is within each unit cell.
■ There are eight corner atoms and six face center atoms.
■ Number of atoms = (6 face center atoms × ½ within cube) + (8 corner atoms × ⅛ within the cube) = 4 atoms within each unit cell.

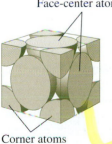

Face-center atoms

Corner atoms

See Exercise 11.

WORKED EXAMPLE 4.3   *Packing Efficiency*

In Worked Example 4.1, it was assumed that the Ti atoms occupied 74% of the space in the solid. Verify that 74% of space is filled with atoms in a face-centered cubic structure.

Plan

- Examine the unit cell for touching atoms.
- Determine the relationship between the unit cell edge dimension and atom radius.
- Use the formula for cube volume: (edge length)^3.
- Determine the volume occupied by atoms; use the sphere volume = $\frac{4}{3}\pi r^3$, where r is the atom radius.
- Packing efficiency = (atom volume)/(cube volume).

Implementation

- Examine Figure 4.13. The face center diagonal atoms just touch.
- Four atomic radii span the face diagonal. The diagonal = $4r$, where r is the cation radius. If a is the edge length, then the diagonal is $\sqrt{2}\,a \Rightarrow 4r = \sqrt{2}\,a.$

- Volume of the unit cube = $\left(\dfrac{4r}{\sqrt{2}}\right)^3 = \dfrac{4^3}{2\sqrt{2}}r^3 = \dfrac{32}{\sqrt{2}}r^3$

- Volume occupied by cations = $4 \times \dfrac{4}{3}\pi r^3$

(Factor of 4 due to four atoms in a unit cell; see Worked Example 4.2)

- Packing efficiency = $\dfrac{4 \times \dfrac{4}{3}\pi r^3}{\dfrac{32}{\sqrt{2}}r^3}$ $100\% = 0.74 \times 100\% = 74\%$

Thus 74% of the space is occupied by hard-sphere model atoms for face-centered cubic packing. This is the maximum packing density that can be created for identical spherical objects and is also the packing efficiency for an HCP lattice.

See Exercises 13, 14.

Hexagonal Close Pack The hexagonal close pack structure name derives from its close pack planes. The difference between the HCP and FCC structures is the stacking of the third layer of atoms, which results in ABC stacking for face-centered cubic and ABA for hexagonal close pack (Figure 4.15). The HCP unit cell is not a cube, but rather a hexagonal cylinder (Figure 4.6d). Since the third layer is shifted in FCC structures relative to HCP structures, 74% of the space is filled in both.

Body-Centered Cubic Lattice
A few metals crystallize in a body-centered cubic (BCC) lattice. Metals that adopt this structure include Fe, Cr, V, Mo, and W. The name body-centered cubic, like

Figure 4.15 The ABC stacking structure of a face-centered cubic lattice and the ABA stacking of a hexagonal close pack lattice shows the relative shifting of the third layer in the two lattices.

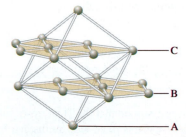

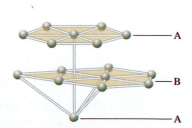

Figure 4.16 The body-centered cubic unit cell shown as (a) a ball-and-stick model, (b) a space-filling model, and (c) a truncated cube model. The space-filling model shows that body diagonal atoms just touch.

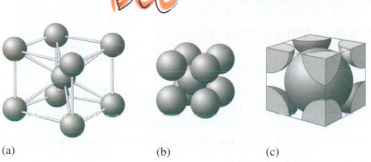

(a) (b) (c)

face-centered cubic, derives from the arrangement of cations on a cube. In this case, one cation is at each corner and one is in the cube center—the "body center" (Figure 4.16).

Atoms at the corners just touch the body center atom. In the BCC structure, each atom has eight nearest neighbors, giving a coordination number of 8 (Table 4.1). The eight nearest neighbors of the body center atom are the corner atoms, and the eight nearest neighbors of a corner atom are the body center atoms of the eight surrounding unit cells. The looser packing results in just 68% of the space being filled with spheres in a BCC lattice (see Exercise 13).

Electron Configuration and Crystal Structure

The crystal structure adopted by a particular metallic solid is the result of a balance among several factors.

- In the absence of mutual repulsion of cations, metals would adopt one of the close pack structures. In their solid form, the noble gases, all of which are neutral atoms with filled s and p valence subshells, exhibit the packing expected of spherical objects—FCC and HCP.
- Cations with filled shells and subshells maximize contact between the charged cations and the electron sea. The alkali metals are an example. Loss of the one valence electron leaves an alkali metal isoelectronic with the previous noble gas, but the remaining cation is positively charged. All of the alkali metals crystallize in a body-centered cubic lattice.
- Electrons confined in the region between neighboring atoms restrict the geometrical relationships among the atoms.

A balance among these three factors determines the crystal structure and also affects the physical properties.

	Table 4.1 Comparison of Common Crystal Structures		
Structure		**Nearest Neighbors (Coordination Number)**	**Packing Efficiency**
BCC		8	68%
FCC		12	74%
HCP		12	74%

4.2 Material Properties

Deformation

Stress, σ Applied load per unit area, F/A_o.

Macroscopically, the first stage of bending, called the elastic stage, is measured in an engineering stress-strain plot. The **stress,** symbolized by the Greek letter sigma, σ, is the force, F, applied per unit area, A_o:

(4.1)
$$\sigma = \frac{F}{A_o}$$

Strain, ε Elongation per unit length, $\Delta\ell/\ell_o$.

The **strain,** given by the Greek letter epsilon, ε, is the elongation, $\Delta\ell$, per unit length, ℓ_o:

(4.2)
$$\varepsilon = \frac{\Delta\ell}{\ell_o}$$

A typical stress-stain plot is shown in Figure 4.17. The stress-strain curve has two regions: a linear portion and a curved portion. The elastic region is the linear portion of the stress-strain plot and the slope is the **elastic modulus,** E:

Elastic modulus Constant of proportionality between stress and strain, linear portion of the stress-strain plot. Elastic modulus, $E = \sigma/\varepsilon$

(4.3)
$$\sigma = E\varepsilon$$

Within the elastic region, the sample returns to its original size and shape upon removing the load. A stiff material has a large elastic modulus, resisting deformation. There is a direct link between the elastic modulus and the electronic structure of the atoms in the solid. Exceeding the elastic limit, which involves moving beyond the linear portion, takes the material into the plastic region. In the plastic region, the material undergoes a permanent change of shape. When processing metals by drawing them into wires or pounding them into sheets, it is desirable to be in the plastic region. In contrast, for applications such as the structural components of a building or the body of a car, it is important to design the structure so that the response remains in the elastic region.

Elastic Response and Bonding

The first stage of bending or deformation is the elastic stage, which is characterized by the elastic modulus. The elastic modulus of fourth-period metals (Figure 4.18) first increases and then decreases from left to right across the period. This is a characteristic

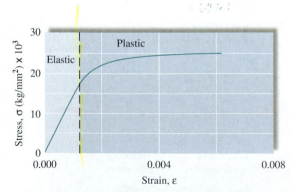

Figure 4.17 A typical stress-strain plot. The linear portion is the elastic region, and the curved portion is the plastic region.

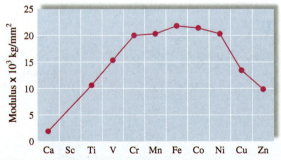

Figure 4.18 The elastic modulus for transition metals and calcium first increases and then decreases across the fourth period.

Before: After:

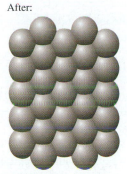

Figure 4.19 The atomic-scale result of application of a stretching force. Bonds along the length are elongated, while bonds across the width are compressed.

periodic variation, suggesting that the electronic structure of the atoms is linked to their elastic response.

On the atomic scale, the macroscopic elastic deformation corresponds to a small change of the spatial relationship between an atom and its neighbors (Figure 4.19) such as a slight compression or elongation of the bonds or a slight change in the angle formed by three adjacent atoms, known as the bond angle. Stretching or bending bonds while keeping the same set of neighbors stores the bending energy in the crystal. Removing the distorting force releases the stored energy, thereby returning the atoms to their original positions and the object to its original shape (Figure 4.20). An atom rigidly bonded to its neighbors is characterized by inflexible bond angles and bond lengths, leading to a large elastic modulus. Conversely, connections that are soft and angles that are unrestricted lead to easy movement of the atoms. The result is a low elastic modulus—a lot of bending or stretching for a little applied force. For comparison, a typical elastic modulus for rubber is three to four *orders of magnitude* lower than that of metals. As a consequence, it takes relatively little force to stretch rubber.

Note the relatively high elastic modulus of copper, a metal normally thought of as relatively flexible. However, flexibility is distinct from elasticity. Metal flexibility is associated with the permanent deformation of the plastic stage. In contrast, elasticity is associated with restoration of the original shape and length. One application where this high elastic modulus of Cu is important is in power lines (Worked Example 4.4). If the elastic modulus of copper was small, power lines would stretch much more when loaded with the ice and snow that occur in northern winters. A calcium wire, for instance, would stretch nearly six times as much as a copper wire.

WORKED EXAMPLE 4.4   *Sagging Power Lines*

Northern winters often feature ice and snow storms. An ice storm can load a power line with as much as 1 in of ice. What is the load, in grams per foot, for a power line? If the utility poles are spaced at 40-ft intervals, what is the load on the copper wire? (Assume that the load divides evenly between the two poles and the mass of ice is equivalent to a single mass at the end of the 20-ft wire.) If the copper wire is a 12-gauge wire, how much does the wire stretch if the response is elastic? The limit to elastic response is typically a strain of 0.1%. How much ice can the wire carry?

Data

Density of ice = 0.92 g/cm^3
12-gauge wire diameter: 2.0 mm
Elastic modulus of copper, $E = 13.48 \times 10^3$ kg/mm^2

Plan

- Determine the ice volume; use $V = \pi r^2 h$. Convert inches to millimeters.
- Use mass = volume × density.
- Determine total load: load per foot times length.
- Determine the cross-sectional area of Cu wire: πr^2, r = wire radius.
- Rearrange Equation (4.3), $\sigma = \varepsilon E$; σ = load per area; $\varepsilon = \Delta \ell / \ell_\text{o}$.
- Determine the load limit from the load that produces a strain of 0.1% ($\varepsilon = 0.001$); use Equation (4.3).

Figure 4.20 The elastic response of a metal takes its name from the more familiar and extensive elastic response of materials like that of a rubber band.

Implementation

- Volume of a cylinder = $\pi r^2 h$. The ice cylinder is essentially solid (except for the wire). r = 2.54 cm, h = 12 in $\times$ 2.54 cm/in.

$$\text{Volume per foot} = \pi \times (2.54 \text{ cm})^2 \left(12 \text{ in} \times \frac{2.54 \text{ cm}}{\text{in}} \right) = 618 \text{ cm}^3$$

- Load per foot = density $\times$ volume = 0.92 g/cm^3 $\times$ 618 cm^3 = 569 g/ft
- Total load = 569 g/ft $\times$ 20 ft = 11400 g
- Cross-sectional area = $\pi r^2 = \pi (1 \text{ mm})^2 = 3.14 \text{ mm}^2$
- Rearranging Equation (4.3),

$$\Delta \ell = \frac{\sigma \ell_o}{E} = \frac{11.4 \text{ kg}}{3.14 \text{ mm}^2} \times \frac{20 \text{ ft}}{13.48 \times 10^3 \text{ kg/mm}^2}$$

$$= 0.00538 = 0.0646 \text{ in} = \text{stretch}$$

- Load capacity: load/area = $\sigma \times E$; load = $\sigma \times E \times$ area.

$$\text{Load} = 0.001 \times (13.48 \times 10^3 \text{ kg/mm}^2) \times (3.14 \text{ mm}^2) = 42.3 \text{ kg}$$

The stretch seems quite minor, but the load that produces a permanent deformation is less than four times larger. For this reason small power lines are often twisted with a steel cable to provide greater mechanical strength.

See Exercise 28.

The energy stored in a compressed or stretched bond can be illustrated by plotting bond energy versus separation between atoms (Figure 4.21). In the absence of an applied force, the atoms settle to the bottom of the energy-separation curve. Either decreasing or increasing the separation increases the energy in the bond. Materials with a large elastic modulus have a narrow steep energy-separation curve, while softer materials have a wider curve. Elastic modulus data (Figure 4.18) suggest that the energy-separation curve gets deeper and narrower from left to right at the start of the fourth period. At Mn, this march pauses, so that the curves for Cr through Ni are very similar. At the end of the fourth-period transition metals, the energy-separation curve again flattens out and gets wider. This trend is connected with the valence electron configuration (Table 4.2).

The s valence electrons are the first to ionize from a Group IA, Group IIA, or transition element atom, so the s valence electrons contribute to the electron sea. Removal of the valence electrons from a Group IA or IIA element leaves only filled subshells; hence there is no preferred bonding direction. Like a collection of magnetic marbles, the atoms shift easily; the energy-separation curve is shallow and wide, and the elastic modulus of all Group IA and IIA metals is very low.

CONCEPT QUESTIONS   What is the valence configuration of Cr? Why do you think the bonds in Cr metal are less flexible than those in V metal? ■

Bonds involving d-orbital electrons are more localized, supplementing the nondirectional bonding from the cation–electron sea attraction. The directed bonds involving d-orbital electrons are more rigid. Therefore, materials with d-orbital bonding are less flexible. Transition elements all contain d-orbital valence electrons, and all of them have greater elastic moduli than the alkali metals or alkaline earth elements. At the start of the transition series, the elastic modulus increases steadily along with the number of d-orbital electrons. This increasing trend is arrested in the middle of the series (Mn, Fe, Co), then the elastic modulus begins to decrease. The electron configuration of Mn

Table 4.2   The Valence Electron Configuration of Fourth-Period Metals

These configurations feature filling of the fourth-shell s orbital and the third-shell d orbitals.

K	Ca	Sc	Ti	V	Cr	Mn	Fe	Co	Ni	Cu	Zn
$4s^1$	$4s^2$	$4s^23d^1$	$4s^23d^2$	$4s^23d^3$	$4s^13d^5$	$4s^23d^5$	$4s^23d^6$	$4s^23d^7$	$4s^23d^8$	$4s^13d^{10}$	$4s^23d^{10}$

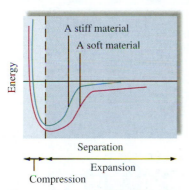

Figure 4.21 The energy of a typical bond. In the absence of an applied force, the atoms settle to the bottom of the energy curve. Compressing the bonds squeezes the nuclei together, so these cores repel each other. Pulling them apart reduces their mutual attraction, which reduces the stability and increases the energy. When the force is removed, the atom rolls down to the low energy separation, releasing the stored energy.

Slip system Combination of a favored slide direction and a close pack plane.

features a half-filled d subshell. Beginning with this half-filled configuration, the increased nuclear charge of Fe through Zn pulls the d-orbital electrons toward the nucleus and makes them less available for bonding interaction with neighboring atoms. Accordingly, the directed bonding decreases from Fe through Zn, as does the elastic modulus. In particular, for Zn the set of five d orbitals contains ten electrons and is filled. This filled subshell constitutes a spherical configuration. The bonds between Zn atoms are nondirectional, requiring little force to move an atom away from its neighbors. As a result, the elastic modulus of Zn is the lowest among the fourth period transition elements.

Yield Point and Plastic Response

Typically, when the strain reaches about 0.1%, the material is pushed beyond the elastic response. On the stress-strain plot (Figure 4.17), the limit to elastic behavior occurs where the straight-line initial portion curves. At this curve, plastic deformation begins and a larger-than-individual-bond view is required.

On the atomic scale, plastic deformation means that the atoms do not return to their original positions upon removal of the force. Instead, bonds between the atoms and their original neighbors are broken and are replaced by bonds with new neighboring atoms (Figure 4.22). Energy is required to slide atoms over each other. However, less energy is required to slide atoms along the rows of a close pack plane (Figure 4.23a) than along other rows (Figure 4.23b). This situation is analogous to gliding along a valley rather than going over mountains. In a hexagonal plane, there are three valley directions, or three directions in which the rows are tightly packed (Figure 4.24a). In contrast, a BCC structure has tightly packed rows in only two directions, and these fit less tightly together than those in an HCP or FCC lattice. The combination of a favored slide direction and a close pack plane is called a **slip system.** Face-centered cubic lattices have eight close pack planes in the unit cell (Figure 4.25). With three slip directions in

Figure 4.22 Plastic distortion occurs when the strain becomes too great and the atoms shift, so that they have new neighbors.

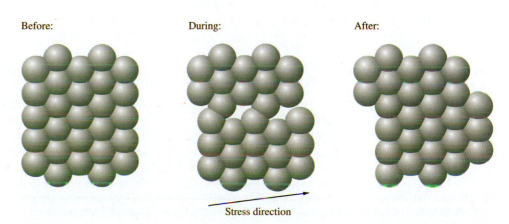

Before: During: After:

Stress direction

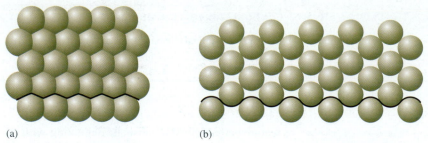

(a)   (b)

Figure 4.23 The dark line represents the variation in energy as two rows of atoms slide over each other. In a close pack structure (a), the variation is quite smooth. In a body-centered cubic structure (b), the larger spacing between atoms means that there is a larger energy barrier to get from one position to the next. Note the larger open spaces in the BCC structure.

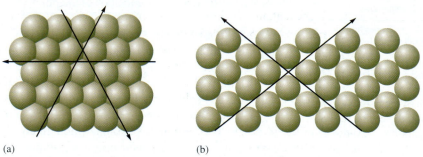

(a)   (b)

Figure 4.24 (a) A close pack plane (FCC or HCP) has three directions in which there are closely packed rows of atoms. (b) A BCC lattice has less densely packed planes and only two directions in which the rows are tightly packed.

The six slip planes of a BCC lattice:

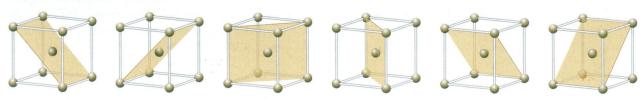

The eight slip planes of a FCC lattice
(two shown in each unit cell):

The two slip planes in an HCP lattice
(the top and bottom planes are
only one half in the the unit cell):

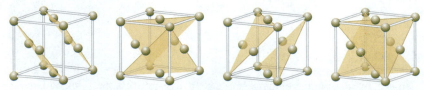

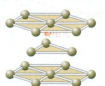

Figure 4.25 Slip planes of the BCC, FCC, and HCP crystal lattices. The BCC lattice is not as tightly packed, so there is more resistance to slip in a BCC lattice than in an FCC or HCP lattice.

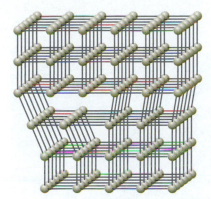

Figure 4.26 A line defect in a simple cubic lattice.

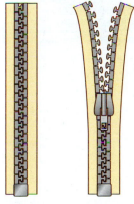

Figure 4.27 It takes a great deal of effort to pull all the teeth of a zipper apart at once. Pulling apart a pair of teeth at a time requires much less effort.

Defect Sites with atoms out of place

Point defect Missing atom or a single impurity atom in an otherwise regular array.

Line defect Row of adjacent atoms shifted slightly from the regular array position.

Grain boundary Region where two adjacent crystalites, each of which is a regular array, are out of register with each other.

Stress hardening Making a metallic solid less flexible by introducing many grain boundaries, often by drawing the metal through a dye, and compressing it between rollers, or pounding it.

each plane, there are 24 slip systems in an FCC lattice. In contrast, an HCP unit cell has only two close pack planes, and thus six slip systems. A BCC lattice has no close pack planes, and more energy is required to slip along the six slip planes.

It would take a very large force for all the bonds along the rows of atoms to let go at once—a much larger force than is actually required. To understand how the planes slide without sliding all at once, it is important to note that structures like those in Figure 4.2 are idealized structures. An actual piece of metal has numerous places where an atom or collection of atoms is out of place. Sites with atoms out of place are called **defects.** Defects, specifically line defects where a row of atoms is misaligned with respect to the next row (Figure 4.26), play a central role in plastic deformation. As an analogy, consider trying to separate a zipper (Figure 4.27) by pulling all the teeth at one time. A significant tug is required to pull apart all the teeth at once. However, if a "defect" is created by pulling one pair of teeth apart, then much less force is required.

Defects in the perfect crystal (simulated in Figure 4.28) can exist as **point defects,** which involve a single atom missing from an otherwise regular array (a vacancy) or a single atom out of position (Figure 4.29a); as **line defects,** which involve a row of adjacent atoms all shifted slightly from the regular array position (Figure 4.29b); or as grain boundaries. **Grain boundaries** are spaces where two adjacent regions, each of which is a regular array, are out of register with each other (Figure 4.29c). The most important of these possible defects for deformation of elemental metals is the line defect. A line defect assists movement of a plane of atoms over one another (Figure 4.30) by shifting the position of the defect. An analogy is a ripple in a rug. Moving a rug by moving a ripple through it is easier than moving the rug against the frictional force of the entire rug on the floor. In a solid, the defect passes through the metal lattice, going from row to row.

The flexibility within the plastic region depends on line defects and the number of slip systems. For example, copper deforms plastically more easily than either iron or titanium in part due to the larger number of slip systems in the copper FCC lattice. The correlation between melting point and flexibility is a result of the energy required to form linear defects. Formation of line defects requires energy because atoms along the line defect have fewer nearest neighbors than those in the remainder of the crystal. Similarly, energy is required to enable the free-flowing motion of a molten metal. Hence there is a correlation between flexibility and melting point.

The rippling of a line defect through the crystalline metal lattice continues until the line reaches the surface of the piece or until the defect reaches a grain boundary. Because the crystal lattices on either side of a grain boundary are out of registry with each other, it is difficult for the defect to cross the grain boundary (Figure 4.31). This is analogous to coming to a ravine: Crossing is problematic. When the defect reaches the grain boundary, either further motion is stopped or a new defect is started. The additional energy required is reflected in a stiffer piece.

Hardening a Metal

Because defect propagation is inhibited at grain boundaries, one way to make a piece stiffer is to process the metal in a way that creates many grain boundaries. Stressing a piece by drawing the metal through a dye and compressing it between rollers are common methods for creating grain boundaries. This approach is called **stress hardening.**

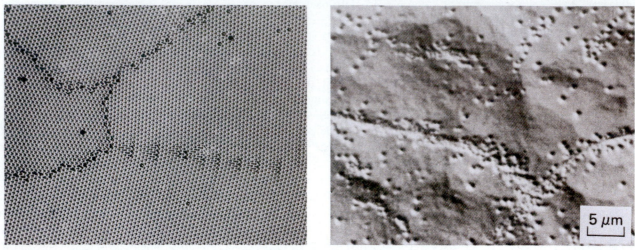

Figure 4.28 (a) Soap bubbles illustrate defects in a lattice. Missing bubbles are analogous to a vacancy, a row of bubbles out of position is similar to a line defect or dislocation, and the interface between two regular arrays skewed with respect to each other is like a grain boundary. (b) Defect sites in copper are observed using a scanning tunneling microscope after the surface has been etched with acid. Note the isolated pits due to either vacancies or line defects emerging from the surface and the string of pits due to a grain boundary.

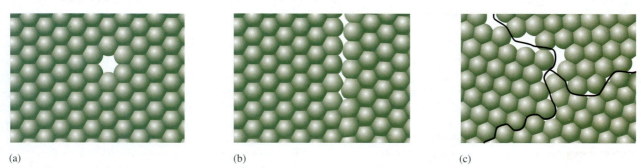

(a)   (b)   (c)

Figure 4.29 (a) A single missing atom in an otherwise regular array is an example of a kind of point defect called a vacancy defect. (b) A row of atoms where the regular crystalline array is disrupted is a line defect. (c) An area where two crystalline arrays meet but are out of register with each other is a grain boundary. The grain boundaries are highlighted with a dark line.

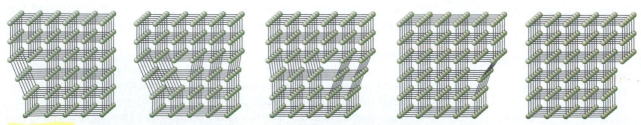

Figure 4.30 Successive "snapshots" of a line defect rippling through a crystal lattice.

Figure 4.31 As the defect ripples through the solid lattice, it is difficult for it to cross the gap at the grain boundary. Grain boundaries act like walls stopping the movement of defects.

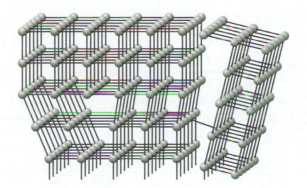

 APPLY IT Obtain a piece of 14-gauge copper wire commonly used in household circuits (14 gauge is the heavy-duty wire). Bend the wire, noting the effort required. Next, pound along the length of the wire, turning it along its axis so that it does not get too flat. (A rock or a hammer can serve as a pounding tool.) Compare the force required to bend the wire before and after pounding it. Describe what has happened within the copper wire, mentioning any other observations you make.

CONCEPT QUESTION   How would you restore flexibility to a stress-hardened copper wire? ■

Electron Mobility: Resistivity

Electrons are relatively loose in the electron sea, consequently metals are conductors. Of course, all metals do not conduct equally well. The copper used in household circuits and the gold used in microelectronics are among the best conductors known. In contrast, manganese is the least conductive of the metals.

Imagine connecting a metal wire with its itinerate electron sea to a source of electrons. To move into the metal, the electron must find an orbital—a wave—that is not already fully occupied. It is tempting to think of the added electron as swimming through the metal wire much like a fish swimming through water, but it is more accurate to think of conduction as analogous to the "wave" at a sporting event. The "wave" starts at some location, then propagates around the stadium, not by people moving through the stadium, but rather by successive people standing up and sitting down. In the same way, electron flow from one end of the wire to the other is achieved by a small movement of electrons down the wire. Thus an electron added at one end of the wire wiggles an electron next to the end of the wire, which wiggles one next to it, and so on, until an electron is ejected out the other end. The ease with which this process occurs determines the conductivity of a material.

Returning to the human "wave" analogy, if a group of people refuse to stand up, the wave has difficulty continuing—it encounters a resistance. For metals, this resistance is quantified by the **resistivity,** symbolized by the Greek letter rho, ρ. Mathematically, resistivity is the inverse of conductivity and is linked to heating of the metal when an electric current passes. Heating can cause a fire, which explains the banning of aluminum for domestic wiring. The resistance to charge flow depends on

Resistivity Tendency not to conduct. Inverse of conductivity. Resistance × (cross sectional area)/length. Typical units: $\mu\Omega$ cm.

the number of charge carriers (in this case, electrons) and the mobility, μ, of those carriers:

(4.4)

$$\rho = \frac{1}{n\mu}$$

In the human "wave" analogy, resistivity is inversely related to the number of people at the event (no crowd, no wave) and their willingness to stand up.

The number of charge carriers (that is, the number of electrons) corresponds to the number of electrons in the electron sea. In the transition series, the number of charge carriers increases steadily from the beginning to the end, resulting in a general decrease of the resistivity. Within this general picture, elements with a half-filled or a completely filled d subshell potentially have a larger resistivity due to lower electron mobility either because additional electrons in these elements must pair in an already occupied orbital or because they must go into another, higher-energy subshell. In this way, mobility is linked to the electron configuration. The electron configuration of copper is $4s^1 3d^{10}$, so copper has many electrons in the electron sea. That is, the electron density is high. With one vacancy in the $4s$ orbital, an electron can easily travel from atom to atom. Thus copper has a large number of charge carriers, and those carriers are very mobile.

In contrast, the electron configuration of manganese is $4s^2 3d^5$. The resistivity (Figure 4.32) of Mn, 144 $\mu\Omega \cdot$ cm ($\mu\Omega = 10^{-6}\ \Omega$) is much greater than the resistivity of the other fourth period transition elements, suggesting that Mn's electrons are much less mobile. This reduced mobility is related to the electron configuration. Due to the half-filled d subshell, adding an electron involves pairing in an already occupied orbital. Pairing requires energy, and using energy results in decreased mobility. Decreased mobility corresponds to an increase in resistance. Pairing in the Cu $4s$ orbital does not require nearly as much energy because the $4s$ orbital is far less confined than the $3d$ orbitals.

The resistivity drops after Mn, because succeeding elements contain electrons in addition to the half-filled subshell, and these additional electrons are mobile. The rise in resistivity at Mn is also observed at Tc and Re (Figure 4.33 and Table 4.3), which

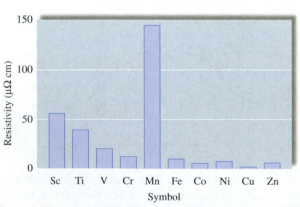

Figure 4.32 Resistivity of the first transition series elements decreases smoothly from the start of the series to the end, with the notable exception of Mn. The electron configuration of Mn features a half-filled d subshell.

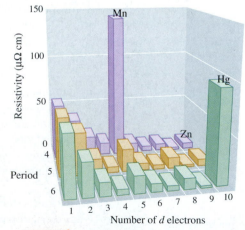

Figure 4.33 Resistivity shows a general decrease across all three transition series. Within this general trend, details of the electron configuration result in exceptionally high resistivity for Mn and Hg.

Table 4.3 Resistivity of Transition Elements ($\mu\Omega \cdot$ cm)

Period										
4	Sc	Ti	V	Cr	Mn	Fe	Co	Ni	Cu	Zn
	56.2	39	20.2	12.7	144	9.98	5.6	7.20	1.725	6.06
5	Y	Zr	Nb	Mo	Tc	Ru	Rh	Pd	Ag	Cd
	25.0	43.3	15.2	5.52	23	7.1	4.3	10.8	1.629	6.8
6	La	Hf	Ta	W	Re	Os	Ir	Pt	Au	Hg
	61.5	34.0	13.5	5.44	17.2	8.1	4.7	10.8	2.271	94.1

share similar electronic configurations. The rise is smaller for Tc and Re, however, indicating that as the shell for the *d* orbitals becomes larger, the energetic premium for pairing is smaller.

CONCEPT QUESTION What fourth-shell subshells are unoccupied in Zn? ∎

The electron configuration of Zn is $[Ar]3d^{10}4s^2$; the next subshell to be occupied is the $4p$ subshell. The resistivity data indicate that in Zn little energy separates the $4p$ subshell from the valence electrons. Since the $4p$ orbitals are empty, they easily accommodate an electron.

At the end of each transition series, a rise in resistivity occurs. For Zn in the fourth period and Cd in the fifth period, this rise is quite small. In the sixth period, the resistivity of Hg ($[Xe]4f^{14}5d^{10}6s^2$) is quite large, comparable to that of Mn. In addition to the filled set of *d* and *s* orbitals, Hg has a filled set of *f* orbitals. The extra positive charge accompanying the 14 *f*-orbital electrons increases the energy separation between the filled *d* and empty *p* orbitals, making conduction relatively difficult.

WORKED EXAMPLE 4.5 *Competition Between Copper and Aluminum*

The data in Table 4.3 indicate why copper is an excellent choice for household wiring. At one point in time, aluminum replaced copper. The resistivity of Al (2.733 $\mu\Omega \cdot$ cm compared with 1.725 $\mu\Omega \cdot$ cm for Cu) seemed to qualify it as a good replacement. Because Al is less dense (2.6989 g/cm^3) than Cu (8.96 g/cm^3), shipping costs were lower for Al. As a result of these factors, a conversion to Al wiring occurred. Use these data to explain why use of Al led to a problem with fires.

Plan

■ Resistance is linked to heating.

Implementation

■ Resistance to electron flow means that energy is lost as the current passes through the wire. This energy loss primarily appears as heat. The resistivity of Al is 1.6 times that of Cu, so more heat is generated, heating materials near the wire and potentially starting a fire. Despite the 3.3 times greater density of Cu, the disadvantages related to the fire risk outweigh the transportation cost savings.

Further comment: As a wire heats up, the resistance increases, magnifying the heat generated. If the heating is great enough, the wire can melt.

See Exercises 33, 34.

CONCEPT QUESTION Using density, melting point, and resistivity data, explain why tungsten is used as an incandescent light bulb filament. Which metal would you choose for a fuse? ∎

4.3   Phase Transitions

Phase A physical state of matter described by a characteristic relationship between the atoms or molecules; examples include solid, liquid, and gas, semi-conducting or metallic tin.

Phase transition Change from one phase of a substance to another phase.

Addition of heat to a material can change the relationships among the atoms. For example, at the melting point, the relatively fixed arrangement in the solid changes to a dizzying dance around each atom. The solid and the molten metal are referred to as **phases,** and the change from phase to phase is called a **phase transition.** Each phase is described by a characteristic spatial relationship among the atoms. For example, in the molten or liquid phase, the atoms are randomly jostling and constantly shifting their nearest neighbors. The transition from the solid to the liquid phase is analogous to the end of class: Just before the end, students sit in chairs often in regularly-spaced rows; at the end of class, there is much more motion and randomness as the class makes its way out the door.

Phase transitions also occur within the solid phase. As the atoms change their bonding and relationships to their nearest neighbors, the physical properties change as well. One dramatic example is the phase transition seen in tin. In one phase, tin is a typical metal—malleable, ductile, and conductive (resistivity, 11.5 $\mu\Omega \cdot$ cm). Below 13.6 °C, the bonding changes, tin's coordination number changes from 8 to 4, and its resistivity increases about six orders of magnitude. In the low-temperature phase, tin is a brittle material. Different forms of an elemental material are called allotropes. The difference in mechanical properties for the two allotropes of tin is responsible for what is referred to as "tin pest" or "tin plague," which can be seen in the extensive damage incurred by organ pipes fashioned of tin in European cathedrals, which are often constantly cold. Only the slow rate of this transition has saved the pipes from total destruction.

Solid Phases of Tin

Diamond structure Tetrahedral arrangement of atoms with every atom identically bonded to four other atoms.

In the low-temperature phase, called α-Sn, each Sn atom has four nearest neighbors in a tetrahedral arrangement (Figure 4.34). This structure is called the **diamond structure** because it is the same as that of the gem diamond. The diamond structure is common among semiconductors. Diamond-structured tin is a semiconductor.

CONCEPT QUESTION   What is the coordination number for Sn in α-Sn?   ■

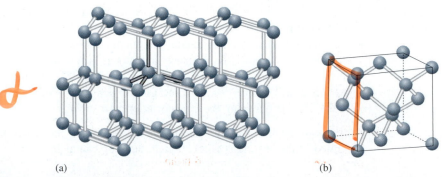

(a)   (b)

Figure 4.34 (a) Below 13.6 °C, the stable phase of tin, called α-Sn, has a tetrahedral arrangement of the atoms in which each Sn atom is surrounded by four others. This is the same crystal structure as seen in diamond. One atom and its four nearest neighbors are shown with a dark line. (b) The unit cell of the diamond structure is related to the face-centered cubic unit cell by the addition of an atom between every other corner atom and the three adjacent face center atoms.

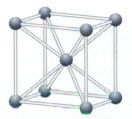

Figure 4.35 Above 13.6 °C, tin is metallic and has eight nearest neighbors in a tetragonal crystal. In this variation on a body-centered cubic unit cell, the distance between the front and back faces is compressed (318.19 pm) compared to height and width (583.18 pm). The bonds to the body center atom (434.5 pm) are relatively long.

Tetragonal solid Crystalline solid characterized by a unit cell with 90° angles and length in two directions being equal. The third length is unique.

Phase diagram A graphical representation of the conditions for stability of phases of a system.

Phase boundary Line separating two phases in a phase diagram.

The greatly reduced electron mobility in α-Sn suggests that the electrons are significantly localized between pairs of atoms. As a fifth-row element, tin is a large atom. The directed bonds are relatively weak, so little force is required to stretch the bonds or bend the angles beyond the elastic limit. At the same time, the diamond structure lacks slip systems, so this phase of Sn is brittle.

Above 13.6 °C, α-Sn transforms: The atomic layers flatten and tin becomes more densely packed, adopting a **tetragonal lattice** structure (Figure 4.35). This form is called β-Sn. The tetragonal unit cell is like a body-centered cell that is somewhat flattened in one direction.

BCC

CONCEPT QUESTION   How many atoms are in a tetragonal unit cell? ■

Bonding in β-Sn is less directed than bonding in α-Sn, and β-Sn is metallic. Metallic β-Sn is significantly more dense (7.31 g/cm³) than α-Sn (5.75 g/cm³). The transformation from β-Sn to α-Sn and the mechanical differences between these two phases cause objects made from room-temperature β-Sn to slowly expand at temperatures less than 13.6 °C due to the lower density of α-Sn. This expansion and the brittleness of α-Sn result in disintegration of the tin into a powder at temperatures less than 13.6 °C; the "tin pest" referred to above. These two phases of tin are an excellent example of the difference in mechanical properties resulting from electrons being localized between pairs of atoms (α-Sn) and the itinerant electrons of metallic bonding (β-Sn).

The phase transitions in tin can be represented on a temperature line (Figure 4.36). This temperature line diagrams the phase transitions at atmospheric pressure. How does this picture change as pressure increases? On both the macroscopic and the atomic scales, increased pressure compresses the material. At the atomic scale, compression results in a shortening of bond distances and adjustment of bond angles to pack more atoms in a given volume: In other words, the density increases. For tin, the β or metallic phase is more dense than the α or semiconducting phase. Hence, at higher pressures, the temperature line of Figure 4.36 shifts so that the transition from β- to α-tin occurs at a lower temperature. Similarly, metallic tin is more dense than molten tin, so increased pressure raises the temperature at which tin melts. A convenient way of representing the stable phases is on a two-dimensional phase diagram (Figure 4.37). A **phase diagram** plots the range of variables—temperature and pressure in this case—over which a given phase is stable. Increased pressure favors the β phase of tin, so the line separating α-tin from β-tin slopes backward. Similarly, increased pressure stabilizes the β phase of tin relative to the liquid, thus the line separating molten and β-tin slopes forward.

In general, the line separating two phases is called the **phase boundary.** The phase boundary line slopes toward the less dense phase, giving a larger area for the more dense phase as pressure increases.

CONCEPT QUESTION   Which is more dense, liquid water or ice? Sketch a temperature line for water at atmospheric pressure. Sketch the pressure-temperature phase diagram for water at elevated pressures. ■

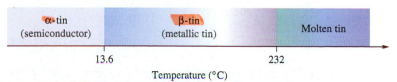

Figure 4.36 The phases of tin plotted on a temperature line illustrate the solid state transition from α-tin to β-tin at 13.6 °C as well as the transition from metallic tin to molten tin at 232 °C. These transitions all occur at one atmosphere pressure.

Figure 4.37 Adding a pressure axis to the temperature line generates a phase diagram: a plot of the temperature and pressure range over which a given phase is stable.

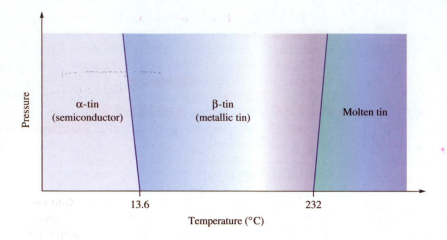

4.4 Alloys

As long ago as 3500 B.C., the Bronze Age began with the discovery that a solution of copper and tin (bronze) has mechanical strength superior to that of either copper or tin alone. Bronze is an example of an **alloy,** a solid solution of two or more elements having the characteristic properties of a metal. The mechanical properties of an alloy reflect how the atoms of each element enter the crystal lattice of the other. Atoms of one element can substitute for the other, and the resulting mix is a **substitutional alloy.** Bronze is an example of a substitutional alloy. Bronze is primarily copper with tin atoms replacing some of the copper atoms, thereby altering the mechanical properties associated with pure copper. When the added atoms are very small, they can be added to the crystalline lattice in the spaces between the hard-sphere atoms. The spaces between the atoms are called **interstitial spaces.** One of the most economically important interstitial alloys is steel. The composition of steel varies, but it always contains primarily iron with a small amount of carbon. Table 4.4 lists several alloys and their uses.

Alloy A homogeneous solid solution of two or more elements having metallic properties.

Substitutional alloy Formation of a solid solution by atoms of one element substituting for the other.

Interstitial spaces Spaces between atoms.

Table 4.4 Selected Alloys

Use	Composition	Key Property
Fuses	50% Bi, 25% Pb, 12.5% Sn, 12.5% Cd	MP 70 °C
Brass	Primarily Cu with up to 40% Zn	Harder than Cu or Zn
Bronze	Cu/Sn	Harder than Cu or Sn
Stainless steel	80.6% Fe, 0.4% C, 18% Cr, 1% Ni	Resists corrosion
Cables, nails, chains	Mild steel, 0.2% C	Strong
Girders, rails	Medium steel 0.2–0.6% C	Stiff
Cutting tools, springs	High-carbon steel, >1% C	Holds an edge, hard
Solder	67% Pb, 33% Sn	MP 275 °C
Plumber solder	95% Sn, 5% Sb	Low MP, no Pb
Flatware	92.5% Ag, 7.5% Cu	Sterling silver
Pewter	85% Sn, 7% Cu, 6% Bi, 2% Sb	Soft luster
Dental fillings	70% Ag, 18% Sn, 10% Cu, 2% Hg	Easily worked
Jewelry	58% Au, 42% Cu	Soft gold color
Jewelry	25% Au, 75% Ag	White gold

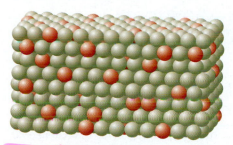

Figure 4.38 Two elements form a substitutional alloy over a large compositional range if they are similar in size, crystal structure, and electronegativity.

Hume-Rothery rules Rules stating that to form a substitutional alloy over the entire composition range, two elements must differ by less than 15% in their atomic radius, have the same crystal structure, and have similar electronegativity.

Homogeneous solution Solution with uniform physical and chemical characteristics.

Heterogeneous mixture A mixture in which the individual components lie in distinct macroscopic regions.

Solubility limit The amount of a substance that can be dissolved in a given amount of solvent.

Substitutional Alloys

For one type of atom to freely substitute for another in a lattice (Figure 4.38), the two elements must have very similar characteristics. A set of "rules" called the **Hume-Rothery rules** are used to evaluate this similarity. To form a *substitutional alloy over the entire composition range*, the two elements must

- Differ by less than 15% in their atomic radius,
- Have the same crystal structure, and
- Have similar electronegativity.

Elements that satisfy the Hume-Rothery rules form a homogeneous, solid solution throughout the composition range. A **homogeneous solution** has uniform physical and chemical characteristics. Sugar dissolved in coffee is a homogenous, single-phase solution, for example. Sugar at the bottom of the cup of coffee along with the sweetened coffee is an example of a two-phase mixture. The sugar–coffee solution is one phase and the solid sugar is a second phase. A mixture of more than one phase is a **heterogeneous mixture.**

For metals, if one or more of the Hume-Rothery rules are not satisfied, the two elements usually form a substitutional alloy containing only a limited amount of one element in the other. This limit is called the **solubility limit.** The term "solubility limit" applies more broadly to any mixture that does not form a solution with uniform properties. Coffee with sugar on the bottom has reached the solubility limit for sugar in coffee, for example.

CONCEPT QUESTIONS What is the melting point of Cu? Of Ni? Predict the melting point of a solution of 10% Cu and 90% Ni. ∎

The Hume-Rothery rules are well satisfied by a mixture of Cu and Ni (Table 4.5), and Cu and Ni form a substitutional solid solution over the entire composition range. Just as a solution of sugar in coffee differs from coffee alone, so the properties of Cu-Ni solutions differ from those of either Cu or Ni alone.

Consider a mixture composed of 50% Cu and 50% Ni. Cu melts at 1085 °C and Ni melts at 1455 °C, so both are molten at this temperature and form a homogeneous molten solution. What happens to this solution as it cools? What might the temperature line (Figure 4.39) look like? The melting point of Ni is nearly 400 °C higher than that of Cu. Hence, although the first solid to appear as the molten solution cools is a homogeneous solution of Cu and Ni, that solid contains more Ni than Cu. This solid exists along with the molten solution; this is a two-phase mixture. When the entire mass has solidified, it forms a homogeneous solid solution—a single phase. Each composition has its own temperature line. Plotting temperature vertically and composition horizontally produces a phase diagram—in this case, a plot of the phase as a function of temperature

Table 4.5 Copper–Nickel Solutions

Hume-Rothery characteristics for Cu and Ni show that these two elements are very similar.

	Radius (pm)	Crystal Structure	Electronegativity
Cu	117	FCC	1.90
Ni	115	FCC	1.91
Difference	2%		0.5%

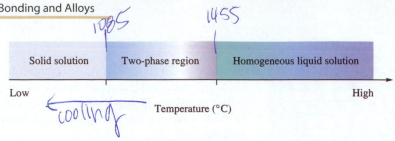

Solid solution   Two-phase region   Homogeneous liquid solution

Low   High

Temperature (°C)

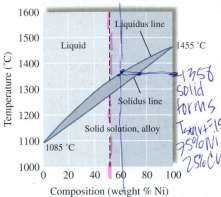

Two-phase, liquid plus solid

Figure 4.40 The copper–nickel phase diagram indicates that copper and nickel form a homogeneous solid solution over the entire composition range.

and composition (Figure 4.40). For two-component systems like Cu and Ni, the composition axis can be plotted either as percent by mass or as mole percent.

CONCEPT QUESTION Figure 4.40 shows the phase diagram for copper and nickel with the 50% Cu, 50% Ni line dotted in. At what temperature does a 50/50 mixture begin to form a solid? ■

When a molten solution of Cu and Ni cools, the first solid formed is always richer in Ni than the molten solution. Slow cooling produces a homogeneous alloy with grain boundaries where the solid particles first formed grow together (Figure 4.41). Practical cooling rates are too fast to produce a completely homogeneous solid and movement of atoms in the solid is very slow. As a consequence, the solid produced has a layered structure, like an M&M candy (Figure 4.41).

The phase diagram can be used to determine the composition of the layers. Within the solid–liquid region, two phases are present. These phases do not have the same composition, however. For example, a 50/50 Cu-Ni solution begins to form a solid at 1325 °C. A horizontal line at 1325 °C is in the liquid area of the phase diagram for compositions from zero to nearly 50% nickel. This line is only in the solid area from 64% to 100% nickel. Thus, at 1325 °C, any solid must have at least 64% nickel. A 50/50 mixture cooled to 1325 °C consists of molten metal that is essentially 50% Cu and 50% Ni with a small amount of solid that is 64% Ni and 36% Cu. Similarly, for any temperature where the composition line is in the two-phase region, the liquid composition is given by the intersection of the constant temperature line and the liquid boundary line. The solid composition is given by the intersection with the solid boundary line.

WORKED EXAMPLE 4.6   *Reading the Phase Diagram*

Phase diagrams are enormously helpful in determining whether a homogenous phase can form from a given mixture or at what temperature the mixture will solidify. Suppose a mixture of 60% Ni and 40% Cu by weight is heated to 1500 °C and then cooled. At what temperature does solid begin to form? What is the composition of this solid? At what temperature does the entire mass solidify?

Plan

■ Locate the 60% by weight nickel line. Use the phase diagram to read the temperature where the 60% composition line crosses into the two-phase region.

■ Solid composition is read from the intersection of the temperature and the solid boundary line.

■ The complete solidification temperature is the temperature at which the 60% composition line crosses into the solid area.

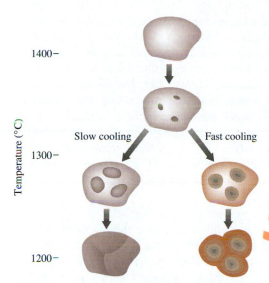

1400—

Temperature (°C)

Slow cooling   Fast cooling

1300—

1200—

Figure 4.41 The cooling rate determines the morphology of the solid formed. Slow cooling produces a homogeneous solid with grain boundaries. Fast cooling produces an inhomogeneous solid consisting of layered regions enriched in nickel in the middle and enriched with copper on the outer surface, with grain boundaries between regions.

Implementation

- The 60% nickel composition line crosses the liquid boundary at 1350 °C. Solid begins to form at this temperature.
- The composition of the solid is 75% nickel and 25% copper. The solid is relatively richer in nickel because copper and nickel readily substitute for each other in the solid, but nickel melts at a higher temperature.
- The 60% composition line crosses into the solid area at 1320 °C.

See Exercises 53–56.

The mechanical properties of an alloy are affected by both the composition and the preparation method, primarily the cooling rate. Since the elastic modulus is determined by the local bond response to an applied load, the elastic modulus is given by the majority element. The plastic response, however, is greatly affected by the formation of an alloy. Plastic flow is arrested at grain boundaries. Even during slow cooling, many crystals grow. When these crystals grow together, a grain boundary forms. Fast cooling produces a layered structure that further inhibits movement of the close pack planes. As a result, the linear portion of the stress-strain curve extends to heavier loads for an alloy. This is the basis for alloys being less deformable than the elemental materials from which they are made. The electrical resistance is increased because the electron interacts differently with each element present. For example, a 50/50 Cu-Ni alloy has a resistivity of 50.05 $\mu\Omega \cdot$ cm compared with 1.725 $\mu\Omega \cdot$ cm for Cu and 7.20 $\mu\Omega \cdot$ cm for Ni. The increased resistance explains why substituting an alloy for copper in power lines will not help carry the ice load in northern winters—much of the power would dissipate as heat.

CONCEPT QUESTIONS   What happens if two metals with poorly matched Hume-Rothery parameters are mixed? If only a small amount of one metal is mixed into a second, what effect does this have on the mechanical properties of the alloy? ■

Most pairs of elements are not as similar as Cu and Ni, so they form a solid solution over only a limited compositional range. Historically, one of the most important solid solutions of this type is bronze; an entire age (3000–500 B.C.) is named for it. Copper, known from antiquity, is easily worked, making it a superior material compared to stone for tools and implements. The plasticity of copper can be a disadvantage because articles fashioned from it are not very durable. The discovery that this durability could be greatly improved by incorporating other elements, particularly tin, into copper was a major advance.

CONCEPT QUESTIONS   Apply the Hume-Rothery rules to Cu and Sn. Do you expect Cu and Sn to form an alloy over the entire composition range? Why or why not? ■

Bronze is not a precise term, but rather refers to a variety of copper-based alloys. The most common alloy consists of 90% Cu and 10% Sn by weight. Tin atoms are 20% larger than copper atoms, so substitution of Sn for Cu creates a strain. Hence Sn has limited solubility in Cu, and the phase diagram is quite complex.

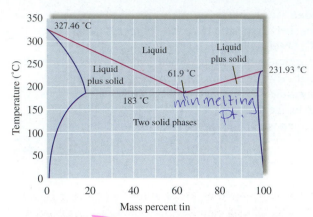

Figure 4.42 The Sn-Pb phase diagram is an example of the type of behavior found among elements with a poor Hume-Rothery match. Several mixtures of tin and lead melt at a lower temperature than either elemental material, a characteristic that is important in application as solder.

A mixture of tin and lead used to solder electrical circuits is also a mixture with limited solubility. This characteristic has a profound and very desirable effect on the alloy's melting point. Lead atoms are only 5% larger than tin atoms, but the electronegativity of lead is greater than that of tin and the elements' crystal structures are significantly different. Each element disrupts the crystallization of the other, so a mixture solidifies at a lower temperature than either metal alone (Figure 4.42). This property enables the solder to melt without melting the wires in the circuit.

At room temperature a Sn-Pb mixture is a heterogeneous solid. The solid region near the 100% Pb axis is an alloy that is mostly Pb with a little Sn; the solid region near the 100% Sn axis is mostly Sn with a little Pb. The area between these regions is a two-phase region consisting of a heterogeneous mixture of the tin-rich and the lead-rich solid phases. A minimum in melting point is reached at a composition of 61.3% Sn. This alloy is a technologically important composition because there is no two-phase region consisting of molten metal and solid: The homogenous melt solidifies at a constant temperature into the two solid phases. The low melting point contributes to the importance of Sn-Pb as a solder material. Uniform solidification makes the heterogeneous solid more regular, improving conduction through the soldered connection.

Interstitial Alloys

One of the most important limited-solubility alloys is steel, which is a mixture of Fe and C. Fe crystallizes in a BCC lattice. In all lattices, spaces exist between the metal atoms, called interstitial spaces (e.g., BCC lattice; Figure 4.43). In some lattices, these spaces are quite large and can accommodate a guest atom if the guest is small enough. Fitting into the interstitial space, the guest atom is termed an **interstitial atom.** Steel takes advantage of incorporation of small atoms in the interstitial spaces of the BCC iron lattice, with atoms of carbon occupying these interstitial spaces to form a solution of Fe and C.

Interstitial atom Guest atom that fits into interstitial space.

Figure 4.43 (a) Two unit cells of the iron lattice are shown with the interstitial site indicated. Due to the more open structure of the BCC lattice, the interstitial sites are larger than for either an FCC or HCP lattice. (b) In steel, carbon occupies some of the interstitial sites.

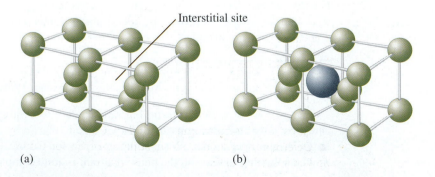

(a)   (b)

WORKED EXAMPLE 4.7   *BCC Interstitial Spaces*

Steel is a solution of Fe and C, with C atoms fitting into the interstitial spaces in the BCC Fe lattice. These interstitial C atoms significantly alter the mechanical and chemical properties of Fe. Is the interstitial space (illustrated in Figure 4.43) large enough to accommodate carbon without distorting the structure?

Data

Radius of Fe = 117 pm
Radius of C = 77 pm

Plan

- Determine the dimensions of the unit cube.
- Determine where carbon sits with respect to the unit cube.
- Each corner atom extends a distance of 117 pm along the edge and face diagonal. Use geometry to determine the hole size compared with the host atoms.

Implementation

- The radius of the Fe atom is 117 pm, so the diagonal of the unit cube is 4×117 pm = 468 pm. The cube edge is 468 pm/$\sqrt{3}$ = 270 pm, and the face diagonal is 270 pm $\times \sqrt{2}$ = 381 pm.
- There are two potential sites for carbon: in the center of a face of the unit cube and along an edge. These sites are equivalent.
- In the face center position (shown in Figure 4.43), the distance to a corner atom is half the face diagonal, or 190 pm. The corner Fe atom extends 117 pm into that space, so the hole has a 73-pm radius. The distance to the body center atoms is half the cube edge, or 135 pm. The body center atom extends 117 pm, leaving a 18-pm radius hole open. Thus the interstitial carbon has an ellipsoidal space with two 73-pm major axes and an 18-pm minor axis. The radius of carbon is 77 pm—a bit squeezed in all directions.

This nonisotropic straining of the iron lattice due to interstitial carbon causes the BCC iron lattice to become distorted into a tetragonal lattice. This distortion inhibits plastic flow and makes steel stronger than iron.

See Exercises 60–62.

Carbon is a little large to fit into the hole in the BCC Fe lattice (Worked Example 4.7). Carbon thus creates a strain in the lattice that has profound implications for both the properties of steel and the processing conditions required for successful steel production. Incorporating carbon into the Fe lattice strains the BCC unit cell causing it to lengthen to a body-centered tetragonal structure, much like that of Sn metal.

WORKED EXAMPLE 4.8   *Carbon in Steel*

Carbon atoms are too large to fit easily into the interstitial sites in the iron lattice. A steel used in the manufacture of a bicycle frame has a carbon content of 0.43% by weight. What fraction of the possible holes is occupied by carbon?

Plan

- Convert 0.43% by weight to atomic percent.
- Determine the number of interstitial sites per iron atom.
- Determine the ratio.

Implementation

■ 0.43 weight % means 0.43 g C per 100 g steel, or 0.43 g C and 99.57 g Fe. The atomic percent is

$$\frac{\dfrac{0.43 \text{ g C}}{12 \text{ g C/mol}} \times 100\%}{\dfrac{0.43 \text{ g C}}{12 \text{ g C/mol}} + \dfrac{99.57 \text{ g Fe}}{56 \text{ g Fe/mol}}} = 2 \text{ atomic \%}$$

This is 1 carbon atom for every 50 iron atoms.

■ A BCC unit cell contains two iron atoms. There are six face center sites for carbon, each of which is shared by two unit cells, or there sites per unit cell. There are twelve edge sites shared by four unit cells, for three more. Thus there are six sites per two iron atoms.

■ Fifty iron atoms can potentially hold 150 carbon atoms, so the fraction of occupied sites is 1/150 or 0.7%.

The carbon atoms must create a significant strain on the lattice to distort the crystal structure of iron with such a low site occupancy.

See Exercises 62, 64, 80, 81.

Austenite A cubic form of carbon steel.

Martensite A tetragonal form of carbon steel formed when austenite is cooled rapidly.

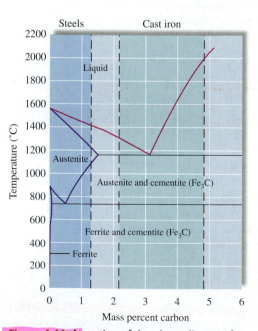

Figure 4.44 A portion of the phase diagram for iron and carbon. Carbon is not large enough to substitute for iron in the lattice, but is too large to fit easily into the interstitial spaces. The result is limited solubility and a complex phase diagram, only part of which is shown here.

Iron with the BCC structure can dissolve about 2% carbon, but is the stable phase only at high temperatures, 700–1400 °C. This cubic phase is called **austenite** in honor of W. C. Roberts-Austin, an early pioneer in the study of steel. Ordinary cooling of austenite results in separation into two phases: an almost pure iron, called ferrite, and solid Fe_3C, called cementite (Figure 4.44). The localized bonding in cementite makes it a very brittle solid. As nearly pure iron, ferrite is relatively soft. The combination of cementite and ferrite produces a gritty, plastic material. Avoiding the phase separation to make steel requires rapid cooling, called quenching. Quenching "freezes" in the random distribution of carbon and iron. At lower temperatures, the lattice becomes tetragonal, stretching in one direction. This is called the **martensite** phase. The martensite phase is less regular than the austenite phase. The terminology, with "austenite" referring to a cubic high-temperature phase and "martensite" referring to a less regular, low-temperature phase, is borrowed from steel and applied to numerous other alloys.

Martensite is not a stable phase of steel. It is generated only by quenching austenite steel. As a result, martensite always contains at least some austenite. The mixed phases make a steel that is both pliable and strong. An example is a material called Damascus steel (Figure 4.45), named for the Syrian capital where the steel was forged. Damascus steel and other high-carbon steels are very pliable at high temperatures and very strong at room temperature. The sharp edge on a 300-year-old sword made from this material, has endured through countless battles. Over hundreds of years, the procedure for making Damascus steel was lost. However, fabricating a cutting tool that holds its edge well is of great economic interest to the tool and die industry. Accordingly, a number of investigators have tried to replicate Damascus steel. Professor Oleg Sherby and specialists at Lawrence Livermore Laboratory have developed a method to incorporate 1.8% carbon in steel, producing a superplastic steel

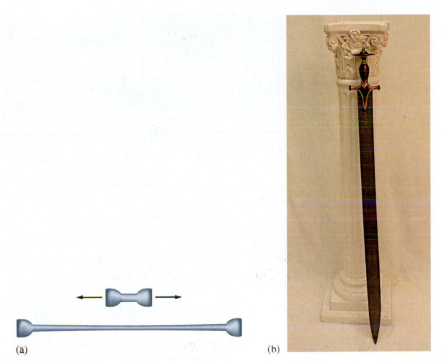

(a)   (b)

Figure 4.45 (a) Superplastic steel can be stretched to 11 times its original length without cracking or breaking. A piece, originally 1 in long, can be pulled at 900 °C to 11 in long. Ordinary steel fails when pulled to twice its original length. (b) A modern, high-carbon steel sword, fabricated to duplicate Damascus steel, holds its edge well. Knowledge of the process for making Damascus steel was lost in the nineteenth century.

(Figure 4.45) that can be stretched at high temperature to 11 times its original length without cracking or breaking. Normally, steel can stretch to only twice its original length before failing. Developing new steel alloys for challenging applications involves understanding atomic-level interactions, phase diagrams, and the effect of processing conditions.

Nitinol

Many useful alloys were discovered by accident while searching for the solution to a different problem. One example is a nickel–titanium alloy called nitinol. In the early 1960s a technician working at the Naval Ordnance Laboratory in White Oak, Maryland, was trying to develop an improved alloy for a missile heat shield. Nickel–titanium alloys had attracted attention as potential heat shield materials because they are very ductile, are resistant to oxidation at high temperatures, and, due to the Ti content, have a lower density than steel. While working with a rod of the Ni-Ti alloy, the technician dropped the rod. It produced a dull thud. When the technician dropped a rod that had become warm due to friction while machining, however, the rod responded with a ringing ping. These differing sounds indicate that the bonding between atoms in the cold rod are more flexible than those in the warmed rod. The story of nitinol vividly illustrates the value of staying alert to the macroscopic manifestations of potentially interesting atomic-level effects.

A fascinating discovery followed the sound observation. To demonstrate the flexibility of the Ni-Ti alloy, an engineer named William Buehler bent a piece of the alloy

Figure 4.46 Looking like a sculpture, this device is called a Simon Nitinol Filter or, more picturesquely, a bird's nest filter. It is inserted into the vein of thrombosis patients to prevent blood clots from traveling to the brain, resulting in stroke. The filter takes advantage of nitinol's temperature response to unfold after insertion. The biocompatibility of titanium and nickle enables the device to remain in place for a long term.

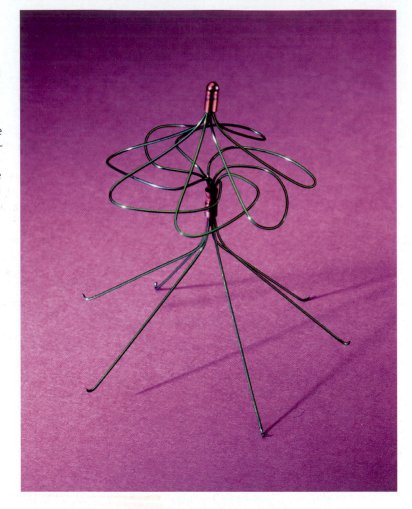

into a corrugated shape. The director of Buehler's engineering group, David S. Muzzey, heated the corrugated wire with a lighter, and the piece of alloy straightened out in a most dramatic fashion. The term "shape memory" was coined, and the new alloy was named nitinol for the two elements Ni and Ti plus Naval Ordnance Laboratory (NOL). The hot–cold cycling can be repeated almost indefinitely and is used to animate the object pictured in Figure 4.46.

Nitinol Crystal Structure

The shape memory effect in nitinol is linked to the crystal structure of the alloy in the two phases and the phase transition between them. The high-temperature phase is called austenite in accord with the terminology of steel. At high temperatures the unit cell is a BCC structure (Figure 4.47) with eight Ti atoms at the corners and Ni in the body center.

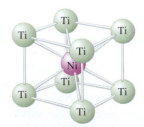

Figure 4.47 The nitinol unit cell showing structure of the high-temperature austenite phase. It could equivalently be shown with the positions of Ni and Ti reversed.

CONCEPT QUESTIONS Nitinol alloys with an atomic composition near 1:1 show the shape memory effect. How many Ti atoms are in the unit cell? How many Ni atoms? Is the structure different if the roles of Ti and Ni are interchanged? ■

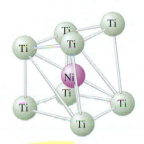

Figure 4.48 The low-temperature phase of nitinol is characterized by a skewed unit cell, compressed along some directions more than others. This irregular shape gives it the name martensite in accord with the terminology from steel.

The low-temperature phase of nitinol (Figure 4.48) is skewed and compressed compared with austenite. Again, in accordance with the terminology from steel, the low-temperature phase is referred to as martensite.

Austenitic nitinol is a hard material that is difficult to deform. In contrast, martensite is quite soft and is easily bent by hand. These differing flexibilities are a direct result of the shape of the unit cell in each phase. The cubic unit cells of austenite stack like sugar cubes. Elastic and plastic deformation occur as discussed earlier. Since Ni and Ti differ in terms of their Hume-Rothery parameters, sliding of planes is inhibited and austenitic nitinol is an inflexible material with a large elastic constant.

The distorted unit cells of martensite stack together much like parallelograms. A large variety of macroscopic shapes are accommodated by different combinations for the tilt of the parallelograms. All tilt choices, however, relate back to the same parent square (Figure 4.49). In three dimensions there are 24 martensite configurations and a number of paths that return martensite to austenite. However, only one restores the atomic ordering with Ni next to Ti and Ti next to Ni. Other paths end up with Ni-Ni neighbors and Ti-Ti neighbors. These configurations all have higher energy. Energy guides the return to the unique starting austenite configuration.

The 24 equivalent configurations for the low-temperature phase give it an abundance of possible macroscopic shapes and correspondingly an easy shift from one shape to the other in response to an applied stress. The result is great flexibility.

The easy transformations among equivalent configurations also explain the dull thud of the low-temperature, martensite form. In the low-temperature form, the atoms merely slip into another of the 24 equivalent configurations rather than transmitting sound energy to the next cell—the sound is muffled. In contrast, the cubic cell of the high-temperature, austenite form lacks geometric flexibility, and the atoms are rigidly coupled. Any disturbance leads to a rippling movement, creating a sound wave. Energy is readily transmitted instead of dissipated.

CONCEPT QUESTION   Nitinol is used in temperature-control switches for coffee makers, hydraulic tube connectors, staples to clamp broken bones, artery stints, and eyeglass frames. How are the two phases of nitinol and the transition between them used in these applications? ∎

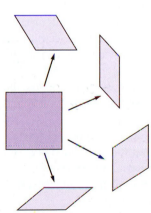

Figure 4.49 Compressing and tilting a square produces a parallelogram. Four distinct parallelograms can be produced from different combinations of compression and tilting of the parent square. All parallelograms relate back to the same square.

Pressure-Induced Phase Transition

A number of medical applications of nitinol rely on the density difference between the martensite and austenite phases of nitinol. Martensite is 0.5% more dense than austenite. Applying pressure pushes atoms closer together, driving nitinol toward the more dense martensite form. It does not take great pressure to convert austenite to martensite at room temperature.

One application of nitinol that takes advantage of this phenomenon is orthodontic braces. With the teeth out of position, there is pressure on the nitinol wire. This pressure squeezes the atoms together, converting the structure into the more dense, martensite phase. As the teeth move due to the gentle pressure of the more flexible martensite phase, the pressure is relieved. The brace then reverts to the less flexible austenite phase, which has been programmed to pull the teeth into the desired positions. The result is a continuous, gentle pressure on the teeth and a superior outcome. If you wore braces in the past decade, they were likely made of nitinol, especially if your orthodontist suggested drinking a cold beverage to counteract the discomfort caused by your braces. The cold temperature induces a transition to the softer form, which puts less pressure on the teeth, causing less pain.

Checklist for Review

KEY TERMS

elastic deformation (p. 119)
plastic deformation (p. 119)
electron sea (p. 121)
metallic bonding (p. 121)
crystal lattice (p. 122)
unit cell (p. 122)
body-centered cubic, BCC (p. 122)
face-centered cubic, FCC (p. 122)
hexagonal close pack, HCP (p. 122)
hard-sphere model (p. 125)
coordination number (p. 125)

close pack (p. 126)
stress (p. 130)
strain (p. 130)
elastic modulus (p. 130)
slip system (p. 133)
defect (p. 135)
point defect (p. 135)
line defect (p. 135)
grain boundary (p. 135)
stress hardening (p. 135)
resistivity (p. 137)
phase (p. 140)
phase transition (p. 140)
diamond structure (p. 140)
tetragonal solid (p. 141)
phase diagram (p. 141)
phase boundary (p. 141)
alloy (p. 142)
substitutional alloy (p. 142)

interstitial spaces (p. 142)
Hume-Rothery rules (p. 143)
homogeneous solution (p. 143)
heterogeneous mixture (p. 143)
solubility limit (p. 143)
interstitial atom (p. 146)
austenite (p. 148)
martensite (p. 148)

KEY EQUATIONS

Stress: $\sigma = \dfrac{F}{A_o}$

Strain: $\varepsilon = \dfrac{\Delta \ell}{\ell_o}$

Elastic modulus: $\sigma = E\varepsilon$

Chapter Summary

The macroscopic material properties of metallic solids and alloys are a result of atomic-level interactions among the atoms. Metallic elements bind their valence electrons loosely and therefore give them up easily. When packed tightly in a crystal, metallic elements can be viewed as consisting of a lattice of cations in a communal electron sea. The electron sea permeates the cation lattice, binding the cations together due to the coulombic attraction to this sea. The loosely held electrons in the sea are supplemented by somewhat more tightly held, directed, d-orbital valence electrons. In the fourth period, the level of d-orbital electron involvement rises to a maximum in the Fe-Co-Ni trio and declines as the d-orbital electrons become drawn closer to the nucleus.

The electron sea model accounts for macroscopic as well as microscopic characteristics of metals: elastic and plastic deformation, density variations, conductivity, and alloy formation. Alloy formation modifies the macroscopic material properties of metals by inhibiting plastic flow, thereby extending the elastic response to much higher loads.

KEY IDEA

The macroscopic properties of materials follow from their atomic-level structure: the electron configuration of the constituent atoms and the crystal structure of the lattice.

OPERATIONAL SKILLS

- Calculate density from atomic size (Worked Example 4.1).
- Determine the number of atoms in a unit cell and the packing density (Worked Examples 4.2, 4.3).
- Use the elastic modulus to determine strain from stress, and vice versa (Worked Example 4.4).
- Evaluate the suitability of metallic materials for applications based on density, flexibility, and resistivity (Worked Example 4.5).
- Read phase diagrams to determine the limits of solubility (Worked Example 4.6).
- Determine the size of interstitial holes in various lattices (Worked Example 4.7).

Exercises

A blue exercise number indicates that the answer to that exercise appears at the back of the book.

■ SKILL BUILDING EXERCISES

4.1 Bonding in Metals

1. Calculate the effective radii of Na and Mg in their respective solids. Compare these with their respective quantum mechanical radii.

2. Calculate the effective radii of K and Ca in their respective solids. Compare these with their respective quantum mechanical radii.

3. The density of copper is 8.96 g/cm³. Calculate the molar density of copper. What is the effective size of Cu in solid copper?

4. For many material applications, the *mass* density is an important parameter, particularly in the transportation industry. For understanding interatomic interactions and bonding, the *atomic* density is more revealing than the mass density. Thus, it is important to be able to convert between the two. The mass density of iron is 7.874 g/cm³. Calculate the molar density of iron. What is the effective size of Fe in solid iron?

5. Calculate the molar densities of Na and Mg. Compare them to the molar density of Al. What is the ratio of the molar density of Al to that of Na? Of Mg to that of Na? Comment on the trend.

6. Assign an electronic configuration to Cr^{3+}. How many *d*-orbital electrons are there?

7. Why do Zn, Cd, and Hg have such low melting points compared with Cr, Mo, and W?

8. Account for each of the following observations:
   a. Zn has a lower melting point than Ca.
   b. Cd has a higher ionization energy than Sr.
   c. The radius of Ca is larger than the radius of Zn.
   d. The melting point decreases from Zn to Cd to Hg.
   e. The radius of Cd is about the same as the radius of Hg.
   f. The radius of Ag is about the same as the radius of Au.

9. Account for the higher melting points of Cu, Ag, and Au compared with K, Rb, and Cs.

10. Account for the greater atomic density of Zn compared with Ca, Cd compared with Sr, and Hg compared with Ba. Discuss why the valence configuration might suggest that the pairs should be similar.

4.1 Crystal Structure of Metals

11. Simple cubic structures have one atom at each of the eight corners of a cube. Calculate the packing efficiency for simple cubic packing. What is the coordination number in simple cubic packing?

12. The element polonium was discovered by Marie Curie and named after her homeland, Poland. Polonium crystallizes in a simple cubic structure and has a density of 9.32 g/cm³. Determine the dimension of the unit cell for polonium.

13. Prove that the packing efficiency for body-centered cubic packing is 68%.

14. Determine the packing density for β-tin.

15. Given the density of tungsten (19.3 g/cm³) and the crystal structure (BCC), calculate the size of the tungsten unit cell and the radius of the tungsten atom.

16. Palladium is a relatively rare catalytic metal that crystallizes in an FCC lattice. The atomic mass of Pd is 106.42. Calculate the mass of a unit cell. Given the density (12.02 g/cm³), calculate the size of the unit cell and the radius of a Pd atom. The surface is the catalytically active portion of a palladium particle. Calculate the number of palladium atoms in 1 cm² of surface assuming that the surface consists of a simple truncation of the bulk crystal structure.

4.1 Close Pack Lattices

17. Is the diamond structure a close pack lattice? Why or why not?

18. Determine the size of the interstitial spaces in the diamond structure. Determine the packing density for a diamond structure.

4.1 Body-Centered Cubic Lattice

19. What is the coordination number of an atom in the body-centered cubic structure? Potassium crystallizes in the BCC structure with a unit cell edge of 533.3 pm and a density of 0.862 g/cm³. Determine the atomic mass of potassium.

20. Few metals crystallize in the BCC structure. Of those that do, which metals are candidates for forming substitutional alloys with iron?

4.2 Material Properties: Deformation

21. Describe elastic and plastic distortion on the atomic and macroscopic levels.

22. Draw a schematic illustration of what is happening on the atomic level as a copper wire is pounded.

23. What are the valence configurations of V, Cr, and Mn? Explain why the elastic modulus of Cr is higher than that of V or Mn.

24. Tungsten is a brittle metal. Silver is a very ductile metal. Sketch the stress-strain curves for these two metals. Which part of the curve reflects the elastic modulus? Which part of the curve reflects the ductility and brittleness for these two metals?

25. Gold is an extremely malleable material that can be made into very thin sheets called gold leaf. One gram of gold leaf can cover 1 m². The density of gold is 19.3 g/cm³. How many atoms thick is the gold leaf?

26. Gold is a very conductive and ductile metal. One gram of gold can be pulled into a wire 2 km long. The density of gold is 19.3 g/cm³. What is the diameter of this gold wire? The resistivity of gold is 2.271 $\mu\Omega \cdot$ cm. What is the resistance per centimeter of this gold wire?

27. In the absence of an applied stress, the iron atoms are separated by 246.0 pm. Under a stress of 79.79 g/mm², the distance increases to 246.9 pm. Calculate the elastic modulus.

28. Carbon is too large to fit into the interstitial spaces in the BCC iron lattice. Assume that carbon is a hard-sphere atom. By how much must the iron unit cell expand to accommodate carbon? The expansion can be thought of as a local strain. Calculate the strain if it is confined to one unit cell. To how many unit cells must the strain extend to reduce the strain to less than 0.1%, the typical limit for linear behavior on the stress-strain curve?

4.2 Material Properties

29. Explain why the mass densities of the sixth-row transition elements are much larger than those of the fifth-row elements, yet their atomic densities are nearly the same.

30. The seventh-row transition elements have been discovered, but few of their properties have been determined because they have only a fleeting existence. Predict the density of element 107, bohrium, named for Neils Bohr in recognition of his contribution to our understanding of the electronic structure of the atom.

4.2 Electron Mobility: Resistivity

31. Account for the following: Cu, Ag, and Au are excellent conductors, having the lowest resistivities of the transition elements.

32. Explain the relatively high resistivity of Mn. Use properties of atomic orbitals to support your explanation.

33. In a metal wire, resistance leads to heating of the wire. (Heating is the power dissipated from passing a current, I: power $= I^2R$.) Resistance $= \rho \times \ell/A$, where ℓ is the length and A is the cross-sectional area. The maximum safe current in a 12-gauge copper wire (0.21-cm diameter) is 20 amp. Determine the power per meter generated by the 12-gauge wire with this current load. The resistivity of Al is three times that of Cu. How much more power is dissipated in an Al wire of the same diameter? How much larger diameter must the Al wire be to generate the same power as the 12-gauge copper wire?

34. The layered structure of graphite was discussed in Chapter 2. It has been observed that graphite conducts very well within a sheet, but poorly between sheets. Account for this observation.

4.4 Alloys

35. Sterling silver is 92.5% Ag and 7.5% Cu by weight. Calculate the atomic percent of each element in sterling silver.

36. Cu and Zn form an alloy called brass. Calculate the number of Zn atoms in 1 cm³ of a 1 wt % Zn alloy. Evaluate the Hume-Rothery parameters for brass, and discuss the potential for complete solubility or the basis for limited solubility as applicable. The density of brass is 8.56 g/cm³.

37. By itself, gold is too soft to be used in jewelry, but when alloyed with Ag it is quite hard. The traditional way of specifying the amount of gold in jewelry is by the karat. One karat means that the jewelry is 1 part in 24 gold by weight. The usual gold content for jewelry is 14 karat. What is the gold to silver atom ratio in a 14-karat ring? The density of gold is 19.3 g/cm³, while that of silver is 10.50 g/cm³. Both crystallize in an FCC structure. What are the dimensions of the gold unit cell, and what is the radius of a gold atom? What are the dimensions of the silver unit cell, and what is the radius of a silver atom? What is the host (majority) atom in a 14-karat ring? Does the guest atom stress the crystal structure of the host?

38. Brass is an alloy of Cu and Zn. Do these elements form a substitutional alloy?

39. Given the following information:

Element	Radius (pm)	Crystal Structure	Electronegativity
Ni	115	FCC	1.8
C	77		
H	32		
O	73		
Ti	132	HCP	1.54
Ag	134	FCC	1.93
Al	118	FCC	1.61
Cu	117	FCC	1.90
Co	116	HCP	1.88
Cr	118	BCC	1.66
Fe	117	BCC	1.83
Pt	130	FCC	2.28
Zn	125	HCP	1.65
Ir	127	FCC	2.2

Which elements are likely to form solid solutions, substitutional impurities, or interstitial impurities in Ni? Of those listed, which other two metals are most likely to form a solid solution?

40. Copper, silver, and gold are all coinage metals. Evaluate the potential for formation of a substitutional alloy for these binary combinations: Cu-Ag, Cu-Au, and Ag-Au.

41. Which elements are candidates for formation of substitutional alloys with Fe? With Pd?

42. Evaluate each of the following pairs for solubility over the entire composition range: Cr and Fe, Ti and Fe, W and Nb. Indicate which of the Hume-Rothery rules are not met in each case where complete solubility is not supported.

4.4 Substitutional Alloys

43. A copper–silver brazing (similar to soldering) is used to join tubes in bicycle frames because it is much harder than Sn-Pb solder. Evaluate copper and silver for their potential to form an alloy. A mixture that is 28.1% by weight Cu melts at 780 °C. Compare this melting point with those of copper and silver. Are the melting point data consistent with your evaluation of the potential for alloy formation? Why or why not?

44. Stainless steel often incorporates elements in addition to iron and carbon. For example, 18–8 stainless steel is 18 wt % Cr and 8 wt % Ni. Cr and Ni substitute for Fe in the BCC lattice.

a. Calculate the atomic percent of each of these elements in 18–8 stainless steel.

b. Does substitution of Cr make the interstitial space larger or smaller?

c. Does substitution of Ni make the interstitial space larger or smaller?

d. Do you expect carbon to be found in unit cells with Ni, with Cr, or with neither? Justify your answer.

45. Ni forms a substitutional alloy with Cu over the entire compositional range. Calculate the number of Ni atoms per cm^3 for an alloy that is 5% Ni and 95% Cu by weight. The density of Ni is 8.902 g/cm^3 and the density of Cu is 8.96 g/cm^3.

46. Silver and gold form an alloy over the entire composition range. Sketch a phase diagram for these two elements for the temperature range 900 °C to 1100 °C.

47. Lead crystallizes in a FCC crystal structure. The density of lead is 11.35 g/cm^3.

a. Calculate the size of a lead atom in elemental lead.

b. Metallic tin crystallizes in a body-centered tetragonal structure. The dimensions of the unit cell are 583.18 pm × 583.18 pm × 318.19 pm. In the tin-rich phase of solder, where will the lead atoms be found (body center or corner) to minimize stress on the lattice?

48. What is the solidification temperature of a molten Cu-Ni solution that is 20% by weight Ni? What is the Ni content of the first solid formed?

49. Describe the physical state of a 1:1 Cu-Ni solution at 1400 °C, 1300 °C, and 1200 °C.

50. Solder consists of Sn and Pb. Evaluate whether these two elements form a substitutional alloy.

51. Which atom in each of the following pairs do you expect to have the larger electronegativity:

a. Cu or Zn

b. K or Ca

c. Cd or Ag

d. Cs or Ba

Justify your answers and then check them.

52. The Hume-Rothery rules give a set of guidelines for alloy formation. Why does a large or small atom (relative to the host) interfere with formation of an alloy? What effect does substitution of an atom with a larger electronegativity have on the solid that prevents alloy formation? How does having a different crystal structure interfere with alloy formation?

53. A mixture of silver and copper is used to braze the tubes of bicycles together. The phase diagram for Ag-Cu is shown in Figure 4.50. Describe the changes that occur as a mixture of 30% Ag and 70% Cu is cooled from 1000 °C to room temperature.

a. At what temperature does the first solid appear?

b. Does the first solid contain more or less copper than the original solution?

c. At what temperature does the last portion of the mixture solidify?

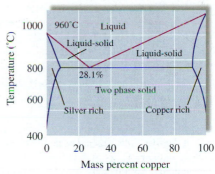

Figure 4.50 The silver–copper phase diagram.

54. Sterling silver consists of silver with 7.5% by weight Cu. Use the phase diagram in Figure 4.50 to describe the changes in a molten mixture of silver with 7.5% copper when cooled from 1000 °C to room temperature. How does this description depend on the cooling rate?

55. A mixture of 7.5 wt % Cu and 92.5 wt % Ag destined for sterling silver is heated to 1000 °C and cooled slowly to 790 °C. Describe the system at 790 °C. Once the mixture has reached 790 °C, it is cooled rapidly to room temperature. Describe the structure of this solid. The resulting sterling silver can be polished to a higher luster than if it is cooled slowly. Explain this observation.

56. A portion of the Ni-Ti phase diagram is shown in Figure 4.51. At what temperature does a mixture that is 50% Ti just begin to solidify?

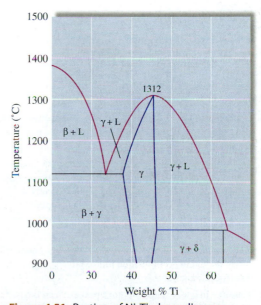

Figure 4.51 Portion of Ni-Ti phase diagram.

4.4 Interstitial Alloys

57. Carbon forms an interstitial solution with Fe in steel. The maximum weight percent carbon is about 2%. Calculate the atomic percent carbon at the maximum solubility.

58. What is the chemical composition of steel? Describe the lattice structure. What effect does C have on the mechanical properties of steel?

59. Carbon atoms are too large to fit into the interstitial spaces in the BCC iron lattice. Do carbon atoms fit into the spaces in the BCC tungsten lattice? How would incorporation of carbon atoms affect the mechanical properties of tungsten?

60. A face-centered cubic lattice contains an interstitial space in the body center. For a host element of 200-pm radius, calculate the radius of the interstitial space. Do any elements fit in this space?

61. Are the interstitial spaces in the diamond lattice larger or smaller than those in the FCC and BCC lattices? Justify your answer and estimate the size of the interstitial spaces in a diamond lattice.

62. Worked Example 4.7 concluded that C atoms are too big to fit into the holes in the Fe lattice. Do N, O, or F atoms fit into this space?

63. Chromium crystallizes in a BCC crystal structure with a lattice parameter—the length of the unit cell—of 291 pm. Calculate the diameter of the largest atom that can fit into the interstitial space without distorting the lattice.

64. Assume that a carbon atom is 77 pm in diameter. How does the iron lattice become distorted to accommodate carbon? Is the distortion uniform in all three directions?

4.4 Nitinol

65. Evaluate Ni and Ti according to the Hume-Rothery rules. Do you expect these two elements to form an alloy over a significant concentration range?

66. Explain how nitinol's temperature response is used during insertion of the Simon Nitinol filter illustrated in Figure 4.46.

67. Explain how nitinol's temperature response is used during insertion of the Simon Nitinol filter illustrated in Figure 4.46.

68. Nitinol has already found numerous other practical applications: safety valves in showers, eyeglasses that can be restored to their original shape, artery expanders, and orthodontic appliances, for example. Recently it has been proposed as a way to straighten the spines of scoliosis victims. Suggest other possible applications for nitinol.

69. Two solid phases of tin were described in this chapter. Which phase do you expect to be the more stable at high pressure? Justify your answer.

70. Describe how the shape memory property can be used in forming an artery stent from nitinol.

■ CONCEPTUAL EXERCISES

71. Inclusion of carbon in the interstitial spaces in the BCC lattice of iron creates a local stress that alters the crystal structure.
a. What element, if any, is small enough to fit into the interstitial spaces without stressing the lattice?
b. Discuss the advantages or disadvantages of replacing the body center atom with a larger atom to create the same stress in the lattice.

72. The resistivity of an alloy is greater than the resistivity of a pure element. Account for this observation using the electron sea model for metallic bonding.

73. Draw a diagram indicating the relationship among the following terms: elastic modulus, electron configuration, electron sea, hard-sphere model, resistivity, density. All terms should be interconnected.

74. Draw a diagram indicating the relationship among the following terms: plastic deformation, unit cell, slip system, elastic modulus, close pack, grain boundary, dislocation. All terms should be interconnected.

☀ APPLIED EXERCISES

75. Discuss how the resistivity and melting point of tungsten make it suitable for use as the filament in an electric light bulb.

76. The operating mechanism of automatic fire sprinklers incorporates a solder link holding a plug over the orifice to a pressurized water line. At the target temperature, the solder melts, releasing the cap and allowing water to flow through a deflector and spray the area. Paper spontaneously begins to burn at 450 °F. Compare this temperature to the melting point of tin. Discuss the advantages of using solder rather than tin as the meltable material to trigger the sprinkler spray. Suppose that one wants to replace the lead in the solder. What characteristics are required in the replacement?

■ INTEGRATIVE EXERCISES

77. Look up the melting points for the first transition series elements. Rationalize these data on the basis of one or more of the following: density, electron configuration, and conductivity.

78. The melting and boiling points of Zn, Cd, and Hg are relatively low.
a. Plot the melting point of the transition elements versus group number.
b. Rationalize the low melting and boiling points of Zn, Cd, and Hg. Cite elastic modulus, density, and electron configuration data to support your arguments.

79. Compare and contrast the response to a bending force in a piece of tin and a piece of martensitic nitinol.

80. What evidence can you cite for *d* orbitals becoming part of the core for elements in the second half of the transition series?

81. In Worked Example 4.7, it was concluded that the interstitial space for a carbon atom in an iron lattice is not spherical. If the carbon atom in the iron lattice is spherical with a diameter of 70 pm, the strain produced from putting carbon in the lattice is not isotropic. Embed the two unit cells containing the carbon atom into the middle of a four by three unit cell layer (Figure 4.52a) and sandwich this cluster between two more 12-cell layers (Figure 4.52b). Calculate the strain in the two unique directions if the extra space to accommodate the carbon atom is spread over the 36 cells of this cluster.

82. (Refer to Exercise 81.) Over how many unit cells must the strain caused by the carbon atom be spread to remain within the elastic region? If the strain is allowed to be plastic in the larger strain region, but elastic in the lower strain region, over how many cells must the strain be spread?

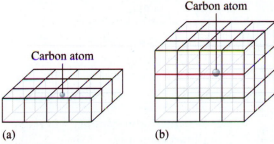

(a)   (b)

Figure 4.52 (a) Schematic of a carbon atom located at the adjoining faces of two BCC unit cells and surrounded by 10 additional unit cells. (b) The 12-unit-cell layer sandwiched between two additional 12-unit-cell layers.

Chapter 5
Chemical Bonding and Modern Electronics

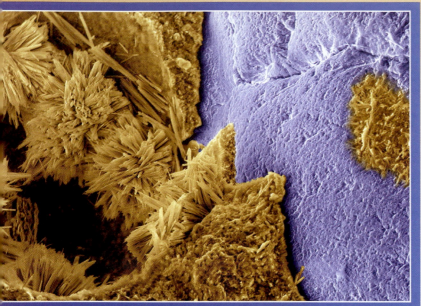

A micrograph of concrete reveals the needles and plate-like crystals that give this major building material its strength and other properties. At an even smaller scale, concrete's properties depend on the ionic and covalent bonds among its atoms.

CONCEPTUAL FOCUS

- Model the behavior of electrons as atoms meet — this is the key to understanding chemistry.
- Use atomic-level interactions to predict the electronic properties of semiconductors and conductors.

SKILL DEVELOPMENT OBJECTIVES

- Correlate interactions and density in covalent materials (Worked Examples 5.1 and 5.2).
- Link interactions and band gap in semiconductors (Worked Examples 5.3 and 5.4).
- Alter the conductivity with doping (Worked Examples 5.3, 5.5, and 5.6).

Many ages of civilization are labeled according to the dominating material in use—the Stone Age, the Copper-Stone or Chalcolithic Age, the Bronze Age, and the Iron Age. In many ways, our current age, the information age, could be aptly labeled the Electronic Materials Age, for electronics dominate our culture. Think of the common items you rely on: calculators, computers, cell phones, CD and DVD players. The materials that these items use differ greatly in their electrical conduction properties, ranging from conductive gold and copper, to semiconductor logic circuits and power indicator lights, to the insulators that prevent short circuits and shocks to the operator. These various conductivity properties are linked to the atomic-level bonding in these materials. Similarly, the mechanical properties—the elastic modulus and brittleness—as well as the melting point are determined by the bonding.

Covalent bond A shared pair of electrons.

Bonding is broadly classified as one of three types: metallic, covalent, or ionic (Figure 5.1). Metallic bonding was explored in Chapter 4, and covalent and ionic bonding are investigated in this chapter. A **covalent bond** is characterized by a directed connection to the nearest neighbor and sharing of valence electrons. An **ionic bond** is characterized not by sharing, but rather by transfer of one or more electrons from one atom to the other. Most often, the bond between nonmetallic elements is a covalent bond, while ionic bonds occur when a nonmetal and a metal combine (Figure 5.2). The dividing line between the three types of bonding is not a sharp demarcation, but rather a gradual shift from one to the other (Figure 5.1). The type of bonding, however, has significant implications for the mechanical and electrical properties of the material.

Covalent bonding features charge carriers (electrons) directed between atoms, locking the atoms into position. For covalently bonded solids, the result is a hard material with a large elastic modulus and little plastic deformation prior to catastrophic failure (breakage). In contrast, metallic bonding delocalizes the charge carriers so that the atoms can shift their positions within the electron sea. Metallic bonding yields a pliable material with a smaller elastic modulus and plastic flow, making the material ductile and malleable.

The distinction between metallic and covalent bonding was overlooked by two corporations, bankrupting one and requiring a governmental bailout to save the other. Rolls Royce decided to use a graphite composite for the compressor blades in an aircraft gas-turbine engine instead of titanium. (Graphite is less dense than titanium, making lighter-weight blades that are less costly to move. Graphite is a covalently bonded material; titanium is a typical metal.) After all the expensive plant design, tooling, and product assembly were completed, the engine had to pass one final, critical test before delivery to companies such as Lockheed Aircraft. The test is called the "bird test" because it simulates flying into a flock of birds, a common scenario during take-off. Due to the directional, covalent bonds in graphite, the brittle blades were stripped out of the engine. Although the test saved many lives, the economic fallout bankrupted Rolls Royce and, but for a U.S. government bailout, would have bankrupted Lockheed as well since the company could not deliver aircraft to its customers. These major repercussions were all due to the minute bonds between atoms and the localization of covalent bonds versus the itinerate metallic bonds.

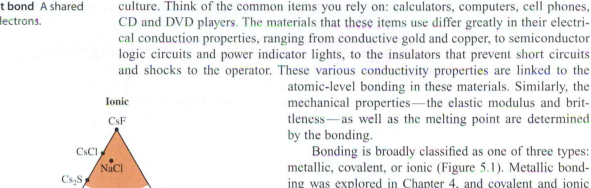

Figure 5.1 The relationships among the three extreme bonding types: covalent, ionic, and metallic. The boundary between the types is not a sharp line, but rather there is a continuum of bonding types. While the bonding in some materials matches one of the extremes, many materials employ a blend of the three bonding types.

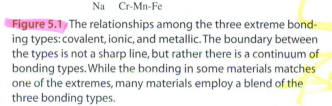

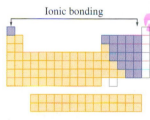

☐ Metallic bonding
☐ Covalent bonding

Figure 5.2 Metallic bonding dominates in bonds among elements on the left side of the periodic table. Covalent bonds are prevalent in bonds between elements on the right side of the periodic table. Ionic bonds are formed when elements on the left side join with those on the right side of the periodic table.

Ionic bond Bond in which one or more electrons transfer from an element with low electronegativity to one with a high electronegativity; the resulting ions hold together due to the attraction between an anion and a cation.

A milder version of the consequences of mixing a little directed, *d*-orbital bonding with the nondirectional, metallic bonding is seen in the variation of the elastic modulus, from left to right, across the transition metals. A small admixture of covalency adds stiffness to the bonds and increases the elastic modulus. Millions of research dollars are spent annually to tailor materials with specific properties, and understanding the type of bonding is one important component of that process.

Conduction—the transfer of charge through a solid—is very sensitive to the type of bonding, the electron configuration of the atoms, and the crystal structure of the material. Conduction is the focus of this chapter. Specifically, the aim is to link the color emitted by a semiconductor device—for example, the power indicator light on a computer monitor—to the atomic-level bonding in the device. This task begins with an exploration of bonding between two atoms, extends to solids, and concludes with an examination of the color of indicator lights and the lasers used to read CDs, DVDs, and the media of the future.

5.1   Introduction to Nonmetallic Bonding

Imagine the encounter between atoms of elements near the opposite sides of the periodic table (Figure 5.3), such as Na and Cl. Elements on the left side of the periodic table have loosely bound, valence electrons—in the case of sodium, a single valence electron. On the right side of the periodic table, elements have *s* and *p* valence subshells that are nearly filled. Specifically, the valence shell of chlorine is one electron short of a filled s^2p^6 configuration and strongly attracts an additional electron. The loosely held electron from Na transfers to Cl (Figure 5.4a), filling the vacancy in chlorine's outer shell and forming a Cl^- ion (isoelectronic with argon), and leaving Na with a neon core and a positive charge (Na^+) (Figure 5.4b). Similar to opposite poles of a magnet, these oppositely charged ions attract each other via a coulombic interaction—the hallmark of ionic bonding. Solids composed of ions are held together as securely as a stack of magnets arrayed north-south-north- . . . . For Na^+ and Cl^-, the result is the familiar white crystalline material—salt (Figure 5.4c). Crystals of NaCl (Figure 5.4d) can be produced by dissolving salt in water and letting the water evaporate. The shape of the macroscopic crystal reflects the cubic arrangement of ions in the crystal.

CONCEPT QUESTION   What interaction occurs between two atoms, both of which strongly attract an additional electron? ∎

A chlorine atom is one electron short of a closed shell configuration and therefore strongly attracts an additional electron. Now imagine two chlorine atoms approaching each other (Figure 5.5). Because the electrons incompletely shield the core charge, as the two atoms approach, the effective positive charge of each atom exerts an attraction for the negatively charged electrons of the other. Because the outer shell is not filled, there is space for an additional electron. (This configuration is the origin of the electron affinity.) One way that the chlorine atom can achieve the illusion of filling that space is by borrowing an electron from the other chlorine atom. Because the two chlorine atoms are completely alike, the second atom also borrows an electron from the first. It is as if each chlorine atom tries to fill the vacancy in

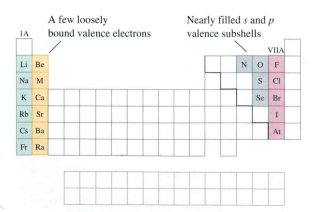

Figure 5.3 Elements on the left side of the periodic table have a few loosely held valence electrons. Those on the right side of the table have nearly filled valence *s* and *p* subshells.

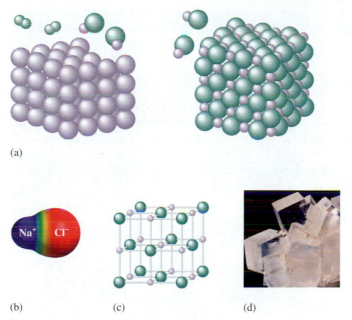

(a)

(b)   (c)   (d)

Figure 5.4 (a) The reaction of metallic sodium with covalently bonded chlorine to form ionic sodium chloride. In the process, sodium atoms (electron configuration: [Ne]$3s^1$) lose one valence electron. Note the significant *decrease* in size from atomic sodium (223 pm) to the sodium ion (102 pm) in sodium chloride. The electron lost by a sodium atom is gained by a chlorine atom (electron configuration: [Ne]$3s^1 3p^5$). Note the *increase* in size of the chlorine atom (97 pm) in molecular chlorine to the chloride ion (184 pm) in sodium chloride. (b) The charge around a NaCl molecule. Red represents the extreme of negative charge, and blue indicates the extreme of positive charge. In-between charges follow the colors of the rainbow. Note the abrupt change from blue to red in NaCl. (c) A ball-and-stick model of the unit cell of solid NaCl, commonly known as salt. Every Na$^+$ ion is surrounded by six Cl$^-$ ions, and vice versa. (d) A single crystal of sodium chloride is cubic, reflecting the arrangement of ions in the crystal.

its own *p* subshell with part of the valence electron density from the other chlorine atom. These two borrowed electrons constitute a *shared pair* of electrons. The electron wave occupied by the shared pair literally envelopes both of the cores, resulting in an overall lowering of the energy. This energy lowering drives the reaction, the joining of the two atoms, and provides the force that keeps the two atoms together. This kind of bond—a shared electron pair—is a covalent bond. The "co-" part refers to something shared, a partner, while "-valent" acknowledges the involvement of the valence electrons.

The octet rule and Lewis dot structures provide a procedure for determining combining rations. One limitation of a Lewis dot structure, however, is that it does not distinguish between ionic and covalent bonding. The distinction between ionic and covalent bonds is important for material properties because covalent bonds have a property not shared by either ionic or metallic bonds: They are *directional* due to sharing of electrons with a specific neighboring atom in a directed orbital. This directional character is in contrast to the totally nondirectional, electrostatic attraction holding metals in a solid or binding ions together. Thus, the degree of covalent character is important for material properties.

Purely ionic bonds, where the valence electron completely transfers from one atom to the other, and completely covalent compounds, which are characterized by equal

Figure 5.5 The encounter between two chlorine atoms is simulated by illustrating the changes in electron density (left column) and morphing the separate *p* orbitals into a shared orbital (right column). The two chlorine cores are represented by the turquoise spheres (dark spheres in the *p* orbitals). At a large separation (450 pm), neither atom is -affected by the presence of the other. As the two chlorine atoms approach, the electron in the partially filled *p* orbital of each atom is attracted to the other atom due to the incompletely shielded nuclear charge and the space available for an additional electron in the half-filled orbital. At a separation of 204 pm, the *p* waves from the two atoms combine to form a wave that is shared by both atoms. The resulting electron density envelopes both atoms and binds them together.

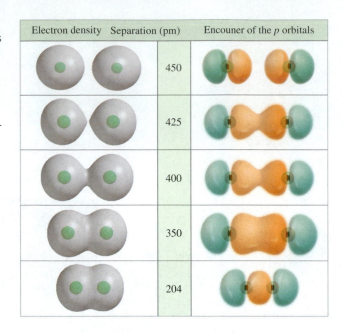

Electron density	Separation (pm)	Encouner of the *p* orbitals
	450	
	425	
	400	
	350	
	204	

sharing between the atoms, are rare. Most bonds lie somewhere along the continuum between these extremes; exactly where depends on the relative tendency of the two bonding atoms to attract electrons. The tendency to attract electrons is indicated by the atoms' electronegativity. The difference in electronegativity (Table 5.1) between two bonded elements indicates whether the electrons are evenly distributed and the bond is covalent, or whether they are weighted to one side and the bond is ionic. For example, the electronegativity difference between Na and Cl is 2.23 and the bond 70% ionic. Thus, there is little "sharing" of the electron between Na and Cl.

The uneven distribution of electron density can be shown in several ways. An **electrostatic charge density** surface (Figure 5.6) indicates the electrical charge at various points on the surface with colors: Red indicates the most negative charge and blue represents the most positive, with in-between charges following the colors

Electrostatic charge density Represents the charge on the molecular surface with colors: red for negative, green for neutral, and blue for positive.

Table 5.1 Percent Ionic Character in a Single Chemical Bond

Difference in electro-negativity	0.1	0.2	0.3	0.4	0.5	0.6	0.7	0.8	0.9	1.0	1.1	1.2	1.3	1.4	1.5	1.6	1.7	1.8	1.9	2.0	2.1	2.2	2.3	2.4	2.5	2.6	2.7	2.8	2.9	3.0
Percent ionic character	0.5	1	2	4	6	9	12	15	19	22	26	30	34	39	43	47	51	55	59	63	67	70	74	76	79	82	84	86	88	89

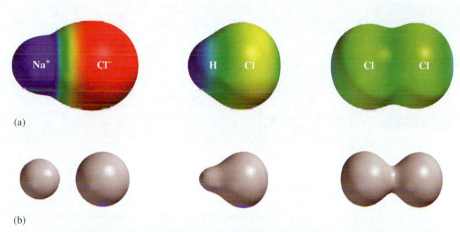

(a)

(b)

Figure 5.6 (a) The electrostatic charge density surfaces for NaCl, HCl, and Cl$_2$ show the differences between ionic (NaCl), mainly covalent (HCl), and purely covalent (Cl$_2$) bonds. In NaCl, the positive charge region (blue) of the Na$^+$ ion is juxtaposed to the negative charge region (red) of the Cl$^-$ ion. Note the abrupt change from extreme positive to extreme negative charge. In contrast, the shift from the positive region around hydrogen in HCl to the mildly negative region around chlorine is more gradual. In purely covalent Cl$_2$, there is little charge variation and the density is symmetrically distributed. (b) The electron density surface presents another picture for the ionic compound NaCl compared with covalent HCl and Cl$_2$. The electron density surface for ionic NaCl shows the electrons of the Na$^+$ ion and the Cl$^-$ ions as two separated spheres. Covalent HCl has an electron density surface that envelopes both the hydrogen atom and the chlorine atom. The electron density in homonuclear Cl$_2$ also envelopes both atoms. Covalent bonds are characterized by a significant electron density connecting the two atoms.

Electron density surface
Encompasses 90% of the electron density – the sum of the square of the wave amplitude.

of the rainbow. Neutral is indicated by green. A second representation, an **electron density surface** (Figure 5.6) encloses 90% of the electron density. The electron density surface shows the largest contrast between ionic and covalent bonding. For NaCl, the electron density occurs in two separate, spherical regions around the Na$^+$ and Cl$^-$ ions. In contrast, for covalent HCl and Cl$_2$, the electron density envelopes the two atoms in the molecule. Ionic NaCl is held together by the electrostatic attraction between opposite charges, while HCl and Cl$_2$ are bound by the enveloping electrons.

CONCEPT QUESTION   Determine the percent ionic character for HCl. Which Group IA element forms the most ionic bond with chlorine? ■

Covalent bonds are prevalent, particularly in compounds containing carbon. Most of the molecules that make up living bodies, for example, are covalent, as is the greenhouse gas, CO$_2$ (Figure 5.7). The electronegativities of carbon and oxygen differ by only 0.89, and the electron density is quite evenly distributed.

CONCEPT QUESTION   How many valence electrons does carbon contribute to CO$_2$? Draw the Lewis dot structure of CO$_2$. ■

To determine the number of bonds in a molecule (i.e., the number of shared electron pairs):

■ Determine the number of electrons needed for each atom to satisfy the octet rule (two for hydrogen).
■ Count the total number of valence electrons available.
■ One-half the difference is the number of electron pairs that need to be shared.

Figure 5.7 CO_2, a molecule that is at the center of the global warming controversy, is an example of a covalent molecule. The electron density wraps around both oxygen atoms and the carbon atom. Carbon is slightly less electronegative than oxygen, so a positive electrostatic charge resides on carbon.

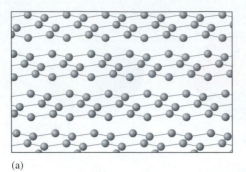
(a)

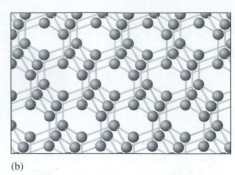
(b)

Figure 5.8 (a) The structure of graphite consists of layers of carbon atoms in a hexagonal arrangement. Each carbon atom is covalently bonded to three other carbon atoms. The separation between the layers is nearly 2.5 times the separation between carbon atoms within a layer, reflecting the difference between localized electron pairs covalently bonding the atoms within a layer and the looser, delocalized attraction between layers. (b) The structure of diamond consists of an extended array of carbon atoms, each covalently bonded to four other carbon atoms. All bonds in diamond are equivalent.

For CO_2 the counting is as follows. Twenty-four electrons are needed by the three atoms. Carbon supplies four valence electrons and the oxygen atoms supply six each, for a total of 16. The difference, eight electrons, represents four pairs, or four bonds. Bonding pairs can also be shown as a line connecting the atoms. Thus, two representations for CO_2 are

$$\ddot{O}::C::\ddot{O} \qquad \text{or} \qquad \ddot{O}=C=\ddot{O}$$

Among substances involving a large number of atoms (e.g., a solid), the bonding significantly affects the material's conductivity. Two forms of carbon, diamond and graphite, illustrate this sensitivity. Graphite consists of layers of carbon atoms forming an hexagonal structure (Figure 5.8a). Within each layer, each carbon atom is covalently bonded by a pair of electrons to three other carbon atoms. The three bonds to the nearest neighbors consist of a pair of localized electrons and use three of the four valence electrons in carbon. The fourth valence electron from the neighboring carbon atoms join in an electron cloud above and below the plane in an extended wave similar to the extended electron states of a metal. The electrons in this cloud are very mobile, and graphite is a conductor. Diamond has a very different structure (Figure 5.8b). Each carbon atom in diamond is covalently bonded to four other carbon atoms, forming an extended network of covalent bonds. The four covalent bonds use all four valence electrons in carbon. Localized between the carbon atoms, the electrons in these bonds are not mobile and hence diamond is an insulator. In addition, the bonds are very inflexible, which makes diamond the hardest material known.

Diatomic Molecules

To understand the contrasting bonding in diamond versus graphite, we begin by investigating the changes in the atomic orbitals of two atoms as they approach each other. As the two atoms approach (Figure 5.9), the electron cloud of each is drawn toward the nucleus of the other atom due to incomplete shielding of the nuclear charge at close distances.

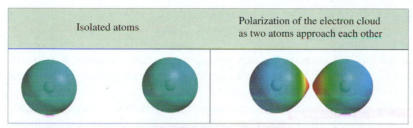

Isolated atoms	Polarization of the electron cloud as two atoms approach each other

Figure 5.9 Interaction between two atoms showing the polarization due to the coulombic attraction of each electron cloud for the nucleus of the approaching atom. At large separation, the electrostatic charge is slightly positive due to incomplete shielding of the nuclear charge that is uniform around the atom. As the two atoms approach each other, attraction due to the incomplete shielding pulls the electron clouds toward the internuclear space. Notice the concentration of electron density between the atoms as they approach each other.

In this encounter, the wave nature, including the phase, of the electron orbital has a profound effect on the interaction between atoms. Picture two water waves coming together (Figure 5.10). If both waves crest where they meet, the result is a higher crest. If one crests and the other troughs, the waves wipe each other out. So it goes with electron waves. If the two electron waves are in phase (crests meet), they reinforce each other, resulting in an enhanced amplitude between the atoms (Figure 5.11, left column). An electron in this wave has a large density between the two atoms and is attracted by both atoms. This attraction both lowers the energy of the electron and binds the atoms together.

Conversely, if the waves in the approaching atoms are out of phase, they mutually cancel each other out (destructive interference), resulting in diminished amplitude—a node—between the atoms (Figure 5.11, right column). An electron in this wave has the greatest density in the region outside the space between the two atoms. The attraction between either atom and the electron is reduced compared with the isolated atom.

The combined waves resulting from either constructive or destructive interference are called **molecular orbitals.** An electron in a constructive-interference molecular orbital interacts not only with the original nucleus but also with the nucleus of the approaching atom. This extra coulombic attraction results in a lower-energy molecular orbital compared with the atomic orbital in the free atom. Conversely, a destructive-interference molecular orbital is characterized by a node between the nuclei. Owing to diminished in-

Molecular orbital An electron orbital spread over a molecule; the square gives the probability of finding the electron at a particular location.

Figure 5.10 Like two water waves, electron waves in two different atoms interact as the atoms come together. (a) Two in-phase waves, or two in-phase orbitals on the separate atoms, reinforce each other, giving rise to an enhanced wave amplitude. (b) Two out-of-phase waves, or two out-of-phase orbitals on the two separate atoms, cancel each other out as they come together, giving rise to a node between them.

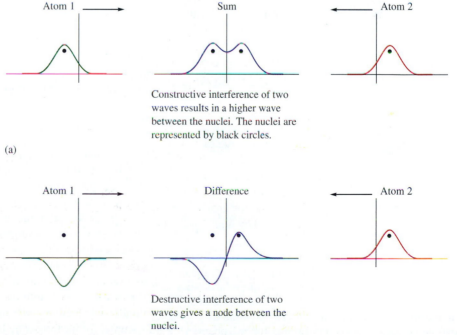

(a)

Atom 1   Sum   Atom 2

Constructive interference of two waves results in a higher wave between the nuclei. The nuclei are represented by black circles.

(b)

Atom 1   Difference   Atom 2

Destructive interference of two waves gives a node between the nuclei.

teraction with either nucleus, an electron in such an orbital is of *higher* energy than the orbital in the free atom. These energies are represented in an energy-level diagram (Figure 5.12), much as for atoms. The orbital energies are represented by lines; energy increases vertically. The energy-level diagram for two ground-state hydrogen atoms is relatively simple. The two 1s orbitals are shown on each side and the two molecular orbitals appear in the middle.

The lower-energy molecular orbital, characterized by having a large amplitude between the two nuclei, is called a **bonding molecular orbital** and is denoted by the Greek letter sigma, σ. A σ orbital is symmetric about the internuclear axis and receives this name in analogy to the highly symmetric atomic *s* orbital (σ is a Greek "s"). The higher-energy orbital, the orbital with a node between the nuclei, is called an **antibonding molecular orbital** and is denoted as σ^* ("sigma star"). The σ and σ^* orbitals are molecular orbitals that electrons can potentially occupy. This is similar to the hydrogen atom, which has 1s, 2s, 2p, . . . orbitals. Hydrogen's single electron can occupy any one of these orbitals. In the ground or lowest-energy state, hydrogen's electron occupies the 1s orbital. The number of atomic orbitals considered when constructing the molecular orbitals depends on the atoms in the molecule. In the hydrogen example, only the 1s orbitals of each atom are included when forming molecular hydrogen. A kind of conservation is observed with orbitals: The number of molecular orbitals produced is *always equal* to the total number of atomic orbitals included. The energy-level diagram for H_2 (Figure 5.12) has two atomic orbitals, one for each of two atoms, and there are two molecular orbitals, one bonding and one antibonding.

Bonding molecular orbital A molecular orbital that, when occupied, lowers the energy of the molecule, characterized by a large electron density between two nuclei.

Antibonding molecular orbital A molecular orbital that, when occupied, raises the energy of the molecule and weakens the bonding. Higher-energy orbital with a node between the nuclei.

CONCEPT QUESTION Hydrogen atoms have an infinite number of orbitals. What is the justification for considering only the 1s orbital? ■

If the two atoms coming together have only one valence orbital each, as is the case with H, He, Li, or Be, then this picture describes the formation of the relevant molecular orbitals: H_2, He_2, Li_2, or Be_2. To describe the bonding in the molecule, it is necessary to fill electrons in the molecular orbitals. Filling molecular orbitals is very similar to filling atomic orbitals: Electrons occupy the most stable orbital available. However, not all electrons can occupy the same orbital. The Pauli exclusion principle applies to molecular orbitals just as atomic orbitals. Thus, a maximum of two electrons (spins opposed) can occupy the same molecular orbital. Electrons fill, two per orbital, the molecular orbital of lowest energy first.

The task that remains is to determine how many electrons to fill in. Electrons are not lost when two atoms come together, they are merely rearranged from the atomic orbitals to the molecular orbitals. Hence electrons are conserved when the molecule forms, and the number of electrons that go into the molecular orbitals is the total number of valence electrons in the separate atoms. For example, for molecular hydrogen there are two electrons, one from each hydrogen atom. Two electrons go into the

In-phase orbitals	(pm)	Out-of-phase orbitals
	600	
	350	
	250	
	74.3	

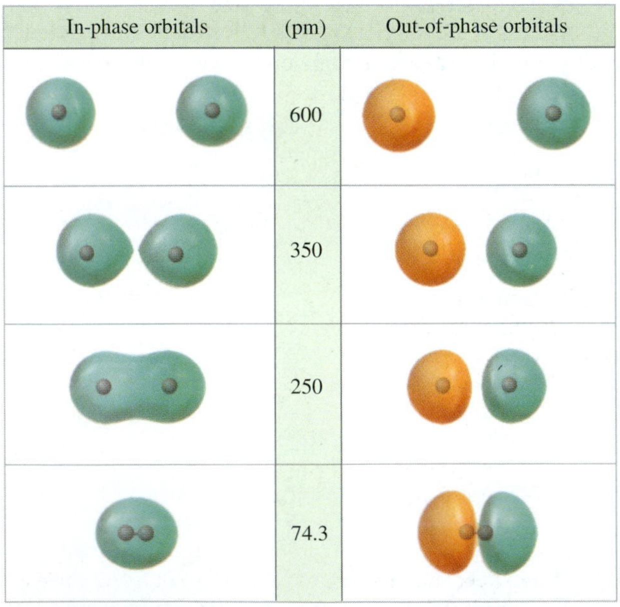

Figure 5.11 (Left column) Two hydrogen atoms approaching with 1s orbitals in phase result in an enhanced amplitude in the internuclear space. (Right column) Out-of-phase orbitals in the two approaching hydrogen atoms cancel each other out in the internuclear space, resulting in diminished amplitude – a node – between the atoms. The two phases are represented as turquoise and yellow.

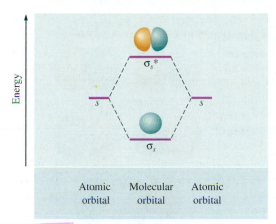

Figure 5.12 The molecular orbital energy-level diagram for formation of a diatomic molecule from two atoms, each with a single *s* orbital, shows the energy of the separate atomic orbitals on each side of the diagram. The molecular orbital energies are shown in the middle. Dashed lines connect the molecular orbital with the atomic orbitals that combine to form that molecular orbital. The bonding and antibonding orbitals are shown for reference.

Bond order An indication of bonding strength. It is $1/2$ the number of bonding electrons minus $1/2$ the number of antibonding electrons.

molecular orbitals (Figure 5.13). The σ, bonding orbital is filled; the σ^*, antibonding orbital is empty. A bond consists of a pair of electrons, so the hydrogen molecule is said to be held together by a *single bond*—a single pair. The lowering of energy upon forming the diatomic hydrogen molecule drives the reaction. Under ambient conditions, elemental hydrogen exists exclusively as H_2, not as separate hydrogen atoms.

Consider the next element, He. The valence orbital in He, like that in H, is a $1s$ orbital, so the energy-level diagram is the same for He as for H (Figure 5.12). Electron occupancy, however, is different. Each He atom brings two electrons into the molecule, for a total of four electrons. These four electrons fill both the bonding and the antibonding orbitals (Figure 5.14). As the name implies, the electrons in an antibonding orbital work against those in the bonding orbital. With an equal number of bonding and antibonding electrons, no net bonding occurs. This model therefore predicts that molecular He_2 does not exist. Indeed, He_2 is not a stable molecule. Except in extreme conditions, helium exists as a single atom, a monatomic gas.

Formally, the net number of bonding pairs is called the **bond order.** It is defined as

(5.1) $$\text{Bond order} = \tfrac{1}{2} \times [(\text{number bonding electrons}) - (\text{number antibonding electrons})]$$

Using this formula, the bond order for H_2 is 1, as concluded above, and that for He_2 the bond order is zero.

CONCEPT QUESTIONS   What is the bond order for Li_2? For Be_2? ∎

Elements beyond Be have valence *p* orbitals in addition to the valence *s* orbital. Forming molecular orbitals from *p* orbitals is somewhat more involved than forming

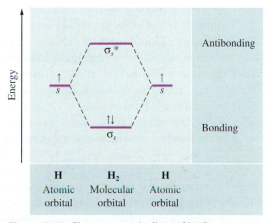

Figure 5.13 Electrons are indicated in the energy-level diagram as arrows. Those on each side are the valence electrons for each of the constituent atoms. The number of electrons in the molecular orbitals is equal to the total number of valence electrons in the separate atoms. This example describes the bonding in hydrogen.

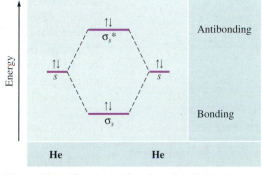

Figure 5.14 The He_2 molecular-orbital energy-level diagram features a filled bonding and a filled antibonding molecular orbital resulting in an unstable dimer.

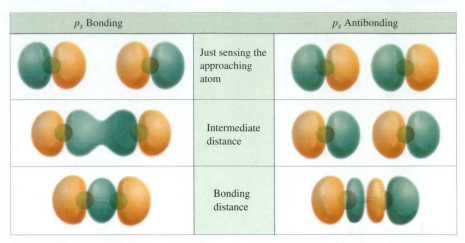

p_z Bonding		p_z Antibonding
	Just sensing the approaching atom	
	Intermediate distance	
	Bonding distance	

Figure 5.15 Formation of the σ_{p_z} orbitals from two p_z atomic orbitals. The atomic cores are located at the node of the p_z orbital. (Left column) As the two atoms approach each other, each in-phase p_z orbital begins to elongate toward the approaching atom. At an intermediate distance, the overlapping p orbital lobes reinforce each other. At bonding distance, the overlapping lobes of the p_z orbitals form a σ_{p_z} bonding orbital that fills in the space between the two nuclei. (Right column) As the two out-of-phase orbitals approach each other, they begin to cancel each other out. Each orbital is repelled away from the internuclear space. At bonding distance, there is a node—an antibond—between the atoms.

molecular orbitals from atomic s orbitals. There are three p orbitals, and in an isolated atom these are all equivalent. However, when two atoms come together, the p orbitals are no longer equivalent: One p orbital faces along the internuclear axis toward the incoming atom, while the other two are perpendicular to the internuclear axis. By convention, the internuclear axis defines the z-axis, so the p_z orbital faces the incoming atom, the p_x and p_y orbitals do not. The p_z orbital is cylindrically symmetric with respect to the internuclear axis, as is the s orbital. As a result, the combination of p_z orbitals is also labeled as a σ molecular orbital. The p orbitals have a node through the nucleus that is retained upon formation of the molecular orbitals, and this node affects the energy of the resulting molecular orbital. To distinguish the molecular orbitals formed by atomic p orbitals from those formed by atomic s orbitals, the molecular orbitals are denoted as σ_{p_z} and σ_s, respectively. As two atoms approach each other (Figure 5.15) the lobes of the two p_z orbitals between the two atoms overlap. If they are in phase, the result is reinforcement between the atoms and a σ_{p_z} bonding molecular orbital. Out-of-phase atomic orbitals interfere destructively, yielding a $\sigma_{p_z}*$ antibonding orbital.

The other two atomic p orbitals, the p_x and p_y orbitals, have a nodal plane that contains the internuclear axis. Forming molecular orbitals out of atomic p_x and p_y orbitals retains this node, so the resulting molecular orbital changes phase across the internuclear axis, much as the atomic p orbitals change phase across the nucleus. In reference to this change of phase, the molecular orbitals are labeled with the Greek letter pi, π (π is the Greek "p"). The p_x and p_y atomic orbitals are equivalent to each other, and both are perpendicular to the internuclear axis. The in-phase combination produces a bonding molecular orbital (Figure 5.16), and the out-of-phase combination produces a node between the atoms, an antibonding molecular orbital.

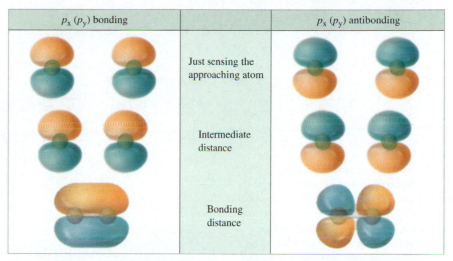

Figure 5.16 In-phase atomic p_x orbitals produce a bonding interaction. The molecular orbital is labeled a π_{p_x} orbital. (There is an equivalent combination of p_y atomic orbitals to produce a π_{p_y} molecular orbital.) The out-of-phase p_x orbitals produce a node between the atoms—an antibonding combination. The antibonding molecular orbital is labeled a π_{p_x} molecular orbital. (There is an equivalent π_{p_y} molecular orbital from interaction of the out-of-phase p_y atomic orbitals.)

To generate a molecular-orbital energy-level diagram, these orbitals need to be put in energy order, from low energy to high energy.

CONCEPT QUESTION   How should the two $1s$ electrons be treated when constructing the molecular-orbital energy-level diagram for second-row diatomic molecules? ■

Consider molecular oxygen, O_2. The portion of the energy-level diagram arising from the $2s$ atomic orbitals in oxygen is the same as that arising from the $1s$ orbitals of hydrogen (Figure 5.17). To complete the molecular-orbital energy-level diagram for O_2, the atomic p orbitals and the molecular σ_{p_z}, π_{p_x}, and π_{p_y}, molecular orbitals need to be added to the energy-level diagram. In the isolated atom, the three p orbitals are all the same energy; when combined in the molecular orbitals, however, they are not. The σ_{p_z} orbital is located directly along the axis between the two oxygen atoms (Figure 5.15), while the lobes of the π_{p_x} and π_{p_y} orbitals are located on the sides of the internuclear axis (Figure 5.16). The more direct connection of the σ_{p_z} orbital stabilizes it and results in a lower energy than that for the π_{p_x} and π_{p_y} orbitals (Figure 5.18). This more direct interaction of the σ_{p_z} orbital also means that the $\sigma_{p_z}*$, antibonding orbital is higher in energy than the $\pi_{p_x}*$ and $\pi_{p_y}*$ antibonding orbitals.

Combining the molecular orbitals derived from the atomic p orbitals with those derived from the atomic s orbitals generates the molecular-orbital energy-level diagram for molecular oxygen (Figure 5.19). Due to the extra nodes in the σ_{p_z}, π_{p_x}, and π_{p_y} orbitals, all of these are less stable, higher-energy orbitals than the σ_s orbital. The electron occupancy of these orbitals determines the bond order and other characteristics of molecular oxygen. Because only the

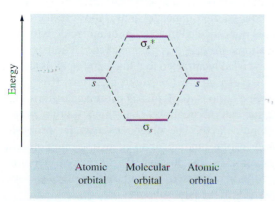

Figure 5.17 The portion of the molecular-orbital energy-level diagram due to the $2s$ orbitals in oxygen.

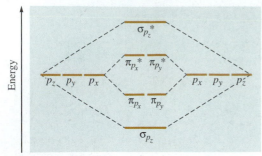

Figure 5.18 The molecular-orbital energy-level diagram resulting from interaction of atomic p orbitals, showing the greater stabilization of the σ_{p_z} orbital compared with the π_{p_x} and π_{p_y} orbitals. The stronger interaction between atomic p_z orbitals also means that the antibonding orbital, $\sigma_{p_z}^*$, is less stable than the $\pi_{p_x}^*$ and $\pi_{p_y}^*$ orbitals. The p_x and p_y orbitals are equivalent, and so are the molecular orbitals related to them.

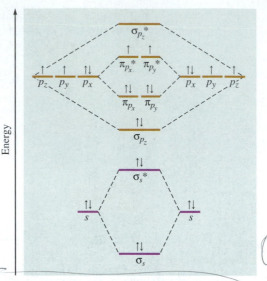

Figure 5.19 The molecular-orbital energy-level diagram for O_2 shows the greater stabilization of the σ_s and σ_s^* orbitals compared with the σ_{p_z} bonding orbital.

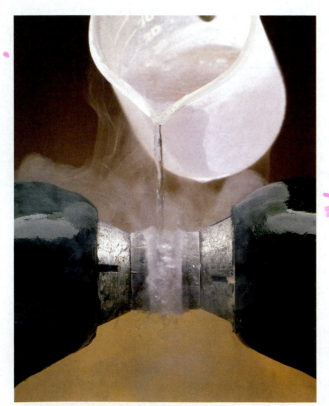

Figure 5.20 Due to the unpaired electrons in molecular oxygen, it can be caught between the poles of a strong magnet.

second-shell atomic orbitals were included to generate the molecular orbitals, only the six second-shell valence electrons are filled in the molecular-orbital energy-level diagram (Figure 5.19).

CONCEPT QUESTION What is the bond order in O_2? ∎

The electron occupancy in this molecular-orbital diagram accounts for an important property of molecular oxygen: its magnetic character. A spinning electron generates a magnetic field due to the charge of the electron. The magnetic fields of two paired electrons essentially cancel each other due to their opposite spins. However, the magnetic field is not canceled in an atom or molecule with unpaired electrons; instead, it is paramagnetic. Paramagnetic materials are weakly attracted to a magnetic field. Because molecular oxygen has two unpaired electrons, it is paramagnetic. One curious result of this paramagnetism is that liquid oxygen can be caught between the poles of a strong magnet (Figure 5.20). The magnetic character probably plays a role in the binding of oxygen to hemoglobin, a process that makes life possible. The unpaired electrons also affect the interaction of oxygen with molecules such as NO and NO_2 in the atmosphere.

The mild attraction of paramagnetic materials to a magnetic field stands in contrast to the weak repulsion of

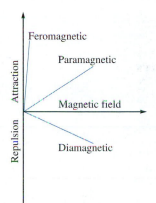

Figure 5.21 The responses of different materials to a magnetic field reveal pairing of electrons in the bonds. Diamagnetic materials have all electrons paired and are weakly repelled by a magnetic field. Paramagnetic materials have unpaired electrons and are attracted to the field. Compared with diamagnetic and paramagnetic materials, ferromagnetic materials are much more strongly attracted to a magnetic field.

diamagnetic materials by a magnetic field (Figure 5.21). Both diamagnetic and paramagnetic effects are insignificant compared to the strong attraction of ferromagnetic materials for a magnetic field. Ferromagnetism results from a large-scale alignment of the magnetic fields of the components (atoms or molecules). Relatively rare, but technologically important, elements, alloys, and compounds are ferromagnetic. Of the elements, only iron, cobalt, nickel, and the rare earth elements are ferromagnetic. Several compounds with oxygen, called oxides, are ferromagnetic; these are often found in important devices. For example, CrO_2 and Fe_2O_3 are used in recording tapes, FeCo is used in "metal" tapes, and FeP is used in video recording tapes. These materials can both be "written" (magnetic fields aligned) and read (the alignment sensed) with magnetic probes, called read heads.

Not all second-period diatomic molecules have the same molecular-orbital energy-level diagram as molecular oxygen. The ordering of the σ_{p_z} and the (π_{p_x}, π_{p_y}) pair may switch. This switch provides insight into the effect of confining electrons. Notice that the σ_{p_z} and the σ_s orbitals have similar symmetry with respect to the internuclear axis (Figure 5.22). That is, electrons in these two orbitals are located in very similar regions of space. Electrons repel each other, so the result of this confinement is that the σ_s orbital is stabilized and the σ_{p_z} orbital is destabilized relative to the ordering shown in Figure 5.19. Stabilization of the σ_s orbital has little effect on the molecule, since it is already the most stable of the bonding orbitals. However, destabilization of the σ_{p_z} orbital can affect the bonds formed. The extent of destabilization depends both on the energy separation of the original atomic 2s and 2p orbitals and on the bond length. The energy separation of the atomic s and p orbitals increases from left to right across the period. This separation tends to diminish the destabilization of the σ_{p_z} orbital as the σ_s orbital is drawn closer to the two nuclei.

The bond length demonstrates a different trend (Figure 5.23). The length is linked to the bond order. Considering the second-period homonuclear diatomic molecules, the bond order increases from a single bond for B_2 to a double bond for C_2 to a triple bond for N_2. Indeed, the N≡N triple bond is the strongest elemental bond known. As the antibonding orbitals fill, the bond order decreases from the triple bond of N_2 to a double bond for O_2 to a single bond for F_2. The bond length is inversely related to the bond-order: Bond length decreases from B_2 through N_2, then increases again through F_2.

Thus, from boron through nitrogen, the effect of the increased energy separation of the atomic orbitals is counterbalanced by increasing confinement to keep the σ_{p_z} orbital higher in energy than the (π_{p_x}, π_{p_y}) pair (Figure 5.24). After nitrogen, the bond length increases, removing this counterbalance, and the σ_{p_z} orbital becomes lower in energy than the (π_{p_x}, π_{p_y}) pair. The molecular-orbital energy-level diagram therefore looks like that of O_2 (Figure 5.19). A similar trend is not observed in the third period and beyond because the bond lengths are all longer.

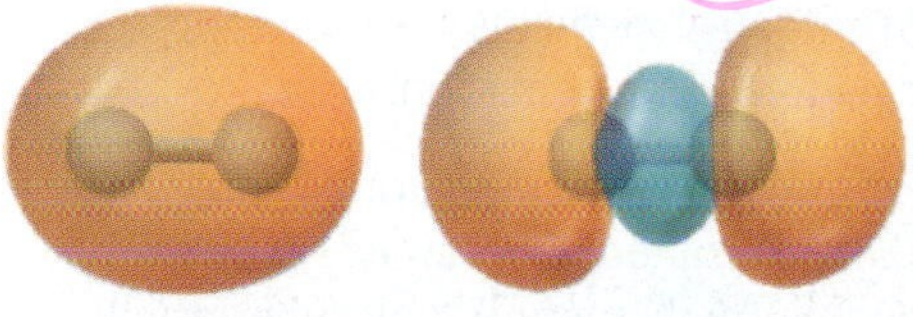

CONCEPT QUESTION   The molecule C_2 is an important species in the interstellar medium. Is C_2 paramagnetic or diamagnetic? ∎

Figure 5.22 The σ_s bonding orbital and the σ_{p_z} orbital are located in a similar region of space and have the same symmetry with respect to the molecule. Electrons in these two orbitals repel each other, destabilizing — that is, raising the energy of — the less stable σ_{p_z} orbital. The extent of this destabilization depends on the energy difference between the atomic s- and p-orbital electrons and the size of the internuclear region (the bond length). The orbitals shown are the σ_s and the σ_{p_z} molecular orbitals for O_2 at the bonding distance.

Heteronuclear diatomic molecules differ from homonuclear molecules in two important respects. First, the atomic orbitals differ in energy, with the more electronegative element having more stable (i.e., lower-energy) orbitals (Figure 5.25). Second, due to the greater stability of the atomic orbitals of the more electronegative element, the bonding orbitals are spatially localized nearer to the more electronegative element (Figure 5.26). The molecular-orbital energy-level

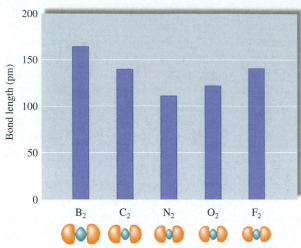

Figure 5.23 The bond length of second-period p-block homonuclear diatomic molecules decreases as the bond order increases from B through N, then increases again as the antibonding orbitals fill. The σ_{p_z} bonding orbital is also pictured. The increasing atomic s- and p-orbital energy separation from B through C stabilizes the σ_{p_z} molecular orbital, but the stabilization is counterbalanced by the decreasing bond length that confines the σ_s and the σ_{p_z} molecular orbital to a small volume. The result is that the σ_{p_z} molecular orbital is less stable than the (π_{p_x}, π_{p_y}) pair. Increasing the bond length from N_2 through F_2 diminishes the counterbalance, and the σ_{p_z} molecular orbital becomes more stable than the (π_{p_x}, π_{p_y}) pair. Along with the increased stabilization, the size of the interatomic region in the σ_{p_z} molecular orbital increases.

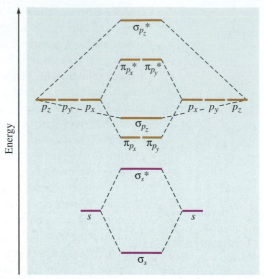

Figure 5.24 The molecular-orbital energy-level diagram for second-period diatomic molecules involving Li, Be, B, or C features a destabilized σ_{p_z} orbital, with the σ_{p_z} orbital being less stable than the (π_{p_x}, π_{p_y}) pair.

diagram shows the shift of electron density toward the more electronegative element by placing the energy of the bonding orbitals closer to the energy of the atomic orbitals of the more electronegative element. The energy of the antibonding orbitals is closer to the energy of the atomic orbitals of the less electronegative elements.

Summarizing the general features of formation of molecular orbitals:

- Orbitals are conserved. The number of molecular orbitals formed equals the total number of atomic orbitals considered. A pair of atomic orbitals generates one bonding (low-energy) and one antibonding (high-energy) molecular orbital.
- Electrons are conserved. For every electron in the atomic orbitals, there is one electron in a molecular orbital.
- Electrons fill molecular orbitals, two with opposite spin per orbital, filling the lowest-energy orbital first. For orbitals of equal energy, follow Hund's rule: half fill the orbitals with parallel spin electrons, then pair any additional electrons.

More Than Two Atoms

In the electron sea model of metals, atoms become ions by donating electrons to the electron sea. The orbitals of the electron sea consist of multiple waves extending over a long range in the solid. Each wave envelopes a large number of atoms. Every atom in the solid contributes atomic orbitals to the extended orbitals as well as electrons to

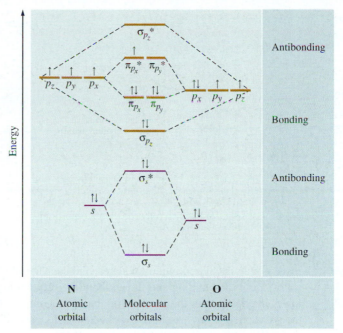

Figure 5.25 The molecular-orbital energy-level diagram for NO illustrates the effect of differing electronegativities on bonding. The bonding molecular orbitals are closer in energy to those of the more electronegative element, and the antibonding orbitals are closer to those of the less electronegative element.

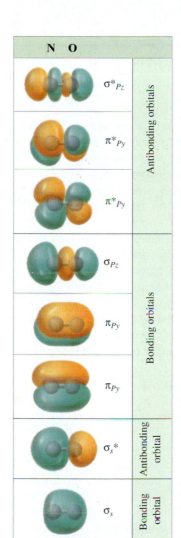

Figure 5.26 The molecular orbitals of NO show that the portion of the orbital in the vicinity of the oxygen atom is more compact and more concentrated than that near to the nitrogen, reflecting the greater electronegativity of oxygen.

the sea. To describe the electronic properties of materials such as metals, the diatomic molecular-orbital model needs to be extended to many atoms—something on the order of a mole of atoms. Constructing orbitals for solid materials follows the same procedure as constructing molecular orbitals, but is somewhat more involved due to the large number of atoms involved. The first step is to determine the atomic orbitals to be included. The following discussion illustrates the need to include all atomic orbitals that are close in energy. Due to the large numbers involved, it is helpful to use a flowchart (Figure 5.27) to track orbitals and electrons.

Consider Be, a typical metal, and follow the flowchart from left to right. Forming the metallic solid from a set of isolated atoms begins by noting the electron configuration: $1s^2 2s^2$. To make the example more concrete, think about assembling the solid from one mole, N_A, of Be atoms (Figure 5.27, the atomic panel on the left). The occupied valence atomic orbitals (in this case, just the atomic $2s$ orbital) certainly need to be included because they are where the electrons for the electron sea originate. A Be atom contains one atomic $2s$ orbital. One mole of Be atoms thus contains one mole of atomic $2s$ orbitals. Orbitals are conserved, so blending these atomic orbitals into molecular orbitals generates one mole of molecular orbitals. These molecular orbitals split into two sets (Figure 5.27, the solid panel to the right): One-half are bonding orbitals and one-half are antibonding. Thus there are $\frac{1}{2}N_A$ bonding orbitals and $\frac{1}{2}N_A$ antibonding orbitals.

With a large number of orbitals, the energy difference between successive orbitals within the bonding set is very small. Similarly, the energy difference between successive orbitals within the antibonding set is small. The small energy difference

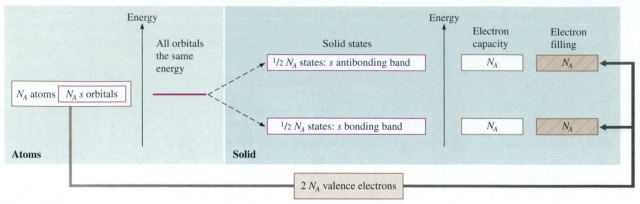

Figure 5.27 Processing flow from atoms to solid for Be with only the valence *s* orbital included.

Band A closely spaced set of energy levels, usually in a solid material. Small-energy difference between successive orbitals within a bonding set.

results in a virtual continuum of states, and each set of orbitals is said to form a **band** of states. Both the bonding and the antibonding orbitals form a band. The bonding band is lower in energy than the antibonding band. In contrast, the N_A atomic orbitals are all of the same energy, but are shown as a rectangle to aid in illustration.

All orbitals—in atomic, diatomic, or larger molecular—can hold a maximum of two electrons. With $\frac{1}{2}N_A$ states, each with the capacity to hold two electrons, the capacity of each band is therefore one mole of electrons.

The number of electrons to fill into these orbitals (i.e., bands) is determined by the number of electrons in the Be atom's $2s$ orbital. Since Be has two electrons in the $2s$ orbital, one mole of Be atoms contains two moles of electrons (Figure 5.27, purple boxes). Each band has the capacity to hold only one mole of electrons, so both the bonding and the antibonding bands are filled. Filling is indicated as cross-hatching in the band-energy diagram (Figure 5.27). With the bonding *and* antibonding orbitals filled, the bond order is zero, suggesting that the solid is not stable. In reality, however, Be forms a stable, metallic solid. This indicates that more orbitals are involved in the bonding of the solid.

CONCEPT QUESTION   What orbitals are in the second shell of Be? ■

Every atom has electron states—orbitals—that are unoccupied. Like vacant apartments, these unfilled states can accommodate electrons. The energy of the atomic orbitals is modified when an atom interacts with other atoms. For example, the atomic *p* orbitals, which are all of the same energy in the isolated atom, have different energies as one orbital faces the second atom and the other two do not. A similar modification occurs when a large number of atoms come together. Specifically, for Be, there are three unfilled *p* orbitals in the same shell as the filled *s* orbital. In an isolated Be atom, these $2p$ orbitals are higher in energy than the $2s$ orbital and are unoccupied. However, the *p* orbitals are not much higher in energy—this is why the σ_s orbital and the σ_{p_z} orbital interact. When a Be atom encounters another Be atom, both the orbitals and the energy separation change. Elemental Be is a stable metal, so the *p* orbitals in the solid, which also form a band, must participate in the metal bonding. Further, for the *p* orbitals of the solid to be occupied by electrons in preference to the antibonding *s* orbitals of the solid, the *p band of the solid must be lower in energy than the s antibonding band* because electrons go into the lowest-energy state available.

There are three times as many *p* orbitals as *s* orbitals, so the bonding band formed from the atomic *p* orbitals (Figure 5.28) consists of three times as many molecular

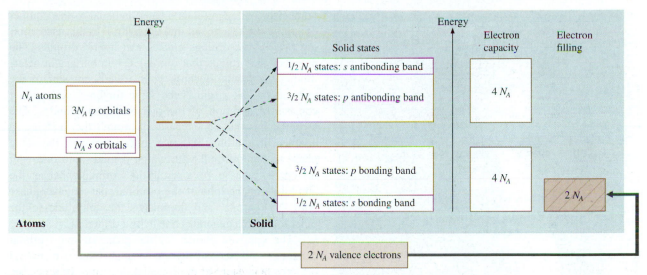

Figure 5.28 Processing flow from a mole of atoms to the solid for Be, including the valence 2*p* orbitals.

orbitals as the *s* bonding band. Similarly, the *p* band of the solid can hold three times as many electrons as the *s* band. Including the *p* band results in a bond order that is nonzero. Consequently, the solid is stable.

CONCEPT QUESTION   To conduct electricity, electrons entering a metal sample must find an unoccupied or half-filled orbital — the Pauli exclusion principle in action. What orbitals are available for conduction in solid Be? ■

Hybrid Orbitals

Hybrid orbital An orbital consisting of a combination of atomic orbitals.

The band structure (Figure 5.28) applies to all elements that have valence *s* and *p* orbitals, which includes the important class of semiconductors that adopt the diamond structure. Each carbon atom in diamond (Figure 5.29) is surrounded by four other carbon atoms, all of which are at the same distance. Forming four bonds requires four molecular orbitals, and four molecular orbitals require four atomic orbitals. However, there are only three atomic *p* orbitals. Further, the angle between the bonds is 109.5° while the *p* orbitals are separated by 90°. Indeed, both the number of equivalent bonds and the angle between the bonds suggest that the orbitals connecting carbon atoms in diamond are different from those in an isolated carbon atom.

Pondering this issue, Linus Pauling (Figure 5.30), a Nobel Prize winner in chemistry, proposed that in the process of forming bonds, the valence orbitals in an atom mix together. Just as two ripples on a pond form a different shape when they converge, so the valence orbitals form a shape that is a melding of the separate shapes. Pauling dubbed the result a **hybrid orbital.**

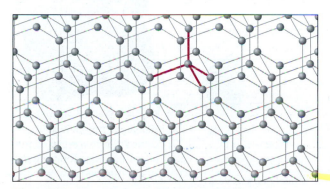

Figure 5.29 The diamond lattice features carbon atoms surrounded by four nearest neighbors in a tetrahedral arrangement.

As an example, consider the combination of an *s* and a single *p* orbital to form a mixed orbital, an *sp* hybrid (Figure 5.31). As with formation of molecular orbitals, the

number of hybrid orbitals created is equal to the number of atomic orbitals mixed into the hybrid. In this case, mixing *two* atomic orbitals creates *two* hybrid orbitals. The phase of the orbitals that mix is very important in determining the result. Because both orbitals are part of the same atom, the origin is the same for both, and mixing consists of addition of the two orbital amplitudes. The *s* orbital has the same phase all around the nucleus. In contrast, a *p* orbital has two lobes: one with the same phase as the *s* orbital with which it mixes and one with the opposite phase. As a result, the *s* orbital enhances the amplitude of the lobe of the *p* orbital that matches phase with the *s* orbital and diminishes the amplitude of the other lobe. This generates one hybrid orbital.

The opposite choice for the phase of the *s* orbital generates the second hybrid orbital. The two hybrid orbitals differ only in their spatial direction. In this case, the orbitals are separated by 180°. In general, the angle between hybrid orbitals can be predicted based on electrostatic repulsion. Negatively charged electrons occupy the orbitals and are most stable when the separation between negative regions is maximized. For two orbitals, the electron separation is maximized if the orbitals are on opposite sides of the nucleus, again predicting a 180° separation for the hybrid orbitals. Each bond requires at least one hybrid orbital, so an atom with two *sp* hybrid orbitals can bond to two other atoms. The three atoms—the atom containing the hybrid orbitals and the two to which it bonds—then form a straight line with the bonds separated by 180°. The result is a linear molecule.

To form four equivalent orbitals, as in diamond, requires the *s* and all three *p* orbitals. To maximize the sep-

Figure 5.30 Linus Pauling received the 1954 Nobel Prize in chemistry for his pioneering work on the nature of the chemical bond.

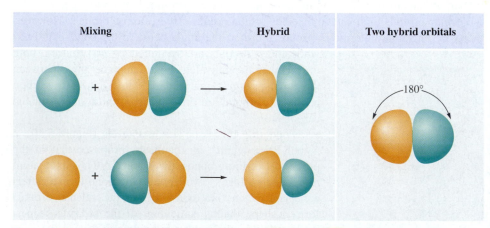

Figure 5.31 Formation of a hybrid orbital from *s* and *p* orbitals. The orbital phase is very important. The lobe of the *p* orbital matching that of the *s* orbital is augmented, while the out-of-phase lobe is diminished. Both combinations occur in the mixing process to produce two *sp* hybrid orbitals. The two *sp* hybrid orbitals are oriented 180° apart.

Figure 5.32 The electron density of the *sp*³ hybrid orbitals forms a tetrahedral shape featuring four regions of electron density.

aration of electrons in these hybrid orbitals, and provide bonding to four other atoms, the hybrid orbitals point to the corners of a tetrahedron (Figure 5.32), which explains the angle of 109.5° (instead of 90°).

APPLY IT   Take four round balloons, blow them up, and tie them into a cluster with the ends tightly intertwined. Observe the spatial arrangement of the balloons.

A hybrid of one s and three p orbitals is found frequently in compounds containing carbon, and is called an sp^3 — pronounced "s p three," hybrid in reference to the single s and three p orbitals.

CONCEPT QUESTIONS   What is the valence electron configuration of C? Which orbitals hybridize to form *three* bonds to C? What is the angle between these three hybrid orbitals? ■

Elemental Semiconductors and Insulators

Consider a large number, say N_A, carbon atoms in diamond (Figure 5.33). Each atom forms a set of four sp^3 hybrid orbitals, so there are $4 \times N_A$ hybrid orbitals. These hybrid orbitals mix to form two bands of allowed states: a bonding band with $2 \times N_A$ states and an antibonding band with $2 \times N_A$ states. Each band, bonding and antibonding, can hold two electrons per state, or $4 \times N_A$ electrons. How many valence electrons does a sample of N_A carbon atoms have? There are four valence electrons in each carbon atom — exactly enough to fill the bonding band leaving the antibonding band empty.

The analogous energy-level diagram for a solid represents the band of states with a rectangle. Electron filling is represented by cross-hatching the rectangle to the level of filling (Figure 5.34). The vertical axis is energy: the bonding band is lower in energy than the antibonding band. Electrons enter into these states, filling the lowest-energy one first. Thus, electrons go into the bonding band prior to the antibonding band. For diamond and other solids where the valence electrons just fill the bonding band, the bonding band is called the **valence band.** With all valence band states filled, an electron added to the solid, as in conduction, must enter the higher-energy, antibonding band.

Valence band The band of states filled with valence electrons in a semiconductor.

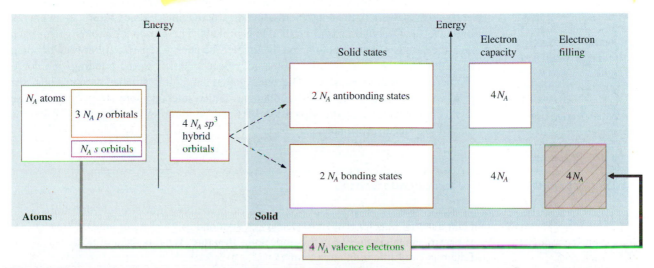

Figure 5.33   Processing flow from atoms to covalent solid for diamond.

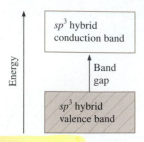

Energy

| sp^3 hybrid conduction band |

Band gap

| sp^3 hybrid valence band |

Figure 5.34 The bands in diamond.

Conduction band An incompletely filled band in a solid. The empty or nearly empty band of states next-highest in energy to the filled band in a semiconductor.

Band gap A range of energies for which there are no allowed electronic states in a solid. Energy difference between the valence band and the conduction band in a semiconductor.

Table 5.2 The Relationships Among the Color, Energy, and Wavelength of Visible Light

Color	Infrared	Red	Yellow	Green	Cyan	Blue	Magenta	Ultraviolet
Energy (eV)	1.65	1.91	2.22	2.53	2.76	2.89	3.10	3.54
Wavelength (nm)	750	650	560	490	450	430	400	350

This antibonding band is called the **conduction band.** Solids with a filled valence band and an empty conduction band are either insulators or semiconductors. The distinction between insulator and semiconductor depends on the energy difference between the valence band and the conduction band, an energy referred to as the **band gap.** The band-gap energy is the bonding/antibonding energy separation in the solid. For diamond this band gap is large, so diamond is an excellent electrical insulator.

Why is Diamond Transparent, Silicon Opaque?

Diamond is a transparent and colorless gem. The next larger member of Group IVA, silicon, is opaque and either black or very reflective if highly polished. Why are these two elements from the same group so different? A solid appears clear and colorless because it transmits all colors of visible light, absorbing no visible photons. To absorb no visible photons, the energy of a photon in the visible region must be insufficient to lift an electron from the low-energy valence band to the higher-energy conduction band. The minimum energy required to lift an electron from the valence band to the conduction band—the band gap—is an important characteristic of many solids. The band-gap energy in diamond, 5.4 eV, is beyond the energy of visible light (Table 5.2).

Si and C are members of the same group, so the same band diagram and the same filling applies to silicon as applies to diamond. The opacity of Si indicates that all visible light is absorbed. The band-gap energy must, therefore, be less than that of visible light. In Si, this band-gap energy is 1.11 eV.

Why is the band gap for Si so much smaller than that for diamond? Think about how the band gap arose. When atoms are far apart, the electron orbitals (the waves) do not overlap (do not interact). No interaction means no energy separation between the in-phase and out-of-phase combinations. When the atoms get close enough that their electron waves overlap, the constructive and destructive interference results in an energy difference for the bonding and antibonding orbitals. The closer the atoms get to each other, the stronger the interaction and the greater the energy separation between bonding and antibonding orbitals. Smaller atoms get closer together, so the bonding–antibonding energy separation is larger: The band gap is larger for smaller atoms. Thus, the orbitals for two carbon atoms in the diamond lattice interact more strongly than those of silicon. Generation of the separation between bonding and antibonding orbitals is referred to as the **overlap rule.**

Overlap rule The energy separation between two electronic states that are the same or close in energy in separate atoms depends on the spatial overlap of the two orbitals.

WORKED EXAMPLE 5.1 *Interactions and Density*

The atomic density of a material reveals information about how strongly the atoms in that material interact. The mass density of diamond is 3.513 g/cm^3 and of Si is 2.33 g/cm^3. Both crystallize in the same solid structure. What is the atomic density of diamond and silicon in atoms/nm^3?

Plan

- Unit analysis: $\dfrac{\text{g}}{\text{cm}^3} \times \dfrac{\text{mol}}{\text{g}} \times \dfrac{\text{atoms}}{\text{mol}} \times \left(\dfrac{\text{cm}}{\text{nm}}\right)^3 = \dfrac{\text{atoms}}{\text{nm}^3}$

$\Rightarrow$ Divide by atomic mass, multiply by N_A, and convert the volume units.

Implementation

- C: $\dfrac{3.513 \text{ g}}{\text{cm}^3} \times \dfrac{1 \text{ mol}}{12.011 \text{ g}} \times \dfrac{6.022 \times 10^{23} \text{ atoms}}{\text{mol}} \times \left(\dfrac{1 \text{ cm}}{10^7 \text{ nm}}\right)^3 = 176 \text{ atoms/nm}^3$

  Si: $\dfrac{2.33 \text{ g}}{\text{cm}^3} \times \dfrac{6.022 \times 10^{23} \text{ atoms}}{\text{mol}} \times \dfrac{1 \text{ mol}}{28.0855 \text{ g}} \times \left(\dfrac{1 \text{ cm}}{10^7 \text{ nm}}\right)^3 = 50.0 \text{ atoms/nm}^3$

The atomic density of Si is only about one-third that of diamond. Because the solid structure is the same for both materials, the lower density of solid Si compared to diamond indicates that Si atoms are farther apart. Hence the electron waves have a much weaker interaction—a smaller overlap—and this weaker interaction results in a smaller band gap.

See Exercise 34.

Worked Example 5.1 shows that the atomic density of diamond is much greater than that of Si, consistent with a stronger interaction for smaller atoms. Succeeding elements in Group IVA are progressively larger, and both Ge and α-Sn crystallize in a diamond structure. These larger atoms have correspondingly smaller band gaps (Figure 5.35). The band gap is directly related to the strength of interaction.

CONCEPT QUESTION The density of α-Sn is 5.769 g/cm^3. Compare the atomic density of α-Sn to that of Si. Is the relative atomic density consistent with the band gaps of these two materials? ■

Mixed-Valence Semiconductors

Look at the power indicator on your computer monitor. The green color is due to the light emitted by a semiconductor composed of Ga and P. Locate Ga and P in the periodic table. Ga is a Group IIIA element with three valence electrons, and P is a Group VA element with five valence electrons. The mixed material, GaP (Figure 5.36), forms a mixed-valence solid. In GaP, four P atoms surround each Ga atom and four Ga atoms surround each P atom. This variation on the diamond structure is called **zinc blende.**

Zinc blende Variation on the diamond structure composed of two elements; every atom of one type is surrounded by four atoms of the second type.

Both Ga and P have s and p valence-electron orbitals, and all four bonds to each atom in this mixed solid are equivalent. Both Ga and P are sp^3 hybridized. The band structure is the same as that of diamond. Filling electrons into these bands is slightly different, however, due to the different valences for the two elements involved (Figure 5.37). Ga has three valence electrons and P has five, so the electrons from Ga fill $\frac{3}{8}$ of the valence band and the electrons from P fill the remaining $\frac{5}{8}$. With an equal number of Ga and P atoms, the valence band is just filled and the conduction band is empty. In the final analysis, GaP differs from diamond and silicon only in the energy of the band gap.

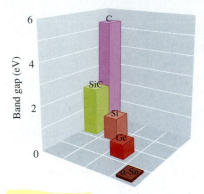

Figure 5.35 Band gap in Group IVA diamond-structure solids. The visible region of the spectrum corresponds to 1.65 – 3.54 eV. Going down the group, the atoms get larger, the separation is larger, the interaction is weaker, and the band gap is smaller.

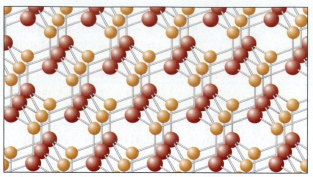

Figure 5.36 The mixed semiconductor GaP consists of a diamond structure in which half of the atoms are Ga (the larger atoms) and the other half are P (the smaller atoms). Each Ga atom is surrounded by four P atoms, and every P atom is surrounded by four Ga atoms. This is called a zinc blende structure.

CONCEPT QUESTION Locate C, Si, Ga, As, and P on the periodic table. In which group is each element located? What is the relationship of the positions of P and As in the periodic table? Group and order these elements according to the number of valence electrons. ∎

The green color of a computer monitor power indicator light is related to the band gap in GaP semiconductors. Imagine an electron being lifted from the valence band to the conduction band, leaving a space behind in the valence band. This space

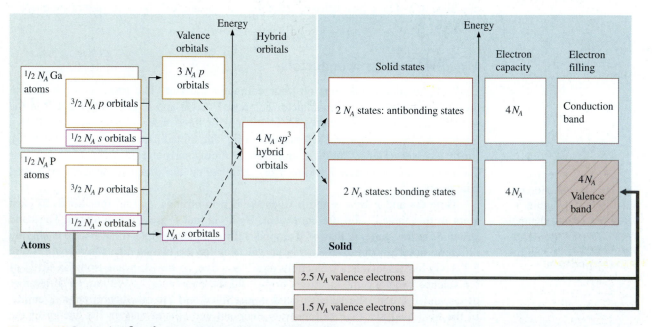

Figure 5.37 Processing flow from atoms to covalent solid for GaP.

Hole, h^+ Space in the valence band of a semiconductor.

is called a **hole.** Removing an electron from one of the atoms leaves that atom with a positive charge, so it is as though the electron has been replaced with a positive charge on the atom. Holes are often labeled as h^+ to indicate this positive charge. Having a hole in the valence band and an electron in the conduction band is an excited state for the GaP solid. When the electron returns to the valence band, the hole is filled and an energy equal to the band-gap energy is given off. This energy emission is very similar to that seen with the hydrogen atom: After being excited (e.g., by an electric discharge), the hydrogen atoms return to the ground state, emitting energy in the form of light in the process. These emission energies provide a way to probe the structure of the electronic states in hydrogen and other atoms. The energy of the light emitted by the semiconductor when it returns to the ground state is similarly revealing about the energy of the band gap. The green color of GaP indicates that this gap is between 2.23 and 2.5 eV.

WORKED EXAMPLE 5.2   *Band Gaps in III–V Semiconductors*

In Group IVA's diamond-structure solids, the smaller elements interact more strongly and thus have a larger band gap. Extending this logic to mixed-valence semiconductors, predict whether the band gap in GaAs is larger or smaller than that of GaP. Is the band gap in AlP larger or smaller than that in GaP?

Plan

The band gap depends on interaction, and interaction depends on size.
- Determine the size of As compared with P.
- Determine the size of Al compared with Ga.

Implementation

- P is in the third period, and As is in the fourth period. Therefore, As is larger. Larger atoms interact less strongly, so the band gap in GaAs is smaller than that in GaP.
- Al is in the third period, and Ga in the fourth period. Therefore, Al is smaller than Ga and the band gap in AlP is larger than that in GaP.

As shown in Figure 5.38, the band gaps in III–V semiconductors are inversely related to the sizes of the elements in the solid.

See Exercises 37–46.

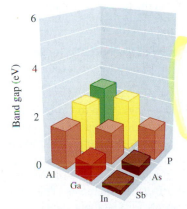

Larger atoms do not interact as strongly as smaller atoms, so the band gap in GaAs (1.35 eV; 919 nm) is lower in energy than that in GaP (2.24 eV). While GaP emits green light, the electromagnetic radiation emitted from GaAs is lower in energy than the visible region of the spectrum. More precisely, it is in the infrared region.

CONCEPT QUESTION   What happens to the band-gap energy when only some of the P atoms in GaP are replaced with As atoms? ■

Figure 5.38 The variation in the band gaps in binary Group III–V semiconductors reflects the dependence of band-gap energy on atomic interaction, which is inversely related to size.

The band gap of the mixed solid SiC (Figure 5.35) is intermediate between that of Si and that of C, reflecting an averaging of the interaction strength. Similarly, Ga with a mixture of As and P has a band gap that is intermediate between the infrared band gap of GaAs and the green band gap of GaP. For example, a semiconductor composed of 40% P and 60% As emits red light. The notation for this composition is $GaP_{0.40}As_{0.60}$. The fractional subscripts emphasize the 1:1 ratio of Ga to the Group VA elements.

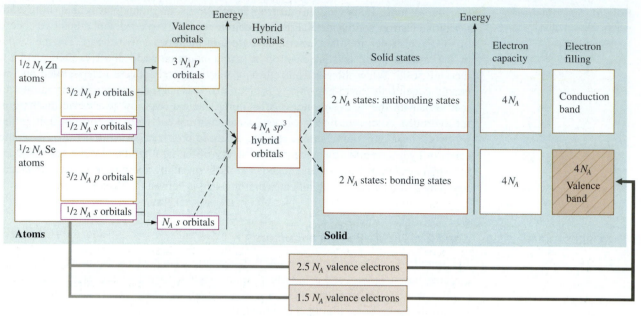

Figure 5.39 Processing flow from atoms to covalent solid for ZnSe.

CONCEPT QUESTION A $GaP_{0.40}As_{0.60}$ semiconductor emits red light, and a $GaP_{1.00}As_{0.00}$ semiconductor emits green light. Predict the color emitted by a semiconductor with the composition $GaP_{0.65}As_{0.35}$. ∎

Group IVA elements have four valence electrons; Group IIIA and VA elements have three and five valence electrons, respectively. Thus, the average number of valence electrons is four. This pattern suggests a chemical version of chopsticks. As long as the binary combination has an average of four valence electrons per atom and crystallizes in the zinc blende structure, the solid has a filled valence band and empty conduction band. These solids are semiconductors. An example of a Group II–VI combination is ZnSe (Figure 5.39).

CONCEPT QUESTIONS What is the valence configuration of Zn? Of Se? What are the valence orbitals in each? The solid, ZnSe, is a semiconductor. What color light does ZnSe emit? ∎

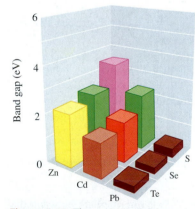

Figure 5.40 The variation in the band gaps in binary II–VI semiconductors reflects the link between band gap and electron confinement. The bonding electrons are more confined for smaller atoms and for combinations with a greater electronegativity difference.

The chopsticks tune works in materials ranging from ZnSe to GaAs to Ge. The band gap for Ge is 0.67 eV, deep in the infrared region (Figure 5.35). GaAs also has a band gap in the infrared region but its band-gap energy, 1.35 eV, is higher than that of Ge. The band gap of ZnSe, the next combination in the chopsticks theme, is 2.58 eV (Figure 5.40), which is in the green region of the visible spectrum. Ge, Ga, As, Zn, and Se are all fourth-period elements. The average size of Ga and As (126 pm and 120 pm, respectively) is about the same as the size of Ge (122 pm). Similarly, the average size of Zn and Se (126 pm and 116 pm, respectively) is nearly the same as the size of Ge. Thus, the bond distances in Ge, GaAs, and ZnSe are all essentially the same. With approximately the same approach distance, the effect of separation on interaction is about the same. As a result, understanding the band-gap energy variation from Ge to GaAs to ZnSe requires a parameter beyond size.

The other parameter that affects atomic interaction and therefore band-gap energy is electron confinement. Electron confinement is related to the electronegativity difference between the two elements involved. In bonding, electronegativity difference is seen as a shift from covalent to ionic bonding. A homonuclear molecule has no electronegativity difference between the atoms, and the electron pair envelopes both nuclei equally. In ionic bonding, the electron pair is localized on the more electronegative element. Between these extremes, in a heteronuclear molecule, the more electronegative element takes a larger share of the electron pair, thus localizing and confining the electron pair to the region nearer to the more electronegative element. This confinement-attraction to the electronegative element makes the bonding orbital more stable—lower in energy—than a less confined electron pair. Similarly, the confinement raises the energy of the antibonding orbital relative to a less confined bonding pair. In a solid, a larger energy difference between the bonding and antibonding orbitals translates into a larger band gap.

The electronegativity difference between Ga and As is 0.37, and the difference between Zn and Se is more than twice that, 0.9. Consistent with confining the bonding electrons to a smaller space, which produces to a larger band gap, the band-gap energy in ZnSe is 2.58 eV, almost blue light and nearly double the energy of GaAs. Predicting relative band gaps is based on two factors:

- The band-gap energy increases with interaction, and interaction increases as the size of the component elements decreases.
- The band-gap energy increases as the electronegativity difference between the component elements increases.

Diodes

Conductivity in semiconductors is very sensitive to minute levels of defects such as substitutional impurities. For example, pure Si is a semiconductor. The ideal model for this solid features every atom on its prescribed site (Figure 5.41). Reality is not

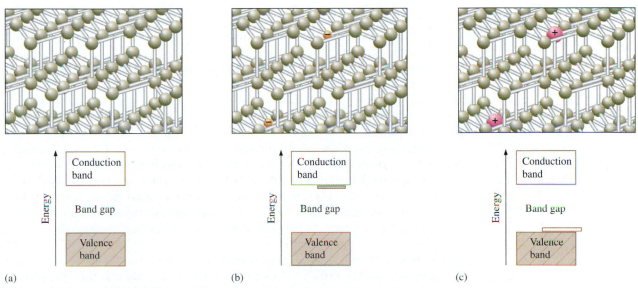

(a)   (b)   (c)

Figure 5.41   (a) Pure silicon features a filled conduction band and an empty valence band. (b) Doping with P adds electrons creating an *n*-type semiconductor. (c) Substitution of Al for Si leaves the valence band incompletely filled, creating a *p*-type semiconductor.

Doping Addition of a low concentration of a second substance into an otherwise pure solid material.

n-type semiconductor
A semiconductor with conduction characterized by the motion of electrons.

p-type semiconductor
A semiconductor with conductivity characterized by the movement of positively charged holes.

p–n junction diode Region of transition from a semiconductor with vacancies in the valence band to one with electrons in the conduction band.

so ideal. In particular, any sample of Si will have a few substitutional impurities. Substitutional impurities are referred to as *dopants,* and deliberate addition of substitutional atoms is called **doping.** Suppose that a few P atoms replace a few Si atoms (Figure 5.41). P is about the same size and electronegativity as Si, so the lattice maintains the diamond structure with the attendant band structure. In particular, the band gap is unaltered. However, P brings with it one more valence electron than Si. This additional electron does not participate in the bonding, so it remains localized near the P atom and its energy is near the conduction-band energy. Only a small added energy, easily supplied by thermal energy, is required for this loosely bound electron to enter the conduction band. Once in the conduction band, the electron is free to move if a voltage is applied across the solid. Every P atom brings one "extra" electron, and this impure Si shows conductivity in proportion to the number of P atoms substituted for Si atoms. With doping levels on the order of 10^{-3} to 10^{-10} atomic percent, the doped material remains a semiconductor. When doping introduces an extra electron, conduction is by electron movement and produces an **n-type semiconductor.**

A Si sample can also very easily have some Al impurities. Indeed, both Si and Al are very abundant on earth. Al has only three valence electrons, so when Al substitutes for Si, one of the bonds between the neighboring Si atoms and Al is missing an electron. It takes little energy for an electron to jump from a nearby Si–Si bond, filling in for the missing electron and leaving the nearby Si atom with a positive charge—a hole. This positive charge is free to move through the solid if a voltage is applied. Conduction is by movement of this *positive* charge, and the material is called a **p-type semiconductor.**

CONCEPT QUESTION Semiconductor-grade Si is now routinely produced that is 99.9999999 atomic percent pure. The density of Si is 2.33 g/cm³. How many impurity atoms per cubic centimeter are found in semiconductor-grade Si? ■

One of the major challenges to the early semiconductor industry was maintaining sufficient cleanliness that the amount of impurity atoms and, therefore, conduction could be carefully controlled. Modern devices feature an impurity atom for every 10^3 to 10^{10} host atoms—very high purity, indeed. Nonetheless, there are 10^{13} to 10^{20} impurity atoms per cubic centimeter.

A mixed-valence semiconductor offers a ready method for addition of electrons or holes. For example, in GaP, if the mixing is not perfect, and a few extra P atoms are included, then there are a few extra electrons and the material is an *n*-type semiconductor. Conversely, a few extra Ga atoms leave the valence band incompletely filled, giving a *p*-type semiconductor. Whether created by imperfect mixing, by doping, or by incomplete purification, juxtaposing a *p*-type semiconductor with an *n*-type semiconductor creates a **p–n junction,** also known as a **diode.** Electrons in the *n* side have a tendency to fill lower-energy holes on the *p* side, a phenomenon called electron drift. Without an added field, the drift does not continue for long, as charge builds up on each side. However, if an external circuit is attached across the junction, and the external circuit creates a potential that also moves the electrons toward the *p* side, then the junction conducts (Figure 5.42).

The energy of an electron changes near the *p–n* junction. On the *n* side, the electron energy is that of the conduction band. On the *p* side, it is the valence-band energy. An electron that crosses the junction thus loses the energy difference at the junction. Alternatively, the electron can be thought of as combining with the

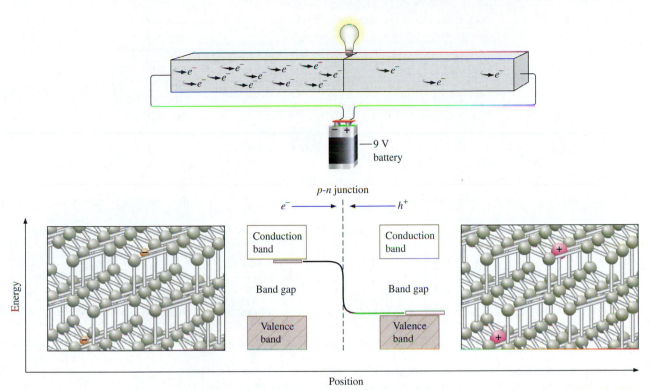

Figure 5.42 Forward bias $p - n$ junction. On the n-type side, the electron is in the higher-energy conduction band. On the p-type side, the electron can occupy the lower-energy valence band because it has vacancies. At the junction, the electron emits the energy difference, which appears as light.

hole, h^+, on the p side. The negative electron annihilates the positive hole, releasing energy

(5.2) $e^- + h^+ \longrightarrow$ energy

If the energy is emitted as light,

(5.3)   Energy $= h\nu = hc/\lambda$

An external circuit injects electrons into the n-type side, pushing electrons to the junction. As electrons cross the $p-n$ junction, the emitted energy appears as light. Pushing electrons toward lower energy is termed **forward bias.**

Forward bias Application of a potential to a p-n junction with the negative pole on the n side and the positive pole on the p side. The conducting configuration.

CONCEPT QUESTION   What happens when the battery leads are reversed? ∎

Reverse bias Application of a potential to a p-n junction with the negative pole on the p side and the positive pole on the n side. The nonconducting arrangement.

Suppose that the external potential causes the electron to drift in the opposite direction. As the electron disturbance ripples up to the junction, it runs into a large energy barrier, electron movement is stopped, and no light is emitted. This is called **reverse bias** (Figure 5.43).

These $p-n$ junctions are present in many modern electronic instruments, such as computers, calculators, cars with computer-controlled electronic ignition systems, and solar energy panels. They are found anywhere that one-way conduction is required.

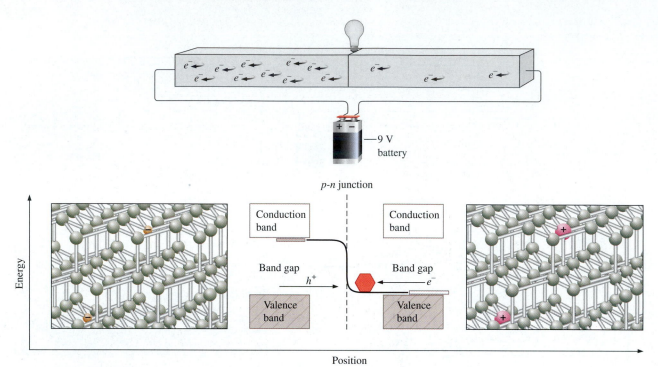

Figure 5.43 Reverse bias $p-n$ junction. On the p side, the electron energy is that of the valence band. At the junction with the n side, the electron needs additional energy equal to the band-gap energy to travel to the other end of the diode. If the battery supplies this amount of energy, the diode short circuits and burns out.

WORKED EXAMPLE 5.3 *Doping*

A semiconductor material needs to have elements added to it to make it somewhat conducting. In addition to Al, what elements could be used to dope elemental Si to make it a *p*-type semiconductor? In addition to P, what elements could be used to create an *n*-type semiconductor?

Plan

■ *p*-Type doping requires an element that is electron-poor compared to Si. These elements are found to the left of Si in the periodic table.

■ *n*-Type doping requires an element that is electron-rich compared to Si. These elements are found to the right of Si in the periodic table.

Implementation

■ For *p*-type semiconductors, the elements to the left of Si are Groups IA, IIA, and IIIA, and the transition elements. Numerous candidates exist, including Ga, B, Cu, and Zn.

■ An element with fewer electrons than Si is required for an *n*-type material. There are fewer choices for *n*-type doping than for *p*-type doping; however, any element to the right of Si in the periodic table is a candidate. These elements are in Groups VA, VIA, and VIIA, and include N, Br, and Sb.

See Exercises 48, 51 – 55.

WORKED EXAMPLE 5.4   *Tuning the Band Gap*

For a new computer application, you are asked to develop a semiconductor with a wider band gap than Si. Owing to economic factors, the new material must contain more Si than any other element. How would you design such a material?

Plan

To be a semiconductor, other components must come from Group IVA or 1 : 1 mole ratios from Groups III and V, or Groups II and VI.

■ A larger band gap requires a smaller size.
■ A larger band gap requires a larger electronegativity difference.

Implementation

■ First consider a pure element. To have a wider band gap than Si, the element must be smaller and in Group IV. Hence, only mixing with carbon will succeed.

■ Among binary mixtures, to ultimately form a ternary material with Si, choose elements in Groups III–V or II–VI. Since the band gap is to be larger, the elements must come from either the second or third period. The choices are Al with P or N, or B with P or N for Group III–V mixing. For Group II–VI mixing, the choices for the Group IIA elements are Mg or Be; for the Group VIA elements, they are O or S. Several fourth-period, transition element oxides are also potential choices due to the electronegativity difference between oxygen and the metal.

See Exercises 39 – 46.

A Blue Diode

Given a large number of elements from which to choose, there is great flexibility for tuning the band gap to a desired energy. For example, a current materials issue concerns making a reliable, inexpensive blue diode. To appreciate the motivation for making a blue diode, consider the following. CD players are read with infrared diode lasers that have an emission wavelength of 780 nm. The storage capacity of a CD is limited by this wavelength, because it determines the spacing between tracks as well as the spacing between pits on a track. CDs have a storage capacity of 650 MB (megabytes), enough to hold two sides of a music album with a little space left over. DVDs are read with a red diode laser, which has a wavelength of 650 nm. The storage capacity of a DVD is 4700 MB, enough to hold a short movie. A blue diode laser with a wavelength in the 400–450 nm range could read a device with a capacity of about 16,000 MB—enough to hold the entire *Star Wars* trilogy with the sound track.

How can a blue diode—one with a band gap of about 3 eV or a wavelength of 415 nm—be made? One strategy is to combine a Group IIIA element with nitrogen. Nitrogen is a very small, highly electronegative Group VA element, so it is a candidate for confining the bonding electrons and forming a large-band-gap semiconductor. Indeed, the band gap of GaN is just about right: 3.34 eV (Figure 5.44). This material is currently the subject of much attention as a potential commercial blue diode. The remaining unresolved issues appear to be chemical issues in the processing of the semiconductor—getting a uniform mixture of Ga and N that results in the zinc blende structure with few enough defects to operate

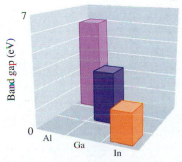

Figure 5.44 Band gap in binary nitride semiconductors.

reliably for thousands of hours. This requirement poses a challenge because nitrogen is sufficiently small and electronegative that it has a strong propensity to have a coordination number of only 3, so it tends to distort the crystal structure. Further, the N≡N triple bond is so favorable that N_2 tends to extrude from the material. The processing problems have yet to be solved.

CONCEPT QUESTIONS   Propose other two- or three-component diodes with the potential to emit in the blue region of the spectrum. Which periods and which groups contain the prime candidates? Are there potential processing issues connected with some otherwise viable alternatives? ∎

5.2   Resistivity and Conductors Versus Semiconductors

Conductivity is the flow of charge. In a metal, the most mobile charged particle is the electron. Imagine connecting a metal wire to a source of electrons. To enter the metal, the electron must find an orbital that is not already fully occupied (Pauli exclusion principle). In s-block elements, an additional electron is easily accommodated in the partly filled s and p bonding band. As a consequence, the s-block elements are all conductors. Group IVA elements have a filled valence band and an empty conduction band. The energy separation between these bands makes these elemental solids either semiconductors or insulators.

Figure 5.45 The range of resistivity of materials varies widely, reflecting the bonding in these materials. The resistivity of silicon is extremely sensitive to minute levels of impurities. (Inset) The resistivity of metals is sensitive to occupancy of the d orbitals.

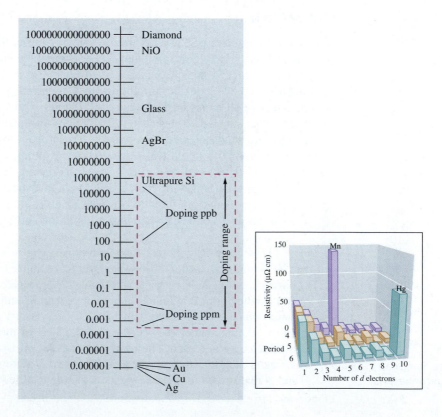

Band energy separation is reflected in the material's resistivity (Figure 5.45). Resistivity is very sensitive to doping, ranging over eight orders of magnitude.

The transition elements have a somewhat more complicated band structure because they contain *d*-orbital valence electrons. The ease with which an electron moves through the metal depends on the number of states readily available for added electrons. The general band structure of metals results in states that are readily available for additional electrons, and the conductivity is high. However, not all metals are equally conductive. In particular, the resistivity—the inverse of the conductivity—of Hg is relatively high for a metal (Figure 5.45, inset). The high resistivity of Hg indicates that the energy overlap between the filled 5*d* band and the unoccupied 6*p* band is not optimal, making it somewhat more difficult for an electron to move through Hg. All metals, however, are much more conductive than the semiconductors. Metal resistivity ranges from about 2 to 144 $\mu\Omega \cdot$ cm. In contrast, semiconductors like Si or Ge have resistivities in the range of tens of $\Omega \cdot$ cm.

CONCEPT QUESTION   What is the electron configuration of Mn? Propose a band structure that accounts for the relatively high resistivity of Mn. ∎

5.3   Oxide Conductors and Semiconductors

Six metal monoxides—TiO, VO, MnO, FeO, CoO, and NiO—illustrate the effect of band structure and filling on conduction. These solids crystallize in a NaCl structure. That is, each metal atom is surrounded by six oxygen atoms and every oxygen atom is similarly surrounded by six metal atoms (Figure 5.46). Oxygen is one of the most electronegative elements, so the metal–oxygen bonds are essentially ionic. Despite these

Figure 5.46 The metal oxides TiO, VO, MnO, FeO, CoO, and NiO all crystallize in a simple cubic, NaCl structure. It is shown in a space-filling model, with oxygen atoms colored red and metal atoms colored green.

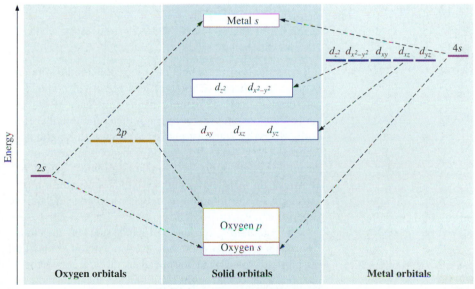

Figure 5.47 The band structure of transition metal oxides for metals on the left-hand side of the transition metal series involve the *d* orbitals of the metal.

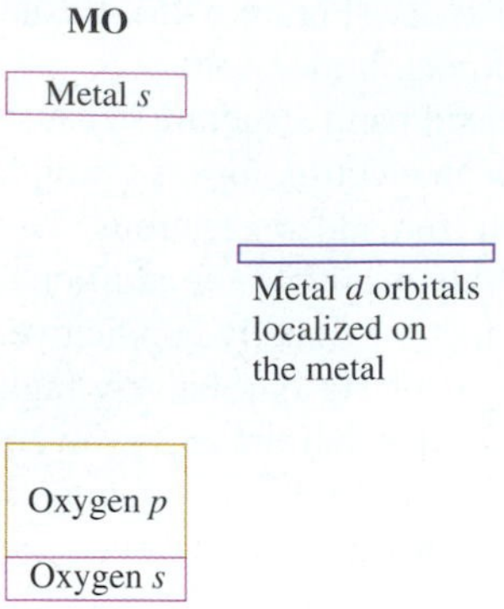

MO

Metal *s*

Metal *d* orbitals localized on the metal

Oxygen *p*

Oxygen *s*

Figure 5.48 Transition elements on the right-hand side of the transition series have a sufficient charge on the nucleus to pull the *d* orbitals to the core, making them unavailable both for bonding and for conduction. The band structure of these metals often does not show the *d* orbitals explicitly. Bonding and conduction involve the metal *s*-valence orbitals and the oxygen *s* and *p* orbitals.

similarities, TiO and VO are metallic conductors, while the remainder of these metal oxides are semiconductors. The key to these contrasting characteristics is found in the nature of the *d* orbitals in these metals, the band structure of the solid, and the band occupancy.

The ionic character of the bonds is reflected in the band structure of the solid (Figures 5.47 and 5.48). If the oxygen and metal orbitals mixed, they would form a bonding and an antibonding combination. However, because oxygen is a more electronegative element than the metal, the bonding combination is a nearly pure oxygen *s* band of states; the antibonding combination is essentially a metal *s* band. This is symbolized in Figure 5.47 by the light dashed arrow, indicating a light contribution to the low-energy *s* band by the metal. This band is customarily labeled as an oxygen *s* band. Similarly, the higher-energy orbital is labeled as a metal *s* band. The oxygen *p* orbitals do not mix with the metal orbitals.

In an isolated metal atom, the five *d* orbitals are all the same energy. However, in the solid, the environment around these orbitals is no longer the same. Two *d* orbitals $d_{x^2-y^2}$, d_{z^2} are directed at the oxygen atoms; the other three *d* orbitals (d_{xy}, d_{xz}, d_{yz}) are pointed between oxygen atoms. Electrons in the orbitals directed at the oxygen atoms are localized in the vicinity of the large electron density of the oxygen atoms. In contrast, electrons in the in-between orbitals are localized away from the oxygen atoms. Electron–electron repulsion results in a higher energy for the set directed at the oxygen atoms.

The filling in of the valence electrons depends on the metal. For TiO, there are six valence electrons per oxygen atom and four per titanium. These ten electrons per TiO unit fill the oxygen *s* band (two electrons) and the oxygen *p* band (six electrons). The remaining two electrons only partly fill the lower-energy *d* band. As a result, TiO is a conductor. For VO, there are 11 electrons to fill in, again leaving the *d* band partly filled. VO is also a conductor.

In contrast to TiO and VO, the oxides MnO, FeO, CoO, and NiO are all semiconductors. This shift in properties is linked to the *d* orbitals. Moving from left to right across the transition series, the *d* orbitals are increasingly pulled into the core and become less available for bonding.

CONCEPT QUESTION Which electrons ionize when Mn, Fe, Co, and Ni form their respective +2 cations? ∎

As the *d* orbitals become more confined to the core, they extend for shorter distances and therefore interact less effectively. The *d* states transform from extended states that overlap forming bands into part of the core (Figure 5.48). Electrons in core states cannot travel through the solid and, therefore, cannot conduct. The metal *s* orbitals extend farthest and so overlap with neighboring atoms—in this case, oxygen atoms. Thus, the metal *s* band does extend over the solid and is able to accept electrons for conduction. The four oxides MnO, FeO, CoO, and NiO are semiconductors because the oxygen 2*s* and 2*p* bands are filled but the metal 4*s* band is empty. The strength of the interaction between oxygen and the metal determines this band-gap energy.

In summary, on the left-hand side of the fourth-period transition series, the *d* orbitals extend far enough to overlap with other metal orbitals and form bands. If a band is partially filled, then the metal oxide is a conductor. If the metal *d* orbitals are more confined, as on the right-hand side of the transition series, then the *d* orbitals do not form bands. Because the oxygen 2*p* band is filled and the metal 4*s* band is empty, these oxides are semiconductors.

WORKED EXAMPLE 5.5   *Conductivity of Oxides*

Transition metal oxides show a tremendous variation in conductivity, ranging from conductors to semiconductors to insulators. Predict the conductivity properties of CdO, NbO, and ZnO.

Plan

Conductivity depends on the availability of the metal *d* orbitals and filling of the band.
- Metals to the left-hand side have *d* orbitals that participate in bonding. For metals on the right-hand side, the *d* orbitals are part of the core.
- Conductors have a partially filled band. Semiconductors have filled and empty bands.

Implementation

- Cd and ZnO are on the right-hand side and Nb is on the left-hand side of the transition series.
- CdO and ZnO have filled oxygen bands and an empty metal *s* band. CdO is a semiconductor and ZnO is an insulator. NbO has filled oxygen bands but the metal *d* orbital band is only partially filled. NbO is a conductor.

See Exercises 63, 64.

WORKED EXAMPLE 5.6   *Forming an Oxide p–n Junction*

With rather large band gaps, oxide semiconductors are good candidates for making a *p–n* junction with little leakage of current under reverse bias. Suggest a strategy for making both *p*- and *n*-type semiconductors from a metal oxide.

Plan

- Making a *p*-type semiconductor involves removing some of the electrons from the valence band.
- Making an *n*-type semiconductor involves adding electrons to the conduction band.

Implementation

- The effects of missing oxygen atoms are that (1) the number of states in the oxygen *p* band is reduced and (2) metal 4*s* electrons cannot transfer into the band due to the missing states. Leaving out some of the oxygen results in the metal 4*s* band having electrons in it, making the material an *n*-type semiconductor. Making the semiconductor with a shortage of oxygen makes it an *n*-type semiconductor.
- A shortage of metal atoms results in incomplete filling of the oxygen *p* band, forming a *p*-type semiconductor.

Because oxygen is very electronegative, the uneven sharing not only results in an alteration of the band structure of Group II–VI semiconductors, but also means that a shortage of the Group VIA element results in conduction by the opposite-charge carrier from other Group II–VI semiconductors. A CdS semiconductor with a shortage of S results in a *p*-type semiconductor. (Exercise for the student: Verify this.)

See Exercise 46.

Checklist for Review

KEY TERMS

covalent bond (p. 159)
ionic bond (p. 160)
electrostatic charge
   density (p. 162)
electron density
   surface (p. 163)
molecular orbital
   (p. 165)
bonding molecular
   orbital (p. 166)
antibonding molecular
   orbital (p. 166)
bond order (p. 167)

band (p. 174)
hybrid orbital (p. 175)
valence band (p. 177)
conduction band
   (p. 178)
band gap (p. 178)
overlap rule (p. 178)
zinc blende (p. 179)
hole, h^+ (p. 181)
doping (p. 184)
n-type semiconductor
   (p. 184)

p-type semiconductor
   (p. 184)
$p-n$ junction, diode
   (p. 184)
forward bias (p. 185)
reverse bias (p. 185)

KEY EQUATIONS

Bond order = $\frac{1}{2} \times$ [(number bonding electrons)
− (number antibonding electrons)]

$e^- + h^+ \longrightarrow$ energy

Energy = $hv = hc/\lambda$

Chapter Summary

As two atoms come together, the interaction depends on two fundamental atomic properties: the hold on the valence electrons (measured by the ionization energy) and the attraction for additional electrons (measured by the electron affinity). These properties are combined in the electronegativity. A large electronegativity difference results in transfer of one or more electrons from the atom with a low electronegativity value to the atom with a high electronegativity value, producing an ionic bond. A small electronegativity difference leaves the electron pair attracted to both atoms, occupying a molecular orbital that envelopes both atoms and forming a covalent bond.

The directionality of covalent bonds makes them fundamentally distinct from ionic and metallic bonds. A material with covalent bonds does not yield easily to a distorting force, but shatters when the bonds become distorted.

Semiconductors form when the bonding or valence band is filled and the antibonding or conduction band is empty. The band gap between these two bands is directly related to the strength of the interactions between atoms in the solid. Doping a semiconductor produces a p-type (electron-poor) or an n-type (electron-rich) semiconductor. Juxtaposing these produces a $p-n$ junction. Metals feature a partly filled band.

KEY IDEA

The wave nature of the electron, atom size, and electronegativity all play key roles in determining the outcome of an encounter between two atoms.

CONCEPTS YOU SHOULD UNDERSTAND

■ Small atoms interact strongly producing a large separation between bonding and antibonding bands. Conductors have partially filled bands. Semiconductors have filled and empty bands.

OPERATIONAL SKILLS

- Predict ionic versus covalent bonding and correlate interactions with density (Worked Examples 5.1 and 5.2).
- Predict relative band-gap energies (Worked Examples 5.3 and 5.4).

- Choose elements for n- and p-type doping and alter a material to increase/decrease the band gap (Worked Examples 5.5 through 5.6).

Exercises

A blue exercise number indicates that the answer to that exercise appears at the back of the book.

■ SKILL BUILDING EXERCISES

5.1 Bonding

1. The following pairs of elements form a chemical bond. Circle the element that contains the majority of the electron density. Explain the basis for your answer.
   a. H and O
   b. Al and N
   c. In and Sb
   d. Zn and Se

2. The following pairs of elements form a chemical bond. Classify the bond as ionic or covalent. Explain the basis for your answer.
   a. Be and O
   b. H and F
   c. Si and C
   d. Ga and P

3. For each of the following, circle the pair of elements with the greater difference in electronegativity:
   a. (H and C) or (H and O)
   b. (Ga and P) or (Ga and As)
   c. (Ga and As) or (Zn and Se)
   d. (Al and N) or (Ga and N)

4. In each of the following, circle the more covalent molecule.
   a. H_2O or H_2S
   b. HCl or CsCl
   c. NH_3 or AsH_3
   d. CH_4 or GeH_4

5. Arrange the elements C, H, O, and F in order of increasing electronegativity. If two elements are nearly equal, indicate this with a $\cong$ symbol.

6. Arrange the elements C, Si, and Ge in order of increasing electronegativity. If two elements are nearly equal, indicate this with a $\cong$ symbol.

7. Ionic materials tend to form a solid at room temperature, whereas small covalent molecules tend to form either liquids or gases. Explain these tendencies based on the bonding in these materials.

8. Ionic solids tend to be brittle, whereas metals are ductile and malleable. Explain this difference based on the bonding in these materials.

9. Classify each of the following solids as ionic, metallic, or covalent:
   a. KF   b. I_2 c. dry ice, CO_2 d. sand, SiO_2

10. Describe what you expect to happen when the following materials are hit with a hammer:
    a. a crystal of salt, NaCl
    b. a cube of gold
    c. a crystal of iodine
    d. diamond

11. NaCl is an ionic solid. MgO crystallizes in the same structure as NaCl. What is the coordination number of Na in NaCl? How many atoms are in the unit cell of MgO? Determine the percent ionic character for an NaCl bond and an MgO bond. The density of NaCl is 2.17 g/cm^3 and that of MgO is 3.6 g/cm^3. How many NaCl units per cm^3? How many MgO units per cm^3? Which has the greater unit density, NaCl or MgO? Why is the density greater?

12. Explain the difference between the orbitals that electrons occupy in an atom and the orbitals that electrons occupy in a molecule. How do they differ from the orbitals occupied by electrons in a solid?

13. Air conditioners still contain chlorofluorocarbons such as CF_2Cl_2, although these materials are being phased out. Draw the Lewis dot structure of CF_2Cl_2. Based on this structure, explain why CF_2Cl_2 is so stable.

14. An eximer laser is based on formation of a molecule that is stable only in an excited state. One example is XeF. Draw a molecular-orbital energy-level diagram for XeF. What is the bond order of XeF? Suggest an excited-state configuration for XeF that is stable.

15. Draw the Lewis dot structure of NF_3. Explain why nitrogen has a propensity to have a coordination number of 3.

16. A number of molecules do not satisfy the octet rule. One of them is NO_2, a molecule responsible for the brown haze in polluted city air. Why does NO_2 not satisfy the octet rule? At night, NO_2 dimerizes to form N_2O_4. Does N_2O_4 satisfy the octet rule?

5.1 Diatomic Molecules

17. How many nodes are in a σ_s molecular orbital? In a $\sigma_s{}^*$ orbital? In a σ_{p_z} orbital? In a $\sigma_{p_z}{}^*$ orbital? In a π orbital? In a π^* orbital? Arrange these orbitals according to the number of nodes. Based on nodes, arrange the orbitals in terms of their energy.

18. What is the bond order of the dimer Na_2?

19. Draw the molecular-orbital energy-level diagram for LiO. What is the bond order?

20. What is the bond order of the heteronuclear diatomic molecule LiBe?

21. Draw the molecular-orbital energy-level diagram for CO. What is the bond order?

22. Draw the molecular-orbital energy-level diagram for BeC. How is this diagram modified for the ion BeC^-? For the ion BeC^+? Which of the three species BeC, BeC^-, or BeC^+ is the most stable, and why? Classify these species as diamagnetic or paramagnetic.

23. Molecular and ionic oxygen species play a role in atmospheric, biological, and oxide semiconductor catalyzed processes. These species include O_2, the peroxide ion $O_2{}^-$, and the superoxide ion $O_2{}^{2-}$. Evaluate the stability of these species by drawing the molecular-orbital diagram, filling in the electrons for each oxygen species. Predict the order of stability for these diatomic species: molecular oxygen, O_2; peroxide ion, $O_2{}^-$; and superoxide ion, $O_2{}^{2-}$. Which are paramagnetic?

24. Predict the order of stability for the following diatomic species: NO, NO^-, and NO^+. Which are paramagnetic?

25. Determine the bond order of each of the following:
    a. NO   b. $O_2{}^{2-}$ c. CN^- d. CO

26. Which of the following diatomic species has the same electron configuration as CN^-?
    a. CO   b. N_2 c. CO^{2+} d. NO^-

27. Sketch the bonding orbital for a combination of an s orbital on one atom and a p orbital on another. Sketch the antibonding combination. Do you expect the bonding combination to be higher or lower in energy than a similar $s-s$ combination? Why?

28. What is the bond order in SiC^-?

5.1 More Than Two Atoms

29. Describe the evidence in support of the band picture for the electronic states of a solid.

30. Explain the differences between the band structure of a conductor, a semiconductor, and an insulator.

31. How many valence electrons per atom are required to fill the following hybrid bands: s bonding, sp bonding, sp^2 bonding, sp^3 bonding, s antibonding, and sp^3 antibonding?

32. Explain what experimental evidence indicates that the bands in solid Li must consist of sp^3 hybrid bands rather than separated s and p^3 bands.

5.1 Hybrid Orbitals

33. Complete the following table:

Number of equivalent bonds	3		2
Hybridization		sp^3	

34. Compare the hybridization of carbon in methane, CH_4, with that of carbon in diamond and graphite.

5.1 Elemental Semiconductors and Insulators

35. Arrange the following in order of size: Al, As, Ga, In, P. In each of the following pairs, select the more electronegative element: (Se or Te), (Al or In), (Sn or Cd), (O or N).

36. Calculate the atomic density of Ge. Use this result to predict whether the band gap in Ge should be larger or smaller than that in Si: Ge density = 5.323 g/cm^3, Si density = 2.33 g/cm^3.

37. Explain why diamond is colorless. Explain why diamond with impurities such as N or B is colored.

38. The element arsenic forms an extended, network solid. Is As a conductor, a semiconductor, or an insulator? Justify your choice.

5.1 Mixed-Valence Semiconductors

39. Select the element in each of the following pairs with a higher electronegativity: (Al or P), (Se or Zn), (Ga or P), (P or Sb).

40. ZnS is a semiconductor with a band gap of 3.54 eV. Draw an energy-level diagram for this material, labeling the atomic orbitals, electron filling, valence band, conduction band, and band gap.
    a. If some of the sulfur is replaced with phosphorus, the conductivity of this material changes. Draw an energy-level diagram for the doped material indicating the filling of electrons. Is this material an n- or p-type semiconductor?
    b. If you were to make a $p-n$ junction out of ZnS, what would you dope the other half with? Explain your choice.
    c. What color is a diode made with this $p-n$ junction—that is, what wavelength is 3.54 eV?

41. Assume that in Exercise 40 you determined that light emitted from the ZnS diode would not be visible. What other element or elements could you partner with Zn rather than sulfur to make a visible diode?

42. Based on our work in this chapter, what color do you predict for an InAs diode? Explain your answer.

43. Using data and models in this chapter, explain the following color variations:

Composition	Emission Color
$GaP_{0.40}As_{0.60}$	red
$GaP_{0.65}As_{0.35}$	orange
$GaP_{0.85}As_{0.15}$	yellow
$GaP_{1.00}As_{0.00}$	green

44. Predict the band-gap energy for a semiconductor of the composition $Al_{0.5}Ga_{0.5}N$. Explain your prediction.

45. GaAs is a black solid. Explain this observation.

46. The element As is very toxic. Suggest elements to replace As in a $GaP_{0.40}As_{0.60}$ diode while maintaining the red color.

47. GaN is the current material of choice for a blue diode. Due to the large size difference between Ga and N, solid GaN has many defects that limit the useful life of the GaN diodes. Suggest ternary combinations starting with AlN to make a blue diode. Justify your choice.

48. The semiconductor GaN is the current material of choice for a blue diode. Is GaN an ionic or covalent semiconductor? Which element should be short of the 1:1 mole ratio to form a p-type semiconductor? Justify your choice.

49. Extremely pure silicon has a purity of 99.99999999 atomic percent. How many silicon atoms per gram have been replaced by impurity atoms? How many per cubic centimeter?

50. Addition of phosphorus to silicon changes the properties of silicon. Do the following quantities increase, decrease, or stay the same? Explain your reasoning.
    a. the hole concentration
    b. the conductivity
    c. the electron concentration

51. A $GaP_{0.40}As_{0.60}$ diode emits light at 650 nm. Calculate the energy of that emission (in eV). Assume that the laser can reliably read features at four times the wavelength along a track, and the spacing between tracks is 10 times the wavelength. How many features (zeros and ones) can be written on a square centimeter of disk space?

52. A YAG laser emits light at 1064 nm. Calculate the energy of the emission (in eV). Is the emission in the visible region of the spectrum? If not, is it higher or lower in energy than the visible region? A YAG laser also emits light at 532 nm and 266 nm. What is the relationship among the energies of the 266-nm, 532-nm, and 1064-nm emission lines? Assume that the laser can reliably read features at four times the wavelength along a track, and the spacing between tracks is 10 times the wavelength. How much more data can be read by a 532-nm or 266-nm beam compared with a 1064-nm beam?

53. Pure Si is a semiconductor. Si may be doped to make it a p-type semiconductor. Which of the following are candidates for doping Si to make it a p-type semiconductor: Ga, In, P, Sb, Ge, and Zn? Explain your choice(s).

54. Classify the following as an insulator, a semiconductor, or an n- or p-type semiconductor.

	Insulator	Semi-conductor	n-Type Semiconductor	p-Type Semiconductor
Diamond				
$Ga_{1.0}As_{1.1}$				
$Ga_{1.0}P_{0.95}$				
CaS				
Si				

55. Classify the following doped semiconductors as n- or p-type semiconductors:

Semiconductor	Dopant	Type	Semiconductor	Dopant	Type
Si	Al		AlP	S	
Ge	P		ZnSe	Sb	
GaAs	Cd		CdTe	In	

56. Solid Ge is a semiconductor. If a few percent of the Ge atoms are replaced by In, the conductivity changes. Draw a band diagram both for pure Ge and for In-doped Ge, indicating the electrons. Clearly label the valence band, conduction band, band gap, and electron occupancy. Explain the conductivity of each material (pure Ge and In-doped Ge).

57. How would you make a $p-n$ junction beginning with pure Si? Which side should be connected to the negative battery terminal to forward-bias this diode? A diode is used in transformers to convert AC power into DC power. Explain the function of the diode in the transformer.

58. A diode is connected as shown in Figure 5.49.
    a. Which part of the diode is the n side? Which is the p side?
    b. On which side are holes located?
    c. Where does electron hole recombination occur?

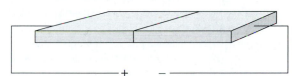

Figure 5.49 A diode wired for conduction.

5.2 Resistivity and Conductors Versus Semiconductors

59. Which of the following has the highest electrical conductivity? The lowest? Explain your answers.
    a. Si b. Ge c. Ag d. Hg

60. Which of the following has the highest electrical resistivity? Explain your reasoning.
    a. NaCl(s) b. diamond c. Rb(s)

61. Silicon is a semiconductor, but addition of very small amounts of impurities can change its resistivity by several orders of magnitude. In contrast, forming a metal alloy has only a minor effect on the resistivity. For example, copper with as much as 3 atomic percent Ni has a resistivity of only five times that of pure copper. Explain these different responses of resistivity to impurities.

62. Diamonds with impurities are colored. Explain the basis for this statement showing any band diagrams you use.

63. Hg has a relatively high resistivity for a metal. This characteristic is not shared by Zn or Cd, smaller members of Group IIB. How do the band diagrams of Zn and Cd differ from that of Hg? Use the electron configuration to justify the differences.

64. Mn has a high resistivity relative to other metals. The subsequent members of Mn's group, Tc and Re, do not share this characteristic. Propose band diagrams for these three elements to explain this difference.

5.3 Oxide Conductors and Semiconductors

65. All of the semiconductors studied in this chapter have 1 : 1 stoichiometry. This question focuses on extension of concepts to non-1 : 1 materials. Predict whether TiO_2 is a semiconductor. Explain your reasoning.

66. CaO is a component of chalk that crystallizes in the NaCl structure. Is CaO a conductor, a semiconductor, or an insulator? Explain your reasoning.

■ CONCEPTUAL EXERCISES

67. Draw a diagram indicating the relationships among the following terms: band gap, bonding orbital, antibonding orbital, bands, electronegativity, size, ionic bond. All terms should be interconnected.

68. Draw a diagram indicating the relationships among the following terms: hybrid orbital, atomic orbital, octet rule, molecular orbital, bond order, electron density. All terms should be interconnected.

69. Draw a diagram indicating the relationships among the following terms: doping, hole, *p*-type semiconductor, *n*-type semiconductor, diode, forward bias. All terms should be interconnected.

70. Draw a diagram indicating the relationships among the following terms: electronegativity, covalent bond, ionic bond, metallic bond, band gap, semiconductor. All terms should be interconnected.

❀ APPLIED EXERCISES

71. Compare magnetic media with semiconductor materials for information storage density. Magnetic particles are typically 100 nm long and 20 nm wide. They are embedded in a polymer binder taking up about 40% of the volume of the magnetic layer. The gap between the tape and the read heads is typically 5 μm.72

72. Solar cells consist of a layer of *n*-type silicon laid on top of a layer of *p*-type silicon. These two layers are sandwiched between a conductive metallic backing and an electrode grid on the front. In the absence of light, this arrangement constitutes an open circuit. When sunlight strikes the surface, current flows. Which side faces the conductive metal backing? Explain why the sandwich must be constructed with the proper side to the metal backing. How much of the solar spectrum can be utilized?

■ INTEGRATIVE EXERCISES

73. Two guiding principles in this chapter are (1) atoms tend to acquire eight valence electrons in their compounds, and (2) two elements form a 1 : 1 semiconductor when the atoms have an average of four electrons per atom. Are these two statements consistent? Predict the conductivity of Al_2O_3.

74. One of the challenges in making a blue diode arises due to the propensity for nitrogen to have 3 as its coordination number. Explain this propensity based on a Lewis dot diagram. What effect does nitrogen's coordination number have on the solid structure of GaN? What are the relative sizes of nitrogen and gallium? What effect does the size differential have on the crystal structure of GaN?

Chapter 6
Shape and Intermolecular Interactions

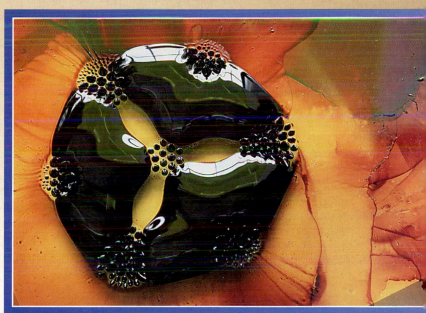

A ferrofluid forms spikes along the magnetic field lines. A soap-like substance called a surfactant separates the magnetic particles. The spike pattern is determined by intermolecular interactions that result from an applied magnetic field and the shape of the various molecules.

CONCEPTUAL FOCUS

- Develop tools for predicting the geometrical shape of a molecule.
- Analyze interactions among molecules based on the shape and on the electron distribution within the molecule.

SKILL DEVELOPMENT OBJECTIVES

- Determine the Lewis structure for a molecule including electron-rich and electron-poor molecules (Worked Examples 6.1 – 6.5).
- Predict the shapes of molecules (Worked Examples 6.6, 6.7, and 6.9).
- Identify polar molecules, calculate charge transfer (Worked Examples, 6.8, 6.11, and 6.12).
- Identify types of intermolecular interactions, calculate interaction energy (Worked Examples 6.10, 6.12, and 6.13).

Water is arguably one of the most important molecules on earth. Biological systems are about two-thirds water by mass, and three-fourths of the earth's surface is covered by water. Despite its small molecular mass (18 grams per mole), water is a liquid at room temperature. In contrast, the halogens are gaseous through chlorine (molecular mass, 71 grams per mole), and the Group VIIIA elements are gaseous all the way through radon (molecular mass, 222 grams per mole) (Figure 6.1). Among the halogens, bromine (mass, 160 grams per mole) is a liquid and iodine (mass, 254 grams per mole) is a solid. The physical form of these and other substances depends on the interaction between the molecular or atomic building blocks. Furthermore, water assumes its prominent place due to the special interaction between a water molecule and other molecules—an interaction that results from the shape of the water molecule. Interactions between water and biomolecules such as DNA and proteins determine the macromolecular shapes of these biomolecules. Their macromolecular shapes make life possible.

Polarizability Ease with which the centers of positive and negative charge can be displaced from each other.

Intermolecular interactions depend on two properties of the interacting molecules: the *geometrical shape* and the **polarizability,** or the ease with which the molecule's electron cloud becomes distorted. The shape and polarizability also determine how well two liquids mix, a property called **miscibility.** Ethanol (often referred to simply as alcohol) and water readily mix in all proportions; ethanol and water are said to be totally miscible. In contrast, oil and water hardly mix at all and are said to be immiscible.

Miscible Describes two liquids forming a uniform mixture.

6.1   Shape

Lewis Dot Structures, Formal Charge

Predicting the shape of a molecule begins with generating the Lewis dot structure. To write a Lewis dot structure, it is necessary to know which atoms are bonded together to form the molecular skeleton. If the skeleton is not known in advance, several guidelines help.

Figure 6.1 Water is a liquid at room temperature. At the same temperature, chlorine is a gas despite a much greater molecular mass. Similarly bromine (molecular mass, 160 g/mol) forms a liquid, while radon (molecular mass, 222 g/mol) forms a gas.

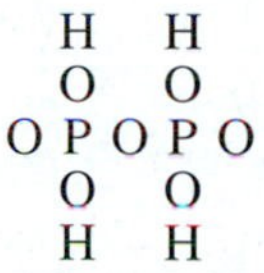

```
H   H
O   O
O P O P O
O   O
H   H
```

Figure 6.2 The skeleton of $H_4P_2O_7$ is an example of a symmetrical starting point for determining the geometrical shape of a molecule.

- The least electronegative element, except for hydrogen, is usually the central element. For example, in CO_2, carbon is less electronegative than oxygen. In $H_4P_2O_7$, phosphorus is less electronegative than oxygen. In HCN, carbon is less electronegative than nitrogen. (Hydrogen is never a central element because it is almost always bonded to only one other atom.)
- Assemble a relatively symmetrical skeleton. For example, CO_2 is O—C—O rather than O—O—C, and $H_4P_2O_7$ has a symmetrical phosphorus–oxygen skeleton (Figure 6.2).

CONCEPT QUESTION What are the skeletons for methane, CH_4, ammonia, NH_3, and carbon tetrachloride, CCl_4? ∎

- Generally, oxygen atoms do not bond to each other. Exceptions are O_2, O_3, and peroxides. The exceptions usually reveal themselves by being reactive or unstable.

CONCEPT QUESTION Carbonic acid, H_2CO_3, is the substance responsible for carbonation in soft drinks. What is the skeletal structure of carbonic acid? ∎

- For molecules with multiple central atoms, such as C_6H_6 and $H_4P_2O_7$, choose a symmetrical structure. This guideline will fail most often for compounds with multiple carbon atoms. For these compounds, the carbon skeleton is often determined experimentally.

CONCEPT QUESTION Propane, C_3H_8, is used in compressed gas tanks for barbeques. What is the skeletal structure for propane? ∎

The Lewis dot structure of a molecule is generated by distributing the electrons around the skeleton. To begin, determine the number of bonds required. Bonds consist of a pair of electrons, and bonding electrons contribute to the octet for both atoms in the bond. This sharing leads to a guideline for determining the number of bonds in a molecule:

- The number of electrons that need to be shared in bonds is determined by counting the total number of electrons needed if no sharing takes place, then counting the number available. The difference is the number that need to be shared. The number of bonds is one-half the number of shared electrons.

(6.1) Number of bonds $= \frac{1}{2}$ (electrons needed − electrons available)

Shared electrons are indicated by dots between the two bonded atoms. The remaining electrons are distributed eight to an atom, starting with the most electronegative atoms. This often results in a number of unshared pairs of electrons. An electron pair not involved in a bond is called a **lone pair.**

Lone pair A pair of electrons not involved in bonding.

WORKED EXAMPLE 6.1 *Combustion-Generated Carbon Dioxide*

Carbon dioxide is generated from the combustion of fossil fuels and is a major greenhouse gas. Greenhouse gases trap infrared radiation. Interaction of CO_2 with infrared radiation is affected by the geometrical structure of CO_2. Generate the Lewis dot structure for CO_2. How many bonds connect the carbon atom to each of the oxygen atoms?

Plan
- Determine the number of electrons needed.
- Count the number of electrons available.
- Use Equation (6.1): Number of bonds $= \frac{1}{2}$ (electrons needed − electrons available).

Implementation
- Electrons needed = (3 atoms) $\times$ (8 electrons/atom) = 24 electrons
- Electrons available = (4 valence electrons from C) + (6 valence electrons from O) $\times$ (2 O atoms) = 16 electrons
- Bonds = $\frac{1}{2}$ (24 − 16) = 4. There are two bonds between the carbon atom and each oxygen atom.

See Exercise 1.

WORKED EXAMPLE 6.2   *Hydrogen Needs Only Two Electrons*

The valence shell of hydrogen is fully occupied by two electrons. As a result, hydrogen needs only two electrons in its Lewis dot structure. Ammonia, NH_3, is an example of an important molecule that contains hydrogen. How many bonds are in ammonia? How many lone pairs are on nitrogen?

Plan
- Determine the number of electrons needed.
- Count the number of electrons available.
- Use Equation (6.1): Number of bonds = $\frac{1}{2}$ (electrons needed − electrons available).
- A lone pair consists of two electrons not involved in a bond.

Implementation
- Electrons needed = (1 atom) $\times$ (8 electrons/atom) + (3 hydrogen atoms) $\times$ (2 electrons/hydrogen) = 14 electrons
- Electrons available = (5 valence electrons from N) + (1 valence electron from H) $\times$ (3 H atoms) = 8 electrons
- Bonds = $\frac{1}{2}$(14 − 8) = 3. There is a single bond between the nitrogen atom and each hydrogen atom.
- Of the eight electrons around nitrogen, six are involved in bonds to the three hydrogen atoms. The remaining two electrons constitute one lone pair.

See Exercise 2.

Resonance structures
Blending of two or more Lewis structures that differ only in the arrangement of the electrons.

Sometimes these guidelines result in symmetrically equivalent atoms having different bonding. Symmetrically equivalent atoms are atoms that are the same until the bonding electrons are distributed. Then two or more structures satisfy all the guidelines and differ only in arrangement of the electrons. These equivalent structures are called **resonance structures.** A double-headed arrow, $\leftrightarrow$, is used to connect resonance structures. Worked Example 6.3 focuses on determining resonance structures.

WORKED EXAMPLE 6.3   *Electrons in Carbonate*

Carbonic acid puts the fizz into soft drinks. In water, carbonic acid generates several substances, including the carbonate ion, CO_3^{2-}. The carbonate ion has several equivalent resonance structures. Determine the number of bonds in the carbonate ion. How many equivalent resonance structures are there for the carbonate ion?

Plan

- Determine the number of bonds.
- Rearrange electrons to attain equivalent structures.

Implementation

- There are 32 electrons needed. Available electrons are the 22 (= 6 × 3 + 4) valence electrons plus 2 additional electrons for the negative charges, giving a total of 24 electrons. Thus, there are four bonds.
- The resonance structures are

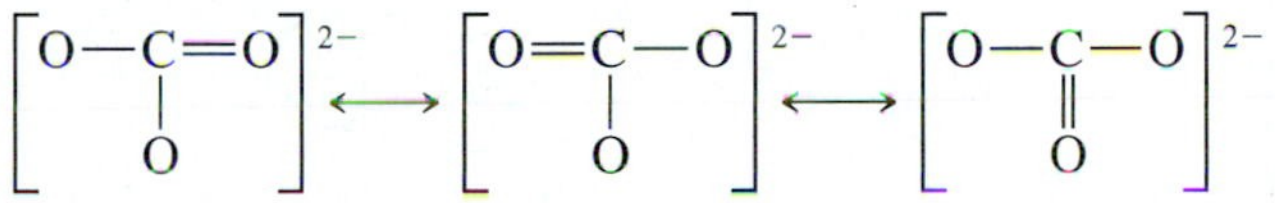

$$\left[\begin{array}{c} O-C=O \\ | \\ O \end{array} \right]^{2-} \longleftrightarrow \left[\begin{array}{c} O=C-O \\ | \\ O \end{array} \right]^{2-} \longleftrightarrow \left[\begin{array}{c} O-C-O \\ \| \\ O \end{array} \right]^{2-}$$

See Exercises 5, 6.

Formal charge Difference between the number of valence electrons for an element and the number assigned the element in a compound: Number of valence electrons − number of lone pair electrons − ½ number of shared electrons.

The final procedure that helps in determining a Lewis dot structure is the assignment of **formal charges.** The formal charge of an element in a compound is determined by assigning the electrons in any bond democratically—that is, equally to both atoms in the bond. Count the number of electrons assigned to the atom, one electron for each bond and two for each lone pair. Subtract this number from the number of valence electrons. This is the formal charge.

(6.2)   Formal charge = (assigned electrons − valence electrons)

The sum of the formal charges is zero for a neutral molecule and is equal to the ion charge for an ion. The preferred structure has formal charges near zero for all atoms. If there is a negative formal charge, it should reside on the more electronegative element, while a positive formal charge should reside on an electropositive element. Two like formal charges should not lie on neighboring atoms. Formal charges are helpful for choosing among Lewis dot structures.

WORKED EXAMPLE 6.4   *Choosing the Best Structure*

Lithium ion batteries contain thionyl chloride, $SOCl_2$. Several Lewis dot structures can be drawn for thionyl chloride, including these three:

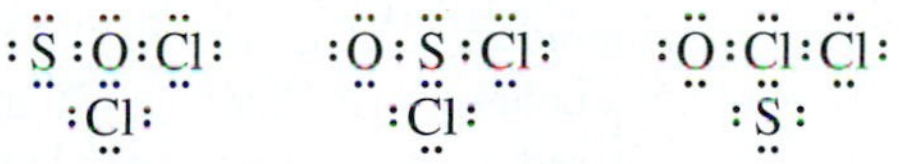

$$\begin{array}{ccc} :\!\overset{..}{S}\!:\!\overset{..}{O}\!:\!\overset{..}{Cl}\!: & :\!\overset{..}{O}\!:\!\overset{..}{S}\!:\!\overset{..}{Cl}\!: & :\!\overset{..}{O}\!:\!\overset{..}{Cl}\!:\!\overset{..}{Cl}\!: \\ :\!\overset{..}{Cl}\!: & :\!\overset{..}{Cl}\!: & :\!\overset{..}{S}\!: \\ (a) & (b) & (c) \end{array}$$

Each of these has a very different atomic skeleton. Which structure is the "best" structure for thionyl chloride?

Plan

- Determine the formal charge of each element.
- Choose the "best" structure based on minimum charge and a negative charge on oxygen.

Implementation

■

Atom	Valence Electrons	−	Electrons Assigned	=	Formal Charge
S in (a)	6		7		−1
S in (b)	6		5		+1
S in (c)	6		7		−1
O (a)	6		5		+1
O (b)	6		7		−1
O (c)	6		7		−1
Cl (a)	7		7		0
Cl (b)	7		7		0
Center Cl in (c)	7		5		+2
Right Cl in (c)	7		7		0

(a) Formal charges are −1 for S, +1 for O, and 0 for Cl (sum = 0).
(b) Formal charges are +1 for S, −1 for O, and 0 for Cl (sum = 0).
(c) Formal charges are −1 for S, −1 for O +2 for the center Cl and O for the right Cl (sum= 0).

■ Structure (c) is ruled out due to the multiple charge. Between (a) and (b), the best structure is (b) because the negative charge resides on oxygen, which is the most electronegative of the three elements.

See Exercises 7, 8.

Limitations of the Octet Rule

The octet rule really should be called the octet guideline because several exceptions to it exist. These include:

■ Many compounds of nitrogen have an odd number of electrons, for example NO and NO_2. The Lewis dot structure of NO_2 is

$$:\ddot{O}:\dot{N}::\ddot{O}:$$

Nitrogen is less electronegative than oxygen, so each oxygen atom is assigned an octet. The odd electron is unpaired and associated with the nitrogen atom.

■ Compounds in which the central element must have more than eight electrons are exceptions. They usually involve elements in the third row and beyond. Examples include PF_5, ICl_3, and XeF_2. These elements simply have more than eight electrons associated with them, called an expanded octet, although the electrons are still paired. A sign for an expanded octet is a requirement for fewer shared pairs than bonds. Bonds are required to hold elements together, so to provide the needed bonds, the octet is expanded (see Worked Example 6.5).

■ Small elements with few valence electrons operate under electron austerity and often have fewer than eight electrons. Examples include compounds of beryllium, where Be often has only four electrons, and compounds of boron, where B often has only six electrons. If there are not enough electrons to provide an octet, the compound exists with an electron shortage, called an incomplete octet (see Worked Example 6.5). A signal of a potential incomplete octet is a negative formal charge on an element with low electronegativity.

WORKED EXAMPLE 6.5   *Electron Riches and Electron Austerity*

The halogens all have seven electrons in their valence shells. Only one more electron is needed to make an octet. However, the larger halogens often form compounds in which the large halogen has more than eight electrons around it. One example is the highly disinfecting compound, ICl_3. Draw the Lewis dot structure of ICl_3.

The opposite of electron riches is electron austerity. The small elements with few valence electrons are insufficiently electronegative to attract electrons from electronegative elements, so in some of their compounds these elements have fewer than eight electrons. Draw the Lewis dot structure of BCl_3.

Plan

■ Count the number of required electrons.
■ Determine the number of valence electrons.
■ Determine the required number of shared electrons.
■ Compare shared electron pairs to required bonds.
■ Check the formal charge.

Implementation

ICl_3:

■ Four atoms with eight electrons each require 32 electrons.
■ Seven valence electrons per atom provide a total of 28 electrons.
■ The difference indicates the need for two shared pairs.
■ Two shared pairs make only two bonds, but three bonds are required to bond iodine to three chlorine atoms. An expanded octet is required.

$$:\ddot{C}l:I:\ddot{C}l:$$
$$:\ddot{C}l:$$

■ Each element has a formal charge of zero.

BCl_3:

■ Four atoms with eight electrons each require 32 electrons.
■ Three electrons come from B, and seven electrons come from each of three Cl atoms, gives 21 electrons from chlorine atoms.
■ The total number of electrons is 24. Four shared pairs of electrons satisfy the needs.
■ Four shared pairs result in a double bond between B and one Cl.
■ The formal charge on chlorine with a double bond is $+1$. The formal charge on boron is -1. Chlorine is much more electronegative than boron. Boron gets only six electrons.

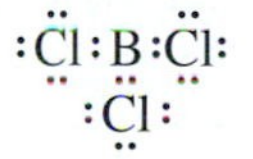

$$:\ddot{C}l:B:\ddot{C}l:$$
$$:\ddot{C}l:$$

Electron-rich (ICl_3) and electron-poor (BCl_3) molecules tend to be reactive.

See Exercises 13, 14.

VSEPR Method

By itself, the Lewis dot structure indicates nothing about the three-dimensional shape of the molecule. This three-dimensional shape is extremely important, however. For example, the three-atom molecule H_2O is a liquid at room temperature whereas the

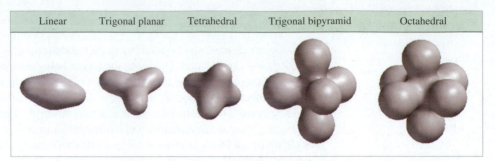

| Linear | Trigonal planar | Tetrahedral | Trigonal bipyramid | Octahedral |

Figure 6.3 The electronic geometry around an atom depends on the number of regions of electron density around the atom. With two regions the electronic geometry is linear, with three regions it is trigonal planar, with four tetrahedral, with five trigonal bipyramidal, and with six octahedral. In these representations, the atomic core, the nucleus plus the core electrons, is found at the exact center of the shape.

Valence shell electron-pair repulsion (VSEPR) Method for determining the shape of a molecule or molecular fragment. All valence pairs repel to establish the electronic geometry. Bonded electrons determine the molecular shape.

three-atom molecule CO_2 is a gas. Both molecules feature an atom flanked by two other identical atoms. But just imagine how the world would change if the water of the oceans was transformed into a dense gas like carbon dioxide.

The geometric shape of water, carbon dioxide, and other molecules is determined by using the Lewis dot representation in conjunction with a method called the **valence shell electron-pair repulsion (VSEPR)** method. In its simplest form, VSEPR is a direct result of Coulomb's law. Electrons are negatively charged and repel each other while being attracted to the positively charged core. Therefore, electron pairs and electrons in bonds around the atom core stay as far away from each other as possible while surrounding the atom. All electrons bonding two atoms together constitute one region of electron density. An atom surrounded by two regions of electron density has a local linear geometry; three regions is trigonal planar; four tetrahedral; five trigonal bipyramidal; and six octahedral (Figure 6.3). Nearly every molecular shape begins with one of these five electronic geometries.

To determine the molecular structure, start by counting the number of electron pairs around the core and distributing them as far apart as possible. This gives the *electronic* geometry.

If every region of electron density represents bonding electrons, then the *molecular* geometry is the same as the *electronic* geometry. The examples illustrated in Figure 6.4 are all symmetric molecules. For example, CH_4 features a central carbon atom bonded to four hydrogen atoms. Carbon has the same geometrical shape when bonded to unlike atoms (Figure 6.5). The tetrahedral geometry is a very common geometry for carbon because each of the four valence electrons for carbon often pairs with an electron from another element to form four bonding pairs, netting eight electrons around carbon.

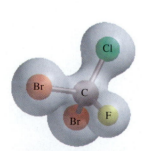

Figure 6.5 Carbon bonded to any four elements has a tetrahedral shape. The molecule illustrated here is $CFClBr_2$.

Figure 6.4 Molecules in which every region of electron density represents a bond have the same *molecular geometry* as *electronic geometry*. The molecules illustrated are all symmetric molecules — molecules in which every bond involves the central element and the same second element.

If some of the regions of electron density represent lone pairs, then the electronic geometry differs from the molecular geometry. As the mass of an electron is much less than the mass of the nucleus, lone pairs are essentially space holders. Only atoms bonded to the central atom determine the molecular geometry. Two molecules in which lone pairs play an extremely important role are ammonia (Worked Example 6.6) and water (Worked Example 6.7).

CONCEPT QUESTIONS Draw the Lewis dot structure of ammonia, NH_3. How many electron pairs are involved in bonds? How many electrons represent lone pairs? Draw the Lewis dot structure of water, H_2O. How many electron pairs are involved in bonds? How many electrons represent lone pairs? ■

WORKED EXAMPLE 6.6   *Ammonia and Protein Building Blocks*

Ammonia, NH_3, is one of the few molecules found in the atmosphere that is basic. Replacing one of the hydrogen atoms of ammonia with a carbon–hydrogen group, such as CH_3 or C_2H_5, forms a substance known as an amine. Amines are one of the structural motifs in the molecules that assemble to form proteins, so ammonia and its relatives are fundamental molecules in both atmospheric chemistry and in life. Determine the molecular geometry of ammonia and the geometry around the nitrogen atom of methyl amine, H_3CNH_2.

Plan

- Determine the molecular skeleton.
- Draw the Lewis dot structure.
- Determine the electronic geometry.
- Determine the atomic arrangement.

Implementation

- Because hydrogen normally forms only one bond, it is almost never a central atom. In ammonia, nitrogen is the central atom and the three hydrogen atoms surround it. For methyl amine, the nitrogen and carbon atoms are bonded to each other, and the hydrogen atoms surround them.

                                       H

H N H   H N C H

H   H H

Ammonia   *Amine*

- To generate the Lewis dot structure of ammonia, count the required electrons, 14, and the valence electrons, 8. Then determine the number of bonds: $\frac{1}{2}(14 - 8) = 3$. Each hydrogen atom is bonded to nitrogen by one pair, and there is an electron pair leftover. To generate the structure of methyl amine, count the required electrons, 26, and the valence electrons, 14. The number of bonds is $\frac{1}{2}(26 - 14) = 6$. Each hydrogen atom is bonded by an electron pair to nitrogen or carbon, and nitrogen and carbon are bonded by an electron pair. This leaves one lone pair. Putting the lone pair on the nitrogen results in a formal charge of zero for all elements, so this is the correct Lewis dot structure.

                                   H

H:N̈:H   H:N̈:C̈:H

Ḧ   Ḧ Ḧ

■ In both ammonia and methyl amine, there are four electron pairs around the nitrogen atom. Hence the *electronic* geometry is tetrahedral. Both structures are shown below with the lone pair indicated in red.

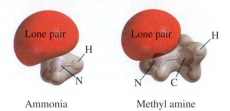

Ammonia   Methyl amine

■ The *molecular* geometry is determined from the arrangement of the *atoms*. Notice that the negatively charged lone pair repels the three bonding pairs, and each of the bonding pairs repel the others and the lone pair. Showing just the atomic skeleton indicates that the geometry around the nitrogen atom is pyramidal in both cases.

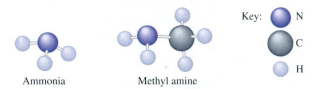

Ammonia   Methyl amine

The lone pair represents a volume of negative charge that results in a strong interaction with positively charged areas of other molecules. The Lewis dot structures and VSEPR shapes are important tools for predicting interactions between molecules.

See Exercise 17.

WORKED EXAMPLE 6.7   *Water*

Water, H_2O, is an extremely important molecule in shaping our world. What is the shape of the water molecule?

Plan
■ Draw the Lewis dot structure.
■ Determine the electronic geometry.
■ Determine the atomic arrangement.

Implementation
■ Supplying every atom in water with 8 electrons (2 electrons for hydrogen) requires 12 electrons. There are 8 valence electrons from the atoms that make up water, so there are two bonds: one between each hydrogen atom and the central oxygen atom.

$$H \; :\overset{\cdot\cdot}{\underset{\cdot\cdot}{O}}: \; H$$

■ There are four electron pairs around the oxygen atom, so the *electronic* geometry is tetrahedral. Note that the clouds of the two lone pairs merge into one large region of high electron density, shown in red.

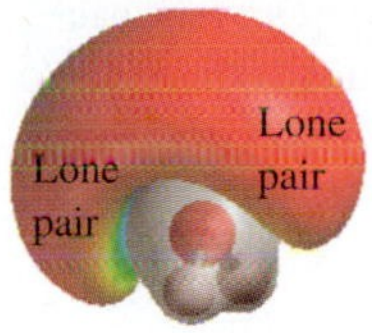

■ The *molecular* geometry is determined from the arrangement of the *atoms*. Each electron pair—the two bonding pairs and the two lone pairs—repels all other electron pairs around the oxygen atom. Showing just the atomic skeleton indicates that the *molecular* geometry is bent planar.

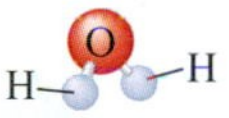

The bent structure of the water molecule makes the lone pairs particularly accessible, allowing water to readily interact with other molecules. Interactions with the lone pair not only affect the physical properties of water but also strongly affect the interaction between water and other molecules, including the proteins, carbohydrates, and DNA that regulate many aspects of living organisms.

See Exercise 18.

The notion of electronic geometry leading to molecular geometry is a general one. Table 6.1 lists the molecular geometries that result from various numbers of lone pairs being associated with the five electronic geometries.

Table 6.1 Examples of the VSEPR Structures of Molecules

Electron Pairs	Electronic Geometry	Lone Pairs	Molecular Geometry	Example
2	Linear 	0	Linear B—A—B	BeF_2 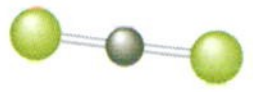
3	Trigonal planar 	0	Trigonal planar 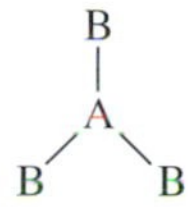	BCl_3 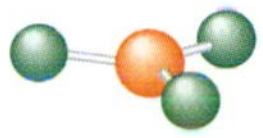

(Continued)

Table 6.1 Examples of the VSEPR Structures of Molecules *(Continued)*

Electron Pairs	Electronic Geometry	Lone Pairs	Molecular Geometry	Example
	One lone pair	1	Bent	SO_2 O_3
4	Tetrahedral	0	Tetrahedral	CH_4
	One lone pair	1	Pyramidal	NH_3
	Two lone pairs	2	Bent planar	H_2O
5	Trigonal bipyramid	0	Trigonal bipyramidal	PF_5
	One lone pair	1	Seesaw	SF_4
	Two lone pairs	2	T-shaped	ICl_3 ClF_3

Electron Pairs	Electronic Geometry	Lone Pairs	Molecular Geometry	Example
	Three lone pairs	3	Linear	XeF$_2$
6	Octahedral	0	Octahedral	SF$_6$
	One lone pair	1	Square pyramidal	IF$_5$ BrF$_5$
	Two lone pairs	2	Square planar	XeF$_4$

Consequences of Shape: Dipole Moment

For both water and ammonia, the existence of the lone pairs not only affects the molecular geometry but also profoundly affects interactions between these and other molecules. The lone pairs are a region of negative charge density. Since molecules are neutral overall, a corresponding region of positive charge density must exist. For water, the positive region is associated with the hydrogen atoms. Separated positive and negative charges constitute a **dipole.**

Dipole Result of separation of the centers of positive and negative charges.

Two methods for representing the charge distribution are illustrated in Figure 6.6. A charge density potential plot represents negatively charged regions as red and positively charged regions as blue, with intermediate colors representing regions of lower charge density. An alternative representation labels the positive ends of the molecule with δ^+ (δ is the Greek letter delta) and the negative ends with δ^-.

CONCEPT QUESTIONS   The ammonia molecule consists of a region of positive charge and a region of negative charge. Where is the negative region? Label it with δ^-. Where is the positive region? Label it with δ^+. ∎

Whenever two bonded atoms differ in electronegativity, the electron density shades toward the more electronegative element, making that region slightly negatively charged

Figure 6.6 Two views of the water molecule show the charge distribution around the molecule. In a charge density potential plot, the negative region is denoted by red and the positive region by blue. Green is neutral. The region near the oxygen atom is negatively charged due to the lone pairs and the greater electronegativity of oxygen. Conversely, the region around the hydrogen atoms is positively charged. Symbolically, the negative region near the oxygen atom is denoted as δ^- and the positively charged region near the hydrogen atoms is denoted as δ^+.

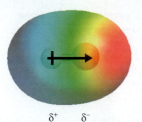
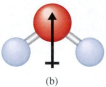

Figure 6.7 There is a greater electron density (red) on the oxygen end of the OH bond than on the hydrogen end (blue) due to the greater electronegativity of oxygen compared with hydrogen. The result is a separation of negative and positive charge—a dipole. The dipole is depicted by an arrow pointing from the positive pole toward the negative pole.

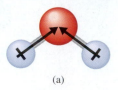

(a)

(b)

Figure 6.8 (a) Each OH bond in water has a dipole due to the greater electronegativity of oxygen compared with hydrogen. (b) Since water is bent, the vector sum of the dipoles is nonzero and the water molecule has a permanent dipole.

and the region of the less electronegative element positively charged. The bond is said to be polar. In water, the OH bond is polar (Figure 6.7). A polar bond is indicated with a vector arrow pointing from the positive pole to the negative pole. The strength, p, of the bond dipole is a function of the magnitude of the charges, q, and the separation, r, between them.

(6.3) $$p = qr$$

For a molecule containing polar bonds to have a molecular or permanent dipole, the vector sum of the dipoles of the individual bonds must be nonzero. Since water is bent, the vector sum of the dipoles is nonzero (Figure 6.8) and water has a permanent dipole.

In contrast to water, methane (CH_4) has no permanent dipole (Figure 6.9). The Lewis dot structure of methane indicates that four electron pairs surround the carbon, so the electronic geometry around carbon is tetrahedral, just as the oxygen in water is tetrahedral. All four electron pairs around carbon in methane are bonded to hydrogen. Carbon is more electronegative than hydrogen (2.55 versus 2.20), so the CH bonds are all polar. However, the vector sum of the four dipoles is zero. Methane is nonpolar.

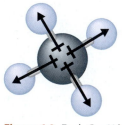

Figure 6.9 Each C—H bond in methane is polar, but the vector sum of the bond dipoles is zero. Methane does not have a permanent dipole.

WORKED EXAMPLE 6.8   *The Importance of Shape*

The shape of a molecule determines if it has a permanent dipole. Does either CO_2 or SO_2 have a permanent dipole moment?

Plan

■ Determine the shape using the Lewis dot structure and VSEPR method.
■ Establish the bond dipoles.
■ Determine the vector sum of the bond dipoles.

Implementation

■ The Lewis dot structures are

$$\ddot{O}::C::\ddot{O} \qquad \ddot{O}::\ddot{S}:\ddot{O}:$$

$$CO_2 \qquad\qquad SO_2$$

The carbon atom in CO_2 is surrounded by two regions of electron density—two bonds—so CO_2 is linear. The sulfur atom in SO_2 is surrounded by three regions of electron density, two bonds and one lone pair, so SO_2 is bent.

■ Both carbon and sulfur are less electronegative than oxygen, so the dipole points from the central atom to the oxygen atom in both cases.

$$\overset{\longleftarrow+\longrightarrow}{O=C=O} \qquad \overset{S}{O \quad O}$$

■ The bond dipoles in CO_2 exactly cancel, so CO_2 is nonpolar. Since SO_2 is bent, the resultant of the bond dipoles is a permanent dipole.

$$\overset{S}{\underset{O \downarrow O}{}}$$

See Exercises 25–28.

The water molecule has two characteristics that make it unusual: The bond dipole is particularly strong, and the two positively charged regions of the hydrogen atoms are matched by the two negatively charged regions of the lone pairs. As a consequence, water forms an extensive network containing chains and rings of molecules. This network is responsible for many of water's unusual properties. For example, the range of temperature over which water is a liquid is much larger than for the analogous molecules H_2S, H_2Te, and H_2Se (Figure 6.10). Solid water forms ice—a fairly open structure that is responsible for ice being less dense than liquid water. So icebergs float—a fact realized in disastrous fashion by the ship *Titanic*.

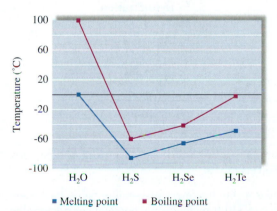

Figure 6.10 The temperature range for the liquid form of H_2O is much larger than that for H_2S, H_2Se, or H_2Te, all of which have the same bent structure. The larger liquid range of water is due to the larger dipole in the OH bond compared with the SH, SeH, or TeH bonds.

Valence Bond Theory

In water, hydrogen and oxygen are joined by a single electron pair—a sigma bond—symbolized with the Greek letter sigma, σ. A σ bond consists of an electron pair localized in an orbital that is directed between the two atomic cores. In the case of water, this σ bond is made from the oxygen atomic s and p orbitals. But the s orbital is nondirectional and the three p orbitals are located along the three Cartesian axes with the lobes separated by 90°. How do these atomic orbitals become the molecular orbitals of water with an H—O—H bond angle of 104.5°? Or consider CO_2, which is a linear molecule. The carbon atomic p orbital contains one electron. Accepting an electron from oxygen to form a two-electron bond fills this p orbital, so how does carbon bond to the second oxygen?

Questions such as these propelled development of a theory called the valence bond theory. Valence bond theory mixes or

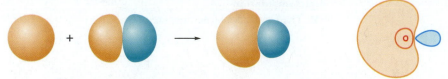

Figure 6.11 An *s* orbital and a *p* orbital centered on the same atom hybridize, enhancing the *p*-orbital lobe that is in phase with the *s* orbital and diminishing the other lobe. The contour plot shows the much larger density of the in-phase lobe compared with the out-of-phase lobe.

hybridizes enough atomic orbitals to form a half-filled molecular orbital for each bond and a filled molecular orbital for each lone pair. For example, forming CO_2 requires generating two half-filled orbitals on carbon. Water requires two half-filled orbitals for the bonds between oxygen and the two hydrogen atoms plus two filled orbitals for the two lone pairs, giving a total of four orbitals.

CONCEPT QUESTION How many orbitals are needed for nitrogen in ammonia? ■

There is a conservation of orbitals: If two molecular orbitals are required, then two atomic orbitals must be hybridized to generate those two orbitals. Carbon in CO_2 has two half-filled *p* orbitals, but the *p* orbitals are located 90° from each other; 180° is required for a linear configuration. The 180° configuration is generated by mixing one valence *p* orbital with the valence *s* orbital (Figure 6.11). The resultant orbital is dubbed an *sp* hybrid. Two half-filled *sp* hybrid orbitals are required for the two σ bonds between carbon and the two oxygen atoms in CO_2.

The hybridization process can also be shown in box diagram notation (Figure 6.12). Each *sp* hybrid is directed toward an oxygen atom to form a σ bond that is filled with one electron from carbon and one from oxygen.

For molecules with no lone pairs, determining the number of valence orbitals that must be hybridized is a matter of counting the number of bonds required.

Number of Bonds	Hybridization
2	*sp*
3	sp^2
4	sp^3
5	sp^2d^2
6	sp^3d^2

CONCEPT QUESTION In many of its compounds, carbon bonds to four other atoms, such as in CCl_4. What is the hybridization of carbon in CCl_4? ■

The treatment of lone pairs is similar to the treatment of bonding pairs. Each lone pair occupies a hybridized orbital. As in the VSEPR method, only bonding pairs determine the molecular geometry. In water, oxygen has two bonding pairs and two lone pairs, so it is sp^3 hybridized. The angle between the two O—H bonds is accounted for by this hybridization. The general result is that the lone pairs are somewhat less localized than the bonding pairs and take up somewhat more space. In a regular tetrahedron, the bond angles are 109.5°. In water, however, the H—O—H angle is 104.5°. Orbitals that are

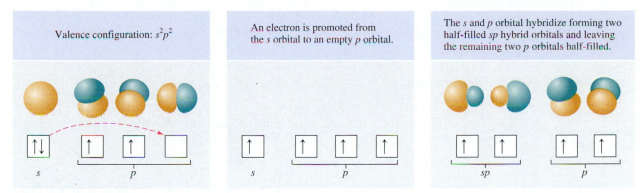

Figure 6.12 A box diagram flowchart shows the formation of an *sp* hybrid orbital. An electron is promoted from the filled *s* orbital to an empty *p* orbital, and the *s* and *p* orbitals mix. The remaining two *p* orbitals are unaffected.

not hybridized are either singly occupied or empty. The valence bond analogue of Table 6.1 is shown in Table 6.2.

CONCEPT QUESTION   One component of acid rain is sulfuric acid, H_2SO_4, which results from oxidation of SO_2 in the atmosphere. What is the hybridization of sulfur in sulfuric acid? ∎

In CO_2, carbon's hybridization is *sp*, accommodating the two bonding pairs. Each bonding pair accounts for two electrons, but there are four electrons shared between the carbon atom and each oxygen atom. The other electron pair forms a π bond. This π bond results from merging an atomic *p* orbital on the carbon atom and an atomic *p* orbital on the oxygen atom (Figure 6.13). The resulting π electron cloud is located

Table 6.2 Hybridization, Electronic Geometry, and Molecular Geometry

Electron Pairs	Electronic Geometry	Hybridization	Lone Pairs	Molecular Geometry	Examples
3	Trigonal planar	sp^2	0	Trigonal planar	BCl_3
			1	Angular	O_3, NO_2^-
4	Tetrahedral	sp^3	0	Tetrahedral	CH_4, CCl_4
			1	Pyramidal	NH_3, SO_3^{2-}
			2	Angular	H_2O
5	Trigonal bipyramidal	sp^3d	0	Trigonal bipyramidal	PF_5
			1	Seesaw (lone pair on equator)	SF_4
			2	T-shaped (both lone pairs on equator)	ICl_3, ClF_3
			3	Linear	XeF_2
6	Octahedral	sp^3d^2	0	Octahedral	$[Cu(NH_3)_6]^{2+}$
			1	Square pyramidal	IF_5, BrF_5
			2	Square planar	XeF_4

above and below the internuclear axis. Bonding between carbon and oxygen thus consists of a σ bond and a π bond—a double bond.

WORKED EXAMPLE 6.9 *Versatile Carbon*

A seemingly infinite variety of molecules consist of only carbon and hydrogen. Bonding between carbon atoms ranges from simple σ bonds to triple bonds. Some carbon–carbon bonding motifs lead to very stable structures. One of these structures is found in benzene, a molecule that consists of six carbon atoms arranged in a hexagonal ring with a single hydrogen atom bonded to each carbon atom. Determine the geometry at each carbon atom in benzene. Starting with the six carbon atoms far apart, indicate what the electron cloud looks like as the six carbon atoms move to bonding distance from each other.

Plan
- Generate the Lewis dot structure.
- Identify half-filled orbitals that can form π bonds.

Implementation
- To generate the Lewis dot structure of benzene, first draw the hexagonal skeleton, and then arrange the six hydrogen atoms around the carbon ring. Sixty electrons are required to provide eight electrons for each carbon atom and two electrons for each hydrogen atom. Only 30 valence electrons are available, indicating 15 bonds. Six bonds are required to connect the hydrogen atoms to the carbon atoms, and another six bonds connect the six carbon atoms in a ring. Three bonds remain. Distribute these between every other carbon pair.

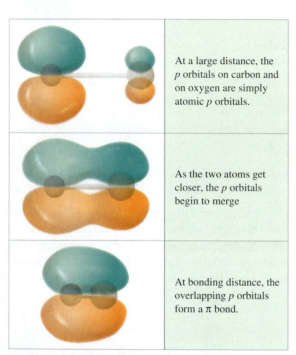

	At a large distance, the *p* orbitals on carbon and on oxygen are simply atomic *p* orbitals.
	As the two atoms get closer, the *p* orbitals begin to merge
	At bonding distance, the overlapping *p* orbitals form a π bond.

Figure 6.13 Formation of the π bond between a carbon atom and an oxygen atom shows development of the molecular orbital that consists of electron density above and below the internuclear axis.

$$\begin{array}{c} H \\ \ddot{C} \\ H:C \quad\quad C:H \\ H:C \quad\quad C:H \\ \ddot{C} \\ H \end{array}$$

This is an example of a resonance structure, as a similar structure differs only in that the double bonds are shifted by one atom.

- The double bonds indicate that there is a singly occupied *p* orbital on every carbon atom. The expanded skeleton looks like (a). After bringing the atoms to bonding distance, it looks like (b).

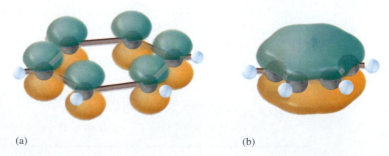

(a)   (b)

Due to the resonance and overlapping p orbitals, this ring motif, called a phenyl ring if one of the hydrogen atoms is replaced by something else, is quite stable. Because it resists heating, it is used in flame retardant materials. The resonance also makes the ring quite rigid, imparting stiffness to molecules that incorporate it.

See Exercises 43, 44.

6.2   Intermolecular Interactions

Ion–Ion Interactions

Intermolecular interactions are a result of the mutual attraction between regions of positive charge and regions of negative charge. The strongest such interaction occurs between two ions. Walking across the carpet on a dry winter day and reaching for a door knob brings first-hand experience of the mutual attraction between positive and negative charges. Attraction and repulsion between particles implies that a force acts between the particles. This force is related to an energy. For charged particles, the attraction and repulsion are described by Coulomb's law:

(6.4)
$$E = \frac{kq_1q_2}{r}$$

The interaction energy, E, is *proportional to the charges q_1 and q_2* and is *inversely proportional to the distance* between them. The proportionality constant, k, is $9.0000 \times 10^9 \ N \cdot m^2/C^2$. If charges q_1 and q_2 are the same sign, then the energy is positive and the two particles repel each other. If the charges have opposite signs, then the energy is negative and the two particles attract each other. Interaction between two ions, called **ion–ion interaction,** both extends farther and is stronger than any other intermolecular interaction.

Ion–ion interaction Interaction between two ions.

APPLY IT Obtain two balloons and blow them up. Rub one against your clothing and observe what happens when you let go of the balloon. Tie a string 18–20 in. long to each balloon. Rub both against your clothing, then hold the free ends of the strings in your hand. Observe the interaction between the two balloons. What do you observe?

WORKED EXAMPLE 6.10   *Attraction*

The charge on an electron is quite small, 1.6×10^{-19} C. The distance between charged particles is also small. For example, common table salt is composed of Na^+ and Cl^- ions separated by 236 pm. Determine the electrostatic energy for one mole of NaCl molecules, assuming a 100% ionic bond. Discuss the relationship between the energy of one mole of NaCl molecules and one mole of solid salt.

Plan

■ Determine the energy of one molecule. Use Equation (6.4): $E = \dfrac{kq_1q_2}{r}$.

■ Scale up to one mole.

■ Consider the relationship of a mole of molecules to a solid consisting of a mole of Na^+ and Cl^- ions.

Implementation

- $E = \dfrac{-(9.00 \times 10^9 \, \text{N} \cdot \text{m}^2/\text{C}^2)(1.60 \times 10^{-19} \, \text{C})(-1.60 \times 10^{-19} \, \text{C})}{(2.36 \times 10^{-10} \, \text{m})}$

  $= -9.76 \times 10^{-19} \, \text{N} \cdot \text{m}$

  $= -9.76 \times 10^{-19} \, \text{J}$

Note: The charge on an electron is used because a negative ion has an excess of one electron. Conversely, a positive ion has an excess of one proton. The charge on a proton is the same as that on an electron. The energy is *negative* indicating an attraction between the positive and negative ions.

- $-(9.76 \times 10^{-19} \, \text{J/molecule}) \times (6.022 \times 10^{23} \, \text{molecules/mol}) = -588 \, \text{kJ/mol}$

- In a molecule of NaCl, the two ions interact only with each other. Solid salt consists of a simple cubic lattice in which every Na^+ ion is surrounded by six Cl^- ions and every Cl^- ion is surrounded by six Na^+ ions. These additional attractions increase the attractive energy, so that the solid salt is bound by more energy than is the single NaCl molecule.

The coulombic attractive energy increases for smaller ions or for two doubly charged ions.

See Exercises 51, 53.

Ion–Dipole Interactions

Ion–dipole interaction
Interaction whenever an ion interacts with a polar molecule.

Most molecules are not charged, but rather are neutral. For neutral molecules, the shape of the molecule profoundly affects its interactions with other molecules. Molecules having a permanent dipole have regions that are positively charged and regions that are negatively charged. The intermolecular interactions between polar molecules and ionic materials result from these positive and negative regions of the polar molecule interacting with the ion charges.

Imagine the interaction between a positively charged sodium ion and several water molecules (Figure 6.14). There is an attraction between the lone pairs of the water molecules and the positive charge of the sodium ion. Clustering water around the sodium ion spreads the positive charge over the water–sodium ion cluster, so the charge density of the cluster is closer to neutral. Similarly, a negative ion such as Cl^- has an attraction for the hydrogen atom side of the water molecule (Figure 6.15). Because the chloride anion is larger than the sodium cation, more water molecules surround the chloride ion and the negative charge is spread over a larger volume. Salty ocean water is a major example of the interaction between NaCl and water.

CONCEPT QUESTIONS Use the δ^+, δ^- notation to illustrate the interaction between a water molecule and a sodium ion. What part of the water molecule is attracted to a Cl^- ion? ■

Figure 6.14 The positively charged sodium ion interacts with the excess electron density around the oxygen atom in water. On average, four water molecules with oxygen atom ends pointing toward the sodium ion surround a sodium ion in water.

The interaction between water and salt is an example of an **ion–dipole interaction.** An ion–dipole interaction occurs whenever an ionic substance interacts with a polar molecule. Such interactions are weaker than ion–ion interactions, typically having an energy of

15–20 kJ/mol. Because a dipolar molecule is neutral, the ion–dipole interaction also decreases more rapidly with distance than does the coulombic interaction between ions. The interaction energy is

(6.5)
$$E \propto -\frac{|z|\mu}{r^2}$$

where z is the charge on the ion and the Greek letter mu, μ, represents the dipole moment of the molecule. Dipole moments are measured in a unit called a debye, named for the Dutch chemist Peter Debye, who pioneered the study of molecular dipoles in the early 1900s. The debye, abbreviated D, is equal to 3.335641×10^{-30} C·m. Measurement of the dipole moment and bond length in simple molecules results in a quantitative measure of charge transferred from one element to the other (see Worked Example 6.11).

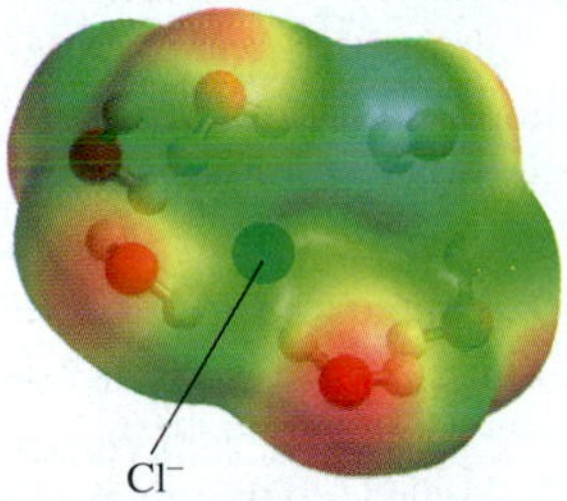

WORKED EXAMPLE 6.11   *Charge Separation*

Hydrogen is less electronegative than all of the halogens. Since hydrogen forms a simple diatomic molecule with any halogen, measuring the dipole moment of the hydrogen halides indicates the extent of transfer of the single electron from hydrogen to the halogen. Determine the amount of charge transferred from hydrogen to each of the halogens: F, Cl, Br, and I. Give the answer in fractions of an electron.

Data

   Dipole moment, μ: HF, 1.91 D; HCl, 1.08 D; HBr, 0.80 D; HI, 0.42 D
   Bond length: HF, 91.6 pm; HCl, 127 pm; HBr, 141 pm; HI, 161 pm

Plan

■ Use the definition of debye, D = 3.335641×10^{-30} C·m. Divide by bond length.

Implementation

■ HF charge separation $= \dfrac{(1.91 \text{ D}) \times (3.335641 \times 10^{-30} \text{ C·m/D})}{(91.6 \text{ pm}) \times (1 \text{ m}/10^{12} \text{ pm})(1.60 \times 10^{-19} \text{ C/electron})}$

   $= 0.435$ electron

The same calculation yields 0.177 electron for HCl, 0.118 electron for HBr, and 0.0544 electron for HI.
Electron transfer to the halogen decreases down the series, consistent with the decline in electronegativity.

See Exercise 59.

Figure 6.15 Surrounding a negatively charged chloride ion with water molecules results in the water molecules orienting with the hydrogen atoms pointing toward the chloride ion. Compared with a sodium ion, a chloride ion is larger, so more water molecules surround the chloride ion.

Cl^-

Hydrogen Bonding

Oxygen is sufficiently electronegative that the hydrogen atoms in water are relatively bare of electrons.

CONCEPT QUESTION   The OH bond in water has a dipole moment of 1.52 D and the bond length is 95.75 pm. What fraction of the electron from hydrogen is transferred to oxygen in water? ■

Dipoles act much like small magnets, and the opposite poles attract each other. With both a very concentrated positive charge—the electron-stripped hydrogen atoms—and

Figure 6.16 The strong dipole of the OH bond results in a strong attraction of the hydrogen end of a water molecule for a lone pair on the oxygen of another water molecule. Further association results in chains and rings of molecules, creating an extensive network.

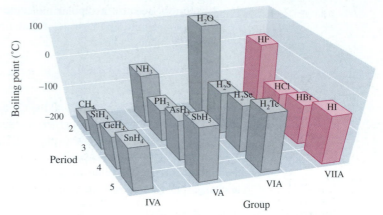

Figure 6.17 The boiling points of stable *p*-block hydrides show a regular variation down a group, with the notable exceptions of NH_3, H_2O, and HF. These three hydrides have anomalously high boiling points.

a very concentrated negative region—the lone pairs on the oxygen atom—the attraction between water molecules is particularly strong (Figure 6.16). Pairing two water molecules still leaves three hydrogen atoms and three lone pairs without interaction, however. These remain concentrated regions of charge, so additional water molecules may interact with this pair. This leads to chains and rings of interacting molecules, forming an extensive intermolecular network. Due to the extensive network, considerable energy is required to convert liquid water into water vapor. Indeed, the boiling point of water is relatively high (Figure 6.17).

An inspection of the boiling points of the stable *p*-block hydrides (Figure 6.17) indicates three hydrides with unusually high boiling points: NH_3, H_2O, and HF. To boil, enough energy must be acquired by the molecules in the liquid so that they can break free of their neighbors. For water, ammonia, and hydrogen fluoride, the energy required is large relative to other *p*-block hydrides. Bond polarity and shape hold the key to this difference.

CONCEPT QUESTION Determine the difference in electronegativity between hydrogen and four other elements: carbon, nitrogen, oxygen, and fluorine. What is the relationship between the electronegativity differences and the boiling points of the hydrides of carbon, nitrogen, oxygen, and fluorine? ∎

A density potential plot (Figure 6.18) of the second-period hydrides illustrates the effect of shape and bond polarity. The color map in Figure 6.18 is consistent across the four hydrides. The surface for methane is green, indicating that the surface is near neutral. In contrast, the surfaces for NH_3, H_2O, and HF show distinct positive (blue) and negative (red) regions. These positive and negative regions are a direct result of the large electronegativity difference between H with its single electron and N, O, and F. Like water, ammonia and hydrogen fluoride form an extensive, strongly interacting network. The large amount of energy required to cause these substances to boil is a reflection of the strength of this interaction.

Larger molecules that incorporate nitrogen or oxygen bonded to hydrogen show similarly strong interactions. Due to the importance of this interaction and resulting bond in determining the properties of water and biomolecules, the bond is given a special name: a **hydrogen bond.** Hydrogen bonding occurs whenever hydrogen is bonded to nitrogen, oxygen, or fluorine and is the strongest, nonionic intermolecular interaction known. Typical hydrogen bond energies are on the order of 20 kJ/mol.

Hydrogen bond A strong intermolecular interaction that occurs when hydrogen is between two very electronegative atoms (nitrogen, oxygen, or flourine).

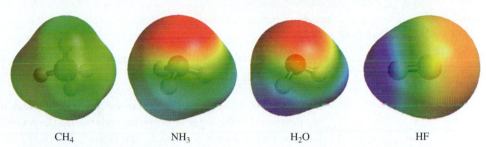

CH$_4$ · · · NH$_3$ · · · H$_2$O · · · HF

Figure 6.18 A density potential plot of the second-period, *p*-block hydrides shows the greater polarity of NH$_3$, H$_2$O, and HF as compared with CH$_4$. Polarity results from the molecular shape and the larger electronegativity difference between hydrogen and nitrogen, oxygen, or fluorine compared with hydrogen and carbon.

Dipole – Dipole Interactions

Dipole – dipole interaction
Interaction between molecules having a permanent dipole when hydrogen bonding is not involved.

The hydrogen-bonding interaction is an example of a larger class of interactions called **dipole – dipole interactions.** The term "dipole – dipole interaction" generally refers to interaction between molecules having a permanent dipole when hydrogen bonding is *not* involved. One example is SO$_2$ in water (Figure 6.19). This interaction involves the sulfur of SO$_2$ and the oxygen of water; thus it is a dipole – dipole interaction.

The interaction between dipoles is represented in two ways (Figure 6.20). Dipole – dipole interaction occurs because of the attraction of the opposite poles: negative and positive. In the case of H$_2$O, the oxygen is electron-rich, leading to a negative pole. As a manifestation of the famous statement that all is relative, although sulfur is generally an electronegative element, in SO$_2$ sulfur is the positive pole of the dipole because it is less electronegative than oxygen. Thus the interaction between water and sulfur dioxide occurs by alignment of the dipoles: the oxygen end of H$_2$O to the sulfur end of SO$_2$ (Figure 6.20a).

An alternate representation uses a curved arrow. The tail of the arrow originates at the electron-rich portion of the interaction, and the head of the arrow points to the electron-poor region (Figure 6.20b).

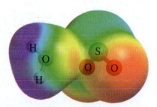

Figure 6.19 The intermolecular interaction of SO$_2$ with water is an example of a dipole – dipole interaction. The isolated SO$_2$ molecule has a dipole due to its bent geometry. Since sulfur and oxygen are both electronegative elements, the oxygen atoms in SO$_2$ are closer to neutral than they are in water. Upon interacting with water — with the negative end of water attacking the somewhat positively charged sulfur — the oxygen atoms acquire greater electron density.

Figure 6.20 Two representations of the interaction between SO$_2$ and H$_2$O. (a) The negative end of the water dipole is attracted to the positive end of the sulfur dioxide dipole. (b) The large electron density on the oxygen atom of water is attracted to the positively charged region near the sulfur atom of sulfur dioxide.

WORKED EXAMPLE 6.12   *Freons*

Chlorofluorocarbons (CFCs) are compounds consisting of carbon, chlorine, and fluorine. One of these is Freon-12: CF_2Cl_2. CFCs are now being phased out due to their deleterious effects on the ozone layer. They are being replaced with compounds that also include hydrogen (HCFCs). The C—H bond is vulnerable to attack in the troposphere, so HCFCs are destroyed in the lower atmosphere and thus do not reach the stratosphere. Determine the structure of Freon-12, CF_2Cl_2, and Freon-22, $HCClF_2$. Does either of these molecules have a permanent dipole? What type of interaction is present in liquid Freon-12? In liquid Freon-22? Draw a representation of the interaction using the curved arrow notation.

Plan

- Draw the Lewis dot structure.
- Evaluate bond polarity and determine the vector sum of the dipoles.
- Look for potential hydrogen bonds or dipoles.
- Interaction occurs from the negative pole to the positive pole.

Implementation

-

$$:\ddot{C}l:$$
$$:\ddot{F}:\overset{..}{C}:\ddot{F}:$$
$$:\ddot{C}l:$$

$$H$$
$$:\ddot{F}:\overset{..}{C}:\ddot{F}:$$
$$:\ddot{C}l:$$

CF_2Cl_2 $HCClF_2$

- The electronegativities are H, 2.20; C, 2.55; Cl, 3.16; and F, 3.98.

    In CF_2Cl_2 all bond dipoles point from the carbon atom to the halogen. The C—F bond dipole is stronger than the C—Cl bond dipole, so CF_2Cl_2 has a permanent dipole.
    In $HCClF_2$, the H—C bond dipole points from the hydrogen toward the carbon. $HCClF_2$ therefore has a significant dipole.

- Freon-12 does not have the potential for hydrogen bonding as it lacks hydrogen. Freon-22 does have a hydrogen atom, but it is not bonded to any of the very electronegative elements N, O, or F. Therefore Freon-22 also does not hydrogen bond. Both Freon-12 and Freon-22 interact via dipole–dipole interactions.
- In the dipole–dipole notation, the dipole of CF_2Cl_2 points from the midpoint of the chlorine atoms to the midpoint of the fluorine atoms. The dipole of $HCClF_2$ is slightly tilted, because the hydrogen atom is less electronegative than the chlorine atom. In the curved arrow notation, the electron-rich region between the fluorine atoms is attracted to the electron-poor region between the chlorine atoms or between the chlorine and the hydrogen atom.

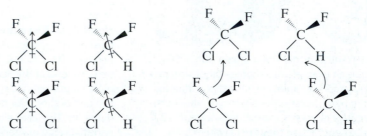

The dipole–dipole interactions among the CFCs and HCFCs are responsible for their ability to provide refrigeration.

See Exercises 67, 68.

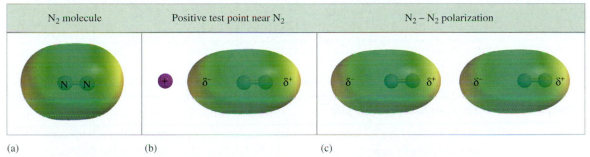

N₂ molecule	Positive test point near N₂	N₂ – N₂ polarization

(a)   (b)   (c)

Figure 6.21 (a) Like all homonuclear diatomic molecules, N_2 is nonpolar. (b) Bringing a positive test charge near the N_2 molecule attracts the negatively charged electrons and repels the positively charged cores. This results in a partial negative charge near the test point and a balancing partial positive away from the test point. (c) Due to the dynamic nature of the electron cloud, occasionally it will be shifted slightly relative to the atomic cores, giving rise to an instantaneous dipole. The positive end of the dipole attracts electrons in the neighboring molecule.

Induced dipole An instantaneous dipole created when the motion of an electron cloud causes the electron density to shift slightly with respect to the atomic cores of the molecule.

Induced dipole – induced dipole An intermolecular interaction resulting from an induced dipole inducing a dipole in a neighboring molecule. The resulting interaction force is called *London force*.

London force An intermolecular force resulting from induced dipoles on neighboring molecules. Also called an *induced dipole-induced dipole* force.

London or Dispersion Forces

Molecules that have no dipole and are not ionic can still have sufficient intermolecular forces to form a liquid. The German American physicist Fritz London first explained this intermolecular force, which is electrical in nature and results from the constant motion of the electron cloud (Figure 6.21). Occasionally the motion of the electron cloud results in the electron density being shifted slightly with respect to the atomic cores of the molecule. This shift creates an instantaneous dipole, called an **induced dipole.** The constant motion of the electron cloud means that the electrons respond rapidly to any change in the local environment, including a neighboring induced dipole. Like a line of dominoes, the induced dipoles ripple through the liquid, holding it together with an intermolecular force known as an **induced dipole – induced dipole** force. This intermolecular force is also called a **London force** in recognition of Fritz London's contribution toward the explanation of this force.

The strength of the London force depends on how loosely the electrons are held. The ease of movement of the electron cloud is characterized by the polarizability of the molecule. As a general guideline, larger atoms are more polarizable than smaller atoms. A larger polarizability is reflected in a larger temperature range for the liquid phase (Figure 6.22).

Figure 6.22 The Group IVA hydrides are all nonpolar molecules, interacting through induced dipole – induced dipole forces. The temperature range over which these hydrides are liquid increases as the size of the central element increases, reflecting a larger polarizability for the larger atoms.

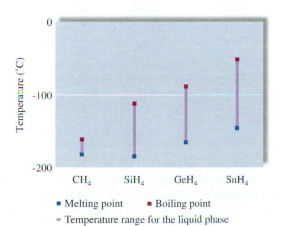

- Melting point   ■ Boiling point
- Temperature range for the liquid phase

The polarizability of a molecule is also responsible for the mixing or solubility of nonpolar molecules in polar liquids — for example, oxygen in water. Although the solubility of oxygen in water is not large, it is essential for aquatic life.

WORKED EXAMPLE 6.13   *Oxygen and Aquatic Life*

Dissolved oxygen in water is essential to aquatic life. Oxygen's solubility in water at room temperature is 2.293×10^{-5} mole fraction. How many oxygen molecules are in a cubic centimeter of water at room temperature? Compare the result with the number of oxygen molecules in air at room temperature. (The earth's atmosphere is 21% oxygen.)

Plan
- Determine moles of water per cm^3 at 298 K (25 °C). Use density.
- Calculate number of oxygen molecules. Use the mole fraction.
- Determine molecules per cm^3 in gas at 298 K. Use the gas law.
- Calculate oxygen molecules per cm^3. Use the mole fraction.

Implementation
- The density of water at 25 °C is 1 g/cm^3. Its molecular weight is 18 g/mol.

$$\text{Water molecules per } cm^3 = \frac{1.0 \text{ g}}{cm^3} \times \frac{1 \text{ mol}}{18.0 \text{ g}} \times \frac{6.022 \times 10^{23} \text{ molecules}}{\text{mol}}$$

$$= 3.3 \times 10^{22} \text{ molecules/}cm^3$$

- To a good approximation, one O_2 molecule takes the place of one water molecule, so the number of oxygen molecules = $(3.3 \times 10^{22} \text{ molecules/}cm^3) \times (2.293 \times 10^{-5}) = 7.6 \times 10^{17}$ oxygen molecules in 1 cm^3 of water.
- Gas density is 1 mol/22.4 L at STP. At 298 K, the density is

$$\frac{1 \text{ mol}}{22.4 \text{ L}} \times \frac{273 \text{ K}}{298 \text{ K}} \times \frac{6.022 \times 10^{23} \text{ molecules}}{\text{mol}} \times \frac{1 \text{ L}}{1000 \text{ } cm^3}$$

$$= 2.5 \times 10^{19} \text{ molecules/}cm^3$$

- If 21% of air is oxygen, the number of oxygen molecules is $(2.5 \times 10^{19} \text{ molecules/}cm^3) \times (0.21) = 5.3 \times 10^{18}$ molecules of oxygen per 1 cm^3 of air. There is about an order of magnitude more oxygen molecules in air than in water.

Aquatic life has adjusted to the lower oxygen concentration by filtering and concentrating the oxygen in water. However, pollution that reduces the amount of oxygen in water can prove fatal for fish and other life in ponds, rivers, and oceans.

See Exercise 71.

Mixing — that is, miscibility — of two liquids can be predicted based on an analysis of intermolecular interactions. Generally, if the intermolecular interactions in two liquids are of comparable strength, then the two liquids are miscible. This is often referred to as the "like dissolves like" principle. If the interactions in two liquids are not comparable, then the stronger interactions would be replaced by weaker ones, resulting in an overall decrease in interaction energy when the liquids are mixed.

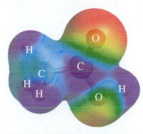

Figure 6.23 Vinegar is a solution of acetic acid in water. Due to the —OH group, the intermolecular interaction between acetic acid and water is a hydrogen-bonding interaction.

Table 6.3 Intermolecular Interaction Energy (in order of decreasing strength)

Interaction	Typical Energy (kJ/mol)	Distance Dependence	Species
ion–ion	250	r^{-1}	ions only
ion–dipole	15–20	r^{-2}	ions and polar molecules
hydrogen bond	20	r^{-3}	N, O, or F bonded to H with shared H
dipole–dipole	2	r^{-3}	polar molecules
dipole–induced dipole	2	r^{-6}	polar molecule with nonpolar molecule
London force: induced dipole–induced dipole	2	r^{-6}	nonpolar portions of molecules

Think about making salad dressing out of vinegar and oil. Vinegar is a solution of acetic acid (Figure 6.23) in water. Acetic acid contains an —OH group, so it forms a hydrogen bond with water molecules. Acetic acid is miscible with water. Chemically speaking, oil is not a precise term, because it consists of a mixture of compounds. All the compounds in oil, however, are hydrocarbons. Hydrocarbons are nonpolar molecules, interacting through London forces. The weaker London forces between water and oil cannot compete with the strong hydrogen-bonding interactions in water. Thus oil, like oxygen, is not very miscible with water.

Relative Strength of Intermolecular Forces

All substances interact via London forces, which are the weakest of the intermolecular forces. London forces are nonetheless important, giving rise to a stable liquid phase in nonpolar substances such as gasoline and oil at room temperature and in liquid nitrogen or noble gases at lower temperatures. Table 6.3 lists the various intermolecular interactions, their relative strengths, and the spatial extension.

Substances are classified according to the strongest intermolecular interactions present. For example, water is classified as a hydrogen-bonded substance even though the mobility of its electrons give rise to induced dipoles. The induced dipole–induced dipole interaction is less strong than the dipole–dipole interaction, and the dipole–dipole interaction is less strong than the hydrogen-bond interaction.

CONCEPT QUESTION   What are the intermolecular interactions present in each of the following: SO_2, CO_2, CCl_3F, PH_3, and H_3CNH_2? ∎

The temperature range over which a substance is a liquid is greatly affected by the type of intermolecular interactions present. Molten salts are typically liquid over more than 600 °C (Table 6.4). In recent years, room-temperature ionic liquids (Figure 6.24) have gained attention because of their potential for use in low-emission manufacturing processes. In addition to having a wide temperature range for stability of the liquid phase, ionic liquids have a very low vapor pressure, so little material escapes into the atmosphere and plant emissions are reduced.

$$H_3C-N\underset{\underset{\smile}{}}{\overset{\frown}{}}\overset{+}{N}-CH_3$$

Figure 6.24 A typical ionic liquid being considered for low-emission manufacturing processes, 1-methyl-3-methylimidazolium (DMIM), is shown in two representations. The chloride salt of DMIM is liquid over a temperature range of more than 200 ºC.

Table 6.4 Ionic Substances

These substances interact via strong ionic–ionic interactions and are liquid over a wide temperature range.

Substance	Melting Point (°C)	Boiling Point (°C)	Range (°C)
KI	981	1323	642
NaBr	747	1390	643
KF	858	1502	644
NaI	660	1304	644
NaCl	800	1465	665
$MgCl_2$	714	1412	698
NaF	996	1704	708
MgO	2826	3600	774
KOH	406	1327	921
NaOH	323	1388	1065
$CaBr_2$	742	1815	1073
CaF_2	1418	2533	1115
$CaCl_2$	775	1935	1161

6.3 Consequences: Surface Tension

On a calm day, the surface of a lake is mirror smooth. Water readily soaks into cotton fabric, but water beads into small, spherical drops on a waxed surface. These phenomena are a result of intermolecular forces.

Imagine molecules of water deep in the lake. Every molecule is surrounded on all sides by neighboring water molecules that interact with it via hydrogen bonds. Now think of a molecule on the surface of the lake. There are water molecules to the left, right, front, back, and below it, but there are no molecules above it. As a result, molecules on the surface have an unbalanced force pulling them into the liquid (Figure 6.25).

Figure 6.25 A drop of liquid is held together by surface tension. Molecules in the interior of the drop interact with neighboring molecules on all sides. At the surface, some of those interactions are missing. Molecules on the surface are therefore pulled inward due to the unbalanced force.

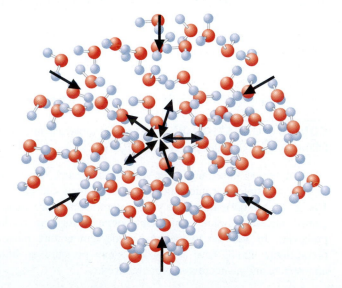

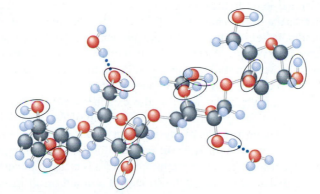

This unbalanced force results in **surface tension,** a force that resists an increase in area of the surface. In the absence of any other force, such as gravity, liquids tend to form spherical drops, as this shape minimizes the surface area.

Liquids with strong intermolecular forces therefore have a high surface tension. The units of surface tension are millinewtons per meter ($mN \cdot m^{-1}$) or equivalently $dyn \cdot cm^{-1}$. Water has a high surface tension, $72\ mN \cdot m^{-1}$, due to the extensive hydrogen-bonding network. Disrupting the network—for example, by replacing one hydrogen atom in water with a CH_3 group to form methanol (Figure 6.26)—results in a lower surface tension, $22.07\ mN \cdot m^{-1}$. Indeed, most liquids have surface tensions that are one-third to one-half that of water.

The interaction of liquids and solids also depends on intermolecular interactions. Cotton fiber is known for its ability to absorb water—a property that results from the —OH groups on the cellulose fibers (Figure 6.27). The hydrogen-bonding interactions between water molecules and cellulose result in water wetting the fiber. Materials that are wet by water are termed **hydrophilic,** which literally means "water loving." In contrast, water-proof fabrics are coated with hydrocarbons. The strong hydrogen-bonding interactions among water molecules pull the water molecules together. The weaker dipole–induced dipole interactions between water molecules and hydrocarbons on the water-proof fabric surface cannot compete with the hydrogen-bonding interactions, so water beads on the surface. Materials that resist water are termed **hydrophobic,** literally "water hating."

Figure 6.26 Methanol can be thought of as water with one hydrogen atom replaced by a CH_3 group. Methanol has mixed intermolecular interactions: The OH portion interacts via hydrogen bonding, but the CH_3 end interacts via London forces.

Surface tension A force that resists an increase in the area of the surface.

Hydrophilic Water loving: Readily absorbing water.

Hydrophobic Water hating: Repelling, tending not to combine with water.

Figure 6.27 Cotton fiber contains strands of cellulose from the structural components of the seed pod of the cotton plant. Cellulose contains many —OH groups and these are available for hydrogen bond interactions. Two water molecules are shown illustrating the hydrogen bonding interaction between water and cellulose. As anyone who has worn jeans out in the rain or snow knows, cotton absorbs water readily. Vinyl surfaces are very hydrophobic, so that water beads up on the surface rather than soaking in.

Checklist for Review

KEY TERMS

polarizability (p. 198)
miscible (p. 198)
lone pair (p. 199)
resonance structures
    (p. 200)
formal charge (p. 201)
valence shell electron-pair
    repulsion (VSEPR)
    (p. 204)
dipole (p. 209)
ion–ion interaction
    (p. 215)

ion–dipole interaction
    (p. 216)
hydrogen bond (p. 218)
dipole–dipole interaction
    (p. 219)
induced dipole (p. 221)
induced dipole–induced
    dipole (p. 221)
London force (p. 221)
surface tension (p. 225)
hydrophilic (p. 225)
hydrophobic (p. 225)

KEY EQUATIONS

Number of bonds = ½(electrons needed
                        − electrons available)

Formal charge = (assigned electrons
                        − valence electrons)

Coulombic energy, $E = \dfrac{kq_1q_2}{r}$

Ion–dipole energy, $E \propto \dfrac{|z|\mu}{r^2}$

Chapter Summary

Molecules stick together into larger structures, implying the existence of a mutually attractive force. The attractive forces are electrical in nature. The strongest intermolecular force is the ion–ion interaction that exists between charged species. Molten salts and ionic liquids are examples. Next in strength is the dipole–dipole interaction that results from unbalanced positively and negatively charged regions. The strongest of the dipole–dipole forces is hydrogen bonding. Hydrogen bonding results when hydrogen with its single electron is bonded to a very electronegative element: N, O, or F. The strongly electronegative element pulls electron density from the hydrogen atom, leaving it with a concentrated region of positive charge.

Intermolecular interactions also exist between nonpolar molecules, as evidenced by formation of a liquid. Instantaneous shifting of the electron cloud results in a transient dipole. The positive end of this transient dipole attracts the electron cloud of neighboring molecules, resulting in a dipole moment for the neighboring molecule. The intermolecular force resulting from the transient dipoles is called a London or dispersion force.

Unbalanced forces at the surface of a liquid result in surface tension. Surface tension is responsible for the beading of water on a waxy surface. When the interactions between the molecules of a liquid and a solid surface are comparable in strength to the interactions within the liquid, the liquid wets the surface.

KEY IDEA

Interactions among molecules are electrical in nature, so they depend on the electron distribution in the molecule. Among nonionic molecules, the shape is very important in determining whether a molecule has a dipole.

CONCEPTS YOU SHOULD UNDERSTAND

- The valence shell electron-pair repulsion method and valence bond theory are built on a foundation that includes the Lewis dot structure. The shape of a molecule and the electron distribution in a molecule are predicted with these tools.

OPERATIONAL SKILLS

- Determine the number of bonds in a molecule (Worked Examples 6.1, 6.2, and 6.9).
- Recognize resonance structures (Worked Example 6.3).
- Choose the best Lewis dot structure (Worked Examples 6.4 and 6.5).
- Determine electronic geometry and molecular geometry (Worked Examples 6.6 and 6.7).
- Identify type of intermolecular interaction (Worked Examples 6.8, 6.10, 6.11–6.13).

Exercises

A blue exercise number indicates that the answer to that exercise appears at the back of the book.

■ SKILL BUILDING EXERCISES

6.1 Shape: Lewis Dot Structures, Formal Charge

1. Draw Lewis dot structures for ONF, NF_3, $GeCl_4$, and SCl_2.

2. Draw Lewis dot structures for CH_4, C_2H_4, and C_2H_2.

3. Solid $CaSO_4$ is an ionic material consisting of Ca^{2+} ions and SO_4^{2-} ions. Draw the Lewis dot structure of SO_4^{2-}.

4. A common method for waterproofing a fabric consists of treating the fibers with a chemical that reacts with the OH groups on the fiber to remove them. One such chemical is trimethylchlorosilane, $(CH_3)_3SiCl$. Draw the Lewis dot structure of trimethylchlorosilane.

5. Ozone is both a desirable and a deleterious molecule. In the stratosphere, ozone protects land life from harmful short-wave ultraviolet radiation. In the troposphere, ozone damages delicate tissue, particularly in the lungs. Both the positive and the negative effects are due to the structure of ozone, which results in a lack of stability. Draw all possible resonance structures for ozone, O_3. Compare the bonding in ozone to that in diatomic oxygen, O_2. Suggest the basis for ozone's lack of stability.

6. Nitrates are commonly used in meat preservatives and in explosives. Draw all possible resonance structures for the nitrate ion, NO_3^-.

7. Draw at least three different Lewis dot structures for CO_2. Choose the best structure from among these.

8. Three different Lewis dot structures are shown for CS_2. Two of these can be classified as resonance structures. Identify them. Choose the best structure for CS_2.

$$:\ddot{S}:C:::S:\qquad :\ddot{S}::C::\ddot{S}:\qquad :S:::C:\ddot{S}:$$
$$\quad\;\text{A}\qquad\qquad\quad\text{B}\qquad\qquad\quad\text{C}$$

9. The gas phosgene, $COCl_2$, is extremely toxic. Draw at least two structures for phosgene and select the best structure.

10. Should hydrogen cyanide, a highly toxic molecule, be assigned the structure HCN or HNC?

11. The thiocyanate ion, SCN^-, is used in a chemical test for iron. In the presence of iron, SCN^- forms a blood red solution. Draw at least three Lewis dot structures for the thiocyanate ion. Choose the best of these structures and indicate why it is the best structure.

12. The molecule N_2O has been referred to as the Teflon of atmospheric chemistry because it is so unreactive that it passes largely untouched through much of the atmosphere. Draw at least three Lewis dot structures for N_2O and choose the best structure. Explain your choice.

6.1 Limitations of the Octet Rule

13. Nitrates, NO_3^-, are common ingredients in explosives. One product of the explosion is gaseous NO_2. NO_2 dimerizes easily, forming N_2O_4. Draw Lewis dot structures for NO_3^-, NO_2, and N_2O_4. Indicate why NO_2 dimerizes to N_2O_4.

14. Draw the Lewis dot structures of PF_5, XeF_2, $B(OH)_3$, and B_2H_6. Are any of these exceptions to the octet rule? For any that are exceptions to the octet rule, indicate in what way they violate the octet rule.

15. Draw the Lewis dot structures of NH_3 and BF_3. These molecules react to form H_3NBF_3. On the basis of the Lewis dot structures, suggest why this reaction occurs.

16. Draw the Lewis dot structures of $B(OH)_3$ and $B(OH)_4^-$. On the basis of these Lewis dot structures, suggest why $B(OH)_3$ reacts with water to yield $B(OH)_4^-$ and H_3O^+.

6.1 VSEPR Method

17. Indicate which of the following molecules or ions have lone pairs on the central atom: PCl_3, PCl_5, BF_3, AsF_3. Draw the Lewis dot structure of each to support your claim.

18. Indicate which of the following triatomic molecules are bent: N_2O, CO_2, NO_2, Cl_2O. Draw the Lewis dot structure of each to support your claim.

19. Peroxide, H_2O_2, is used as a disinfectant. What is the shape of the peroxide molecule? Predict the relative strength of the O—O bond in peroxide and in molecular oxygen.

20. Indicate the geometry of the following: H_3BO_3, O_3, H_3CNO_2 (at both the carbon and the nitrogen).

21. The methyl radical, CH_3, is a short-lived species that is important in understanding the elementary steps of a reaction. The shape of the methyl radical plays a fundamental role in deconvoluting the elementary steps. Draw the Lewis dot structure of the methyl radical. What is its geometrical shape?

22. The acetate ion, $H_3CCO_2^-$, is an important component of vinegar. What is the shape around each of the carbon atoms in the acetate ion? What are the expected H—C—H and O—C—O bond angles?

23. The shape of a molecule affects the encounter of that molecule with other molecules. As discussed in the chapter, the shape determines whether the molecule has a dipole moment. In addition, the shape affects the approach distance—the exposure of regions of high charge density. What is the geometry at each of the nitrogen atoms in DMIM, an ionic liquid anion? What is the shape of the pentagonal ring in DMIM?

$$H_3C-N\overset{\displaystyle\frown}{\underset{}{\quad}}\overset{+}{N}-CH_3$$
DMIM

24. Molecular shape affects packing in the solid state. Identify all sp^3 and sp^2 hybridized carbon atoms in coronene. Predict the shape of coronene.

Coronene

6.1 Consequences of Shape: Dipole Moment

25. What is the dipole moment of the molecule acetylene, C_2H_2?

26. Nitrogen is a fundamental element in formation of proteins and other structures in biological tissues. Protein structure is complex due to the large sizes of protein molecules, and local interactions near the nitrogen atoms play a role in determining that structure. As a simplified analogue, determine whether the following molecules or ions have a dipole moment: $N(CH_3)_3$, $NH_2CH_2CH_2COOH$ at the terminal nitrogen or terminal carbon, $N(CH_3)_4{}^+HCOO^-$.

27. Solubility of a solute in a solvent is highly dependent on the dipole moment of the solvent and the electron distribution in the solute. Determine which of the following solvents have a dipole moment: benzene, C_6H_6; acetone, H_3CCOCH_3; carbon disulfide, CS_2; ethanol, H_5C_2OH; acetic acid, H_3CCOOH.

28. Determine which of the following molecules have a permanent dipole: C_6F_6, SiF_4, PF_3, IF_5, AsF_5.

29. Draw the structures of water and ammonia. Label each with the δ^+ and δ^- notation.

30. The O—H bond length in water is 96 pm and the H—O—H bond angle is 104.5°. If the O—H bond were completely ionic, there would be a +1 charge on each hydrogen atom and a −2 charge on the oxygen atom. In this case, what would the dipole moment of water be? Compare this with the actual value of 1.854 D.

31. The carbon monoxide molecule, CO, has a very small dipole moment of 0.110 D and a bond length of 113 pm. Determine the amount of electron transfer from carbon to oxygen.

32. Lithium forms a gaseous molecule with hydrogen, LiH. The bond length is 159 pm and the dipole moment is 5.884 D. What is the extent of electron transfer from Li to H in LiH? Compare the answer to that for LiF ($\mu = 3.326$ D, bond length = 156 pm). Label LiF and LiH with δ^+ and δ^- notation.

33. The OH radical is extremely important in cleaning pollutants in the troposphere. The bond length in OH is 97 pm and its dipole moment is 1.668 D. What is the extent of electron transfer from H to O in OH? Compare the extent of electron transfer in OH to that in H_2O. (The O—H bond length in water is 96 pm, the H—O—H bond angle is 104.5°, and $\mu = 1.854$ D.)

34. HF is used to etch silicon in the semiconductor industry. HF is also a dangerous material, penetrating into the skin with little evidence until it reaches aqueous solutions like blood. It then causes an extreme burning sensation. Death follows quickly unless the problem is treated immediately. These characteristics are all dependent on the electron distribution in HF. The dipole moment of HF is 1.826 D and the bond length is 91.7 pm. Determine the extent of electron transfer from H to F in HF. Compare this to the extent of electron transfer in HI ($\mu = 0.0448$ D and bond length = 161 pm).

6.1 Valence Bond Theory

35. The central elements in CH_4, NH_3, and H_2O are all sp^3 hybridized, but only methane has no dipole moment. Explain this observation.

36. What is the hybridization of carbon in the following molecules: C_2H_6, C_2H_4, and C_2H_2?

37. Formic acid, HCOOH, is the substance that results in the stinging sensation accompanying the sting of a bee or bite of an ant. What is the hybridization of carbon in formic acid?

38. What is the hybridization of nitrogen in methyl isocyanide, H_3CNC? What is the hybridization of nitrogen in trimethyl amine, $(CH_3)_3N$?

39. Nitric acid, HNO_3, is an important chemical in the polymer industry. What is the hybridization at the nitrogen in nitric acid?

40. Give an example of a molecule that does not contain carbon, but that has sp^3 hybridization.

41. Transition elements near the beginning of all three transition series form a number of compounds, and these often use the d orbitals in addition to the valence s orbital. What is the hybridization in $TiCl_4$?

42. The transition elements form a number of compounds that are referred to as complex ions, and these often use the d orbitals in addition to the valence s and p orbitals. What is the hybridization in $FeFl_6{}^{3-}$?

43. What is the hybridization of the central oxygen in ozone, O_3? Compare that with the hybridization of sulfur in SO_2.

44. The molecule naphthalene is the main ingredient of mothballs. The structure of naphthalene can be drawn as shown here. Do all of the carbon atoms have the same hybridization? How many electrons are in the π cloud of naphthalene? Draw at least one additional resonance structure for naphthalene.

Naphthalene

45. The molecule TRIS is used in making gas-permeable, hard contact lenses. The shape of TRIS plays an important role in forming the pores that enable gas molecules to pass easily. Identify all sp^2 and sp^3 hybridized carbon atoms in TRIS. What hybridization is found at the silicon atoms?

TRIS

46. Caffeine is the stimulant found in coffee, tea, and many soft drinks. The structure of caffeine is shown here. Identify all sp^2 hybridized atoms in caffeine.

Caffeine

6.2 Intermolecular Interactions: Ion–Ion

47. Simple salts are held together primarily by ion–ion interactions. The melting and boiling points of these salts reflect the distance between the ions and the ion charge. Make a plot of the melting and boiling points of the sodium halide salts. Is there a correlation between halide ion size and melting point? Is there a correlation between ion size and boiling point? The largest of the halides, astatine, is radioactive, so many properties of its compounds have not been determined. From an estimate of the radius of the As⁻ ion (240 pm) predict the approximate melting and boiling point of NaAs.

Salt	Melting Point (°C)	Boiling Point (°C)
NaBr	747	1390
NaCl	800	1465
NaF	996	1704
NaI	660	1304

48. Both NaCl and CaO crystallize in the same lattice structure. Account for the following: The melting point of CaO is 2899 °C while that of NaCl is 800 °C.

49. Molten salts are generally good conductors of electricity due to the charge carried by the ions. In contrast, the conductivity of covalent materials is rather poor. Which of the following is the best conductor of electricity when melted? Which is the poorest conductor? Substances: HCl, NaCl, CCl_4, ICl.

50. When dissolved in water, ionic materials dissociate into their constituent ions and are excellent conductors of electricity. The following substances are all soluble in water: NaBr; Na_2S; acetone, CH_3—O—CH_3; methylene chloride, CH_2Cl_2. Choose the ones that are expected to be good conductors.

51. Assuming that each of the following molecules is 100% ionic, calculate the electrostatic energy of KCl, MgO, CaO, and NH_4Cl. Compare your results with that for NaCl in Worked Example 8.10. Bond length: KCl, 267 pm; MgO, 175 pm; CaO, 182 pm; NH_4Cl (where NH_4 is treated as a point charge at the nitrogen atom), 303 pm.

52. CaO is an ionic solid derived from lime, $CaCO_3$, by heating. The internuclear distance in CaO is 182 pm. Determine the electrostatic energy of a gas-phase CaO molecule. Draw a cartoon of the interaction between a water molecule and a CaO molecule. Indicate the interaction between a CaO molecule and a CO_2 molecule.

53. The lattice energy of an ionic salt is well approximated by the electrostatic energy of a gas-phase molecule times the Madelung constant, M. The constant M is the sum of a series of pairwise interactions. For example, for NaCl, the lattice is a simple cubic lattice with alternating Na⁺ and Cl⁻ ions (Figure 6.28). The first term in the electrostatic series is that due to nearest neighbors: $6 \times \dfrac{q_{Na^+}q_{Cl^-}}{r}$, where the factor of 6 is due to the six nearest neighbors. The next term is that due to repulsion of like charges: $\frac{1}{2}\left[8 \times \dfrac{q_{Na^+}q_{Na^+}}{\sqrt{2}r} + 8 \times \dfrac{q_{Cl^-}q_{Cl^-}}{\sqrt{2}r} \right]$; eight like charges surround each ion and factor of $\frac{1}{2}$ is due to double counting. What is the next term in the series? The series converges to 1.747 for NaCl. Write an expression for the lattice energy of NaCl per mole of NaCl.

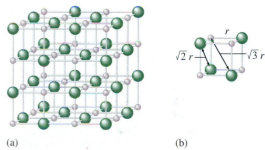

(a)   (b)

Figure 6.28 (a) The sodium chloride lattice consists of a simple cubic lattice of alternating Na⁺ and Cl⁻ ions. (b) A cube of the NaCl lattice illustrates the electrostatic interactions.

54. MgO crystallizes in the same lattice structure as NaCl. The crystal lattice energy of NaCl is 790 kJ/mol and that of MgO is 3791 kJ/mol. Suggest the basis for the larger lattice energy for MgO. The Na—Cl distance in solid sodium chloride is 270 pm. What is the Mg—O separation in solid magnesium oxide? (*Hint:* See Exercise 53.)

55. The lattice energy of sodium bromide, NaBr, is 754 kJ/mol, while that of NaCl is 790 kJ/mol. The Na—Cl distance in solid sodium chloride is 270 pm. What is the Na—Br separation in solid sodium bromide? (*Hint:* See Exercise 53.)

56. NaCl and CaS crystallize in the same cubic structure. The lattice energy is 790 kJ/mol, and the Na—Cl distance is 270 pm. In CaS, the separation between the Ca^{2+} cation and the S^{2-} anion is 289 pm. Determine the lattice energy of CaS. (*Hint:* See Exercise 53.)

57. Ionic liquids are receiving attention due their potential as solvents of low volatility. An example of an ionic liquid is 1-methyl-3-methylimidzolium, abbreviated as DMIM. The melting point of DMIM salts decreases with increasing anion size. Explain the basis for this trend. The melting point of the chloride salt is 87 °C, while the melting point of NaCl is 800 °C. Suggest a basis for the lower melting point of DMIM-Cl compared with NaCl.

H_3C—N$\overset{+}{\underset{\smile}{}}$N—$CH_3$

DMIM

58. Dust particles develop an electrostatic charge and are attracted to any surface that also develops a charge. This phenomenon is commonly observed as the dust that accumulates on a computer monitor or a television screen. On furniture, the charge is often the result of friction—hence the advice against dry dusting. One common ingredient in furniture polish is a siloxane polymer. Another common polish is lemon oil, the main ingredient of which is limonene. Explain how these two common dusting agents prevent dust buildup.

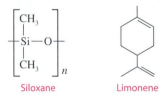

Siloxane   Limonene

6.2 Ion–Dipole Interactions

59. A dipole moment results from separation of charge. Determine the extent of charge separation in each of the following: CO, ClF, NaCl, CsCl, OH.

Substance	Dipole Moment (D)	Separation (pm)
CO	0.12	182.2
ClF	0.88	162.8
NaCl	9.0	236.0
CsCl	10.42	290.6
OH	1.668	97

60. Ionic materials dissolve in water due to the attraction between the dipole of water (1.854 D) and the charge on the ion. Rank the following ions in order of interaction energy with water, from strongest to weakest: Ca^{2+}, K^+, Li^+, Mg^{2+}, Na^+, Zn^{2+}.

6.2 Hydrogen Bonding

61. Two molecules are shown below: a polyester and cellulose. The polyester is similar to the fiber in polyester clothing. Cellulose is a major component of cotton fibers. Based on these structures, which absorbs water better, a cotton towel or a polyester towel? Explain your answer by indicating the potential for hydrogen bonding in each of the molecules.

62. Molecules used to make soft contact lenses result in a soft lens due to their ability to hydrogen-bond with water. Two such molecules are HEMA and DHPMA. Indicate all locations where hydrogen bonding can occur.

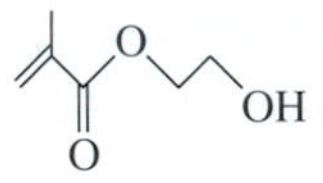

HEMA 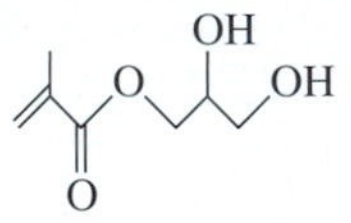   DHPMA

63. Which of the following can form hydrogen bonds?

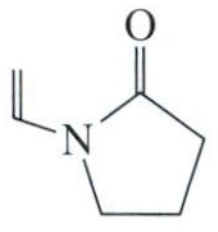

A 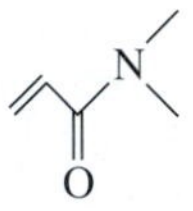   B 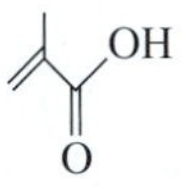   C

64. The molecules $B(OH)_3$ and B_2H_6 are examples of electron austerity and thus are exceptions to the octet rule. The interaction of each of these molecules with water is dominated by this electron austerity. Draw the Lewis dot structure for each of these molecules. Indicate the interaction of each molecule with water using curved arrow notation.

6.2 Dipole–Dipole Interactions

65. Which of the following interact by dipole–dipole forces in the liquid form: CCl_4, SO_3, NO_2, or N_2O_5? Support your answer with the geometrical structure.

66. Liquid methylene chloride, H_2CCl_2, is used as a degreasing agent in fabrication of metal parts. Indicate the structure of this molecule and the intermolecular interaction among methylene chloride molecules.

67. Ethylene glycol, CH_2OHCH_2OH, is a common ingredient of antifreeze. Use curved arrow notation to indicate the interaction between an ethylene glycol molecule and a water molecule.

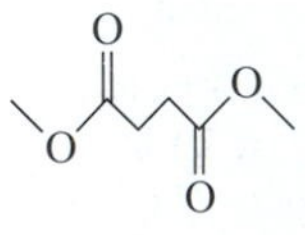

A polyester

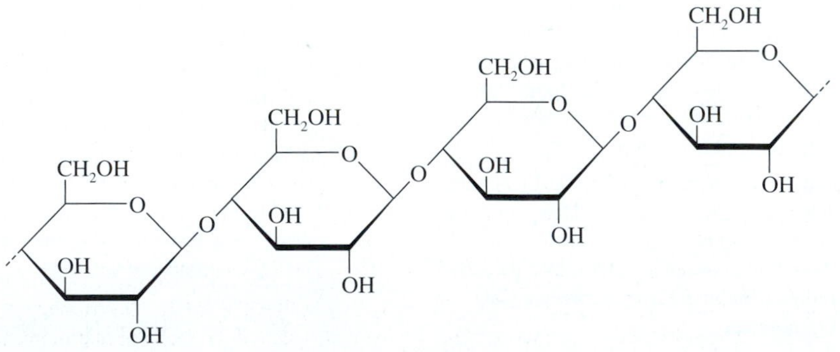

Cellulose

68. Use the dipole moment and internuclear separation data given here to determine the extent of electron transfer in NO, CO, and SO. Comment on any trend you observe.

	CO	NO	SO
Dipole moment (D)	0.110	0.159	1.55
r		182	175	208

69. The dipole moment of water is 1.854 D, the O—H distance is 97 pm, and the H—O—H bond angle is 104.5°. Calculate the extent of charge transfer from H to O in water. The dipole moment in H_2S is 0.97 D, the H—S bond distance is 133 pm, and the bond angle is 92.2°. Compare the extent of charge transfer in H_2S with that in H_2O.

70. Label the following molecules with the δ^+ and δ^- notation: H_3CCN, PCl_3, F_2O, PCl_5. Justify your labels.

6.2 London or Dispersion Forces

71. Given the following solubility data for gases in water, calculate the number of molecules of each gas dissolved per cubic centimeter of water. Comment on any trend you observe.

Solublity (mole fraction)				
He	Ne	Ar	Kr	Xe
7.04×10^{-6}	8.40×10^{-6}	2.45×10^{-6}	5.04×10^{-6}	9.05×10^{-6}

72. Plot the boiling point of the following substances versus period in the periodic table: He, Ne, Ar, Kr, Xe, Rn. Comment on any trend you observe.

73. Discuss the basis for the statement that among substances with similar structure, the boiling point increases linearly with molecular mass.

6.2 Relative Strength of Intermolecular Forces

74. Hard contact lenses are made from a polymer based on methyl methacrylate. What type of intermolecular interaction is present between methyl methacrylate molecules? Justify your answers.

Methyl methacrylate

75. Hard contact lenses with good gas permeability were formulated to allow the cornea of the eye to receive oxygen from the air. The cornea needs a direct supply of oxygen because it has no blood vessels. Two molecules used to make gas-permeable hard contact lenses are TRIS, illustrated in Exercise 45, and methyl methacrylate, illustrated in Exercise 74. Identify the sites, if any, with potential for intermolecular interactions via each of the following: hydrogen bonding, dipole–dipole interactions, and London dispersion forces.

76. The molecule HFIM is a component of gas-permeable contact lenses. Is HFIM a candidate for formation of hydrogen bonds? Why or why not?

HFIM

6.3 Consequences: Surface Tension

77. A windshield glass treatment uses CFC-113 to make the glass water-repellent. Does CFC-113 have a dipole moment? What intermolecular force operates in liquid CFC-113? What intermolecular force operates between water and CFC-113?

$$FCl_2C—CF_2Cl$$
CFC-113

78. Use the following data to explain how the breathable, water-repellent fabric Gore-Tex® functions. Gore-Tex consists of a Teflon film over a synthetic fiber with extremely small pores—about one-and-a-half billion per square centimeter. Each pore is less than 0.5 μm in diameter. Rain drops of less than 1 μm in diameter are unstable and vaporize. Water does not wet Teflon.

■ CONCEPTUAL EXERCISES

79. Draw a diagram indicating the relationships among the following terms: lone pair, electronic geometry, molecular geometry, VSEPR, dipole, valence bond, hybridization. All terms should have at least one connection.

80. Draw a diagram indicating the relationships among the following terms: ion–ion interactions, hydrogen bonding, induced dipole, London force, polarizability, hydrophilic and hydrophobic. All terms should have at least one connection.

APPLIED EXERCISES

81. The active ingredient of the commercial formulation known as Rain-X® is a long-chain molecule with the structure shown below, where n is some large integer. The manufacturer claims that Rain-X improves vision in rain, prevents the buildup of frost, ice, and road grime. Does the active ingredient have a permanent dipole moment? Indicate the interaction between the active ingredient in Rain-X and water molecules. Suggest how Rain-X works considering the answers to the previous questions.

82. In theater productions, a dramatic effect is sometimes added with the use of fog. The art of creating an artificial fog is based on the science of intermolecular interactions. The machines that create fog use a combination of compounds

A and B, shown below. Does either A or B have a permanent dipole? Is either A or B a candidate to form hydrogen bonds? Explain the interaction between A and water, and between B and water. Neither A nor B is very volatile, so they cannot create fog either alone or in combination. However, when combined with water and heated, a fine fog is created. Explain the action of A and B in creating a fog.

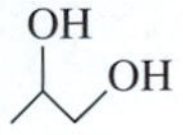

A 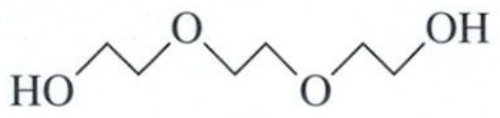   B

■ INTEGRATIVE EXERCISES

83. Bubbles in a soft drink are spherical, as are the bubbles in a glass of beer or champagne. These bubbles are primarily composed of carbon dioxide. Discuss why these bubbles are spherical. Consider the following in your answer: What is the major constituent of soda, beer, or champagne? What forces are involved in generating the spherical shape?

84. Molecular shape plays an important role in the successful manufacture of contact lenses. The original hard contact lenses were based on methyl methacrylate, MMA (panel A). Adding molecules that pack less efficiently makes pores in the lens, allowing oxygen to penetrate; the result is called breathable lenses. The molecules abbreviated TRIS and HFIM (panel A) are two examples. A soft lens is based on a molecule that absorbs water. Analyze the shapes of TRIS

and HFIM, indicating the origin of the poor packing. Of the molecules in panel B, which might be used as the basis of a soft contact lens? What is the site for hydrogen bonding, if any, for molecules in panel B?

Panel A

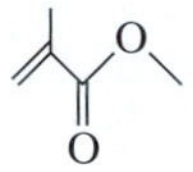

MMA 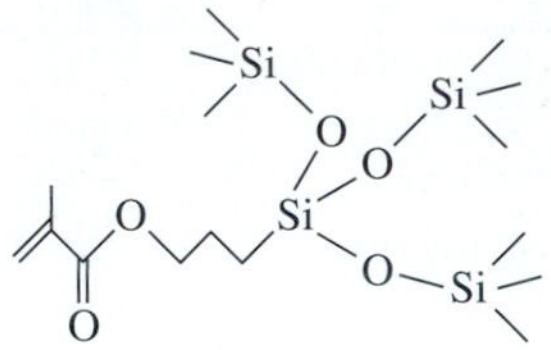   TRIS 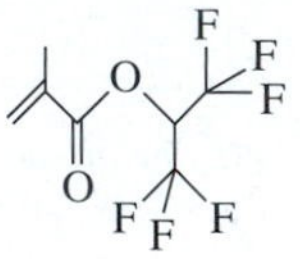   HFIM

Panel B

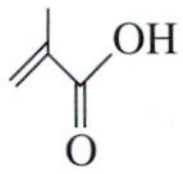

MAA 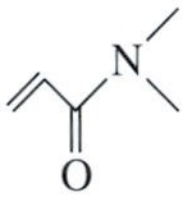   DMA 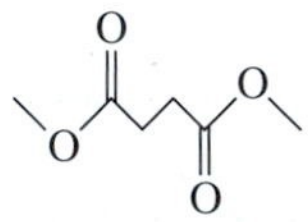   EDA

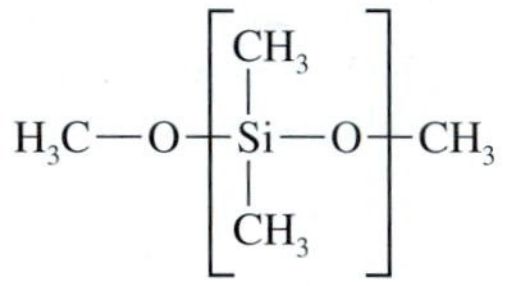

TMSM

Chapter 7
Thermodynamics and the Direction of Change

Thermal energy and instabilities in the atmosphere spawn tornadoes, hurricanes, and cyclones. These storms can occur in either hot weather or over heated ocean water from which the storm draws its energy.

CONCEPTUAL FOCUS

- Track energy in reactions and develop criteria for reactions that can occur.
- Relate energy to heat flow and work done.

SKILL DEVELOPMENT OBJECTIVES

- Determine heat absorbed or evolved in a process from the heat involved in standard processes (Worked Examples 7.1, 7.3, 7.6).
- Calculate work done by a reaction (Worked Example 7.2).
- Evaluate interaction energy from heat (Worked Examples 7.4, 7.5, 7.7 – 7.9).
- Predict reaction direction (Worked Examples 7.10 – 7.12).

E nergy is released when two atoms form a chemical bond. Rearranging bonds as molecules or compounds react absorbs or releases energy, reflecting the changing interactions. This energy flow makes the world go 'round, enabling life as we know it. The relationship of energy, and specifically heat energy, to chemical change is important for understanding chemical interactions and is of practical importance for real-world applications. For example, the heat evolved in constructing a dam out of concrete is sufficient to crack the dam unless the heat is taken away by water circulating in pipes installed in the massive structure (Figure 7.1a). The process responsible for much of the heat involved in the setting of concrete is the interaction of mortar, primarily $Ca(OH)_2$, with water (Figure 7.1b and 7.1c).

While energy flow is important, many of the benefits of modern chemistry result from the ability to rearrange bonds in a predictable way—to transform readily available, inexpensive substances into new, highly desirable ones. Examples include forming stainless steel from easily corroded iron, pharmaceuticals from simple chemicals, or polymers from petroleum stock. The ability to predict the outcome of the interaction between molecules is based on two quantities: the heat evolved or absorbed by a reaction and the energy dispersal.

Reactions that occur at constant pressure are very common in chemistry. Concrete sets, you digest your food, batteries drive charge flow—all of these are processes that occur under the constant pressure of the earth's atmosphere. Due to the prevalence of constant-pressure processes, the heat evolved in this case is given a specific symbol, **ΔH,** and is referred to as the **heat of reaction,** or the **enthalpy change** for the system. The heat of reaction reveals important information about interactions among the reactants, the products, and shifting interactions as a reaction progresses. As indicated in the case studies that follow, the interactions alone do not indicate whether a reaction can occur. To predict the direction of a reaction, information about the energy dispersal is needed. This topic is addressed in the later sections of the chapter.

Heat of reaction, ΔH Heat produced during constant-pressure processes. Also known as *enthalpy change*.

Enthalpy change, ΔH Heat transferred for constant-pressure processes. Also known as *heat of reaction*.

(a)

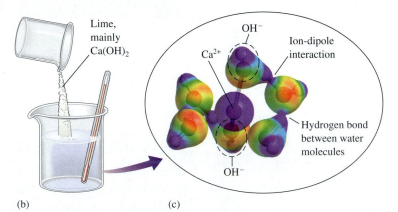

Lime, mainly $Ca(OH)_2$

OH$^-$

Ca^{2+}

Ion-dipole interaction

Hydrogen bond between water molecules

OH$^-$

(b)   (c)

Figure 7.1 (a) The heat produced as concrete sets, such as in a large dam, can be sufficient to crack the structure unless cooling coils with circulating water remove the heat. (b) A significant portion of that heat is attributable to reaction between lime, $Ca(OH)_2$, and water. (c) A molecular-level view of the interactions between $Ca(OH)_2$ and water. Note the very positive potential of the Ca^{2+} ion.

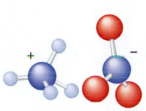

7.1 Case Study: Cold Packs, Hot Packs, and Welding

Injuries in sporting accidents are often treated with an "instant" cold pack. Such a pack consists of a tough outer pouch containing water (often with a blue dye to signify cold) and an inner pouch containing a salt. The term *salt* is used in the generalized chemical sense, meaning a substance consisting of oppositely charged ions held together by coulombic forces. Magnifying the cold-pack salt by a hundred million times reveals that it consists of two ions (Figure 7.2), both of which are made up of several atoms. One ion is NH_4^+, called the ammonium ion; the other is NO_3^-, called the nitrate ion. When the inner pouch of the cold pack is broken (Figure 7.3), water from the   outer pouch mixes with the salt, separating the oppositely charged ions and surrounding them with water molecules. The cold sensation accompanying this process indicates that heat is drawn into the pack from the area around it—including the injured area.

While sprains and pulled muscles are often treated with cold therapy, heat is usually recommended for muscle aches or spasms. One method for producing the heat is to use a hot pack. Like cold packs, hot packs consist of an outer pouch containing water and an inner pouch containing a salt. In this case, the salt is often $CaCl_2$, a solid consisting of Ca^{2+} and Cl^- ions. For this salt, forming the solution results in heat flowing out, just as heat flows out when wood or other fuel is burned.

The amount of heat produced in a chemical reaction can be quite substantial. For example, in remote areas welding of railroad rails is done using the reaction of metallic aluminum with iron oxide:

(7.1) $$Fe_2O_3(s) + 2Al(s) \longrightarrow 2Fe(l) + Al_2O_3(s)$$

The reaction in Equation (7.1) generates so much heat that the elemental iron produced is molten (Figure 7.4)! The molten iron flows around the rails, and the rails conduct the heat away, thereby solidifying the iron and fusing the rails together. Because all of the oxygen needed to form Al_2O_3 comes from the Fe_2O_3, once started the reaction is self-sustaining and can take place in the absence of added oxygen. It can even be used to weld underwater. This reaction is referred to as the thermite reaction.

In the following sections, these three reactions are examined in more detail, connecting the heat flow with differences in interactions among the reactants and those among the products.

Figure 7.2 The term "salt" is often used in chemistry to denote a solid consisting of oppositely charged ions. The ions are held together by coulombic attraction between the positively charged ion and the negatively charged ion. The salt illustrated here consists of NH_4^+ ions and NO_3^- ions. Both of these ions contain several elements covalently bonded together. The charge indicates that the group of ions has collectively either lost (+) or gained (−) an electron. The group NH_4^+ is called an ammonium ion, and the group NO_3^- is called a nitrate ion. Both are commonly found in the environment, in the foods we eat, and in our bodies.

Figure 7.3 When the inner pouch is broken, the salt and water mix, forming a solution and drawing heat in from around the cold pack.

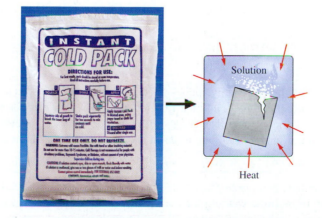

Figure 7.4 Finely divided aluminum is added to powdered iron oxide, and the reaction is started by lighting a strip of magnesium. Once started the reaction is self-sustaining and very spectacular, generating enough heat to raise the temperature of the mixture to nearly 2000 °C. This reaction should not be tried at home; only properly trained personnel should run it.

APPLY IT Go to a sporting goods store or a ski shop and obtain a product called a "hand warmer." Two varieties of hand warmer exist: One contains a solution and is reusable; the other contains a solid and is designed for one-time use. For the solution variety of hand warmer, activate the warmer. Describe your observations. How is the product restored to its original solution state? What are the components of the solid hand warmer? Propose a mechanism for the solid hand warmer.

7.2 The First Law of Thermodynamics: Heat and Energy in Chemical Reactions

Conservation of energy applies to chemical processes just as it does to any other process. For example, light is emitted from an atom when an electron in an excited state falls to a lower energy state: The energy difference appears as a photon. The principle of conservation of energy is so important in chemical reactions that it is given the status of a law.

The first law of thermodynamics: *The energy of an isolated system is constant.*

Here the term "isolated system" refers to everything affected by the reaction. For example, in thermite welding the reactants (Al and Fe_2O_3), the products (Fe and Al_2O_3), plus the rails and surrounding air all constitute the isolated system.

Caution: Thermodynamics uses much of the same language employed in everyday life, but uses it in a very precise way. For example, it is common to link heat and temperature. But think about a mixture of ice and water: Adding heat to the mixture shifts the proportions of ice and water, but the temperature remains of 0 °C. *Heating* a container of ice water does not warm it; it does not change the temperature.

Internal energy A property of a system that can be changed by a flow of work, heat or both.

Heat, *q* Energy transferred in a constant volume process.

Work, *w* Force acting over a distance. For processes with gases: $dw = -PdV$.

In a chemical reaction, *energy* refers specifically to the **internal energy**—the sum of the potential energy due to interactions between molecules, the kinetic energy due to the motion of the molecules, and the chemical energy stored in the bonds. It is often useful to subdivide the isolated system, which according to the first law has constant energy, into the reaction of interest and everything else—the surroundings. Only rarely is the energy change in the reaction sufficient to appear in the surroundings as light. Fireflies and excited atoms are examples of such exceptions. More typically, energy is observed as a combination of **heat** and a volume change. Heat is given the symbol *q*. The volume change in a constant-pressure process is referred to as pressure-volume **work** and given the symbol, *w*. Work is discussed further later in this chapter. Heat and work are the currencies with which energy is extracted from a system and transferred to the surroundings. Focusing on the system of interest leads to an alternate statement of the first law:

> The first law of thermodynamics: *The change in internal energy of a system is equal to the heat transferred to the surroundings plus the work done on the surroundings.*

(7.2) $\Delta E = q + w$

The change in energy of the system *plus* surroundings is zero: Energy lost by the system is gained by the surroundings. For example, combustion of gasoline in an automobile engine drives the pistons—the volume changes—and heats the car.

Heat

Many important chemical processes occur at atmospheric pressure, such as the freezing of a lake in winter, digestion of food, the response of the body to a drug, or rusting of an automobile body. For this important class of reactions, that part of the energy change in the reaction that appears as heat in the surroundings is called enthalpy, from the Greek *thalpein* meaning "to heat" (Figure 7.5). Energy releasing reactions heat the surroundings—the products have a lower enthalpy than the reactants ($\Delta H < 0$)—and are termed **exothermic** (from the Greek *exo,* meaning "out" or "outside"). The hot-pack reaction is exothermic. Conversely, energy-absorbing reactions cool the area around the reaction—the products have a higher enthalpy than the reactants ($\Delta H > 0$)—and are called **endothermic** (from the Greek *endon,* meaning "within").

Exothermic A process in which heat is released.

Endothermic A process in which heat is absorbed.

Figure 7.5 The enthalpy landscape. Energy flowing out during a reaction appears as heat. Reactions requiring energy input absorb energy, cooling the area around the reaction.

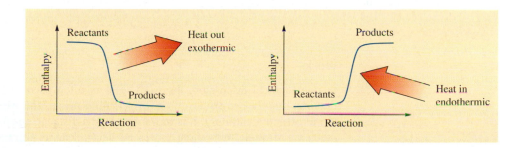

Heat capacity, C_p The ratio of the heat supplied to the temperature rise per unit mass or mole.

If the cold pack is put into a container of water rather than against an injury, the water temperature decreases. Measuring the temperature decrease determines how much heat is extracted from the water. The **heat capacity, C_p,** of any substance is the amount of heat required to raise the temperature by one degree K. The heat capacity of water is 4.18 J/K · g, so the heat absorbed in the cold-pack reaction can be determined by multiplying the heat capacity of water by the temperature drop and the known mass of water.

(7.3)
$$\Delta H = C_p \, \Delta T$$

The heat capacity is a measure of the energy storage capacity of a material. The heat capacity of water is relatively high, nearly six times that of rock (~0.7 J/K · g) and helps to moderate the temperature on earth. During the day, energy reaching the surface from the sun is largely captured by water, keeping the temperature on the planet from becoming searing hot. Conversely, at night heat is released as water cools, keeping the temperature from becoming frigid.

CONCEPT QUESTION Why do temperatures in the desert become very cold at night despite being very hot during the day? ■

WORKED EXAMPLE 7.1 *Heating with Gas*

A comment about units: In previous chapters the energy unit electron-volt (eV) was used to describe processes involving single atoms, such as removing one electron from an atom in the gas phase. An electron-volt is a very small unit of energy. A larger-magnitude unit, joules per mole (J/mol), is used when describing the energy for processes involving moles of atoms or molecules. Chemical reactions usually involve on the order of a mole of atoms or molecules, so J/mol is the more suitable unit. One electron-volt per atom for a mole of atoms corresponds to 96,485 J/mol.

An apparatus is designed that burns gas at 1 atm pressure. All of the heat generated is captured by a large vessel of water. Octane, one of the components of gasoline, is burned in this vessel and 2.28 g (0.02 mol) raises the temperature of 5.00 L of water by 5.230 °C. How much heat is generated per mole of octane burned?

Plan
- C_p(water) = 4.184 J/K · g; need grams of water in 5.00 L
- Use $\Delta T = 5.203$ °C = 5.230 K and Equation (7.3), $\Delta H = C_p \, \Delta T$
- Scale up to 1 mol

Implementation
- 5.00 L = 5.00×10^3 mL. Density of water is 1 g/mL, so 5.00×10^3 mL = 5.00×10^3 g
- $\Delta H = (4.184 \text{ J/K} \cdot \text{g})(5.230 \text{ K})(5.00 \times 10^3 \text{ g}) = 1.09 \times 10^5 \text{ J} = 109 \text{ kJ}$
- 109 kJ/0.02 mol = 5.47×10^3 kJ/mol

The heat absorbed by the water is equal to the heat extracted from the octane, so $\Delta H_{\text{combustion}}$(octane) $= -5.47 \times 10^3$ kJ/mol. Octane is one component of gasoline used in automobiles, and 0.02 mol of it is about 3 mL (molecular weight, 114.23 g/mol; density, 0.6986 g/mL) or a little over ½ teaspoon. So ½ teaspoon is sufficient to heat 350 mL of water from room temperature to 90 °C—a good temperature for making a mug of tea!

See Exercises 1–26.

Work

In nontechnical discussion, the word *work* is often used to describe any activity that requires muscle or mental effort. In scientific usage, the word *work* has a somewhat more restrictive meaning: It is movement over a distance against a force. No matter how much we struggle, if nothing moves no work is done. In chemical systems, the most common form of work is pressure-volume work. For example, a car engine burns gasoline converting the internal energy into work to propel the car down the road and into heat that warms the engine and interior of the car. The work associated with combustion of the gasoline occurs in the cylinders (Figure 7.6). Pushing the pistons up expands the volume of the cylinders and extracts energy from the gasoline. The force on the piston is applied over the area of the piston: Force per unit area is pressure, the pressure of pressure-volume work. The volume of pressure-volume work is the volume change for the cylinder, the distance that the piston moves times the area. Mathematically, the work is

(7.4) $$w = -\text{force} \times \text{distance} = -\left(\frac{\text{force}}{\text{area}}\right) \times (\text{area} \times \text{distance})$$

$$= -(\text{pressure} \times \text{volume change}) = -P_{\text{applied}} \Delta V$$

Pushing the piston outward against the applied pressure (P_{applied}) decreases the energy of the reacting system, so by convention, it has a negative sign. Worked Example 7.2 illustrates how work is calculated.

WORKED EXAMPLE 7.2   *The Work of Driving a Car*

Force

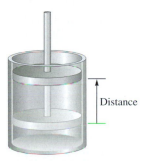

(a)

Distance

(b)

Figure 7.6 (a) Schematic of a car piston showing the force against which the piston moves. (b) Work is done as the piston moves from the initial position to the final position due to combustion of the fuel.

A typical automobile features a 3.0-L, six-cylinder engine. Six cylinders means that there are six pistons in the engine. The 3.0 L refers to the displacement of those pistons when the engine runs. Suppose the car is running down the highway and the engine is turning at 2500 rpm (revolutions per minute). The pistons fire 2500 times per minute, expanding by 3.0 L each time. Assume that the pistons are pushing against an effective pressure of 1.5 atm, how much work is done by the combustion of gasoline per minute? If the car cruises at 60 mi/h and gets 30 mi to the gallon, how much fuel is consumed in 1 min? Assume the fuel is all octane, determine the work produced per mole and compare it with heating determined in Worked Example 7.1.

Plan
- Determine the total volume change per minute.
- Use Equation (7.4), $w = -P_{\text{applied}} \Delta V$.
- Convert to energy units, kilojoules (1 L · atm = 101.3 J).
- Determine gallons/minute, convert to liters (1 gal = 3.7854 L).
- Use density (0.6986 g/mol) and molecular weight (114.23 g/mol) to convert volume to moles; scale up.

Implementation
- Volume change per minute, $\Delta V = (3.0 \text{ L/revolution}) \times (2500 \text{ revolutions/min}) = 7.5 \times 10^3 \text{ L/min}$
- $w = -P_{\text{applied}} \Delta V = -(1.5 \text{ atm}) \times (7.5 \times 10^3 \text{ L/min}) = -1.1 \times 10^4 \text{ L} \cdot \text{atm/min}$
- $w = (-1.1 \times 10^4 \text{ L} \cdot \text{atm/min}) \times (1.01 \times 10^2 \text{ J/L} \cdot \text{atm}) = -1.1 \times 10^6 \text{ J/min}$
  $= -1.1 \times 10^3 \text{ kJ/min}$
- Fuel consumption in L/min:

$$\frac{60 \text{ mi}}{\text{h}} \times \frac{1 \text{ gal}}{30 \text{ mi}} \times \frac{1 \text{ h}}{60 \text{ min}} \times \frac{3.7854 \text{ L}}{\text{gal}} = 0.126 \text{ L/min} = 126 \text{ mL/min}$$

- $126 \text{ mL/min} \times \dfrac{0.6986 \text{ g}}{\text{mL}} \times \dfrac{\text{mol}}{114.23 \text{ g}} = 0.771 \text{ mol/min}$

$$\frac{1.1 \times 10^3 \text{ kJ/min}}{0.771 \text{ mol/min}} = 1.5 \times 10^3 \text{ kJ/mol}$$

In Worked Example 7.1, it was determined that combustion of octane produced 5.47×10^3 kJ/mol in heat. Work extracted is roughly one-fourth the heat produced. (*Note:* Converting all of the internal energy to work requires the car engine to be 100% efficient. Real engines are never 100% efficient.)

See Exercises 27–34.

Hess's Law

Hess's law The energy change in any process is the same irrespective of the path followed.

Using heat energy to probe interactions relies on an important result: The enthalpy change for any process, ΔH, depends only on the amount of matter and the nature of the starting and ending substances, *not* on the nature of the process. In particular, the enthalpy change for a process is the same whether that process occurs in one step or in a series of steps. This principle, which was formulated by G. H. Hess in 1840, is known as **Hess's law.**

> Hess's law: The net enthalpy change of a process is the algebraic sum of the enthalpy changes for the several steps.

State function A property that depends only on the state of the system: It is independent of the pathway to arrive at that state.

Enthalpy is said to be a **state function.** That is, enthalpy depends only on the present condition of the system, not on how it arrived at that condition. An immediate consequence of Hess's law is that the enthalpy change for a reaction, ΔH, is equal in magnitude but opposite in sign for the reaction in the reverse direction.

CONCEPT QUESTION Energy, called the latent heat of fusion, ΔH_{fus}, must be added to ice to melt it into liquid water. How does spraying a citrus crop with water prevent the fruit from being ruined by frost? ∎

Hess's law is used to subdivide a process into smaller steps so that the enthalpy change can be determined from known enthalpies.

WORKED EXAMPLE 7.3   *Snowstorms and Heat*

Water in clouds often becomes supercooled—that is, remains liquid at a temperature below 0 °C—before forming ice. When liquid water freezes, energy in the form of heat is released. Compare the energy released per gram of water from a cloud of supercooled raindrops at -10 °C with that from a cloud of drops that freeze at 0 °C.

Data
ΔH_{fus}(water, 0 °C) = 333.6 J/g
Heat capacity (ice) = 2 J/g·K
Heat capacity (water) = 4.2 J/g·K

Plan
∎ Devise a path from water to ice at -10 °C that passes through water → ice at 0 °C.

Know	**Want**
Heat released for	Heat released for
water $\xrightarrow{(0\ °C)}$ ice	water $\xrightarrow{(-10\ °C)}$ ice

∎ Determine the heat for the added steps.

Implementation
∎ Path

Water $(-10$ °C$) \longrightarrow$ water $(0$ °C$) \longrightarrow$ ice $(0$ °C$) \longrightarrow$ ice $(-10$ °C$)$

■ The first added step, water $(-10\ °C) \rightarrow$ water $(0\ °C)$, requires water to be warmed by 10 °C (equal to 10 K). The second added step, ice $(0\ °C) \rightarrow$ ice $(-10\ °C)$, cools ice by 10 °C. Warming water by 10 °C requires input of $(4.2\ J/g \cdot K \times 10\ K) = 42\ J/g$. The cooling of ice involves $\Delta H = [2\ J/g \cdot K \times (-10\ K)] = -20\ J/g$. At $-10\ °C$:

$$\text{water}\ (-10\ °C) \xrightarrow{+42\ J/g} \text{water}\ (0\ °C) \xrightarrow{-333.6\ J/g} \text{ice}\ (0\ °C) \xrightarrow{-20\ J/g} \text{ice}\ (-10\ °C)$$

At 0 °C:

$$\text{water}\ (0\ °C) \xrightarrow{-333.6\ J/g} \text{ice}\ (0\ °C)$$

Freezing of supercooled water at $-10\ °C$ releases 22 J/g less heat than freezing at 0 °C does. A cloud at 0 °C will not completely form snow because heat released upon freezing of part of the cloud warms the remainder. Warming decreases as temperature does, so a $-10\ °C$ cloud will completely form snow.

See Exercises 39–46.

The key to using a known process to determine the enthalpy of an unknown process is to link the known process to the unknown one by one or more steps where the enthalpy can be determined (Worked Example 7.3). Like using stepping stones to cross a stream, as long as the series of steps leads from one side to the other, the enthalpy change for the process can be determined from the individual steps. Breaking a process into a series of steps reveals information about the interactions taking place. For example, the difference between the interaction of iron and aluminum with oxygen can be determined from the thermite reaction. Writing the thermite reaction as the transformation of one mole of Fe_2O_3 into one mole of Al_2O_3 requires two moles of Al and produces two moles of Fe. Along with the iron and aluminum oxide, 847.6 kJ of heat is produced.

(7.5) $Fe_2O_3(s) + 2Al(s) \longrightarrow 2Fe(s) + Al_2O_3(s)$ -847.6 kJ

The negative sign indicates released heat.

Oxygen is transferred from iron to aluminum in the thermite reaction. Its transfer is accompanied by evolution of heat, suggesting that oxygen interacts more strongly with aluminum than with iron. This conclusion can be quantified by splitting the transfer into separate iron and aluminum reactions (Figure 7.7).

Iron is much less electronegative (1.83) than oxygen (3.44). As a consequence, oxygen has a much stronger attraction for additional electrons than does iron. Oxygen's attraction for additional electrons results in a strong interaction between iron and oxygen—an interaction that is responsible for rusting of iron in the environment. Removing oxygen from iron therefore requires enthalpy input. This enthalpy has been determined to be 822.2 kJ/mol Fe_2O_3. Removed from iron, oxygen exists as a diatomic molecule and iron as solid iron:

(7.6) $Fe_2O_3(s) \longrightarrow 2Fe(s) + 1.5O_2(g)$ 822.2 kJ/mol

Aluminum is similarly less electronegative (1.61) than oxygen. Indeed, it is less electronegative than iron and much more enthalpy input is required to remove oxygen from aluminum: 1669.8 kJ/mol Al_2O_3.

(7.7) $Al_2O_3(s) \longrightarrow 2Al(s) + 1.5O_2(g)$ 1669.8 kJ

Figure 7.7 The elements involved in the thermite reaction are Al (purple), Fe (green), and O (red). Enthalpy relationships in the reaction are shown symbolically. Energy input is required to remove oxygen from iron, and energy is released when oxygen combines with aluminum. The exothermicity of the reaction is determined by a balance between the strength of the interaction between Fe, Al, and O_2 and the solid lattices formed.

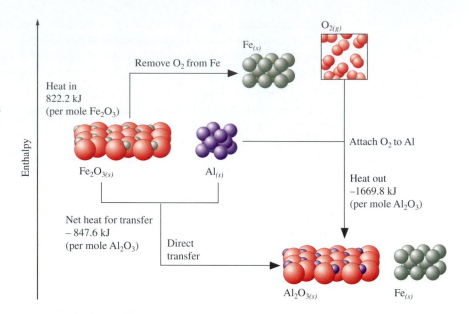

There is a reversibility to this process: Enthalpy input is required to remove oxygen from aluminum, so enthalpy is released when oxygen is attached to aluminum:

(7.8) $$2Al(s) + 1.5O_2(g) \longrightarrow Al_2O_3(s) \qquad -1669.8 \text{ kJ}$$

The large exothermicity indicates that aluminum attracts oxygen very strongly.

The net result of the thermite reaction is a transfer of oxygen from iron to aluminum. Like profit or loss, the net enthalpy (Figure 7.7) is the difference between the enthalpy that must be invested to remove oxygen from iron and the enthalpy that is released when the oxygen is attached to aluminum. Hess's law indicates that it does not matter whether the reaction actually proceeds by first removing the oxygen from iron and then attaching it to aluminum or by a direct transfer. The net enthalpy is the same.

CONCEPT QUESTION The thermite reaction is sufficiently exothermic that the iron produced is *molten*. However, Equation (7.5) gives enthalpy data for the production of *solid* iron. How can Equation (7.5) be used to determine the exothermicity of the thermite reaction with molten Fe as a product? ■

Enthalpy data can also be used to determine the strength of the interactions. Turning the $Fe_2O_3(s)$ dissociation reaction around, formation of Fe_2O_3 from its elements is exothermic ($\Delta H = -822.2$ kJ). The enthalpy involved is about half the value for formation of Al_2O_3 ($\Delta H = -1669.8$ kJ) from its elements. Oxygen is bound to aluminum in $Al_2O_3(s)$ much more strongly than it is bound to iron in $Fe_2O_3(s)$. This is one of the powerful uses of Hess's law.

WORKED EXAMPLE 7.4 *Relative Bond Strengths*

The band-gap energy in mixed-valence semiconductors depends on the strength of interactions between the elements. Both Fe_2O_3 and Al_2O_3 are semiconductors with the same metal-to-oxygen ratio and the same crystal structure. Predict which has the larger band-gap energy. Explain your reasoning.

Plan

■ The magnitude of the band-gap energy is determined by interactions. Stronger interactions result in larger band gaps.

Implementation

■ Elements that interact more strongly form semiconductors with wider band gaps. Because aluminum interacts more strongly with oxygen than iron does, the band gap is expected to be larger.

The actual band gap is 9.5 eV for Al_2O_3, this material is quite a good insulator. In contrast, the band gap in Fe_2O_3 is 2.34 eV, in the yellow region of the visible spectrum.

See Exercises 47, 48, 56.

Hess's Law and Enthalpy of Formation

Imagine measuring the reaction enthalpy for the millions of possible reactions under various conditions, a daunting task. Fortunately, due to Hess's law the enthalpy change for every reaction need not be measured. Instead, chemists have defined a standard reaction to form any given molecule. The associated enthalpy is called the **standard enthalpy of formation.** The standard reaction involves forming the compound from its constituent elements in their most stable state at 1 atm pressure and the specified temperature, usually 25 °C. The notation for the enthalpy of formation is ΔH_f°, where the subscript "f" indicates the formation reaction, and the superscript ° indicates reactants and products in their respective standard states. The most stable form of several elements is listed in Table 7.1.

Enthalpy of formation, ΔH_f° Enthalpy change in the reaction consisting of forming a substance from its elements in their most stable form at 25 °C and one atmosphere pressure.

CONCEPT QUESTIONS   What is the reaction for formation of an element in its most stable form at atmospheric pressure and 25 °C? What is ΔH_f° for an element? ■

Because matter is conserved, in a balanced reaction the number of moles of each element is the same among the reactants as it is among the products. Thus the formation reactions provide a standard path from reactants to products.

The reaction enthalpy (branch A) is equal to the sum of branches B and C. Branch B is the reverse of formation of the reactants, so

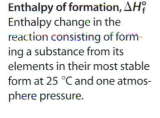

(7.9) $\Delta H(\text{reaction}) = \sum \Delta H_f^\circ(\text{products}) - \sum \Delta H_f^\circ(\text{reactants})$

Worked Example 7.5 illustrates use of ΔH_f° data.

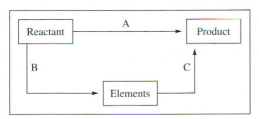

Table 7.1 Stable Forms of Elements at 1 atm Pressure and 25 °C	
Element	**Stable Form**
hydrogen	$H_2(g)$
oxygen	$O_2(g)$
carbon	graphite
nitrogen	$N_2(g)$
chlorine	$Cl_2(g)$
fluorine	$F_2(g)$
bromine	$Br_2(l)$
iodine	$I_2(s)$
metals	solid
sulfur	$S_8(s)$

WORKED EXAMPLE 7.5   *Home Heating*

The precise composition of natural gas is dependent on the source and the season, but the principal component of this fuel is always methane (CH_4). Combustion consists of combining natural gas with O_2 to produce CO_2 and H_2O along with heat. How much heat is produced per mole of methane?

Data

Substance	CH_4	$CO_2(g)$	$H_2O(l)$
ΔH_f°(kJ/mol)	−74.6	−393.5	−285.8

Plan

- Write the balanced combustion reaction.
- Sketch the steps from reactants to products through the elements.
- Change signs for reactions that are reversed. For reactions involving more than one mole, ΔH_f° is multiplied appropriately.

Implementation

- The balanced reaction includes enough oxygen to completely convert CH_4 to CO_2 and H_2O: $CH_4(g) + 2O_2(g) \longrightarrow CO_2(g) + H_2O(l)$

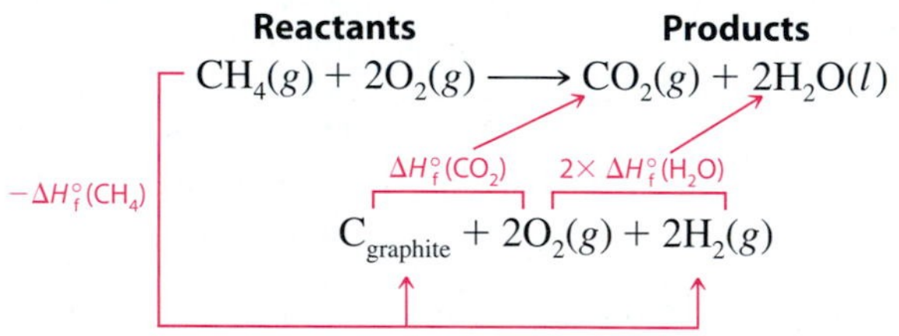

-

CH$_4$ is broken into its elements: carbon as graphite and hydrogen as H_2 gas. Oxygen is already in elemental form. The products are formed by combining O_2 gas with graphite to form CO_2, and O_2 gas with H_2 gas to form water.

- Breaking methane into graphite and hydrogen is the reverse for formation of $CH_4(g)$.

$$\Delta H = -\Delta H_f^{\circ}(CH_4) + \Delta H_f^{\circ}(CO_2) + 2 \times \Delta H_f^{\circ}(H_2O)$$
$$= -(-74.6 \text{ kJ}) + (-393.5 \text{ kJ}) + 2 \times (-285.8 \text{ kJ})$$
$$= -890 \text{ kJ}$$

The reaction is exothermic in accordance with production of heat.

See Exercises 39 – 46.

Extensive Any property that depends on the amount of material.

Intensive Any property that does not depend on the sample size.

In determining the enthalpy of the reaction (e.g., in Worked Example 7.5), the enthalpy of formation is multiplied by the stoichiometric coefficient because the stoichiometric coefficient specifies how many moles of the substance are involved. Any property that depends on the amount of material is termed an **extensive** property. Heat, volume, and mass are extensive properties, for example. Any property that does not depend on the sample size is termed **intensive.** Temperature, melting point, and density are examples of intensive properties.

WORKED EXAMPLE 7.6   *Hess's Law and Reaction Enthalpy*

Digestion of sugar is an important source of energy for the body. This oxidation reaction produces the same products as combustion.

$$C_{12}H_{22}O_{11}(s) + 12O_2(g) \longrightarrow 12CO_2(g) + 11H_2O(l)$$

The combustion reaction is exothermic by 5639.7 kJ/mol. Determine the heat of formation of sugar.

Plan

- Exothermic reaction $\Rightarrow \Delta H$ is negative.
- ΔH(digestion of sugar) $= 12 \times \Delta H_f^\circ(CO_2) + 11 \times \Delta H_f^\circ(H_2O)$
  $- \Delta H_f^\circ(C_{12}H_{22}O_{11}(s))$
  $\Longrightarrow \Delta H_f^\circ(C_{12}H_{22}O_{11}(s)) = 12 \times \Delta H_f^\circ(CO_2) + 11 \times \Delta H_f^\circ(H_2O)$
  $- \Delta H$(digestion of sugar)

Implementation

- ΔH(combustion) $= -5639.7$ kJ
- $\Delta H_f^\circ(C_{12}H_{22}O_{11}(s)) = 12 \text{ mol} \times (-393.5 \text{ kJ/mol}) + 11 \text{ mol} \times (-285.8 \text{ kJ/mol})$
  $- (-5639.7 \text{ kJ}) = -2226.1$ kJ

Note that there is 4.185 J/cal, and a food calorie is one kcal. Thus, 1 mol of sugar delivers $-(5.6397 \times 10^6 \text{ J}) \div (4.185 \text{ J/cal}) \times (1 \text{ food calorie}/1000 \text{ cal}) = -1348$ food calories. The negative sign indicates that the reaction releases energy, and this energy is used by the body as fuel.

See Exercises 39–46.

Crystal lattice energy,
$\Delta H_{\text{x'tal lattice}}$ Energy released when ions in the gas phase combine forming a crystalline lattice.

Hydration enthalpy,
$\Delta H_{\text{hydration}}$ The energy involved when an ion in the gas phase is surrounded by water molecules to form an aqueous solution.

(a)

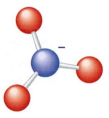

(b)

Figure 7.8 Some ions consist of several atoms bound together with a net charge. (a) The ammonium ion consists of a nitrogen atom surrounded by four hydrogen atoms. The cluster has lost one electron for a net positive charge. (b) The nitrate ion consists of a nitrogen atom surrounded by three oxygen atoms. The cluster has gained an additional electron for a net negative charge.

Hess's Law and Solution Formation

Ionic solids are very common in nature. Many are formed from binary combinations of Group IA or IIA metals with either halogens or oxygen. For example, table salt is a combination of Na^+ and Cl^-. Other ionic salts contain polyatomic ions—atoms bound together in which the bound group has a net charge. The ammonium and nitrate ions involved in the cold-pack reaction are examples of polyatomic ions. The ammonium ion (Figure 7.8a) consists of a nitrogen atom and four hydrogen atoms bound into a cluster from which a single electron is removed to give a polyatomic ion with a positive charge. The nitrate ion (Figure 7.8b) similarly consists of a nitrogen atom plus three oxygen atoms bound into a cluster that has gained one electron to form a negative ion.

Dissolving salt in water is sometimes accompanied by absorption of heat, sometimes by release of heat. To understand these contrasting results, let us examine the process. On the reactant side, the crystalline solid and liquid water are separated. On the product side, water molecules surround the separated ions. Focusing on the reactant, imagine assembling the solid crystalline lattice from the separated ions in the gas phase (illustrated for NaCl in Figure 7.9). Attraction between the ions results in a release of energy when the ions form the crystalline solid. Conversely, energy input is required to break the lattice into its ions. This energy is called the **crystal lattice energy** $\Delta H_{\text{x'tal lattice}}$. The crystal lattice energy of sodium chloride is 769 kJ/mol and that of ammonium nitrate is 661 kJ/mol.

Starting with the same ions in the gas phase, imagine surrounding the ions with water molecules (Figure 7.10), a process called **hydration.** The associated enthalpy is referred to as the **hydration enthalpy.** Energy released on hydration is a direct result of the attraction between water molecules and the charged ions in the salt (Figure 7.11). The hydration enthalpy of Na^+Cl^- is -765 kJ/mol and that of $NH_4^+NO_3^-$ is -635.3 kJ/mol.

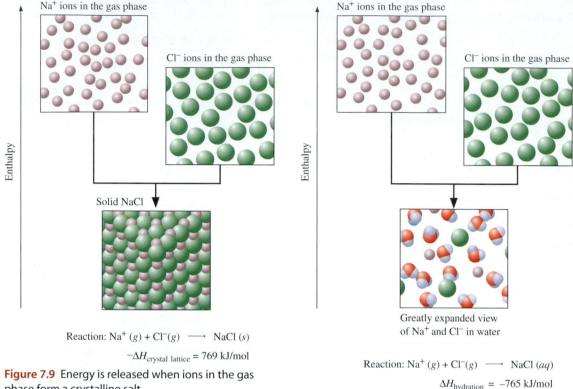

Reaction: Na$^+$ (g) + Cl$^-$(g) $\longrightarrow$ NaCl (s)

$-\Delta H_{\text{crystal lattice}}$ = 769 kJ/mol

Figure 7.9 Energy is released when ions in the gas phase form a crystalline salt.

Greatly expanded view of Na$^+$ and Cl$^-$ in water

Reaction: Na$^+$ (g) + Cl$^-$(g) $\longrightarrow$ NaCl (aq)

$\Delta H_{\text{hydration}}$ = −765 kJ/mol

Figure 7.10 Energy is released when ions in the gas phase are surrounded by water molecules.

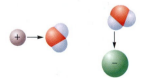

Figure 7.11 The electron distribution around the water molecule is not uniform. There is a partial negative charge near the oxygen atom and a partial positive charge near the hydrogen atoms. Positive ions are attracted to the oxygen side of water, and the hydrogen side is attracted to the negative ions.

The enthalpy involved in dissolving a salt (as opposed to the gas-phase ions) reflects the balance between the enthalpy evolved for assembling the crystalline lattice and the enthalpy evolved from hydrating the ions. For the hot-pack reactions, there is a net release of enthalpy (Figure 7.12) because the magnitude of the crystal lattice energy is smaller than the hydration enthalpy. In Hess's law terms, the uphill investment in breaking up the crystal lattice is smaller than the downhill payoff in hydrating the ions. The net result is release of heat, an exothermic solution formation.

The opposite is true for NH$_4$NO$_3$ in the cold-pack reaction (Figure 7.13). Separating the crystal lattice into its component ions is more endothermic than hydrating the NH$_4^+$ and NO$_3^-$ ions is exothermic. The uphill investment is larger than the downhill payoff, so solution formation is endothermic.

Crystal Lattice Enthalpy Breaking a crystal lattice into separated ions in the gas phase is always endothermic due to the strong interaction between the oppositely charged ions in the solid. Indeed, the crystal lattice enthalpy can be calculated based on coulombic interactions, which are attractive between oppositely charged ions and repulsive between ions of the same charge. The result is

(7.10)
$$\Delta H_{\text{x'tal lattice}} = M \frac{NZ^+ \times Z^- e^2}{R}$$

where M is a constant called the Madelung constant that depends on the lattice geometry, N is Avogadro's number, Z^+ and Z^- are the charges on the ions, and e is the charge

Figure 7.12 The process of making a solution of $CaCl_2$ is exothermic by (2268 kJ/mol -2350 kJ/mol) $= -82$ kJ/mol. The ions are held together in their lattice less strongly than the separated ions interact with water.

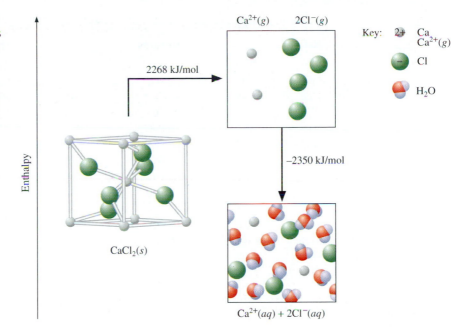

on the electron. R is the spacing between the center of the positive and negative charges and is equal to the sum of the radii of the negative and the positive ions. Equation (7.10) indicates that the lattice enthalpy *decreases* with increasing lattice spacing. Small ions—those from elements in the first few periods of the periodic table—are closer to the oppositely charged ion than the larger ions from latter periods. For example, anion size increases in the series F^-, Cl^-, and Br^-. As a consequence, the separation between a positive ion such as Li^+ and F^- is smaller than that between Li^+ and Br^- (Figure 7.14). Because coulombic interaction decreases as separation increases, the crystal lattice enthalpy is smaller for LiBr than for LiF (Table 7.2).

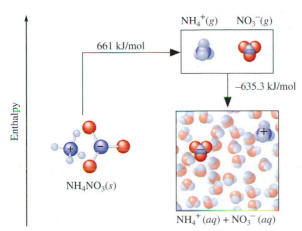

Figure 7.13 The enthalpy change in the formation of ammonium nitrate solution indicates that the ammonium and nitrate ions are bound together in the crystal lattice of the solid more strongly than the two ions interact with water.

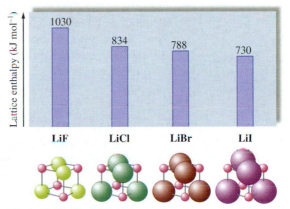

Figure 7.14 The unit cell becomes progressively larger in the series. LiF, LiCl, LiBr, LiI. The interaction between the negatively charged anion and the positively charged Li^+ ion decreases as the size of the anion increases.

Table 7.2 Crystal Lattice Enthalpy for Several Ionic Salts (kJ/mol)

Halides		Nitrates		Hydroxides	
LiF	1030	$LiNO_3$	848	LiOH	1021
LiCl	834	$NaNO_3$	755	NaOH	887
LiBr	788	KNO_3	685	KOH	789
LiI	730	$AgNO_3$	820	AgOH	918
NaF	910	$Mg(NO_3)_2$	2481	$Mg(OH)_2$	2870
NaCl	769	$Ca(NO_3)_2$	2268	$Ca(OH)_2$	2506
NaBr	732	$Zn(NO_3)_2$	2376	$Zn(OH)_2$	2795
KF	808	$Cd(NO_3)_2$	2238	$Al(OH)_3$	5627
KCl	701	NH_4NO_3	661		
KBr	671				
AgCl	910	**Oxides**		**Sulfates**	
AgBr	897	MgO	3356	$(NH_4)_2SO_4$	1766
$MgCl_2$	2477	CaO	3414	$CaSO_4$	2489
$CaCl_2$	2268	ZnO	4142	$BaSO_4$	2469
$AlCl_3$	5376	SnO	3652	Cs_2SO_4	1596

CONCEPT QUESTION Graph the crystal lattice enthalpy for NaF, NaCl, and NaBr as a function of $1/R$, where $R = r_{Na^+} + r_{x^-}$. (Ion size data are given in Figure 3.14 in Chapter 3.) How well does Equation (7.10) fit the data? ∎

Coulombic attraction increases with charge, so the lattice enthalpy also *increases* with ion charge. The crystal lattice enthalpy of salts consisting of highly charged ions is larger than that of single-charged ions.

WORKED EXAMPLE 7.7 *Lattice Enthalpy*

Although ionic crystals are not made from ions in the gas phase, the crystal lattice enthalpy—the enthalpy required to break the salt into the constituent ions in the gas phase—reveals the strength of interaction unclouded by other interactions. The crystal lattice enthalpy of NaCl is 769 kJ/mol. MgO crystallizes in the same lattice structure as NaCl. Is the crystal lattice enthalpy of MgO larger or smaller than that of NaCl? Why? Estimate the crystal lattice enthalpy of MgO based on the NaCl data and compare it with the data in Table 7.2.

Plan

■ Use Equation (7.10): $\Delta H_{x'tal\ lattice} = M \dfrac{NZ^+ \times Z^- e^2}{R}$

Compare the effect of the charges for NaCl and MgO.
Compare the effect of the size for NaCl and MgO.
■ Calculate and compare the result with the data in Table 7.2.

Implementation

■ The charges on the ions in MgO are Mg (+2) and O (−2); on NaCl the charges are Na (+1) and Cl (−1). Doubling the charge multiplies the lattice enthalpy by a factor of 4. Due to the charge, the lattice enthalpy of MgO is expected to be 4 × (769 kJ/mol) = 3076 kJ/mol. The ion sizes are Na^+ (102 pm), Cl^- (184 pm), Mg^{2+} (72 pm), and O^{2-} (140 pm). The ratio of sizes is (102 + 184)/(72 + 140) = (286/212) = 1.35, so the

Figure 7.15 The hydration enthalpy of several elemental ions.

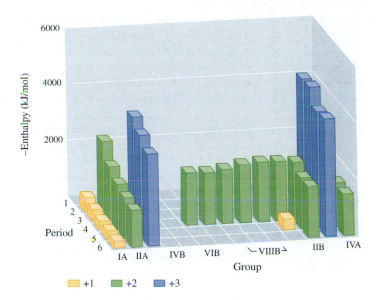

crystal lattice enthalpy of MgO is expected to be 35% greater due to the smaller size of the two ions involved.

■ Combining size and charge, the crystal lattice enthalpy of MgO is predicted to be (769 kJ/mol) × 4 × 1.35 = 4150 kJ/mol based on that of NaCl. The actual value is 3356 kJ/mol, suggesting that the actual charges on the ions in MgO are somewhat smaller and the ions somewhat larger than the +2 and −2 charges imply.

See Exercises 47, 48.

Ion Hydration Ion hydration enthalpy is always negative, because heat is always evolved when an ion is surrounded by water molecules. Ion hydration enthalpy is observed to vary linearly with the ratio of ion charge to ion radius (Figure 7.15). This relationship is explained by considering both the number of water molecules that fit around an ion and the force of attraction between the ion and water. Hydration enthalpy is classified into three categories: The magnitude is smallest for +1 ions, is larger for +2 ions, and is largest for +3 ions (Figure 7.15 and Table 7.3).

Table 7.3 Hydration Enthalpy for Several Monatomic Ions (kJ/mol)

Li$^+$	Be^{2+}											B	C	N	O	F$^-$	Ne
−515	−2487															−506	
Na$^+$	Mg^{2+}											Al^{3+}	Si	P	S	Cl$^-$	Ar
−405	−1922											−4660				−364	
K$^+$	Ca^{2+}	Sc^{3+}	Ti	V	Cr^{2+}	Mn^{2+}	Fe^{2+}	Co^{2+}	Ni^{2+}	Cu^{2+}	Zn^{2+}	Ga^{3+}	Ge	As	Se	Br$^-$	Kr
−321	−1592	−3960			−1850	−1845	−1920	−2054	−2106	−2100	−2044	−4685				−337	
Rb$^+$	Sr^{2+}	Y^{3+}	Zr	Nb	Mo	Tc	Ru	Rh	Pd	Ag$^+$	Cd^{2+}	In^{3+}	Sn^{2+}	Sb	Te	I$^-$	Xe
−296	−1445	−3620								−375	−1806	−4109	−1554			−296	
Cs$^+$	Ba^{2+}	La^{3+}	Hf	Ta	W	Re	Os	Ir	Pt	Au	Hg^{2+}	Tl^{3+}	Pb^{2+}	Bi	Po	At	Rn
−263	−1304	−3283									−1823	−4184	−1480				

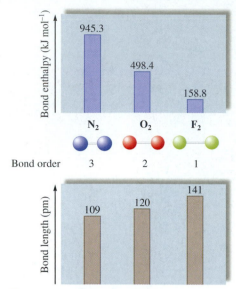

Figure 7.16 The atomization energy for a diatomic molecule is a direct measure of the strength of the homonuclear bond. Among the second-row elements, the bond strength decreases from the triple bond of nitrogen to the double bond of oxygen to the single bond of fluorine. The bond length varies in the opposite way, increasing along the series.

Enthalpy of atomization
Energy required to transform a sample into atoms in the gas phase.

Hess's Law and Enthalpy of Atomization

The **enthalpy of atomization** is the energy input necessary to convert a substance into its component atoms in the gas phase. It is a pure measure of the total bond energy of a substance. For example, the bond order for the second-row elements nitrogen, oxygen, and fluorine decreases from 3 to 2 to 1. These are all diatomic molecules, so the atomization enthalpy is equal to the bond enthalpy and the bond enthalpy decreases along with the bond order (Figure 7.16).

CONCEPT QUESTIONS   The halogens chlorine, bromine, and iodine all form single-bonded, diatomic molecules. The atomization enthalpy decreases from 242.6 kJ/mol for Cl_2 to 192.8 kJ/mol for Br_2 to 151.1 kJ/mol for I_2. How are bond lengths expected to change in this series? Advance challenge: The F—F bond is the shortest among the halogens, yet fluorine's atomization enthalpy is nearly as low as that of I_2. Can you propose an explanation for the very low atomization enthalpy of F_2? (Hint: Think of electron crowding.) ∎

Atomization enthalpies are always positive because energy is stored in chemical bonds. For solid substances, the enthalpy of atomization also includes a contribution from the interaction between the molecules or atoms of the solid that is responsible for holding the solid together. An examination of atomization enthalpy for the metallic elements (Figure 7.17 and Table 7.4) reveals that atomization enthalpy is related to the electron configuration. Atomization enthalpy is least for Group IA metals, increases to a maximum in the middle of the transition series, and then decreases to the end of the transition elements. The tendency toward a maximum in the middle of the transition series is interrupted at the s^2d^5 configuration—a filled s subshell and a half-filled d subshell. The atomization enthalpy for p-block metals is higher than that for elements at the end of the transition series. The p-block

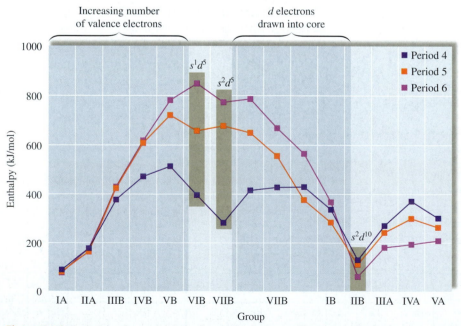

Figure 7.17 Enthalpy of atomization of metallic elements illustrating the trend across a period.

Table 7.4 Atomization Enthalpy per Atom for Several Elements at 25 °C (kJ/mol)

H 218.0																	He 0
Li 159.3	Be 324.0											B 565.0	C 716.7	N 472.6	O 249.2	F 79.38	Ne 0
Na 107.5	Mg 147.1											Al 330	Si 450	P 316.7	S 277.2	Cl 121.3	Ar 0
K 89.0	Ca 177.8	Sc 377.8	Ti 473.0	V 514.2	Cr 396.6	Mn 283.3	Fe 416.3	Co 428.4	Ni 430.1	Cu 337.4	Zn 130.4	Ga 272.0	Ge 372.0	As 302.5	Se 227.1	Br 96.4	Kr 0
Rb 80.9	Sr 163.6	Y 424.7	Zr 608.8	Nb 721.3	Mo 658.1	Tc 678.0	Ru 650.1	Rh 556	Pd 376.6	Ag 284.9	Cd 111.8	In 243.3	Sn 301.2	Sb 264.4	Te 196.6	I 75.2	Xe 0
Cs 76.5	Ba 177.8	La 431.0	Hf 619.2	Ta 782.0	W 849.4	Re 774	Os 787	Ir 669	Pt 565.7	Au 368.2	Hg 61.4	Tl 182.2	Pb 195.2	Bi 209.6	Po	At	Rn 0

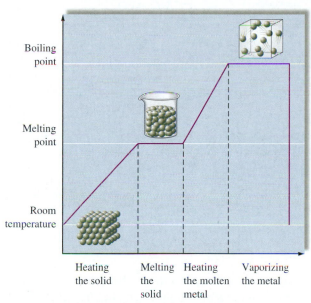

Figure 7.18 A sequence of steps for atomizing a solid. First the solid is warmed to the melting point. The temperature rise halts until the solid is melted. The liquid is then warmed until it boils. The temperature rise stops until all atoms of the metal are in the gas phase. Finally, the gas-phase atoms are cooled to room temperature.

Latent heat of fusion The enthalpy change that occurs upon melting a solid at its melting point at a pressure of one atmosphere.

metals also feature more directed, covalent bonds than do the transition metals. The lowest atomization enthalpy of the elements occurs for the noble gases (Group VIIIA). With their filled valence s and p subshells, atoms of these elements have little attraction for other atoms. Indeed, the noble gases are all monatomic gases at room temperature. Already single atoms, their atomization enthalpy is zero.

The atomization enthalpy for a metal can be thought of as follows. Imagine heating a metallic solid from room temperature to its melting point (Figure 7.18). At first the solid gets warm and the enthalpy input is equal to the heat capacity multiplied by the temperature increase. (The heat capacity varies somewhat with temperature, but is nearly constant.) When the melting point is reached, additional heat input does not raise the temperature. Instead, the solid melts. The enthalpy input required to melt the solid is called the **latent heat of fusion.** Once all the metal is molten, continued heating raises the temperature. The temperature rise is determined by the heat capacity of the liquid metal and the enthalpy input. Finally, the boiling point is reached. Again, the temperature does not rise further until all of the liquid metal is vaporized. The heat required to vaporize the metal at its boiling point is called the **latent heat of vaporization.**

If the now-hot, vaporized metal were cooled to room temperature without condensing or solidifying the metal—a thought experiment—the process would result in atomization of the metal. The total enthalpy of the cycle (heating, melting, heating, boiling, and cooling to room temperature) is the enthalpy of atomization.

WORKED EXAMPLE 7.8   *Interaction Energy*

For a solid material, the atomization enthalpy is the total of the bond energies and the energy of interaction between the molecules in the solid. Determine the atomization enthalpy of Fe_2O_3 and of Al_2O_3. Comment on the difference between them and the exothermicity of the thermite reaction.

Latent heat of vaporization
The enthalpy change that accompanies vaporizing one mole of a liquid at a pressure of one atmosphere.

Plan

■ Devise a path from atoms to the solid using known enthalpies: ΔH_f° and $\Delta H_{atomization}$.

Implementation

■ Path: atoms → elements in most stable form → solids
$\Delta H_f^\circ(Al_2O_3) = -1669.8$ kJ/mol; $\Delta H_f^\circ(Fe_2O_3) = -822.2$ kJ/mol
Elements involved are iron, aluminum, and oxygen.
$\Delta H_{atomization}(Al) = 330.0$ kJ/mol
$\Delta H_{atomization}(Fe) = 416.3$ kJ/mol
$\Delta H_{atomization}(O_2) = 249.18$ kJ/mol

| 2 mol Al atoms
3 mol O atoms | → | 2 mol solid Al
1.5 mol O_2 gas | → | 1 mol solid Al_2O_3 |

$$2 \times -\Delta H_{atomization}(Al)$$
$$3 \times -\Delta H_{atomization}(O)$$

$$\Delta H_f^\circ(Al_2O_3)$$

$$\Delta H_{atomization}(Al_2O_3) = -\Delta H_f^\circ(Al_2O_3) + 2 \times \Delta H_{atomization}(Al) + 3 \times \Delta H_{atomization}(O)$$
$$= +1669.8 \text{ kJ/mol} + 2 \text{ mol} \times (330.0 \text{ kJ/mol}) + 3 \text{ mol}$$
$$\times (249.18 \text{ kJ/mol})$$
$$= 3077 \text{ kJ per mol } Al_2O_3 \text{ formed}$$

The similar calculation for Fe_2O_3 yields 2402 kJ per mol Fe_2O_3 formed. It takes a little more heat to atomize iron, but the interaction of iron with oxygen is weaker than that between aluminum and oxygen. The strong interaction between aluminum and oxygen drives the thermite reaction.

See Exercises 47, 48, 56.

Entropy, S Thermodynamic function that measures the energy dispersal in a system.

7.3 The Second Law of Thermodynamics: The Way Forward

A comparison of the hot-pack and cold-pack reactions shows that chemical reactions can occur with either a net increase or a net decrease in the internal energy. It is evident that energy alone does not predict the direction of a reaction. Instead, another property is needed to make this determination. To imagine this property, think about opening a container of coffee. Soon the aroma of the coffee beans spreads throughout the room. It would violate experience for the aroma to spontaneously return to the container! Similarly, a drop of ink in a container of water soon colors the entire volume of water. In both of these cases, the molecules that carry the aroma of coffee and the dye that carries the color of ink tend to spread throughout all the space available. This tendency to disperse is encapsulated in a property called entropy. **Entropy,** as suggested by Ludwig Boltzmann in 1877, is determined by the number of different microscopic arrangements that lead to the same macroscopic observable situation (Figure 7.19). Entropy is often dubbed "disorder." A messy room, a disordered room, is said to have high entropy because there are many ways to have a messy room. In contrast, when everything has a place and is put in that place, there is only one way to arrange the items in the room. This is a low-entropy state.

CONCEPT QUESTIONS Which has greater disorder: a crystalline, metallic solid or a molten, liquid metal? Liquid water or water vapor (Figure 7.20)? ■

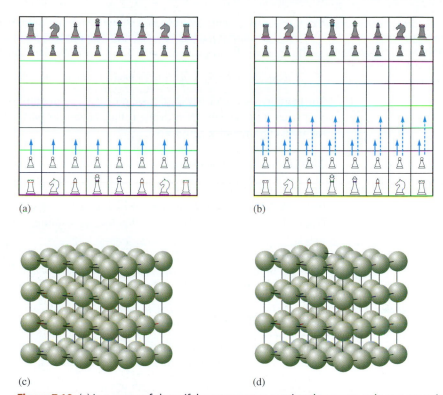

(a)   (b)

(c)   (d)

Figure 7.19 (a) In a game of chess, if the pawns were restricted to move only one space in the first move and no other pieces could move, there are 8 possible first moves. (b) Allowing the pawns to move one or two places doubles the number of possible first moves to 16. The second set of rules produces greater entropy. (c) There is only one way to arrange a perfect crystal with all atoms at their lattice sites. This is a zero-entropy configuration; there are no choices. (d) There are $4 \times 4 \times 4 = 64$ possibilities for having only one atom missing from the perfect cubic crystal. The entropy is much higher. The entropy of a system is proportional to the number of possible arrangements.

Experience tells us that it takes effort to attain the low-entropy state. Just as a ball tends to roll downhill, so the tendency to more equivalent configurations—high entropy—helps to drive a reaction. This tendency is embodied in the second law of thermodynamics.

The second law of thermodynamics: *The entropy of an isolated system increases in a spontaneous process.*

Figure 7.20 (a) With liquid water at 25 °C, there are many different ways the water molecules can be arranged and fill the container to the same volume. (b) Water vapor at the same temperature as liquid water can be distributed in many more ways due to the vapor filling the entire volume available to it.

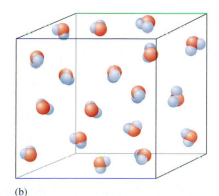

(a)   (b)

Spontaneous process A process that occurs naturally without application of an external force.

Here, "isolated system" refers to everything affected by a process. In particular, the universe is isolated because it includes everything. A **spontaneous process** is one that occurs without being driven by an external force. Thus the second law can also be stated as follows:

The universe tends to a state of maximum entropy.

For the universe to tend toward a state of maximum entropy, any change must necessarily either increase entropy or leave it unchanged. In particular, no process can decrease the total entropy of the universe.

The Universe in Two Parts

Because the universe is quite large, the task of analyzing a possible reaction is made tractable by dividing the universe, which consists of everything that is affected by the reaction, into two parts: the system and the surroundings. Symbolizing entropy as S,

(7.11) $$\Delta S_{universe} = \Delta S_{system} + \Delta S_{surroundings} \geq 0$$

Entropy, like enthalpy, depends only on the condition of the system, not on the system history. Entropy is also a state function.

Entropy of the System

CONCEPT QUESTION   Which has greater entropy, liquid water or ice? Reconcile the freezing of water with the second law of thermodynamics. ∎

Third law of thermodynamics The entropy of a perfect crystal at 0 K is zero.

Entropy and enthalpy are both state functions, but they differ in a fundamental way—their respective zero-value states. Enthalpy cannot be measured absolutely. Instead, the convention is that the elements in their most stable form at atmospheric pressure and 25 °C have an enthalpy value of zero (much like the zero-energy state for the electron in an atom is defined to be the ionized state). Entropy is different. Boltzmann's link between entropy and the number of microscopic arrangements available to a system leads to the **third law of thermodynamics:** *A perfect crystal has zero entropy at absolute zero,* or $S_{system} = 0$ at 0 K. "Perfect" means that all particles in the crystal are flawlessly aligned (there are no defects) and all particles are in their minimum energy state. Warming the crystal above 0 K imparts movement to the particles so that they are no longer flawlessly aligned and the entropy increases from zero. Entropy values ($S°$) are therefore absolute values, while enthalpy values ($\Delta H_f^°$) are relative.

The hot-pack reaction illustrates entropy change in a spontaneous process (Figure 7.21a). The hot pack starts with calcium chloride in a solid, crystalline array in one pouch and liquid water in a second pouch. The product or ending material consists of Ca^{2+} and Cl^- ions surrounded by water molecules (Figure 7.21b). Because entropy is a property of the system (a state function), the change in *system* entropy can be determined by the difference in entropy for the products and that of the reactants (Table 7.5).

Table 7.5 Entropy Values: Hot-Pack Reaction

Substance	$S°$ (J/mol · K)
$CaCl_2(s)$	108.4
$Ca^{2+}(aq)$	−53.1
$Cl^-(aq)$	56.5

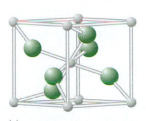 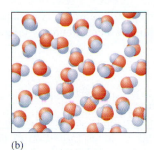 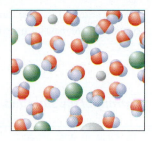

(a)   (b)

Figure 7.21 (a) A schematic of the $CaCl_2$ crystal illustrates the regular array of a crystalline solid. Liquid water has a far less regular structure. (b) When the ions Ca^{2+} and Cl^- are added to water, the small size and large charge of the Ca^{2+} ion draws water molecules close to it, so that the water has fewer available configurations. The net result is a loss of entropy when $CaCl_2$ dissolves in water.

(7.12) $$CaCl_2(s) \longrightarrow Ca^{2+}(aq) + 2Cl^-(aq)$$

$$\Delta S = S°(Ca^{2+}(aq)) + 2 \times S°(Cl^-(aq)) - S°(CaCl_2(s))$$
$$= (-53.1 + 2 \times 56.5 - 108.4) \text{ kJ/K}$$
$$= -48.5 \text{ J/K}$$

There is a loss of entropy upon dissolving one mole of $CaCl_2$ in water. A significant part of the reason for this loss of entropy is the ordering of water molecules around the small, doubly charged calcium ion.

WORKED EXAMPLE 7.9   *Solution Formation and Entropy*

Like the hot pack, the cold pack contains a salt and water. The cold-pack salt is ammonium nitrate, NH_4NO_3. In water, NH_4NO_3 breaks up into NH_4^+ and NO_3^- ions. Determine the entropy change upon solution formation. Compare the result with that for the hot-pack $CaCl_2$ discussed earlier. The data are provided in Table 7.6.

Plan
- Write the reaction.
- $\Delta S = S°$ (products) $- S°$ (reactants).

Implementation
- NH_4NO_3 (*x'tal*) $\rightarrow NH_4^+$ $(aq) + NO_3^-$ (aq)
- $\Delta S = S°$ (products) $- S°$ (reactants)
  $= (113.4 \text{ J/mol} \cdot \text{K}) \times 1 \text{ mol} + (146.4 \text{ J/mol} \cdot \text{K}) \times 1 \text{ mol} - (151.1 \text{ J/mol} \cdot \text{K})$
  $\times 1 \text{ mol} = 108.7 \text{ J/K}$

In contrast to the case of the $CaCl_2$ solution, the entropy change in making a NH_4NO_3 solution is positive. Both ions in NH_4NO_3 are relatively large, and both carry only a single charge. Water molecules are not tightly bound to either ion, so the solution

Table 7.6 Entropy Values: Cold-Pack Reaction

Substance	$S°$ (J/mol · K)
$NH_4NO_3(s)$	151.1
$NH_4^+(aq)$	113.4
$NO_3^-(aq)$	146.4

retains much of the flexibility of liquid water. The solution has more available configurations than the crystal plus liquid water, so entropy increases.

See Exercises 58, 59.

Entropy of the Surroundings

The second law of thermodynamics asserts that the entropy of the universe must increase in any process. This includes formation of the hot-pack and cold-pack solutions. As determined earlier, making a $CaCl_2$ solution entails a decrease of entropy for the system. Solution formation is spontaneous—it occurs without added energy or entropy—so the decrease in system entropy for the hot pack must be balanced by an equal or greater increase in the entropy of the surroundings.

CONCEPT QUESTIONS What is the minimum entropy increase for the surroundings for the hot-pack reaction? What is the minimum entropy increase for the surroundings for the cold-pack reaction? ■

The only quantity exchanged between the hot pack and the surroundings is heat. Heat flow to the surroundings increases the movement of particles in the surroundings. More movement corresponds to an increase in the number of ways the surroundings can be arranged—in other words, an entropy increase. Quantitatively, the increase in disorder is proportional to the heat added to the surroundings. For a constant pressure process, heat is enthalpy.

(7.13) $$\Delta S_{surroundings} = \Delta H_{surroundings}/T$$

Heat *added to* the surroundings is heat *extracted from* the system.

(7.14) $$\Delta H_{surroundings} = -\Delta H_{system}$$

Thus

(7.15) $$\Delta S_{surroundings} = -\Delta H_{system}/T$$

The entropy increase is proportional to $1/T$, where T is the absolute temperature.

In the hot-pack reaction, the system (the hot pack) entropy decreases by 48.5 J/K · mol. Thus the entropy of the surroundings must increase by at least 48.5 J/K per mol $CaCl_2$. The hot-pack reaction is exothermic by -82 kJ/mol (Figure 7.12), and the entropy increase of the surroundings occurs at room temperature, 298 K. So

(7.16) $\Delta S_{surroundings} = -\Delta H_{system}/T = -(-82 \text{ kJ/mol})/298 \text{ K} = -275 \text{ J/mol} \cdot \text{K}$

The surroundings increase in entropy more than five times as much as required by the second law of thermodynamics. The entropy *loss* by the system is compensated for by heat leaving the system, resulting in an even larger *gain* in entropy for the surroundings.

CONCEPT QUESTIONS Does the entropy of the surroundings increase or decrease for the cold-pack reaction? Calculate the entropy change and check for compliance with the second law of thermodynamics. ■

Gibbs Free Energy

The statement that the entropy of the surroundings results from heat exchange with the system can be incorporated into the second law, $\Delta S_{universe} \geq 0$:

(7.17) $\Delta S_{universe} = \Delta S_{system} + \Delta S_{surroundings} = \Delta S_{system} - \Delta H_{system}/T \geq 0$

This expression provides a criterion, based solely on properties of the system, that predicts whether a process can occur in the absence of external energy input. It is often somewhat easier to think of processes running downhill, like a ball coming to rest at the bottom of a bowl. Equation (7.14) is therefore rearranged by multiplying by $-T$ to arrive at

Free energy, G The energy that is free to do work at constant temperature and pressure. Determines the direction of spontaneous change, $\Delta G < 0$ is spontaneous for reactions at constant temperature and pressure. Also known as *Gibbs free energy*.

(7.18)
$$\Delta H_{system} - T\Delta S_{system} \leq 0$$

Everything in Equation (7.18) refers to the system, so the subscript "system" may be dropped. However, to avoid confusion about entropy increasing, remember that ΔS refers specifically to *system* entropy. The second law is about *total entropy*, which includes the surroundings—the ΔH part, too.

Predicting spontaneous processes is very important for chemists, so Equation (7.17) defines an important function called the **free energy.** Free energy is also called the **Gibbs free energy** and is labeled G in honor of Josiah Willard Gibbs, an American physicist who proposed this criterion for a spontaneous process in 1877. His proposal for identifying spontaneous processes is a consequence of the second law of thermodynamics: For a process at constant pressure, the Gibbs free energy tends to a minimum,

Gibbs free energy, G A thermodynamic function equal to the enthalpy (H) minus the product of the entropy (S) and the Kelvin temperature (T); $G = H - TS$. Also known as *free energy*.

(7.19)
$$\Delta G = \Delta H - T\Delta S \leq 0$$

where all functions refer to the system.

WORKED EXAMPLE 7.10   *Etching Glass*

Glass including quartz (SiO_2) is attacked by aqueous HF solution, which etches the surface and can dissolve the glass entirely. Explain why HF etches glass and determine whether HCl does as well. Table 7.7 contains the relevant data.

Plan
- Write the reactions.
- Determine ΔG for each process.
- Determine ΔH and ΔS for each reaction.

Implementation
-
$$SiO_2(s) + 4HF(aq) \longrightarrow SiF_4(g) + 2H_2O(l)$$
$$SiO_2(s) + 4HCl(aq) \longrightarrow SiCl_4(g) + 2H_2O(l)$$
- Gibbs free energy
For an HF etch:

$$\Delta G = \Delta G_f^\circ (SiF_4) + 2 \times \Delta G_f^\circ (H_2O) - 4 \times \Delta G_f^\circ (HF) - \Delta G_f^\circ (SiO_2)$$
$$= [-1572.65 + 2(-237.129) - 4(-278.9) - (-856.4)] \text{ kJ}$$
$$= -74.9 \text{ kJ}$$

Table 7.7 Data for Worked Example 7.10

Substance	ΔH_f° (kJ/mol)	ΔG_f° (kJ/mol)	S° (J/mol · K)
$SiF_4(g)$	−1614.94	−1572.65	282.49
$SiCl_4(g)$	−657.01	−616.98	330.73
$SiO_2(s, \text{quartz})$	−910.94	−856.64	41.84
$HCl(aq)$	−167.159	−131.228	56.5
$HF(aq)$	−332.63	−278.9	−13.8
$H_2O(l)$	−285.8	−237.129	69.91

For an HCl etch:

$$\Delta G = \Delta G_f^\circ \, (\text{SiCl}_4) + 2 \times \Delta G_f^\circ \, (\text{H}_2\text{O}) - 4 \times \Delta G_f^\circ \, (\text{HCl}) - \Delta G_f^\circ \, (\text{SiO}_2)$$
$$= [-616.98 + 2(-237.129) - 4(-131.228) - (-856.4)] \text{ kJ}$$
$$= 290.1 \text{ kJ}$$

■ Enthalpy

For an HF etch:

$$\Delta H = \Delta H_f^\circ \, (\text{SiF}_4) + 2 \times \Delta H_f^\circ \, (\text{H}_2\text{O}) - 4 \times \Delta H_f^\circ \, (\text{HF}) - \Delta H_f^\circ \, (\text{SiO}_2)$$
$$= [-1614.94 + 2(-285.8) - 4(-332.63) - (-910.94)] \text{ kJ}$$
$$= 54.92 \text{ kJ}$$

For an HCl etch:

$$\Delta H = \Delta H_f^\circ \, (\text{SiCl}_4) + 2 \times \Delta H_f^\circ \, (\text{H}_2\text{O}) - 4 \times \Delta H_f^\circ \, (\text{HCl}) - \Delta H_f^\circ \, (\text{SiO}_2)$$
$$= [-657.01 + 2(-285.8) - 4(-167.159) - (-910.94)] \text{ kJ}$$
$$= 351.0 \text{ kJ}$$

Entropy

For an HF etch:

$$\Delta S = S^\circ \, (\text{SiF}_4) + 2 \times S^\circ \, (\text{H}_2\text{O}) - 4 \times S^\circ \, (\text{HF}) - S^\circ \, (\text{SiO}_2)$$
$$= [282.49 + 2(69.91) - 4(-13.8) - (41.84)] \times (298.15 \text{ K}) \text{ J}$$
$$= 130 \text{ kJ}$$

For an HCl etch:

$$\Delta S = S^\circ \, (\text{SiCl}_4) + 2 \times S^\circ \, (\text{H}_2\text{O}) - 4 \times S^\circ \, (\text{HCl}) - S^\circ \, (\text{SiO}_2)$$
$$= [330.73 + 2(69.91) - 4(56.5) - (41.84)] \times (298.15 \text{ K}) \text{ J}$$
$$= 60.4 \text{ kJ}$$

The HF etching reaction is spontaneous as indicated by the negative ΔG value. The system entropy increases significantly (436 J/K · mol), which offsets the endothermic reaction ($\Delta H = 55$ kJ). In contrast, the HCl system entropy increase is somewhat smaller (203 J/K · mol) and is overwhelmed by the large endothermicity of the reaction (351 kJ/mol). As a result, HCl can be stored safely in glass containers for long periods of time.

See Exercises 61–64.

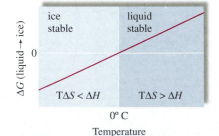

Figure 7.22 Temperature affects the balance between entropy ($T\Delta S$) and enthalpy (ΔH). At low temperatures, the exothermicity overwhelms the local entropy reduction and ice is stable. At higher temperatures, the entropy loss on formation of ice overwhelms the exothermicity and the reverse reaction is spontaneous; water is stable.

Temperature and Spontaneity

The sign of ΔG provides an extremely powerful tool for predicting the spontaneous direction of a reaction. In particular, the two components of ΔG provide insight into whether the reaction is driven by changing interactions (as measured by ΔH) or by increasing disorder (as measured by ΔS). For some reactions, both components act in concert. For many important reactions, however, temperature determines the balance between entropy and enthalpy to alter the spontaneous direction. For example, the stable phase of H_2O at room temperature is liquid water, while at -10 °C the stable phase is solid ice. The stable phase switches from liquid to solid. The freezing of water can be expressed as a reaction:

(7.20) $$H_2O(l) \longrightarrow H_2O(s)$$

For this reaction, ΔG shifts from a positive value at room temperature (liquid stable) to a negative value (ice stable) at -10 °C (Figure 7.22). Examining the two components of ΔG points to the origin of this shift. Heat must be removed

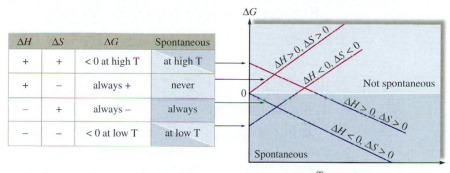

ΔH	ΔS	ΔG	Spontaneous
+	+	< 0 at high T	at high T
+	−	always +	never
−	+	always −	always
−	−	< 0 at low T	at low T

Figure 7.23 ΔG must be negative for a constant-pressure process to be spontaneous. If the enthalpy change and the entropy change are of the same sign, the two components counterbalance each other and the temperature determines spontaneity.

from water to form ice, so ΔH is negative. A negative value of ΔH tends to make ΔG negative, favoring ice. Entropy provides a counterbalancing component of ΔG. Ice is more ordered than water, so ΔS is negative. A negative value of ΔS tends to make ΔG positive since ΔS is multiplied by $-T$ in ΔG, which favors liquid water. There is a balance between ΔH and ΔS for freezing water. At temperatures below 0 °C, ΔG for the freezing of water is negative, above 0 °C, ΔG is positive.

This balance between ΔH and ΔS exists whenever ΔH and ΔS have the same sign (Figure 7.23).

WORKED EXAMPLE 7.11   *The Turnaround Temperature*

Combustion of gasoline in an automobile is an exothermic process, raising the temperature in the engine to a very high level. NO and NO_2 are among the substances produced. NO_2 is responsible for the brown haze that hovers over major cities. If the engine is kept very cool, NO is produced rather than NO_2. How cool must the engine be to avoid production of NO_2? Table 7.8 provides the relevant data.

Plan
- Write the reaction for production of $NO_2(g)$ from $NO(g)$.
- Determine ΔH and ΔS.
- $T = \Delta H / \Delta S$.

Implementation
- $2NO(g) + O_2(g) \longrightarrow 2NO_2(g)$
- $\Delta H = 2 \times (33.2 - 91.3) \text{ kJ} = -116.2 \text{ kJ}$
  $\Delta S = [2 \times (240.1 - 210.8) - 205.2] \text{ J/K} = -144.6 \text{ J/K}$
- $T = \dfrac{\Delta H}{\Delta S} = \dfrac{(-116.2 \text{ kJ}) \times (1000 \text{ kJ/J})}{-144.6 \text{ J/K}} = 803.6 \text{ K}$

Table 7.8 Data for Worked Example 7.11			
Substance	ΔH_f° (kJ/mol)	ΔG_f° (kJ/mol)	S° (J/mol · K)
NO(g)	91.3	87.6	210.8
$NO_2(g)$	33.2	51.3	240.1
$O_2(g)$	0	0	205.2

If the engine is kept to temperatures less than about 500 °C, NO will be produced rather than NO_2.

See Exercises 65 – 68.

7.4   Coupled Reactions

Many chemical processes involve more than one simultaneous reaction. For example, iron oxide, Fe_2O_3, is the familiar substance rust. Left to itself, Fe_2O_3 does not decompose to metallic iron and elemental oxygen. However, coupling the decomposition of Fe_2O_3 with the production of Al_2O_3 results in the violent thermite reaction. In the language of Gibbs free energy, ΔG for the decomposition of Fe_2O_3 is positive. But ΔG for the *simultaneous* decomposition of Fe_2O_3 and production of Al_2O_3 is negative: It is a spontaneous process. This is similar to circus performers or children on a teeter-totter (Figure 7.24). Neither the child nor the circus performer spontaneously goes up in the air, but coupling another action does launch the person upward.

ATP

Reactions in living organisms are governed by the same constraints that apply to reactions in the laboratory: They are spontaneous if ΔG is negative, and they are not spontaneous if ΔG is positive. There is no "free lunch"; organisms must get the energy for synthesis of complex molecules such as proteins, DNA, and RNA from energy-releasing reactions. The origin of the energy-releasing reactions is the "burning" of foods—sugars and fats—which converts the food to CO_2 and H_2O by combining the "fuel" with oxygen in a process called **oxidation.** The cells in our bodies consume molecules that are rich in carbon but lacking in oxygen, take these molecules apart, and then combine the pieces with oxygen. In the process, the cells use the released energy to

Oxidation Removal of one or more electrons.

Figure 7.24 Processes that require free energy input can occur if they are coupled to another process so that the overall process is spontaneous. (a) A circus performer cannot fly through the air spontaneously. With the help of a partner imparting energy to a springboard, however, the performer easily launches upward. (b) A child on a teeter-totter does not spontaneously bounce up in the air, but if another playmate on the other end drops down, the child rises easily.

maintain the organism. If the energy were released all at once in an uncontrolled fashion, the organism would simply burn up. Instead, nature breaks the oxidation into smaller steps, shuttling energy to where it is needed. The key energy-shuttling reaction common to all living systems is the conversion of adenosine triphosphate (ATP) to adenosine diphosphate (ADP) (Figure 7.25). ATP/ADP conversion shuttles energy from the steps of the combustion to bootstrap energetically unfavorable processes such as protein synthesis. Several enzymes are involved, carefully regulating the rate of energy release.

ATP features three phosphate—phosphorus and oxygen—groups attached to an adenosine sugar molecule (Figure 7.25). The phosphate end of ATP features four negative charges in close proximity, making the bonds relatively weak and the phosphate group easily removed with addition of water.

(7.21) $ATP^{4-} + H_2O \rightleftharpoons ADP^{3-} + HPO_4^{2-} + H^+$ $\Delta G = -34.5 \text{ kJ}$

Addition and removal of water occurs every few seconds in an active cell, turning over the entire pool of ATP in about a minute. It has been estimated that if ATP were not regenerated, a person would have to consume his or her entire body weight in ATP each day. ATP hydrolysis is a relatively low-energy reaction compared with the combustion of sucrose, the primary energy source:

(7.22) $C_6H_{12}O_6 + 6O_2 \longrightarrow 6CO_2 + 6H_2O$ $\Delta G = -2872 \text{ kJ}$

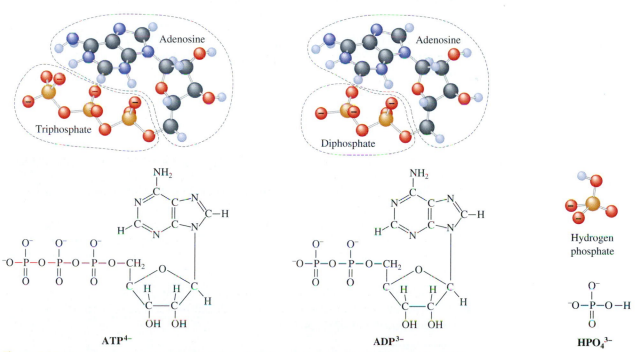

Figure 7.25 The energy engine of the body is controlled by (a) adenosine triphosphate, ATP^{4-}. Due to the large number of negative charges close together, this is a high-energy molecule. The energy can be lowered by hydrolysis of one phosphate group, decreasing the number of closely spaced negative charges by one to generate adenosine diphosphate, ADP (b), phosphate (c), HPO_4^{2-}, and H_3O^+. In the metabolic breakdown of glucose, the phosphate group is transferred to glucose. Color code: P (yellow), carbon (black), oxygen (red), nitrogen (blue), hydrogen (light blue).

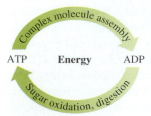

Figure 7.27 Energy is recycled in the cells of living organisms. Energy from exoenergetic reactions converts ADP to ATP by incorporating a phosphate ion (HPO_4^{2-}) from the cellular fluid. When energy is needed for complex molecule buildup, the energy is released via hydrolysis of ATP, converting it to ADP and releasing a phosphate ion into the cellular fluid.

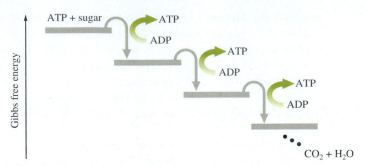

Figure 7.26 Digestion of sugar (food) releases energy that is used to convert ADP to ATP. The energy stored in ATP is used to synthesize complex molecules in the body, regenerating ADP in the process. Thus goes the cycle of life.

If the oxidation of sugar occurred in one step, it would be very difficult to harness more than a small fraction of the energy released. The body has cleverly devised a series of smaller steps. One is the phosphorylation of glucose:

(7.23) Glucose + HPO_4^{2-} + H^+ ⟶ [glucose phosphate]$^-$ + H_2O $\Delta G = 13.8$ kJ

CONCEPT QUESTION Does coupling phosphorylation of glucose, Equation (7.23), with the hydrolysis of ATP, Equation (7.21), result in a spontaneous process? ∎

Each of the smaller sucrose oxidation steps releases enough free energy to regenerate ATP from ADP (Figure 7.26). Complex molecule assembly uses the energy stored in ATP, returning it to ADP and ensuring that the cycle continues as long as the organism is alive (Figure 7.27).

WORKED EXAMPLE 7.12   *Solar Power*

Plants also use the ATP-ADP pair to shuttle energy. The major difference is that animal organisms utilize oxidation of fats and sugars as the energy source for cycling of ATP, whereas plants use sunlight for this purpose. Chlorophyll in plants appears green, so it absorbs light in the red and blue regions of the visible spectrum. Assume that the conversion is 100% efficient. How many ADP molecules are converted to ATP from absorption of one mole of blue (430 nm) photons? From the absorption of one mole of red (650 nm) photons?

Plan
- Determine the energy of one mole of photons, $E = N\,h\nu = N\,hc/\lambda$.
- Divide the photon energy by the energy needed to convert one ADP to ATP, as shown in Equation (7.21), 34.5 kJ.

Implementation

$$\blacksquare \quad E = \frac{6.022 \times 10^{23} \text{ particles}}{\text{mol}} \times 6.6262 \times 10^{-34} \text{ J} \cdot \text{s} \times \frac{2.9979 \times 10^8 \text{ m/s}}{430 \text{ nm}}$$

$$\times \frac{\text{nm}}{10^{-9} \text{ m}} = 278 \text{ kJ/mol for blue photons}$$

$$E = 184 \text{ kJ/mol for red photons}$$

- There are about eight turnovers per mole of blue photons, and a little over five turnovers per mole of red photons.

See Exercises 79, 80.

Checklist for Review

KEY TERMS

heat of reaction, enthalpy
  change, ΔH (p. 234)
internal energy (p. 237)
heat, q (p. 237)
work, w (p. 237)
exothermic (p. 237)
endothermic (p. 237)
heat capacity, C_p (p. 238)
Hess's law (p. 240)
state function (p. 240)
enthalpy of formation,
  ΔH_f° (p. 243)
extensive (p. 244)
intensive (p. 244)
crystal lattice energy,
  $\Delta H_{\text{x'tal lattice}}$ (p. 245)
hydration enthalpy,
  $\Delta H_{\text{hydration}}$ (p. 245)

enthalpy of atomization
  (p. 250)
latent heat of fusion
  (p. 251)
latent heat of
  vaporization (p. 252)
entropy, S (p. 252)
spontaneous process
  (p. 254)
third law of
  thermodynamics
  (p. 254)
free energy, Gibbs
  free energy, G (p. 257)
oxidation (p. 260)

KEY EQUATIONS

$\Delta E = q + w$

$\Delta H = C_p \Delta T$ constant pressure, no phase change, no reaction

$w = -P_{\text{applied}} \Delta V$

$\Delta H = \sum \Delta H_f^\circ \text{ (products)} - \sum \Delta H_f^\circ \text{ (reactants)}$

$\Delta H_{\text{x'tal lattice}} = M \dfrac{NZ^+ \times Z^- e^2}{R}$

$\Delta S_{\text{universe}} \geq 0$

$\Delta G = \Delta H - T\Delta S$

Chapter Summary

In this chapter, criteria are developed to determine when a reaction can occur. The fundamental quantity that determines the direction of a reaction is entropy. Specifically, the entropy (S) of the universe increases in a spontaneous process. Measuring the entropy of the universe poses a challenge. Meeting this challenge consists of dividing the universe into two parts: the system and the surroundings. Heat is exchanged between the system and the surroundings. Because most chemical reactions occur under the constant pressure of the earth's atmosphere, the entropy change of the surroundings is entirely due to heat (H) evolved from the system, and the entropy change of the surroundings is $-\Delta H/T$. Adding the entropy change of the surroundings to the entropy change of the system yields the entropy change for the entire universe. The Gibbs free energy (G) is defined to quantify the change in entropy of the universe for a chemical reaction:

$$\Delta G = \Delta H - T\Delta S$$

The Gibbs free energy decreases in a spontaneous reaction; the reaction proceeds if ΔG is negative and the reverse reaction proceeds if ΔG is positive. This is a central criterion for chemists.

The energy exchange between the system and the universe occurs because of the difference between the enthalpy of the products and that of the reactants. Enthalpy provides information about chemical interactions. Heat is released from the system if the interactions among the products are weaker than those among the reactants. In this case, the reaction is exothermic and ΔH is negative. A negative ΔH along with increased system entropy (positive ΔS) ensures a spontaneous reaction. Conversely, a positive ΔH along with decreased system entropy results in a nonspontaneous reaction. Reactions with the same sign for ΔH and ΔS switch direction with temperature.

KEY IDEA

Entropy determines which reactions can precede. The heat evolved in a process probes changing interactions among reactants and products.

CONCEPTS YOU SHOULD UNDERSTAND

- First law of thermodynamics: The energy change in a reaction is manifested as heat flow and work done.

OPERATIONAL SKILLS

- Calculate heat evolved from heat captured by water (Worked Example 7.1).
- Calculate work from volume change (Worked Example 7.2).

- Use Hess's law to calculate enthalpy, entropy, and free energy (Worked Examples 7.3, 7.6).
- Use enthalpy to determine relative bond energies (Worked Examples 7.4, 7.5, 7.7–7.9).
- Estimate lattice enthalpy from ion size and charge (Worked Example 7.7).
- Use atomization energy to determine bond energies (Worked Example 7.8).
- Evaluate reactions for spontaneity (Worked Examples 7.10–7.12).

Exercises

A blue exercise number indicates that the answer to that exercise appears at the back of the book. Thermodynamic data needed in these exercises are found in the appendix.

■ SKILL BUILDING EXERCISES

7.2 Heat

1. Is condensation of 10 g of water exothermic or endothermic? What is the relationship of the condensation enthalpy for 20 g of water to that for 10 g of water?

2. Is $O_2 \rightarrow 2O$ exothermic or endothermic? Explain.

3. Predict which of the following are exothermic.
   a. formation of H_2 from 2H
   b. formation of water from H_2 and O_2
   c. sublimation of CO_2
   d. evaporation of alcohol

4. The term "cold cream" aptly describes the cooling sensation that accompanies application of these skin care products. Although the exact formulation is a proprietary secret of each manufacturer, typical cold creams consists of 20% by weight alcohols (mainly isopropanol). The vapor pressure of ethanol (a typical alcohol) is 59 torr at 25 °C; by comparison, water vapor has a vapor pressure of 24 torr at the same temperature. Explain why cold cream is cold. How does cold cream compare with perspiration?

5. During freeze warnings in Florida, farmers spray their crops with warm water to protect them from frost. What chemical principles are growers using to protect their harvest? The following data may be useful: $\Delta H_{fusion}(H_2O) = 6.01$ kJ/mol, $C_p(H_2O) = 4.184$ J/g·K. Assume a typical fruit tree contains 5 L of sugar water, the freezing point of which is −4 °C. If

the temperature is predicted to drop to −6 °C, and typical well water is 13 °C, how much water must be sprayed on each tree to protect it?

6. A box of baking soda is often recommended as a handy fire extinguisher in the kitchen. Three key components are needed to sustain a fire: fuel, an oxidizing agent, and heat. The baking soda ($NaHCO_3$) hint is based on removing two of these required elements. Explain how baking soda works to extinguish a fire. The following data may be helpful: the decomposition reaction for $NaHCO_3$ is

$$2NaHCO_3(s) \longrightarrow Na_2CO_3(s) + H_2O(g) + CO_2(g)$$
$$\Delta H = 135.6 \text{ kJ/mol}$$

7. The reaction of Mg metal with CO_2 is

$$2Mg(s) + CO_2(g) \longrightarrow 2MgO(s) + C(graphite)$$

   a. Determine the enthalpy for the reaction of Mg metal with CO_2.
   b. Explain why a CO_2 fire extinguisher will not put out a Mg metal fire.

8. Aluminum has a heat capacity that is more than twice that of copper. Identical masses of aluminum and copper, both at 0 °C, are simultaneously dropped into identical cans of hot water. When each has come to equilibrium, is
   a. The temperature of the aluminum block higher than that of the copper block?
   b. The temperature of the copper block higher than that of the aluminum block?
   c. The temperature of both blocks the same?

9. Determine which of the following statements are false and explain why they are false.
   a. As a material freezes, it transfers heat to the environment and decreases in temperature.
   b. As a material freezes, it absorbs heat from the environment to maintain the same temperature.
   c. As a material freezes, it neither absorbs nor releases heat because the temperature remains the same.

10. The heat capacity of iron is 0.449 J/g · K. A blacksmith is hammering a 250-g horseshoe that has been heated to 1200 K. To cool the horseshoe, the blacksmith tosses it into a bucket containing 25 L of water at 25 °C. When the horseshoe has cooled, what is the temperature of the horseshoe and the water?

11. a. What is the relationship of ΔH for condensation of 20 g of water to that for 10 g of water?
    b. What is the relationship of ΔH for evaporation of 20 g of water to that for condensation of 20 g of water?

12. Fermentation consists of conversion of fructose to ethanol and CO_2.
    a. Is this an exothermic or endothermic process?
    b. Suppose that you are in charge of a winery and need to keep the bottles at a constant temperature during fermentation. How much water is required for each liter of alcohol produced? Assume that the water starts at room temperature (25 °C) and cannot get warmer than 30 °C.

13. Iron is a more electronegative metal than is aluminum, so when Fe_2O_3 is combined with aluminum metal a violent reaction occurs, forming iron metal and Al_2O_3. Suppose that 2 mol of Al (54 g) and 1 mol of Fe_2O_3 (160 g) are combined. All the heat released in the reaction is used to heat some water from room temperature (25 °C) to 90 °C. How much water is heated?

14. Liquid hydrazine (H_2NNH_2) is often used as solid rocket fuel. Calculate the heat produced when it is burned producing $N_2 + H_2O$. Data: $\Delta H_f(H_2NNH_2) = 50.6$ J/mol.

15. For which process is ΔH greater: condensation of supercooled steam or condensation of steam at 100 °C? Both produce water at 90 °C.

16. Explain why steam at 100 °C gives a more severe burn than water at 100 °C.

17. A popular cereal, Cheerios, derives most of its energy from its 22 g of carbohydrates (primarily sucrose, $C_{12}H_{22}O_{11}$). Calculate the number of calories contained in the carbohydrates. The label claims 110 Cal per serving. How many calories are from fat and the little protein in the cereal? (1 food calorie = 1000 cal.)

18. Average daily calorie need is based on a resting metabolic rate for an average adult of 80 watts (1 W = 1 J/s). How many calories per day must the average adult consume to maintain a resting rate? (One food calorie = one chemica kcal). Vigorous exercise can raise the caloric requirement by a factor of 7.5. How many more calories are needed by a person who exercises 30 minutes each day?

19. The total energy consumption in the United States is 10^6 kJ per person per day. How many liters of octane are required to produce this energy, assuming a typical 30% efficiency in using energy? (The density of octane is 0.6989 g/cm³.)

20. Reasonable efficiency for converting chemical energy to electrical energy via combustion is 30%. Calculate how many pounds of coal must be burned to run your reading lamp for two hours of studying for a chemistry exam. A typical reading lamp has a 75-W bulb. Assume that coal is pure graphite. (1 W = 1 J/s.)

21. Determine the combustion enthalpy per gram for the following fuels: methane, octane, methanol, and ethanol. Which fuel is most efficient per gram? Postulate what type of fuel is most efficient.

22. Determine which fuel, methane or propane, produces more heat per gram. Explain why.

23. A fuel cell typically converts methanol (H_3COH) to formaldehyde (H_2CO). Compare the energy produced by a methanol fuel cell versus combustion of methanol.

24. Acetylene is formed industrially by partial oxidation of propane:

    $$C_3H_8(g) + 2O_2(g) \longrightarrow C_2H_2(g) + CO(g) + 3H_2O(l)$$

    a. Calculate the heat of this reaction.
    b. The propane starting material comes from liquefied natural gas (95% propane and 5% methane). Methane does not form acetylene, but it does oxidize. Calculate the heat produced by methane combustion to CO and H_2O.
    c. What is the percent extra heat due to the methane side reaction?

25. a. Calculate ΔH for the conversion of O_2 to ozone.
    b. In the atmosphere, ozone occurs primarily in the lower stratosphere, the layer of the atmosphere just above the troposphere. The temperature of the lower stratosphere is higher than that of the troposphere. Explain this temperature profile using your data in (a).

26. You are in charge of constructing a dam that is 7 m thick at the base and tapers to 3 m thick at a height of 15 m for a dam that is 30 m wide. The density of concrete is 2.6 g/cm³, and the hydration enthalpy is about 20 kJ/g. This dam is to be cooled by water to keep the temperature between 25 °C and 45 °C. What volume of water is needed if the water starts at 13 °C?

7.2 Work

27. Calculate the work required to blow up a 10-cm diameter, spherical balloon against the pressure of the atmosphere.

28. Assuming all volume change is due to gases, calculate the work for combustion of 1 mol of octane producing gaseous water. Determine the work per liter of octane (density = 0.6986 g/cm³). Compare your answer to the work calculated in Worked Example 7.2. Comment on the difference.

29. Assuming all volume change is due to gases, calculate the work for combustion of 1 mol of methane and 1 mol

of methanol. What can you conclude about the desirability of incorporating oxygen into molecules to be used for fuel?

30. Assuming all volume change is due to gases, calculate the work produced during combustion of rocket fuel (liquid H_2NNH_2) to produce $N_2 + H_2O(g)$.

31. For an average adult, an average breath is $\frac{3}{4}$ pint of air and resting breathing involves about 20 breaths per minute. Calculate the work done each day in breathing.

32. A 3-cm^3 syringe is plugged on the end so that the air inside cannot escape. The gas in the syringe is compressed by rapidly depressing the plunger so that no heat is transferred to the surrounding air. The temperature inside the syringe is observed to rise by 7.3 °C. Assume that the pressure applied to the piston is a constant 1.2 atm. (The heat capacity of air is 1.03 J/g·K; the average molecular weight of air is 28.8 g.)
    a. What is the volume change of the gas?
    b. How much work was done on the gas?
    c. Determine the change in internal energy of the air in the syringe.

33. Baked goods "rise" due to gases produced in a chemical reaction during baking (baking soda or yeast) or due to liquid water turning into stream during baking (steam leavening, as in puff pastries). Calculate the work done by a typical sheet of puff pastry (8 in by 11 in by 0.75 in) expanding to four times its original volume when baked at 400 °F. Explain why the pastry is more successful if the oven is preheated.

34. Stretching a rubber band or bungee cord requires energy input. The work required changes with elongation as given by $w = \frac{1}{2} kx^2$, where k is the restoration constant for the rubber and x is the elongation. A rubber cord used in bungee jumping is 20 m long, doubles its length in a jump, and has a restoration constant of 11 N/m. How much work does the jumper do on the cord during a jump? (Neglect bounces — calculate only the work for the first stretch.) Compare the work done during the first meter of a jump to that in the last meter.

7.2 Heat and Work

35. When drinking a glass of ice water, the temperature of the water quickly rises from 0 °C to body temperature (37 °C). How many calories does the body expend in warming the water? (Glass volume: 8 oz or about 250 mL.) Energy is also expended in climbing the stairs. How much energy is expended in carrying a 20-lb bag of groceries up three flights of stairs (approximately 10 m)? Which is the more effective means for losing weight: drinking ice water or carrying groceries up the stairs? (Gravitational acceleration is 9.8 m/s^2.)

36. A circus performer hauls his 110-lb partner to the top of the trapeze (20 m). How much energy does the performer expend? How many grams of glucose must the performer consume to acquire this energy, assuming conversion of glucose to muscle energy is 30% efficient?

37. For an average adult, an average breath is $\frac{3}{4}$ pint of air and resting breathing involves about 20 breaths per minute. How many grams of sucrose must the average adult consume to provide the energy for breathing assuming that the conversion of sucrose to work is 50% efficient? (See Exercise 31 for the work done.)

38. Light sticks produce their glow as a result of an energy-releasing chemical reaction. The color of the light is determined by the reaction energy and the structure of the dye molecule that actually emits the light. The reaction energy must at least equal the emitted light energy. The light intensity is determined by the efficiency of transferring energy from the original energetic molecule to a dye molecule in solution.
    a. Determine the minimum reaction energy for a light stick that emits yellow-green light at 486 nm.
    b. Determine the minimum reaction energy for a light stick that emits light in the red region at 680 nm.

7.2 Hess's Law and Enthalpy of Formation

39. Calculate ΔH for the digestion (oxidation) of sugar, $C_{12}H_{22}O_{11}$, versus alcohol, C_2H_5OH (ethanol). Which provides more calories per gram? A typical alcoholic drink contains 2 oz of alcohol and a typical candy bar contains 6 oz of sugar. A person is going on a diet to lose weight. Which should he or she give up: a daily nightcap or the afternoon candy bar snack?

40. Iron is produced by roasting — heating iron oxide, Fe_2O_3, with elemental carbon produced by heating wood in an oxygen-poor atmosphere. Calculate the enthalpy for production of iron.

$$Fe_2O_3(s) + 3C(graphite) \longrightarrow 2Fe(s) + 3CO(g)$$

41. Given the enthalpy of formation of FeO (-266.5 kJ/mol) and Fe_2O_3 (-824.2 kJ/mol) at 25 °C, determine the enthalpy for the reaction

$$2FeO(s) + \tfrac{1}{2}O_2(g) \longrightarrow Fe_2O_3(s)$$

42. Look up the enthalpy of formation of the following series of compounds: HF, HCl, HBr, and HI. Comment on any trend for this series.

43. Calculate the heat of combustion for methane (CH_4), ethane (C_2H_6), propane (C_3H_8), and butane (C_4H_{10}) — all are fuels used in heating. Discuss any general trends for this series of hydrocarbons in terms of heat content per carbon, per gram, and per mole.

44. Calculate the heat of combustion for methane (CH_4), methanol (CH_3OH), propane (C_3H_8), propanol (C_3H_7OH), propaldehyde (C_3H_6O), and propanoic acid (C_2H_5COOH). Determine both the heat per mole and the heat per gram. Are there any general trends for this series of substances?

45. Calculate the heat produced in the following reaction:

$$3MnO_2(s) + 4Al(s) \longrightarrow 3Mn(s) + 2Al_2O_3(s)$$

Is this process endothermic or exothermic?

46. Use enthalpy of formation data to determine the latent heat of vaporization for propane, C_3H_8.

7.2 Hess's Law and Solution Formation: Crystal Lattice Enthalpy

47. The band gap in CdS is 2.42 eV in the yellow region of the visible spectrum. The band gap in CdTe is 1.5 eV deep in the red region of the spectrum. The formation reactions of both CdS and CdTe are exothermic. Predict which is more exothermic, and support your prediction.

48. CaO crystallizes in the same crystal lattice as NaCl. Calculate the crystal lattice enthalpy of CaO based on that of NaCl. The ion size for Ca^{2+} is 100 pm. Compare your result to the data listed in Table 7.2 and comment on any difference between your result and the data. (Ionic radii: Na^+, 102 pm; Ca^{2+}, 100 pm; Cl^-, 181 pm; and O^{2-}, 140 pm.)

7.2 Hess's Law and Solution Formation: Ion Hydration

49. Plot the hydration enthalpy for the Group IA metal ions in one series and the Group IIA metal ions in a second series. What is the relationship between ion size and hydration enthalpy?

50. Make a plot of hydration enthalpy for the following series of ions: Na^+, Mg^{2+} and Al^{3+}. Justify your choice of variable for the horizontal axis. Predict the hydration enthalpy for Cr^{6+}.

51. $BaSO_4$ is often used as a contrast agent for diagnostic X rays of lower gastrointestinal tract problems. $BaSO_4$ is a very insoluble material, which is fortunate because Ba^{2+} is a toxic ion. Is the dissolution of $BaSO_4$ endothermic or exothermic? Does the enthalpy of solution contribute to the lack of solubility?

52. AgCl is a classic example of the exception to the general rule that chloride salts are soluble. Does the enthalpy of solution contribute to the insolubility of AgCl?

7.2 Hess's Law and Enthalpy of Atomization

53. Compare the atomization enthalpy for $CaCl_2$ and NaCl. Comment on the difference.

54. Plot the atomization enthalpy versus atomic number for the fourth-period Group IA, Group IIA, and transition elements. Rationalize the trend.

55. Plot the atomization enthalpy for the Group IA elements. Explain the trend and predict an approximate value for the unstable element, Fr.

56. Consider the following reaction:

$$Cl(g) + CH_4(g) \longrightarrow HCl(g) + CH_3(g)$$

a. Determine ΔH for this reaction.
b. Determine $\Delta H_{\text{atomization}}$ for $CH_3(g)$.
c. What is the average energy of a CH bond in CH_4? In CH_3? Compare these two results.

7.3 Entropy

57. Describe the entropy changes for the following processes: melting a solid, boiling a liquid, subliming a solid, and condensing a gas.

58. Calculate the solution entropy for the series NaCl, $MgCl_2$, and $AlCl_3$. Comment on the trend you observe.

59. Patients often report a warm, flushed feeling as one of the side effects of being injected with a contrast agent prior to undergoing a CAT scan. Contrast agents are ionic substances that have a high iodine content. One example is iotalamic acid, which has the molecular formula $C_{11}H_9I_3N_2O_4$. The high iodine content is important for absorption of X rays to enhance the image.
a. Calculate the mass percent iodine in iotalamic acid.
b. Injection of iotalamic acid and the charge-balancing cation (often Na^+) raises the ionic strength of the blood. To restore the normal ionic strength, water is drawn out of nearby tissue in a process called osmosis. Explain osmosis using entropy arguments.
c. Due to the added blood volume, extra energy must be expended to circulate the blood. Explain the warm sensation felt by many patients.

60. When a stretched rubber band is relaxed, it cools. (Prove this to yourself by stretching a rubber band against your lip and note the temperature. Let the rubber band relax and again note the temperature.) What happens to the entropy of the rubber band when it relaxes? Draw a cartoon of the long molecules of rubber in the rubber band in both the stretched and the relaxed states.

7.3 Gibbs Free Energy

61. The Goldschmidt process is used industrially to produce elemental metals.

$$2Al(s) + Cr_2O_3(s) \longrightarrow Al_2O_3(s) + 2Cr(?)$$

a. Calculate the heat produced per gram of metallic chromium.
b. Do you expect this process to be spontaneous? Why or why not?
c. Will the chromium produced be solid or liquid? (*Hint:* Use heat capacity and heat produced to determine whether it will melt.)

62. Titanium is a strong, yet lightweight metal. Determine whether Ti forms an oxide at 25 °C.

63. Two oxides of nitrogen are prevalent in polluted city air: NO_2 and N_2O_4. NO_2 is the form responsible for the brown haze observed on bad pollution days. Which nitrogen oxide is more stable at room temperature?

64. Ethanol (C_2H_5OH) is the alcohol found in alcoholic beverages including wine. Acetic acid (CH_3COOH, vinegar) is a product of oxidation of alcohol by oxygen. Is this oxidation spontaneous at room temperature?

7.3 Temperature and Spontaneity

65. One step in the production of metallic iron is reduction of iron oxide with carbon.

$$2Fe_2O_3(s) + 3C(graphite) \longrightarrow 4Fe(s) + 3CO_2(g)$$

a. Evaluate the spontaneity of this reaction at room temperature.
b. How does heating the reaction change the conclusion?

66. Some metal oxides are very stable, whereas others decompose to yield the metal and oxygen when heated. Determine

which of the following are stable when heated. For those that are unstable, determine the decomposition temperature.
a. MgO
b. CaO
c. ZnO
d. CuO

67. A number of different arrangements of $C_4H_{10}O$ are possible. Two of them are 2-butanol and diethyl ether.
a. Determine the latent heat of vaporization for each.
b. Determine the entropy change for the condensation of each.
c. What is the boiling point of each?

68. HgO can be heated to generate molecular oxygen. At what temperature does HgO decompose?

7.4 Coupled Reactions

69. Oxidation of one glucose molecule leads to the regeneration of 38 molecules of ATP. Show that this coupling makes synthesis of ATP spontaneous. (The entropy of glucose can be neglected.) What fraction of the glucose oxidation energy is used in ATP synthesis?

70. Hydrolysis of ATP produces ADP and releases energy:

$$ATP^{4-} + H_2O \rightleftharpoons ADP^{3-} + HPO_4^{2-} + H^+$$
$$\Delta G = -30.5 \text{ kJ}$$

The initial step in the metabolic breakdown of glucose is the addition of a phosphate group (called phosphorylation):

$$\text{Glucose} + HPO_4^{2-} + H^+ \rightleftharpoons [\text{glucose phosphate}]^-$$
$$+H_2O \qquad \Delta G = 13.8 \text{ kJ.}$$

Show that coupling the phosphorylation of glucose with hydrolysis of ATP results in a spontaneous reaction.

71. Copper forms an oxide, Cu_2O. The standard free energy of formation of Cu_2O is -140 kJ/mol at 375 K, so heating does not liberate copper from oxygen. The oxidation of carbon to CO is spontaneous at 375 K. The standard free energy of formation of CO is -143.8 kJ/mol. Show that coupling the reduction of copper with the oxidation of carbon results in a spontaneous reaction.

72. The oxide of chromium, Cr_2O_3 is very stable and does not decompose even at very high temperatures. However, when heated with aluminum, chromium is liberated from the oxide and may be recovered as chromium metal.
a. Show that Cr_2O_3 is stable with respect to decomposition into Cr and O_2.
b. Show that coupling reduction of Cr_2O_3 with oxidation of aluminum produces a spontaneous reaction.

■ CONCEPTUAL EXERCISES

73. Draw a diagram indicating the relationship among the following terms: ΔH, endothermic, exothermic, enthalpy of formation, hydration enthalpy, atomization enthalpy, latent heat of fusion, latent heat of vaporization. All terms should have at least one connection.

74. Draw a diagram indicating the relationship among the following terms: internal energy, heat, work, entropy, third law

of thermodynamics, enthalpy. All terms should have at least one connection.

75. The atomization enthalpy of graphite is 716.7 kJ/mol, that of O_2 is 498.4 kJ/mol, and that of CO_2 is 1608 kJ/mol. Determine the enthalpy of formation of CO_2. Draw a diagram indicating the relationship of the atomization enthalpy to the formation enthalpy. Explain why enthalpy of atomization is always positive.

76. The main component of natural gas is methane, CH_4. The combustion of methane produces CO_2 and H_2O.
a. Write the balanced reaction for the combustion of methane.
b. Use enthalpy of formation data to determine the enthalpy of combustion of methane.
c. Use enthalpy of atomization data to determine the enthalpy of combustion of methane.
d. Draw a diagram indicating the relationship of the calculation in (b) to that in (c).

■ APPLIED EXERCISES

77. Nestlé is test-marketing a self-heating can of Nescafé. Inside the can is a small container of lime (CaO) and water. When a person pushes a button on the can, the two mix and generate enough heat to raise the temperature of the contents to 40 °C in a few minutes. Discuss the feasibility of such a scheme. What are the limitations of such a device? The heat of hydration of CaO is approximately 600 kJ/mol. Assume the drink contains 250 g and the heat capacity is the same as water, the major constituent.

78. Natural gas is primarily methane, CH_4, and is used to heat homes. A typical home is approximately 2000 ft^2 and the ceilings are 8 ft high. How many grams of methane are required to raise the temperature in the home from 40 °F to 70 °F? How many grams of CO_2 does this reaction produce? The following data may be helpful: The heat capacity of air is 1.01 J/g · K and the enthalpy of combustion of methane is -890.8 kJ/mol. Assume that the molecular weight of air is the same as nitrogen, its major component.

■ INTEGRATIVE EXERCISES

79. Living organisms obtain their energy from the oxidation of glucose, capturing the energy produced from oxidation by coupling it to the production of ATP. This process is not 100% efficient; some of the energy is "lost" as heat keeping organisms warm. Assume that only half of the energy released from the oxidation of glucose is stored in ATP. Calculate how many grams of glucose are required to maintain a 140-lb body at normal body temperature: 98.6 °F. Assume that the heat capacity of the body is the same as water (its major component) and that glucose oxidation raises the temperature from an ambient temperature of 65 °F.

80. Plants convert solar energy into various forms of sugar, including sucrose. Animals, in turn, use sucrose as the source

of energy to maintain their bodies, so life on earth traces its energy source to the sun. The light-harvesting molecule is chlorophyll, which absorbs in the blue (450 nm) and red (680 nm) regions of the spectrum.

a. Determine the enthalpy required to synthesize sucrose from CO_2 and H_2O.

b. How many blue photons are required to provide the energy to synthesize one molecule of sucrose from CO_2 and H_2O?

c. How many red photons are required for the same synthesis?

d. It has been estimated that more than 4.2×10^{17} kJ is stored annually by photosynthesis. How many tons of carbon are stored as carbohydrates each year?

Chapter 8
Equilibrium: A Dynamic Steady State

A balance between the pulling forces of the opposing teams keeps the tug-of-war in equilibrium.

CONCEPTUAL FOCUS

- Construct a dynamic model of equilibrium consisting of continuous forward and reverse reactions.
- Connect molecular-level interactions and thermodynamics with equilibrium.

SKILL DEVELOPMENT OBJECTIVES

- Determine concentrations at equilibrium (Worked Examples 8.1, 8.3–8.10).
- Relate pressure and concentration equilibrium constants to each other (Worked Example 8.2).
- Determine the Gibbs free energy from concentrations and equilibrium from the standard Gibbs free energy (Worked Example 8.11).
- Change equilibrium concentrations by adding reactants or products or by changing pressure or temperature (Worked Example 8.12).

Look around you and you will see that most substances, such as a cup of coffee saturated with sugar or the dressing on a salad, appear not to change. However, within the calm macroscopic view, continuous activity is occurring (Figure 8.1). Sugar molecules are constantly going into solution while others are rejoining the solid at the bottom of the cup. The vinegar in the salad dressing has even more activity. Vinegar primarily consists of acetic acid and water and these components interact continuously: Intact acetic acid molecules interact with water and occasionally a fragment joins with water breaking the acetic acid molecule apart, even as other fragments reform intact acetic acid molecules. All of this action is linked to molecular-level interactions; the focus of this chapter is developing the connections between the action and the interactions.

8.1   Gas-Phase Reactions

Look at the air around you. Everything appears calm. However, magnifying the air by 10 million times reveals a buzz of activity. The dominant molecule in air is molecular nitrogen, a very stable molecule featuring a triple bond. Next in abundance is molecular oxygen, a molecule that is extremely important to life. Nitrogen, oxygen, and a host of other molecules are continually colliding, changing direction, and slowing down or speeding up. Among the scarcer species is NO_2, the molecule that gives polluted city air its brown haze. As NO_2 molecules fly about, occasionally two NO_2 molecules collide. Even less frequently, two colliding NO_2 molecules fuse together forming a single molecule of N_2O_4.

CONCEPT QUESTION   Write the Lewis dot diagrams for NO_2 and N_2O_4. Why does NO_2 dimerize to form N_2O_4? ■

Isolating NO_2 molecules in a sealed container results in a container filled with a dark brown gas. Joining two NO_2 molecules together to form N_2O_4 (Figure 8.2) lightens the color of the gas because N_2O_4 is colorless. However, such a container would not become totally colorless. Instead, some brown color would remain. When no further color change occurs, there is a mixture of NO_2 and N_2O_4 in the container. Although the color remains constant, NO_2 molecules continue to bond together to form N_2O_4, and at the same time N_2O_4 molecules dissociate to regenerate two NO_2 molecules.

CONCEPT QUESTION   Which of the following has the highest entropy: a 1-L container filled with 0.04 mol of nitrogen atoms and 0.08 mol of oxygen atoms arranged as 0.04 mol of NO_2, 0.02 mol of N_2O_4, or a mixture of NO_2 and N_2O_4? ■

There are many ways to arrange 2.5×10^{22} (0.04 mol) NO_2 molecules in a 1-L container and just as many ways of arranging 2.5×10^{22} N_2O_4 molecules in the same container. However, there are even more ways of arranging a mixture of 2.5×10^{22} molecules containing both NO_2 and N_2O_4. Forming a mixture increases the entropy.

Figure 8.1   Equilibrium is dynamic: The rate in one direction is balanced by the rate in the other. Although a macroscopic change is not observed, a lot of action is occurring.

Figure 8.2 In the dizzying dance of molecules in the gas phase, occasionally two NO_2 molecules collide and join together to form a molecule of N_2O_4 (stages simulated on right-hand side). NO_2 is responsible for the brown haze over polluted cities. When dimerized as N_2O_4, it is colorless.

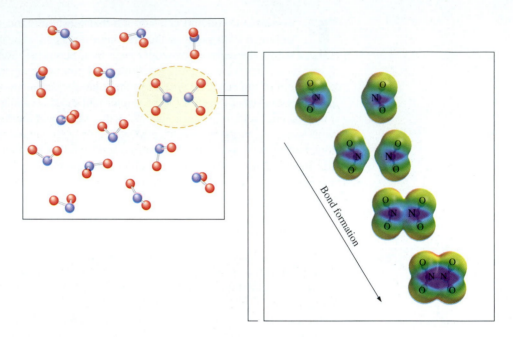

The entropy is *maximized* and the Gibbs free energy is *minimized* for a mixture. For any chemical reaction in a closed container, the mixture has the lowest Gibbs free energy and is called the *equilibrium state,* or simply **equilibrium.**

Equilibrium Lowest Gibbs free energy for a chemical reaction in a closed container. Also called *equilibrium state.*

For the NO_2/N_2O_4 mixture, the stability gained by joining two NO_2 molecules to give each nitrogen an octet is balanced by the increased entropy from having a mixture of NO_2 and N_2O_4. The competition between increased entropy and increased stability determines the proportion of NO_2 versus N_2O_4.

Equilibrium is attained when opposing reactions occur at equal rates. Look at the two fishermen in Figure 8.1. Despite the flurry of activity, neither boat's water level gets reduced. The rate at which products form reactants equals the rate at which reactants form products. To indicate that the reaction proceeds in both directions, it is written with two reaction arrows:

(8.1) $$2NO_2 g \rightleftharpoons N_2O_4 g$$

or more generally

(8.2) $$a\text{A} + b\text{B} \rightleftharpoons p\text{P} + s\text{S}$$

where A, B, P, and S indicate the reactants and products and *a*, *b*, *p*, and *s* are the stoichiometric coefficients in the balanced reaction.

Equilibrium constant, K_c Ratio of concentrations. Specifically, the product of the product concentrations divided by the product of the reactant concentrations. More exactly, it is the product of the product *activities* divided by the product of the reactant *activities.*

The salient question is, What determines the distribution among A, B, P, and S at equilibrium? How is entropy balanced against interactions? The answer to the distribution question was formulated in the early to mid-1800s by Claude Berthollet, a scholar working for Napoleon. Designating the concentration of a chemical species by enclosing its symbol in square brackets, at equilibrium the ratio

(8.3) $$\frac{[\text{P}]^p\,[\text{S}]^s}{[\text{A}]^a\,[\text{B}]^b} = K_c$$

is a constant and is known as the **equilibrium constant, K_c.** Note that K, carries a subscript "c" to indicate that it refers to concentration in molarity units. Every reaction has its own value for K_c.

Applying Equation (8.3) to the dimerization of NO_2, Equation (8.1), results in

(8.4)
$$\frac{[N_2O_4]}{[NO_2]^2} = K_c$$

The value of K_c is 164.8 at 25 °C and 1 atm pressure. Note that the equilibrium constant *has no units*. (The technical reason for this is discussed later in the chapter.)

WORKED EXAMPLE 8.1   *Brown Haze*

As mentioned earlier, NO_2 is the molecule responsible for the brown haze over polluted cities. In a 1-L vessel containing only NO_2 and N_2O_4 at room temperature (25.0 °C) and atmospheric pressure, what fraction of the molecules are NO_2 and what fraction are N_2O_4? ($K_c = 164.8$.)

Plan
- Use $PV = nRT$ to determine the number of moles in 1 L.
- Use $K_c = 164.8$ to solve for concentrations (two equations and two unknowns).
- Use the definition mole fraction NO_2 = moles NO_2/total moles.

Implementation
- $P = 1$ atm, $T = 298.1$ K, $V = 1$ L
  The total number of moles is $n = PV/RT = 0.0409$ mol
  $\Rightarrow n(N_2O_4) + n(NO_2) = 0.0409$ mol
  Since the total volume is 1 L, $[N_2O_4] + [NO_2] = 0.04089$ mol/L.

- $$K_c = [N_2O_4]/[NO_2]^2 = 164.8 \implies [N_2O_4] = 164.8\,[NO_2]^2$$
  $$\implies 164.8\,[NO_2]^2 + [NO_2] = 0.0409$$
  $$\implies [NO_2] = 0.0130 \text{ mol/L}, [N_2O_4] = 0.0279 \text{ mol/L}$$

- Mole fraction $NO_2 = 0.0310/0.0409 = 0.318$; mole fraction $N_2O_4 = 0.682$

The brown color comes from the 31.8% of the mixture that is NO_2.

See Exercise 1.

For gas-phase reactions such as the dimerization of NO_2, it is often more convenient to express the equilibrium constant with concentrations being replaced by partial pressures. Conversion between concentration and partial pressure involves the gas law. For moderate pressures, the ideal gas law, $PV = nRT$, applies.

(8.5)
$$p = \left(\frac{n}{V}\right)RT = cRT$$

Partial pressure (in atmospheres) is equal to concentration times the product of the gas-law constant, R, and the absolute temperature, T.

(8.6)
$$K_c = \frac{[N_2O_4]}{[NO_2]^2} = \frac{p_{N_2O_4}/RT}{(p_{NO_2}/RT)^2} = \frac{p_{N_2O_4}}{(p_{NO_2})^2}\,RT = K_p\,RT$$

K_p is 6.74 for the dimerization of NO_2. In general, Equation (8.3) results in a factor of $(1/RT)$ for each mole of gas in the reaction. As a result

(8.7)
$$K_c = K_p\,(1/RT)^{\Delta n}$$

where Δn is the change in the number of moles of gas.

WORKED EXAMPLE 8.2   *Making Fertilizer*

Nitrogen, one of the essential elements for living organisms, is made into bioavailable nitrogen from atmospheric nitrogen by an enzyme found in some plants. This process is known as nitrogen fixation. Animals lack the ability to fix nitrogen. To meet the needs of the growing human population, a process was therefore developed to "fix" nitrogen, thereby converting it into ammonia to fertilize crops and increase the supply of bioavailable nitrogen. The process for "fixing" nitrogen is known as the Haber-Bosch process in honor of the German chemist Fritz Haber, who developed the process, and Karl Bosch, the engineer who developed the equipment needed for industrial production of ammonia. The reaction is

$$N_2 + 3H_2 \rightleftharpoons 2NH_3$$

The value of K_c is 9.60 at 300 °C. What is the value of K_p?

Plan

- Determine the number of moles of gas involved in the reaction.
- Use Equation (8.7): $K_c = K_p(1/RT)^{\Delta n}$.

Implementation

- There are two moles of gas on the product side ($2NH_3$) and four on the reactant side (N_2 plus $3H_2$).
- $K_c = K_p \times (1/RT)^{-2} = K_p \times (RT)^2 \Rightarrow K_p = K_c/(RT)^2$
  At 300 °C (573 K), $RT = 47.0$ L·atm/mol $\Rightarrow K_p = 9.60/(47.0)^2 = 4.34 \times 10^{-3}$

Forty billion pounds of ammonia are manufactured annually in the United States, mostly by the Haber-Bosch process.

See Exercises 1, 2.

8.2   Heterogeneous Equilibria

Homogeneous equilibria
An equilibrium system where all reactants and products are in the same phase.

Heterogeneous equilibria
Equilibria involving reactants and/or products in more than one phase.

Many equilibria, like the gas-phase equilibria discussed in Section 8.1, occur in only one phase. These are called **homogeneous equilibria.** Many other processes involve substances in separate phases. These are called **heterogeneous equilibria.** For example, limestone forms from atmospheric CO_2 and solid CaO, metals corrode by combining solid metal with atmospheric oxygen, and solid salts dissolve in liquid water to give ocean water. The reaction describing the formation of limestone is

(8.8) $$CaO(s) + CO_2(g) \rightleftharpoons CaCO_3(s)$$

How should the concentration of the solid be treated in this equilibrium? The density of a pure liquid or solid is a constant at any given temperature and pressure, and changes little over moderate temperature or pressure ranges. Due to this constancy, pure liquids and solids are not included in the equilibrium constant. It is as though the constant density is incorporated into the equilibrium constant. The expression for the equilibrium constant for formation of $CaCO_3$ is thus

(8.9) $$K_c = \frac{1}{[CO_2]} = \frac{RT}{p_{CO_2}} = K_p \times RT$$

Although the two solids CaO and $CaCO_3$ do not appear in the equilibrium constant, both must be present for equilibrium to be established. If both solids are present, CaO and $CaCO_3$ control the partial pressure of CO_2 in the gas phase. It is believed that limestone stores much of the CO_2 present in the early earth's atmosphere.

CONCEPT QUESTION Write the expression for the equilibrium constant K_c for the corrosion of tin (Sn) forming tin oxide (SnO_2). What is the relationship between K_c and K_p for the corrosion of tin? ∎

Several important equilibria involve dissolving gases in water. For example, combustion of sulfur-containing coal produces SO_2. SO_2 dissolves in water.

(8.10)
$$SO_2(g) + H_2O(l) \rightleftharpoons H_2SO_3(aq)$$

Subsequent processes yield solutions that contribute to acid rain around coal-fired power plants. Water is a pure liquid, so the equilibrium constant expression for dissolving SO_2 in water is

(8.11)
$$K_c = \frac{[H_2SO_3]}{[SO_2]} = \frac{RT[H_2SO_3]}{p_{SO_2}}$$

Note that although water is present in Equation (8.10), it does not appear in the equilibrium constant, Equation (8.11). The partial pressure of SO_2 and the aqueous concentration of H_2SO_3 are linked by the equilibrium.

CONCEPT QUESTION CO_2 dissolves in water to form H_2CO_3, a process responsible both for storage of CO_2 and for making natural water slightly acidic. What is the reaction for dissolving CO_2 in water? Write the expression for the equilibrium constants K_c and K_p for this process. ∎

For aqueous solutions, water is referred to as the solvent and substances dissolved in water are called solutes. Solutions involving other nearly pure liquids are similar. The nearly pure liquid is the **solvent** and any substance dissolved in the liquid is a **solute.** For example, methanol gas (CH_3OH) dissolves in acetic acid (CH_3COOH) to form methyl acetate (CH_3COOCH_3) and water:

Solvent The dissolving medium in a solution.

Solute Substance dissolved in a liquid to form a solution.

(8.12) $$CH_3OH(g) + CH_3COOH(l) \rightleftharpoons CH_3COOCH_3(sln) + H_2O(sln)$$

where the notation "sln" indicates that the products are dissolved in acetic acid. The expression for the equilibrium constant is

(8.13) $$K_c = \frac{[CH_3COOCH_3]\,[H_2O]}{[CH_3OH]} = \frac{RT[CH_3COOCH_3][H_2O]}{p_{CH_3OH}}$$

Because water is not the solvent, the concentration of water *does* appear in the equilibrium constant expression, whereas the solvent, *acetic acid, does not.*

8.3 Variations on Equilibrium Constants

In keeping with the many types of reactions, many variations on the equilibrium constant exist. Table 8.1 lists the names of several types of equilibrium constants, the associated reactions, and some specific examples. The symbol for the equilibrium constant is K. Subscripts associate the equilibrium constant with a class of reactions. For example, the equilibrium constant for acid ionization is K_a.

Table 8.1 Variations on K

Symbol	Name	Aqueous Solution Examples	K
K_a	Acid ionization constant	$CH_2COOH(aq) + H_2O(l) \rightleftharpoons H_3O^+(aq) + CHCOO^-(aq)$	1.8×10^{-5}
		$HF(aq) + H_2O(l) \rightleftharpoons H_3O^+(aq) + F^-(aq)$	7.2×10^{-4}
K_1	Successive ionization constants for	$H_2SO_4(aq) + H_2O(l) \rightleftharpoons H_3O^+(aq) + HSO_4^-(aq)$	K_1 = very large (~100)
K_2	polyprotic acids	$HSO_4^-(aq) + H_2O(l) \rightleftharpoons H_3O^+(aq) + SO_4^{2-}(aq)$	$K_2 = 1.2 \times 10^{-2}$
K_b	Base ionization constant	$NH_3(aq) + H_2O(l) \rightleftharpoons NH_4^+(aq) + OH^-(aq)$	1.8×10^{-5}
		$CH_3NH_2(aq) + H_2O(l) \rightleftharpoons CH_3NH_3^+(aq) + OH^-(aq)$	5.0×10^{-4}
K_{sp}	Solubility constant	$AgCl(s) \rightleftharpoons Ag^+(aq) + Cl^-(aq)$	1.8×10^{-10}
		$Au(OH)_3(s) \rightleftharpoons Au^{+3}(aq) + 3OH^-(aq)$	1.0×10^{-53}
K_d	Dissociation constant for a complex ion	$[Ag(NH_3)_2]^+(aq) \rightleftharpoons Ag^+(aq) + 2NH_3(aq)$	6.3×10^{-8}
		$[CuCl_2]^-(aq) \rightleftharpoons Cu^+(aq) + 2Cl^-(aq)$	1.0×10^{-5}
		$[Co(NH_3)_6]^{+3}(aq) \rightleftharpoons Co^{+3}(aq) + 6NH_3(aq)$	2.2×10^{-34}
K_f	Formation constant for a complex ion	$Ag^+(aq) + 2NH_3(aq) \rightleftharpoons [Ag(NH_3)_2]^+(aq)$	$1.6 \times 10^{+7}$
		$Cu^+(aq) + 2Cl^-(aq) \rightleftharpoons [CuCl_2]^-(aq)$	$1.0 \times 10^{+5}$
		$Co^{+3}(aq) + 6NH_3(aq) \rightleftharpoons [Co(NH_3)_6]^{+3}(aq)$	$4.5 \times 10^{+33}$

The key to decoding and using these variations on the equilibrium constant is knowing what chemical reaction the constant corresponds to. Remember:

> *An equilibrium constant*
> *by any other name is still*
> *an equilibrium constant.*

8.4 Aqueous Solutions

Solutions with water as the solvent receive the most attention because water is essential for cycling ions, molecules, and nutrients throughout the global sphere. Indeed, three-fourths of the earth's surface is covered with water and 70% of the human body's mass is water. Under ambient conditions, nearly every surface has a coating of water, and this coating is a key ingredient in corrosion and wear. Corrosion and wear generate ions from solids, continuing the cycling of substances through the environment.

Acids and Bases

Hydronium ion The H_3O^+ ion; a hydrated proton.

Hydroxide ion The OH^- ion.

Consider a glass of water. On the surface, it appears static. Beneath the surface, however, water molecules are jostling each other continuously. Furthermore, for every half billion water molecules, a hydrogen core transfers from one water molecule and attaches to the lone pair of another, resulting in two charged ions: H_3O^+, called the **hydronium ion,** and OH^-, called the **hydroxide ion** (Figure 8.3). Although once in a half billion might seem like an insignificant number, consider that there are about 6×10^{13} (60 trillion) such ions in a cubic centimeter of water. Nothing is constant in the glass of water save for this continuous jiggling and exchanging of protons like hot potatoes. At any instant, there are 60 trillion H_3O^+ ions and 60 trillion OH^- ions; the numbers stay constant but the specific water molecules that have been transformed change continuously.

How does this picture change when something else is added to water? For example, adding a teaspoon of vinegar to a glass of water most definitely alters its taste. On the

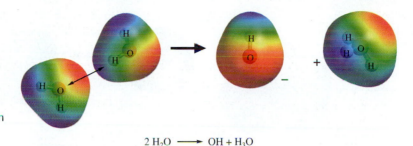

$$2\,H_2O \longrightarrow OH + H_3O$$

Figure 8.3 Water molecules are constantly jostling one another. Occasionally, a proton transfers from one water molecule to another, resulting in an H_3O^+ ion and an OH^- ion. The potential color range is shifted one unit positive for the H_3O^+ ion (i.e., green corresponds to a potential of $+1$) and one unit negative for the OH^- ion (green corresponds to a potential of -1).

The equilibrium constant
has no units. Technically, the
K_c equilibrium constant
involves concentration relative to an ideal one molar
solution. Like an ideal gas, an
ideal one molar solution is
one where the solute molecules do not interact with
one another. The ratio of the
concentration to the ideal
one molar concentration
removes the units and is
called the **activity.** The
activity of pure liquids and
solids is 1. For most purposes, it is sufficient to
remember the following
points: (1) for dilute solutions, the numbers that go
into the equilibrium constant are numerically equal
to the concentration in
moles per liter but have no
units; (2) for pure or nearly
pure liquids and solids, the
activity is 1; and (3) for gases,
it is either the concentration
for K_c or the partial pressure
in atmospheres for K_p.

Acid (Arrhenius definition)
An acid is a substance that
contains H and produces H^+
in aqueous solutions.

Activity The effective
concentration or pressure of
a substance: For the solvent,
activity equals one; for a
solid, activity equals one.

**Ion-product constant for
water, K_w** The equilibrium
constant for the autoionization of water;
$K_w = [H^+][OH^-]$. At 25 °C,
K_w equals 1.0×10^{-14}

molecular level, vinegar belongs to a class of substances known as acids. The simplest definition of an acid is the Arrhenius definition: An **acid** is a molecule that ionizes in water to generate H^+ ions. The H^+ ion is a bare proton, so it is extremely small and does not exist isolated in aqueous solution. Instead, H^+ is always associated with at least one water molecule. Hence it is often written as H_3O^+. Addition of H^+ from acetic acid, the acidic component of vinegar, upsets the equality between the number of H_3O^+ and OH^- ions in solution. However, while the relative concentration of H^+ and OH^- varies, their product is a constant.

$$(8.14) \qquad [H_3O^+][OH^-] = 1.0 \times 10^{-14} \equiv K_w \qquad \text{25 °C}$$

In pure water, $[H_3O^+]$ is the same as $[OH^-]$, and both are 1.0×10^{-7} M. In acidic solutions, $[H_3O^+]$ is larger than $[OH^-]$. In basic solutions, $[OH^-]$ is the larger quantity. K_w is called the ionization constant or the **ion-product constant for water.** Like gas-phase equilibria, the dynamic solution reaction is represented with an equilibrium equation.

$$(8.15) \qquad 2H_2O(l) \rightleftharpoons H_3O^+(aq) + OH^-(aq)$$

Before adding the acetic acid, one liter of water contains 1.0×10^{-7} moles of H_3O^+ ions and 1.0×10^{-7} moles of OH^- ions: It is a "neutral" solution. After addition of the acid, the H_3O^+ ion concentration increases *and the OH^- ion concentration decreases* to maintain a constant product.

Vinegar is an approximately 1 M acetic acid solution. If every acetic acid molecule dissociated, an additional mole of H^+ ions would be released per liter of solution. Acetic acid, however, is a weak acid. Weak acids incompletely ionize in water. For acetic acid, most of the molecules remain intact; only 0.4% ionize, yielding approximately 4×10^{-3} M H_3O^+. This is a considerably higher concentration than is found in pure water but it is also considerably less than the concentration of added acetic acid.

On the molecular level, acetic acid is CH_3COOH (Figure 8.4) and the CH_3COO^- ion is called an acetate ion. When acetic acid is put in water, there is a competition between the acetate ion and water for the acid's proton. Transferring the proton to water generates H_3O^+. The balance between molecular acetic acid and acetate ion plus hydronium ion is given by the equilibrium constant

$$(8.16) \qquad K_a = \frac{[CH_3COO^-][H_3O^+]}{[CH_3COOH]} = 1.8 \times 10^{-5}$$

Figure 8.4 Illustration of the equilibrium between molecular acidic acid and its associated ions in aqueous solution. The acetic acid proton transfers to a water molecule.

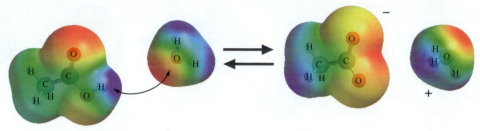

$$CH_3COOH(aq) + H_2O(l) \rightleftharpoons CH_3COO^-(aq) + H_3O^+(aq)$$

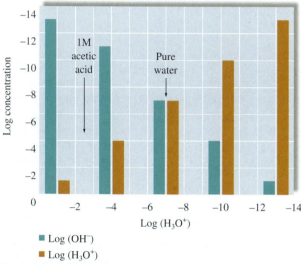

■ Log (OH⁻)
■ Log (H₃O⁺)

Figure 8.5 Increasing H_3O^+ concentration is accompanied by decreasing OH^- concentration. The product $[H_3O^+] \times [OH^-]$ is a constant. Because these concentrations vary over so many orders of magnitude, concentration is *plotted on a log scale.*

Acid ionization constant, K_a
The equilibrium constant for ionization of an acid.

pH scale A convenient way to represent solution acidity; $pH = -\log[H^+]$ (log base 10).

K_a is called the **acid ionization constant.** K_a for acetic acid is a small number, indicating that there are many more intact acetic acid molecules than acetate ions and hydronium ions in solution.

Describing Acids and Bases: The pH Scale

There are many more H_3O^+ ions than OH^- ions in the 1 M acetic solution. Nonetheless, the water ionization constant remains at 1.0×10^{-14}, so the hydroxide ion concentration is suppressed: $[OH^-] = 1.0 \times 10^{-14}/[H_3O^+] = 2.5 \times 10^{-12}\ M$. Handling exponential notation can be awkward, so the concentrations of H_3O^+ and OH^- ions are often expressed on a logarithmic scale (Figure 8.5). The log concentration scale is known as the **pH scale.** The term "pH" comes from the French *puissance d'hydrogène,* meaning "power of hydrogen." The pH scale is related to molar concentration of H_3O^+ by

(8.17) $$pH = -\log[H_3O^+]$$

Pure water has a pH of 7, and the 1 M acetic acid solution has a pH of 2.4. Any solution with a pH less than 7 is said to be acidic and any solution with a pH greater than 7 is called basic. The OH^- ion concentration is often expressed on a pOH scale: $pOH = -\log[OH^-]$. From the properties of logarithms, $pH + pOH = 14$. In analogy with pH, various other "p" terms are also used. The "p" stands for the negative logarithm.

CONCEPT QUESTION What is the pOH of a 0.01 M acetic acid solution? ■

WORKED EXAMPLE 8.3   *Basic Solutions*

Ammonia is one of the few molecules in the atmosphere that forms a basic solution in water. Although the concentration of ammonia in the atmosphere is quite low except near areas with a large population of water fowl, ammonia nonetheless plays an important role in neutralizing acids and generating solid particulates that serve as cloud condensation nuclei. Determine the pH of a solution that is 0.01 M in $NH_3(aq)$. $K_b(NH_3) = 1.8 \times 10^{-5}$.

Plan

- Write the equation for the reaction.
- Write the equilibrium constant expression.
- Define the unknown: moles NH_3 ionized in 1 L of water $= x$.
- $pOH = -\log[OH^-]$.

Implementation

- $NH_3(aq) + H_2O(l) \rightleftharpoons NH_4^+(aq) + OH^-(aq)$

- $K_b = \dfrac{[NH_4^+][OH^-]}{[NH_3]}$

- The reaction stoichiometry indicates that the numbers of moles of NH_4^+ and moles of OH^- are equal, and both are equal to the number of moles of aqueous ammonia that ionize. Thus, $x = [NH_4^+] = [OH^-]$ and $[NH_3] = 0.01 - x$.

$$K_b = \frac{[NH_4^+][OH^-]}{[NH_3]} = \frac{(x)(x)}{(0.01 - x)} = \frac{x^2}{(0.01 - x)} = 1.8 \times 10^{-5}$$

Since $K_b \ll 1$, x is expected to be much less than 0.01; $x^2 \cong 1.8 \times 10^{-7}$, so $x = 4.2 \times 10^{-4}$. Note that within the significant figures of the problem, 4.2×10^{-4} is negligible compared with 0.01.

- $[OH^-] = 4.2 \times 10^{-4}\,M \Rightarrow pOH = 3.4 \Rightarrow pH = 10.6$, a basic solution.

Ammonia-based cleaning products are approximately 1 M solutions of ammonia and have a correspondingly higher pH.

See Exercises 14–18.

APPLY IT Boil two leaves of red cabbage in two cups of water for 5 minutes. Let cool and pour the juice into three clear plastic cups, labeled 1, 2, and 3. Note the color of the solution. Cup 3 is a reference solution. Add $^1/_4$ teaspoon of vinegar to cup 1 and $^1/_4$ teaspoon of household ammonia to a cup 2. Record the color change. Add drops of vinegar to cup 2 until the color matches that of cup 3. Add $^1/_8$ teaspoon of baking soda to cup 2 and 3: Are the colors of these solutions the same after addition of baking soda? What would you add to cup 1 to restore the original color? (*Note* if the solution is very dark, dilute the cabbage juice before addition of vinegar, ammonia, or baking soda.) (Solutions may be poured down the drain after dilution. Do not consume the solutions. Dispose of the plastic cups in a solid waste container.)

Solubility

Solubility The amount of a substance that dissolves in a given volume of solvent at a given temperature.

Solubility is an example of a heterogeneous equilibrium between a solid and its ions in solution. Consider the familiar material salt, NaCl. A teaspoon of salt dissolves completely in a cup of water. However, a teaspoon of salt only partially dissolves in a teaspoon of water. The coulombic attraction between Na^+ and Cl^- is the same in both cases, but the amounts of water and ions are very different. With only a teaspoon of water, the ion concentration is much greater. The solution has a threshold level of ions; when this level is exceeded, a solid forms. The level is determined by the product of the ion concentrations and is called the **solubility product**. For regular salt (NaCl), the solubility product is very high. Thus NaCl is generally thought of as a soluble salt. Other salts, such as $CaCO_3$ in seashells or $Ca_3(PO_4)_2$ in bone, have a much smaller solubility product.

Solubility product The equilibrium product of the concentrations of the ions in a salt raised to their stoichiometric coefficeints.

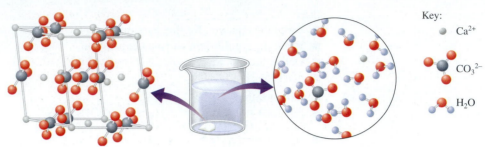

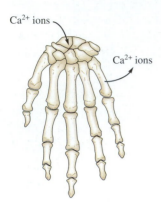

Figure 8.6 Calcium in bones and bodily fluids is constantly being exchanged at millions of building sites. The constant rebuilding repairs micro cracks and imperfections, allows the organism to grow, and aids in healing of fractured bones.

Figure 8.7 The equilibrium between solid $CaCO_3$ and Ca^{2+} plus CO_3^{2-} ions in solution.

Solubility Guidelines

Due to the importance of aqueous solutions, a set of general guidelines has been formulated for predicting solubility of ionic salts.

- *Always soluble* cations: Li^+, Na^+, K^+, Rb^+, Cs^+, and NH_4^+
- *Generally soluble* anions: NO_3^-, CH_3COO^-, ClO_3^-, ClO_4^-, and SO_4^{2-}
  Exceptions:

  > Insoluble: halides and pseudo-halides (CN^-, SCN^-) of Ag^+, Hg_2^{2+}, and Pb^{2+}
  > Insoluble: sulfates (SO_4^{2-}) of Pb^{2+}, Ba^{2+}, and Hg^{2+}
  > Sparingly soluble: sulfates (SO_4^{2-}) of Ca^{2+} and Ag^+

- *Generally insoluble* anions: OH^-, CO_3^{2-}, PO_4^{3-}, AsO_4^{3-}, and S^{2-}
  Exception:

  > Soluble: heavier Group IIA metal OH^-

Using Solubility Products

Bone (Figure 8.6) is often thought of as inert, but in living organisms it is constantly being degraded and restored at millions of remodeling sites. This ongoing remodeling also maintains a steady level of calcium in the blood—calcium that regulates many body functions. The dynamic exchange between calcium in bone and calcium in solution is critical for healing micro cracks and imperfections and allows the bone structure to grow as the organism does. Sea shells also contain calcium in the form of $CaCO_3$, and the exchange between the shell and seawater is similar to that between bone and body fluids.

Picture what happens in a beaker with water and solid $CaCO_3$ (Figure 8.7). When in contact with water, calcium ions (Ca^{2+}) and carbonate ions (CO_3^{2-}) ions are constantly coming together to join the solid and departing from the solid to be surrounded by water molecules. In the solution, the water molecules come between the oppositely charged ions. This dynamic exchange is represented by the equilibrium

$$\text{ions in solid} \rightleftharpoons \text{ions in solution}$$

(8.18)
$$CaCO_3(s) \rightleftharpoons Ca^{2+}(aq) + CO_3^{2-}(aq)$$

The similar equation for $Ca_3(PO_4)_2$ in bone is

(8.19)
$$Ca_3(PO_4)_2(s) \rightleftharpoons 3Ca^{2+}(aq) + 2PO_4^{3-}(aq)$$

CONCEPT QUESTION   How do the equilibrium constant expressions for solubility of $CaCO_3$ and for the solubility of $Ca_3(PO_4)_2$ differ? ∎

Saturated solution A solution containing ions in contact with the solid salt of the ions.

A solution containing ions in contact with the solid salt of the ions is called a **saturated solution.** A saturated solution of $CaCO_3$ contains about a five times larger concentration of Ca^{2+} ions than a saturated $Ca_3(PO_4)_2$ solution (see Worked Example 8.4).

WORKED EXAMPLE 8.4   *A Matter of Solubility: Sea Shells and Bones*

The cycling of calcium ions through the biosphere is linked to the inorganic calcium cycle via calcium ions in solution. Determine the calcium ion concentration for a solution in contact with bone, $Ca_3(PO_4)_2$, and with sea shells, $CaCO_3$.

Data

$$K_{sp}(CaCO_3) = 4.8 \times 10^{-9}$$
$$K_{sp}(Ca_3(PO_4)_2) = 1.0 \times 10^{-25}$$

Plan

- Write the equation for the reaction.
- Write the equilibrium constant expression.
- Define the unknown: x = moles solid dissolved in 1 L of water. Solve for x.

Implementation

For $CaCO_3$:

- $CaCO_3(s) \rightleftharpoons Ca^{2+}(aq) + CO_3^{2-}(aq)$
- $K_{sp}(CaCO_3) = [Ca^{2+}][CO_3^{2-}]$; note that $CaCO_3(s)$ does not appear.
- x moles dissolves in 1 L $\Longrightarrow [Ca^{2+}] = x$ and $[CO_3^{2-}] = x$

$$K_{sp}(CaCO_3) = (x)(x) = x^2 = 4.8 \times 10^{-9}$$
$$\Longrightarrow x = 6.9 \times 10^{-5} \Longrightarrow [Ca^{2+}] = 6.9 \times 10^{-5}\ M$$

For $Ca_3(PO_4)_2$:

- $Ca_3(PO_4)_2(s) \rightleftharpoons 3Ca^{2+}(aq) + 2PO_4^{3-}(aq)$
- $K_{sp}(Ca_3(PO_4)_2) = [Ca^{2+}]^3 [PO_4^{3-}]^2$
- x moles dissolve in 1 L $\Longrightarrow [Ca^{2+}] = 3x$ and $[PO_4^{3-}] = 2x$

$$K_{sp}(Ca_3(PO_4)_2) = (3x)^3(2x)^2 = 108x^5 = 1.0 \times 10^{-25}$$
$$\Longrightarrow x = 3.9 \times 10^{-6} \Longrightarrow [Ca^{2+}] = 1.2 \times 10^{-5}\ M$$

The calcium ion concentration in a solution in contact with bone is about one-fifth the calcium ion concentration in a solution in contact with sea shells.

See Exercises 20–23.

WORKED EXAMPLE 8.5   *Equilibrium Constants and Interactions*

When exposed to moisture, the iron oxides FeO and Fe_2O_3 hydrate to form $Fe(OH)_2$ and $Fe(OH)_3$. Which of these hydrated oxide coats has greater solubility in water?

Plan

- Write the equations for the reactions.
- Write the equilibrium constant expression.
- Use K_{sp} to determine solubility.

Implementation

- $Fe(OH)_2(s) \rightleftharpoons Fe^{2+}(aq) + 2OH^-(aq)$; $Fe(OH)_3(s) \rightleftharpoons Fe^{3+}(aq) + 3OH^-(aq)$
- $K_{sp}(Fe(OH)_2) = [Fe^{2+}][OH^-]^2$; $K_{sp}(Fe(OH)_3) = [Fe^{3+}][OH^-]^3$

■ For $Fe(OH)_2$:

$$K_{sp} = 7.9 \times 10^{-15} = x(2x)^2 \Longrightarrow x = 1.3 \times 10^{-5} \Longrightarrow \text{molar solubility}$$
$$= 1.3 \times 10^{-5}\ M$$

⚠ A frequent error in equilibrium calculation is forgetting either the stoichiometry ($2x$) or the exponential factor (power of 2).

For $Fe(OH)_3$:

$$K_{sp} = 6.3 \times 10^{-38} = x(3x)^3 \Longrightarrow x = 2.2 \times 10^{-10} \Longrightarrow \text{molar solubility}$$
$$= 2.2 \times 10^{-10}\ M$$

The molar solubility of the Fe^{3+} hydroxide is much smaller than that of the Fe^{2+} hydroxide, a direct reflection of the greater coulombic attraction between the $+3$ ion and the OH^- than between the $+2$ ion and the OH^- ion.

See Exercises 22, 23.

Linked Equilibria

Common Ion Effect

If all the Ca^{2+} and PO_4^{3-} ions in body fluids came from bone, then the calcium ion concentration would be $1.2 \times 10^{-5}\ M$ and the phosphate ion would be $7.8 \times 10^{-6}\ M$. However, not all Ca^{2+} in blood and other body fluids comes from dissolution of bone. Some calcium is supplied in the diet, and other calcium ions come from various proteins that regulate body functions. Once in solution, all calcium ions are the same, so the reactions that supply calcium ions are linked by the common aqueous ion concentration. For bone, because the solubility product is a constant, a higher calcium concentration suppresses the solubility of bone. For example, suppose that the additional sources of calcium raise the calcium ion concentration to $2\ mM$ (millimolar), a concentration typical of body fluids. How soluble is bone in this solution? With calcium concentration regulated by other systems, the solubility of bone is linked to the phosphate ion concentration.

(8.20)
$$K_{sp}(Ca_3(PO_4)_2) = [Ca^{2+}]^3\,[PO_4^{3-}]^2 = (2 \times 10^{-3})^3[PO_4^{3-}]^2$$
$$= 1.0 \times 10^{-25} \qquad [Ca^{2+}]\ \text{fixed}$$
$$\Longrightarrow [PO_4^{3-}]^2 = 1.25 \times 10^{-17} \Longrightarrow [PO_4^{3-}] = 3.5 \times 10^{-9}$$

Dissolving $Ca_3(PO_4)_2$ supplies two PO_4^{3-} ions per mole, so the molar solubility of bone in a $2\ mM$ calcium ion solution is 1.75×10^{-9} moles per liter. This is nearly four orders of magnitude less than in water—good news for bones in the body! Reducing the solubility of the salt by introducing one of the ions from another source is called the **common ion effect.**

Common ion effect An ion appearing in more than one simultaneous equilibrium.

In the body, Ca^{2+} ion is also used as a signal transmitter for many functions, including muscle contraction to make the heart beat faster in response to stress. Since this is a vital function, the proteins (Figure 8.8) that transmit the Ca^{2+} ions bind them very tightly and draw Ca^{2+} ions from the bones if Ca^{2+} is not supplied in the diet. That's why you were admonished to drink your milk as a child.

CONCEPT QUESTIONS The typical calcium ion concentration of seawater is $1.4\ mM$. Are sea shells more or less soluble in seawater than pure water? What is the molar solubility of sea shells in seawater? ■

Spectator Ions

For some applications, it important to control the concentration of components of a mixture to achieve a specified result. For example, barium salts are often administered to patients to

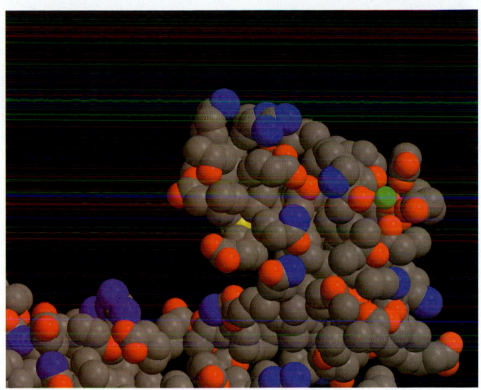

Spectator ion An ion that does not participate in a reaction. It is present to maintain overall electrical neutrality.

Figure 8.8 The protein calmodulin binds Ca^{2+} in the body, thereby regulating the level of Ca^{2+} ion in bodily fluids. The positive Ca^{2+} is tightly bound in the protein by several oxygen atoms (three are designated here) and somewhat less negative nitrogen atoms (one denoted as NH_2 here). This strong binding results in Ca^{2+} being drawn out of bone if it is not supplied in the diet.

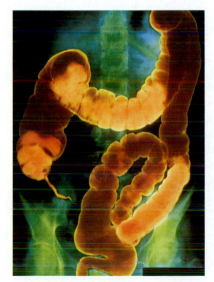

Figure 8.9 Barium atoms in $BaSO_4$ absorb X rays well.

aid in viewing the gastrointestinal tract because barium absorbs X rays well (Figure 8.9). However, barium is toxic: It can replace calcium both in bones and in the proteins that carry messages through the body. Hence, it is desirable to limit the amount of barium ion in the system. One approach is to administer barium as the insoluble salt, $BaSO_4$. However, the barium ion concentration can be limited even further by mixing the $BaSO_4$ with Na_2SO_4. Sodium salts are soluble, so

(8.21) $$Na_2SO_4(s) \longrightarrow 2Na^+(aq) + SO_4^{2-}(aq)$$

Note the one-way reaction arrow in Equation (8.21). Like all reactions in a closed system, this reaction is an equilibrium. However, the concentrations are so far to the right that for all practical purposes the reaction goes to completion. The solubility of Na_2SO_4 in water is 281.1 g per kilogram of water, a concentration of 2 M. In contrast, a saturated solution of $BaSO_4$ has a concentration of 1.04×10^{-5} M. Sodium sulfate, like all sodium salts, is considered soluble.

Sodium ion does not actively participate in the reaction between barium and sulfate (SO_4^{2-}). Instead, sodium acts as a counter ion to SO_4^{2-} to maintain electrical neutrality. Na^+ is termed a **spectator ion.** The presence of the Na_2SO_4 suppresses the Ba^{2+} ion in solution (Worked Example 8.6).

In general, spectator ions are identified from the solubility guidelines. Soluble combinations in a mixture merely maintain electrical neutrality. For the purposes of determining concentrations, they can and should be ignored (Worked Example 8.6).

WORKED EXAMPLE 8.6 *Limiting Toxic Ions*

Because Ba^{2+} is a toxic substance, Na_2SO_4 is often added to the solution of $BaSO_4$ when examining the gastrointestinal tract with X rays. Determine the efficacy of this strategy. Calculate the Ba^{2+} concentration in a saturated solution of $BaSO_4$. Compare this result with the Ba^{2+} concentration in a saturated solution with 1 mM Na_2SO_4. $K_{sp}(BaSO_4) = 1.08 \times 10^{-10}$.

Plan

■ Write the equation for the reaction of Ba^{2+} and $SO_4{}^{2-}$.
■ Write the expression for the equilibrium constant, K_{sp}.
■ In the saturated solution, $[Ba^{2+}] = [SO_4{}^{2-}]$.
■ Na_2SO_4 is soluble $\Rightarrow [SO_4{}^{2-}] = mM$. Calculate $[Ba^{2+}]$ from K_{sp}.

Implementation

■ $BaSO_4(s) \rightleftharpoons Ba^{2+}(aq) + SO_4{}^{2-}(aq)$
■ $K_{sp} = [Ba^{2+}][SO_4{}^{2-}]$
■ In the saturated solution:

$$[Ba^{2+}] = [SO_4{}^{2-}] \Longrightarrow K_{sp} = [Ba^{2+}] \times [SO_4{}^{2-}] = [Ba^{2+}]^2$$
$$\Longrightarrow [Ba^{2+}] = \sqrt{1.08 \times 10^{-10}} = 1.04 \times 10^{-5}\ M$$

■ In the Na_2SO_4 solution:

$$[SO_4{}^{2-}] = mM \Longrightarrow K_{sp} = [Ba^{2+}][SO_4{}^{2-}] = [Ba^{2+}] \times 0.001$$
$$\Longrightarrow [Ba^{2+}] = 1.08 \times 10^{-7}\ M$$

The addition of the Na_2SO_4, even at the millimolar level, has suppressed the Ba^{2+} ion concentration by two orders of magnitude. This represents an excellent outcome for the patient, as neither $SO_4{}^{2-}$ nor Na^+ is toxic. Addition of Na_2SO_4 is very effective.

See Exercises 25, 26.

WORKED EXAMPLE 8.7 *Making Solids*

It is possible to form a saturated solution beginning with no solid. Consider $BaSO_4$, which is used for gastrointestinal X-ray diagnosis. A 0.5 mM solution of $Ba(NO_3)_2$ is a crystal-clear solution, because $Ba(NO_3)_2$ is a soluble salt. Similarly, 1.0 mM Na_2SO_4 will completely dissolve in 1 L of water. However, a solid forms when 1.0 millimole of Na_2SO_4 is added to 1 L of 0.5 mM $Ba(NO_3)_2$. What is the composition of the solid? What is the concentration of barium in the saturated solution?

Plan

■ Consider the combination of anions and cations. Which are insoluble?
■ Write the equation for the reaction, and determine the initial concentrations.
■ Is the concentration change to attain equilibrium large? If so, which reactant is in short supply and limits the reaction from going to completion? Determine the new initial concentrations from allowing the reaction to go to completion.
■ Determine the change needed to establish equilibrium.
■ Calculate $[Ba^{2+}]$ from K_{sp}.

Implementation

■ The cations present are Na^+ and Ba^{2+}; the anions are $NO_3{}^-$ and $SO_4{}^{2-}$. Sodium salts are soluble and Ba^{2+} is soluble with $NO_3{}^-$. $BaSO_4$ is insoluble, however.

- $Ba^{2+}(aq) + SO_4^{2-}(aq) \rightleftharpoons BaSO_4(s)$. This reaction is the reverse of the solubility reaction, so rewrite it as $BaSO_4(s) \rightleftharpoons Ba^{2+}(aq) + SO_4^{2-}(aq)$. The initial concentrations are

$$[Ba^{2+}] = 0.5 \ mM$$
$$[SO_4^{2-}] = 1.0 \ mM$$

- Both of these concentrations will be significantly reduced at equilibrium because the product of the initial concentrations is 5×10^{-7} while K_{sp} is 1.08×10^{-10}. The concentration of Ba^{2+} is less than that of SO_4^{2-} and they combine in a $1:1$ ratio, so the limiting reactant is Ba^{2+}. If all the Ba^{2+} reacts,

$$[Ba^{2+}] = 0$$
$$[SO_4^{2-}] = 1.0 \ mM - 0.5 \ mM = 0.5 \ mM$$

- $$BaSO_4(s) \rightleftharpoons Ba^{2+}(aq) + SO_4^{2-}(aq)$$

Initial:	0	0.0005 M
Change to equilibrium:	$+x$	$+x$
Final:	x	$(0.0005 + x) \ M$

- $K_{sp} = [Ba^{2+}][SO_4^{2-}] = (x)(0.0005 + x) = 1.08 \times 10^{-10}$. x is expected to be small compared with $0.0005 \Rightarrow 0.0005x = 1.08 \times 10^{-10} \Rightarrow x = 2 \times 10^{-7}$, or $[Ba^{2+}] = 2 \times 10^{-7} \ M$. Check: $2 \times 10^{-7} \ M$ is about three orders of magnitude smaller than 0.5 mM.

On completion of the X-ray series, the body can be flushed of residual Ba^{2+} ions by administering additional Na_2SO_4.

See Exercises 28, 29, 37, 38.

Buffers

Maintaining a nearly constant pH is important for many systems, particularly biological systems. The pH of blood is closely regulated to 7.4 ± 0.5 despite a flux of acidic and basic substances. A solution that resists a change in pH upon addition of small amounts of acid or base is termed a **buffered solution** or simply a **buffer**. Buffers resist changes of pH because they contain both an acidic species to consume added base and a basic species to consume added acid. For low pH, this dual character is achieved by combining a weak acid with a salt containing the anion of the acid. The weak acid supplies H^+ to neutralize any added base. The anion combines with H^+, forming the acid, and thereby limits the H_3O^+ concentration. Because the anion combines with H^+, it is called the **conjugate base**. Acetic acid, discussed previously, is an example of a weak acid. The conjugate base is the acetate ion.

Buffered solution A solution that resists a change in its pH when either hydroxide ions or protons are added.

Buffer A solution containing a weak acid and its salt (or a weak base and its salt) in approximately equal concentrations.

Conjugate base What remains of an acid molecule after a proton ionizes.

CONCEPT QUESTION Carbonic acid, H_2CO_3, is a weak acid. What is the conjugate base of carbonic acid? ■

CONCEPT QUESTION Acetic acid is a weak acid. Is a solution consisting of 1 M $NaCH_3COO$ acidic or basic? Justify your answer. ■

The negative charge on the acetate ion (Figure 8.10) is attracted to the positive end of water, the hydrogen end. Transferring the proton from water to acetate generates a molecule of acetic acid; what is left of the water is the hydroxide ion.

The chemical reaction is

(8.22) $\quad CH_3COO^-(aq) + H_2O(l) \rightleftharpoons CH_3COOH(aq) + OH^-(aq)$

Figure 8.10 Acetate ion interacts with water by the electrostatic attraction between the acetate negative charge and the positive end of the water molecule. Transfer of a proton from water to acetate generates acetic acid and a hydroxide ion.

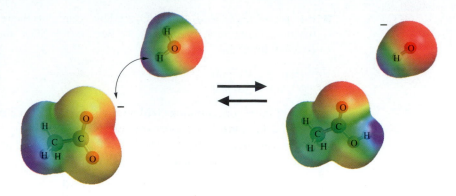

The expression for the corresponding equilibrium constant is

(8.23) $$K = \frac{[CH_3COOH][OH^-]}{[CH_3COO^-]}$$

Notice that Equation (8.23) is similar to $(K_a)^{-1}$ for acetic acid, though it lacks the hydronium ion and adds a hydroxide ion. A standard algebraic technique—multiply by 1 in the form of $[H_3O^+]/[H_3O^+]$—turns Equation (8.23) into a more familiar expression:

(8.24) $$K = \frac{[CH_3COOH][OH^-]}{[CH_3COO^-]} \frac{[H_3O^+]}{[H_3O^+]} = \frac{[CH_3COOH]}{[CH_3COO^-][H_3O^+]} \frac{[H_3O^+][OH^-]}{1}$$
$$= (K_a)^{-1} K_w = (1.8 \times 10^{-5})^{-1}(1.0 \times 10^{-14}) = 5.5 \times 10^{-10}$$

The pOH of this solution is 4.6 and the pH = 9.4: It is a fairly basic solution. Acetate ion is the conjugate base of acetic acid.

WORKED EXAMPLE 8.8   *pH of a Weak Acid*

Acetic acid is a weak acid, so it is a candidate to form a buffer. Determine the pH of a 1 *M* solution of acetic acid. $K_a = 1.8 \times 10^{-5}$.

Plan

- Write the equation for the reaction.
- Set up the initial concentrations.
- Set up the change needed to establish equilibrium.
- Write an expression for the unknown equilibrium concentrations.
- Write an expression for the equilibrium constant.
- Solve for $[H_3O^+]$.
- pH = $-\log [H_3O^+]$.

Implementation

- $CH_3COOH(aq) + H_2O(l) \rightleftharpoons CH_3COO^-(aq) + H_3O^+(aq)$
- $CH_3COOH(aq) + H_2O(l) \Longrightarrow CH_3COO^-(aq) + H_3O^+(aq)$ $K_a = 1.8 \times 10^{-5}$

  Initial: 1 *M*

- To establish equilibrium, some of the acetic acid must ionize to generate CH_3COO^- and H_3O^+:

  $CH_3COOH(aq) + H_2O(l) \Longrightarrow CH_3COO^-(aq) + H_3O^+(aq)$ $K_a = 1.8 \times 10^{-5}$

  Change: $-x$ $+x$ $+x$

Since acetic acid is an acid, the 10^{-7} M [H_3O^+] from pure water is negligible.

■ Unknown equilibrium concentrations:

[CH_3COOH] = $(1 - x)$ M
[CH_3COO^-] = x M
[H_3O^+] = x M

■ $K = \dfrac{[CH_3COO^-][H_3O^+]}{[CH_3COOH]} = \dfrac{x^2}{1 - x} = 1.8 \times 10^{-5}$

■ Since the equilibrium constant for this reaction is a small number,

$$x \ll 1 \Longrightarrow x \cong \sqrt{1.8 \times 10^{-5}} = 4.2 \times 10^{-3}$$

■ pH $= -\log(4.2 \times 10^{-3}) = 2.4$

Check: Compared with 1, 4.2×10^{-3} is negligible within the two significant digits of the problem.
If the solution was 1 M in a strong acid such as HCl, the pH $= -\log(1) = 0$, far more acidic than the pH of 2.4 for acetic acid. Acetic acid is a weak acid.

See Exercises 47–50.

CONCEPT QUESTIONS A 1 M acetic acid solution has a pH of 2.4, and a 1 M sodium acetate solution has a pH of 9.4. What is the pH of a solution made by combining 500 mL of 1 M acetic acid with 500 mL of 1 M NaCH$_3$COO? Will this solution be acidic, basic, or nearly neutral? ■

If a weak acid and its salt are in nearly equal concentrations, the solution pH reflects the counterbalancing acidic and basic influences. For example, in a solution with both acetic acid and acetate ion, acetic acid neutralizes any added base and acetate ion neutralizes added acid. In a solution that contains 0.5 M acetic acid and 0.5 M acetate ion, both the reactant, CH_3COOH, and one product, CH_3COO^-, of the acid ionization

(8.25) $CH_3COOH(aq) + H_2O(l) \rightleftharpoons CH_3COO^-(aq) + H_3O^+(aq)$

are initially present in large concentration.

(8.26) $CH_3COOH(aq) + H_2O(l) \rightleftharpoons CH_3COO^-(aq) + H_3O^+(aq)$

Initial: 0.5 M   0.5 M

To establish equilibrium, some of the acid must ionize to generate the H_3O^+ required by equilibrium.

(8.27) $CH_3COOH(aq) + H_2O(l) \rightleftharpoons CH_3COO^-(aq) + H_3O^+(aq)$

Change to equilibrium: $-x$ $+x$ $+x$
Final: $0.5 - x$ $0.5 + x$ x

Substituting these unknown concentrations into the equilibrium constant gives

(8.28) $K_a = \dfrac{[CH_3COO^-][H_3O^+]}{[CH_3COOH]} = \dfrac{(0.5 + x)x}{0.5 - x} = 1.8 \times 10^{-5}$

If x is small ($x \ll 1$), then $0.5 \pm x \approx x$ and Equation (8.28) becomes $x \approx 1.8 \times 10^{-5}$. Check: 1.8×10^{-5} is, indeed, negligible compared with 1 within the significant digits given. *Notice that the hydrogen ion concentration is simply equal to the acid ionization constant.* So a solution with equal acetic acid and acetic ion concentrations has a pH of 4.7. In general, if the acid and salt concentrations are nearly equal, then

(8.29)
$$\frac{[\text{salt}]}{[\text{acid}]}[\text{H}_3\text{O}^+] = K_a$$

Addition of H_3O^+ or OH^- shifts the balance between acid and salt, but changes pH very little. Equation (8.29) may look familiar to students of biology. Take the $-\log$ of both sides:

(8.30)
$$\log\left(\frac{[\text{acid}]}{[\text{salt}]}\right) + \text{pH} = \text{p}K_a$$

Henderson-Hasselbalch equation Equation relating the pH of a buffer to the pK_a of a weak acid (or pK_b of a weak base) and the acid and salt concentrations.

(8.31)
$$\text{pH} = \text{p}K_a + \log\left(\frac{[\text{salt}]}{[\text{acid}]}\right)$$

This formula (8.31) is often used to determine how to make a buffer: choose a weak acid with a pK_a near to the desired pH and use the appropriate [salt]/[acid] ratio. Equation (8.31) is called the **Henderson Hasselbalch** equation. There is an exactly analogous base equation.

For high pH, the dual character is attained by combining a weak base with its **conjugate acid.**

Conjugate acid The species formed when a proton is added to a base.

(8.32)
$$\text{pOH} = \text{p}K_b + \log\left(\frac{[\text{salt}]}{[\text{base}]}\right)$$

For example, a base buffer can be made with NH_3 and NH_4Cl. NH_4^+ is the conjugate acid of the base ammonia.

WORKED EXAMPLE 8.9   *Buffering Blood*

To keep the body functioning properly, blood is buffered at a pH of $7.3-7.5$. Suggest the buffer system that is responsible for maintaining this pH level. What [Salt]/[Acid] ratio is required?

Plan

- Use $\text{pH} = \text{p}K_a + \log\left(\dfrac{[\text{salt}]}{[\text{acid}]}\right)$, find p$K_a$ between 7 and 8

- Modify pH by [salt]/[acid]

Implementation

- Table 1 in Appendix A lists K_a values for several acids. A K_a on the order of 10^{-7} or 10^{-8} will generate a buffer in the correct range. The following acids are in the correct range: K_2 of arsenic acid, K_1 of carbonic acid, K_3 of citric acid, K_2 of chromic acid, K_a of hypochlorous acid, K_2 of phosphoric acid, K_2 of phosphorous acid, K_2 of selenous acid, K_2 of sulfuric acid, or K_2 of tellurous acid.

- Some of these acids are quite toxic, so they are not appropriate candidates. Potential remaining candidates include K_1 of carbonic acid, K_3 of citric acid, K_2 of phosphoric acid, or K_2 of phosphorous acid. The most likely candidates are carbonic acid (H_2CO_3) or phosphoric acid (H_3PO_4), as they are involved in other physiological processes.

$$K_1 (\text{H}_2\text{CO}_3) = 4.3 \times 10^{-7}, \text{p}K_1 = 6.4$$
$$\Longrightarrow \log([\text{salt}]/[\text{acid}]) = 1.0 \Longrightarrow [\text{salt}] = 10\,[\text{acid}]$$
$$\Longrightarrow [\text{HCO}_3^-] = 10\,[\text{H}_2\text{CO}_3]$$

$$K_2\ (H_3PO_4) = 6.23 \times 10^{-8},\ pK_2 = 7.2$$
$$\Longrightarrow \log\ ([salt]/[acid]) = 0.2 \Longrightarrow [salt] = 1.6\ [acid]$$
$$\Longrightarrow [HPO_4{}^{2-}] = 1.6\ [H_2PO_4{}^-]$$

Although the phosphoric acid system is closer, the actual buffer system in blood is the carbonic acid system. Its use is related to the heterogeneous equilibrium of $CO_2(g)$ with water and is connected with breathing and metabolism.

See Exercises 53, 54.

Multiple Equilibria

It is believed that the level of CO_2 in the prebiotic atmosphere of the earth was very high—a necessary condition for the earth to achieve temperatures high enough to initiate biological activity. Over time, the very high CO_2 concentrations were reduced by processes that also occur in the modern atmosphere (Figure 8.11). For example, CO_2 from the air dissolves in rainwater, forming carbonic acid, H_2CO_3. This acid dissolves rock that contains calcium, and both the calcium ion and dissolved CO_2 (as $HCO_3{}^-$ or $CO_3{}^{2-}$) find their way to the sea. Marine organisms incorporate the calcium and the CO_2 into their chalky shells in the form of $CaCO_3$. When these organisms die, the $CaCO_3$ forms limestone. As tectonic activity pushes the limestone on the sea floor under the continental crust, it is heated and CO_2 is released, leaving calcium-containing rock. The gas eventually enters the atmosphere during volcanic activity. This cycle repeats itself in a kind of global recycling program. A totally nonbiological portion of this cycle is responsible for formation of stalactites and stalagmites in limestone caverns (Figure 8.12). Multiple, linked equilibria make the world go round.

Figure 8.11 Calcium and carbon dioxide cycle through the environment in linked biological and inorganic cycles via multiple, interlinked equilibria.

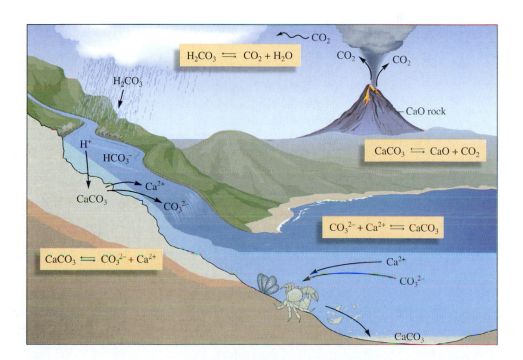

Figure 8.12 Carlsbad Caverns in New Mexico contains a colorful display of stalactites and stalagmites. The color variations are due to metal ions — mainly Fe^{+2} and Fe^{+3} dissolved in the $CaCO_3$ matrix.

CONCEPT QUESTIONS What is the solubility of $CaCO_3$ in seawater (pH 7)? What is the solubility of $CaCO_3$ in rainwater of pH 6? What partial pressure of CO_2 is needed to acidify rainwater to pH 6? ■

8.5 Eight Steps to Conquering Variations on *K*

Determining the concentration of substances involved in an equilibrium is important in many practical settings, with questions ranging from whether scale ($CaCO_3$) will form on a boiler and subsequently clog pipes connected to the boiler, to what concentrations of specific ions are hazardous, to whether plaque forms on arteries. The procedure for determining concentration for all equilibrium problems includes certain common elements. At its heart, the procedure is an algebraic exercise. The common steps are summarized here.

1. Identify the chemical reaction occurring. Identify and delete spectator ions.

CONCEPT QUESTION What are the spectator ions in Worked Example 8.7? ■

2. Write the equilibrium constant expression for the reaction.

Concept Questions in this section have a more drill-and-practice, check-up slant than other sections.

CONCEPT QUESTION Write the reaction that occurs when 0.01 mol of NH_4Cl is added to 1 L of water. What is the expression for the equilibrium constant for this reaction? ■

3. If the equilibrium constant expression does not correspond to a standard form, then write it as a combination of standard forms to algebraically determine a value for the equilibrium constant. See Equations (8.22) to (8.24).

CONCEPT QUESTION The base ionization constant for NH_4OH is 1.8×10^{-5}. What is the equilibrium constant for the following reaction?

$$NH_4^+ \, (aq) + 2H_2O(l) \rightleftharpoons NH_4OH(aq) + H_3O^+ \, (aq) \quad \blacksquare$$

If the exponent of the equilibrium constant is a positive number, the equilibrium constant is large. The opposite is true if the exponent is a negative number.

4. Record the initial concentrations given in the problem.
5. The goal is to simplify the calculation. **Stop,** and look at the concentrations and equilibrium constant. If the equilibrium constant is large and the initial reactant concentrations are much larger than the concentrations of the products, most of the reactants will become products at equilibrium. The calculation is greatly simplified by shifting reactants to products and treating the problem as a limiting reactant problem that goes to completion. Conversely, if the equilibrium constant is very small and the initial product concentrations are relatively large, then simplify the calculation by shifting products to reactants.

CONCEPT QUESTION What is the limiting reactant when a 0.5 *M* NH₄Cl solution is added to a 1 *M* NaOH solution? ■

6. The limiting reactant(s) in Step 5 have a starting concentration equal to zero after the shift. At equilibrium the concentration of the limiting reactant is not zero, but it is small. Stoichiometry determines how all other concentrations change and the changes are small. Herein lays the simplification.
7. **Stop.** If the concentration change is small, it can be neglected when added to or subtracted from a larger concentration. Simplify and solve the algebraic equation. [See Equation (8.28) and discussion.]

CONCEPT QUESTION Does the pH change when solid $Pb(OH)_2$ is added to a 0.5 *M* NaOH solution? $K_{sp}(Pb(OH)_2) = 2.8 \times 10^{-16}$. ■

8. **Review.** Is the concentration change small compared with the larger ones? Is the answer sensible?

Frequently Asked Questions

Do equilibrium reactions go to completion? For example, can lead (Pb^{2+}) be eliminated from drinking water by forming a solid?

Equilibrium reactions never go to completion. Hence there will always be some lead ions in solution if this method is chosen.

Can Pb^{2+} be reduced below 10 ppb (parts per billion) with CO_3^{2-}?

Data
$K_{sp}(PbCO_3) = 1.3 \times 10^{-13}$

Plan
■ Determine the molar concentration of a 10 ppb (parts per billion by weight) solution.
■ Determine the CO_3^{2-} concentration needed from K_{sp}.
■ If this is a practical $CO_3^{2-}(aq)$ concentration, then the answer is yes.

Implementation
■ One liter of water weighs 1 kg; 10 ppb by weight is therefore 10×10^{-9} kg $= 10 \times 10^{-6}$ g $Pb^{2+}(aq)$. Dividing by the atomic mass of Pb gives 4.8×10^{-8} *M* $Pb^{2+}(aq)$.
■ Since $K_{sp} = [Pb^{2+}][CO_3^{2-}]$, if $[Pb^{2+}] = 4.8 \times 10^{-8}$ *M*, then $[CO_3^{2-}] = 1.3 \times 10^{-13}/4.8 \times 10^{-8}$ or 2.7×10^{-6} *M*. This is the carbonate concentration required.

▪ This is an attainable concentration. For example, it can be achieved by dissolving $CaCO_3$: $K_{sp} = 1.0 \times 10^{-8}$. A saturated $CaCO_3$ solution has 1.0×10^{-4} M CO_3^{2-}, a higher concentration than required. Thus, the lead can be reduced to 10 ppb with $CO_3^{2-}(aq)$.

Are reactant and product concentrations equal at equilibrium? What does equilibrium mean?

Reactant and product concentrations are not necessarily equal. Instead, *equilibrium* means a state of balance between opposing forces or actions. In this case, the opposing actions are formation of product and formation of reactant. At equilibrium, these rates are equal.

How can reactant and product concentrations differ if the rates of reactant and product formation are equal?

If the equilibrium favors products, a smaller fraction of the product molecules transform to reactants than the reverse. The larger number of product molecules persists. A mouse pulling strongly can balance an elephant pulling weakly.

Are equilibrium concentrations determined by stoichiometric ratios? For example, in the dissociation of $Pb(OH)_2$, are the Pb^{2+} and OH^- ions always present in a $1:2$ ratio?

Pb^{2+} and OH^- are often not present in the $1:2$ stoichiometric ratio. *If* the solid lead hydroxide is the only source of Pb^{2+} and OH^- ions, meaning that the solution is made by adding solid $Pb(OH)_2$ to water, then $[Pb^{2+}]$ will be half $[OH^-]$. However, the answer is no if the sources of sthe Pb^{2+} and OH^- ions are independent or if there is a second source for either—for example, if Pb^{2+} comes from $Pb(NO_3)_2$ and OH^- from NaOH. In this case, the ratio of *precipitated ions* is $1:2$ as stoichiometry dictates. The ions left in solution are the *original concentrations minus the precipitated ions* and these will not be equal.

8.6   Reaction Direction and the Thermodynamic Connection

The Reaction Quotient

Reactants and products are often combined in nonequilibrium concentrations. The reaction proceeds in the direction required to alter the concentrations until reactants and products are in equilibrium. For example, if an insoluble salt such as $CaCO_3$ is added to water, the initial salt ion (Ca^{2+} and CO_3^{2-}) concentrations are zero. Salt therefore dissolves until the ion concentrations reach equilibrium values. Addition of $Ca(NO_3)_2$ to the saturated solution increases the calcium ion concentration. The product of the calcium and carbonate ion concentrations exceeds K_{sp}. Solid $CaCO_3$ precipitates until the product is again equal to the solubility product. In general, determining how concentrations change utilizes the **reaction quotient, Q.** For the reaction

> **Reaction quotient, Q** The ratio of the product of product concentrations raised to their stoichiometric coefficients to the product of reactant concentrations raised to their stiochiometric coefficients.

(8.33)
$$aA + bB \rightleftharpoons pP + sS$$

the reaction quotient is

(8.34)
$$Q = \frac{[P]^p [S]^s}{[A]^a [B]^b}$$

The terms used to write Q are the same as are used to write the equilibrium constant. The distinction between Q and K is that K is a fixed constant and the concentrations in K

$Q > K$	$Q = K$	$Q < K$
$a\text{A} + b\text{B} \rightleftharpoons c\text{C} + d\text{D}$	$a\text{A} + b\text{B} \rightleftharpoons c\text{C} + d\text{D}$	$a\text{A} + b\text{B} \rightleftharpoons c\text{C} + d\text{D}$
A larger number of product molecules are transformed into reactants than reactants into products to bring concentrations to the equilibrium balance.	The forward and reverse transformations are equal at equilibrium.	A larger number of reactant molecules are transformed into products than products into reactant to bring concentrations to the equilibrium balance.

Figure 8.13 Change of concentrations to attain equilibrium. The longer arrow symbolizes the enhanced reaction.

are the *equilibrium* concentrations. In contrast, Q is not fixed. Its value changes until $Q = K$, at which point the system is in equilibrium. For example, at the moment when one mole of acetic acid is placed in water, no acetate ions are in solution and the H_3O^+ concentration is 10^{-7} M. At this instant,

(8.35)
$$Q = \frac{[CH_3COO^-][H_3O^+]}{[CH_3COOH]} = \frac{(0) \times (1 \times 10^{-7})}{1} = 0$$

and 0 is less than 1.8×10^{-7}. As the acetic acid molecules ionize, the H_3O^+ and CH_3COO^- concentrations increase. After much activity, the concentrations reach the equilibrium values. Forward and reverse reactions continue at the balanced rates required at equilibrium with *no net change of concentration.*

Q is used to determine what happens when reactants and products are mixed such that the reaction quotient is not equal to K (Figure 8.13). If $Q > K$, the numerator is relatively too big. Products become transformed into reactants, simultaneously decreasing the numerator and increasing the denominator. Conversely, if $Q < K$, reactants transform into products until $Q = K$.

WORKED EXAMPLE 8.10   *Photography and Silver*

Silver is used in photographic processes. Because silver is an expensive metal, it is desirable to recover as much silver as possible from processing solutions. A solution contains 13 μM ($\mu = 10^{-6}$) Ag^+. If 58.5 mg of NaCl is added per liter of solution, will AgCl precipitate? The solubility limit of NaCl slightly exceeds 6 M. How much silver remains in solution for a 6 M Cl^- (aq) solution?

Data
$K_{sp}(\text{AgCl}) = 1.8 \times 10^{-10}$

Plan
- Determine $[Cl^-]$.
- Is Q larger than K?
- Use K_{sp}.

Implementation
- 58.5 mg NaCl is 0.00100 mol NaCl, and NaCl is a soluble salt, so $[Cl^-] = 0.00100$ M.
- $Q = [Ag^+][Cl^-] = (1.3 \times 10^{-5}) \times (1.00 \times 10^{-3}) = 1.3 \times 10^{-8} > K$. AgCl will precipitate.

■ With $[Cl^-] = 6\ M$, the silver ion is greatly suppressed.

$$[Ag^+] = \frac{1.8 \times 10^{-10}}{6}\ M \Longrightarrow [Ag^+] = 3 \times 10^{-11}\ M$$

Notice that it is not possible to recover *all* of the silver from solution via precipitation.

See Exercises 59, 60.

Gibbs Free Energy and *K*

The sign of the Gibbs free energy value indicates whether a reaction is spontaneous in the forward direction ($\Delta G° < 0$) or in the reverse direction ($\Delta G° > 0$). This suggests a connection between the Gibbs free energy and the equilibrium constant. That connection is given by

(8.36) $$\Delta G° = -RT \ln K$$

where R is the gas law constant, T is the absolute temperature, and ln is the base for the natural logarithm. A negative value of $\Delta G°$ means that the reaction is spontaneous in the forward direction, so the product concentrations increase and reactant concentrations decrease, leading to K greater than 1 and ln K positive. K greater than 1 means that products dominate the mixture, consistent with the forward reaction. Conversely, a spontaneous reverse reaction ($\Delta G° > 0$) reduces product concentrations and increases reactant concentrations, So $K < 1$ and ln K is negative.

Equation (8.36) is often used to determine equilibrium constants from standard thermodynamic data (Worked Example 8.11).

WORKED EXAMPLE 8.11 *Acid Rain*

Coal-fired power plants generate oxides of sulfur from the sulfur in the coal. Both SO_2 and SO_3 are produced in this process. Scrubbing SO_3 from the exhaust is much easier than scrubbing SO_2 because SO_3 reacts readily with water, forming nonvolatile H_2SO_4. Determine the balance between SO_2 and SO_3 produced at 1500 °C, a typical combustion temperature, assuming that these sulfur oxides are in equilibrium with oxygen.

Plan
■ Write the equation for the conversion of SO_2 to SO_3.
■ Determine $\Delta G°$ from $\Delta H - T\Delta S$ (thermodynamic data appear in the appendix).
■ Solve Equation (10.36) for K.

Implementation
■ $2SO_2(g) + O_2(g) \rightleftharpoons 2SO_3(g)$
■ $\Delta H_f°$: for $SO_2(g) = -296.8$ kJ/mol
  for $O_2(g) = 0$ kJ/mol
  for $SO_3(g) = -395.7$ kJ/mol
$\Delta H° = 2 \times (-395.7\ \text{kJ/mol}) - 2 \times (-296.8\ \text{kJ/mol}) - 1 \times (0\ \text{kJ/mol})$
  $= -197.8$ kJ/mol
Similarly $\Delta S°$: for $SO_2(g) = 248.2$ J/K · mol
  for $O_2(g) = 205.2$ J/K · mol
  for $SO_3(g) = 256.8$ J/K · mol
$\Delta S° = 2 \times (256.8\ \text{J/K·mol}) - 2 \times (248.2\ \text{J/K·mol}) - 1 \times (205.2\ \text{J/K·mol})$
  $= -188.0$ J/K · mol

Figure 8.14 Change of concentrations to attain equilibrium. The longer arrow symbolizes the enhanced reaction.

$Q > K_{eq}$	$Q = K_{eq}$	$Q < K_{eq}$
$aA + bB \rightleftharpoons cC + dD$	$aA + bB \rightleftharpoons cC + dD$	$aA + bB \rightleftharpoons cC + dD$
$\Delta G > 0$	$\Delta G = 0$	$\Delta G < 0$

$$\Delta G° = \Delta H° - T\Delta S° = -197.8 \text{ kJ/mol} - (1773 \text{ K}) \times (-0.1880 \text{ kJ/K} \cdot \text{mol})$$
$$= 135.5 \text{ kJ/mol}$$

$$\blacksquare \quad \ln K = -\frac{\Delta G°}{RT} = -\frac{135.5 \times 10^3 \text{ J/mol}}{(8.3145 \text{ J/mol} \cdot \text{K})(1773 \text{ K})} = -9.193$$

$$\implies K = 1.017 \times 10^{-4}$$

At combustion temperature, the equilibrium favors SO_2 over SO_3. This result makes it a challenge to scrub sulfur oxides out of smoke stacks.

See Exercises 1, 2.

The connection between the Gibbs free energy and the equilibrium constant can be extended to nonequilibrium concentrations with a slight alteration of Equation (8.36):

(8.37) $$\Delta G = \Delta G° + RT \ln Q$$

At equilibrium, the reaction goes in both directions, so ΔG is zero. At equilibrium, Q is equal to K and Equation (8.37) becomes the same as Equation (8.36).

When Q is larger than K, ΔG is positive and the reaction runs in the reverse direction, reducing products and increasing reactants until Q is equal to K. Figure 8.14 summarizes the relationships among Q, K, and ΔG.

At equilibrium, the net reaction runs neither forward nor backward and $\Delta G = 0$. Since $G = \Delta H - T\Delta S$, $\Delta H = T\Delta S$. Interactions, ΔH, are balanced by entropy, $T\Delta S$.

8.7   Factors That Affect Equilibrium

Le Châtelier's Principle

One of the remarkable properties of a system at equilibrium is that after a disturbance, it returns to equilibrium. Many predator–prey examples illustrate the same phenomenon. Consider the wolves and rabbits on the western plains of the United States. If a virus sweeps through the rabbit population, reducing the number of rabbits, there is insufficient food for the wolves and some of them also die. When the virus passes, the rabbits multiply. The plentiful food supply results in survival of more wolves, and their numbers increase. In chemical terms, a stress applied to the rabbit population results in a response in the wolf population that reduces the stress on the rabbit population. In a chemical reaction, this tendency to return to the equilibrium state is embodied in **Le Châtelier's principle:**

Le Châtelier's principle If a stress is applied to a system at equilibrium, the concentrations change in a direction that tends to reduce the stress.

> When a stress is applied to a system at equilibrium, the system responds in the way that best reduces the stress.

Three common types of stresses are addition of one of the components of the reaction, change in temperature, and change in pressure.

Concentration

WORKED EXAMPLE 8.12   *Upsetting Equilibrium: Concentration*

A 1 M acetic acid solution has a pH of 2.4. Describe the changes in this solution when 1 M $NaCH_3COO$ is added to it.

Plan

- Write the equation for the reaction.
- Identify ions participating in the reaction.
- Determine whether reactants or products are added.
- Evaluate Q.

Implementation

- $CH_3COOH(aq) + H_2O(l) \rightleftharpoons CH_3COO^-(aq) + H_3O^+(aq)$
- $NaCH_3COO$ is soluble, forming CH_3COO^- and Na^+. The Na^+ ion does not participate in the reaction; it is a spectator ion.
- Acetate ion is a product of acetic acid ionization. Equilibrium is stressed.
- $Q > K_{eq}$.

With $Q > K_{eq}$, the acetate ion captures H^+ from H_3O^+, reducing the H_3O^+ concentration, and the acid concentration increases. The reduced H_3O^+ concentration results in an increased pH.

See Exercises 75, 76.

Addition of a reactant or a product does not change the *value* of the equilibrium constant, K. Instead, when equilibrium is reestablished, the concentrations of the components are altered relative to the concentrations before the disturbance. The ratio of product concentrations to reactant concentrations returns to the same value.

Temperature

Raising the temperature introduces more heat energy into an equilibrium system. Consider a reaction that is exothermic. Including heat in the reaction, the equilibrium looks like

(8.38) $aA + bB \rightleftharpoons pP + sS + \text{heat}$

Adding heat increases the amount of product in this reaction. The system responds to this stress by diminishing the amounts of the other products and increasing the amounts of the reactants. This changes the value of the equilibrium constant. Addition of heat is *the one stress that alters the value of the equilibrium constant.*

As an example, consider the dimerization of brown NO_2 to colorless N_2O_4. This dimerization is exothermic by 57.20 kJ/mol N_2O_4 produced. How does an increase in temperature at midday or a decrease at night affect the haze? The reaction is exothermic, so heat is a product:

(8.39) $2NO_2(g) \rightleftharpoons N_2O_4(g) + \text{heat}$ $\Delta H = -57.20$ kJ

Heating the equilibrium mixture adds heat to the product side, putting a stress on the equilibrium. To relieve this stress, the reverse reaction, dissociation of N_2O_4, increases relative to the forward reaction, dimerization of NO_2, resulting in a net increase of brown NO_2. The haze (Figure 8.15) is thus murkier at midday.

The opposite happens at night. As the evening air cools, it removes heat—a product—from the equilibrium. The system relieves this stress by increasing the

Figure 8.15 City haze.

forward reaction; the dimer concentration therefore increases and the haze clears. At 25 °C, K_c for Equation (8.39) is 164.8; at 100 °C, its value diminishes to 4.72.

CONCEPT QUESTION The cold-pack reaction, involving solvation of NH_4NO_3, is endothermic. How does the solubility of NH_4NO_3 on a warm summer day compare to that in the dead of winter? ■

Pressure

For many systems, pressure has little effect on the equilibrium because most phases are not very compressible. In other words, increased pressure does not stress the reaction. Two exceptions to this guideline are reactions involving gases and reactions at extreme pressures. An example where extreme pressure has an effect is the reaction

(8.40)   Graphite $\rightleftharpoons$ diamond

Graphite and diamond are two allotropes of pure carbon. At ordinary pressures, the more stable form is graphite (Figure 8.16). However, under the extreme pressures found deep inside the earth, diamond becomes the more stable phase because diamond is denser than graphite. This is an illustration of Le Châtelier's principle: The system, graphite, is put under stress by the pressure, and the system shrinks to relieve the pressure by converting to more dense diamond.

A second example is the conversion of ice to liquid water:

(8.41)   Ice $\rightleftharpoons$ water

In this case, liquid water is the more dense form: Ice cubes float in soft drinks, iced tea, and lemonade. Ice melts when a weighted wire is laid over it (Figure 8.17). The wire passes through the water below it. Without the pressure of the wire, the water refreezes. In the process, the wire passes through the block, leaving the block intact.

Similarly, equilibria involving gases respond to favor the side with fewer gas-phase molecules. The equilibrium constant remains the same, but the equilibrium balance is

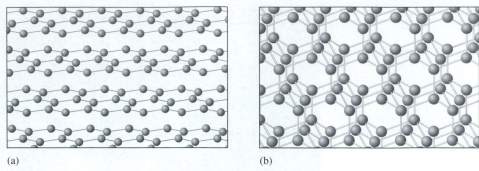

(a)   **(b)**

Figure 8.16 Graphite (a) has a more open structure than diamond (b). Putting pressure on graphite squeezes the layers together, transforming it into diamond.

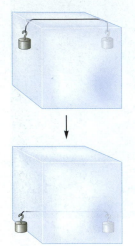

Figure 8.17 Ice is less dense than water and melts under the pressure of a weighted wire. The water refreezes as the wire passes through the block, leaving the block of ice intact.

upset when the pressure increases because the concentrations of all gases increase when the volume decreases. For example, the reaction

(8.42) $$2NO_2(g) \rightleftharpoons N_2O_4(g)$$

favors the formation of N_2O_4 at higher pressures. For this reaction

(8.43) $$Q = \frac{[N_2O_4]}{[NO_2]^2} = \frac{(n_{N_2O_4})/V}{(n_{NO_2})^2/V^2} = \frac{(n_{N_2O_4}) \times V}{(n_{NO_2})^2}$$

At equilibrium, $Q = K_c$. As the volume decreases, pressure increases, and $Q < K_c$. In response, more product is formed to restore $Q = K_c$. Conversely, a volume expansion results in the formation of more reactant. K_c remains constant throughout this process.

CONCEPT QUESTION The Haber-Bosch process for formation of ammonia,

$$N_2 + 3H_2 \rightleftharpoons 2NH_3$$

uses high temperature and high pressure. Does high pressure enhance formation of ammonia? ■

Checklist for Review

KEY TERMS

equilibrium (p. 272)
equilibrium constant, K_c (p. 272)
homogeneous equilibria (p. 274)
heterogeneous equilibria (p. 274)
solvent (p. 275)
solute (p. 275)
hydronium ion (p. 276)
hydroxide ion (p. 276)
acid (p. 277)
activity (p. 277)
ion-product constant for water, K_w (p. 277)
acid ionization constant, K_a (p. 278)

pH scale (p. 278)
solubility (p. 279)
solubility product (p. 279)
saturated solution (p. 281)
common ion effect (p. 282)
spectator ion (p. 283)
buffered solution (p. 285)
buffer (p. 285)
conjugate base (p. 285)
Henderson-Hasselbalch equation (p. 288)
conjugate acid (p. 288)
reaction quotient, Q (p. 292)
Le Châtelier's principle (p. 295)

KEY EQUATIONS

Equilibrium constant for reaction
$aA + bB \rightleftharpoons pP + sS$

$$K_c = \frac{[P]^p [S]^s}{[A]^a [B]^b}$$

$$\Delta G = \Delta G° + RT \ln Q$$

$$K_c = K_p (1/RT)^{\Delta n}$$

$$pH = -\log [H_3O^+]$$

$$pH = pK_a + \log ([salt]/[acid])$$

$$pOH = pK_b + \log ([salt]/[base])$$

$$Q = \frac{[P]^p [S]^s}{[A]^a [B]^b}$$

Chapter Summary

Many reactions generate a mixture of reactants and products because a mixture has greater entropy than either reactants or products alone. Equilibrium reactions are described by equilibrium constant:

$$K_c = \frac{[P]^p [S]^s}{[A]^a [B]^b}$$

for the reaction

$$aA + bB \rightleftharpoons pP + sS$$

where square brackets denote concentration in moles/liter for solutes, concentration for gases (or partial pressure in atmospheres for K_p), and a value of 1 for the solvent or any precipitate.

Equilibrium is a dynamic process in which the rates of the forward and reverse reactions are equal. Removal of a reactant or a product upsets this balance. The system responds to this imbalance by increasing concentrations on the diminished side. This is Le Châtelier's principle:

> When a stress is applied to a system at equilibrium, the system responds in the way that best reduces the stress.

One consequence of this principle is that the value of the equilibrium constant depends only on temperature. Raising the temperature decreases the equilibrium constant for an exothermic reaction and increases it for an endothermic reaction.

KEY IDEA

Equilibrium is characterized by balanced change: Reactants are continuously being transformed into products and products into reactants. These transformations are balanced in such a way that the ratio of product concentrations multiplied together to the concentrations of the reactants multiplied together is a constant.

CONCEPTS YOU SHOULD UNDERSTAND

■ Equilibrium is determined by a balance between interactions and entropy: $\Delta G° = \Delta H° - T\Delta S° = -RT \ln K$.

OPERATIONAL SKILLS

■ Write the equilibrium constant for any reaction (Worked Examples 8.1–8.10).

■ Calculate concentrations given starting concentrations and the equilibrium constant for a specified reaction (Worked Examples 8.1, 8.5–8.10).
■ Calculate pH and choose buffers to regulate pH (Worked Examples 8.9, 8.10).
■ Identify and eliminate spectator ions (Worked Examples 8.6, 8.7, 8.10).
■ Calculate K from $\Delta G°$ or $\Delta G°$ from K, and ΔG from Q (Worked Example 8.1).
■ Determine the response of an equilibrium system to a stress (Worked Example 8.12).

Exercises

A blue exercise number indicates that the answer to that exercise appears at the back of the book. Data on acid ionization, base ionization, complex ion formation constants, and solubility products are given in the appendix.

■ SKILL BUILDING EXERCISES

8.1 Gas-Phase Reactions

1. Coal-fired power stations emit SO_2 from oxidation of the sulfur in the coal. SO_2 can be oxidized to yield SO_3

$$2SO_2(g) + O_2(g) \rightleftharpoons 2SO_3(g)$$

Data: K_c is 4.08×10^{-3}. The partial pressure of O_2 in the atmosphere is 0.20 atm.
a. Write an expression for the equilibrium constant for oxidation of SO_2 to SO_3.
b. What is the value of K_p?
c. What fraction of the sulfur atoms are present as SO_2 at 25 °C and 1 atm pressure?

2. Carbon monoxide, CO, reacts with chlorine gas, Cl_2, to yield gaseous $COCl_2$.
   a. Write an expression for the equilibrium constant for reaction of CO with Cl_2.
   b. The value of K_c is 4.57×10^9. What is the value of K_p?

8.1 The Equilibrium Constant

3. Write an expression for the equilibrium constant for each of the following reactions.
   a. $H_2CO_3(aq) + H_2O(l) \rightleftharpoons H_3O^+(aq) + HCO_3^-(aq)$
   b. $CdS(s) \rightleftharpoons Cd^{2+}(aq) + S^{2-}(aq)$
   c. $3NO_2(g) + 3H_2O(l) \rightleftharpoons 2H_3O^+(aq) + 2NO_3^-(aq) + NO(g)$

4. Write an expression for the equilibrium constant for each of the following reactions.
   a. $PbS(s) \rightleftharpoons Pb^{2+}(aq) + S^{2-}(aq)$
   b. $2SO_2(g) + O_2(g) \rightleftharpoons 2SO_3(g)$
   c. $HNO_2(aq) + H_2O(l) \rightleftharpoons H_3O^+(aq) + NO_2^-(aq)$

8.2 Heterogeneous Equilibria

5. The Statue of Liberty is green due to a coating of $CuCO_3$ and $Cu(OH)_2$ on the copper cladding. Both materials are insoluble salts with a small solubility product. Write the expression for K_{sp} for each of these salts.

6. Write the expression for the equilibrium constant for the following reactions.
   a. $Fe_2S_3(s) \rightleftharpoons 2Fe^{3+}(aq) + 3S^{2-}(aq)$
   b. $CdS(s) \rightleftharpoons Cd^{2+}(aq) + S^{2-}(aq)$

7. At 25 °C, the mole fraction of molecular oxygen in water is 2.301×10^{-5}. The atmosphere is 22 mole percent oxygen. Compare the number of oxygen molecules per cm^3 in water to the concentration of molecular oxygen in the atmosphere.

8. CO_2 is the substance that gives soda its sparkle as well as its crisp, acidic bite. With a 1-atm partial pressure of CO_2, the mole fraction solubility of CO_2 in water is 6.15×10^{-4}. Determine the equilibrium constant for the reaction:

$$CO_2(g) \rightleftharpoons CO_2(aq)$$

8.4 Acids and Bases

9. What is the $HOOCCOO^-$ ion concentration in a solution made by addition of 0.01 mol HOOCCOONa to 1 L of water?

10. HF is a weak acid. Indicate the reaction that occurs when 0.01 mol of NaF is added to 1 L of water. Write the equilibrium constant for this reaction.

11. If 100 mL of 0.1 M NH_4Cl is added to 1 L of water, what is the OH^- ion concentration?

12. Carbonated beverages get their effervescence from dissolved CO_2 gas. The bite comes from the acidic nature of the beverage resulting from the reaction

$$H_2CO_3(aq) + H_2O(l) \rightleftharpoons HCO_3^-(aq) + H_3O^+(aq)$$

Determine the pH of a solution that is 6.15×10^{-4} M in H_2CO_3.

13. Plop, plop, fizz, fizz, oh what a relief it is—the fizz of Alka-Seltzer. Alka-Seltzer tablets consist of 1916 mg of sodium bicarbonate ($NaHCO_3$), 1000 mg of citric acid, and some pain medication.
   a. Determine the pH when the citric acid of an Alka-Seltzer tablet is dissolved in 50 mL of water.
   b. Write the equation for the reaction that occurs when $NaHCO_3$ is added to the solution in part (a).
   c. The mole fraction solubility of CO_2 in water of pH = 7 is 6.15×10^{-4}. Write equations for the set of reactions that lead to the fizz of Alka-Seltzer.

14. Triethylamine, $(C_2H_5)_3N$, is a weak base related to ammonia. $K_b = 1.02 \times 10^{-3}$. What is the pH of a 0.01 M solution of triethylamine?

8.4 Describing Acids and Bases: The pH Scale

15. You have a 1 M solution of the strong acid HCl and the weak acid HF. Compare the pH of these two solutions.

16. What is the pH of a 0.1 M solution of CH_3COONa?

17. At what pH does $Zn(OH)_2$ just begin to precipitate from a 0.20 M solution of $ZnCl_2$?

18. The pK_a of tartaric acid, $HOOC—(CHOH)_2—COOH$, the major acidic component of wine, is 2.98. Determine the pH of a 0.5 M tartaric acid solution.

8.4 Solubility

19. Paintings discolor over time, usually darkening. Although some of this discoloration is due to environmental factors leading to formation of black PbS, some is due to the combination of pigments used in the original painting. Thus the seeds of destruction were sown when the painting was created. Indicate the reaction that occurs from a mixture of the following:
   a. white lead, $2PbCO_3 \cdot Pb(OH)_2$, and ultramarine, $NaAlS_2$
   b. chrome yellow, $PbCrO_4$, and ultramarine, $NaAlS_2$

20. PbS can be used to form a nanoparticulate device called a quantum dot. PbS is insoluble, but lead is toxic. Therefore it is desirable to limit the amount of lead left in solution.
   a. If equal volumes of equimolar concentrations of Pb^{2+} and S^{2-} are mixed, what is the lead ion concentration at equilibrium?
   b. If equal volumes of 0.01 M S^{2-} and 10^{-4} M Pb^{2+} are combined, how much lead is left in solution?

21. CdS is an insoluble material that can be used to form quantum dots (see Exercise 20). Because Cd^{2+} is toxic, it is desirable to limit the Cd^{2+} left in solution to nanomolar concentrations (10^{-9} M). What $[S^{2-}]$ is required to achieve this result if equal volumes of Cd^{2+} and S^{2-} solutions are used?

22. Calculate the molar solubility of $Sn(OH)_2$ and $Sn(OH)_4$. Comment on any trend that you observe.

23. Calculate the molar solubility of $PbCl_2$, $PbBr_2$, and PbI_2. Comment on any trend you observe.

24. Suppose 0.001 mol $Sn(NO_3)_2$ is dissolved in 1 L of 0.001 M NaI solution. Indicate the spectator ion(s). In what form is most of the Sn found, $Sn^{2+}(aq)$ or SnI_2?

25. Suppose 0.001 mol $Sn(NO_3)_2$ is dissolved in 1 L of 0.001 M Na_2S solution. Indicate the spectator ion(s), if any. When equilibrium is reached, are there more moles of $Sn^{2+}(aq)$ or SnS?

26. A solid forms when 0.001 mol $AgNO_3$ is dissolved in a solution containing 0.001 mol NaCl and 0.001 mol NaI. What is the solid?

27. Write the equation for the reaction that occurs when 0.02 mol $CuSO_4$ is added to 1 L of 0.1 M NaOH solution. Write an expression for the equilibrium constant for this reaction.

28. What is the Ag^+ ion concentration left in solution if AgCl is precipitated by adding enough HCl to a solution of $AgNO_3$ to make the final Cl^- ion concentration 0.1 M?

29. In a saturated solution of AgCl, the concentrations of the Ag^+ and Cl^- ions are each 1.25×10^{-5} M. What will the Ag^+ ion concentration be if sufficient NaCl is added to the solution to increase the Cl^- ion concentration 100-fold?

30. $AgNO_3$ is added to a solution containing 0.001 mol Cl^- and 0.001 mol Br^- per liter. What are the concentrations of Cl^- and Br^- remaining when AgCl just begins to precipitate?

31. What is the solubility of $Cu(OH)_2$ in a solution that is *buffered* at pH = 5? How much $Cu(OH)_2$ dissolves in this solution?

32. Aluminum hydroxide, $Al(OH)_3$, is insoluble. Determine the molar solubility of aluminum hydroxide in neutral rain.

33. PbS can form a quantum dot solution. Determine the amount of Pb^{2+} and S^{2-} left in solution if equal volumes of 0.06 M $Pb(NO_3)_2$ and 0.06 M Na_2S are combined.

34. CdS can form a quantum dot solution. Determine the amount of Cd^{2+} and S^{2-} left in solution if equal volumes of 0.06 M $Cd(NO_3)_2$ and 0.06 M Na_2S are combined.

35. Grapes are one of a few fruits that contain a significant concentration of tartaric acid, $HOOC—(CHOH)_2—COOH$. Tartaric acid is a weak acid that ionizes into bitartarate ion:

$$HOOC—(CHOH)_2—COOH(aq) + H_2O(l) \rightleftharpoons$$
$$HOOC—(CHOH)_2—COO^-(aq) + H_3O^+(aq)$$

The bitartarate ion combines with potassium, also found in high concentration in grape juice, to yield potassium bitartarate. The solubility of potassium bitartarate in water is 1 g per 162 mL of water. In alcohol, potassium bitartarate is far less soluble, requiring 8820 mL of alcohol to dissolve 1 g.
   a. Determine K_{sp} for potassium bitartarate in water.
   b. Determine K_{sp} for potassium bitartarate in ethanol.
   c. Based on these data, suggest an explanation for formation of "wine crystals"—crystals of potassium bitartarate.

36. Zinc is used to hot-dip nails that will be used in exterior applications to prevent their rusting. In part, this application depends on the insolubility of $Zn(OH)_2$ formed from hydration of ZnO. How much $Zn(OH)_2$ dissolves in 1 L of solution?

8.4 Common Ion Effect

37. The solubility of Ag_2CrO_4 is 0.0030 g per 100 mL.
   a. Calculate the solubility product.
   b. Determine the solubility in 0.01 M $AgNO_3$.
   c. Determine the solubility in 0.01 M K_2CrO_4.

38. What would happen if solid AgCl were added to a saturated solution of AgI?

8.4 Spectator Ions

39. Suppose that PbS is to be synthesized out of $Pb(NO_3)_2$ and K_2S. Of the ions present when these solids are added to water, which are spectators?

40. In each of the following reactions, identify any spectator ions present.
   a. A 0.05 M $CuCl_2$ solution is mixed with a 1 M NaOH solution.
   b. A 0.001 M $AgNO_3$ solution is mixed with a 0.1 M HCl solution.

41. Identify the spectator ion(s) in the following problem: Determine the pH of a solution made by addition of 0.2 mol $NaCOOCH_3$ to a 0.1 M CH_3COOH solution.

42. Identify the spectator ion(s) in the following problem: At what pH does $Zn(OH)_2$ just begin to precipitate in a 0.20 M $ZnCl_2$ solution?

43. Identify the spectator ion(s) and write the net ionic equation for the following.
   a. A solution of 0.01 M $Cd(NO_3)_2$ is added to a solution of 0.001 M Na_2S.
   b. 0.02 M $AgNO_3$ is combined with 0.05 M KI and 0.02 M NaBr.
   c. 1 M HCl dissolves $Cu(OH)_2$.
   d. $Fe(OH)_2$ and $Fe(OH)_3$ dissolve in 1 M HCl.
   e. A solution of 0.03 M $Cu(NO_3)_2$ is combined with a solution of 1 M NaOH.

44. Identify the spectator ion(s) and write the net ionic equation for the following.
   a. NH_4Cl is added to water.
   b. A solution of HCl is added to a solution of NaOH.
   c. A solution of NH_4OH is added to a solution of HCl.

45. Identify the spectator ion(s) and write the net ionic equation for the following.
   a. Acetic acid (CH_3COOH) is added to water.
   b. Formic acid (HCOOH) is added to water.

46. Identify the spectator ions when the following solutions are combined.
   a. 0.01 M Na_2SO_4 and 0.001 M $Ba(NO_3)_2$
   b. 0.001 M $NaCrO_4$ and 0.05 M $PbCl_2$
   c. 0.1 M NH_3 and 0.01 M $Cu(NO_3)_2$

8.4 Buffers

47. HF is a weak acid. What would you combine with HF to form a buffer? What is the approximate pH of your buffer?

48. Indicate the reaction that occurs when 0.01 mol NaF is added to 1 L of water. What is the pH of the resulting solution?

49. Suppose 100 mL of 0.1 M $C_6H_5NH_3Cl$ is added to 1 L of water. What is the OH^- concentration?

50. Suppose 100 mL of 0.1 M NH_4Cl is added to 150 mL of 0.1 M NH_4OH. What is the pH of the resulting solution?

51. Suppose 0.1 mol of solid NaOH is added to 1 L of 0.125 M acetic acid, CH_3COOH. What is the pH of the resulting solution?

52. What is the solubility of $Cu(OH)_2$ in a solution buffered at a pH of 5?

53. Seawater has a pH of 7.8–8.3. A tropical fish aquarium must be buffered at this pH. Suggest a buffer system to use. What [salt]/[acid] concentration is required?

54. Suggest a buffer system that could maintain a tropical fish aquarium's pH at 9 (see Exercise 53).

55. What is the OH^- concentration when 100 mL of 0.1 M Na_2HPO_4 solution is added to 150 mL of 0.1 M NaH_2PO_4?

56. Suppose 0.1 mol of solid NaOH is added to 1 L of 0.125 M citric acid, $C_3H_5O(COOH)_3$. What is the pH?

57. Suppose 0.01 mol of solid HCOONa is added to 1 L of 0.1 M HCOOH. What is the $HCOO^-$ ion concentration? What is the pH?

58. The pK_a of tartaric acid, $HOOC$—$(CHOH)_2$—$COOH$, the major acidic component of wine, is 2.98.
    a. What is the pH of a 0.01 M solution of tartaric acid?
    b. What is the pH of a 0.01 M solution of potassium bitartarate, $KOOC$—$(CHOH)_2$—$COOH$?
    c. Give a recipe for formation of a buffer based on tartaric acid that has a pH of 4.

8.4 Multiple Equilibria

59. The green patina on a copper object is a coating of $CuCO_3$ and $Cu(OH)_2$. This patina can be removed with a variety of household substances, including vinegar (acetic acid) or lemon juice (citric acid). Indicate the equilibria involved. Which is more readily removed, the $CuCO_3$ or the $Cu(OH)_2$? For example, which is more soluble in 1 M acetic acid?

60. A solution is made by adding 0.01 mol $CdSO_4$ (soluble) and 0.1 mol Na_2S to 1 L of a pH = 10 buffered solution. Write equations for any chemical reactions that occur. What equilibrium determines the $Cd^{2+}(aq)$ concentration? Write the equilibrium constant expression for it.

61. The primary buffer system for blood is the H_2CO_3 and HCO_3^- system, which maintains the pH of the blood at 7.4. What problem would a person have if his or her blood became too acidic? Too basic?

62. Marble used in statues and monuments consists of $CaCO_3$. Weathering or erosion of these statues and monuments is accelerated by acidic precipitation due to pollution from power plants, manufacturing, auto exhaust, and other sources.
    a. What is the molar solubility of marble in neutral water?
    b. How much more soluble is marble in rainwater with a pH of 4.4?

63. Hydrangeas are a variety of flower that can vary in color from blue to pink. Although hydrangea color is often taken as a signature of the pH of the soil in which the plant grows, the color is actually due to the concentration of aluminum ions. In neutral or slightly acidic soil, aluminum is unavailable because it forms an insoluble hydroxide, $Al(OH)_3$. As the pH is lowered to 5, sufficient aluminum ion is available that the blue color develops.
    a. Determine the solubility of $Al(OH)_3$ in a pH = 7 solution.
    b. Determine the solubility of $Al(OH)_3$ at pH = 5.

64. Disappearing ink is made from an acid–base indicator and a basic solution. One combination is the indicator thymolph-

thalein, abbreviated In, and a sodium hydroxide solution. In a strong base, the indicator is a deep blue color typical of ink. Carbon dioxide from the air combines with the basic solution, neutralizing the base and converting the indicator to a colorless form.

$$HIn \text{ (colorless)} \Longleftrightarrow H^+ + In^- \text{ (deep blue)}$$

pK_a for thymolphthalein is 9.9.
    a. What concentration of NaOH is required to produce the blue ink?
    b. Write equations for the set of reactions, starting with CO_2 in the air, that are responsible for the disappearing part of disappearing ink.
    c. Once dried, the ink color can be restored by spraying the paper with aqueous ammonia. Write equations for the reactions responsible for restoration of the color.

8.6 The Reaction Quotient

65. Will $CaSO_4$ precipitate from a solution made by combining 50 mL of 0.001 M $CaCl_2$ with 50 mL of 0.001 M Na_2SO_4? Justify your answer.

66. The equilibrium constant, K_c, for the reaction

$$H_2(g) + I_2(g) \Longleftrightarrow 2HI(g)$$

is 794. A 5.0-L vessel is filled with 0.06 mol H_2, 0.09 mol I_2, and 0.05 mol HI. Is this mixture at equilibrium? If not, in which direction will the reaction proceed? What are the equilibrium concentrations of each of the gases?

8.6 Gibbs Free Energy and *K*

67. The industrial synthesis of methanol, CH_3OH, uses the following reaction.

$$CO(g) + 2H_2(g) \Longleftrightarrow CH_3OH(g)$$

    a. Calculate the equilibrium constant for this reaction at 25 °C.
    b. Is the equilibrium constant larger or smaller at 100 °C? Justify your answer.

68. Aniline, $C_6H_5NH_2$ is a weak base related to ammonia. It reacts with water as follows:

$$C_6H_5NH_2(aq) + H_2O(l) \Longleftrightarrow C_6H_5NH_4^+(aq) + OH^-(aq)$$
$$K_b = 4.27 \times 10^{-10}$$

    a. Calculate $\Delta G°$ for the base ionization of aniline.
    b. Calculate $\Delta G°$ for the base ionization of ammonia.
    c. Compare the strength of interaction of aniline and ammonia with water.

8.7 Le Châtelier's Principle

69. Which of the following actions changes the value of an equilibrium constant?
    a. addition of a reactant
    b. increasing the pressure
    c. addition of spectator ions
    d. raising the temperature

70. Account for the following: Water boils at a lower temperature on a mountain top than at sea level.

71. Account for the following: A deep-sea diver must surface slowly or bubbles of nitrogen will form in the blood, causing

pain or even death. If the gas tanks in the diving apparatus are filled with a helium–oxygen mix, this problem is less severe than if the tank is filled with a nitrogen–oxygen mix. The mole fraction solubility of N_2 in water is 1.177×10^{-5}, that of oxygen in water is 2.301×10^{-5}, and that of He in water is 0.708×10^{-5}.

72. Account for the following: Ice melts under the pressure of a skater's blade.

73. The following reaction is endothermic:

$$PCl_3(g) + Cl_2(g) \Longleftrightarrow PCl_5(g)$$

What is the effect of each of the following?
a. increasing the total pressure
b. addition of PCl_5
c. increasing the volume
d. increasing the temperature

74. The following reaction is exothermic:

$$3Fe(s) + 4H_2O(l) \Longleftrightarrow Fe_3O_4(s) + 4H_2(g)$$

What is the effect of each of the following?
a. addition of Fe
b. increasing the temperature
c. removal of $H_2(g)$
d. addition of water

8.7 Concentration

75. The synthesis of nitric acid begins with oxidation of ammonia.

$$4NH_3(g) + 5O_2(g) \Longleftrightarrow 4NO(g) + 6H_2O(g)$$

a. Is the yield for this reaction improved by increasing the vessel size?
b. How is the production of NO affected by increasing the partial pressure of water?
c. How is the production of NO affected by increasing the partial pressure of oxygen?

76. Nail polish remover is ethyl acetate, $CH_3COOC_2H_5$. Ethyl acetate is made by the reaction

$$CH_3COOH(sln) + C_2H_5OH(l) \Longleftrightarrow CH_3COOC_2H_5(sln) \\ + H_2O(sln).$$

How is the yield of ethyl acetate affected by each of the following?
a. addition of C_2H_5OH
b. removal of water
c. addition of CH_3COOH

8.7 Temperature

77. True or false: Raising the temperature increases the value of K for an endothermic reaction. Explain your reasoning.

78. How will an increase in temperature affect each of the following reactions?
a. $H_2(g) + Cl_2(g) \Longleftrightarrow 2HCl(g) \qquad \Delta H = +92$ kJ
b. $H_2(g) + I_2(g) \Longleftrightarrow 2HI(g) \qquad \Delta H = -25$ kJ

79. At room temperature, the equilibrium constant for the reaction

$$N_2O_4(g) \Longleftrightarrow 2NO_2(g)$$

is 6.07×10^{-3}. At 35 °C the equilibrium constant is 1.26×10^{-2}. Is the reaction exothermic or endothermic? Justify your choice.

80. When CO_2 dissolves in water, the following series of reactions occur:

$$CO_2(g) + H_2O(l) \Longleftrightarrow CO_2(aq)$$
$$CO_2(aq) + H_2O(l) \Longleftrightarrow H_2CO_3(aq)$$
$$H_2CO_3(aq) + H_2O(l) \Longleftrightarrow H_3O^+(aq) + HCO_3^-(aq)$$

a. Warming a CO_2-saturated solution releases CO_2 into the air. Are the reactions exothermic or endothermic?
b. With a partial pressure of CO_2 equal to 1 atm, the mole fraction of CO_2 in water is 6.15×10^{-4}. What is the equilibrium constant for the first reaction?
c. The pH is 3.8 at 25 °C. Determine the equilibrium constant for the second reaction.

81. Obtain a balloon or wide rubber band and perform the following exercise. Grasping the balloon firmly with your hands, place the taut balloon against your lip. Pull rapidly and report your observation. Based on this observation, what would happen if a weight was suspended at the end of a stretched balloon and the balloon was heated? Will the weight lift or drop? Justify your answer based on your observation.

82. Red wine occasionally is observed to have crystals of potassium bitartarate, $KOOC—(CHOH)_2—COOH$, on the bottom of the bottle. To prevent formation of these crystals, some wine makers cool the wine for several days or weeks and filter off the crystals prior to bottling. The solubility of potassium bitartarate in water is 1 g/162 mL of water at room temperature and 1 g/16 mL at 100 °C. Is the heat of solution of potassium bitartarate exothermic or endothermic? Justify your answer with the data given.

8.7 Pressure

83. The reaction in the Haber-Bosch process for making ammonia is

$$N_2(g) + 3H_2(g) \Longleftrightarrow 2NH_3(g)$$

The goal of the industrial synthesis is to maximize the NH_3 yield. Write the equilibrium constant for this reaction. Should this reaction be run at high pressure or low pressure?

84. Write the expression for the equilibrium constant for the following reactions and indicate how product yield is affected by an increase in pressure.
a. $2SO_3(g) \rightleftarrows 2SO_2(g) + O_2(g)$
b. $2O_3(g) \rightleftarrows 3O_2(g)$
c. $NO_2(g) + SO_2(g) \rightleftarrows NO(g) + SO_3(g)$

85. How will an increase in pressure affect each of the following gas-phase reactions?
a. $H_2(g) + I_2(g) \rightleftarrows 2HI(g)$
b. $4NH_3(g) + 5O_2(g) \rightleftarrows 4NO(g) + 6H_2O$ (steam)
c. $2H_2(g) + O_2(g) \rightleftarrows 2H_2O$ (steam)

86. The fluid in the capsule between bones such as those in the fingers is a pale yellow, viscous fluid containing dissolved CO_2, N_2, and O_2. Explain the popping sound when you "crack" your knuckles.

■ CONCEPTUAL EXERCISES

Exercises 87 and 88 are designed to simulate arrival at equilibrium through a series of steps. The instructions indicate how to get from one step to the next. Record the number of each type of molecule present at each step. Relate the number of molecules of each type at equilibrium to the prescription for converting molecules at each step.

87. Examine a prototypical reaction $A \rightleftharpoons B$.
    a. Start with 50 molecules of A in 1 L. The reaction prescription is as follows. At each step, 0.1 of the A molecules present react to yield B molecules, and 0.005 of the B molecules present react to form A molecules. What are the equilibrium concentrations? How many cycles does it take to arrive at equilibrium?
    b. Run the reaction again starting with 50 B molecules and no A molecules. How do the equilibrium concentrations differ? How many cycles does it take to get to equilibrium? Compare the result with that from part (a).

88. In this simulation, the reaction generates two products:

$$A \rightleftharpoons B + C$$

At each step, let one-tenth of the A molecules be transformed into B and C. Because B and C need to come together to form A, the fraction that is transformed involves the product of the concentration of B times that of C. Set the fraction of $[B] \times [C]$, molecules that are transformed into A molecules to 0.025. Start with 100 molecules of A and 10 molecules of C. What are the equilibrium concentrations? How many cycles does it take to arrive at equilibrium?

89. Three cations labeled A, B, and C are mixed with four anions labeled W, X, Y, and Z with the following observations:
    i. A solution of AW is mixed with a solution of BX and a precipitate forms.
    ii. A solution of CW is mixed with a solution of BX and a precipitate forms.
    iii. A solution of AW is mixed with a solution of BY but no precipitate forms.
    iv. A solution of AZ is mixed with a solution of BX and a precipitate forms.
    a. Write a reaction equation for each observation.
    b. Classify the cations as always soluble or sometimes soluble.
    c. Classify the anions as generally soluble or generally insoluble.

90. Describe the sequence of events when salt is added to water. Use each of the following terms: solubility, saturated solution, ionic compound, solvent, solute, equilibrium, dissolution, precipitation.

☀ APPLIED EXERCISES

91. Over time, automatic coffee makers, steam irons, faucet heads, and shower heads accumulate deposits from hard water. Hard water contains dissolved calcium ions that combine with carbonate ions generated by dissolving CO_2 from the air in water. These deposits can be cleaned by soaking them in vinegar, which contains acetic acid, CH_3COOH.
    a. Write the series of reactions that lead from $CO_2(g)$ to dissolved carbonate ion, CO_3^{2-}
    b. Acetic acid is a weak acid. Write the equation for the acid ionization reaction.
    c. Indicate the reaction responsible for dissolving the $CaCO_3$ deposits.

92. Egg shells consist of $CaCO_3$. Placing an egg in vinegar overnight leaves the egg with only the inner membrane intact. What chemical reactions are responsible for the disappearance of the egg shell? When preparing eggs for dying, it is recommended that a small amount of vinegar be added to the dye solution. Explain why this step helps with coloring the egg.

Chapter 9

Electrochemistry: Batteries, Corrosion, Fuel Cells, and Membrane Potentials

A nickel is half electroplated with copper. Electrons from the electroplating circuit and Cu^{2+} cations in solution deposit elemental copper on the coin.

CONCEPTUAL FOCUS

- Connect the force that drives electrochemical reactions to the potential to gain or lose an electron.

SKILL DEVELOPMENT OBJECTIVES

- Use electronegativity to locate electrons in a compound and assign oxidation numbers to track electrons (Worked Examples 9.1–9.3).
- Use oxidation numbers to predict formulas and identify elements that gain or lose electrons in a reaction (Worked Examples 9.4, 9.5).
- Use the potential for electron loss to determine battery voltage (Worked Examples 9.6, 9.7).
- Trace electron flow to design fuel cells and corrosion protection (Worked Examples 9.8–9.10).
- Calculate stress in an oxide coat, evaluate stability of coat (Worked Example 9.11).
- Connect the cell potential to the Gibbs free energy and the equilibrium constant, K (Worked Examples 9.12–9.15).

B atteries have become commonplace items in today's society—from the tiny units powering digital watches to the larger versions driving laptop computers to their much larger cousins supplying the energy to run electric buses. In 2004, batteries represented a $17 billion global market.

Batteries remain the subject of intensive research, belying their rather mature origins, which date back to the voltaic pile invented by Alessandro Volta in 1800. Research continues in this area because battery technology has not kept pace with the exponential rate of development found in the remainder of the electronics industry. Indeed, batteries remain the heaviest, least efficient component of nearly all portable electronic devices. Current research focuses on developing batteries that generate more power for longer periods of time in a smaller, lighter-weight package. To paraphrase Emerson, "Build a better battery and the world will beat a path to your door!"

All batteries have the same basic components: a positive terminal and a negative terminal separated by a membrane and immersed in an ionic solution or paste that conducts either electrons or positive ions. These same basic components are found in fuel cells, devices that depend on a continuous supply of reactants to generate power.

Oxidation Removal of one or more electrons.

The processes taking place in a battery are similar to those occurring in another familiar phenomenon: corrosion of a metal (Figure 9.1), commonly referred to as rusting. In corrosion, the neutral metal atom loses some of its electrons either to become a positive ion or to form a compound, often with oxygen. Losing one or more electrons is broadly termed **oxidation,** a name that originated with burning of a substance in oxygen but is now employed more generally to refer to loss of electron density by an atom in a reaction. Matter is conserved in chemical reactions, so when one atom loses an electron, another atom must gain an electron. This electron gain is broadly termed **reduction.** Oxidation and reduction *always* occur in pairs, like bookends. The whole reaction consisting of the oxidation half and reduction half is termed a **redox** reaction. Charged particles or, more specifically, the imbalance of charged particles across cell membrane signals muscles to contract and extend, enabling physical activity such as running to occur (Figure 9.1).

Reduction Addition of electrons.

Redox Contraction of oxidation and reduction.

Figure 9.1 Electrochemistry is the engine behind all of the items pictured here: corrosion of ships and cars, and the signaling pathways that control muscles for the runner to run.

9.1   Case Study: Utility Poles, Nails, and Tin Cans

Left exposed to a moist atmosphere iron, an important structural material, combines with oxygen to become the familiar red rust. Yet iron, the major component of steel, is used in numerous applications, including underground pipes, utility poles, nails, and metal food cans. How is the inevitable corrosion kept at bay?

Steel (iron) utility poles and pipes are often connected to a magnesium stake by an electrical conductor (Figure 9.2). The Mg stake disintegrates and needs to be replaced periodically; this suggests that magnesium corrodes—loses its electrons—more easily than iron, protecting iron from degradation. This is indeed the case, and Mg is said to be more active than Fe. The most common atmospheric substance removing the electrons in corrosion is oxygen. Oxygen is one of the most electronegative elements, so it tends to attract electrons when oxygen interacts with other elements. When oxygen interacts with either the iron in a steel pole or the magnesium stake, electrons are supplied by the magnesium stake since Mg is more electropositive (the opposite of electronegative) than iron. The electrical conductor connecting the steel pole to the magnesium stake is critical, because it forms a pathway to convey electrons to and through the iron in the pole to oxygen.

Nails used in exterior applications (Figure 9.3) are usually galvanized—that is, dipped in zinc. Like Mg, Zn is a more active metal than Fe. In this case, the Zn contacts Fe directly, so a separate conductor is not necessary. The Zn coat protects the iron in the nail from corrosion as long as the Zn lasts, preventing the exterior siding and shingles on wooden buildings from being spoiled with unsightly, rusty streaks.

Metal food cans are sometimes referred to as "tin cans" due to the tin that coats the steel body (Figure 9.4). Tin is a less active metal than iron, so it protects the iron only if there are no breaks in the coating. Once the coating is scratched or cracked through, however, iron "protects" the tin, and the can rusts rapidly as demonstrated by speedy rusting of crushed, discarded cans left outdoors. This rapid deterioration is instrumental in the breakdown of the cans in garbage dumps.

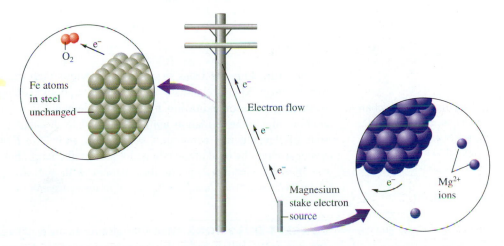

Figure 9.2 A steel utility pole is connected to a magnesium stake. The magnesium feeds electrons to the iron in the steel pole via the conducting connector, preventing oxidation of the iron.

O₂

Fe atoms
in steel
unchanged

Electron flow

Magnesium
stake electron
source

Mg²⁺
ions

Figure 9.3 Galvanized nails or pails utilize the greater activity of Zn compared with Fe to protect the iron from corrosion.

Figure 9.4 Metal food cans consist of Sn-coated iron. Sn is relatively resistant to corrosion, even when it is in contact with acidic food. Due to the greater structural strength of Fe, the body of the can consists of an Fe-based steel. Sn protects the can from corrosion by forming a barrier between the Fe and O_2.

To recap, Mg and Zn are more active than Fe, so they prevent oxidation of Fe by supplying electrons to O more readily than does Fe. Sn is less active than Fe, providing electrons less readily than Fe; it protects only if the Sn barrier has no break. These four elements can be ordered in a activity series:

Mg (most active), Zn, Fe, Sn (least active)

The relative activity of each metal is related to its electron configuration. Mg, the most active of the four metals, is a Group IIA element and has only two electrons in its valence shell. Mg easily loses these two electrons to attain the electron configuration of Ne, thereby forming a +2 cation, Mg^{2+}.

CONCEPT QUESTION Write the electron configurations of Zn, Fe, and Sn. Which electrons are lost when these elements form +2 ions? ■

Tracking electron flow is the key to unraveling redox reactions. Although electron flow away from Mg to O_2 to form the +2 cation is apparent, other cases, such as Fe in FeO or Fe_2O_3, are not as readily deciphered. The next section develops a bookkeeping method for these less obvious cases.

9.2   Oxidation Numbers

Oxidation state The charge on an element resulting from electrons being assigned to the more electronegative element in a bond.

To aid in tracking electrons in cases less obvious than the loss of two electrons from Mg to form Mg^{2+}, chemists have created a bookkeeping procedure to assign an **oxidation state,** or **oxidation number,** to elements in a compound. For simple ions, such as Mg^{2+}, the oxidation number is equal to the charge on the ion. The oxidation state of elements in a compound, such as iron in FeO or Fe_2O_3, is assigned using electronegativity. The oxidation state is not an actual charge, but merely reflects the shift in the bonding electrons toward the element with the stronger attraction. In general, *shared electrons in a bond are assigned to the more electronegative element involved in the bond.*

CONCEPT QUESTIONS What is the electronegativity difference between iron and oxygen? Would you assign the shared electrons in FeO to iron or to oxygen? ■

Oxidation number Charge on an atom due to the difference between the number of valence electrons and the number of electrons resulting from assigning all electrons in a bond to the more electronegative element in the bond.

Because oxygen is nearly the most electronegative element in the periodic table, electrons in the bond between oxygen and another element are assigned to oxygen. Oxygen forms two bonds to complete its octet, so assigning electrons by electronegativity gives eight valence shell electrons to oxygen (one shared electron from each bond, plus oxygen's six valence electrons). With eight electrons in the valence shell, oxygen has a -2 charge. The oxidation number of -2 reflects this charge.

Similarly, in most common compounds, hydrogen is less electronegative than the element to which it is bonded, so it is assigned no electrons and an oxidation number of $+1$. These general observations are summarized in two guidelines for assigning oxidation numbers:

- The oxidation number of oxygen is usually -2.
- The oxidation number of hydrogen is usually $+1$.

Exceptions to these two guidelines follow from an examination of electronegativities. Oxygen is assigned a positive oxidation number when it is combined with fluorine because fluorine is the one element that is more electronegative than oxygen. In peroxides, HOOH, oxygen has an oxidation number of -1. Oxygen is assigned both electrons in the bond between oxygen and hydrogen. However, the two electrons in the O—O bond are divided evenly between the two oxygen atoms because both are equally electronegative. Thus, in peroxide, oxygen has seven electrons and a -1 charge. Hydrogen is assigned a negative oxidation number when it is combined with low-electronegativity metals, such as the alkali metals (Group IA) and the alkaline earth metals (Group IIA). Any element in pure form—that is, not combined with another element and with no charge—has an oxidation number of zero. Electrons are shared equally among atoms in this case, so there is no net gain or loss of electrons.

- The oxidation numbers of alkali metals are $+1$ and of alkaline earth metals are $+2$.
- The oxidation number of an element in its elemental form is zero.

Two additional rules ensure that the total number of electrons is accounted for:

- The oxidation numbers in neutral compounds must sum to zero.
- The oxidation number of a simple ion, such as Cl^-, is equal to the charge on the ion. For more complex ions, the oxidation numbers must sum to the ion charge.

The first rule provides a practical check that the total number of valence electrons in a compound is equal to the total number brought into the compound by its constituent elements. The latter rule accounts for electrons lost or gained by an ion.

WORKED EXAMPLE 9.1   *Electrons in Water*

Water is an important participant in many redox reactions, particularly in the environment: "Water, water, everywhere . . .". Assign oxidation numbers to H and O in water.

Plan

- Count the valence electrons.
- Oxygen is more electronegative than hydrogen, so assign eight valence electrons to oxygen. Assign the remaining electrons to hydrogen.
- The net charge is the oxidation number.

Implementation

- Valence electrons: oxygen (6) plus one for each of two hydrogen atoms totals 8.
- Oxygen gets all 8. Hydrogen gets none.
- Net charge on oxygen: gained two valence electrons, oxidation number is -2. Net charge on each hydrogen: lost one valence electron, oxidation number is $+1$.

See Exercises 1–4.

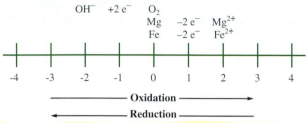

Figure 9.5 The oxidation–reduction number line. Oxidation corresponds to moving to the right and reduction to moving to the left: Right Oxidation Left Reduction (ROLR).

Formally, loss of electrons corresponds to oxidation and gain of electrons corresponds to reduction (LEO GER). It is helpful to put oxidation numbers on a number line (Figure 9.5) to aid in visualizing electron gain and loss. Oxidation corresponds to an increase in oxidation number, reduction to a decrease. On the oxidation line, moving to the right corresponds to oxidation and moving to the left is reduction: right oxidation left reduction (ROLR).

 APPLY IT Obtain some galvanized tacks or nails at a hardware store, some tincture of iodine from a drugstore, and some household bleach. Put the tacks or nails in a clear container and nearly cover them with the iodine solution. Observe the color of the solution. Let stand for about 30 min. What happened to the solution color? Pour the solution into a second clear container. Add a few drops of the bleach. How does the solution color change? Zn ions in aqueous solution make a clear and colorless solution. I^- ions impart a brown-purple color. Household bleach contains Cl^- ions, which impart a pale-yellow color. Use the color information to describe the processes in solution.

Periodic Trends

The connection between the valence configuration, electrons required to form an octet, and electronegativity results in a periodic pattern for oxidation states (Figure 9.6). For example, the halogens require only one more electron to attain a filled s and a filled p valence subshell. Hence the common oxidation state for the halogens is -1. In some reactions, the larger halogens lose electrons—for example, to more electronegative fluorine or oxygen—and have a positive oxidation number. Indeed, larger elements tend to have a variety of oxidation states. Note the larger variety of oxidation states among the transition elements in Figure 9.6.

WORKED EXAMPLE 9.2 *Identifying Electrons Lost to Form Ions*

Elements near the right side of the periodic table acquire electrons to complete the valence shell and become negative ions. Metallic elements tend to lose, rather than gain electrons. To evaluate the stability of metal ions, it is helpful to identify the electrons that are lost in forming the ion. Write the valence configuration of Zn, and identify the electrons lost to form the $+2$ ion. Explain why the only common oxidation state of zinc is $+2$.

Common oxidation states of the elements

IA	IIA	IIIB	IVB	VB	VIB	VIIB	VIIIB	VIIIB	VIIIB	IB	IIB	IIIA	IVA	VA	VIA	VIIA	VIIIA
H 1																	He 0
Li 1	Be 2											B 3	C ±4, 2	N ±3, +5, 4, 2	O −2	F −1	Ne 0
Na 1	Mg 2											Al 3	Si 2, 4	P ±3, 4, 5	S ±2, 4, 6	Cl ±1, 3, 5, 7	Ar 0
K 1	Ca 2	Sc 3	Ti 3, **4**	V 2, 3, **4**, 5	Cr 2, 3, 6	Mn **2, 3**, 4, 6, 7	Fe **2, 3**	Co 2, 3	Ni 2, 3	Cu 1, 2	Zn 2	Ga 3	Ge 4	As ±3, 5	Se −2, 4, 6	Br ±1, 5, 7	Kr 0
Rb 1	Sr 2	Y 3	Zr 4	Nb 3, **5**	Mo 2, 3, 4, 5, 6	Tc 7	Ru 2, 3, 4, 6, 8	Rh 2, 3, 4	Pd 2, 4	Ag 1	Cd 2	In 3	Sn **4**, 2	Sb ±3, 5	Te ±2, 4, 6	I ±1, 5, 7	Xe 0
Cs 1	Ba 2	La 3	Hf 4	Ta 5	W 2, 3, 4, 5, 6	Re 2, 4, 6, 7	Os 2, 3, 4, 6, 8	Ir 2, 3, 4, 6	Pt 2, 4	Au 1, **3**	Hg 1, **2**	Tl **3**, 1	Pb **4**, 2	Bi **3**, 5	Po 2, 4	At ±1, 3, 5, 7	Rn 0

Figure 9.6 Common oxidation states of the elements. The most common state is indicated in bold. Metals generally have positive oxidation states, consistent with their propensity to form positive ions. Nonmetals, particularly the smaller ones, have negative oxidation numbers. The semimetals are on the boundary between primarily positive oxidation numbers and primarily negative ones. Electron gain and loss reactions are common, particularly among the transition elements.

double-check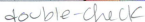

Plan

- Identify the valence electrons.
- The highest-energy electrons are lost to form the positive ion.
- Examine the electron configuration of the ion for closed or half-filled subshells. These are the more stable configurations.

Implementation

- Zinc is in the fourth period, so it has both $3d$ and $4s$ valence electrons. The 12 valence electrons just fill both of these subshells. The valence configuration is $3d^{10}4s^2$.
- The loss of two electrons leaves a $+2$ charge. Transition elements *always* lose s valence electrons first. The loss of the two $4s$ electrons results in a valence configuration of $3d^{10}$—a complete $(n − 1)d$ subshell—for the Zn^{2+} ion.
- Ten electrons in the d subshell constitute a closed shell configuration for the Zn^{2+} ion. This is a stable configuration. The stability of the Zn^{2+} ion electron configuration is consistent with the Zn^{2+} ion losing no more electrons and forming no other oxidation states.

See Exercises 5, 6, 12, 14.

Recognizing Redox Reactions and Predicting Formulas

Reactions involving electron transfer are extremely common. For example, electron transfer is involved in digestion, in the browning of an apple left in air, and in the rusting of tools left outside. Oxidation numbers help with recognition of electron transfer reactions: If the oxidation number of any element changes during a reaction, electron transfer is occurring and the reaction is a redox reaction.

Tracking electrons in a reaction now becomes a matter of following the oxidation number. An element is oxidized when electrons are removed—the oxidation number *increases*—and is reduced when electrons are gained—the oxidation number *decreases*. Thus, in the reaction of iron with oxygen to produce FeO, the oxidation number of Fe goes from 0 to +2: Iron is oxidized.

WORKED EXAMPLE 9.3   *Generation of Smog*

One product in automobile exhaust is NO_2. NO_2 is responsible for the brown haze in polluted city air. At night NO_2 readily dimerizes to colorless N_2O_4. NO_2 also readily dissolves in water to produce nitric acid, HNO_3. Does either dimerization or acid production involve a redox process? Why or why not?

Plan

■ Assign oxidation numbers to the reactants and products.

■ An increase in oxidation number indicates oxidation; a decrease indicates reduction.

Implementation

■ The reactant is NO_2 in both cases. Because the oxidation number of oxygen is -2, that of nitrogen must be $+4$ (the sum of -2, -2 and $+4$ is zero). In N_2O_4, the oxidation number of nitrogen is $+4$. (Four oxygen atoms at -2 is -8. This is balanced by two nitrogen atoms at $+4$.) In HNO_3, the oxidation number of nitrogen is $+5$. (Three oxygen atoms at -2 is -6 plus one hydrogen atom at $+1$ is balanced by one nitrogen atom at $+5$.)

■ From NO_2 to N_2O_4 there is no change of oxidation number. The dimerization does not involve a redox process. From NO_2 to HNO_3, the oxidation number of nitrogen increases—it is oxidized. Thus formation of the acid is a redox process, and at least one more molecule must be involved and must undergo reduction.

Incorporation of NO_2 into water is one of the two major processes responsible for the formation of acid rain. NO is also a product, and production of NO from NO_2 involves reduction of nitrogen from a $+4$ to a $+2$ oxidation state.

See Exercises 3, 4, 17, 18.

WORKED EXAMPLE 9.4   *Oxidation Numbers Predict Formulas*

Oxidation numbers are powerful tools for predicting chemical formulas. Three oxides of sulfur are formed from burning of sulfur-containing coal and subsequent oxidation in the environment. What is the formula for each of the three oxides?

Plan

■ The common oxidation states of sulfur are ± 2, $+4$, and $+6$. The normal oxidation state of oxygen is -2.

■ The algebraic sum of the oxidation numbers in a compound is zero.

Implementation

■ Only the positive oxidation states need to be considered because the oxidation state of oxygen is -2 when combined with any element other than fluorine.

■ The algebraic sum of the oxidation states +2, +4, or +6 with −2 per oxygen atom requires 1, 2, and 3 oxygen atoms, respectively.

Sulfur oxidation state:	+2	+4	+6
Molecular formula:	SO	SO_2	SO_3

See Exercises 15, 16, 19, 20.

9.3   Activity Connected with the Periodic Table

CONCEPT QUESTION Find the electronegativity values for Mg, Zn, Fe, and Sn. What is the correlation between electronegativity values and activity for these elements? ■

The activity of metallic elements is associated with how readily the element loses its electrons. Elements that lose electrons readily are associated with low electronegativity values, so the most active metals are associated with the lowest electronegativity values.

Electronegativity generally decreases in moving toward the lower-left corner of the periodic table, and the most active metals are found at the beginning of each period. The deviations in this general trend reflect the electron shell structures of the atoms. For example, in the fourth period the electronegativity dips slightly at Mn with its half-filled d subshell and again at Zn with its filled d subshell (Figure 9.7). Consistent with this dip, Zn and Mn are more active than Fe. Sn is the least active of the four elements. A Group IVA element, Sn is located near the right side of the periodic table. With a relatively high electronegativity, and hence a greater tendency to draw electrons to it, Sn is the least active toward electron loss of the four metals investigated.

The Activity Series

Activity series A list of metallic elements in order of the tendency to oxidize.

A large variety of metals can be put together into an **activity series,** which shows the relative tendency for an element to give up electrons and become an ion in an action

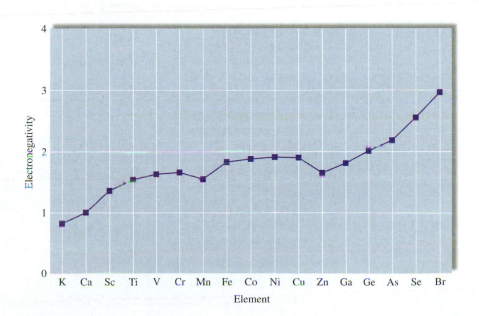

Figure 9.7 Electronegativity generally increases across the fourth period, consistent with the generally rising electronegativity from left to right across the periodic table. As we have often seen, there are some dips in the general trend at Mn and again at Zn, which have half-filled and filled d subshells, respectively. The activity of the metallic elements follows the same general trend.

(Table 9.1). Any element in such a series becomes an ion more readily than those elements below it. For example, Cu appears above Ag on the activity list. If a piece of Cu is placed in a solution containing Ag^+ ions (Figure 9.8), Cu supplies electrons to the Ag^+ ions. As a result neutral, metallic Ag whiskers are deposited on the Cu, and Cu^{2+} ion go into solution, turning the solution blue.

The most active metals—Li, Na, K, and Ca—give up electrons so readily that the valence electrons are transferred to hydrogen in water (Figure 9.9), forming molecular hydrogen. That is, the most active metals *liberate hydrogen from water.* Due to the reactivity of the most active metals with water, samples of these metals are stored under oil to protect them from moisture in the atmosphere.

Metals just below this most active group in Table 9.1 require a little more energy to transfer electrons to hydrogen in water. This added energy is provided by converting water to steam, and the next most active metals liberate hydrogen from steam. Metals just below the group that liberate hydrogen from steam transfer an electron to H^+ ions in acidic solutions, generating H_2 and leaving the metal ion in solution. The least active metals—those found at the bottom of the metal activity list—do not transfer an electron to hydrogen.

Due to the importance of water and hydrogen, the activity with respect to water and acids is also indicated on the metal activity list.

To summarize these points about metal activity:

- Any metal displaces ions of a metal below it from solution.
- The most active metals liberate hydrogen from water or steam.

Caution: Never put the most active or very active metals in acid solution. The reaction would be violent and potentially hazardous. Always test an unknown metal first with cold water, then warm water. Only if both of these tests are negative should the metal be put in acid solution.

Table 9.1 Metal Activity Series

Most active—lose electrons to become ions most readily

Li	
K	Displace hydrogen
Ca	from cold water
Na	
Mg	
Al	
Mn	
Zn	Displace hydrogen
Cr	from steam
Fe	
Cd	
Co	
Ni	Displace hydrogen from
Sn	nonoxidizing acids
Pb	
H	(a nonmetal)
Sb	(a metalloid)
Cu	
Ag	
Hg	
Pt	
Au	

Least active—most likely to retain electrons

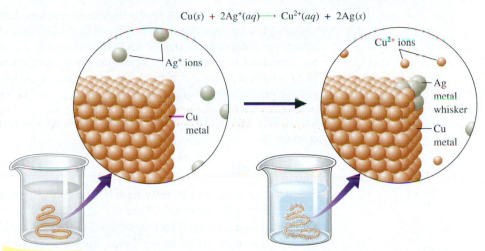

$$Cu(s) + 2Ag^+(aq) \longrightarrow Cu^{2+}(aq) + 2Ag(s)$$

Figure 9.8 A copper wire "tree" soon glitters with silvery whiskers when placed in a solution containing Ag^+ ions. The Ag^+ ions are reduced to metallic Ag. In exchange, the elemental Cu atoms in the wire become Cu^{2+} ions, turning the solution a blue—a characteristic color for Cu^{2+} ion in solution. At the atomic level, Ag^+ ions in solution contact the metallic Cu surface. One electron per Ag^+ ion is transferred, transforming Ag^+ ion into metallic Ag, and leaving the copper surface short of electrons. This transfer creates a Cu^{2+} ion that departs into solution.

Half-reaction One part of an oxidation–reduction reaction. One half reaction represents oxidation, the other represents reduction.

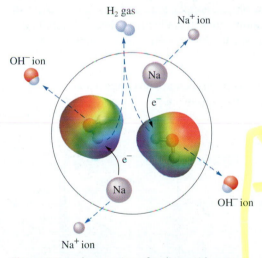

Figure 9.9 In the reaction of sodium with water, each Na atom transfers an electron to a hydrogen atom in water, severing the bond between the hydrogen atom and the oxygen atom. Two hydrogen atoms join to form H_2. The oxygen atom plus the remaining hydrogen atom, along with the electrons from the bond between the departed hydrogen atom and oxygen, form a hydroxide ion (OH^-). The Na atom minus its loosely held electron is a Na^+ ion. The dotted oval indicates the reactants—$2H_2O$ plus 2Na. The products—$2Na^+$, $2OH^-$, plus H_2—are outside the oval.

- The metals of intermediate activity liberate hydrogen from acids.
- The least active metals do not liberate hydrogen and, therefore, do not react with water.

Redox reactions can be represented similarly to other chemical reactions with reactants and products, by placing their chemical symbols on the left and right sides of an arrow, respectively. Consider, for example, the reaction of Cu and Ag^+

(9.1) $$Cu(s) + 2Ag^+(aq) \longrightarrow Cu^{2+}(aq) + 2Ag(s)$$

or sodium with water

(9.2) $$2Na(s) + 2H_2O(l) \longrightarrow 2Na^+(aq) + 2OH^-(aq) + H_2(g)$$

Balanced reactions have the same total charge on both the reactant and the product sides [two in Equation (9.1) and zero in Equation (9.2) as well as the same number of atoms for each element. An alternative representation that shows the electrons more explicitly separates the reaction into an oxidation and a reduction **half-reaction.** Electrons are produced in the oxidation half-reaction. For example,

(9.3) $$Cu(s) \longrightarrow Cu^{2+}(aq) + 2e^-$$

The two electrons produced in Equation (9.3) are never free. Instead, they transfer directly from Cu to Ag^+ but are shown in the equation explicitly to facilitate tracking electrons. Similarly, electrons consumed are reactants in the reduction half-reaction:

(9.4) $$Ag^+(aq) + e^- \longrightarrow Ag(s)$$

Two electrons are produced in the oxidation reaction, Equation (9.3), but only one is consumed in the reduction reaction, Equation (9.4). Since electrons cannot be free in solution, the reduction reaction is doubled to consume two electrons before the reactions are combined to produce the total reaction for the reduction of Ag^+ by Cu, Equation (9.1). Half-reactions are helpful for evaluating activity as well as determining the voltage produced by a battery.

CONCEPT QUESTION What are the two half-reactions for the sodium–water reaction? ∎

Galvanic Cells

Suppose that instead of directly transferring from Cu to Ag^+, the electrons are routed through a wire on their way from Cu to Ag^+ (Figure 9.10). Electrons moving in a wire constitute a current, so this separation generates an electrical current. How can the electron source (Cu) be separated from the electron sink (Ag^+)? Look again at the scenario in the utility pole (Figure 9.2): When oxygen attacks the iron, the electron source, Mg, is separated from the sink, O_2, by a conductive wire. Like dominos falling, the oxygen taps the iron for electrons, the iron taps the conductive wire, and the conductive wire gets the electrons from the Mg stake. A similar separation can be achieved in the Cu/Ag^+ case (Figure 9.10a) by immersing a strip of Ag in a solution containing Ag^+ ions and a strip of copper in a separate solution of Cu^{2+} ions and then connecting the two metal strips—called **electrodes**—with a wire. Silver draws electrons through the wire from the copper strip, liberating Cu^{2+} ions into solution. Metallic silver is deposited on the Ag electrode. As electrons are transferred, charge builds up on both sides: positive charge on the copper side and negative charge on the silver side. Coulombic repulsion soon stops the electron flow. A conductive bridge, called a **salt bridge,** therefore connects the two sides and completes the circuit (Figure 9.10b).

Electrode Location for oxidation or reduction in an electrochemical cell.

Salt bridge A concentrated salt solution in a gel that provides a conducting path between two compartments of an electrochemical cell.

CONCEPT QUESTIONS How does the description change if no copper ions are present? How does it change if no silver ions are present? ∎

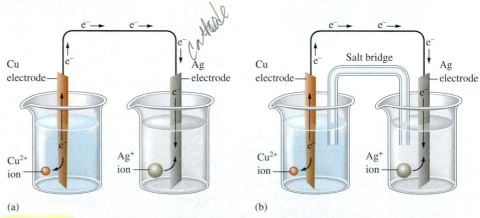

(a)   (b)

Figure 9.10 (a) Separating the source of electrons (Cu) from the sink (Ag^+) and connecting the two with a conducting wire causes the electrons to flow through the wire. The flow of charge soon stops, however, due to a buildup of positive charge on the copper side as Cu^{2+} ions are created and a depletion of positive charge on the silver side as Ag^+ ions are consumed. (b) Addition of a salt bridge enables a compensating charge flow, completing the circuit.

Galvanic cell A device in which chemical energy from a spontaneous redox reaction is changed to electrical energy that can be used to do work (a battery).

Electron source   Electron sink

Anode   Cathode

Oxidation   Reduction

9 V
battery

Figure 9.11 Redox reactions are at the heart of all batteries.

Anode The electrode in a galvanic cell at which oxidation occurs.

Cathode The electrode in a galvanic cell at which reduction occurs.

This device—the two solutions, the metal strips, the wire, and the salt bridge—is called a **galvanic cell,** or battery. The short-hand notation for the copper–silver galvanic cell is

(9.5) $Cu \mid Cu^{2+} \parallel Ag^+ \mid Ag$

Oxidation is indicated first. A single vertical line ($\mid$) indicates a phase boundary, such as between the solution and the copper strip or electrode. The double vertical line ($\parallel$) indicates the salt bridge. It is followed by the reduction reaction. In this convention, electrons travel in the external circuit from left to right and the electrodes are on the ends. The reactions can be read from this notation; for example, Cu oxidizes to Cu^{2+} and Ag^+ reduces to Ag. A $Cu \mid Cu^{2+} \parallel Ag^+ \mid Ag$ cell is not a very practical battery, however, because silver is a very expensive metal and, as we will see, a $Cu \mid Cu^{2+} \parallel Ag^+ \mid Ag$ battery generates a small voltage. A much more practical pair is Cu and Zn.

CONCEPT QUESTION Which metal, Cu or Zn, is the more active? ■

Battery Potential and the Electromotive Series

A typical flashlight C or D battery generates 1.5 volts (V), and a rectangular battery (the kind with snap leads) generates 9 V. These voltages are related to the potential, driving electrons from the material being oxidized to the material being reduced. In either a galvanic cell (Figure 9.10) or a battery (Figure 9.11), the **anode** is where oxidation occurs; it is where electrons are produced and is the negative pole—negative electrode—of the battery. Conversely, the **cathode** is where reduction occurs; it is where electrons are consumed and is the positive electrode of the battery. Connecting the two sides of the battery with a wire results in a flow of electrons from the anode to the cathode, just as electrons move from Cu to Ag in the galvanic cell (Figure 9.10).

WORKED EXAMPLE 9.5   *A Copper–Zinc Galvanic Cell*

Consider a galvanic cell constructed with a copper strip in a solution of $Cu^{2+}(aq)$ ions and a zinc strip in a solution of $Zn^{2+}(aq)$ ions. Which metal is oxidized? Which metal is the anode and which is the cathode? Draw a schematic picture of this cell indicating electron flow. Write the two half-reactions for this galvanic cell.

Plan

- The metal with the lower electronegativity is more active and loses electrons more easily. It is the metal being oxidized.
- The metal being oxidized forms the anode and that being reduced is the cathode. Electrons flow from the anode to the cathode.
- Identify the ion formed by each metal.

Implementation

- Copper is a relatively inactive metal and is less active than zinc. Zinc loses its electrons and is oxidized.
- The zinc electrode is the anode; the copper electrode is the cathode.
- Zn and Cu both commonly form +2 ions. The reactions are thus

$$Zn(s) \longrightarrow Zn^{2+}(aq) + 2e^-$$
(oxidation reaction, anode, negative pole)

$$Cu^{2+}(aq) + 2e^- \longrightarrow Cu(s)$$
(reduction reaction, cathode, positive pole)

When coupled with Zn, Cu is reduced. When coupled with Ag, however, Cu is oxidized. Copper's activity is between that of Ag and that of Zn. It is the *relative* activity that determines whether a metal is oxidized or reduced and the direction of electron flow.

See Exercises 49, 54.

Electrons flow from the negative electrode to the positive electrode because the electrons are more stable—have a lower energy—when they are associated with the substance at the positive pole. The force that pulls electrons from the anode (negative electrode) to the cathode (positive electrode) is due to an electrochemical potential and is variously called the **cell potential (E)**, the **cell voltage (V)**, or the **electromotive force (emf)**. Electromotive *force* is really a historical misnomer because it is not a force, but rather a potential. The potential for taking on electrons is referred to as the **standard reduction potential (E°)**—*standard* because it is the potential at 25 °C with solution species present in 1 *M* concentrations. It is a *reduction* potential because it refers to a reaction in which one or more electrons are *gained*. Gaining electrons *reduces* the oxidation number. Standard potentials for several reactions are listed in Table 9.2.

Notice the similarity between the reduction potential series and the metal activity series. Cu, Ag, and Au are on the bottom of the activity series, and they have the most positive potential for being reduced to the neutral metal—that is, the least tendency to lose one or more electrons to become a positive ion. The most active metals are classified

Cell potential, E Potential that results in a driving force in a galvanic cell that pulls electrons from the reducing agent in one compartment to the oxidizing agent in the other. Also called *electromotive force* or *cell voltage*.

Cell voltage (V) Potential that pulls electrons from the anode to the cathode. Also called *cell potential (E)* or the *electromotive force*.

Electromotive force (emf) (misnomer, a potential) Potential for taking on electrons.

Standard reduction potential (E°) The potential for a reduction reaction to occur at 25 °C with all ions present in 1 *M* concentration. Measured with respect to the reduction of H^+ to H_2.

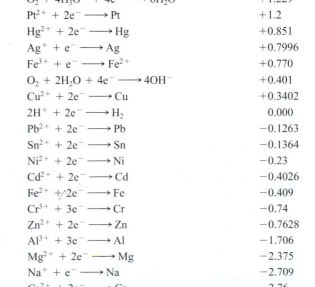

Table 9.2 Standard Reduction Potentials

Reduction Half-Reaction	E° (V)
$Au^{3+} + 3e^- \longrightarrow Au$	+1.42
$O_2 + 4H_3O^+ + 4e^- \longrightarrow 6H_2O$	+1.229
$Pt^{2+} + 2e^- \longrightarrow Pt$	+1.2
$Hg^{2+} + 2e^- \longrightarrow Hg$	+0.851
$Ag^+ + e^- \longrightarrow Ag$	+0.7996
$Fe^{3+} + e^- \longrightarrow Fe^{2+}$	+0.770
$O_2 + 2H_2O + 4e^- \longrightarrow 4OH^-$	+0.401
$Cu^{2+} + 2e^- \longrightarrow Cu$	+0.3402
$2H^+ + 2e^- \longrightarrow H_2$	0.000
$Pb^{2+} + 2e^- \longrightarrow Pb$	−0.1263
$Sn^{2+} + 2e^- \longrightarrow Sn$	−0.1364
$Ni^{2+} + 2e^- \longrightarrow Ni$	−0.23
$Cd^{2+} + 2e^- \longrightarrow Cd$	−0.4026
$Fe^{2+} + 2e^- \longrightarrow Fe$	−0.409
$Cr^{3+} + 3e^- \longrightarrow Cr$	−0.74
$Zn^{2+} + 2e^- \longrightarrow Zn$	−0.7628
$Al^{3+} + 3e^- \longrightarrow Al$	−1.706
$Mg^{2+} + 2e^- \longrightarrow Mg$	−2.375
$Na^+ + e^- \longrightarrow Na$	−2.709
$Ca^{2+} + 2e^- \longrightarrow Ca$	−2.76
$K^+ + e^- \longrightarrow K$	−2.924
$Li^+ + e^- \longrightarrow Li$	−3.045

Increasing activity

Increasing potential to be reduced

as active because they tend to end up as positive ions. In other words, they have a *great tendency to be oxidized,* which means that their *potential for being reduced is negative.* In fact, the metal activity list is essentially an abbreviated, and less detailed, reduction potential list with the element with most negative reduction potential appearing at the top of the metal activity list.

Standard reduction potentials are also used to determine the voltage produced by a battery or galvanic cell. For example, the reduction potential for Cu^{2+} to Cu is $+0.3402$ V and that for Zn^{2+} to Zn is -0.7628. The *reduction* potential for Cu is higher than that for zinc, so Cu is reduced; that is, the copper side is the cathode. The potential or voltage for this cell is the difference in potential for electrons being associated with Cu^{2+} ($+0.3402$ V) and that for being associated with Zn^{2+} (-0.7628 V). The standard voltage of a battery based on the copper–zinc reaction is the difference between them: 1.1030 V.

CONCEPT QUESTION What is the potential of the $Cu|Cu^{2+}||Ag^+|Ag$ cell (Figure 9.10)? ■

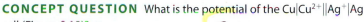

$E_{cathode} - E_{anode} = E_{cell}$

WORKED EXAMPLE 9.6 *The Two-Potato Clock Potential*

At first glance, it might appear that the potato or lemon battery pictured in Figure 9.12 depends on the same reactions as a $Zn|Zn^{2+}||Cu^{2+}|Cu$ galvanic cell and, therefore, has the same potential. However, the Cu^{2+} ion concentration of a potato or lemon is quite low. In addition, the battery functions even in salt water with no Cu^{2+} ions. Thus, there must be another sink for the two electrons from Zn—another oxidant. What other oxidant in Table 9.2 could be responsible for the operation of the clock battery? What potential is produced?

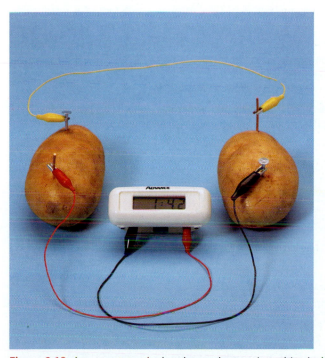

Figure 9.12 A two-potato clock or lemon battery is a whimsical device that derives its power from an electrochemical reaction. The potato or lemon serves as the salt bridge, enabling ions to move and complete the circuit.

Plan

- Consult Table 9.2 to find a potential oxidant.
- The battery voltage is the difference in potentials for the two half-reactions.

Implementation

- In the absence of metal ions, the likely oxidant is O_2, which has two possible half-reactions:

  a. $O_2 + 4H^+ + 4e^- \longrightarrow 2H_2O + 1.229$ V
  b. $O_2 + 2H_2O + 4e^- \longrightarrow 4OH^- + 0.401$ V

- The lemon is acidic so (a) is possible. However, neither the potato nor salt water is acidic, so (b) is the likely half-reaction. The reduction potential for Zn^{2+} is -0.7628 V, so the cell potential with (b) and the reduction of zinc is

$$+0.401 \text{ V} - (-0.7628 \text{ V}) = 1.164 \text{ V}$$

Oxygen is a somewhat stronger oxidant than Cu^{2+}, so the cell potential is somewhat higher with oxygen as an oxidant. Oxygen is often the oxidant in the environment.

 *Eanode − Ecathode*

See Exercises 28, 29, 97, 98.

9.4 Balancing Redox Reactions

Balancing redox reactions can prove challenging because electrons play a critical role yet appear only implicitly. The method used here focuses on the electrons by explicitly counting and balancing them at the beginning of the process. The balancing procedure is illustrated with an example and followed by a summary of the steps involved.

One of the impediments to using methanol (H_3COH) as a substitute for gasoline in internal combustion engines is that the fuel is not always completely oxidized and one of the side products is formaldehyde (H_2CO). What are the oxidation and reduction reactions involved in oxidation of methanol to formaldehyde? Does either half-reaction suggest a method to avoid formaldehyde production?

Step 1. Write the skeleton half-reaction indicating the major reactants and products.
In this case the only identified reactant is methanol (H_3COH) and the only product is formaldehyde (H_2CO).

(9.6) $\qquad\qquad H_3COH \longrightarrow H_2CO$ $\qquad\qquad$ *Not balanced*

Step 2. Assign oxidation numbers to identify the element oxidized and balance that element.

$$\overset{+1\;-2\;-2\;+1}{H_3COH} \longrightarrow \overset{+1\;\;0\;-2}{H_2CO}$$

(9.7) $\qquad\qquad H_3COH \longrightarrow H_2CO$ $\qquad\qquad$ *Not balanced*

CONCEPT QUESTIONS How are the oxidation numbers of carbon in H_3COH and H_2CO determined? Why is the oxidation number of carbon different in these two compounds? ■

Carbon is oxidized. It begins with an oxidation number of -2 in methanol but winds up with an oxidation number of zero in formaldehyde: Two electrons have been lost. With one carbon in both the reactant and the product, carbon is balanced. The next step explicitly shows these electrons.

Step 3. Balance the electrons.

(9.8) $H_3COH \longrightarrow H_2CO + 2e^-$ *Not balanced*

As written, Equation (9.8) is not balanced for charge. There are two options for balancing charge: addition of H_3O^+ if the solution is acidic or addition of OH^- if it is basic. With respect to balancing the reaction, it does not matter which option is chosen. Instead, the choice is dictated by conditions of the reaction. The example is a gas-phase reaction, and neither OH^- nor H_3O^+ will appear in the final reaction. Thus, the added charged substance must also appear in the reduction half-reaction and cancel when the two half-reactions are combined. (This also serves as a check on the accuracy of the preceding steps.)

Step 4. Balance charge. Adding H_3O^+, the charge balance becomes

(9.9) $H_3COH \longrightarrow H_2CO + 2e^- + 2H_3O^+$ *Not balanced*

Equation (9.9) is balanced except for hydrogen and oxygen. You should be able to balance them with water molecules. If this is not possible, an error occurred in an earlier step—go back and check.

Step 5. Balance H and O with H_2O.

(9.10) $H_3COH + 2H_2O \longrightarrow H_2CO + 2e^- + 2H_3O^+$ *Balanced*

Equation (9.10) is a balanced oxidation half-reaction. To describe the oxidation of methanol to formaldehyde, a partner reduction half-reaction is needed to receive the electrons. In this case, the electrons are consumed by oxygen in the air and the reduction half-reaction is balanced in the same way as the oxidation reaction. Reduction of oxygen to water results in the following half-reaction:

(9.11) $O_2 + 4H_3O^+ + 4e^- \longrightarrow 6H_2O$

Note that Equation (9.11) uses H_3O^+, a product in Equation (9.9).

The final step involves combining the two half-reactions to form a complete reaction. The oxidation and reduction reactions are not independent. Electrons produced in the oxidation half-reaction *must be consumed* in the reduction half-reaction. Adding oxidation and reduction reactions must result in cancellation of the electrons.

Step 6. Multiply the half-reactions by the least common multiple of the number of electrons so that electrons produced in the oxidation half-reaction are consumed in the reduction half-reaction. In this case, the methanol oxidation reaction must be multiplied by 2.

$$2H_3COH + 4H_2O \longrightarrow 2H_2CO + 4e^- + 4H_3O^+$$
$$O_2 + 4H_3O^+ + 4e^- \longrightarrow 6H_2O$$

(9.12) $\overline{2H_3COH + O_2 \longrightarrow 2H_2CO + 2H_2O}$

Summarizing the steps for balancing redox reactions:
(Complete Steps 1–5 for each separate half-reaction.)

Step 1. Write a skeleton half-reaction.
Step 2. Assign oxidation numbers, identifying the element being oxidized (reduced), and balance the element being oxidized (reduced).
Step 3. Balance electrons.
Step 4. Balance charge with H_3O^+ or OH^-.
Step 5. Balance hydrogen and oxygen by adding water.
Step 6. Combine the oxidation and reduction half-reactions, multiplying as needed to cancel electrons.

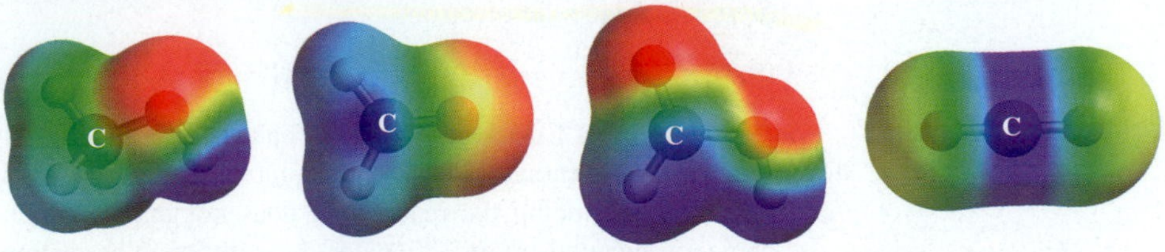

Figure 9.13 Increasing oxidation of carbon from methanol to formaldehyde to formic acid to carbon dioxide shows the increasingly positive (increasingly blue color) potential around the carbon atom. The formal oxidation number increases from -2 to 0 to $+2$ to $+4$, correlating with the increased positive potential at the carbon atom.

CONCEPT QUESTIONS What is the oxidation number of carbon in formic acid, HCOOH, and carbon dioxide, CO_2? Write the oxidation reaction for formaldehyde forming formic acid and the oxidation reaction for formation of CO_2 from formaldehyde. ■

Oxygen can oxidize formaldehyde to formic acid or CO_2 (Figure 9.13). Thus, falling into the formaldehyde trap might be avoided if the oxygen partial pressure is increased.

CONCEPT QUESTIONS How can you determine whether oxidation of formaldehyde to formic acid or CO_2 is a spontaneous process? Locate the necessary data and determine whether formic acid or CO_2 is a more stable product for oxidation of methanol than is formaldehyde. ■

9.5 Applications

Batteries

C and D Cell Batteries

Inside the common 1.5-V flashlight battery (Figure 9.14) is a zinc cup filled with an electrolyte paste. This paste contains ions that serve as the salt bridge. In the center and insulated from the zinc cup is a carbon electrode. In this case, the active metal zinc is oxidized in the half-reaction:

(9.13) $$Zn(s) \longrightarrow Zn^{2+}(aq) + 2e^-$$

The electrons travel through the external circuit, passing through the flashlight bulb and to the carbon electrode where the reduction reaction

(9.14) $$2MnO_2(s) + 2e^- + H_2O(l) \longrightarrow Mn_2O_3(s) + 2OH^-(aq)$$

occurs. This battery works because zinc is an active metal that readily gives up two electrons to become a positive ion. In MnO_2, manganese is in a $+4$ oxidation state. Zinc is sufficiently active to reduce it to a $+3$ oxidation state in Mn_2O_3.

WORKED EXAMPLE 9.7 *Flashlight Reduction Reaction Potential*

The precise reaction sequence in a flashlight dry cell is incompletely known. Nonetheless, an approximate potential for the reduction reaction given in Equation (9.14) can be determined from the $Zn|Zn^{2+}$ potential and the fact that the battery produces 1.5 V. What is the approximate potential for the reduction reaction, Equation (9.14)? Write the cell notation for the flashlight battery.

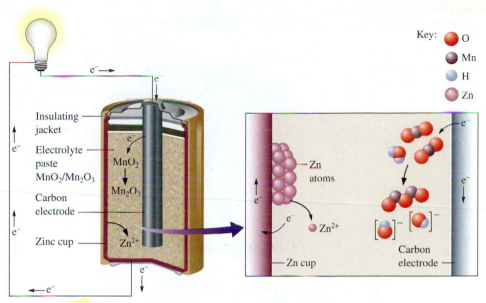

Key: O   Mn   H   Zn

Figure 9.14 A flashlight battery consists of a zinc cup anode and a cathode made of an inert carbon electrode immersed in a paste containing MnO_2 and Mn_2O_3. (right) An atomic-level view of the internal workings of a flashlight battery. Zn is oxidized at the anode, producing electrons that travel through the external circuit to the carbon electrode. There the electrons reduce MnO_2 to Mn_2O_3. OH^- ions drift toward the anode and Zn^{2+} ions toward the cathode to complete the circuit. The cell potential deteriorates as Zn ions are produced.

Plan

- Identify the anode and the cathode reactions.
- Cell potential = reduction potential (cathode reaction) − reduction potential (anode reaction).
- Cell notation: anode∥cathode.

Implementation

- The anode reaction is the oxidation reaction: oxidation of Zn. The reduction potential for Zn is -0.7628 V. Equation (9.14) gives the cathode reaction.
- Cell potential = 1.5 V = potential [Equation (9.14)] − $(-0.7628$ V$) \Rightarrow$ potential [Equation (9.14)] = $(1.5 - 0.7628)$ V = 0.74 V
- Cell notation: $Zn|Zn^{2+}\|MnO_2|Mn_2O_3$

See Exercises 47–52.

Lead Storage Battery

WORKED EXAMPLE 9.8   *Reactions in a Lead Storage Battery*

The lead storage battery in most automobiles consists of an anode where lead is oxidized to $PbSO_4(s)$ and a cathode where $PbO_2(s)$ is reduced to $PbSO_4(s)$ (Figure 9.15). The electrolyte is H_2SO_4. What are the oxidation, reduction, and net reactions for the lead storage battery? Write the notation for a lead storage battery.

Plan

- Balance the oxidation and reduction half-reactions in acidic conditions.
- Multiply the half-reactions by the appropriate integer to cancel electrons and add.
- Cell notation: anode∥cathode.

Figure 9.15 The atomic-level view of the reactions involved in a car battery shows lead's oxidation to $PbSO_4$ at the anode and reduction of PbO_2 to $PbSO_4$ at the cathode. Lead ion is common to both electrodes, enabling the battery to be recharged from the alternator as the engine runs.

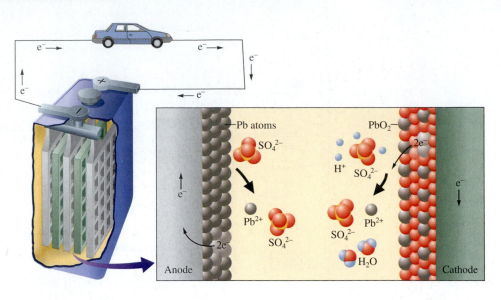

Implementation

■ Anode reaction: $Pb \longrightarrow PbSO_4$

$Pb(s) + SO_4^{2-}(aq) \longrightarrow PbSO_4$	Balance sulfate
$Pb(s) + SO_4^{2-}(aq) \longrightarrow PbSO_4(s) + 2e^-$	Balance electrons

Cathode reaction: $PbO_2 \longrightarrow PbSO_4$

$PbO_2 + SO_4^{2-}(aq) \longrightarrow PbSO_4$	Balance sulfate
$PbO_2 + SO_4^{2-}(aq) + 2e^- \longrightarrow PbSO_4$	Balance electrons
$PbO_2 + SO_4^{2-}(aq) + 2e^- + 4H^+(aq) \longrightarrow PbSO_4$	Balance charge
$PbO_2(s) + SO_4^{2-}(aq) + 2e^- + 4H^+(aq) \longrightarrow$	
$\qquad\qquad PbSO_4(s) + 2H_2O(l)$	Balance H & O

Fuel cell A primary electrochemical cell in which the reactants are continuously fed in while the cell is in use.

■ Two electrons are produced at the anode and two are consumed at the cathode, so adding the two reactions gives the overall reaction:

$$Pb(s) + PbO_2(s) + 2SO_4^{2-}(aq) + 4H^+(aq)$$
$$\longrightarrow 2PbSO_4(s) + 2H_2O(l)$$

■ Cell notation: $Pb|PbSO_4||PbO_2|PbSO_4$

See Exercises 58–61.

Fuel Cells

Fuel cells are very similar to batteries. The distinction is that at least one reactant is constantly fed into the **fuel cell.** For example, the fuel cell used in the manned space program in the *Gemini, Apollo,* and Space Shuttle missions (Figure 9.16) consists of the oxygen reaction,

$$(9.15) \qquad O_2(g) + 2H_2O(l) + 4e^- \longrightarrow 4OH^-(aq)$$

coupled with the hydrogen reaction,

$$(9.16) \qquad H_2(g) + 2OH^-(aq) \longrightarrow 2H_2O(l) + 2e^-$$

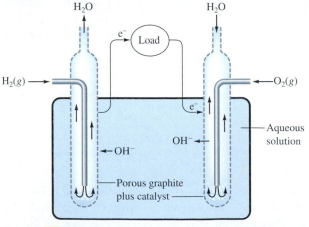

Figure 9.16 A hydrogen–oxygen fuel cell is an environmentally friendly method for producing electrical energy. The only by-product is water. In self-sufficient environments like the Space Shuttle, the water can be used to support life systems.

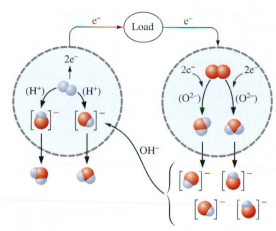

Figure 9.17 In the fuel cell, electron transfer is indirect. Hydrogen passes its electron to the external circuit and, minus its electron, combines with OH⁻ to form water. On the other side of the cell, oxygen accepts the electrons, becoming, in effect, O^{2-}, which attaches to a water molecule. The combination then splits into two OH⁻ ions.

producing the net reaction,

(9.17) $$2H_2(g) + O_2(g) \longrightarrow 2H_2O(l)$$

The direct reaction of H_2 with O_2 to produce water occurs with explosive violence if a mixture of the two gases is simply sparked. This explosion is avoided in a fuel cell by separating the two half-reactions: H_2 gas is pumped into the anode and O_2 gas into the cathode. Electrons produced by hydrogen in the oxidation reaction (Figure 9.16) flow out of the anode and into the cathode, where oxygen is reduced. The rate of energy production is controlled by the hydrogen and oxygen gas pressure. As is frequently the case with fuel cells, ions produced at one electrode drift to the other electrode to complete the circuit. For the hydrogen–oxygen fuel cell, OH⁻ is produced at the cathode (Figure 9.17) and drifts to the anode, where it combines with hydrogen that has lost its electron (effectively H⁺) and forms water.

The environmentally benign product, water, makes the hydrogen fuel cell very attractive. The challenge lies in safe production or transport of the gaseous reactants. The latter issue remains a topic of cutting-edge research.

CONCEPT QUESTIONS   The oxidation of methanol to formaldehyde by oxygen can form the basis of a fuel cell. Draw a schematic for design of a methanol–air fuel cell. Should the solution be acidic (containing H_3O^+) or basic (containing OH⁻)? Which ion completes the circuit? ■

WORKED EXAMPLE 9.9   *Zinc–Air Fuel Cell*

One advantage of fuel cells is that they are lighter in weight than batteries because one or more of the reactants are not incorporated in the original package. An example is the zinc–air cell, in which molecular oxygen is supplied from air. Such a cell does not function until the plastic coating covering the air openings is removed. In just a few minutes, however, it is up to power. At the anode, zinc is oxidized to ZnO. The overall cell reaction is

$$2Zn + O_2 \longrightarrow 2ZnO$$

Determine the anode reaction and the cathode reaction, and suggest a design for the cell.

Plan

- Balance the anode reaction.
- Based on the anode reaction, select an oxygen reduction reaction.
- Identify an ion to complete the circuit.
- Suggest an electrode for the cathode reaction.

Implementation

- The identified participants in the anode reaction are Zn (reactant) and ZnO (product)

Step 1	$Zn \longrightarrow ZnO$	
Step 2	$\overset{0}{Zn} \longrightarrow \overset{+2 -2}{ZnO}$	Identify oxidation numbers
Step 3	$Zn \longrightarrow ZnO + 2e^-$	Balance electrons
Step 4	$Zn + 2OH^- \longrightarrow ZnO + 2e^-$	Balance charge
Step 5	$Zn + 2OH^- \longrightarrow ZnO + 2e^- + H_2O$	Balance H and O

- The anode reaction indicates that OH$^-$ is a good ion to complete the circuit, suggesting a basic paste could be used in the cell.
- The cathode reaction, reduction of oxygen, should produce OH$^-$ to drive the anode reaction: $O_2 + 2H_2O + 4e^- \longrightarrow 4OH^-$.
- The cathode electrode is merely a conduit for the electrons from the anode reaction. A number of metals can be used for the cathode as long as the activity is lower than that of Zn. In effect, Zn protects the cathode from oxidation. Copper is one choice, as are gold and silver.

Zinc–air fuel cells are used to power heart pacemakers and other applications where a compact and portable power supply is needed. Power is controlled by the oxygen supply.

See Exercises 73, 74.

Electrolysis of Water

In batteries and fuel cells, electrons are transferred from a metal or other material with a low reduction potential to one with a higher reduction potential. Electrolysis reverses this spontaneous process and pushes electrons uphill, using electrical energy (Figure 9.18). At the molecular level, water electrolysis is nearly the reverse of the process that occurs in the H_2/O_2 fuel cell. In the fuel cell, the electron sink is oxygen gas, which, along with the added electrons, combines with water to form OH$^-$. In electrolysis, oxygen gas is generated when electrons are pulled out of water. As an electronegative element, oxygen is usually in a -2 oxidation state.

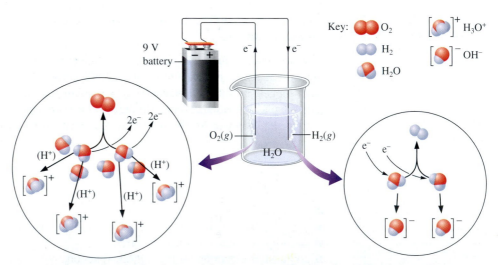

Figure 9.18 A molecular-level view of the electrolysis of water shows the electron flow. In all electrochemical reactions, oxidation occurs at the anode. The anode of an electrolysis apparatus is connected to the positive pole of the battery, attracting electrons and increasing the oxidation number of the substance being electrolyzed. In the case of water, removing electrons generates two H$^+$ ions, each of which immediately associates with a water molecule to form an H_3O^+ ion. With a shortage of electrons, two oxygen atoms join to form O_2. On the other side, electrons are pumped into the cathode. These electrons transfer to water, severing the bond between oxygen and hydrogen. The newly acquired electron becomes associated with hydrogen. Two such liberated hydrogen atoms link to form H_2, also generating OH$^-$ in solution.

By applying electrical energy, however, oxygen is driven uphill to an oxidation state of zero.

By convention, oxidation always occurs at the anode. Thus, in an electrolytic cell containing water, oxygen is generated at the anode. Similarly, hydrogen is produced at the cathode. The combination of electrolysis and a fuel cell has been proposed as a method for energy storage. In such a system, solar energy is used to generate electricity during sunny periods and the electricity is, in turn, used to electrolyze water. The hydrogen and oxygen are stored for use during nonsunny periods. This combination has not seen widespread usage due to safety concerns.

CONCEPT QUESTION   What is produced at the anode and cathode when molten salt (NaCl) is electrolyzed? ◼

Corrosion Prevention

Sacrificial Anode

Rust consumes a significant portion of the gross national product of the United States every year. For example, approximately 25% of the steel produced in the country annually merely replaces steel that has rusted away. One strategy for preventing corrosion uses a sacrificial metal, called a **sacrificial anode.** A sacrificial anode is a metal that has a lower reduction potential (greater activity toward oxidation) than the metal it protects. With a lower reduction potential, the sacrificial metal acts as an anode in a cell formed from the two metals—hence the name *anode.* Because the more active metal is consumed, it is picturesquely termed the *sacrificial* anode.

more — + in table 9.2

CONCEPT QUESTIONS   For the utility pole depicted in Figure 9.2, where is the point of oxidation? Where is the point of reduction? ◼

The location of oxidation and the location of reduction need not be the same. Consider the steel nails depicted in Figure 9.19. Corrosion produces Fe^{2+} ions from metallic Fe in the steel nail. The dark blue color indicates that Fe^{2+} ions are localized near the head and tip of both nails. Additionally, the bent nail has Fe^{2+} localized near the bend. In contrast, the companion reduction reaction produces OH^- ions from molecular oxygen. This reaction occurs everywhere along the length of the nail.

CONCEPT QUESTIONS   Deformation can be divided into stages: elastic, plastic, and fatigue and failure. The bent nail pictured in Figure 9.19 has been stressed to the plastic deformation stage. Plastic deformation begins at defect points. What do you think is the relationship between defects in plastic deformation and corrosion at the bend? Why do the nails also corrode at the head and tip? ◼

Plastic deformation begins at defects—sites where bonds are strained due to atoms being out of place or sites where atoms are entirely missing. Furthermore, the defects propagate as the material is further deformed. What happens to those defects when they reach the edge, the surface, of the material? At the surface, the strained bonds render the surface atoms more prone to oxidation—leading to generation of Fe^{2+} at the bend of the nail (Figure 9.19). For the head and tip, think about how a nail is made. First the metal is drawn into a wire, taking advantage of the ductility of metals. The long wire is cut, at one end into a point for the tip, and at the other end into a blunt cut that is then flattened to create the head. Cutting and flattening produce defects, so the oxidation reaction spreads from points of stress.

Sacrificial anode Metal of lower reduction potential which protect an object made of a metal with a higher reduction potential.

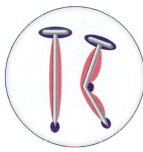

Figure 9.19 Examining the location of corrosion provides insight into the atomic-level events involved in corrosion. In this illustration, the nails are surrounded by a solution containing an indicator that remains colorless except in the presence of OH^- ions, which turn it pink. The surrounding medium also contains $[Fe(CN)_6]^{3-}$ ions. Normally colorless, the $[Fe(CN)_6]^{3-}$ ions turn dark blue in the presence of Fe^{2+} ions. In time, both nails have dark blue patches near the head and tip, and the bent nail has a blue patch at the bend. The length of both nails is surrounded by pink solution, indicating reduction along the entire length of the nails.

In the earth's environment, the prime candidate for reduction is O_2. Oxygen is both quite electronegative and plentiful. Of the reactions listed in Table 9.2, two involve molecular oxygen:

(9.18)
$$O_2(g) + 4H_3O^+(aq) + 4e^- \longrightarrow 6H_2O(l)$$

(9.19)
$$O_2(g) + 2H_2O(l) + 4e^- \longrightarrow 4OH^-(aq)$$

CONCEPT QUESTION Look at Equations (9.18) and (9.19). Which do you think is occurring in Figure 9.19? ■

The indicator incorporated in the solution surrounding the nails (Figure 9.19) turns pink in the presence of OH^- ions. Indeed, nearly the entire length of both nails is surrounded by pink solution. Equation (9.18) includes no OH^- ions, so the reduction reaction must be given by Equation (9.19).

WORKED EXAMPLE 9.10 | *Follow the Electron Flow*

Oxidation and reduction do not have to occur at the same location. Trace the electron flow for the oxidation at the bend of the nail shown in Figure 9.19. What is required for oxidation and reduction to occur at separate sites?

Plan

- Find the sites of oxidation and reduction.
- Electrons flow from the oxidation site to the reduction site.

Implementation

- The appearance of the deep blue color at the bend of the nail indicates that oxidation occurs there. The pink color surrounding the body of the nail indicates that reduction can occur anywhere along the body.
- For electrons to flow from the oxidation site to the reduction site, these two sites must be connected by a conductor (see Figure 9.20 below). Metals are conductors, so they satisfy this requirement.

See Exercises 30, 41, 43, 54.

CONCEPT QUESTION If there was no solution around the nail, it would quickly acquire a positive charge in the vicinity of the bend due to electrons drifting away because of formation of Fe^{2+} ions. In the setup shown in Figure 9.19, the positive charge is neutralized by the $[Fe(CN)_6]^{3-}$ counterions. Why do dry metal objects not corrode despite plentiful O_2 in the air? ■

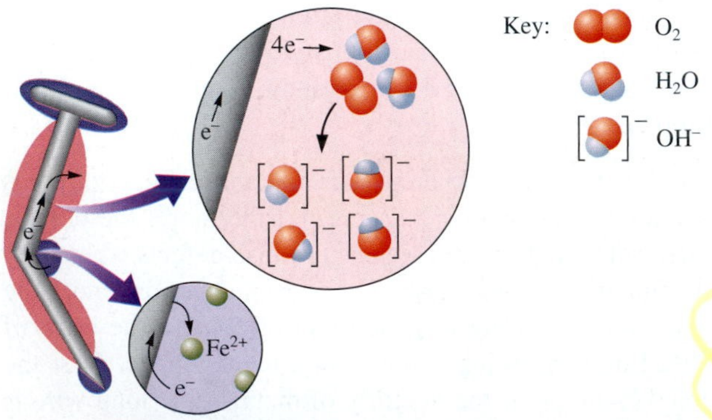

Key: O_2, H_2O, $[OH]^-$

Figure 9.20 The site of oxidation (electron source) must be connected to the site of reduction (electron sink) by a conductive material.

Oxide Coat

Some metals form an oxidized coat that is very tough and protects the underlying metal from further corrosion. Chromium is one of the most common of these elements. Other metals form an oxide coat that flakes off, exposing the underlying metal to further corrosion and ultimately leading to failure of the object. Probably the most familiar example of a flaking oxide is the mixture of FeO and Fe_2O_3: red rust. Incorporation of oxygen in the iron lattice expands the lattice (Figure 9.21), straining the bonds

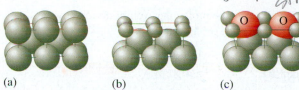

compressive stress

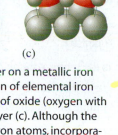

Figure 9.21 Formation of an oxide layer on a metallic iron lattice involves simultaneous conversion of elemental iron (a) into Fe^{2+} ions (b) and incorporation of oxide (oxygen with two additional electrons) in the Fe^{2+} layer (c). Although the Fe^{2+} ions are smaller than the neutral iron atoms, incorporation of the oxide in the lattice causes an expansion of the distance between iron species. In FeO, the iron–iron distance increases by 21%, resulting in a compressive strain as the iron lattice tries to expand but cannot due to bonding to the underlying metal. The compressive strain in Fe_2O_3 is even larger, 29%. In the presence of water, FeO hydrates to form $Fe(OH)_2$, with a P-B ratio of 3.72 indicating a highly strained surface layer. Moisture hastens degradation of rusted objects.

between iron atoms in the oxide layer and iron atoms in the metal resulting in a great compressive stress in the oxide layer. The compressive stress in the surface oxide layer is conveniently evaluated by comparing the density of the metal atoms in the elemental metal to the density in the metal oxide. This ratio is referred to as the **Pilling-Bedworth ratio** or **P-B ratio.** The P-B ratio consists of the volume per metal atom in the oxide divided by the volume per metal atom in the elemental metal:

(9.20) $\quad$ P-B ratio =

$$\frac{\text{(metal oxide molar volume)/(metal atoms per formula unit)}}{\text{metal molar volume}}$$

Imagine the metal atoms remaining in their original positions. The atoms shrink as valence electron density is transferred to oxygen. This newly opened space is occupied by oxygen with its newly acquired electrons. If oxygen fits into the opened space in the metal lattice, then the volume per metal atom is the same in the oxide as in the elemental metal and the P-B ratio is 1. If oxygen is too large to fit into this space, however, the metal lattice tries to expand but cannot due to the bonds to the metal below it. The result is a compressive stress in the oxide layer and a P-B ratio larger than 1. A ratio somewhat larger than 1 is desirable because slight compression of the oxide layer effectively seals the surface. Conversely, if the oxygen is smaller than the space opened by forming the metal ions, then the underlying metal is exposed and oxidation marches through the solid. In this case, the P-B ratio is less than 1 and the oxide coat is porous. The optimal value for the P-B ratio is thus between 1 and 2, corresponding to an oxide film that is nonporous and is protective, because it adheres well to the underlying metal. A P-B ratio larger than 2 indicates that the oxide layer is greatly compressed and will eventually flake off.

①Ideal
if fits perfectly

Pilling-Bedworth ratio, P-B ratio Ratio of the density of metal atoms in the elemental metal to the density of the metal atoms in the metal oxide.

_PB < 1_
⟹ porous surface.

P-B > 1 ⟹ compressive stressives $\quad$ _PB = 1-2 ⟹ OPTIMUM_

_01/26_

WORKED EXAMPLE 9.11 $\quad$ _Assessing Stress in an Oxide_

Rust consists of a mixture of FeO and Fe_2O_3. Which oxide is the source of greater stress in the oxide coat? Rust also hydrates to form $Fe(OH)_2$. How does the stress caused by formation of the hydroxide compare to that from the oxide? Comment on which component is primarily responsible for flaking of rust.

Plan
- Get density data for the various compounds.
- Determine the volume per metal atom in the various compounds.
- Use Equation (9.20):

$$\text{P-B ratio} = \frac{\text{(metal oxide molar volume)/(metal atoms per formula unit)}}{\text{metal molar volume}}.$$

Implementation
- Density:

$$Fe = 7.874 \text{ g/cm}^3 \qquad Fe_2O_3 = 5.24 \text{ g/cm}^3$$
$$FeO = 6.0 \text{ g/cm}^3 \qquad Fe(OH)_2 = 3.4 \text{ g/cm}^3$$

FeO $\quad$ _Fe–Fe → 21%_
distance

_Fe_2O_3_ $\quad$ _Fe↔Fe → 29%_

_$Fe(OH)_3$_ $\quad$ _PB = 3.72_

■ (Illustrated for Fe_2O_3) The molar volume is the molecular weight divided by the density. The molecular weight of Fe_2O_3 is 159.69 g/mol.

$$\text{Molar volume} = \frac{\text{molecular weight}}{\text{density}} = \frac{159.69 \text{ g}}{5.24 \text{ g/cm}^3} = 30.5 \text{ cm}^3/\text{mol}$$

The volume per metal atom is the molar volume divided by the number of metal atoms per formula unit—in this case, 2:

(9.21)   Oxide molar volume per metal atom = 15.2 cm³/mol

■ The molar volume of iron is 7.10 cm³/mol, which is the volume per metal atom. Thus the P-B ratio is

(9.22) $$\text{P-B ratio} = \frac{15.2 \text{ cm}^3/\text{mol}}{7.10 \text{ cm}^3/\text{mol}} = 2.14$$

A similar calculation for FeO results in a P-B ratio of 1.77; for $Fe(OH)_2$, the P-B ratio is 3.72. Thus, Fe_2O_3 is the source of some strain, while the hydroxide is the source of much more. Hydration of the hydroxide, therefore, is primarily responsible for the flaking of rust.

See Exercises 83, 84.

As a surface phenomenon, oxidation is clearly highly dependent on the morphology or structure of the surface, which is in turn greatly affected by the shaping or manufacturing process. If the shaping process creates defects on the surface, such as atoms out of place, missing atoms, or dislocation fronts, atoms at these sites are missing bonding partners. The unshared electrons are then more readily available to electronegative oxygen, and incorporation of oxygen at these sites oxidizes the metal atoms. For example, a bent nail rusts at the head, the tip, and the bend (Figure 9.19) because these areas include a large concentration of defects. Similarly, a dislocation front that runs to the surface forms a line of dangling bonds, strained bonds, or missing atoms. Corrosion follows this line, creating cracks and ultimately leading to mechanical failure of the piece.

Because metals are conductive, the site of this corrosion (the site of oxidation) need not be near the site of reduction. For example, a bolt holding a bridge together can rust despite an intact paint coat covering the head if it is in electrical contact with metal in the remainder of the bridge. Because the area under the head is the most strained, the bolt head may become severed, with a catastrophic result for the bridge.

To prevent or minimize this problem, materials can be incorporated that stop crack propagation by forming covalently bonded, crystalline structures within the metal (high-carbon steel works this way) or by incorporation of metals that form a tough, covalent coat (chrome steel works this way). Post-processing annealing relieves stresses and thus also minimizes corrosion.

9.6   The Thermodynamic Connection

A negative value for the Gibbs free energy indicates that a reaction is spontaneous in the forward direction. Conversely, a positive value indicates that a reaction is spontaneous in the reverse direction. Similarly, the standard electrochemical potential indicates whether the reaction goes forward ($E° > 0$) or in the reverse direction ($E° < 0$) under standard conditions. It is probably not surprising, therefore, to learn that a connection exists between $\Delta G°$ and $E°$. This relationship is given by

(9.23)
$$\Delta G° = -nFE°$$

where n is the number of moles of electrons transferred in the reaction and F is the Faraday constant, named in honor of the nineteenth-century physicist, Michael Faraday, who laid the foundations of our understanding of electricity. The Faraday constant corresponds to the electric charge on one mole of electrons: 96,485 C/mol.

Electrons are the currency of change in an electrochemical reaction. Moving one coulomb of charge, 1/96,485 moles of electrons, between two electrodes that differ by one volt requires one joule of energy. Conversely, spontaneous movement of 1/96,485 moles of electrons between two electrodes with a potential difference of one volt releases one joule of energy to do useful work.

WORKED EXAMPLE 9.12   *The Electrochemistry – Thermodynamic Connection*

Standard cell potentials can be used to determine the free energy change for a chemical reaction. Consider the copper–zinc galvanic cell:

$$Zn \mid Zn^{2+} \parallel Cu^{2+} \mid Cu$$

Determine the potential of this cell and the free energy for the reaction.

Plan
- Write the balanced reaction.
- Determine the half-reactions.
- Determine the cell potential.
- Use Equation (9.23): $\Delta G° = -nFE°$.

Implementation
- $Zn(s) + Cu^{2+}(aq) \rightleftharpoons Zn^{2+}(aq) + Cu(s)$
- The half-reactions are

$$Zn(s) \longrightarrow Zn^{2+}(aq) + 2e^- \qquad E° = 0.7628 \text{ V}$$
$$Cu^{2+}(aq) + 2e^- \longrightarrow Cu(s) \qquad E° = 0.3402 \text{ V}$$

- Cell potential = 0.7628 V + 0.3402 V = 1.103 V.
- $\Delta G° = -nFE°$: $n = 2$ since two electrons are transferred from copper to zinc; $F = 96,485$ C/mol; and $E° = 1.103$ V.

$$\Delta G° = -(2 \text{ mol e}^-) \times (96485 \text{ C/mol e}^-) \times (1.103 \text{ V}) \times (1 \text{ J/C} \cdot \text{V})$$
$$= -212.8 \text{ kJ/mol}$$

Since the Gibbs free energy value is negative, the reaction is spontaneous for reactants and products in their standard states. For the solids, Zn and Cu, the solid is the standard state. For the ions in solution, Cu^{2+} and Zn^{2+}, the standard state is a 1 M solution. In contact with 1 M solutions, then, the reduction of copper by zinc is spontaneous.

See Exercises 85, 86.

Equilibrium and the Dead Battery

We have all had the frustrating experience of having a battery go "dead." From a chemical perspective, what causes a battery to die? The most obvious answer is that electrons are no longer liberated because the anode is consumed. For example, in the two-potato clock the zinc electrode is often consumed. Without zinc to provide electrons, the clock no longer functions. Another frequently encountered example is the magnesium stake used to protect a utility pole. When the stake is consumed, it no longer protects the pole from corrosion. Hence the magnesium stakes need to be replaced on a regular basis. Another cause of a dead battery is that the reaction reaches equilibrium, which means there is no net change and electrons no longer flow.

Electron transfer occurs as part of the drive toward equilibrium. At equilibrium, the forward and reverse reactions occur at the same rate. One consequence of the equal rates is that the number of electrons transferred from reactants to products is equal to the number transferred from products to reactants. The net flow of electrons, like the net flow of atoms, is therefore zero. With no net electron flow, the current is zero and so is the voltage. Indeed, the voltage produced by a battery often differs from the standard value, and the most common reason is that the solution-phase ions are not in their standard state: the ideal 1 M concentration. The relationship between the standard voltage, $E°$, and the voltage, E, is determined by the relationship between the standard Gibbs free energy and the Gibbs free energy:

(9.24) $$\Delta G = \Delta G° + RT \ln Q$$

Substituting Equation (9.23), $\Delta G° = -nFE°$, and the analogous relationship $\Delta G = -nFE$, into Equation (9.24) results in

(9.25) $$-nFE = -nFE° + RT \ln Q$$

or

Nernst equation An equation expressing the cell potential in terms of the concentrations of the substances involved in the cell reaction. $E = E° - (RT/nF) \ln Q$

(9.26) $$E = E° - \frac{RT}{nF} \ln Q$$

Equation (9.26) was developed by the 25-year-old German chemist Walther Nernst in 1889 and is referred to as the **Nernst equation.** At room temperature, the combination $RT/F = 0.025680$ V. Multiplication by 2.303 converts Equation 9.27 into $E = E° - \dfrac{0.0592 \text{ V}}{n} \log Q$. At equilibrium, E is zero and

(9.27) $$E° = \frac{RT}{nF} \ln K$$

WORKED EXAMPLE 9.13   *Cell Potential and the Equilibrium Constant*

The connection between the standard cell potential and the standard Gibbs free energy also provides a method for determining the equilibrium constant for a galvanic cell reaction. The standard Gibbs free energy change for the

$$\text{Zn} \mid \text{Zn}^{2+} \parallel \text{Cu}^{2+} \mid \text{Cu}$$

cell is -212.8 kJ/mol. At room temperature, what is the relationship between the zinc and copper ion concentrations at equilibrium?

Plan
- Write the equation for the reaction.
- Write the expression for K.
- $\Delta G° = -RT \ln K$.

Implementation
- $\text{Zn}(s) + \text{Cu}^{2+}(aq) \rightleftharpoons \text{Zn}^{2+}(aq) + \text{Cu}(s)$
- Since Cu and Zn are both solids, they do not appear in the equilibrium constant.

Thus $K = \dfrac{[Zn^{2+}]}{[Cu^{2+}]}$, and the equilibrium constant gives the concentration ratio for the two ions.

■ $K = \exp\left(-\Delta G^\circ/RT\right) = \exp\left[-\left(\dfrac{(-212.8 \text{ kJ/mol})}{(8.3145 \text{ J/K} \cdot \text{mol})(298 \text{ K})} \times \dfrac{1000 \text{ J}}{\text{kJ}}\right)\right]$

$= 1.99 \times 10^{37}$

The zinc ion concentration is about 37 orders of magnitude larger than the copper ion concentration at equilibrium. If the zinc ion is present in lower concentration, $Q < K$, the reaction proceeds as written, generating a cell potential.

See Exercises 85, 86.

WORKED EXAMPLE 9.14   *Running Down*

The standard potential of the $Zn|Zn^2\|Cu^{2+}|Cu$ galvanic cell is 1.103 V. It is determined that a cell produces 0.67 V. Determine the ratio $[Zn^{2+}]/[Cu^{2+}]$ in this cell.

Plan

■ Write the equation for the cell reaction.

■ Write the expression for the reaction quotient, Q.

■ Use Equation (9.26): $E = E^\circ - \dfrac{RT}{nF}\ln Q$.

Implementation

■ Zinc is oxidized and Cu reduced in the cell reaction.

$$Zn(s) + Cu^{2+}(aq) \rightleftharpoons Zn^{2+}(aq) + Cu(s)$$

■ Cu and Zn are solids, so they do not appear in Q.

$$Q = \dfrac{[Zn^{2+}]}{[Cu^{2+}]}$$

■ $E^\circ = 1.103$ V and $E = 0.67$ V. Rearranging Equation (9.26),

$$Q = \exp\left\{-(E - E^\circ)\dfrac{nF}{RT}\right\} = \exp\left\{-(0.67 - 1.103)\text{ V}\dfrac{(2) \times (96485 \text{ C/mol})}{(8.3145 \text{ J/mol} \cdot \text{K})(298 \text{ K})}\right.$$

$$\left. \times \left(\dfrac{1 \text{ J}}{\text{C} \cdot \text{V}}\right)\right\} = 4.42 \times 10^{14}$$

As more Zn^{2+} ion is produced, the cell potential decreases even more, ultimately falling to zero. At that point, the Zn^{2+} ion concentration has increased to 1.99×10^{37} times the Cu^{2+} ion concentration.

See Exercises 87, 88.

CONCEPT QUESTION As illustrated in Worked Example 9.14, one of the disadvantages of a galvanic cell using aqueous solutions is that the potential falls as soon as the oxidized product begins to appear. Is there a cell design for which such that the potential remains nearly constant until a reactant or product is almost fully consumed? ■

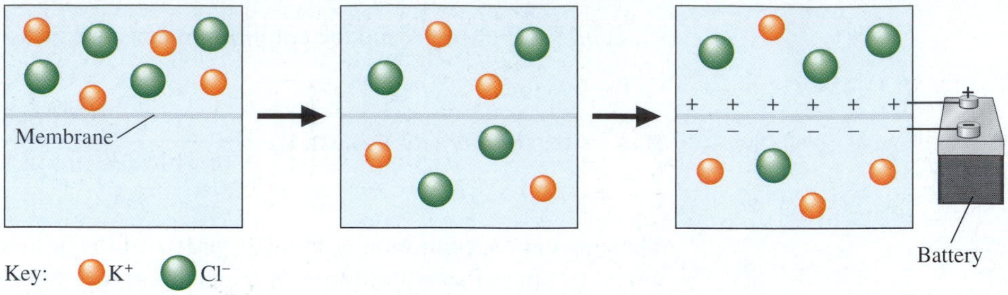

Key: ● K⁺   ● Cl⁻

Figure 9.22 A membrane separates distilled water from a solution containing the ions K^+ and Cl^-. Due to the concentration difference across the membrane, the ions diffuse across the membrane to equalize the concentration. Attaching a battery across the membrane drives positive ions (K^+ in this example) to the negatively charged side and negative ions (Cl^- in this example) to the positive side, so that the concentrations are no longer equal. The imbalance is determined by the applied potential.

Concentration Cells

The relationship between the cell voltage and concentration suggests another method to construct a battery: Bring together two solutions with different concentrations. How can the spontaneous mixing of the solutions be prevented to harness the potential? Imagine a membrane that is permeable to all substances. Starting with ions on one side of the membrane and distilled water on the other (Figure 9.22), entropy results in diffusion across the membrane to equalize the ion concentrations on the two sides. Applying a voltage across the membrane alters this balance, with the negative ions in higher concentration on the positively charged side of the membrane and positive ions in higher concentration on the negative side. Increasing the potential increases the concentration imbalance (Worked Example 9.15). If the potential is suddenly removed, the tendency toward equilibrium provides a force to restore equal concentrations—a force proportional to the removed potential, E. Imbalance of charges is very common in biological systems. Indeed, all cells have a potential across their plasma membranes due to this imbalance, and this membrane potential affects the trafficking of charged substances across the membrane.

WORKED EXAMPLE 9.15   *Driving Ions*

Biological systems use active pumps to create an ion concentration imbalance across a cell membrane. What potential difference is required for the potassium ion concentration inside the cell to be double that outside the cell?

Plan
- Write the equation for transport across the membrane.
- Write an expression for Q.
- Use Equation (9.26): $E = E^\circ - \dfrac{RT}{nF} \ln Q$.

Implementation
- Ion(outside) $\leftrightarrows$ Ion(inside)
- $Q = \dfrac{[\text{Ion (inside)}]}{[\text{Ion (outside)}]}$
- In the absence of an applied potential, concentrations are equal on both sides of the cell; $Q = K = 1$ and $E = E^\circ$. There is no potential difference across the membrane. Applying a potential so that $[\text{Ion(inside)}] = 2 \times [\text{Ion(outside)}]$ leads to $Q = 2$. Potassium is a singly charged ion, so $n = 1$, and

$$E - E° = -\frac{(8.3145 \text{ J/mol}\cdot\text{K})(298 \text{ K})}{(1)(96,485 \text{ C/mol})} \times \frac{1 \text{ C}\cdot\text{V}}{\text{J}} \times \ln(2) = 0.018 \text{ V}$$

A potential difference of just 18 mV creates a factor of 2 difference in ion concentration.

See Exercises 87, 88.

Cell membrane

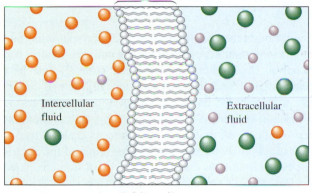

Intercellular fluid

Extracellular fluid

Key: ● Na^+ ● K^+ ● Cl^-

Figure 9.23 As a result of selective transport and permeability, ion concentrations differ from the inside to the outside of cells. Typical concentrations are

	$[Na^+]$	$[K^+]$	$[Cl^-]$	[other]
Inside cell	15 mM	150 mM	10 mM	100 mM
Extracellular	150 mM	5 mM	120 mM	none

Biological membranes do not have miniature batteries connected to them. Instead, the membrane has selective permeability: Some ions can cross while others cannot. The membrane potential is a result of this selective permeability. Intercellular body fluid has approximately the same ion composition as seawater, while the fluid inside the cell is highly concentrated in potassium and depleted in sodium and chloride (Figure 9.23). Additional negative ions inside the cell are too large to penetrate the membrane and other charged substances are embedded in the membrane itself. The 30-fold potassium imbalance leads to a potential of −85 mV at a cell temperature of 17 °C. Imbalance in other ion concentrations mitigates this value so that the measured membrane potential is −70 mV. This small voltage is used in touch-sensitive switches on lamps and other household appliances.

CONCEPT QUESTIONS   How large is the membrane potential generated by the sodium ion imbalance? Is the sodium membrane potential positive or negative? What does the sign signify? ∎

9.7   Metallurgy

Metallury The process of separating a metal from its ore and preparing it for use.

Ore Natural mineral source of a metal: rock.

Metals have a wide range of properties. Although most metals are not found in elemental form in nature, a few are. Some elements are easily separated from other substances found with them, whereas others are very difficult to obtain as pure elements. **Metallurgy** refers to the science of extracting a metal and converting it into a useful form. Metallic elements are often found in an oxidized state, having transferred their valence electrons to other elements. Production of the elemental state of these metals is therefore, a reduction process.

The term **ore** refers to a mineral containing a metal of interest, usually chemically combined with oxygen or sulfur and hence called an oxide or sulfide, respectively. The oxide or sulfide is physically mixed with less useful materials such as sand, soil, or clay. The ore is first separated from the less useful material by density. Most metal sulfides and oxides are more dense than the less useful materials—primarily silicates, metal ions bound with SiO_3^{2-} or SiO_4^{4-}. Once separated, if the metal is a sulfide, it is often first converted to the metal oxide and SO_2 by heating in air, a process called *roasting*. Finally, the metal oxide is reduced with coke, coal that has been heated in an oxygen-poor atmosphere. Heating coke with a metal oxide results in oxidation of the carbon to CO and CO_2 along with reduction of the metal. The energy required in these processes depends on how strongly the metal ion is bound in the ore. The least active metals are loosely bound to oxygen or sulfur, whereas the more active ones are very tightly bound. This chapter's survey of common, useful metals, starts with metals most loosely bound to oxygen.

CONCEPT QUESTION Which metallic elements are unlikely to ever be found as free metals in nature? ■

Coinage Metals: Cu, Ag, and Au

The least active metals are those below hydrogen in the activity series and include copper, silver, and gold.

Copper

Copper has been used by humans for more than 5000 years and still plays an important role in our modern world. The most important copper ores are the sulfides, oxides, and carbonates: chalcocite, Cu_2S; cuprite, Cu_2O; the beautiful green mineral malachite, which is a mixture of $CuCO_3$ and $Cu(OH)_2$; and chalcopyrite, a mixed iron and copper sulfide with the formula $CuFeS_2$. Copper has been mined for several centuries, so the remaining ores are typically less rich in copper. In ancient times, copper was found as a pure metal—boulder of copper—but now the most common source is chalcopyrite.

Several steps are required to extract the 99.9999% pure copper required for electrical wiring from chalcopyrite. The first step utilizes the greater activity of iron compared with copper to extract iron as a silicate ($FeSiO_3$) by heating to 1100 °C.

(9.28) $2CuFeS_2(s) + 3O_2(g) + 2SiO_2(s) \longrightarrow 2CuS(s) + 2SO_2(g) + 2FeSiO_3(s)$

The SiO_2 used in this process usually comes from sand—a readily available material as about 26% of the earth's crust is composed of silicates. At elevated temperatures, CuS is unstable primarily due to the low activity of copper, and CuS readily converts to Cu_2S. Further heating in air separates copper from oxygen and sulfur because Cu is a relatively inactive metal.

(9.29) $2Cu_2S(s) + 3O_2(g) \longrightarrow 2Cu_2O(s) + 2SO_2(g)$

(9.30) $2Cu_2O(s) + Cu_2S(s) \longrightarrow 6Cu(s) + SO_2(g)$

Copper obtained in this way has a high level of impurities. To obtain really pure copper for wiring, it is refined further in an electrolytic process.

Silver

Refuse dumps from the processing of silver found in Asia Minor indicate that silver was separated from lead as early as 3000 B.C. Silver occurs principally as argentite, Ag_2S, and as cerargyrite, AgCl, along with other metals such as lead, zinc, copper, nickel, and gold. One method for obtaining elemental silver is to extract it with cyanide ion, CN^-, with which it forms a strongly bound complex ion, $[Ag(CN)_2]^-$.

(9.31) $AgCl(s) + 2CN^-(aq) \rightleftharpoons [Ag(CN)_2]^-(aq) + Cl^-(aq)$

Elemental silver is obtained from the complex by reaction with a more active metal, often zinc. Zinc transfers its two valence electrons to two silver ions, reducing silver to metallic silver.

(9.32) $2[Ag(CN)_2]^-(aq) + Zn(s) \longrightarrow 2Ag(s) + Zn^{2+}(aq) + 4CN^-(aq)$

Gold

Gold is often found in pure form because it is one of the least active metals. Elemental gold nuggets can be separated from sand by panning—swirling sand and gold in a low-sided pan (Figure 9.24). The sand, primarily silicon oxides, is less dense than gold. Thus the sand swirls over the side of the pan, leaving the denser gold behind. The low activity is the reason for the permanent beauty of gold jewelry and decorations made with gold leaf.

Figure 9.24 Separation of gold from sand and other rock uses the greater density of gold. The less dense materials swish over the side of the pan, leaving dense gold behind.

Because gold is so valuable, it is also recovered from the sludge left behind from refining other metals. Like silver, it can be extracted with CN^- and reduced with Zn.

Structural Metals: Fe and Al

Iron

Iron is discussed here both because it is representative of metals of intermediate activity and because it is one of the more abundant elements in the universe—fourth most abundant on earth after oxygen, silicon, and aluminum. Iron, which is the major component of steel, is an important structural and machine metal.

The most common iron ores are hematite, Fe_2O_3, and magnetite, Fe_3O_4. The latter is a mixed oxide of iron consisting of FeO and Fe_2O_3. Both iron oxides are reduced with coke (carbon), which is partially oxidized to CO and subsequently to CO_2 as the iron oxide is reduced.

$$(9.33) \qquad Fe_2O_3(s) + 3C(s) \longrightarrow 2Fe(l) + 3CO(g)$$

$$(9.34) \qquad Fe_2O_3(s) + 3CO(g) \longrightarrow 2Fe(l) + 3CO_2(g)$$

Limestone, $CaCO_3$, is added to the mixture to react with silicates in the ore. When heated, calcium carbonate decomposes to CaO and CO_2:

$$(9.35) \qquad CaCO_3(s) \longrightarrow CaO(s) + CO_2(g)$$

Calcium is a more active metal than iron, so it reacts with the silicates in preference to iron, producing a material generally referred to as slag:

$$(9.36) \qquad CaO(s) + SiO_2(s) \longrightarrow CaSiO_3(l)$$

Figure 9.25 Iron is produced in a commercial slag furnace.

The slag floats on the surface of the molten iron, which flows out of the slag furnace (Figure 9.25). The slag crust serves the secondary purpose of protecting the molten iron from oxidation by oxygen in the air.

The iron that is produced and used in the manufacture of steel has a high carbon content, resulting in large domains of Fe_3C, which makes it brittle. Removal of part of the carbon along with addition of other metals such as V, W, Ni, and Cr produces a stronger material known as steel.

Aluminum

CONCEPT QUESTIONS Find aluminum on the activity list and in the table of reduction potentials. Will procedures used to refine iron work with aluminum? Why or why not? ■

The debut of aluminum as a useful material occurred only relatively recently in part because of the high activity of aluminum, which makes it difficult to obtain pure aluminum. Aluminum articles are lighter in weight than steel ones because aluminum is less dense than iron, which is the major component of steel. This light weight has led to thousands of uses for aluminum, ranging from structural materials to cooking utensils to airplanes and machine parts. Aluminum is the third most abundant element on earth, constituting about 8% of the crust. However, it is never found free in nature. Aluminum is a somewhat contradictory material in that the major impediment to its earlier widespread usage and one of its most desirable characteristics are closely related: Aluminum readily forms an oxide.

CONCEPT QUESTION The density of aluminum is 2.6989 g/cm³ and the density of Al_2O_3 is 3.97 g/cm³. Is the oxide coat porous, protective, or compressive? ■

Look at the position of Al in the activity list. Metals more active than Al liberate hydrogen from water. Isolating Al therefore requires a different strategy than displacing Al^{3+} ions from solution with a more active metal, because the more active metal would liberate hydrogen from the water solvent and join Al^{3+} as an ion in solution. Heating the oxide with coke or CO will not work because aluminum is too active to be reduced by coke or CO. Isolation of metallic Al is an engineering problem involving simultaneously delivering electrons to Al^{3+} ions and protecting the newly generated metal from exposure to atmospheric oxygen or water.

Oersted first prepared aluminum in 1825 by heating the chloride with potassium amalgam, a solution of potassium and mercury. This strategy for aluminum production utilizes the greater activity of potassium to deliver electrons to aluminum from potassium but does little to protect the newly won aluminum metal from oxygen. As a consequence, the product is not very pure.

Widespread utilization of aluminum awaited an economical method for extracting it from the rocks of the earth. As often happens, a solution to the problem of economical production of pure aluminum was discovered simultaneously, by Hall in the United States and Héroult in France, in 1886. The Hall–Héroult process uses molten cryolite, Na_3AlF_6, as the solvent. Cryolite melts at 1009 °C, a much lower temperature than that needed to melt Al_2O_3 (2045 °C). Due to the strongly electronegative character of fluorine, cryolite dissociate into ions, Na^+ and $[AlF_6]^{3-}$, and conducts electricity well. Electrons are delivered to Al^{3+} in Al_2O_3 dissolved in cryolite by immersing electrodes into the molten bath and turning on the power. The newly formed metallic aluminum is more dense than the molten Na_3AlF_6 and is protected from atmospheric moisture and oxygen by the molten cryolite. The reactions occurring in the electrolysis are complex, but can be summarized as follows:

(9.37)
$$Al^{3+} + 3e^- \longrightarrow Al(l)$$

(9.38) $2O^{2-} \longrightarrow O_2 + 4e^-$

The $[AlF_6]^{3-}$ ion in the solvent ensures that electrons are transferred to aluminum from oxygen in the oxide rather than a component of the solvent because fluorine is more electronegative than oxygen.

   The search for a less energy-intensive procedure for production of aluminum continues. Given the importance of aluminum in modern society, such a procedure would have great economic impact. Because of the high energy requirements for the electrolysis step used in the Hall–Héroult process (Figure 9.26), it is economically profitable to recycle aluminum.

The Most Active Metals: Na and K

CONCEPT QUESTIONS Can aqueous salt solution electrolysis be used to obtain elemental Group IA or IIA metals? Why or why not? ∎

The problem of protecting a newly produced metal from oxidation becomes even more challenging with the most active metals such as sodium or potassium. Both readily react with oxygen and nitrogen in the air and so must be stored under oil or kerosene. Both were originally isolated by Humphrey Davy in 1807 from their hydroxides, potash (KOH) and caustic soda or lye (NaOH), by electrolysis. The most abundant source of potassium is potash, and electrolysis of potash is still the method of choice for production of potassium. Sodium is currently obtained from salt, NaCl. Salt is obtained from a seemingly endless supply by evaporation of seawater. NaCl melts at 810 °C, so the procedure consists of heating salt to 810 °C and electrolyzing the molten salt.

(9.39) $2Na^+Cl^-(l) \xrightarrow{\text{Electrolysis}} 2Na(l) + Cl_2(g)$

The chlorine gas bubbles out of the mixture and is trapped for use in manufacture of pharmaceuticals and plastics, such as polyvinyl chloride (PVC) plumbing.

CONCEPT QUESTION At which electrode is Cl_2 generated? Trace the electron flow in this electrolysis. ∎

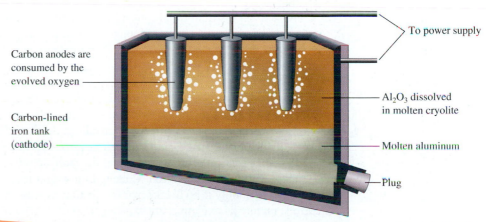

Figure 9.26 Schematic of the apparatus for the electrolytic production of aluminum showing molten aluminum sinking to the bottom of the tank. The molten cryolite protects the newly won aluminum from attack by oxygen in the air.

Checklist for Review

KEY TERMS

oxidation (p. 306)
reduction (p. 306)
redox (p. 306)
oxidation state (p. 308)
oxidation number (p. 309)
activity series (p. 313)
half-reaction (p. 315)
electrode (p. 316)
salt bridge (p. 316)
galvanic cell (p. 317)
anode (p. 317)
cathode (p. 317)
cell potential, E (p. 318)

cell voltage, V (p. 318)
electromotive force,
  emf (p. 318)
standard reduction poten-
  tial, $E°$ (p. 318)
fuel cell (p. 324)
sacrificial anode (p. 327)
Pilling-Bedworth ratio,
  P-B ratio (p. 329)
Nernst equation (p. 332)
metallurgy (p. 335)
ore (p. 335)

KEY EQUATIONS

P-B ratio =
$$\frac{\text{(metal oxide molar volume)/(metal atoms per formula unit)}}{\text{metal molar volume}}$$

$$E = E° \text{ (reduction)} - E° \text{ (oxidation)}$$

$$\Delta G° = -nFE°$$

$$E = E° - \frac{RT}{nF} \ln Q$$

Chapter Summary

Electron transfer reactions are very common. Generally, metallic elements react by giving up electrons—that is, by being oxidized. On the other side of the periodic table, the nonmetallic elements react by taking on additional electrons to attain filled s and p valence subshells. However, a closer look at electron transfer reactions reveals a more complex picture. Electrons tend to be associated with elements of greater electronegativity. Among metals, the tendency to transfer electrons from one element to another is the basis for many modern devices, including batteries and fuel cells. The driving force for electron transfer derives from the potential difference for an electron associated with one element compared with another, known as the standard reduction potential.

An uneven distribution of ions also results in an electrical potential. This electrical potential is found extensively in biological systems. Every cell has a membrane potential of tens of millivolts, for example. These potentials are involved in many signaling processes, including the response of nerve cells.

There is a down side to the electron transfer process known as corrosion, which leads to weakening of structures such as bridges, underground pipes, and utility poles. In environmental corrosion, the oxidant is often molecular oxygen, which is reduced to water or hydroxide. Several methods have been developed for inhibiting the inevitable march of elemental metals toward their oxidized forms. Galvanic protection of steel couples the oxidation of iron with that of the more active metal zinc, with the result that zinc is oxidized rather than iron. Coating iron with a less active metal protects the iron from corrosion only if no breaks mar the coating. Breaks in the coating cause the iron to corrode rapidly as oxidative attack on any part of the coating results in electron transfer from iron to the oxidizer.

Metallurgy is the process of obtaining elemental metals from their ores. The exact method chosen for each metal is related to the activity of that metal, with the least active metals being either present in nature as the elemental metal or easily obtained from the ore in which it is found. The most active metals are never found in elemental form and are difficult to extract from the minerals in which they are found.

KEY IDEA

Electrons flow spontaneously from active metals to those of greater reduction potential. Electrolytic processes reverse this flow by application of electric power.

CONCEPTS YOU SHOULD UNDERSTAND

- In the absence of an applied potential, ions are distributed uniformly. A concentration gradient corresponds to a force driving ions to the region of lower concentration. The potential is determined by the ratio of ion concentrations in different parts of a system.

OPERATIONAL SKILLS

- Assign oxidation numbers to the elements in a compound, and use oxidation numbers to predict formulas (Worked Examples 9.1–9.4).

- Identify redox reactions, the element oxidized, and the element reduced (Worked Examples 9.5, 9.6).
- Determine the voltage produced by a battery or galvanic cell (Worked Example 9.7).
- Identify anode and cathode reactions for batteries, fuel cells, and electrolytic cells. Trace electron and ion flow in these devices (Worked Examples 9.8–9.10).
- Evaluate the oxide coat (Worked Example 9.11).
- Relate the Gibbs free energy to the cell potential (Worked Examples 9.12–9.14).
- Calculate the potential across a cell membrane (Worked Example 9.15).

Exercises

A blue exercise number indicates that the answer to that exercise appears at the back of the book.

■ SKILL BUILDING EXERCISES

9.2 Oxidation Numbers

1. What is the oxidation number of Cl in HOCl?

2. Explain why the oxidation number of oxygen in HOOH is -1 rather than the usual -2. Explain why the oxidation number of hydrogen in NaH and CaH_2 is -1 rather than the usual $+1$.

3. Indicate the oxidation number of N in the following: N_2O, NO, N_2O_3, NO_2, N_2O_4, N_2O_5, NH_3, N_2H_2, HNO_3, HNO_2.

4. What is the oxidation number of oxygen in OF_2?

9.2 Periodic Trends

5. Write the valence configuration for Sn and its $+2$ and $+4$ ions. The $+4$ ion is more common. From the electronic configuration, indicate why the two oxidation states are $+2$ and $+4$ and why the $+4$ state is more common.

6. How many valence electrons does each of the halogens have? How many valence electrons does oxygen have? What oxidation state do you expect for the halogens? For oxygen? Justify your answers.

7. What is the most electronegative element in the second row of the periodic table? What is the least electronegative element in this row?

8. What is the most electronegative element in Group VA? In Group VIA? In Group VIIA?

9. What is the least electronegative element in Group IA?

10. Account for the following properties of Group IA elements.
   a. They are all powerful reducing agents.

   b. They all have the lowest first ionization potential of their respective periods.
   c. They exist in compounds and ions only in the $+1$ ionization state.

11. Account for the following properties of Group IIA elements.
   a. Their compounds are all ionic unless combined with another Group IIA metal.
   b. They exist in compounds and salts primarily as $+2$ ions.

12. What is the electronic configuration of Al? How does this configuration explain the fact that in most aluminum compounds, aluminum is in a $+3$ oxidation state?

13. In which section of the periodic table are the strongest oxidizing agents found? In which section are the strongest reducing agents found? Where are the least active metals found?

14. What is the electronic structure of Fe? Explain why common ionization states of Fe are $+2$ and $+3$.

9.2 Recognizing Redox Reactions and Predicting Formulas

15. Fluorine forms four binary compounds with iodine. List at least three of them.

16. Chlorine forms several binary compounds and ions with oxygen. Indicate the oxidation number of chlorine in each of the following: Cl_2O, ClO^-, ClO_2^-, ClO_3^-, ClO_4^-.

17. Identify which of the following are redox reactions.
   a. When heated, limestone ($CaCO_3$) decomposes to CaO and CO_2.
   b. When heated, HgO decomposes to metallic Hg and O_2 gas.

18. Assign oxidation numbers to each of the elements in the reaction of Br_2 with I^-.

$$Br_2(g) + 2I^-(aq) \longrightarrow I_2(s) + 2Br^-(aq)$$

Identify the electron source and the electron sink.

19. Indicate the formula of the product expected when each of the following is heated in excess oxygen: Zn, Ti, In, Cd, Sr, Cs, Sn, Pb.

20. Indicate the formula expected when each of the following is heated in an atmosphere of excess chlorine: Ge, Ga, Rb, Ti, Cu, Co, V.

9.3 Activity Connected with the Periodic Table

21. Explain why the halogens never occur in nature in elemental form.

22. Which metals are commonly found in elemental form? Why?

23. Gold is often found in elemental form, but sodium never is. Explain why.

24. Explain the following: K and Cu each have a single s valence electron, yet K is much more active than Cu. Ca and Zn each have two s valence electrons and both form $+2$ ions, yet Ca is more active than Zn.

9.3 The Activity Series

25. Indicate the product(s), if any, for the following potential reaction.

$$Ag^+(aq) + Ni(s) \longrightarrow$$

26. Indicate the product(s), if any, for the following potential reaction.

$$Au^{+3}(aq) + Cu(s) \longrightarrow$$

27. Complete and balance the following equation.

$$Fe_3O_4(s) + Al(s) \longrightarrow$$

28. Will Ni^{2+} oxidize Ag? Will it oxidize Al? Give the net ionic equation for any reaction that does occur.

29. What do you expect to happen if a piece of lead is put in a beaker with Sn^{2+}? If a piece of copper is put in a beaker with Ag^+?

30. If copper and steel (Fe) are in contact (as in a domestic water system), will corrosion occur? If so, which metal corrodes?

31. Which of the following can exist in contact with water without reacting with water? For any that react, give the balanced reaction equation.
    a. calcium metal, Ca
    b. aluminum metal, Al
    c. lead metal, Pb

32. Which of the following can exist in contact with an acid solution without reacting? For any that react, give the balanced reaction equation.
    a. tin metal, Sn
    b. aluminum metal, Al
    c. zinc metal, Zn
    d. silver metal, Ag

33. Which do you expect to be a stronger oxidizing agent, Co^{3+} or Cr^{3+}. Explain your answer.

34. Zinc is used to galvanize iron. List three other metals that could also be used to galvanize iron.

35. What would you expect to happen if copper and aluminum wiring are in contact?

36. Brass is an alloy of Cu and Zn in a 2 : 1 ratio. Explain why brass does not develop a green coat as copper does.

37. Stainless steel contains about 10% chromium with iron. Explain why stainless steel is more rust resistant than iron.

38. Do you expect zinc to reduce cadmium ion? Iron to reduce mercury ion? Zinc to reduce lead ion? Magnesium to reduce potassium ion?

39. Monel is an Fe/Cu/Ni alloy. Which of these three metals oxidizes when monel corrodes?

40. Write the possible oxidation and reduction reactions that may occur when a piece of calcium metal is immersed in each of the following.
    a. water
    b. HCl
    c. KOH

41. Underground steel pipes are sometimes protected by connecting them to a block of Mg or Al with an insulated copper wire. Comment on how this system inhibits deterioration of the pipe. Why is the copper connecting wire insulated?

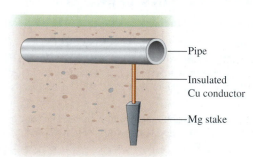

Pipe

Insulated Cu conductor

Mg stake

42. Predict the product(s) of the reaction of an alkali metal and HCl. Predict the product(s) of the reaction of an alkali metal with gaseous Cl_2.

43. Explain why the iron struts inside the Statue of Liberty quickly corroded when the asbestos pads separating them from the copper cladding disintegrated.

44. Predict whether the alkali metals can be prepared by reduction of the oxide with metallic aluminum. Justify your answer.

9.3 Galvanic Cells

45. Explain the difference between an oxidation reaction and a reduction reaction. Which occurs at the anode? Which occurs at the cathode?

46. Dental amalgams are composed of Ag_2Hg_3, Ag_3Sn, and Sn_xHg ($7 \leq x \leq 9$). Consider the following half-reactions:

$$3Hg_2^{2+}(aq) + 4Ag(s) + 6e^- \longrightarrow 2Ag_2Hg_3(s) \quad E° = 0.85 \text{ V}$$
$$Sn^{2+}(aq) + 3Ag(s) + 2e^- \longrightarrow Ag_3Sn(s) \quad E° = -0.05 \text{ V}$$

a. Biting on a piece of aluminum foil produces an uncomfortable sensation. What is the potential of the cell formed by aluminum and the filling?

b. Does touching a filling with a stainless steel fork produce the same sensation as biting on aluminum foil? Why or why not?

9.3 Battery Potential and the Electromotive Series

47. Calculate the value of $E°$ for each of the following.
    a. $I_2(s) + Mg(s) \rightarrow Mg^{2+}(aq) + 2I^-(aq)$
    b. $Ag(s) + Fe^{3+}(aq) \rightarrow Ag^+(aq) + Fe^{2+}(aq)$
    c. $2Mg(s) + O_2(g) + 2H_2O(l) \rightarrow 2Mg(OH)_2(s)$
    d. $Mn(OH)_2(s) + 2Ag + S^{2-} \rightarrow Ag_2S + Mn + 2OH^-$

48. Choose an oxidizing agent for each of the following processes.
    a. $Fe(s) \rightarrow Fe^{3+}(aq) + 3e^-$
    b. $Fe(s) \rightarrow Fe^{2+}(aq) + 2e^-$ but not
       $Fe(s) \rightarrow Fe^{3+}(aq) + 3e^-$
    c. $Fe^{2+}(aq) \rightarrow Fe^{3+}(aq) + e^-$
    d. $2F^-(aq) \rightarrow F_2(g) + 2e^-$

49. A mercury battery is based on the oxidation of zinc and reduction of HgO to Hg. Determine the anode and cathode reactions and the cell potential.

50. In principle, a battery could be made of aluminum metal and chlorine gas. Write the balanced reaction for this battery. Which reaction occurs at the anode? What is the battery potential?

51. A miniature battery consists of Zn, a KOH paste, water, Hg, and HgO. Zinc and KOH are consumed, and Hg and K_2ZnO_2 are formed. Indicate the half-reactions and the overall reaction for this battery.

52. The NiCd battery is based on the following reaction:

    $Cd(s) + NiO_2(s) + 2H_2O(l) \longrightarrow Cd(OH)_2(s)$
    $+ Ni(OH)_2(s)$

    What are the two half-reactions that go into this reaction? What is the potential of this battery?

53. In a mixture of NaI, NaCl, NaBr, and NaF, what oxidizing agent could be used to oxidize iodide to elemental iodine without oxidizing the other species? After iodide is oxidized, what could be used to oxidize only the bromide? Which is next, the chloride or the fluoride? What could be used to oxidize it?

54. Originally, the Statue of Liberty consisted of copper plates over steel ribs. When it was restored, the steel ribs were replaced with stainless steel, and Teflon spacers were installed between the ribs and the copper cladding. Comment on two ways in which this structure will improve the longevity of the statue.

9.4 Balancing Redox Reactions

55. Complete and balance the following reactions.
    a. $Cu(s) + Ag^+(aq) \rightarrow$
    b. $Na(s) + H_2O(l) \rightarrow$
    c. $Ag(s) + Zn^{2+}(aq) \rightarrow$

56. For each of the following redox reactions, determine whether the reaction is balanced. Balance any equations that are not balanced.

    a. $Al(s) + Fe^{2+}(aq) \rightarrow Al^{3+}(aq) + Fe(s)$
    b. $SO_2(g) + O_3(g) \rightarrow SO_3(g) + O_2(g)$
    c. $Zn(s) + Ni^{2+}(aq) \rightarrow Zn^{2+}(aq) + Ni(s)$
    d. $Ca(s) + H_2O(l) \rightarrow Ca(OH)_2(s)$

57. Write the half-reactions for the oxidation of coke (C) to CO and to CO_2 by O_2. Write the overall reaction for each of these half-reactions.

58. Production of lead from the ore galena, PbS, involves several steps. First PbS is oxidized to PbO and SO_2 by molecular oxygen. Write the balanced equation for oxidation of PbS. Which element is oxidized? Which is reduced?

59. After PbO is produced in the reaction described in Exercise 58, the oxide is reduced with coke, elemental carbon, to produce elemental lead and carbon dioxide, CO_2. Write the balanced equation for this process.

60. The dangerously toxic substance phosgene, $COCl_2$, is made by mixing carbon monoxide, CO, with chlorine, Cl_2, in the presence of sunlight. Write the balanced equation for this reaction. Assign oxidation numbers to the elements in the reactants and products.

61. Write the oxidation half-reaction generating the following ions in acid solution: Al^{3+}, Cr^{3+}, and Fe^{3+}. Write the reduction half-reaction for the liberation of H_2 from acid solution. Write the balanced equation for each of these reactions.

62. Write a balanced equation for the reaction of the Group IA metals with water.

63. Write a balanced equation for each of the following reactions.
    a. $Rb(s) + H_2O(l)(\text{cold!}) \rightarrow$
    b. $Ba(s) + H_2O(l)(\text{cold!}) \rightarrow$
    c. $Cu(s) + Fe^{2+}(aq) \rightarrow$

64. Write the oxidation and reduction half-reactions and balance the following equation.

    $Cl_2(g) + Br^-(aq) \longrightarrow$

65. Dropping a piece of K into a beaker of water results in a rapid reaction and a fire. Write the balanced equation for reaction of K with water. Indicate the source of the fire. Write the balanced equation for the fire reaction.

66. Identify the half-reactions in Equation (9.29).

    $2Cu_2S + 3O_2 \longrightarrow 2Cu_2O + 2SO_2$

67. Identify the half-reactions in Equation (9.30).

    $2Cu_2O + Cu_2S \longrightarrow 6Cu + SO_2$

68. Identify the half-reactions in Equation (9.32).

    $2[Ag(CN)_2]^-(aq) + Zn(s) \longrightarrow 2Ag(s) + Zn^{2+}(aq)$
    $+ 4CN^-(aq)$

69. Identify the half-reactions in Equation (9.33).

    $Fe_2O_3(s) + 3C(s) \longrightarrow 2Fe(l) + 3CO(g)$

70. Identify the half-reactions in Equation (9.34).

    $Fe_2O_3(s) + 3CO(g) \longrightarrow 2Fe(l) + 3CO_2(g)$

9.5 Batteries

71. The combination of chlorine gas and zinc has been proposed as the basis for a battery. Write the reactions involved in such a battery. Identify the anode and cathode reactions. What is the potential of this proposed battery?

72. In principle, a battery can be made from aluminum metal and chlorine gas. Write the reaction involved in such a battery. Identify the anode and cathode reactions. What is the potential of this proposed battery? Which has the larger potential, this $Al|Al^{3+}||Cl_2|Cl^-$ battery or a similar $Zn|Zn^{2+}||Cl_2|Cl^-$ battery?

9.5 Fuel Cells

73. To provide nearly instant backup electrical power in case of a power outage, the telephone company uses a device called an aluminum–air battery. Oxidation of aluminum occurs at the anode when the device is filled with water. Suggest the reaction for the cathode. Which ion is a good candidate for completing the circuit? How would you design this system to include the required ions?

74. Design a fuel cell that uses CO produced from incomplete combustion and O_2 as fuel. Sketch the circuit, including the electron flow. What are the two half-reactions involved in the fuel cell? Give the overall reaction.

9.5 Electrolysis

75. Write the electrode reactions for the electrolytic production of magnesium metal from molten $MgCl_2$. Draw a schematic of the electrolytic cell, indicating the electron flow.

76. Write the electrode reactions and trace the flow of electrons in the electrolytic production of fluorine from molten CaF_2.

77. Calcium is obtained commercially by electrolysis of molten $CaCl_2$. How many coulombs are needed to produce 5.0 g of Ca metal? If the circuit supplies 10 A, how long will it take to produce 5.0 g of Ca metal?

78. Commercial production of aluminum uses an electrolytic process. A commercial electrolysis unit operates at 5.0 V. How many kilowatt-hours (kWh) are required to produce 1 lb of aluminum? If electricity costs 9 cents per kWh and the unit is 45% efficient, what is the cost for 1 lb of aluminum?

9.5 Corrosion Prevention

79. Do you expect a piece of cold worked metal to be more or less corrosion resistant than an annealed piece? Explain.

80. Steel may be protected from corrosion by coating it with Zn. Would a Cu coat be similarly effective? Why or why not?

9.5 Sacrificial Anode

81. What are the advantages and disadvantages of using zinc rather than aluminum in galvanizing iron?

82. Asphalt shingles at the edge of a roof are usually supported by an aluminum drip edge, both to support the shingles and to inhibit water from running down the face of the building. Nails used to attach shingles are often galvanized steel (Zn-coated Fe). What happens to the function of Zn when the galvanized nail is driven through the aluminum drip edge?

9.5 Oxide Coat

83. In the *Wizard of Oz*, the tin woodsman asks Dorothy to oil his joints so that he can move again.
   a. Explain why, if the woodsman were really tin, this action would not have been necessary.
   b. If the tin woodsman was composed of the same material as a tin can, oil would help. Explain why.
   c. Tin does form an oxide coat. Evaluate this coat for protection.

84. Stainless steel contains chromium and vanadium as minor components. Which is the anode when either of these metals couples with iron? Evaluate the oxide coat of each of these elements for prevention of rust.

9.6 Equilibrium

85. Silver is a by-product of the extraction of lead from its ore. Calculate the equilibrium constant and $\Delta G°$ for reduction of silver ion by lead.

86. Ammonia is a molecule essential to life on earth. At this time, ammonia is synthesized from atmospheric nitrogen by the very efficient, FeS-containing nitrogenase enzyme. Before this enzyme evolved, it is believed that ammonia may have been produced with the inorganic material, FeS, in the reaction

$$N_2 + 3H_2S + 3FeS \Longrightarrow 2NH_3 + 3FeS_2$$

   a. Determine $\Delta G°$ for this reaction at room temperature (25 °C).
   b. Determine the cell potential, $E°$.
   c. Determine the equilibrium constant.
   d. Is ammonia production increased or decreased at higher temperatures?

9.6 Concentration Cells

87. Nerve cells use sodium and potassium ion pumps extensively, along with selective ion permeability. A nerve cell is subjected to a 20-mV pulse, changing the membrane potential from −70 mV to −50 mV. In response to this pulse, the structure of the membrane changes and it becomes permeable to sodium ions but not to the larger potassium ions. How does the sodium ion concentration difference across the membrane change in response to the lowered potential? (Body temperature is 38 °C.)

88. In plants, illumination of chloroplasts releases H^+, generating a pH gradient that drives ATP synthesis. Illumination drops the pH to about 5 from the resting value of 8. Determine the change in membrane potential triggered by this drop in pH.

9.7 Metallurgy

89. Chromium metal may be prepared from its oxide by the Goldschmidt process:

$$Cr_2O_3(s) + 2Al(l) \longrightarrow 2Cr(l) + Al_2O_3(s)$$

Can metallic zinc be prepared from its oxide by a similar process? Explain your answer. Give the balanced equation for this reaction if it occurs.

90. Three reactions important in the metallurgy of iron are
   a. $FeO(s) + CO(g) \rightarrow Fe(s) + CO_2(g)$
   b. $CaCO_3(s) \rightarrow CaO(s) + CO_2(g)$
   c. $C(s) + CO_2(g) \rightarrow 2CO(g)$

Identify those that are redox reactions. Indicate the oxidation and reduction half-reactions.

91. Write the half-reaction taking place at each electrode when molten KCl is electrolyzed.

92. Suggest a method to obtain Mn from an ore containing Mn_2O_3. Can Mn be reduced with coke? With CO?

■ CONCEPTUAL EXERCISES

93. Draw a diagram indicating the relationship among the following terms: oxidation, reduction, redox, anode, cathode, salt bridge, cell potential. All terms should have at least one connection.

94. Draw a diagram indicating the relationship among the following terms: corrosion prevention, Pilling-Bedworth ratio, sacrificial anode, galvanic protection, tin cans, protective oxide, porous oxide. All terms should have at least one connection.

95. A box of baking soda ($NaHCO_3$) is a handy and effective fire extinguisher in the kitchen. At high temperature, baking soda decomposes, generating CO_2 gas in an endothermic reaction ($\Delta H = 135.6$ kJ/mol). Baking soda is said to be double-acting in extinguishing the fire. Explain this statement.

96. Magnesium is a very active metal that can be oxidized with oxygen to MgO, nitrogen to Mg_3N_2, water to $Mg(OH)_2$ and CO_2 to MgO. Write the redox reaction associated with each of these four oxidations. As a result of these processes, a magnesium fire is particularly hazardous. Explain why water or a CO_2 fire extinguisher should not be used with a magnesium fire. How would you extinguish a magnesium fire?

■ APPLIED EXERCISES

97. Anyone who has encountered the odor of a skunk remembers the pungent fragrance for a long time—even longer if the odor resulted from an encounter between a pet and the skunk. Two compounds responsible for the skunk odor are the thiols H_7C_4SH and $H_{11}C_5SH$. One remedy for the family pet is to bathe it in a mixture of hydrogen peroxide (HOOH) and sodium bicarbonate ($NaHCO_3$). Chemically, the thiols are oxidized to the disulfides $H_7C_4SSC_4H_7$ and $H_{11}C_5SSC_5H_{11}$. What is the oxidizing agent? Write the half-reaction for the two oxidation and the reduction reactions. Write the overall reaction for deodorizing the family pet.

98. If left exposed to air, copper, brass, and bronze items acquire a green patina and less aesthetic black patches. These can be removed by rubbing the object with vinegar and salt. Explain the processes involved in generation of the patina, and explain how the patina and patches are removed. Is either process a redox process? Helpful information: Vinegar contains acetic acid, CH_3COOH; $CuCO_3 \cdot Cu(OH)_2$ is green; $2CuCO_3 \cdot Cu(OH)_2$ is blue; and CuO is black.

■ INTEGRATIVE EXERCISES

99. Methanol has been proposed as the basis for a fuel cell. Determine the equilibrium constant for oxidation of methanol to CO_2 and water at 25 °C. Calculate the potential of a methanol fuel cell. One issue with use of methanol in a fuel cell is production of formaldehyde (CH_2O) from partial oxidation. Formaldehyde has been associated with several health concerns. Determine the relationship between the cell potential from oxidation of methanol to formaldehyde and that from oxidation to CO_2 and water.

100. The spontaneous reaction of O_2 to form water is used extensively in biological systems to drive nonspontaneous processes in the cells. At pH 7, the standard reduction potential for O_2 to water, $E^{\circ\prime}$, is 0.82 V. (Note that this value differs from E° due to the difference in $[H_3O^+]$ relevant to physiological systems; hence the prime on the symbol for the standard reduction potential.) Oxygen reduction is coupled to oxidation of NADH (nicotinamide adenine dinucleotide) to NAD^+. The standard physiological reduction potential for NAD^+ is -0.32 V. Does coupling the reduction of oxygen with the oxidation of NADH produce a spontaneous reaction? Determine $E^{\circ\prime}$, $\Delta G^{\circ\prime}$, and K for this reaction.

Chapter 10

Coordination Chemistry: Gems, Magnetization, Metals, and Membranes

Gemstones from Sri Lanka include diamonds and highly colored gems. Color in gemstones results from transition metal impurities caught in a clear and colorless matrix.

CONCEPTUAL FOCUS

■ Connect color with interaction energy and equilibrium.
■ Connect magnetic properties with electron spin and *d*-orbital interaction.

SKILL DEVELOPMENT OBJECTIVES

■ Interpret color to evaluate relative interaction strength (Worked Example 10.1).
■ Relate the *d*-orbital splitting to the molecular geometry (Worked Example 10.2).
■ Determine metal coordination number from ligands and calculate metal oxidation state from the charge on the complex (Worked Example 10.3).

■ Determine concentration of metal complexes using equilibrium constants and use equilibrium constants to quantitatively evaluate interaction strengths (Worked Example 10.4).
■ Correlate electron pairing with the magnetic properties of transition metal complexes, and particularly those of Fe(II) and Fe(III) (Worked Example 10.5).
■ Explain technological applications of magnetism (Worked Example 10.6).

Gemstones have fascinated humankind throughout the eons. From colorless diamond, cool to the touch but sparkling with an inner fire, to intensely colored ruby and emerald, gems were considered to be mysterious and mystical. Though still alluring, the atomic–molecular interactions key to the beauty of gemstones have been unraveled in recent times (Figure 10.1).

The atomic–molecular interactions in colored gemstones are due to a new kind of bond, one in which both electrons of the bond originate with one of the two atoms involved in the bond. Because it is a shared pair of electrons, the bond is still referred to as a covalent bond. However, to distinguish it from a bond formed from one electron from each atom, this new kind of covalent bond is called a **coordinate-covalent bond.** A metal ion with its concentrated positive charge is surrounded by ions or molecules having lone pairs of electrons. The negative charges of the electrons around the positive metal ion alter the energy of electrons remaining in the metal ion. Unpaired electrons in *d* orbitals result in important magnetic interactions. Direct interaction among *d*-orbital electrons in neighboring atoms leads to the stronger, permanent form of magnetism used as the basis for disk drives, CDs, and audio tapes (Figure 10.1). The same coordinate-covalent interactions are often involved in transport of ions across cell membranes. The specific case of calcium ion transport is examined in this chapter.

Coordinate-covalent bond
A bond consisting of two electrons from one bonding molecule/ion/atom and none from the second.

10.1 Gems

The relationship between light and energy provides a visible probe for the interactions in gemstones. Since the interactions in gemstones have much in common with all coordinate-covalent interactions, the general features of coordinate-covalent bonding are examined in conjunction with gemstone color. The environment around the metal ion is the key to these interactions. In ruby, for example, this metal ion environment is determined by the solid structure of host. Thus both the relationship between color and energy and the solid structure of ruby are needed to explain the color of ruby.

Gems and other coordinate-covalent materials are transparent, but colored. Their transparency is due to light transmitted through the material. Not all colors make it through, however. Absorbed colors are filtered out, leaving the complementary color. Consider ruby. To appear red, green and blue light must be removed from white light; that is, green and blue light are absorbed. Electrons are responsible for this absorption. To absorb blue and green light of energy 2.5–3.0 eV, there must be an electron state 2.5–3.0 eV above the ground state. The composition and structure of ruby hold the key to finding these states.

Figure 10.1 The beauty of gemstones results from the crystal structure and their color from the interplay between light and electrons. Magnetic properties — the basis for audio tapes — are also determined by electrons in materials.

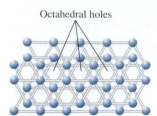

Octahedral holes

Figure 10.2 Two layers of a hexagonal lattice illustrate the spaces between atoms, the *holes*. Atoms in the second layer sit in the triangular hollows of the first layer. Only half of the triangular hollows are occupied.

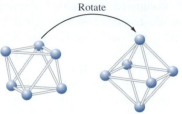

Rotate

Figure 10.3 In the first view of an octahedral solid (left), the solid sits on a triangular face. The second view shows the eight faces: four above the equatorial plane and four below it. The name "octahedron" comes from the eight faces of this solid structure. An octahedron has six vertices, and the six atoms defining this structure sit at the vertices.

Figure 10.4 A tetrahedral solid structure is defined by the four atoms consisting of three atoms in the first layer with the one atom of the second layer sitting in the triangular hollow of the first layer. The name "tetrahedron" is in reference to the four faces of the solid formed with the four vertices.

Holes Spaces between atoms in a hard-sphere model of a solid.

Octahedral holes A cavity in a crystalline lattice that is surrounded by six atoms. Six atoms define the vertices of an octahedral solid.

Tetrahedral holes A cavity in a crystalline lattice that is surrounded by four atoms.

Ruby is composed of an Al_2O_3 solid, called corundum, that has a small concentration, 0.5%, of Cr^{3+} ions. The corundum solid structure consists of a hexagonal array of oxygen atoms with aluminum between the layers. Examining two layers of the hexagonal oxygen lattice reveals two types of spaces, called **holes,** between the layers (Figure 10.2). (Physical holes are spaces in contrast to holes in semiconductors that result from electron shortage and are charged.) One type of hole is surrounded by six atoms: three atoms in the first layer and three in the second. These six atoms define the vertices of an octahedral solid (Figure 10.3) and are referred to as **octahedral holes.** The other type of hole is surrounded by four atoms: three in the first layer and one in the second. These holes are called **tetrahedral holes** due to the four-sided figure defined by these vertices (Figure 10.4). In corundum, aluminum sits in two-thirds of the octahedral holes.

CONCEPT QUESTIONS The unit cell in an HCP array is a hexagonal cylinder spanning three planes. The cylinder is capped by hexagons in the top and bottom planes. The intermediate plane of the unit cell contains a three-atom triangle. How many oxygen atoms are in the HCP unit cell? Locate the octahedral holes. How many octahedral holes are in the unit cell? Show that filling two-thirds of the octahedral holes with aluminum leads to a formula of Al_2O_3 for corundum. ∎

Since oxygen is an electronegative element and aluminum is an electropositive one, corundum can be thought of as an array of oxygen minus two ions, referred to as oxide ions, with aluminum plus three ions in the octahedral holes. Oxygen ions are much larger than aluminum ions, $r(O^{2-}) = 140$ pm versus $r(Al^{3+}) = 53$ pm, so the Al^{3+} ions fit into the spaces between the layers of O^{2-} (Figure 10.5). Chromium also forms a +3 ion whose size, 62 pm, is comparable to that of Al^{3+}. Substituting Cr^{3+} for Al^{3+} puts the Cr^{3+} ion in an octahedral environment (Figure 10.6). Think of the Cr^{3+} ion as sitting at the origin of a coordinate system with the negative oxide ions located on the axes (Figure 10.6).

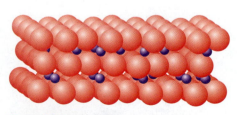

Figure 10.5 Corundum consists of a hexagonal array of oxygen (indicated in red) with aluminum (shown in purple) in holes between the layers.

Corundum, without Cr^{3+}, is a clear, colorless, extremely hard crystal. The color in ruby originates from the addition of the Cr^{3+} ions; the electronic structure of Cr^{3+} is a key ingredient in determining

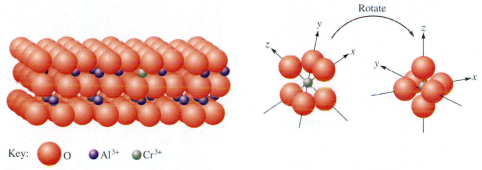

Key: O Al^{3+} Cr^{3+}

Figure 10.6 Substitution of Cr^{3+} (green) for Al^{3+} (purple) in the corundum lattice puts the Cr^{3+} ion in an octahedral environment. The Cr^{3+} ion and its six surrounding oxygen atoms are taken out of the lattice to demonstrate the octahedral environment.

this color. Neutral Cr has the electronic configuration $[Ar]4s^13d^5$. The +3 ion is formed by removing one $4s$ electron and two $3d$ electrons, leaving a configuration of $[Ar]3d^3$. In free space (i.e., with no other atoms close by), an electron occupying any one of the five d orbitals has the same energy (Figure 10.7). In the solid lattice, however, the Cr^{3+} ion is not in free space but rather in an environment surrounded by the six negatively charged oxide ions. These negative charges alter the energy of an electron in any one of these orbitals as follows. Imagine an electron in a d orbital caught in the middle of the six oxide ions. An electron is negatively charged, so it experiences a repulsive force due to the negative oxide charges. The strength of this repulsion is inversely proportional to the distance between the electron in the d orbital and the negatively charged ions (Coulomb's law). Look again at the shape of the five d orbitals (Figure 10.8). With negative charges on each end of each axis, an electron in an orbital on the axes experiences a greater repulsion than an electron between the axes.

Two of the orbitals, $d_{x^2-y^2}$ and d_{z^2}, are directed along the x-, y-, and z-axes, in line with the oxide ions. In contrast, the remaining three d_{xy}, d_{xz}, and d_{xy} orbitals are directed between the axes, so that they have a less direct encounter with the negative ions. An electron in the $d_{x^2-y^2}$ or d_{z^2} orbital experiences a strong repulsion, whereas an electron in the d_{xy}, d_{xz}, or d_{xy} orbital experiences a much weaker repulsion. With diminished repulsion, an electron in any of the d_{xy}, d_{xz}, or d_{yz} orbitals is some-

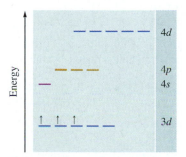

Figure 10.7 The energy-level diagram for a Cr^{3+} ion that has no other atoms or ions nearby shows that all five d orbitals have the same energy.

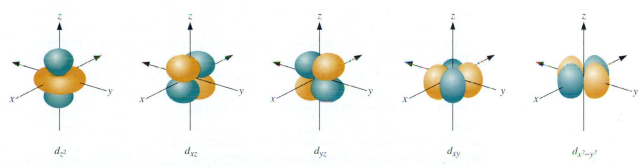

d_{z^2} d_{xz} d_{yz} d_{xy} $d_{x^2-y^2}$

Figure 10.8 The shape and orientation of the five d orbitals relative to the negative oxide ions determines the energy of an electron in those orbitals.

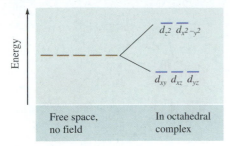

Figure 10.9 The d orbitals are no longer the same energy when the Cr^{3+} ion is in an octahedral environment. The d orbitals split into two sets: The $d_{x^2-y^2}$ and d_{z^2} pair is higher in energy than the d_{xy}, d_{xz}, and d_{xy} set.

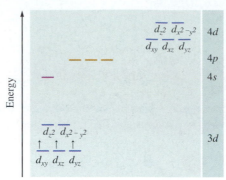

Figure 10.10 The energy-level diagram for Cr^{3+} in an octahedral field with the three valence electrons filled in shows the d_{xy}, d_{xz}, and d_{xy} orbitals with one electron each (Hund's rule) and the higher-energy $d_{x^2-y^2}$ and d_{z^2} orbitals unoccupied.

what lower in energy, more stable, than an electron in the $d_{x^2-y^2}$ or d_{z^2} orbital. As a consequence, the five orbitals split into two energy sets (Figure 10.9).

Electrons fill these orbitals in the usual fashion: two per orbital, lowest energy first. Cr^{3+} has three electrons that go into these orbitals (Figure 10.10). In the octahedral field created by the six surrounding oxide ions, Cr^{3+} has one electron in each of the three lower-lying d orbitals.

Absorption of energy can lift any one of the three $3d_{xz}$, $3d_{yz}$, or $3d_{xy}$ orbital electrons to any empty orbital. The lowest-energy transition is to the d_{z^2} or the $d_{x^2-y^2}$ orbital. This transition imparts the red color to ruby. Furthermore, the color indicates the energy separation between the sets of orbitals. In ruby, green-blue light is absorbed (Figure 10.11) indicating that interaction with the oxide ions separates the two sets of d orbitals by 2.5–2.9 eV.

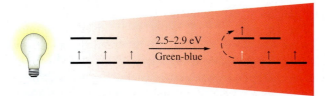

Figure 10.11 White light impinges on the ruby crystal. Green-blue photons, photons of energy 2.5–2.9 eV, lift an electron from the lower-energy set of orbitals to the higher-energy set. As the green-blue photons are filtered out, the light becomes increasingly red.

WORKED EXAMPLE 10.1 *Interactions and Colors*

Beryl is a common mineral with the formula $Be_3Al_2(SiO_3)_6$. The largest beryl crystal found to date comes from Madagascar and measures 8 m long, is 3.5 m in diameter, and weighs 380 tons. The gem aquamarine consists of a beryl host with Fe^{2+} impurities and appears blue. Heliodor is similar to aquamarine, but in this case the iron impurity is in the +3 oxidation state and the gem's color is yellow. In both cases, the iron ions are in an octahedral environment surrounded by six oxygen atoms. What color is absorbed by each of these gems? Rationalize the absorbed color by citing the oxidation state of iron and relating it to Coulomb's law.

Plan
- The color absorbed is the complement of the perceived color.
- More energy absorbed implies a greater interaction.
- Coulomb's law indicates that the interaction energy is directly proportional to the charges and inversely proportional to the separation.

Implementation

- Aquamarine appears blue so the complement, red plus green or yellow light, is absorbed. Heliodor appears yellow, so red and green light are transmitted and blue light is absorbed.
- Blue light is higher in energy than is yellow light. Heliodor absorbs higher-energy photons than does aquamarine.
- Heliodor contains iron in a +3 oxidation state, while aquamarine contains iron in a +2 oxidation state. The higher charge on Fe(III) results in a stronger coulombic interaction with the oxygen atoms that are negatively charged due to the high electronegativity of oxygen.

See Exercises 1, 2.

CONCEPT QUESTIONS   Emerald, like ruby, consists of an Al_2O_3 crystal with Cr^{3+} ions. In the case of emerald, the Al_2O_3 crystal also contains some $Be_3Si_6O_{15}$. The presence of $Be_3Si_6O_{15}$ alters the lattice constant, expanding the size of the octahedron containing the Cr^{3+} ion. What color is absorbed by emerald? Is the energy absorbed by emerald greater than or less than that absorbed by ruby? Is the greater distance between Cr^{3+} and the oxide ions in emerald compared with ruby consistent with the energy absorbed by these two gems? ■

10.2   **Transition Metal Complexes**

APPLY IT   Obtain some household ammonia from the grocery store, a galvanized nail from a hardware store, and some root killer (Roebic, Root Killer K-77) from a garden supply store. Make a solution consisting of 2 match-head sized crystals of the root killer (crushed) and $^1/_4$ cup water in a clear plastic cup. Describe the color of the solution. Pour several tablespoons of ammonia cleaner into the solution. How does the color change? Add the galvanized nail to the solution and let sit for an hour. Observe the color of the solution. Has the appearance of the nail changed? Root killer contains $CuSO_4$. Cu^{2+} ions in aqueous solution impart a characteristic, pale blue color. Cu^{2+} ions interact with ammonia to yield a deep blue solution. Galvanized nails contain a Zn coating. Zn^{2+} ions yield a clear, colorless aqueous solution. Use the color information to describe the processes in solution. (The solution may be diluted and poured down the drain. Do not consume the solution. Dispose of plastic cup in a solid waste container.)

Transition metal complex
A substance consisting of a transition metal ion surrounded by ligands bound with coordinate covalent bonds.

Ligand   A molecule or ion that bonds to a transition metal ion via a coordinate covalent bond.

Ruby is part of a much larger class of colorful materials called **transition metal complexes.** Many gems, including emerald, alexandrite, aquamarine, citrine quartz, blue and green azurite, malachite green, and red garnet, derive their colors from similar transitions. Transition metal complexes come in a variety of geometrical shapes. In every case, the d-orbital energy levels can be determined by coulombic repulsion between electrons in the metal d orbitals and the negatively charged lone pairs of the atom or ion coordinating to the metal. The coordinating substance is called a **ligand,** from the Latin *ligare,* meaning "to bind." The most common numbers of ligands are six, leading to octahedral complexes, and four, leading to tetrahedral complexes. In a tetrahedral complex, the ligands are not directly in line with any of the d orbitals (Figure 10.12). Rather, the distance between the ligands and the d_{xy}, d_{xz}, and d_{yz} set of orbitals is smaller than the distance to either the $d_{x^2-y^2}$ or d_{z^2}

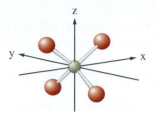

Figure 10.12 In a tetrahedral complex, the ligands are not directly aligned with any of the d orbitals. The ligands are closer to the d_{xy}, d_{xz}, and d_{yz} orbitals than to either of the $d_{x^2-y^2}$ or d_{z^2} orbitals.

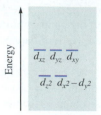

Figure 10.13 The d-orbital energy levels in a tetrahedral complex are split by a smaller energy than in the corresponding octahedral complexes.

orbital. As a result, the d-orbital splitting pattern in a tetrahedral complex (Figure 10.13) is the opposite of that of an octahedral complex. Due to the less direct alignment, the d-orbital splitting in a tetrahedral complex is smaller than that in an octahedral complex. This smaller splitting in tetrahedral environments is sometimes used to determine whether the metal ion occupies the tetrahedral or octahedral holes in a solid lattice. Worked Example 10.2 illustrates how to construct the energy-level diagram for another geometry.

WORKED EXAMPLE 10.2   *Geometry and d-Orbital Energy*

Metal ions are found in many geometries. The d-orbital energy levels can be determined from the geometry. Platinum catalyzes a variety of reactions, including some important medicinal reactions, most famously in treatment of testicular cancer. Platinum complexes often form a square planar configuration, and this square planar geometry is critical to the medicinal function of platinum. A simple complex containing platinum is the square planar complex $[PtCl_4]^{2-}$. Create the d-orbital energy-level diagram for $[PtCl_4]^{2-}$. Explain your answer.

Plan

- Locate the four Cl^- ions.
- Determine the proximity of the d-orbital lobes to the Cl^- ions.
- Separation determines energy.

Implementation

- The first task is to decide where the ligands should be placed. Because four of the five d orbitals are geometrically similar, the unique d-orbital axis (Figure 10.8) determines the unique direction. The d_{z^2} orbital is the unique one; the z-axis is unique, and the x- and y-axes are equivalent. To form a square planar complex, the ligands are placed in the plane formed by the equivalent x- and y-axes. The four ligands are placed on the $\pm x$- and $\pm y$-axes. The z-axis contains no ligands.
- Orbitals with density in the x-y plane are the $d_{x^2-y^2}$, d_{xy}, and d_{z^2} orbitals. Of these, the $d_{x^2-y^2}$ orbital points directly along the axes, the d_{xy} orbital points between the axes, and the d_{z^2} orbital has only a small ring in the x-y plane. The remaining orbitals, d_{xz} and d_{yz}, point above and below the x-y plane.
- The orbital closest to the four Cl^- ion ligands is the $d_{x^2-y^2}$ orbital; it is highest in energy. The next highest-energy orbital is the d_{xy} orbital, followed by the d_{z^2} orbital.

The d_{xz} and d_{yz} orbitals are equivalent, being removed from the ligands to the greatest extent and lowest in energy. The energy-level diagram for a square planar complex is

Complex ion A charged species consising of a metal ion surrounded by ligands.

Complex The combination of a metal ion and the ligands joined to the metal by coordinate covalent bonds.

Note that the energy-level diagram for a square planar complex is more complicated than that for an octahedral or tetrahedral complex. The common theme is that the energy ordering is determined by coulombic interactions between electrons and ligands. Electron filling is determined by the metal ion's electron configuration. In the complex ion $[PtCl_4]^{2-}$, the platinum ion has a charge of $+2$ and an electron configuration of $[Xe]4f^{14}\,5d^8$, so all of the orbitals except $d_{x^2-y^2}$ are filled.

See Exercises 3, 4.

10.3   Terminology

Positive metal ions are formed by loss of electrons, so they are relatively small. Because they consist of a concentrated, positive charge, metal ions are often surrounded by several small molecules or negative ions. The metal ion with its attendant molecules or ions is called a **complex ion** if it is charged or simply a **complex** if it is neutral. For example, the radius of the Al^{3+} ion is 53 pm, much smaller than the radius of a water molecule (180 pm). In aqueous solution, there is a strong attraction between the $+3$ charge and the lone pairs on oxygen in water. The Al^{3+} ion is thus fairly tightly bound to six water molecules (Figure 10.14).

The combination is denoted

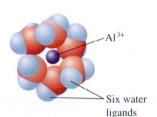
Al^{3+}
Six water ligands

Figure 10.14 There is a strong attraction between an Al^{3+} ion and the lone pairs on the water molecules in aqueous solution. The Al^{3+} ion (purple) is surrounded by six water molecules, binding to the water molecules through a lone pair on oxygen (red), with the hydrogen atoms (light blue) being directed away from the positive Al^{3+} ion.

(10.1) $[Al(H_2O)_6]^{3+}$

The complex is enclosed in square brackets with the charge indicated outside the brackets. In this case, the complex has a $+3$ charge, the result of a $+3$ charge on aluminum and no charge on the six surrounding water molecules. Thus the combination is a complex ion. The six surrounding water molecules are the ligands. All of the ligands have unbonded pairs of electrons that are attracted to the positive metal ion and form a covalent bond. Some ligands have more than one atom with a lone pair. These multi-pronged ligands are called **multidentate,** quite literally "multi-toothed." The specific number of teeth is indicated by the prefix *mono* for one, *bi* for two, *tri* for three, *quadri* for four, *quinque* or *penta* for five, and *hexa* for six. In a more colorful description, the ligand is said to **chelate** to the metal; *chelate* comes from the Greek word "chele," meaning claw.

mono = one
bi = two
tri = three
quadri = four
quinque or *penta* = five
hexa = six

Water is an example of a monodentate ligand. Although water contains two lone pairs, both are associated with a single atom: the oxygen atom. Geometrical constraints prevent both lone pairs from coordinating to the metal ion, so water is monodentate. Ammonia is another example of a monodentate ligand, as are the halogen ions.

Two common multidentate ligands are ethylenediamine (en) and ethylenediamine tetraacetate (EDTA) (Figure 10.15). Ethylenediamine consists of two ammonia molecules, with one hydrogen of each being replaced by a hydrocarbon chain that connects the two

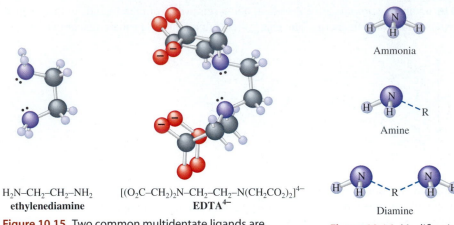

$H_2N–CH_2–CH_2–NH_2$
ethylenediamine

$[(O_2C–CH_2)_2N–CH_2–CH_2–N(CH_2CO_2)_2]^{4-}$
EDTA^{4-}

Figure 10.15 Two common multidentate ligands are ethylenediamine, abbreviated en, and ethylenediamine tetraacetate, abbreviated EDTA. Color code: oxygen (red), nitrogen (dark blue), carbon (black), hydrogen (light blue).

Ammonia

Amine

Diamine

Figure 10.16 Modification of ammonia leads to amines and diamines. "R" here refers to any hydrocarbon chain.

Dentate Literally, "having teeth." Lone pairs in a ligand that coordinate to a metal ion.

Multidentate Multi-toothed; a ligand having more than one lone pair that can form a coordinate covalent bond.

Chelate A polydenate ligand bound to a metal ion.

NH_2 groups (Figure 10.16). Ammonia that has one or more hydrogen atoms replaced by a hydrocarbon chain is called an **amine** group. In ethylenediamine, the hydrocarbon chain links the two amine groups, so the result has a lone pair on each of two nitrogen atoms; thus ethylenediamine is *bidentate*. EDTA also has two amine groups. In EDTA, each of the two remaining hydrogen atoms of the amines is replaced by CH_3COO^-, called an acetate group. The negative charge and lone pairs on oxygen make the acetate group a very good ligand. With four acetate groups and two amine groups, EDTA is *hexadentate*, biting into a metal ion with six lone pairs. EDTA literally wraps around the metal (Worked Example 10.3).

The total number of coordinate-covalent bonds to the metal ion is called the **coordination number.** Thus a metal ion coordinated to a single EDTA ligand has a coordination number of 6. It takes three ethylenediamine ligands or six water molecules to achieve a coordination number of 6.

WORKED EXAMPLE 10.3   *Chelation Therapy*

The Pb^{2+} ion is toxic in the body, and ingesting it can produce a condition known as lead poisoning. Lead poisoning is most often seen in children who ingest paint chips in older buildings. One therapy for lead poisoning is to administer EDTA as a chelating agent. What is the charge on the lead–EDTA complex? What is the coordination number of lead in the complex?

Amine Ammonia that has one or more hydrogen atoms replaced by a hydrocarbon.

Coordination number For complexes, the number of electron pairs donated to a transition metal ion.

Plan
■ Charge is determined by the sum of the metal ion and ligand charges.
■ Coordination number is the number of coordinate covalent bonds.

Implementation
■ The charge on the lead ion is $+2$. The charge on EDTA is -4. Thus the complex has a -2 charge: $[PbEDTA]^{2-}$.
■ EDTA is a hexadentate ligand forming six coordinate-covalent bonds with the lead ion. The coordination number is 6.

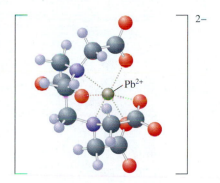

Because the complex $[PbEDTA]^{2-}$ is charged, it is soluble in aqueous solutions, literally washing the harmful lead ions out of the body via the waste stream.

See Exercises 7, 8.

10.4   Interactions

There are three probes for the strength of interaction in coordination complexes: color, the equilibrium constant, and, for metal ions in water, the enthalpy of hydration.

Color

When a metal ion has no ligands around it, an electron has the same energy in all of the five d orbitals. Bringing ligands, with their negatively charged lone pairs, into the vicinity of the metal ion alters the energy equivalence of the orbitals, lowering the energy of those orbitals farther removed from the ligands and raising the energy of those closest to the ligands. The energy separation between the sets of orbitals depends on how strongly the ligand interacts with the metal ion. Consequently, the energy required to lift an electron from one of the lower-energy set of orbitals to the higher-energy set also depends on how strongly the ligands interact with the metal.

For example, a solution of the complex ion $[CoCl_6]^{4-}$ is purple, a solution of $[Co(H_2O)_6]^{2+}$ is pink, and a solution of $[Co(NH_3)_6]^{2+}$ is yellow-orange. Each of these complexes involves Co^{2+} and six ligands forming an octahedral complex. Thus the different colors provide information about the strength of interaction between the cobalt ion and the ligands. The solutions are clear and colored and act as a filter, selectively removing some light colors from the visible spectrum (Figure 10.17).

The energy of the color removed from the spectrum corresponds to the energy separation between the sets of d orbitals: green for $[CoCl_6]^{4-}$, blue-green for $[Co(H_2O)_6]^{2+}$, and blue for $[Co(NH_3)_6]^{2+}$ (Figure 10.17). Moving up the color−energy spectrum, the six chloride ligands separate the d orbitals by the smallest energy of these three complex ions, and the six ammonia ligands separate them by the most. Since the metal ion d orbitals remain the same and the geometry of all three complexes is the same, the different energy separation must be due to the ligands. More specifically, the difference is due to the ligand lone pairs (Figure 10.18). Chloride has four lone pairs and the electrons are delocalized over the entire sphere encompassing the ion. This is a far less directed lone pair than is found in ammonia, which has a single, directed lone pair.

By comparing the color of a large number of complexes with various metal ions, chemists have identified a **spectrochemical series.** It is called a spectrochemical series

Spectrochemical series
A series of ligands with increasing d orbital splitting; increasingly shorter wavelength absorption.

Figure 10.17 The color of transition metal complexes reflects the strength of the interaction between the positively charged metal and the ligands. The observed color is the complement of the color absorbed. The absorbed color is a direct measure of the splitting between the sets of *d* orbitals.

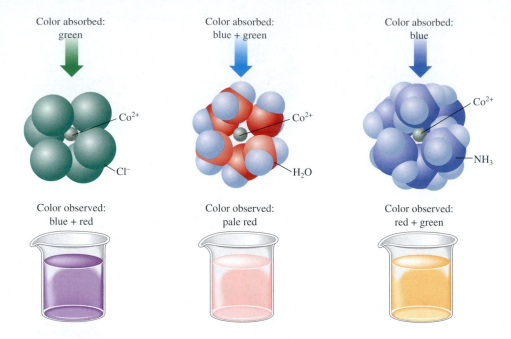

Color absorbed: green

Color absorbed: blue + green

Color absorbed: blue

Color observed: blue + red

Color observed: pale red

Color observed: red + green

because it is based on the spectrum of colors generated by these beautiful complexes. The series is

(10.2) $I^- < Br^- < Cl^- < F^- < OH^- < H_2O < NH_3 < en < CN^- < EDTA$

Varying the charge (the oxidation state) of the metal ion influences the strength of the interaction between the ligand and the metal ion. For example, both Fe^{2+} and Fe^{3+} form octahedral complexes with water. The $[Fe(H_2O)_6]^{2+}$ complex is pale green-blue, whereas the $[Fe(H_2O)_6]^{3+}$ complex is pale blue (Figure 10.19). Pale blue corresponds to a low-intensity absorption of red and green light, and pale green-blue corresponds to a low-intensity absorption of red light. Red light is lower in energy than green light, indicating that the Fe^{2+} ion interacts less strongly with the six water ligands than does the Fe^{3+} ion. The electrostatic interaction is stronger for the more highly charged Fe^{3+} ion.

CONCEPT QUESTION An emerald green solution of $NiSO_4$ contains the complex ion $[Ni(H_2O)_6]^{2+}$. The emerald green solution turns dark blue on addition of aqueous ammonia. On the molecular level, what is the source of the color change? ∎

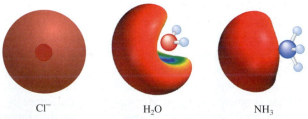

Cl⁻   H₂O   NH₃

Figure 10.18 The strength of the interaction in a coordinate-covalent bond depends on the localization of the lone pairs. In Cl^-, the four lone pairs encompass the entire ion. In water, the two lone pairs are more localized than those in chloride, but less so than the single lone pair in ammonia. Ammonia provides the largest energy separation and has the most directed lone pair.

Equilibrium

The equilibrium constant for a reaction provides a second measure of the relative interaction between a metal ion and a set of ligands. This is very useful information, particularly when complexes have different geometries. For example, consider the octahedral $[Ni(NH_3)_6]^{2+}$ complex and the square planar $[Ni(CN)_4]^{2-}$ complex (Figure 10.20). Splitting of the *d*-orbital electron energies is very different in these two different geometries. It is possible to determine which ligand binds more strongly, however, by using the equilibrium constants.

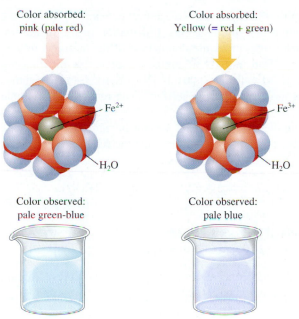

Color absorbed:
pink (pale red)

Fe²⁺

H₂O

Color observed:
pale green-blue

Color absorbed:
Yellow (= red + green)

Fe³⁺

H₂O

Color observed:
pale blue

Figure 10.19 These two iron complexes illustrate the effect of metal ion charge on the energy separation for *d*-orbital electrons. The larger charge leads to a stronger interaction and a larger splitting.

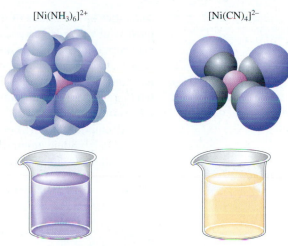

$[Ni(NH_3)_6]^{2+}$

$[Ni(CN)_4]^{2-}$

Figure 10.20 Although the two complexes of Ni^{2+}, that with ammonia (NH_3) and that with cyanide (CN^-), are different colors due to the differing geometries, the color cannot be used directly to discern which ligand interacts more strongly.

Suppose a solution contains $0.001\ M\ Ni^{2+}$, $0.5\ M$ ammonia, and $0.5\ M$ cyanide. In what form is most of the Ni^{2+} found: as $[Ni(H_2O)_6]^{2+}$ due to the large excess of water, or as the $[Ni(NH_3)_6]^{2+}$ or the $[Ni(CN)_4]^{2-}$ complex?

CONCEPT QUESTIONS $[Ni(CN)_4]^{2-}$ has a -2 charge, while $[Ni(NH_3)_6]^{2+}$ has a $+2$ charge. Account for this difference. Which is the limiting reagent in a solution with $0.001\ M\ Ni^{2+}$ and $0.5\ M$ ammonia? In a solution with $0.001\ M\ Ni^{2+}$ and $0.5\ M\ CN^-$? ∎

The logic supporting the use of equilibrium constants to determine interaction strength is as follows. If nickel binds more strongly to ammonia, the amine complex should contain more of the Ni^{2+} than does the cyanide or water complex. Determining which complex contains most of the Ni^{2+} ion requires the equilibrium constant for formation of each of these complex ions. The reaction for formation of the amine complex is

(10.3) $[Ni(H_2O)_6]^{2+}(aq) + 6NH_3(aq) \rightleftharpoons [Ni(NH_3)_6]^{2+}(aq) + 6H_2O(l)$
$$K_f = 4.07 \times 10^8$$

Due to the small diameter of the Ni^{2+} ion, it complexes either with water or with ammonia. The equilibrium constant for formation of a complex ion from its hydrated ion is called the **formation constant** and denoted as K_f. As usual, the equilibrium constant is the product of the product concentrations divided by the product of the reactant concentrations with the solvent excluded.

Formation constant, K_f The equilibrium constant for forming a complex ion from its constituent metal ion and ligands.

(10.4) $$K_f = \frac{[[Ni(NH_3)_6]^{2+}]}{[[Ni(H_2O)_6]^{2+}][NH_3]^6}$$

The order of magnitude of the equilibrium constant, 10^8, indicates that the products are present in much larger concentration than the reactants. Calculation of the nickel ion concentration complexed with water recognizes the dominance of the amine complex. When starting with 0.001 M nickel ion and 0.5 M ammonia, nickel is the limiting reactant. Using the entire 0.001 M Ni^{2+} forms 0.001 M $[Ni(NH_3)_6]^{2+}$ and consumes 0.006 M ammonia (six ammonia molecules per nickel ion). The remaining ammonia concentration is 0.494 M.

At equilibrium, the concentration of $[Ni(H_2O)_6]^{2+}$ cannot be zero. To form some $[Ni(H_2O)_6]^{2+}$, some amine complex must dissociate to provide the required Ni^{2+}. Let x be the unknown concentration of $[Ni(NH_3)_6]^{2+}$ that dissociates:

(10.5) $\quad [Ni(H_2O)_6]^{2+}(aq) + 6NH_3(aq) \rightleftharpoons [Ni(NH_3)_6]^{2+}(aq) + 6H_2O(l)$

$$K_f = 4.07 \times 10^8$$

Start:	0 M	0.494 M	0.001 M
Equilibrium:	x M	$(0.494 + 6x)$ M	$(0.001 - x)$ M

Stoichiometry dictates that x is also the concentration of the water complex. Solving for x, and assuming that x is small compared with 0.001,

(10.6) $\quad K_f = \dfrac{[[Ni(NH_3)_6]^{2+}]}{[[Ni(H_2O)_6]^{2+}][NH_3]^6} = \dfrac{(0.001 - x)}{(x)(0.494 + 6x)^6} \approx \dfrac{(0.001)}{(x)(0.494)^6}$

$$= 4.07 \times 10^8$$

results in $x = 1.7 \times 10^{-10}$ or $[[Ni(H_2O)_6]^{2+}] = 1.7 \times 10^{-10}$ M. Thus very little nickel remains as the water complex despite the presence of more than 100 times as many water molecules as ammonia molecules in the solution.

For the cyanide complex, $[Ni(CN)_4]^{2-}$, the equilibrium constant again favors the products, this time by 31 orders of magnitude. The formation reaction is

(10.7) $\quad [Ni(H_2O)_6]^{2+}(aq) + 4CN^-(aq) \rightleftharpoons [Ni(CN)_4]^{2-}(aq) + 6H_2O(l)$

$$K_f = 1.0 \times 10^{31}$$

Start:	0 M	0.496 M	0.001 M
Equilibrium:	x M	$(0.496 + 4x)$ M	$(0.001 - x)$ M

Solving for the nickel–water complex, $[[Ni(H_2O)_6]^{2+}]$,

(10.8) $\quad K_f = \dfrac{[[Ni(CN_4)]^{2-}]}{[[Ni(H_2O)_6]^{2+}][CN^-]^4} = \dfrac{(0.001 - x)}{(x)(0.496 + 4x)^4} \approx \dfrac{(0.001)}{(x)(0.496)^4}$

$$= 1.0 \times 10^{31}$$

results in $[[Ni(H_2O)_6]^{2+}] = 1.6 \times 10^{-33}$ M. The cyanide complex dominates over water by even more than does the ammonia complex. Thus cyanide binds to nickel more strongly than either water or ammonia.

CONCEPT QUESTIONS Determine the equilibrium constant for the following reaction:

$$[Ni(NH_3)_6]^{2+}(aq) + 4CN^-(aq) \rightleftharpoons [Ni(CN)_4]^{2-}(aq) + 6NH_3(aq)$$

In the competition for Ni^{2+}, which ligand dominates: NH_3 or CN^-? What is the concentration of $[Ni(NH_3)_6]^{2+}$ in a solution that is 0.5 M in NH_3, 0.5 M in CN^-, and 0.001 M in Ni^{2+}? ∎

WORKED EXAMPLE 10.4   *Binding and Oxidation State: Toxicity of Cyanide*

In the body, iron is found as Fe^{2+} in hemoglobin, although it can be oxidized to Fe^{3+}. Once oxidized to Fe^{3+}, the hemoglobin no longer transports oxygen effectively because it fails to unbind the O_2. Compare the ligand binding strength of Fe^{2+} to that of Fe^{3+} by determining (a) the concentration of $[Fe(H_2O)_6]^{2+}$ in a solution that is 0.001 M in Fe^{2+} and 0.006 M in CN^- and (b) the concentration of $[Fe(H_2O)_6]^{3+}$ in a solution that is 0.001 M in Fe^{3+} and 0.006 M in CN^-.

Data

$$K_f([Fe(CN)_6]^{4-}) = 7.7 \times 10^{36}$$
$$K_f([Fe(CN)_6]^{3-}) = 7.7 \times 10^{43}$$

Plan

■ Write the equilibrium equation for the reaction.
■ Write the expression for the equilibrium constant.
■ Determine the unknown hexaaquo complex concentration.

Implementation

■ a. $[Fe(H_2O)_6]^{2+}(aq) + 6CN^-(aq) \rightleftharpoons [Fe(CN)_6]^{4-}(aq) + 6H_2O(l)$
   b. $[Fe(H_2O)_6]^{3+}(aq) + 6CN^-(aq) \rightleftharpoons [Fe(CN)_6]^{3-}(aq) + 6H_2O(l)$

■ a. $K_f([Fe(CN)_6]^{4-}) = \dfrac{[[Fe(CN)_6]^{4-}]}{[[Fe(H_2O)_6]^{2+}][CN^-]^6}$

   b. $K_f([Fe(CN)_6]^{3-}) = \dfrac{[[Fe(CN)_6]^{3-}]}{[[Fe(H_2O)_6]^{3+}][CN^-]^6}$

■ In both cases, the equilibrium constant indicates that products dominate ($K \gg 1$). Because iron and cyanide are present in stoichiometric concentrations, both will be tied up in the cyanide complex. Let x be the concentration of the hexaaquo complex.

a. $[Fe(H_2O)_6]^{2+}(aq) + 6CN^-(aq) \rightleftharpoons [Fe(CN)_6]^{4-}(aq) + 6H_2O(l)$

Initial:	0.001	0.006	
Start:	0	0	0.001 M
Equilibrium: x		$6x$	$0.001 - x$

For $[Fe(H_2O)_6]^{2+}$:

$$K_f([Fe(CN)_6]^{4-}) = \frac{0.001 - x}{x(6x)^6} = 7.7 \times 10^{36} \Longrightarrow x \approx \sqrt[7]{\frac{0.001}{(6)^6(7.7 \times 10^{36})}}$$
$$= 4.3 \times 10^{-7}$$

So, $[Fe(H_2O)_6]^{2+} = 4.3 \times 10^{-7}\ M$.
   b. A similar calculation with the Fe^{3+} ion yields $[[Fe(H_2O)_6]^{3+}] = 4.3 \times 10^{-8}\ M$.

Both concentrations of the hexaaquo complex are quite small, a sign of strong binding by the cyanide ion. The even smaller concentration of the Fe^{3+}–water complex illustrates the stronger binding that results from the larger ion charge. One of the reasons that cyanide is so toxic is this very strong binding to metal ions, particularly to iron in hemoglobin.

See Exercises 41–43.

Thermodynamics

Energy is released when an isolated ion is surrounded by ligands, thereby forming coordinate-covalent bonds. For the important case involving surrounding an ion with

Hydration enthalpy The energy evolved when an ion in the gas phase is surrounded by water molecules to form an aqueous solution.

water molecules, the released energy is called the **hydration enthalpy.** For example, the hydration enthalpy of Li^+ corresponds to the reaction

(10.9) $$Li^+(g) \longrightarrow Li^+(aq) \qquad \Delta H_{hydration} = -515 \text{ kJ/mol}$$

Examining hydration enthalpy for a series of transition metal ions shows the effect of the *d*-orbital splitting. Consider the Cr^{2+} ion in water (Figure 10.21). In free space, unaffected by any other atoms or molecules, an electron in any one of the five *d* orbitals has the same energy. As a result, the four valence electrons of Cr^{2+} can occupy any four of the five *d* orbitals. However, when water surrounds the Cr^{2+} ion, the energy picture changes: Three orbitals are lowered in energy and two are raised in energy. The energy separation increases as the water molecules approach the Cr^{2+} ion more closely. At the bonding distance, one of the higher-energy orbitals is unoccupied.

The empty *d* orbital in the higher-energy set effects the hydration enthalpy of the Cr^{2+} ion. Consider chromium's neighbor in the periodic table, manganese. The Mn^{2+} ion has the electron configuration $[Ar]3d^5$. Every *d* orbital in this ion is occupied by one electron. The average electron energy for the hydrated Mn^{2+} ion is thus higher than the average electron energy for the hydrated Cr^{2+} ion. Due to the higher energy that remains in the ion, somewhat less energy is released when Mn^{2+} hydrates ($\Delta H_{hydration} = -1841$ kJ/mol) than when Cr^{2+} does ($\Delta H_{hydration} = -1904$ kJ/mol). Mn^{2+} is somewhat smaller than Cr^{2+} ($r_{Mn}^{2+} = 67$ pm, $r_{Cr}^{2+} = 77$ pm). Size alone would result in a stronger interaction between Mn^{2+} and water, and hence more energy released on interaction.

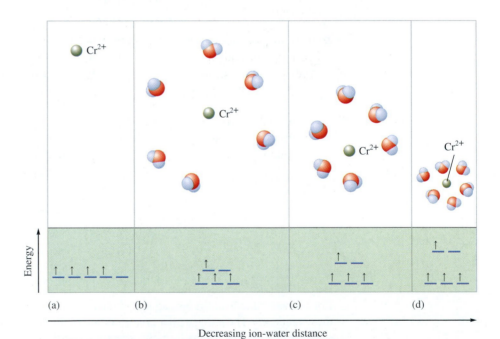

Figure 10.21 The hydration of a metal ion by water is illustrated by the coordination of Cr^{2+} to six water molecules, forming the complex $[Cr(H_2O)_6]^{2+}$. Free of interaction with water molecules, electrons in the five *d* orbitals all have the same energy. Cr^{2+} has an electron configuration of $[Ar]3d^4$. In the isolated Cr^{2+} ion, the four valence electrons go into four of the five *d* orbitals. Surrounding the chromium ion with six water molecules in an octahedral configuration, however, alters the energy of an electron in the *d* orbitals. Three electrons go, one each, into the lower-energy set of orbitals, leaving one electron in the higher-energy pair. As the water molecules close in on the Cr^{2+} ion, the energy separation increases.

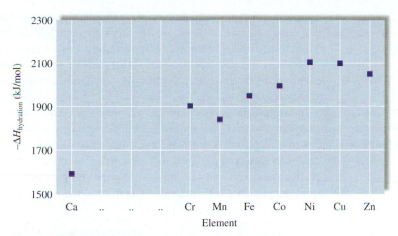

Figure 10.22 The hydration enthalpy of the transition metal +2 ions reflects the splitting of the *d*-orbital electron energy levels. Ca^{2+}, Mn^{2+}, and Zn^{2+}, with empty, half-filled, and filled *d* orbitals, respectively, release no additional energy due to the ligands. With the lower-energy set of orbitals filled and the higher-energy set empty, Ni^{2+} has the greatest additional energy release: nearly 150 kJ/mol higher than the straight-line trend from Ca^{2+} through Zn^{2+}.

Contrary to this expectation, hydration of the larger Cr^{2+} ion releases more energy. The additional energy released due to the empty, higher-energy *d* orbital in Cr^{2+} is significant.

Examining the hydration energy of the fourth-period transition metal +2 ions (Figure 10.22) reinforces the picture of extra energy being released due to splitting of the *d*-orbital electron energies. Elements at the beginning of the series do not form stable +2 ions in water. However, starting with Mn^{2+}, the hydration enthalpy increases for Fe^{2+} with one filled *d* orbital and four half-filled *d* orbitals. The hydration enthalpy increases by another increment for Co^{2+}, which has two filled *d* orbitals in the lower-energy set. Another increment occurs for Ni^{2+}, which has all three lower-energy *d* orbitals filled and the higher-energy set only half-filled. The next ion, Cu^{2+}, has no incremental increase in energy release. The additional *d*-orbital electron in Cu^{2+} is in the higher-energy set of orbitals. Finally, Zn^{2+} has all orbitals filled, and its hydration enthalpy is about 50 kJ/mol less than that of Cu^{2+}.

CONCEPT QUESTION In what way would the graph in Figure 10.22 change if the +2 ions Mn through Zn coordinated to four water molecules in a tetrahedral arrangement rather than six water molecules in an octahedral arrangement? Discuss at least two changes. ∎

10.5   Magnetism

Permanent magnets are used in a variety of devices, ranging from stereo speakers and watch movements, to audio and video tapes, refrigerator door seals, and computer memory devices. The first known magnetic material was loadstone, a form of Fe_3O_4, discovered in antiquity in China. It was observed that a small, floating piece of loadstone always points north in the earth's magnetic field. This observation led to the development of compasses, which were widely used for navigational purposes by 1600. Compasses depend on the magnetic properties of the electron. Today, mass storage of information, indispensable for the electronics industry, uses both the charge and the magnetic properties of the electron. Specifically, the **magneto resistance effect,** a

Magneto resistance effect
A change in resistance that occurs when the data sensor passes near a magnetized region of a magnetic disk.

change in resistance that occurs when the data sensor passes near a magnetized region of a magnetic disk, serves as the basis for the $30 billion hard-drive industry. The magneto resistance effect was discovered by Lord Kelvin in 1856, but its commercialization was delayed by 135 years, until suitable materials were developed. The field is still open to development of stronger magnets that offer reduced size, lighter weight, and lower cost.

Ferromagnetic The ability to be permanently magnetized; having large domains of aligned spins.

Materials that form permanent magnets are called **ferromagnetic.** To form a ferromagnetic material, the atoms must have unpaired electrons. However, merely having unpaired electrons is not sufficient. Interactions among d-orbital electrons on neighboring atoms also play a critical role. Similarly, interactions between d-orbital electrons and the ligands determine the number of unpaired electrons in the d orbitals, resulting in paramagnetic versus diamagnetic materials.

Unpaired d Electrons

The various forms of magnetism—paramagnetism, ferromagnetism, and diamagnetism—all have their roots in the spin of the electron. The full name for electron spin is the spin *magnetic* quantum number, which is so named because of the magnetic field generated by the electron spin. All materials have electrons, and all electrons have spin. However, most materials are only weakly magnetic, indicating no magnetization in the absence of an applied magnetic field and having only a very small magnetization in an applied field.

There are two types of weak magnetization: paramagnetism and diamagnetism. *Paramagnetism* is due to atoms containing unpaired electrons. The unpaired spins tend to align themselves in the presence of an applied magnetic field, so that paramagnetic materials are more stable within the magnetic field than outside it. Paramagnetic materials are, therefore, drawn into a magnetic field. The energy difference inside and outside the field depends on the number of unpaired spins, providing a mechanism to determine the number of unpaired spins. Operationally, a lightweight tube containing the substance of interest is suspended from a balance and the mass determined. The magnetic field is then turned on and the mass determined again. The mass difference

(10.10) $$(m_{\text{field}} - m_{\text{no field}})g = \tfrac{1}{2}\chi\mu_{\text{o}}H^2A$$

depends on the gravitational acceleration g (9.807 m/s^2), the vacuum permeability μ_{o} (1.257 $\times$ 10^{-6} N/A^2), the applied magnetic field H, and the tube cross-sectional area A. The important molecular parameter is χ, the volume magnetic susceptibility. For a given substance, the susceptibility is proportional to $n(n + 2)$, where n is the number of unpaired spins.

The second type of weak magnetization is *diamagnetism.* Substances with no unpaired spins are diamagnetic. With all electrons paired, when a diamagnetic material is put in a magnetic field, the electron spins try to align themselves with the field. Because the spins are paired, however, it takes energy to align spins with the field. As a result, the substance has a lower energy outside the field than inside the field, and diamagnetic materials are repelled by an applied field. Diamagnetic substances lose mass in a magnetic field. The volume magnetic susceptibility is determined in the same manner as that of paramagnetic substances, but χ is negative in this case.

Determining whether a substance is paramagnetic or diamagnetic and identifying the number of unpaired spins gives information on the strength of the ion ligand

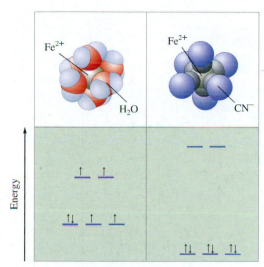

Figure 10.23 Electron filling in the octahedral iron complexes depends on the strength of the interactions with the ligands. A strong ligand such as CN^- results in a large splitting and pairing of electrons in the lower-energy set of orbitals. In contrast, d-orbital energy splitting for a weak ligand such as water is smaller; it therefore costs less energy to put electrons in the higher-energy set than it costs to pair electrons in the lower-energy set. $[Fe(CN)_6]^{4-}$ is diamagnetic, whereas $[Fe(H_2O)_6]^{2+}$ is strongly paramagnetic.

High spin complex A complex in which unpaired electrons occupy the higher-energy set of orbitals rather than pairing in the lower-energy set.

Weak field ligand A ligand that produces only a small splitting between sets of d orbitals in a metal ion; splitting sufficiently small that electrons occupy the higher energy set of orbitals rather than pairing with electrons in the lower energy set of orbitals.

Low spin complex A complex in which electrons pair in the lower-energy set of orbitals rather than unpair by occupying the higher-energy set of orbitals.

interactions or probes a change in oxidation state of a metal ion in a reaction. As an example, consider the two complex ions $[Fe(H_2O)_6]^{2+}$ and $[Fe(CN)_6]^{4-}$. Both are octahedral complexes. The $[Fe(H_2O)_6]^{2+}$ ion is strongly paramagnetic and has four unpaired spins. In contrast, the complex $[Fe(CN)_6]^{4-}$ is diamagnetic and has no unpaired spins. The difference between these complex ions reflects a difference in the strength of interaction with the ligands versus the energy required to pair spins in an orbital. As both complexes are octahedral, the same basic energy-level diagram applies (Figure 10.9). The electron configuration of Fe^{2+} is $[Ar]3d^6$, and the difference between the two complexes lies in how the six d electrons fill the energy levels (Figure 10.23). The CN^- ion is a much stronger ligand than is H_2O, and the six CN^- ligands form a shorter coordinate-covalent bond to the Fe^{2+} ion. The difference in energy between the higher- and lower-energy set of orbitals is larger in the cyanide complex than in the water complex. Indeed, the energy difference between the sets of orbitals in the iron–cyanide complex is larger than the energy required to pair electrons in the same orbital. Consequently, the electrons pair up in the lower-energy set of orbitals. With all electrons paired, the cyanide complex is diamagnetic.

Water is a weaker ligand than is cyanide. In the $[Fe(H_2O)_6]^{2+}$ complex, the splitting energy is much smaller than in the cyanide complex. Indeed, the splitting is smaller than the pairing energy, leaving four unpaired electrons. $[Fe(H_2O)_6]^{2+}$ is thus strongly paramagnetic and strongly attracted to a magnetic field. A complex in which unpaired electrons occupy the higher-energy set of orbitals rather than pairing in the lower-energy set is referred to as a **high spin complex** and the ligand is called a **weak field ligand.** A complex with electrons paired in the lower-energy set of orbitals rather than unpaired in the higher-energy set is called a **low spin complex.** The ligand is called a **strong field ligand.**

One of the most important reactions in which iron changes from high to low spin occurs in hemoglobin, the oxygen-carrying component of red blood. This spin reversal is thought to be responsible for Fe^{2+} remaining as Fe^{2+} rather than oxidizing to Fe^{3+}. In fact, life depends on this switch in spin. Hemoglobin consists of Fe^{2+} coordinated to four nitrogen atoms in a ring called a porphyrin ring (Figure 10.24). A fifth coordination site is occupied by part of the globin protein, which protects the Fe^{2+} ion from excess oxygen that would directly oxidize Fe^{2+} to Fe^{3+}. In the inactive form, the sixth coordination site of iron is occupied by water. When molecular oxygen replaces water in the sixth coordination site, iron could transfer an electron to oxygen, becoming Fe^{+3} and ultimately forming Fe_2O_3, meaning that hemoglobin would become a pile of rust. Fortunately for humans and many animals, rather than transferring an electron to oxygen, two electrons in iron reverse spin, forming a low spin complex.

CONCEPT QUESTIONS Draw the d-orbital energy-level diagram for the two forms of hemoglobin, one with water in the sixth coordination site and one with oxygen in the sixth site. Is either form diamagnetic? Is either form attracted to a magnetic field? How many unpaired spins would iron have if it remained a high spin complex and oxidized to Fe^{3+}? (Although Fe^{2+} in hemoglobin is not in a completely octahedral environment, the energy-level diagram is essentially that of an octahedral complex.) ∎

Figure 10.24 The active center in hemoglobin consists of an Fe^{2+} ion (green) coordinated to four nitrogen atoms (blue) in the porphyrin ring of carbon atoms (black). The fifth coordination site is occupied by part of the protein that protects the iron ion from oxidation to Fe^{3+}. The final coordination site is occupied by either water or oxygen.

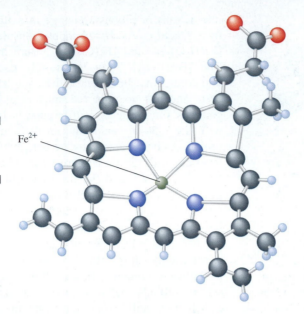

Fe^{2+}

WORKED EXAMPLE 10.5   *Tracing a Reaction with Magnetization*

Strong field ligand A ligand that produces a large splitting between sets of *d* orbitals in a transition metal ion; a splitting sufficiently large that electrons pair in the lower energy set of orbitals rather than occupying the higher energy set.

In hemoglobin, iron binds molecular oxygen in the lungs, where the partial pressure of oxygen is high. As the blood circulates, it enters the muscle tissue, where the oxygen pressure is low and oxygen is released. Carbon monoxide interferes with the oxygen transport function of hemoglobin by binding to iron. A student performs the following experiment to explore the role of carbon monoxide binding to iron. A solution is made with $FeSO_4$, and a magnetic measurement reveals a strongly paramagnetic solution. Carbon monoxide is bubbled through the solution, and the color changes. A magnetic measurement reveals that the new solution is diamagnetic. Puzzled, the student leaves the uncovered sample in the laboratory for several hours. Upon returning to the laboratory, the student observes that the solution has changed color again and is now weakly paramagnetic. Explain the student's observations.

Plan

- Determine the electron configuration of iron in the original sample.
- Unpaired spins lead to a paramagnetic substance, and paired spins to a diamagnetic substance.

Implementation

- The oxidation state of iron in $FeSO_4$ is $+2$. The electron configuration of Fe^{2+} is $[Ar]3d^6$.
- The sample starts out strongly paramagnetic, suggesting the existence of several unpaired spins. In water, Fe^{2+} is complexed to six water molecules as $[Fe(H_2O)_6]^{2+}$. Strong paramagnetism suggests that water is a weak field ligand [energy-level diagram (a) on page 365], leading to four unpaired spins. The color change on bubbling CO through the solution suggests that the water ligands are replaced by CO ligands. To become diamagnetic, CO must be a strong field ligand, forcing pairing of electrons in the three lower-energy orbitals (diagram b). Weak paramagnetism upon standing indicates a reaction

leading to an unpaired electron. Oxidation of the Fe^{2+} to Fe^{3+} is consistent with this magnetism (diagram c). Leaving the sample uncovered exposed the Fe^{2+} to oxygen from the atmosphere, also consistent with oxidation of Fe^{2+} to Fe^{3+}.

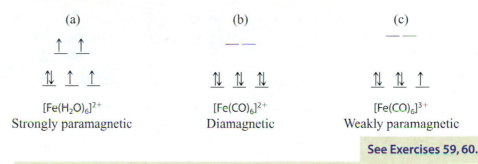

(a)

$[Fe(H_2O)_6]^{2+}$
Strongly paramagnetic

(b)

$[Fe(CO)_6]^{2+}$
Diamagnetic

(c)

$[Fe(CO)_6]^{3+}$
Weakly paramagnetic

See Exercises 59, 60.

Diamagnetic and paramagnetic substances are usually thought of as nonmagnetic because they cannot be permanently magnetized and their interaction with a magnetic field is relatively weak. By comparison, ferromagnetism is a very strong effect. In a magnetic field in which paramagnetic and diamagnetic substances gain or lose some weight, ferromagnetic materials become stuck fast to the poles of the magnet. Ferromagnetism, however, is quite rare. In the whole of the periodic table, only Fe, Co, and Ni are ferromagnetic at room temperature. One of the earliest uses of X-ray crystallography sought to determine whether the crystal structures of these three transition elements held the key for explaining ferromagnetism. They do not. The crystal structures of all three are different: Iron is body-centered cubic, cobalt is hexagonal close pack, and nickel is face-centered cubic. Rather than the crystal structure, the *d*-orbital localization and overlap are the critical factors.

Ferromagnetic materials, like paramagnetic materials, have unpaired electrons. In paramagnetic materials, the spin on one atom is not correlated with its neighbors, and the spins are random (Figure 10.25). When placed in a magnetic field, the spins in paramagnetic materials tend to align themselves with the field; there is a net magnetization. However, upon removal of the field, the spins again become random—nothing locks the spins together. In contrast, in ferromagnetic materials, the *d* orbitals in each atom overlap sufficiently with their neighbors that even after removal of an external field, the magnetic field of the neighboring electrons locks their spins together. At the same time, the *d* orbitals are sufficiently compact that the correlation between spins is sufficiently strong to keep them aligned. A delicate balance exists.

Both *s*- and *p*-orbital electrons are too diffuse to lock. At the start of the first transition series, Sc through Mn, the *d*-orbital electrons are valence electrons and also too diffuse to lock spins (Figure 10.26). At the end of the transition series, the *d*-orbital electrons are drawn deeper into the core and overlap is insufficient for correlation. Only for Fe, Co, and Ni do the *d* orbitals extend far enough for the spins to correlate, yet remain sufficiently

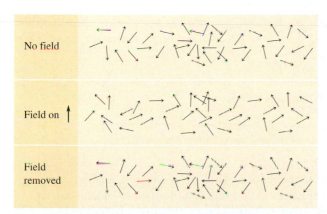

No field	
Field on ↑	
Field removed	

Figure 10.25 In the absence of a magnetic field, paramagnetic materials are characterized by a random distribution of directions for the unpaired spins. Application of a magnetic field tends to align the spins so that the net magnetic moment is nonzero. Here, more spins are pointing in an upward direction than are pointing downward when the field is on. Removing the field restores the random distribution of spins.

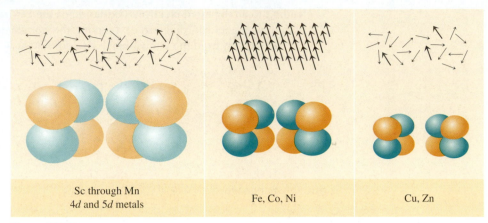

Figure 10.26 Schematic representation of *d* orbitals in the transition elements. Sc through Mn and the 4*d* and 5*d* orbitals are diffuse, providing insufficient correlation among spins on nearby atoms, so these elements are paramagnetic (or diamagnetic for Zn, Cd, and Hg). In Cu and Zn, the *d* orbitals have become part of the core and have insufficient overlap to provide electron spin correlation. Only in Fe, Co, and Ni are the *d* orbitals just right, having unpaired spins and sufficient overlap such that the magnetic field of the electrons becomes aligned with the field of neighboring electrons, creating large domains with aligned spins. Large domains with aligned spins produce a ferromagnetic material.

compact for the locking to be strong enough to maintain the larger domains, ranging from 10^{17} to 10^{21} atoms, required for permanent magnetism. In contrast, the 4*d* and 5*d* orbitals are more diffuse than the 3*d* orbitals and lack the necessary cooperative effect.

A few oxides are ferromagnetic, including Fe_3O_4 and CrO_2. In the oxides, the metal ions are small, so the *d* orbitals do not directly overlap. Instead, the oxide ions act as a bridge locking the electron spins together. The electronegativity of oxygen and the consequent charge on the metal have the effect of reducing the size of the metal's *d* orbitals. The reduced *d*-orbital size results in a shift to the left in the periodic table for ferromagnetic oxides, so that the magnetic oxides are those of Cr, Mn, and Fe.

CONCEPT QUESTIONS   What is the electron configuration of chromium in CrO_2? Would you expect TiO_2 to be magnetic? Why or why not? ∎

WORKED EXAMPLE 10.6   *Fluid Vacuum Seals*

One recent development in vacuum science technology is a material called a ferrofluid, which is used to prevent gas from crossing an atmosphere vacuum barrier. Ferrofluids consist of nanometer-diameter particles of either iron or iron oxide coated with a surfactant—a soap—and suspended in a carrier fluid, usually a hydrocarbon or a fluorocarbon. The surfactant coats the particles to prevent their agglomeration. The carrier fluid's viscosity and vapor pressure are important parameters in practical applications. Ferrofluids are often used to seal shafts in rotating applications, for example. Because the seal is fluid, there is essentially no friction and rotation speeds as high as 10,000 rpm are practical. A schematic of a typical seal is shown on the next page.

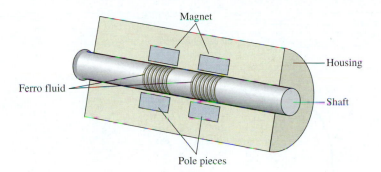

a. Examine the schematic and explain how the fluid acts as a seal.
b. Which element—Fe, Co, or Ni—is the best choice for this application? Explain why.
c. Why is it important to keep the particles separated?

Plan

- Sealing requires preventing gas from crossing the seal.
- Oxygen will inevitably get into the fluid. Consider the magnetic properties of the oxides.
- Consider friction.

Implementation

- The magnetic field pulls the magnetic particles and the fluid associated with it into the grooves in the shaft of the seal. With the seal region filled with fluid, gas molecules from outside cannot cross to the inside.
- The fluid will be exposed to oxygen on the atmosphere side of the seal, and the metal particles will slowly and inevitably oxidize. Iron oxide is magnetic, but neither the oxides of nickel nor those of cobalt are magnetic. If either Ni or Co is used, the performance of the fluid will degrade over time as the metal oxidizes and the magnet ceases to pull the fluid into the seal region. The iron oxide Fe_3O_4 is magnetic, so exposure to oxygen will not degrade the performance of a seal based on iron particles.
- If the particles are large, then, like grains of sand, they may become caught in the mechanism. Catching and grinding in the mechanism will cause friction between the particles and the seal, eventually wearing the seal away.

See Exercise 61.

10.6   Bioconnect

One of the intriguing phenomena in biological systems involves the transport of ions across a cell membrane against a concentration gradient—that is, from a region of lower to a region of higher concentration. Identification of some of the membrane spanning proteins involved in ion transport resulted in the 2003 Nobel Prize for physicians-turned-chemists Peter Agre and Roderick MacKinnon. Flow against a concentration gradient is counter to common experience. Put a drop of ink in a glass of water and the ink spreads out, uniformly coloring the water. The ink flows from a high-concentration region, the original drop, to the low-concentration regions in the remainder of the glass. Recent evidence indicates that one mechanism for flow against a concentration gradient in biological systems involves coordinate-covalent bonds working in conjunction with photo energy.

 One of the ions transported against a concentration gradient is Ca^{2+} (Figure 10.27). Calcium is an essential mineral for both plant and animal life. The ligand involved in calcium ion transport is shown schematically in Figure 10.28 and is known as a

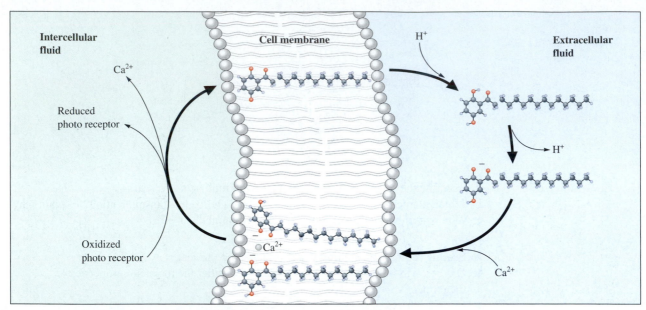

Figure 10.27 Transport of ions across cell membranes against a concentration gradient involves complexing the ion to increase its solubility in the uncharged cell membrane and releasing it inside the cell.

hydroquinone. The proton in the OH group nearest to the hydrocarbon chain ionizes, releasing the proton and generating the negatively charged ion (Figure 10.29). The negative charge is stabilized by the proximity of the two electronegative oxygen atoms. Both oxygen atoms have lone pairs, so this ionized hydroquinone is a bidentate ligand.

Two hydroquinone ion ligands coordinate to a single Ca^{2+} ion (Figure 10.30). Both ligands are bidentate, so the calcium ion has a coordination number of 4 in a nearly tetrahedral structure. Encapsulated in the tetrahedral structure, the solution around the

Figure 10.28 The hydroquinone involved in calcium transport contains a long-chain hydrocarbon. The close proximity of the —OH group to the $=O$ on the long chain is an important factor in hydroquinone's role as a bidentate ligand for Ca^{2+} ions.

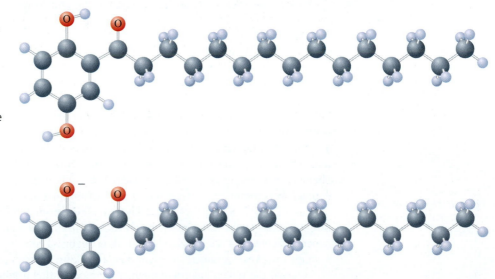

Figure 10.29 Ionization of the hydroquinone (Figure 10.28) leaves a negative charge stabilized by the close proximity of the two oxygen atoms.

Figure 10.30 Two bidentate ionized hydroquinone ligands coordinate to a Ca^{2+} ion to form a neutral complex. The structure around the calcium ion is nearly tetrahedral, so the environment is well shielded from the calcium ion.

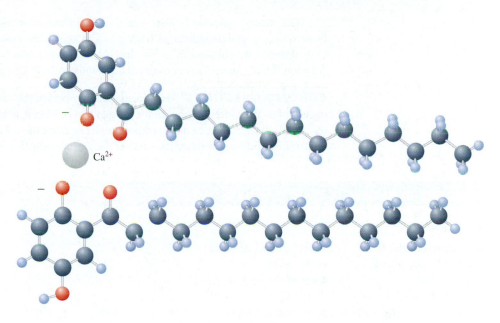

complex is well shielded from the concentrated field of the +2 charge of the calcium ion. The complex is neutral—the sum of the −1 charge on each ligand and the +2 charge on calcium. Just as oils do not dissolve well in water, so neutral hydrocarbon chains are not very soluble in the aqueous intercellular solution but are quite soluble in the cell membrane. Consequently, the neutral calcium ion complex becomes embedded in the cell membrane, taking the calcium ion with it.

When the calcium ion complex arrives at the inner surface of the cell membrane, it encounters a molecule called a *photoreceptor*. A photoreceptor absorbs photons and stores the energy for use by the organism. Photoreceptors are usually colored substances with a preferential absorption in part of the visible spectrum. The most commonly encountered photoreceptor is the green pigment in plants called chlorophyll. Photoreceptors in our eyes enable us to perceive color. In the calcium transport mechanism, the photoreceptor in the cell is oxidized by absorbing a photon. The oxidized form has a higher energy content than the nonactivated photoreceptor, so oxidation stores the photon energy. The oxidized photoreceptor accepts two electrons from each ionized hydroquinone ligand in the complex, oxidizing the ionized hydroquinone to the quinone (Figure 10.31). The quinone is also a bidentate ligand. Because it is neutral, however, the quinone is a much weaker ligand, binding the calcium ion less strongly. The calcium ion is thus released into the cellular solution, increasing the Ca^{2+} ion concentration in the cell.

Figure 10.31 The quinone shown here is formed by reduction of the ionized hydroquinone.

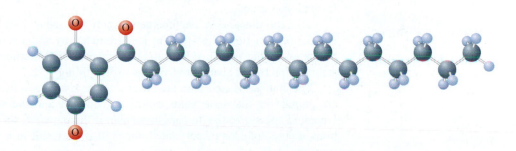

The neutral quinone is also less soluble in the aqueous cellular solution than the hydroquinone, and thus diffuses back into and across the cell membrane. The quinone is then reduced by substances in the intercellular fluid and picks up a proton from the solution. Thus, the cycle is completed and ready to start again.

CONCEPT QUESTIONS Assign oxidation numbers to the carbon attached to the —OH groups of (a) the hydroquinone, (b) the ionized hydroquinone, and (c) the quinone. How many electrons are released in oxidizing the ionized hydroquinone to the quinone? Does the oxidation number of the OH-carbon reflect the electron release? ■

Checklist for Review

KEY TERMS

coordinate-covalent
   bond (p. 347)
holes (p. 348)
octahedral holes (p. 348)
tetrahedral holes (p. 348)
transition metal
   complex (p. 351)
ligand (p. 351)
complex ion (p. 353)
complex (p. 353)
dentate (p. 354)
multidentate (p. 354)
chelate (p. 354)
amine (p. 354)
coordination
   number (p. 354)

spectrochemical series
   (p. 355)
formation constant, K_f
   (p. 357)
hydration enthalpy
   (p. 360)
magneto resistance effect
   (p. 361)
ferromagnetic (p. 362)
high spin complex (p. 363)
weak field ligand (p. 363)
low spin complex (p. 363)
strong field ligand (p. 364)

Chapter Summary

The color of gemstones derives from the interaction of transition metal ions with light: An interaction that results from splitting of the d-orbital energies by ligands that surround the ion. The bond between the metal ion and the surrounding ions or molecules is a coordinate-covalent bond, in which both electrons of the bond come from the surrounding ion or molecule. The color of the complex unit offers a qualitative way to measure the interaction between the coordinating ligand and the metal ion. A quantitative measure of bonding strength is provided by the equilibrium constant for formation of the complex. Those complexes with the strongest interactions have the largest formation constants.

The coordinate-covalent interaction between metal ions and water results in an exothermic reaction when the ion is surrounded by water molecules. The enthalpy of hydration of the transition metals shows a periodic variation, reflecting the d-orbital stabilization by the ligand field and the orbital filling.

The spin associated with the electron gives rise to a magnetic field. When all spins are paired, the magnetic fields cancel one another and the substance is diamagnetic. Unpaired spins give rise to paramagnetism if the unpaired spins are in random orientations. Large domains of correlated unpaired spins result in a ferromagnetic substance.

Correlation of spins depends on characteristics of the *d* orbitals; the *d* orbitals must extend sufficiently far that the *d* orbital on one atom overlaps with the *d* orbital on neighboring atoms but cannot be so diffuse that the electron spins do not lock. Fe, Co, and Ni are the only elements that meet this requirement. Metal oxides meet the correlation requirement because of the presence of an oxygen bridge. The ferromagnetic oxides are those of Cr, Mn, and Fe.

Coordinate-covalent bonds play a role in biological systems. The most well-known role is that played by iron in hemoglobin. The porphyrin ring and surrounding globin protein protect the Fe^{2+} ion from oxidation to Fe^{3+}. When binding molecular oxygen for transport, the Fe^{2+} ion switches from high spin to low spin rather than oxidize. Other molecules coordinate to metal ions and transport them across cell membranes, sometimes against a concentration gradient.

KEY IDEA

As a consequence of coulombic interaction between ligands and metal ions, the *d* orbitals in a metal ion do not all have the same energy. This energy separation gives rise to the color and the magnetic properties of metals and complexes.

CONCEPTS YOU SHOULD UNDERSTAND

- The complement of the color of a complex measures the interaction between the metal ion and its ligands.
- Unpaired electrons give rise to magnetism: paramagnetism if the electron spins on neighboring atoms are randomly oriented, or ferromagnetism if they are correlated.

OPERATIONAL SKILLS

- Connect the color of a transition metal complex to energy absorbed (Worked Example 10.1).
- Relate the *d*-orbital splitting pattern to the geometrical arrangement of the ligands (Worked Example 10.2).
- Determine the metal ion coordination number from the ligands and determine the charge on the complex from the metal ion plus ligand charges (Worked Example 10.3).
- Determine complex ion and hydrated ion concentrations from K_f (Worked Example 10.4).
- Use magnetic data to determine ligand binding strength (Worked Example 10.5).
- Explain the basis for technological applications of magnetism (Worked Example 10.6).

Exercises

A blue exercise number indicates that the answer to that exercise appears at the back of the book.

■ SKILL BUILDING EXERCISES

10.1 Gems

1. The gemstone emerald is green. In contrast, the gemstone ruby is red. Both gems have Cr^{3+} ions substituted for about 0.5% of the aluminum ions so that the Cr^{3+} ions are surrounded by six oxide ions in an octahedral arrangement. The emerald host crystal is beryl: $Be_3Al_2(SiO_3)_6$. The ruby host crystal is corundum: Al_2O_3. Compared with corundum, beryl is softer and less dense: The distance between the oxide ions and aluminum is expanded in beryl. Discuss the color of these gemstones in light of the environment around the Cr^{3+} ions.

2. An aqueous solution of Ni^{2+} is bright emerald green, and a solution of Cu^{2+} is bright blue. Both metal ions have a coordination number of 6 in water. Write the formula for each of these hydrates. Which hydrate absorbs higher-energy photons? Which interaction is stronger, that between Ni^{2+} and water or that between Cu^{2+} and water?

10.2 Transition Metal Complexes

3. Draw the energy-level diagram and fill in electrons for $[Ni(CN)_4]^{2-}$, which is a *square planar* molecule.

4. Draw the energy-level diagram and fill in the electrons for $[Ag(NH_3)_2]^+$, which is a *linear* molecule.

5. Due to its low reactivity, silver is often found as elemental silver in the environment. Nonetheless, it must be separated from other substances found with it. The metallurgy of silver consists of treating the crushed ore with NaCN and aerating it for about two weeks, thereby forming the $[Ag(CN)_2]^-$ complex. The solution is then separated from the remaining solid material. Silver is obtained by treating the solution with metallic zinc.
   a. Write equations for the chemical reactions corresponding to the metallurgy of silver.
   b. What is the oxidation state of silver in $[Ag(CN)_2]^-$?
   c. What is the coordination number of silver in $[Ag(CN)_2]^-$?
   d. What is the geometry of the complex $[Ag(CN)_2]^-$?
   e. Draw an energy-level diagram for the $[Ag(CN)_2]^-$ complex, filling in the electrons.

6. Gold forms a complex with CN^- that is used in the metallurgy of gold. Gold is two coordinate with an oxidation state $+1$.
   a. Write a formula for the gold cyanide complex.
   b. Draw an energy-level diagram for the gold cyanide complex, filling in the electrons.
   c. Do you expect the Au^+ ion to form a colored complex with cyanide? Why or why not?

10.3 Terminology

7. The complex CrO_4^{2-} is bright yellow, and the complex MnO_4^- is an intense purple. What is the oxidation state of the metal in each case? What are the molecular geometry and the coordination number of each of these oxides?

8. Identify the ligand(s) and give the coordination number and oxidation number for the central atom or ion in each of the following:
   a. $[Cr(NH_3)_5Cl]^{2+}$
   b. $[Fe(CN)_6]^{4-}$
   c. $[PtCl_4]^{2-}$
   d. $[Co(en)_3]^{3+}$
   e. $[CuEDTA]^{2-}$

10.3 Color

9. Potters produce a variety of colors in the glaze on a pot both by incorporating various metal ions in the glaze and by choosing an oxidizing or reducing atmosphere in the kiln. Use energy-level diagrams to explain why iron in the glaze produces red to yellow colors in an oxidizing atmosphere, while green colors are produced in a reducing atmosphere.

10. In ceramic glazes, the host matrix is aluminum silicate, Al_2SiO_5. Addition of copper ions imparts a blue color to the glaze, much like the blue of the hexaaquo complex. What is the local geometry around the copper ion?

11. The color of glazes containing Cr^{3+} ions depend heavily on the remaining components of the glaze. Incorporating Zn^{2+} results in a green color. With Pb^{2+} ions, chromium produces an orange or red color. Which ion, Zn^{2+} or Pb^{2+}, produces an oxide lattice with the larger lattice spacing?

12. Fill in the following table:

Complex	Color Observed	Color Absorbed	Wavelength Absorbed (nm)	Energy (eV)
$[CoEDTA]^{2-}$	violet		560	
$[Co(NH_3)_6]^{3+}$		violet	430	
$[Fe(CN)_6]^{3-}$	red		500	

13. Account for the following: All complexes of Ti^{4+} are colorless.

14. In solid corundum, aluminum ions are surrounded by six oxide ions in an octahedral configuration. Substitution of Cr^{3+} results in ruby. The oxide Cr_2O_3 has the same hexagonal structure as ruby, but Cr_2O_3 is green. The unit cell in corundum has lattice constants $a = 0.4763$ nm and $c = 1.3003$ nm; the unit cell of Cr_2O_3 has lattice constants $a = 0.4960$ nm and $c = 1.3599$ nm. Explain these materials' different colors, citing the lattice constants to support your explanation.

15. Incorporation of a small concentration of Cu^{2+} in MgO produces an absorption band with a maximum at 900 nm. Introducing a small concentration of Cu^{2+} into $MgAl_2O_4$ produces two bands with maxima at 1660 nm and 710 nm. What conclusion can you draw about the structure of $MgAl_2O_4$?

16. Silica gel is a common desiccant used to keep components dry in shipping and storage. For this application, it is very useful to have a visual indicator of when the gel has absorbed large amounts of water. A small concentration of Co^{2+} is incorporated in the silica gel to perform this function. When the indicator gel is dry, it is blue. This color turns to pink when moisture is absorbed. Explain this color change.

17. The active center in hemoglobin is Fe^{2+}. When coming from the lungs to the muscles, the complex formed is red. When making the return trip from the muscles to the lungs, the complex is blue. What do these colors indicate about the relative binding strength of the Fe(II)–O_2 complex versus the Fe(II)–CO_2 complex?

18. Is either of the complexes mentioned in Exercise 17 a candidate for being paramagnetic? Why or why not? (The hemoglobin complex is octahedral around the Fe^{2+} center.)

19. Nearly all anions can complex with metal ions. The complex of iron with thiocyanate, NCS^-, results in an intense red complex, $[Fe(NCS)_6]^{3-}$.
   a. What color does the $[Fe(NCS)_6]^{3-}$ complex absorb?
   b. Use energy-level diagrams to illustrate the origin of the color of $[Fe(NCS)_6]^{3-}$.
   c. Is the $[Fe(NCS)_6]^{3-}$ complex a candidate to form a high spin complex? Why or why not?

20. Chromium forms a variety of colored complexes. Listed in Table 10.1 are a series of ammonia and chloride complexes. Interpretation of the color becomes more complicated when two or more different ligands are involved. Use the spectrochemical series to predict the average interaction strength in these complexes.

10.4 Equilibrium

21. Hydroxide ion is a ligand for many metal ions. Zinc forms a four-coordinate complex with hydroxide. Use this information to write reactions corresponding to the following sequence.
   a. Sodium hydroxide is added to a transparent, colorless solution of $Zn(NO_3)_2$.
   b. Following addition of a small amount of sodium hydroxide, a white precipitate forms.
   c. Upon addition of more sodium hydroxide, the solution returns to a transparent, colorless state.

Table 10.1 Complexes of Chromium

Complex	$[Cr(NH_3)_6]^{3+}$	$[Cr(NH_3)_5Cl]^{2+}$	$[Cr(NH_3)_4Cl_2]^+$	$[Cr(NH_3)_3Cl_3]$	$[Cr(NH_3)_2Cl_4]^-$
Color	yellow	purple	green	violet	orange-red

22. Calculate the concentration of the following complex ions (assume the volume does not change on addition of a solid).
    a. $[Cu(H_2O)_6]^{2+}$ in a solution made by adding 0.001 mol $Cu(NO_3)_2$ to 1 L of 0.1 M NH_3
    b. $[Co(H_2O)_6]^{2+}$ in a solution made by adding 0.005 mol $Co(NO_3)_2$ to 1 L of a 0.03 M ethylenediamine solution
    c. $[Ni(H_2O)_4]^{2+}$ in a solution made by adding 0.001 mol $NiSO_4$ to 1 L of a slightly basic 0.01 M KCN solution
    d. $[Cu(NH_3)_4]^{2+}$ in a solution made by combining 950 mL of 0.004 M $[Cu(NH_3)_4]^{2+}$ with 50 mL of 1 M ethylenediamine solution (Assume the volumes are additive.)

23. Account for the following observations: When ammonium hydroxide is added to a Cu(II) solution, a blue precipitate forms. Upon further addition of ammonium hydroxide, a deep blue solution forms.

24. What is the effect of addition of ammonium chloride to the final solution in Exercise 23?

25. When a solution of NH_4OH is added to solid AgCl, the AgCl is partially dissolved. Write the reaction for the equilibrium that is established. If some nitric acid is added to this system, does the amount of solid AgCl increase, decrease, or stay the same? Explain your answer.

26. Suppose 9.9 mg CuCl is added to 1 L of solution containing 0.1 M KCN. Assuming no change in volume when the CuCl is added, what are the concentrations of Cu^+, Cl^-, K^+, and CN^- in the resulting solution?

27. What is the molar concentration of NH_3 needed to convert exactly 50% of the silver ion to the $[Ag(NH_3)_2]^+$ complex if 0.1 mol $AgNO_3$ is added to 1 L of ammonia solution? (Assume that the $AgNO_3$ does not change the volume.)

28. Write the equilibrium that is relevant for the following: A solution contains 0.010 M Ni^{2+} and Zn^{2+}. If KCN is added to this solution until $[CN^-]$ is 1.0 M, how much Ni^{2+} and Zn^{2+} will remain in solution?

29. Write the equilibrium that is relevant for the following: What concentration of NH_3 is required to dissolve completely 0.010 mol AgBr in 1 L of solution?

30. In what form is the Cd^{2+} ion in a solution made by addition of 0.01 mol $CdSO_4$ to 1 L of 1 M ammonia?

31. If 0.1 mol KCN is added to the solution in Exercise 30, in what form is the Cd^{2+} ion?

32. In what form is Fe^{3+} if 1 mol $Fe(NO_3)_3$ and 0.01 mol NaF are dissolved in 1 L of solution?

33. Where is the Fe^{3+} if 1 mol KCN is added to the solution in Exercise 32?

34. Determine the Cd^{2+} concentration when 0.01 mol $CdSO_4$ and 0.1 mol KCN are added to 1 L of 0.1 M ammonia.

35. Set up, but do not solve for, the F^- concentration in a solution made by adding 1 mol of $K_3(AlF_6)$ to 1 L of water.

36. What is the concentration of the Zn^{2+} ion in a solution of 0.1 mol solid $ZnCl_2$ added to 1 L of 0.4 M ammonia?

37. Which solution furnishes the higher concentration of Cd^{2+} ion, a 0.1 M $Cd(NH_3)_4$ Cl_2 solution or a 0.1 M K_2 $Cd(CN)_4$ solution?

38. A solution contains 0.01 mol Cl^- ion and 0.07 mol NH_3 per liter. If 0.01 mol of solid $AgNO_3$ is added to 1 L of this solution, will AgCl precipitate?

39. How much KCN is required to dissolve all the solid in 1 L of solution containing 0.001 mol $NiCO_3(s)$?

40. How many moles of NH_4Cl must be added to 100 mL of 0.1 M NH_4OH solution to prevent precipitation of $Mn(OH)_2$ when this solution is added to 100 mL of a 0.002 M solution of $MnCl_2$? Assume no change in volume on addition of NH_4Cl.

41. Copper is unique among the fourth-period transition elements in forming a +1 ion. Cu^+, like Cu^{2+}, forms a complex with ammonia: Cu^+ forms a two-coordinate complex and Cu^{2+} forms a four-coordinate complex. Write the reaction for formation of each of these complexes from an aqueous solution of the metal ion with ammonia. Determine the concentration of $[Cu(H_2O)_2]^+$ ion in a solution containing 0.001 M CuCl and 0.5 M ammonia. Determine the concentration of $[Cu(H_2O)_4]^{2+}$ in a solution made from 0.001 M $CuCl_2$ and 0.5 M ammonia. Compare the two concentrations and comment on the result.

42. The bidentate ligand ethylenediamine forms an octahedral complex with both Co^{2+} and Co^{3+}. Set up and perform a suitable equilibrium calculation to determine which complex is more tightly bound.

43. Cobalt is commonly found as a +2 and a +3 ion. Both ions form an octahedral complex with ammonia. Calculate the concentration of $[Co(H_2O)_6]^{2+}$ in a solution with 0.001 M Co^{2+} and 0.5 M ammonia. Upon oxidation of Co^{2+} to Co^{3+}, what is the concentration of $[Co(H_2O)_6]^{3+}$ remaining in solution?

44. A solution of $Ni(NO_3)_2$ forms a green precipitate upon addition of a low concentration of aqueous ammonia. Adding even more ammonia dissolves the precipitate, resulting in a deep blue solution. Write the chemical reactions corresponding to these observations.

10.4 Thermodynamics

45. The enthalpy of hydration for an ion is a measure of the strength of interaction between the metal ion and water. The hydration enthalpy of all metal ions is exothermic. The relative exothermicity reveals details about the interaction of water with specific ions.
    a. Write the reaction corresponding to hydration of Mn^{2+} and Ni^{2+} with water.
    b. Discuss why the reactions are exothermic.

46. The exothermicity of reaction warms the solution when a metal ion is hydrated by water. To a first approximation, the exothermicity is proportional to the ion charge to radius ratio. A closer inspection reveals differences between ions. For example, Mn^{2+} and Ni^{2+} are nearly the same size—the radius of Mn^{2+} is 67 pm and that of Ni^{2+} is 69 pm. Despite the slightly larger size of Ni^{2+}, its hydration enthalpy is significantly more negative than that of Mn^{2+}. Compare the temperature rise for addition of 0.01 mol Mn^{2+} in 1 L of water with that for 0.01 mol Ni^{2+} in 1 L of water. Comment on this apparent contradiction.

Data: $\Delta H_{\text{hydration}}$ (Mn^{2+}) = -1841 kJ/mol; $\Delta H_{\text{hydration}}$ (Ni^{2+}) = -2105 kJ/mol; heat capacity of water = 4.184 J/g·K.

47. Hydration of metal ions is an exothermic reaction, forming a complex ion with the surrounding water molecules. The stability of the complex increases with the charge and decreases with the ion radius. The exothermicity can be significant. Determine the temperature rise of water upon hydration of 0.01 mol of Na^+ ($r = 102$ pm), Mg^{2+} ($r = 72$ pm), and Al^{3+} ($r = 53$ pm) in 1 L of water. Data: $\Delta H_{\text{hydration}}$ (Na^+) = -405 kJ/mol; $\Delta H_{\text{hydration}}$ (Mg^{2+}) = -1922 kJ/mol; $\Delta H_{\text{hydration}}$ (Al^{3+}) = -4660 kJ/mol; heat capacity of water = 4.184 J/g·K.

48. Hydration of very small, highly charged ions can result in a coordinate-covalent bond sufficiently strong that the coordinated water molecule ionizes, leaving OH^- coordinated to the ion and generating H_3O^+ in solution. Generally, only multiply charged cations are sufficiently small to result in a detectable change in pH. Determine the pH of 0.01 M solutions of $Al(NO_3)_3$, $Cr(NO_3)_3$, and $Fe(NO_3)_3$. Data: K_a (Al^{3+}) = 1×10^{-5}; K_a (Cr^{3+}) = 1×10^{-4}; K_a (Fe^{3+}) = 6×10^{-3}.

10.5 Magnetism

49. A sample of the solid $Fe_2(SO_4)_3$ is vigorously attracted to a magnet. The solid $FeSO_4$ is also attracted vigorously. In contrast, the solid $K_3Fe(CN)_6$ is attracted less vigorously, and the solid $K_4Fe(CN)_6$ is essentially unaffected by a magnetic field. Explain these observations.

50. $Cr(CO)_6$ is a volatile, diamagnetic, white solid. Describe the arrangement of the electrons in this chromium solid.

51. A student is measuring the magnetic susceptibility of $[Co(NH_3)_6]^{2+}$ with the sample contained in an open tube. Just as she is about to measure the weight gain of the sample, the fire alarm goes off and everyone has to leave the building. Because the formal laboratory period is nearly over when students are allowed back into the building, the student stores the sample in her locker for measurement the next day. Upon returning to the laboratory, the student is dismayed to find that the sample has lost weight in the magnetic field. Explain the student's observations.

52. Suggest why complexes in which the central metal ion has the d^8 electronic configuration are frequently found in square planar geometry.

53. How many unpaired electrons are in the high spin complex, $[FeCl_6]^{3-}$? How many in the low spin complex, $[Fe(CN)_6]^{3-}$?

54. All complexes of V^{3+} have the same number of unpaired electrons irrespective of the ligand. Why?

55. How many unpaired electrons are in each of the following: Cr^{3+}, Cr^{2+}, Mn^{2+}, Fe^{2+}, Co^{3+}, and Co^{2+} for both the weak field and the strong field cases?

56. Indicate the electron configuration of the following ions in an octahedral field: Mn^{2+}, Ni^{2+}, Fe^{2+}, and Fe^{3+} for both the weak field and the strong field cases.

57. Determine the electron configuration (or give the energy-level diagram with electron occupancy) for $[Co(CN)_6]^{3-}$, a low spin complex, and for $[CoF_6]^{3-}$, a high spin complex.

58. The complex $[Fe(CN)_6]^{4-}$ is octahedral and diamagnetic. The complex $[Fe(H_2O)_6]^{2+}$ is octahedral and paramagnetic. Explain these observations using energy-level diagrams. What is the hybridization in each of these complexes?

59. A magnetic measurement indicates that an aqueous solution containing Fe^{2+} is paramagnetic. Upon addition of 3 mol of 1,10-phenanthroline (Figure 10.32), the paramagnetism disappears. Does 1,10-phenanthroline form a tetrahedral or an octahedral complex with Fe^{2+}? Show the energy-level diagrams that you use to support your conclusion.

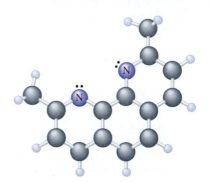

Figure 10.32 The ligand, 1,10-phenanthroline, is a bidentate ligand.

60. A deep blue, clear solution of Ni^{2+} and 0.5 M ammonia becomes cloudy and red upon addition of 2 mol of dimethylglyoxime (Figure 10.33). The deep blue solution is paramagnetic, but this paramagnetism disappears upon addition of dimethylglyoxime. Use these data to determine the coordination number of the nickel dimethylglyoxime complex. Draw the energy-level diagram and fill in the electrons for the nickel–ammonia complex. Draw a second energy-level diagram for the nickel–dimethylglyoxime complex, filling in the electrons.

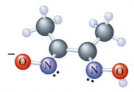

Figure 10.33 The bidentate ligand, dimethylglyoxime anion, has a single negative charge.

61. Canadian and U.S. nickels are the same size and weight although the compositions of the two coins differ. Canadian nickels are pure nickel, whereas U.S. nickels are 70% copper and 30% nickel. Explain how vending machines and pay phones might distinguish between the two coins.

62. Account for the following: Compounds of main-group elements are nearly all diamagnetic. The magnetism of covalent compounds of transition elements depends on the oxidation state of the metal. The magnetic state of transition

metal complexes depends on both the oxidation state of the metal and the nature of ligands binding to the metal ion.

10.6 Bioconnect

63. Cyanide is a particularly toxic substance due to its strong affinity for metal ions located in enzymes that provide the energy needed for cell respiration. Specifically, Fe^{3+} is found in the respiratory enzyme cytochrome oxidase. When the function of cytochrome oxidase is inhibited, the cells die. An antidote for cyanide poisoning consists of administering Co^{2+} ions as Co_2EDTA. The formation constant of Co_2EDTA is 2.0×10^{16}.
  a. What is the charge on Co_2EDTA?
  b. What is the equilibrium constant for replacement of the EDTA ligand by cyanide?

64. The oldest form of oxygen-transporting molecule is found in the blood of mollusks and arthropods that have roamed the earth since before the time of the dinosaurs. At the center of the oxygen-carrying protein are two copper atoms. As a consequence, the blood of horseshoe crabs is blue. Although perfectly capable of transporting oxygen, this copper system is less efficient than the iron system used in hemoglobin. The copper system binds one oxygen molecule in 5000, compared with four oxygen molecules out of 4000 for hemoglobin.
  a. Discuss the interaction of iron and copper with oxygen in their respective oxygen-carrying proteins. Which interaction is stronger?
  b. When oxygen binds to iron in hemoglobin, iron switches from a high spin complex to a low spin complex. Is copper capable of a similar spin switch?

65. In the body of a healthy person, the oxygen partial pressure in the lungs is 110 mm Hg. Atmospheric pressure is 760 mm Hg. Hemoglobin molecules leave the lungs 97% saturated with oxygen. The oxygen pressure in tissues while mildly exercising is 40 mm Hg and hemoglobin unloads 25% of its oxygen. While vigorously exercising, the oxygen pressure falls to 20 mm Hg and hemoglobin unloads 35% more oxygen. Discuss hemoglobin unloading of oxygen in terms of equilibrium. Determine an equilibrium constant for oxygen binding to hemoglobin (a) in the lungs, (b) in mildly exercising tissue, and (c) in vigorously exercising tissue. (*Note:* Due to cooperative binding of oxygen to hemoglobin, the equilibrium constants are not all the same.)

66. The average adult body contains about 5 L of blood. Hemoglobin in the blood contains about 2.1 g of iron. When oxygenated in the lungs, 97% of the iron in hemoglobin is coordinated to an oxygen molecule and is a bright red, low spin complex. After unloading more than half of its oxygen, blood is a blue color and the iron center has a high spin.
  a. Determine the molar concentration of iron in blood.
  b. Determine the concentration of oxygen molecules in oxygenated blood.
  c. Explain the red color of oxygenated blood.
  d. When deoxygenated, the sixth coordination site on iron in hemoglobin is occupied by water. What does the color of deoxygenated blood suggest about the binding of water to iron? Is this color consistent with the spin of the two complexes? Show any energy-level diagrams used in answering this question.

67. Myoglobin is the oxygen-carrying protein found in muscle tissue that imparts the deep red color to red meat. Myoglobin's affinity for oxygen is three to ten times that of hemoglobin—three times when mildly exercising to ten times when vigorously exercising. Set up equilibrium reactions showing the transfer of oxygen from hemoglobin to myoglobin. Use equilibria to discuss the sequence of events that occur when a person goes from mild exercise to vigorous exercise. (See Exercise 65 for relevant data.)

68. Carbon dioxide does not bind to hemoglobin. Instead, it is circulated in the blood to the lungs as dissolved bicarbonate and carbonic acid. Discuss the role played by pH in this transport. Cite relevant equilibrium constants. Data: The normal pH of human blood is 7.4.

69. Carbon monoxide (CO) binds to hemoglobin about 250 times more strongly than does molecular oxygen. If half the iron sites are bound to CO, the person smothers. In the lungs, the oxygen partial pressure is 110 mm Hg. Based on the relative binding strength, determine the lethal partial pressure of CO.

70. In a normal healthy individual, about 1% of the iron sites are occupied by CO. Smokers and individuals in heavily trafficked (highly polluted) areas add 15% more CO to their hemoglobin. Use the data in Exercise 70 to determine the CO partial pressure in a smoker's lungs.

■ CONCEPTUAL EXERCISES

71. Draw a diagram indicating the relationship among the following terms: transition metal complex, chelate, coordination number, spectrochemical series, coordinate-covalent bond, multidentate. All terms should have at least one connection.

72. Draw a diagram indicating the relationship among the following terms: formation constant, spectrochemical series, high spin complex, low spin complex, hydration enthalpy, ligand. All terms should have at least one connection.

⚙ APPLIED EXERCISES

73. Metal ions—particularly copper, iron, and nickel—catalyze the reaction of oxygen with fats in foods, resulting in rancidity. The metal ions come from a variety of sources, including soils and machinery used to process the foods. Comment on the use of $EDTA^{4-}$ to guard against oxidation. Consult the table of complex ion formation constants (Appendix A). Which metal ions are effectively sequestered by $EDTA^{4-}$ in the presence of Ca^{2+} ions in many foodstuffs?

74. Magnetite, Fe_3O_4, is a ferroelectric material. The magnetite structure consists of an FCC lattice of oxygen atoms with iron in both the octahedral and tetrahedral holes. The unit cell consists of eight FCC cubes of oxygen. Eight tetrahedral sites are occupied by Fe^{3+}, eight octahedral sites are occupied by Fe^{3+}, and eight octahedral sites are occupied by Fe^{2+}.
  a. Show that the unit cell is consistent with the formula Fe_3O_4. How many formula units are in a unit cell?
  b. Which is larger, the octahedral hole or the tetrahedral hole?
  c. The magnetization from the octahedral sites is larger than that from the tetrahedral sites. Magnetization from the

tetrahedral sites opposes that due to the octahedral sites, so magnetization can be increased by inserting nonmagnetic atoms in the tetrahedral sites. Suggest ions that are good candidates for increasing the magnetization of magnetite.

■ INTEGRATIVE EXERCISES

75. A low spin tetrahedral complex has not been reported, although many high spin complexes exist. What does this observation tell you about the splitting in tetrahedral complexes? Use geometrical arguments to indicate why this result is expected.

76. Explain the following observation: $[Ni(H_2O)_4]^{2+}$ is tetrahedral and paramagnetic, while $[Ni(CN)_4]^{2-}$ is square planar and diamagnetic. Show any energy-level diagrams used in your explanation.

77. $[Pd(CN)_4]^{2-}$ is a square planar complex. Do you expect this complex ion to be paramagnetic or diamagnetic? Why?

78. The formula for the complex formed by ammonia and Co^{3+} is $[Co(NH_3)_6]^{3+}$, and the complex is paramagnetic. The formula for the complex formed by chloride and Pt^{2+} is $[PtCl_4]^{2-}$ and this complex is diamagnetic. Indicate the geometrical configuration of each complex.

Chapter 11
Polymers

Scientists have long envied the strength and elasticity of spider silk. Dragline silk is about five times as strong as steel, can stretch twice as much as nylon, and is waterproof (as demonstrated by the dew shown in this photo).

CONCEPTUAL FOCUS

- Recognize molecular groups that can lead to bond-forming reactions.
- Connect macromolecular structure with material properties.

SKILL DEVELOPMENT OBJECTIVES

- Apply equilibrium and solubility concepts to acid–base condensation reactions (Worked Examples 11.1–11.3, 11.5).
- Connect material properties with monomer structure (Worked Example 11.4).

- Trace electron flow in conductive polymers (Worked Example 11.6).
- Connect physical and chemical properties of polymers with the monomers and the bonding (Worked Example 11.7).
- Identify the formation mechanism from a polymer's structure, and predict a formation mechanism from the monomer's functional groups (Worked Examples 11.8–11.10, 11.12).
- Recognize structures that generate radical initiators (Worked Example 11.11).

377

Natural and synthetic polymers (Figure 11.1) dominate much of the modern world. Polymers are versatile molecules with properties that range from the extremely hard material of airplane windows, which can withstand a bullet, to the soft and pliable material of food wraps. On the molecular level, a polymer is like an extremely long train of identical linked cars. Indeed, the term **polymer** is a juxtaposition of its parts: *poly,* meaning "many," and *mer,* meaning "units." A polymer is a large molecule with multiple, repeating units.

The story of how polymers with tailored properties were created begins with the work of the German chemist Hermann Staudinger in the early 1920s. Staudinger was the first to realize that polymers are not simply clumps of small molecules, but rather truly large molecules. For his pioneering work and insights, Staudinger received the 1953 Nobel Prize in chemistry. Today synthetic polymers have become embedded in daily life, and it is virtually impossible to open your eyes without encountering polymers with widely differing properties. Even the single polymer, polyethylene, can form the soft flexible material of plastic bags, the waxy material that coats milk cartons, or a fiber whose strength per pound exceeds that of steel. Polymers are indeed versatile molecules.

The disparate properties of polymers stem from a combination of the elemental composition of the units (the **monomers**), intermolecular interactions, and the **macromolecular structure.** Macromolecular structure refers to the spatial relationship of one part of a large molecule to other parts of the same molecule: whether the long chain is stretched straight, curled, pleated, or clumped. The various macromolecular structures dramatically affect how two polymer molecules fit together. Processing conditions strongly influence the macromolecular structure and hence the physical properties of the polymer. Indeed, the disparate properties of polyethylene are a direct result of the distinct conditions under which the polymer is assembled.

Polymer A molecule consisting of many repeat units that are all the same.

Monomer A small molecule that is the building block for a polymer.

Macromolecular structure Spatial relationship of one part of a large molecule to other parts of the same molecule.

(a)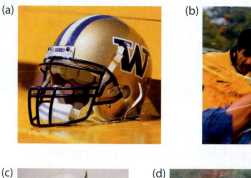

(b)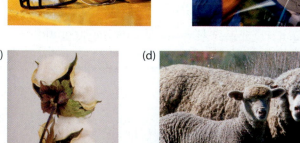

(c)

(d)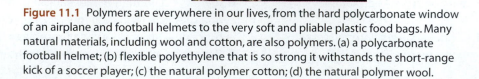

Figure 11.1 Polymers are everywhere in our lives, from the hard polycarbonate window of an airplane and football helmets to the very soft and pliable plastic food bags. Many natural materials, including wool and cotton, are also polymers. (a) a polycarbonate football helmet; (b) flexible polyethylene that is so strong it withstands the short-range kick of a soccer player; (c) the natural polymer cotton; (d) the natural polymer wool.

Polymers are quite large molecules, with molecular weights in the 100,000–300,000 amu range. For macromolecules, masses are conventionally expressed in **daltons,** where one dalton is one amu. With masses in the 100,000–300,000 dalton range, polymers are orders of magnitude more massive than the molecules we have encountered to this point.

Dalton Expression for mass. One dalton is one amu.

CONCEPT QUESTION One of the larger molecules that we have investigated so far is $[PbEDTA]^{2-}$, which is used to treat lead poisoning in children. What is the mass of $[PbEDTA]^{2-}$ in daltons? ∎

The systematic investigation of the composition of natural polymers with the goal of imitating them began at DuPont, in the laboratory of Wallace H. Carothers. The first goal was to make materials imitating the natural polymers silk, rubber, and cellulose — three materials that remain important today. The first synthetic polymer to be developed was nylon, a silk analogue, which was introduced in 1939 at the New York World's Fair. It was an instant success; 4 million pairs of nylon hosiery were sold in the first few hours. Nylon remains an important polymer, accounting for nearly one-third of the polymer tonnage produced in the United States.

Two distinct chemistries are involved in nearly all polymer synthesis. An acid–base condensation reaction is the fundamental chemistry involved in making nylon. The other type of chemistry involves pulling electrons from multiple bonds or strained bonds to form a new bond. Formation of rubbers, for example, features multiple-bond or strained-bond chemistry. The double-bond motif is also present in ethylene, the monomer for polyethylene. Unpairing electrons in a bond to form new bonds is involved in the generation of conducting polymers as well. Both acid–base condensation reactions and electron-unpairing reactions are examined in this chapter.

11.1 Case Studies

Nylon

Nylon, like nearly all polymers, is composed primarily of carbon and hydrogen, a composition known as a **hydrocarbon.** Carbon is unique among the elements in forming long, strong chains of carbon bonded to carbon. Various functional groups can be added to this simple theme. The two functional groups that are relevant to synthesis of nylon are the amine and acid groups. An **amine group** substitutes a hydrocarbon chain for one of the hydrogen atoms of ammonia (Figure 11.2). Incorporation of the hydrocarbon chain has little effect on ammonia as a base (Worked Example 11.1).

Hydrocarbon A compound composed of carbon and hydrogen.

Amine group An organic compound with an $-NH_2$ group.

WORKED EXAMPLE 11.1 *Basic Strength*

Ammonia is a strong base that is often used as a household cleaner. Water and ammonia compete for one of water's protons when in solution. The base ionization reaction for ammonia is

(11.1) $$NH_3(aq) + H_2O(l) \rightleftharpoons NH_4^+(aq) + OH^-(aq)$$

The equilibrium constant is $K_b(NH_3) = 1.8 \times 10^{-5}$. In a 1 M ammonia solution, what is the ratio of solvated ammonia molecules to ammonium ions?

Replacing one or more hydrogen atoms in ammonia with a hydrocarbon chain produces an amine. Making ammonia into an amine modifies the interaction with water.

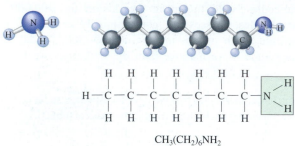

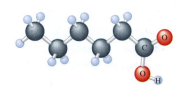

$$CH_3(CH_2)_6NH_2$$

Figure 11.2 Ammonia with one hydrogen atom replaced by a hydrocarbon forms an amine. The number of carbon atoms and the configuration in the hydrocarbon chain can vary widely. This example has a simple seven-carbon chain. The amine is shown in both a ball-and-stick model and a stick-diagram model. The amine group is highlighted in the stick diagram. The amine can also be represented with an in-line formula.

Replacing one hydrogen atom with a methyl group produces methyl amine, CH_3NH_2, the substance responsible for the fishy odor of herring brine. Compare the interaction of methyl amine with water to that of ammonia with water. K_b (CH_3NH_2) $= 4.5 \times 10^{-4}$.

Plan

- Write the expression for the equilibrium constant.
- Determine the relationship between $[NH_3]/[NH_4^+]$ and the equilibrium constant.
- Solve for $[NH_3]/[NH_4^+]$.
- Repeat for NH_2CH_3.
- Compare NH_3 to NH_2CH_3.

Implementation

- $K_b = \dfrac{[NH_4^+][OH^-]}{[NH_3]}$
- Stoichiometry indicates that $[OH^-] = [NH_4^+]$. Let $[NH_4^+] = x\ M$; then $[NH_3] = (1 - x)\ M$. $[NH_3]/[NH_4^+] = (1 - x)/x$.
- $K_b = x^2/(1 - x) = 1.8 \times 10^{-5} \Rightarrow x = 4.2 \times 10^{-3} \Rightarrow [NH_3]/[NH_4^+] = 240$
- For CH_3NH_2: $[CH_3NH_2]/[CH_3NH_3^+] = 47$
- Both ammonia and methyl amine compete favorably with water for the proton, with methyl amine competing slightly more favorably than ammonia.

Acid group The $-COOH$ portion of an organic molecule.

The cleaning power of methyl amine is somewhat greater than that of ammonia due to the greater ionization. However, due to the fishy odor, it never made the grade as a cleaning product!

See Exercises 7, 77, 78.

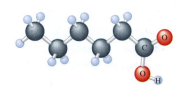

$$CH_3(CH_2)_4COOH$$

Figure 11.3 An organic acid consists of the $-COOH$ group. The number of carbons and the configuration of the hydrocarbon attached to the acid group vary widely. A simple five-carbon chain is illustrated here. The acid is shown as a ball-and-stick model, a stick diagram, and an in-line formula. The acid group is highlighted in the stick diagram.

The **acid group** consists of $-C\underset{\diagdown OH}{\overset{\diagup O}{}}$ (Figure 11.3), where the elements at the open bond can be varied.

CONCEPT QUESTION What is the range of normal oxidation numbers for carbon? Assign an oxidation number to the carbon atom in an acid group. Explain the attraction between the carbon atom of the acid group and a lone pair of electrons. ■

The carbon of an acid functional group is bonded to two oxygen atoms. Oxygen is nearly the most electronegative element in the periodic table. Hence electron density collects around the oxygen atoms at the expense of the carbon atom. This electron density loss leads to two senses in which the $-COOH$ functional group acts as an acid. First, it ionizes in water, transferring a proton to water:

(11.2) $-C\underset{OH(aq)}{\overset{O}{}} + H_2O(l) \rightleftharpoons -C\underset{O^-(aq)}{\overset{O}{}} + H_3O^+(aq)$

Here the functional group acts as an Arrhenius acid. Altering the hydrocarbon chain has little effect on the acidity of the acid.

WORKED EXAMPLE 11.2   *Organic Acids*

The two-carbon acid illustrated in Figure 11.4 is known as acetic acid. Acetic acid is the acidic component of vinegar, which is itself a major component of salad dressing. The tangy taste of salad dressing comes from the physiological response to the acid ionization. What is the degree of ionization—that is, the ratio of molecular H_3CCOOH acid to ionized H_3CCOO^- acid in a solution that is 1 M in acetic acid? ($K_a = 1.8 \times 10^{-5}$)

Replacing the $—CH_3$ group of acetic acid with a hydrogen atom results in formic acid, HCOOH, the active ingredient in ant bites and many insect stings. Compare the degree of ionization of acetic acid with that of formic acid. (K_a (HCOOH) = 1.8×10^{-4})

Plan

- Write the acid ionization reaction.
- Write the expression for the equilibrium constant.
- Determine the relationship between $[H_3CCOOH]/[H_3COO^-]$ and the equilibrium constant.
- Solve for $[H_3CCOOH]/[H_3COO^-]$.
- Repeat for HCOOH.
- Compare acetic acid to formic acid.

Implementation

- $H_3CCOOH(aq) + H_2O(l) \rightleftharpoons H_3COO^-(aq) + H_3O^+(aq)$

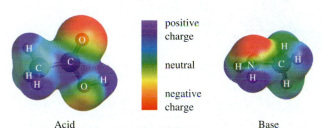

positive charge

neutral

negative charge

Acid   Base

Figure 11.4 A density potential plot of the two-carbon acid, acetic acid, illustrates the positive charge at the acid carbon due to collection of electron density around the electronegative oxygen atoms. The electropositive carbon atom is an electron-pair acceptor. A similar density potential plot for an amine illustrates the negative region of the nitrogen lone pair. The amine is an electron-pair donor.

- $K_a = \dfrac{[H_3COO^-][H_3O^+]}{[H_3CCOOH]}$
- Stoichiometry indicates that $[H_3COO^-] = [H_3O^+]$. Let $[H_3COO^-] = x$ M; then $[H_3CCOOH] = (1 - x)$ M. $[H_3CCOOH]/[H_3COO^-] = (1 - x)/x$.
- $K_a = x^2/(1 - x) = 1.8 \times 10^{-5} \Rightarrow x = 4.2 \times 10^{-3} \Rightarrow [H_3CCOOH]/[H_3COO^-] = 240$
- For HCOOH: $[HCOOH]/[HCOO^-] = 76$
- The acids have a very similar degree of ionization, with the stinging formic acid being slightly more ionized.

The sting of an insect bite is due to the acid ionization. Hence application of baking soda to an insect bite reduces the sting.

See Exercises 7, 77, 78.

The second sense in which the $—C\!\!\begin{smallmatrix}O\\ \\OH\end{smallmatrix}$ group is an acid is in the more general sense developed by Lewis. The acid carbon has lost electron density (Figure 11.4) to the two oxygen atoms bonded to it, leaving the carbon with a partial positive charge (oxidation number = +3). The partial positive charge has an electrostatic attraction for a negative charge, such as that of a lone pair. This is similar to the attraction between two water molecules resulting in the autoionization of water:

(11.3) $2H_2O(l) \rightleftharpoons H_3O^+(aq) + OH^-(aq)$

Lewis acid An electron pair acceptor.

Lewis base An electron pair donor.

The focus on electrons in this interaction leads to the Lewis definition of an acid and a base: A **Lewis acid** is a substance capable of accepting an electron pair, and a **Lewis base** is a substance capable of donating an electron pair. Due to the emphasis on the electrons, the Lewis definitions of an acid and a base are very general.

In forming complex ions, metal ions also act as Lewis acids. There is an important difference between metal ions and the acid group, however: Metal ions have low-lying empty orbitals into which an electron pair can be accepted to form a coordinate-covalent bond. Not so with carbon. It is a second-row element, so the closest energy empty orbital is far removed. Thus the *maximum coordination for carbon is 4.*

For a base such as ammonia to form a bond to the acid carbon, one of carbon's bonding pairs must be released. For example, if the acid carbon releases the OH group, then carbon restores the coordination number to 4. Similar logic applies to nitrogen, which forms a maximum of three bonds plus a lone pair. Transferring a hydrogen atom to the OH group of the acid forms water—a stable molecule. The result of the interaction (Figure 11.5) is formation of the functional group known as an **amide.**

Amide A bond which results from condensation of an amine and an acid.

Joining an acid and an amine, and thereby forming a single amide bond with no other functional groups, makes only a small molecule. A modification is needed to construct a longer chain.

Figure 11.5 Formation of the bond between the acid carbon and the amine nitrogen is accompanied by elimination of —OH from the acid and —H from the amine to form water.

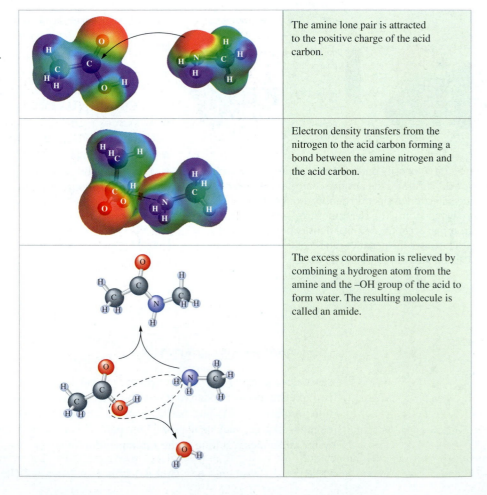

The amine lone pair is attracted to the positive charge of the acid carbon.

Electron density transfers from the nitrogen to the acid carbon forming a bond between the amine nitrogen and the acid carbon.

The excess coordination is relieved by combining a hydrogen atom from the amine and the –OH group of the acid to form water. The resulting molecule is called an amide.

The essential modification, first implemented by Wallace Carothers in 1927, is to modify the hydrocarbon portion of the acid and amine molecules to incorporate a second reactive center (Figure 11.6). With a second amine group, the molecule becomes a *diamine*, and each amine group can react with an acid. Similarly, the acid is made into a *diacid*. Reaction of a diamine with a diacid produces an amide with an additional amine at one end and an additional acid on the other end. Each of these groups can react further, and so on.

Alternately, an acid can be incorporated in the hydrocarbon portion of an amine. The result is an **amino acid.** Proteins are amino acids. Proteins are the main constituents of hair, nails, skin, muscle tendons, and enzymes that catalyze many essential reactions. In proteins, the amide bond is known as a **peptide bond.**

Amino acid A molecule containing both an amine and an acid functional group.

Peptide bond The bond resulting from the condensation reaction between amino acids.

WORKED EXAMPLE 11.3   *Room-Temperature, Bench-Top Synthesis of Nylon*

Practical application of polymer synthesis involves consideration of several issues. Suppose you are given the task of synthesizing a nylon fiber on the bench top as Carothers did at DuPont. The first issue is to prevent the monomers from simply solidifying into a large mass. If your goal is controlled addition of monomers to the growing polymer chain, which of the monomers in Figure 11.6 would you choose? In what way should the monomers in Figure 11.6 be modified? What is the preferable pH?

Plan

- Choose two monomers to separate them from the chain.
- Control monomer addition to the growing chain by using solubility.
- Choose pH such that protonation of the base is prevented.

Implementation

- If a single monomer is chosen (that is, the amino acid), there will be no way to prevent instant reaction. Choose a diamine and a diacid instead.
- Solubility depends on the proportion of hydrocarbon chain to functional groups. Short chains are water-soluble; long chains are soluble in organic solvents. Choose the length of the long-chain hydrocarbon so that the monomer is soluble in an organic solvent. Choose a short hydrocarbon chain for the other monomer. Ethylenediamine is a short-chain diamine that is soluble in water. Choose ethylenediamine for water solubility and a long-chain diacid for solubility in an organic solvent.
- To prevent protonation of the diamine, make the aqueous solution containing the diamine very basic.

A common bench-top synthesis of nylon (Figure 11.7) uses a further modification of the diacid. The —OH group is replaced by a —Cl, the result is called an acid chloride. If the diacid is used, it deprotonates in contact with the basic amine solution and chain growth is inhibited. The diacid chloride does not deprotonate, so HCl is produced rather than water. The HCl is neutralized by the basic diamine solution, producing water and salt.

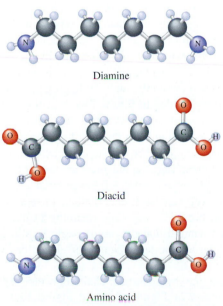

Diamine

Diacid

Amino acid

Figure 11.6 Polymers are made from molecules with two reactive centers. Examples shown here are (from top to bottom) a seven-carbon diamine, an eight-carbon diacid, and a seven-carbon amino acid.

See Exercises 1, 4–6.

A jumble of polymer strands by themselves do not produce a very resilient fiber. Think of a plate of spaghetti: The individual strands readily

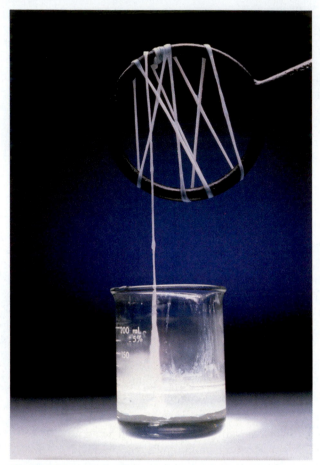

Figure 11.7 Solubility principles are used to control end-on growth of a nylon chain. A short-chain diamine is soluble in an aqueous solvent, and a long-chain diacid is soluble in an organic solvent. Nylon produced at the interface is pulled into a long fiber.

Stick diagram Representation of an organic molecule consisting of lines representing bonds, carbon atoms occur at the junction between lines, hetero atoms are shown explicitly and there are a sufficient number of hydrogen atoms to fill out the four coordination of carbon.

slide over one another so that the whole has little cohesion. However, if the strands have a very sticky sauce, then they hold together. On the molecular level, the analogue of the sticky sauce is the interaction between strands.

CONCEPT QUESTIONS Which elements or groups are necessary for hydrogen bonding? Does nylon contain these elements or groups? ■

The backbone of nylon contains units that can interact via hydrogen bonds: oxygen plus hydrogen bonded to nitrogen. To hydrogen bond, the N—H group must line up with the oxygen. This alignment is accomplished by stretching the fiber as it is drawn so that the polymeric chains are parallel to each other. The amide N—H bonds on one chain interact with the oxygen atom of the amide bond on the next chain to form a hydrogen bond (Figure 11.8). Hydrogen bonds are individually weaker than covalent bonds, but the regular structure of the stretched chains contains the possibility a large number of hydrogen bonds at very frequent intervals. The result is a very strong intermolecular interaction and a very strong fiber.

It is quite cumbersome to draw ball-and-stick models, particularly for large molecules, so chemists have devised abbreviated notations. One notation is the **stick diagram,** which is frequently used for smaller molecules as well. Figure 11.9 shows the stick diagram for nylon made from a six-carbon amino acid. In this notation, carbon atoms are located at the junctions of lines that represent the bonds. Atoms other than carbon and hydrogen, called heteroatoms, are shown explicitly. To illustrate hydrogen bonding in nylon, the hydrogen atoms involved in hydrogen bonding are shown explicitly, with the hydrogen bond being indicated with a dotted line. Hydrogen atoms needed to fill the four coordination of carbon atoms or three coordination of nitrogen are not shown. A more compact notation for polymers (Figure 11.10) shows just the repeated unit. In this notation, the ends of the repeated unit are the bonds that link monomers together. The unit is enclosed in parentheses, with n indicating the number of repetitions. Because polymers are most often produced with a range of values for the number of units, n is often left unspecified.

In addition to being very strong, nylon is a very resilient, elastic fiber. In metals, elasticity results from straining bond angles and bond lengths while retaining nearest-neighbor atoms. If the stress on a metal is high enough that the atoms begin to slide, then the elastic response regime is exceeded. The atomic-level picture of elasticity in polymeric materials is slightly different.

Nylon is a flexible, stretchy fiber because the hydrocarbon chain between the hydrogen-bonded amides is flexible. A comparison of Figure 11.8 and Figure 11.11 illustrates stretched and relaxed nylon fibers. All fibers—cotton, wool, silk, and the synthetic ones—share the characteristic that the chain connecting the two functional ends is somewhat flexible. The functional ends contain polar groups so that interchain hydrogen bonds occur at frequent, regular intervals.

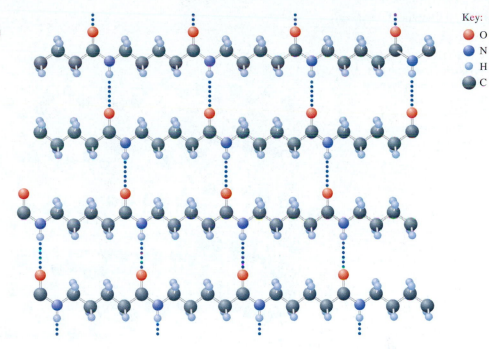
Figure 11.8 Ball-and-stick model of hydrogen bonding in nylon-5.

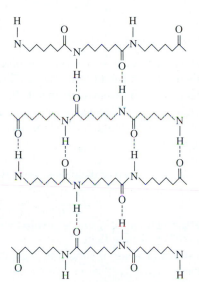

Figure 11.9 An abbreviated notation, called a stick diagram, is often used for organic molecules. Shown here is the stick diagram for nylon made from an amino acid. In this notation, bonds are represented as lines, carbon atoms are located at the junctions between lines, and atoms other than carbon or hydrogen, called heteroatoms, are shown explicitly. Hydrogen atoms involved in hydrogen bonding are also shown explicitly. Hydrogen bonds are indicated with dotted lines.

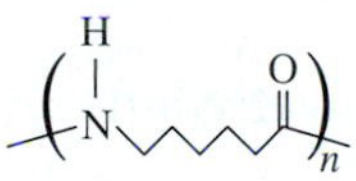

Figure 11.10 A shorthand notation for polymers shows only the repeated unit with the linking bonds at each end of the unit. The repeated unit is enclosed in parentheses, and the subscript n indicates the number of copies of the unit in the polymer. Often n is unspecified and encompasses a range of values.

Figure 11.11 The hydrocarbon chains between the hydrogen-bonded amide groups are flexible. Flexibility in the chain results in a resilient, elastic fiber.

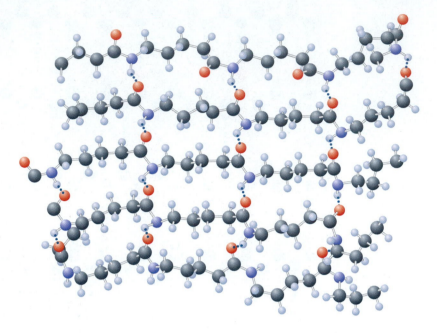

Polyethylene: Structure Property Control

Polyethylene is a prime example of how the material properties of a polymer can be altered by controlling the synthesis conditions. The synthesis conditions determine the arrangement of monomers, affect the average chain length, and control the macromolecular shape. Two parameters, length and shape, determine chain packing. Chain packing affects the material properties and the utility of the polymer. Polyethylene can be found in a variety of forms (Figure 11.12):

■ The waxy coating of a milk carton or the petals and leaves of realistic artificial flowers
■ Clear food wrap
■ Hard bottles used for bleach
■ Sheets of "artificial ice" on skating rinks
■ The world's strongest fiber, having six times the strength of steel on a pound-for-pound basis

High-density polyethylene (HDPE) The form of polyethylene that consists of unbranched chains resulting in efficient packing and a high density.

Low-density polyethylene (LDPE) The form of polyethylene that consists of branched chains resulting in inefficient packing and a low density.

Long, straight chains can pack very efficiently. The chains are analogous to logs floating down a stream. The stream current aligns the logs, often forming a continuous covering on the stream. Efficient packing of the polymer chains makes a material that is strong and dense, called **high-density polyethylene (HDPE).** Melting HDPE and drawing it into a fiber under pressure packs the long chains into a tight bundle and produces a very strong fiber, stronger than steel on a pound-for-pound basis. Fusing these fibers together results in a sheet that is both cloth-like and extraordinarily tear-resistant. This material is used for polypack shipping envelopes.

In contrast, tangled chains with bits branching off do not pack efficiently. Inefficient packing makes a flexible material that has gaps between molecules—this is, a low-density material. This **low-density polyethylene (LDPE)** is the polymer that coats cartons, is used for food wraps, and makes artificial flowers.

Figure 11.12 The many faces of polyethylene are illustrated in the varied products fashioned from it, including flexible plastic food wraps, waxy carton coatings and artificial flowers, rigid bottles, slippery sheets, and extremely strong fibers.

Ethylene The molecule $H_2C \!=\! CH_2$.

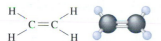

Figure 11.13 Ethylene consists of two carbon atoms joined by a double bond, shown here in a stick diagram and as a ball-and-stick model.

Polyethylene is a polymer made from a single monomer, **ethylene** (Figure 11.13). Ethylene has two carbon atoms, which are linked by a double bond, plus four hydrogen atoms to complete the skeleton. This monomer has only one functionality—the double bond. The polymerization of ethylene takes advantage of this double bond.

The double bond consists of (Figure 11.14) two electrons in a σ-bond molecular orbital and two electrons in a π-bond molecular orbital. The σ bond directly joins the two carbon atoms and is a very stable, low-energy orbital. Not so the π orbital (Figure 11.15). It consists of the sideways overlap of one p orbital from each of the carbon atoms. Sideways overlap results in the π orbital being located above and below the plane of the molecule. With a node through the molecular plane, the π orbital is a high-energy orbital.

Because the electrons in the π orbital are not tightly bound, it takes relatively little energy for the two paired electrons to become unpaired (Figure 11.16), occupying a p orbital in each carbon atom. The double bond opens up, forming two unpaired

Figure 11.14 Bonding in ethylene consists of a σ bond directed between the carbon atoms and a π bond located above and below the plane of the molecule. The σ bond is more stable, lower energy than the π bond.

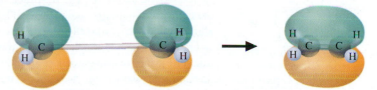

Figure 11.15 The π molecular orbital in ethylene consists of the sideways overlap of two atomic p molecular orbitals. The orbital consists of two lobes: one above and one below the molecular plane. The π molecular orbital has a node through the molecular plane.

Figure 11.16 The π molecular orbital electrons in ethylene can become unpaired, forming a diradical. The unpaired electrons attack at a region of high electron density to form an electron pair, a bond. The result retains two unpaired electrons but is a longer molecule.

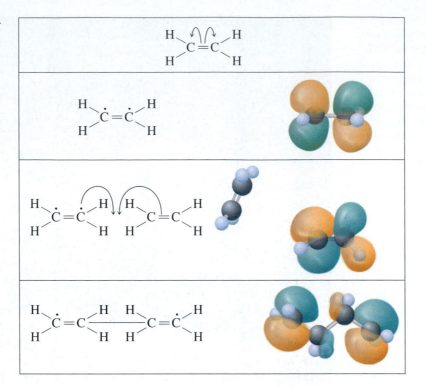

Radical An atom, molecule, or ion with at least one unpaired electron.

electrons. An unpaired electron is called a **radical.** Radicals are involved in many reactions and have recently received much attention in medicine, where the damaging effect of radical attack is warded off with antioxidants. In ethylene, the radicals are very reactive and do not remain unpaired for long. Unpaired electrons are attracted to regions with loosely bound electrons (Figure 11.16). Joining the unpaired electron with a loosely bound electron in another ethylene forms an electron pair, a bond, between the carbon atom with the unpaired electron and that with the loosely bound electron. The result is a longer chain (with four carbons) and propagation of the radical.

Chemists depict the sequence of steps for addition of one ethylene molecule to another with electron-pushing, arrow diagrams (Figure 11.16). The arrow starts at the electron, and its head points to the electron's destination. Proton shifts are similarly shown with arrows: the tail of the arrow at the hydrogen and the lead points to the proton's destination.

The four-carbon chain produced is still a diradical, so the process can continue. Another addition at the end carbon forms a longer chain. But notice that nothing keeps the radical on the end of the chain. A little rearrangement results in the shift of the radical to another position on the chain (Figure 11.17). Electrons easily shift position in a molecule; shifting the H involves only moving the proton. If this rearranged radical reacts, it forms a *branched* chain. Branched-chain growth creates a polymer with inefficient packing, LDPE. Further rearrangement (Figure 11.18) can lead to collapse of the diradical and termination of the chain. To confine the radical to the end of the chain making an unbranched chain and HDPE requires prevention of this radical rearrangement.

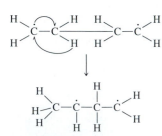

Figure 11.17 Shifting the unpaired electron results in a branched point of attachment for the growing chain.

Low-Density Polyethylene

The most common use of polyethylene is as a plastic wrap. Approximately 100 billion pounds of polyethylene is manufactured every year. Of this amount, 40% is used in plastic wraps.

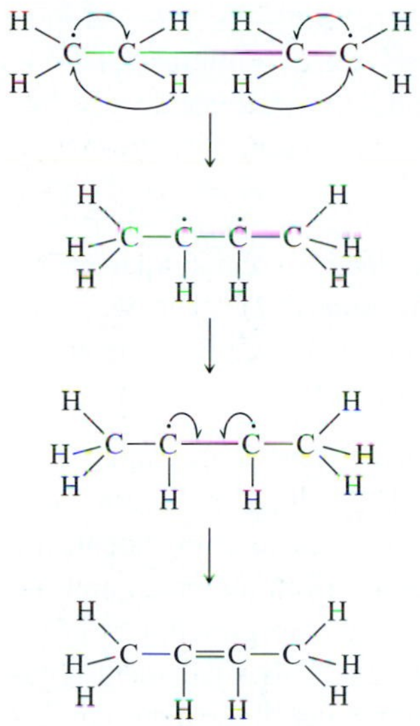

Figure 11.18 A double shift results in collapse of the diradical, formation of a double bond, and termination of chain growth.

Heterogeneous catalysis
Process in which a substance in a different phase (often solid) increases the rate of a reaction without being consumed in the reaction.

One way to generate a radical so as to initiate polyethylene formation is to take a molecule that has a very weak bond and heat it until that bond breaks. An example is benzoyl peroxide (Figure 11.19). Oxygen is a relatively small atom, and the stable form of oxygen features a double bond. As a result, the O—O single bond in benzoyl peroxide is weak. Heating benzoyl peroxide to about 100 °C breaks this bond, forming two radicals. Each radical eliminates CO_2 to form a benzyl radical.

(11.4)

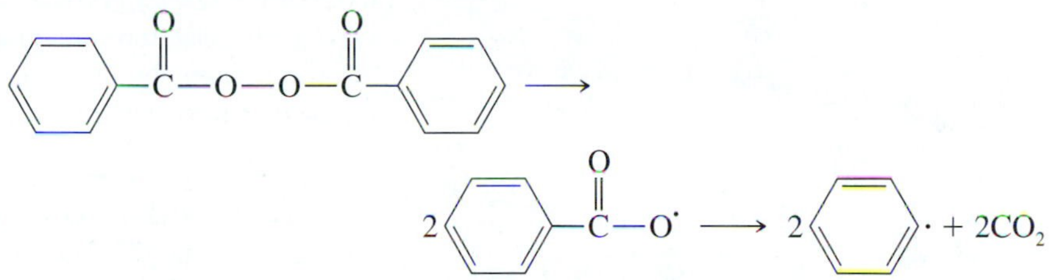

Because the stable molecule CO_2 has been eliminated, the pieces cannot be easily reassembled into benzoyl peroxide. Instead, the benzyl radicals attack the double bond in ethylene, and the reaction is off and running. This method for initiating chain polymerization results in a very fast polymerization. As nothing in the mechanism prevents rearrangement of the position of the radical in the growing polyethylene chain, branches develop (Figure 11.20) and the end product is LDPE.

High-Density Polyethylene

Preventing side-chain formation requires another method for chain extension, one that forces end-on addition. Although the full details of how this method works are not well understood, the salient features have been identified. In the 1950s, two European-based chemists, Karl Ziegler of West Germany and Giulio Natta of Italy, revolutionized the production of alkene polymers by using a solid surface to catalyze the polymerization of alkenes. This development was so significant that Ziegler and Natta won the 1963 Nobel Prize in chemistry for their work.

Because the reaction occurs on a solid surface while the monomers and the growing chain remain in solution, the method is called **heterogeneous catalysis.** It relies on the tendency of small, highly charged metal ions to complex to electron-rich molecules and

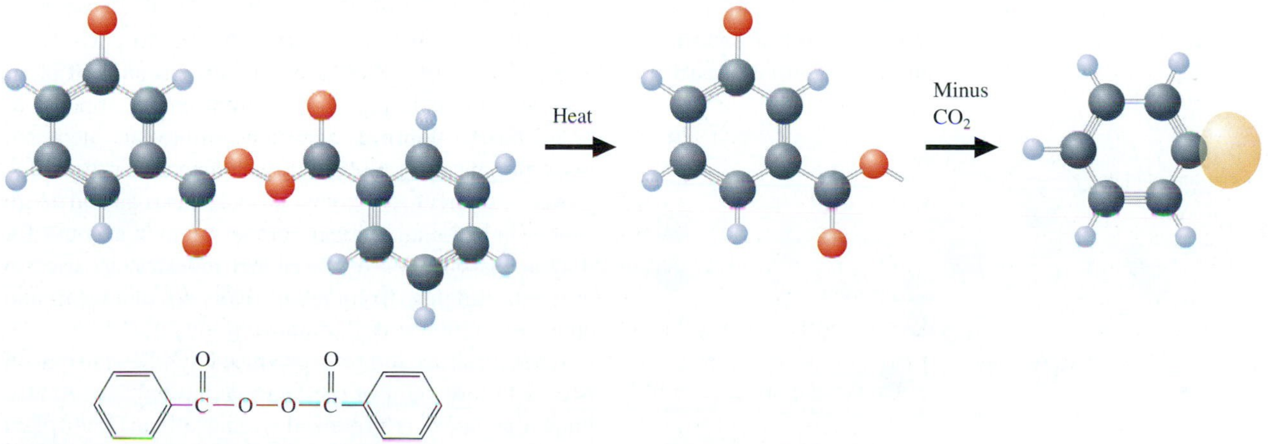

Figure 11.19 Benzoyl peroxide is shown as both a stick diagram and a ball-and-stick model. Due to the weak single O—O bond, when heated the O—O bond in benzoyl peroxide breaks. The stable molecule CO_2 is eliminated and the benzyl radical is generated. The lone electron orbital in the benzyl radical is shown on the right.

Figure 11.20 Branched and jumbled polymer chains do not pack very efficiently, so the material has a low density. With ample space to move, the chains easily respond to a stress, so the material is highly flexible.

Coordination polymerization Polymerization mechanism with addition to a double bond occurring via coordination to a solid catalyst.

ions. The double bond in ethylene is electron-rich, for example. A common implementation of this method uses a mixture of $TiCl_4$ and $Al(C_2H_5)_3$. The mixture of $TiCl_4$ and $Al(C_2H_5)_3$ is a deeply colored solid, where the color indicates that titanium is present in a $+3$ oxidation state.

CONCEPT QUESTIONS   What is the electron configuration of Ti? What are the electron configurations of Ti^{3+} and Ti^{4+}? Explain the following statement: If titanium remained in a $+4$ oxidation state in the catalyst, the catalyst would not be colored. ■

It is believed that the electron-rich ethylene is coordinated to both the Ti^{+3} and the electropositive Al, pinching the coordinated monomer (Figure 11.21). Then, in a concerted rearrangement, the monomer is added onto the growing chain. Both the chain and the monomer are coordinated to the Ti-Al center, so growth can occur only at the end attached to the Ti-Al center. This attachment regulates the site of addition and prevents branching of the chain. In reference to the role of the metal complex in the reaction, the method is called **coordination polymerization.** Not surprisingly, coordination polymerization occurs more slowly than radical polymerization.

Stretching the growing polymer as it forms aligns the long axes of the numerous polymer chains (Figure 11.22), producing a highly ordered material with long, unbranched chains. Because the chains pack regularly, this material has a very high density.

Stretching imparts strength in two ways. First, pulling to a small diameter aligns virtually all the chains with the long axis parallel to the long fiber direction. To tear across the fiber, these very strong covalent bonds must be broken. Second, aligning the chains increases the interactions between the chains. The difference in electronegativity between carbon and hydrogen is quite small, so a C—H bond is not very polar and the interaction, the London dispersion force, is relatively weak. However, when this force accumulates over hundreds of units and works synergistically with very strong covalent bonds, the resulting fiber has a high tensile strength that can rival that of steel.

Rubber

Of the 2.2 million metric tons of synthetic rubber produced in North America every year, the majority is used to make tires. Of course, there are tens of thousands of other uses for this material, ranging from engine mounts to belts, hoses, flooring, adhesives, footwear, electrical insulation, bearing pads to protect against earthquakes, gloves, catheters, condoms, carpet backing, and foam rubber in bedding. Rubber coatings, which account for only a small portion of the rubber industry, are used in products ranging from tennis balls to rain gear and constitute a billion-dollar industry.

Most rubber today is produced synthetically, so it is easy to lose sight of the fact that natural rubber, also known as latex, is produced by more than 2500 plant species. The limited use of natural rubber stems from its low molecular weight. To be useful in commercial products, the molecular weight needs to be in the

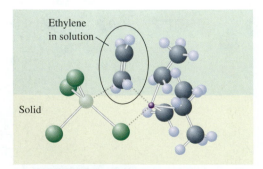

Key:
- Al
- Ti
- Cl
- C
- H

Figure 11.21 The heterogeneous polymerization of ethylene using a mixture of $TiCl_4$ and $Al(C_2H_5)_3$ as the catalyst to open up the double bond of ethylene.

Figure 11.22 Aligned poly-ethylene chains produced by drawing the growing polymer as it forms on the heterogeneous catalyst.

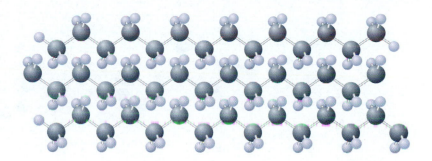

vicinity of 1 million daltons, producing particles about 4.0 microns in diameter. Only a very few plants produce latex in this size range. Early attempts to produce synthetic rubber made a sticky product. To produce a useful rubber, the configuration of the added units needs to be controlled.

CONCEPT QUESTION A common monomer in rubber consists of five carbon atoms and eight hydrogen atoms. How many monomers are required to form a polymer of molecular weight 1 million daltons? ▪

The simplest rubber—the basis of natural latex—begins with the monomer isoprene (Figure 11.23). Isoprene has two double bonds, compared to just one in ethylene. The second double bond makes the polymerization reaction somewhat more involved, but is essential for cross-linking of rubber. As in ethylene, radical attack (Figure 11.24) produces a bond by pairing one of the π-bond electrons with the radical electron. The remaining π-bond electron remains unpaired, propagating the radical. After attack there are three relatively weakly bound electrons in isoprene: the two electrons of the remaining double bond and the unpaired, radical electron. The combination is more stable if the double bond is rearranged to appear in the middle of the isoprene unit (Figure 11.24).

The two ends of the isoprene molecule are not fixed with respect to each other. That is, the ends can rotate freely about the single bond. Electron rearrangement thus produces two distinct atomic configurations around the double bond. One configuration has both the —H and the —CH$_3$ groups on the same side of the double bond (Figure 11.25), called a

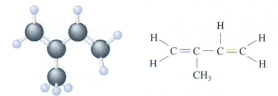

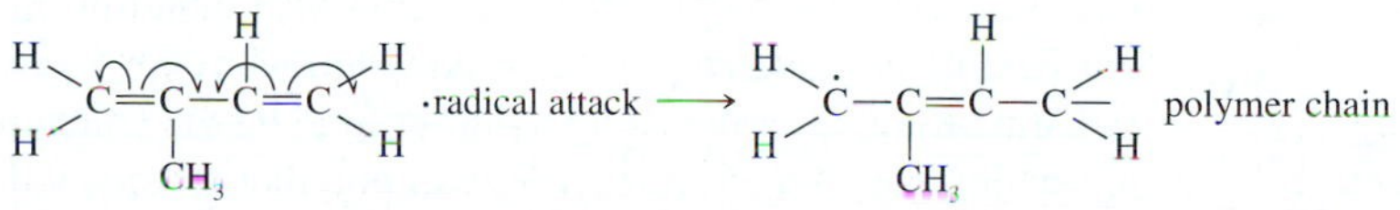

Figure 11.23 The monomer for natural latex is isoprene, shown here as both a ball-and-stick model and a stick diagram.

Figure 11.24 Attack on either double bond of iso-prene produces a radical. The unit is stabilized by shifting electrons so that the remaining double bond appears in the middle of the unit.

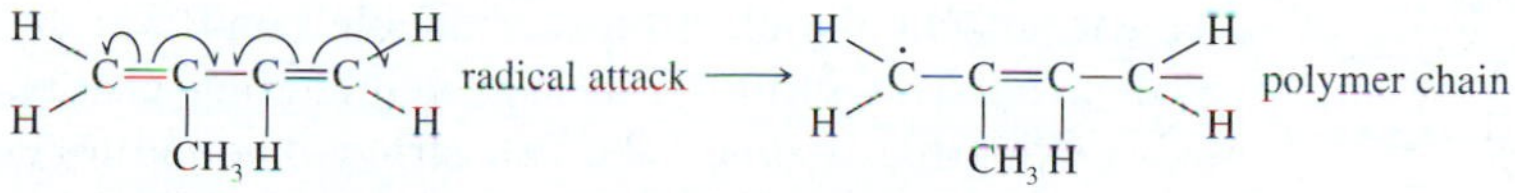

Figure 11.25 Double-bond rearrangement in the polymerization of isoprene can result in the —H and —CH$_3$ groups appearing on the same side of the double bond (shown here), called the *cis* configuration, or on opposite sides (Figure 11.26), called the *trans* configuration.

cis-polyisoprene   *trans*-polyisoprene

Figure 11.26 Natural latex consists of *cis*-polyisoprene. *trans*-Polyisoprene makes a harder material due to denser packing and lower flexibility of the two —CH$_2$— groups linking the double bond.

Cis Arrangement of two functional groups on each end of a double bond in which both groups are on the same side of the double bond with hydrogen atoms on the other side.

Trans Arrangement of two functional groups on each end of a double bond in which the groups are on the opposite side of the double bond.

cis configuration. The second configuration has the —H and —CH$_3$ groups on opposite sides of the double bond (Figure 11.26) and is called a ***trans*** configuration. Arrangement about the double bond affects packing of the polymer chains and hence the physical properties of the polymeric material. Natural latex is produced exclusively in the *cis* configuration. Early attempts to produce synthetic polyisoprene produced a mixture of *cis* and *trans* configurations—and a sticky goop.

Development of the Ziegler-Natta catalysis in the 1950s provided a method for production of 100% *cis*-polyisoprene. Either all *cis*- or all *trans*-polyisoprene produces a useful rubber. In the *cis* form, the CH$_2$ groups linking the double-bonded carbon atoms have greater flexibility in orientation, so *cis*-polyisoprene produces a softer rubber.

CONCEPT QUESTION Examine the schematic of the Ziegler-Natta catalyst (Figure 11.21). How does the Ziegler-Natta catalyst favor the all *cis* configuration of polyisoprene? ▪

Natural rubber is not a particularly resilient material. Think of the many polymer chains as a mass of spaghetti or earthworms. As anyone who has been fishing knows, earthworms can clump into a ball that is easily picked up. Soon, however, individual worms wriggle out of the mass and the whole clump disintegrates. To make a resilient rubber, the strands need to be linked more robustly.

The process of linking polymer strands was discovered long before the molecular nature of polymers was understood. In 1839, the American inventor Charles Goodyear discovered that heating rubber in the presence of sulfur binds the polymers into a tough, but flexible product. Due to the high temperature involved, Goodyear called the process **vulcanization,** after the Roman god of fire and the volcano. On the molecular level, vulcanization consists of connecting two polymer strands via a covalent bond with a small number of linking atoms (Figure 11.27). The covalent connection between the two strands is called a **cross-link.**

Vulcanization A process in which sulfur is added to rubber and the mixture is heated, causing crosslinking of the polymer chains and thus adding strength and resiliency to the rubber.

Cross-link Covalent bonds holding two polymer chains together.

Cross-linking highlights the importance of the remaining double bond: It provides a site for cross-linking. At high temperature sulfur, an electronegative element, attacks the electron-rich double bond. The double bond opens, forming a covalent bond to sulfur. Attack at neighboring chains results in a strong covalent bond linking the chains. Like two strings knotted together, the two polymer chains are attached. The remaining unpaired electron on each chain reacts with more sulfur, linking to yet another chain. The result is a network of connected polymer chains.

Figure 11.27 Vulcanization of rubber links the individual strands with a tough, covalent bond. In this example, *cis*-polyisoprene is cross-linked by addition of a sulfur linkage across the remaining double bond in polyisoprene.

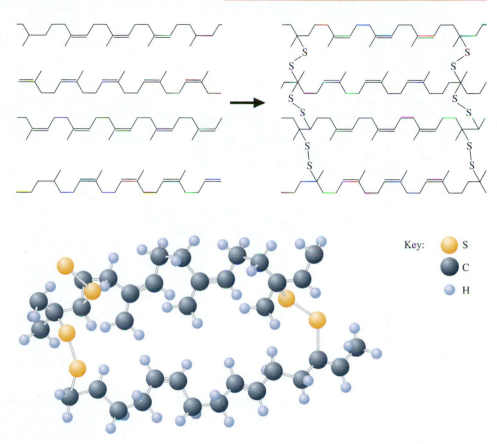

Key: S   C   H

Stresses that reconfigure the chain cannot disrupt the link because covalent bonds are very strong.

The physical properties of rubber are greatly affected by the degree of cross-linking. The portions of the polymer chains between the cross-links are free to slide over one another, to stretch and distort bond angles similar to the elastic response of a metal. The elastic response of a rubber, however, is many times that of a metal. Typical metals exceed the elastic response limit after a strain of only 0.1%. In contrast, typical polymers can elongate by 500% without suffering degradation. The cross-links hold the key to this enormous stretchability (Figure 11.28).

CONCEPT QUESTIONS Polyethylene is a versatile material. Can polyethylene be cross-linked to form a rubber? Why or why not? ▪

The cross-linked sites serve as anchors, keeping the chains from sliding completely away (wriggling out of the mass) from one another, and bringing the object back to its original shape upon removal of the stress. Frequent cross-linking limits the range of free movement, making for a stiffer rubber. Less frequent cross-linking produces a softer rubber.

CONCEPT QUESTION Obtain a rubber band that is at least $^3/_8$ inch wide. Mark a diamond shape on the band in ink. Stretch the rubber band: how has the diamond shape changed? Explain this macroscopic observation in terms of the molecular-level changes in the rubber polymers tagged with the ink. ▪

Figure 11.28 Stretching a polymer rearranges the macromolecular structure. Because the polymer chains are cross-linked with strong covalent bonds, the chains remain associated. As illustrated here, the stretching can be very large. Upon removal of the stress, the chains relax to assume their previous configuration and the object returns to its original shape. Colors: carbon (black), hydrogen (light blue), sulfur (yellow).

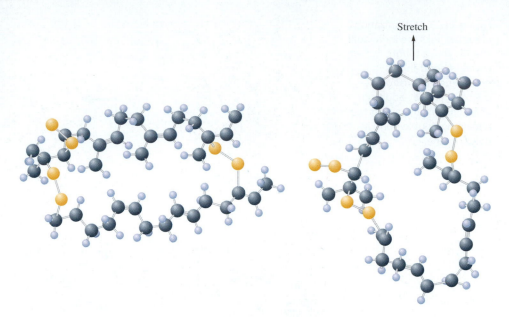

Stretch

WORKED EXAMPLE 11.4   *A Vibration-Damping Rubber*

The term "rubber" generally evokes an image of a bouncy material. To bounce, the chain distortion caused by an impact must be readily restored. On the molecular level, this process corresponds to the unhindered return of the polymer chain to its prestressed configuration. Some applications require a rubber that damps vibrations; pads designed to minimize the damage from earthquakes or supports for vibration-sensitive equipment are but two examples. One vibration-damping rubber is made from norborene,

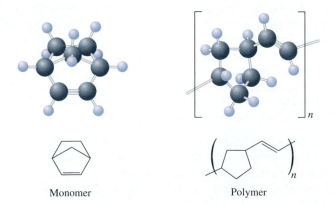

Monomer   Polymer

The vibration-damping characteristic is due to the structure of the polymer chains. How does the molecular structure of the chain contribute to vibration damping?

Plan

■ Examine the structure of the linking chain.

■ Find the flexible and inflexible sections of the chain.

Implementation

- The chain consists of a double bond and a pentagonal ring.
- Neither the double bond nor the pentagonal ring is very flexible. The linking chain is relatively bulky and rigid. Coiling and uncoiling of this chain entails passing these bulky groups past each other. Norborene forms a stiff, vibration-damping rubber due to the atomic-level friction of these bulky groups as they pass each other. Friction dissipates energy, so energy imparted by compressing the rubber (as from an impact with the floor) is partially lost to friction. Dissipated energy is not returned in the bounce, so motion is damped.

See Exercises 18, 32.

APPLY IT   Obtain a balloon with a neck large enough to stretch over your finger. Stretch the balloon over your pointer finger and hold it in place with your thumb. Grasp the balloon with your free hand and stretch rapidly. Does your finger feel warmer or cooler? Release the balloon. What temperature change does your finger detect? In which state, stretched or relaxed, do the long-chain molecules in the balloon interact more strongly? How should the molecules in the balloon be altered to make the balloon stiffer?

Conducting Polymers

Polymers are constituted primarily of carbon and hydrogen, two elements normally thought of as nonconductors. A few polymers, however, are conductors. One of these is called polyaniline. In addition to being a conductor, polyaniline is a colorful material.

Aniline, $C_6H_5NH_2$ (Figure 11.29), can be thought of as a variation on ammonia: Replace one of the hydrogen atoms of ammonia with a six-membered carbon ring called a phenyl ring or **phenyl group.** The phenyl ring has a significant effect on the properties of ammonia (Worked Examples 11.1 and 11.5).

Phenyl group The benzene molecule minus one hydrogen atom: — C_6H_5.

WORKED EXAMPLE 11.5   *Electron Donation: Base Strength*

The colorful character of aniline compounds launched the synthetic dye industry. In 1856, 17-year-old William Henry Perkin was experimenting in his home laboratory while on holiday and accidentally cooked up a black goop that stained cloth a beautiful mauve color. The electrons that give rise to this color also affect interaction of this relative of ammonia with water. Ammonia interacts with water by donating an electron pair to form the complex $H_3N \cdot HOH$, which then dissociates to form NH_4^+ and OH^-. Aniline interacts with water in a similar manner. Write the reaction that occurs between aniline and water, and compare the pH of 1 M solutions of ammonia, methyl amine, and aniline.

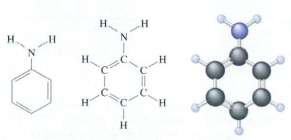

Figure 11.29 Three representations of aniline: a structural diagram, a stick diagram, and a ball-and-stick model.

Data

$K_b(NH_3) = 1.8 \times 10^{-5}$
$K_b(\text{methyl amine}) = 4.5 \times 10^{-4}$
$K_b(\text{aniline}) = 4.3 \times 10^{-10}$

Plan

- Write the base ionization reaction for aniline.
- Write the expression for K_b (aniline).
- Solve for $[OH^-]$, pOH, and pH.

Implementation

- $$\text{aniline}(aq) + H_2O(l) \longrightarrow \text{anilineH}^+(aq) + OH^-(aq)$$

Initial:		1 M	
Change:	x	$-x$	x
Equilibrium:	x	$(1 - x)\,M$	x

- $$K_b = \frac{[\text{aniline H}^+][OH^-]}{[\text{aniline}]} = \frac{(x)^2}{(1 - x)} = 4.3 \times 10^{-10}$$

- Since K_b is quite small, x is small $(x \ll 1) \Rightarrow x^2 = 4.3 \times 10^{-10} \Rightarrow$
  $x = 2.1 \times 10^{-5} \Rightarrow [OH^-] = 2.1 \times 10^{-5} \Rightarrow pOH = 4.7 \Rightarrow pH = 9.3.$

A similar calculation for ammonia results in $[OH^-] = 2.4$ and pH = 11.6. For methyl amine, pOH = 1.7 and pH = 12.3.

The ammonia and methylamine solutions are about $2\frac{1}{2}$ times more basic than the aniline solution. Aniline is less basic, indicating that the presence of the phenyl ring makes the lone pair less available for donation.

A note about aniline dyes: Perkin retired as a millionaire at the age of 37, after which he became a university professor and made numerous contributions to organic chemistry. Colorful chemistry can also be profitable chemistry!

See Exercises 7, 77, 78.

Interaction between the phenyl ring and the nitrogen lone pair modifies the properties of the nitrogen lone pair. The phenyl ring carbon atoms are sp^2 hybridized (Figure 11.30). Each carbon atom in the ring forms three σ bonds: one to a hydrogen atom and two to neighboring carbon atoms. The fourth valence electron in carbon is in a p orbital that is perpendicular to the plane formed by the sp^2 hybrid. In the phenyl ring, all six carbon atoms have an electron in a p orbital, and all are perpendicular to the plane of the ring (Figure 11.30b). These orbitals are sufficiently near to one another that they

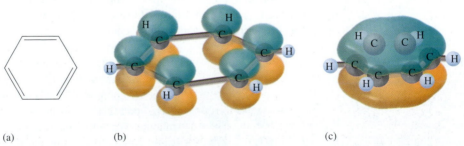

(a)   (b)   (c)

Figure 11.30 (a) A stick diagram of a phenyl ring shows the six carbon atoms joined by alternating single and double bonds. Due to resonance, all bonds are equivalent. (b) The C—C bonds consist of a σ bond and the sideways overlap of an atomic p orbital in each of the carbon atoms. Here the bond has been stretched sufficiently that the separated p orbitals are visible. (c) Bringing the carbon atoms to bonding distance reveals the electron orbital formed by the overlap of the six separate atomic p orbitals.

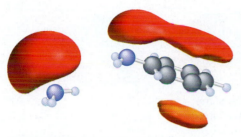

Figure 11.31 Ammonia and aniline are related molecules. In aniline one of the hydrogen atoms of ammonia is replaced with a six-carbon ring. The nitrogen lone pair blends with the electrons of the carbon ring.

overlap, forming an extended electron wave above and below the plane of the ring. With a node containing the plane of the ring, the molecular orbital is called a π orbital.

Due to the proximity of the nitrogen lone pair to the phenyl ring's π orbital, the lone pair melds with the π electron cloud (Figure 11.31), forming an electron wave that extends over the ring and the nitrogen atom. Extending the wave has two important consequences. First, the electron pair is less localized and less concentrated, and thus it interacts less strongly with water. As a result, aniline is a weaker base than is ammonia. Second, because the electron wave extends over the entire ring and the nitrogen atom, when several aniline molecules join to form polyaniline, the new molecule has the potential to conduct electricity.

Polymerization of aniline uses the loosely bound electrons of the melded lone pair and π orbital. The negative charge of these loosely bound electrons is attracted to a region of positive charge or positive potential. When the positive potential is sufficiently large, aniline loses one of the loosely bound electrons, forming a *radical cation* (Figure 11.32). Aniline is also oxidized. A common procedure for polymerizing aniline uses a positively charged, conductive glass for the location of the polymerization (Figure 11.33). The conductive glass forms the positive electrode.

CONCEPT QUESTIONS   Removal of one electron from aniline oxidizes it and produces a radical cation. Why is oxidized aniline a radical? Why is it a cation? ■

Oxidation reduces the number of electrons associated with the nitrogen and changes the nitrogen hybridization from sp^3 to sp^2. The unpaired, remaining nitrogen electron occupies a nitrogen p orbital. This p orbital is coplanar with the phenyl ring's carbon p orbitals, overlapping with the adjacent carbon atom's p orbital. All seven p orbitals (six from the phenyl ring plus the one from nitrogen) form an extended molecular orbital (Figure 11.32). Overlap of the extended molecular orbitals of adjacent aniline units produces an even more extended molecular orbital that spreads over the long polymer chain (Figure 11.34). This configuration is important for the conductivity of polyaniline.

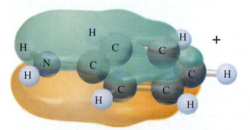

Figure 11.32 Application of a positive potential removes an electron from aniline. With one electron removed, nitrogen becomes sp^2 hybridized, with the odd electron residing in an atomic p orbital. The geometry at the nitrogen atom becomes trigonal planar, and the p orbital strongly overlaps with the phenyl ring's π orbital. The positive charge is localized primarily at the carbon atom opposite the more electronegative NH_2 group.

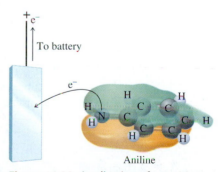

Figure 11.33 Application of a positive potential to a conducting glass attracts the high electron density associated with the nitrogen lone pair. If the potential is high enough, an electron is transferred from aniline to the conducting glass, leaving the aniline cation on the glass.

Figure 11.34 The extended molecular orbital in polyaniline is formed from overlap of the atomic *p* orbitals on the six carbon atoms of the phenyl ring plus the *p* orbital of the nitrogen atom. Extending over the long polymer chain, this molecular orbital has the potential to conduct a current.

Electron conduction is also important in the production of the polymer. Oxidation of aniline produces an unpaired electron, and this unpaired electron initiates polymerization of aniline. To propagate the polymerization and produce a long chain, an additional electron must be removed for each monomer added. If the growing chain does not conduct, aniline would have to be adjacent to the positive electrode to be oxidized and would form an exceedingly thin, one-molecular-layer coating on the glass. To add on to the monomer and form a polymer, the transferred electron must be able to travel through the oxidized aniline to the glass electrode. Similarly, to add on to the free end of the growing polymer chain, the removed electron must be able to travel along the chain to the electrode. Every aniline molecule added to the growing chain is accompanied by removal of an additional electron. Polymerization thus requires application of a steady, positive potential.

As the polymer grows, one of the hydrogen atoms from nitrogen and one of the hydrogen atoms from the carbon ring are removed (Figure 11.35). The accompanying reduction reaction thus involves production of H_2 at the negative electrode.

Conduction consists of a flow of charge. Tracing the charge flow through aniline begins with the repeating four-monomer unit of the conducting form of aniline (Figure 11.36). A positive charge represents a hole—that is, a destination for electrons. Chemists use a double-headed arrow to represent shifting a pair of electrons (Figure 11.36b). Electrons flow in one direction and holes in the other, just as in semiconductors.

The loosely held electrons responsible for conduction in polyaniline also give rise to the color observed (Table 11.1).

The fully reduced form of polyaniline (Figure 11.37) absorbs ultraviolet radiation at 310 nm. This wavelength is beyond the visible region, so the material appears as a clear transparent coating on the glass. The energy separation between the highest-energy occupied set of orbitals on the molecule (called the HOMO, for *highest occupied molecular orbital*) and either the *lowest-energy unoccupied molecular orbital* (LUMO) or the

Figure 11.35 A chain of partially oxidized polyaniline.

Key:

- N
- C
- H

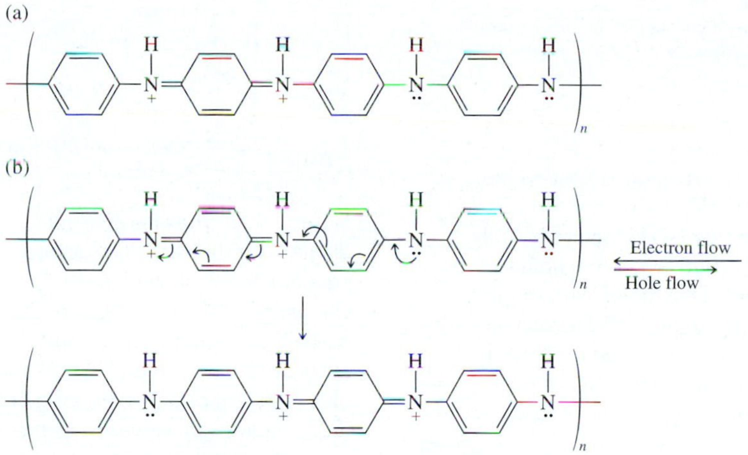

Figure 11.36 (a) The conducting form of polyaniline consists of repeating units, containing four aniline monomers. This form is oxidized by one electron for every two monomers compared with the fully reduced form of aniline. (b) Shifting pairs of electrons to adjacent positions as indicated by the double-headed arrow shifts the positions of the positive charges.

(a)

(b)

Electron flow

Hole flow

lowest-energy partially filled orbital corresponds to light of wavelength 310 nm. This separation is analogous to the band gap in a semiconductor.

Oxidization of fully reduced polyaniline removes electrons from the HOMO of reduced polyaniline, leaving the HOMO only partly occupied (Figure 11.38), and corresponds to the removal of one electron for every two monomers. The absorption in the

Table 11.1 Forms of Polyaniline

Leucoemeraldine, fully reduced, insulating, clear
λ max = 310 nm

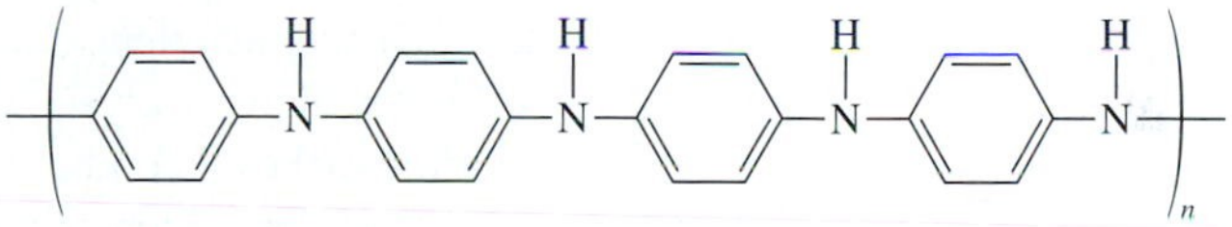

Emeraldine salt, partially oxidized, protonated, conducting, green
λ max = 320, 420, 800 nm

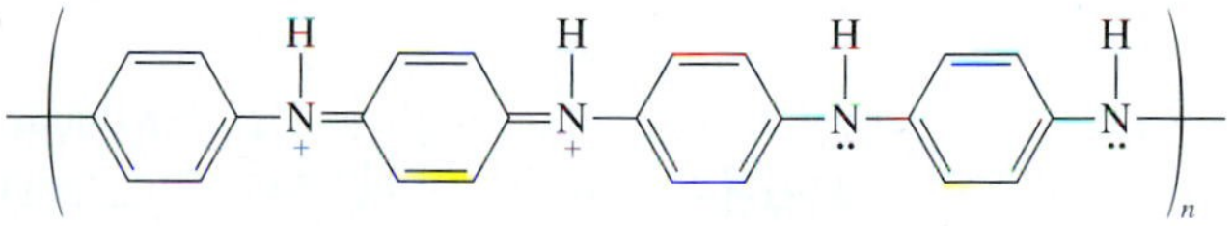

Emeraldine base, partially oxidized, insulating, blue
λ max = 320, 620 nm

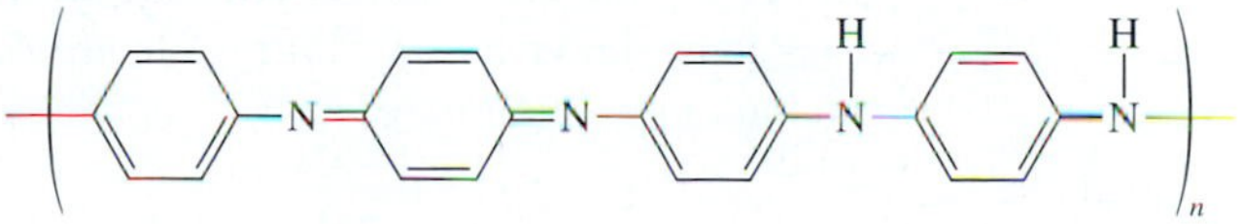

Pernigraniline, fully oxidized, insulating, purple
λ max = 320, 530 nm

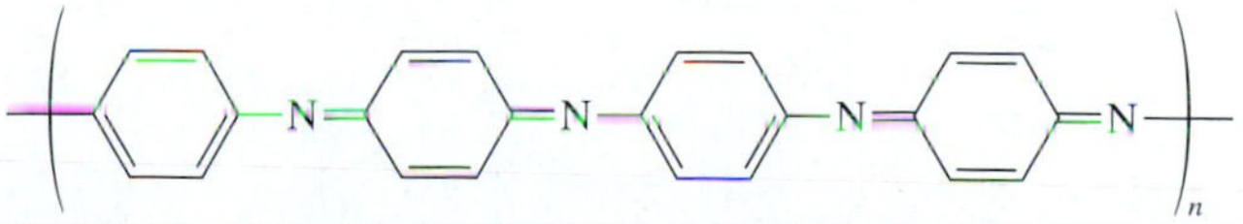

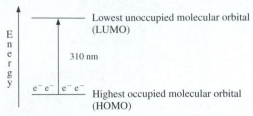

Figure 11.37 The energy-level diagram for fully reduced polyaniline, called leucoemeraldine shows the highest occupied molecular orbital (HOMO) and the lowest unoccupied molecular orbital (LUMO). Electrons are shown for illustrative purposes only. The actual number of electrons depends on the polymer chain length.

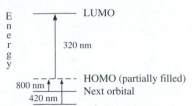

Figure 11.38 The energy-level diagram for the partially oxidized, emeraldine salt shows a partially occupied HOMO. Transitions to the partially occupied HOMO result in absorption of very red (800 nm) and blue (400 nm) photons. Filtering out red and blue light leaves transmitted green photons, so emeraldine appears green.

first oxidized form occurs at 320 nm. Two other bands appear: one at 800 nm, and the other at 420 nm. These bands correspond to transitions from lower electronic orbitals to the now partially occupied HOMO.

WORKED EXAMPLE 11.6   *An All-Hydrocarbon Conductor*

The first conducting polymer made was also chemically the simplest, containing only carbon and hydrogen:

$$\left(\!\!-C\!=\!C\!-\!\!\right)_{n}$$

This polymer is known as polyacetylene due to the monomer, $HC\equiv CH$, acetylene, from which it is made. As shown above, polyacetylene is not a conducting polymer. Doping polyacetylene with iodine makes it conductive. Explain how addition of iodine changes the conductivity of polyacetylene. Draw a schematic of conduction in polyacetylene.

Plan

■ Conduction requires both an electron source and an electron sink. Identify them.

Implementation

■ With alternating single and double bonds, there is no destination for electrons in polyacetylene that would make this material a conductor. Removing some of the π-bond electrons will make such a space. Electronegative elements are excellent removers of electrons. Thus adding some iodine removes electron density from the π bond, generating a hole. The hole left from removal of an electron can be represented as shown in (a). Shifting electrons, as indicated by the arrows in (b), results in the representation (c).

a.

b.

c.

The positive charge has moved left and the negative charge has moved right. This shift continues all along the chain.

The Japanese scientist Hideki Shirakawa discovered conducting polymers in 1977. For this discovery and subsequent development, Shirakawa, along with Alan Mac Diarmid and Alan Heeger, won the 2000 Nobel Prize. A huge advantage of conducting polymers is that they can be fabricated into light, flexible sheets. For purposes of comparison, note that the conductivity of a superb insulator, Teflon, is 10^{-16} $\Omega^{-1} \cdot m^{-1}$. Good conductors, silver and copper, have a conductivity of 10^8 $\Omega^{-1} \cdot m^{-1}$. Iodine-doped polyacetylene has a conductivity of 10^5 $\Omega^{-1} \cdot m^{-1}$.

See Exercises 23 – 26.

WORKED EXAMPLE 11.7   *Flying High: Corrosion Protection of Airplanes*

The hull of an airplane is composed of aluminum and titanium because they are lightweight, yet strong metals. Under the stress of flying, however, both metals corrode. Currently the coating of choice to protect the hull is chromium. Although a dense metal, the thin coating adds little to the mass of the airplane. However, chromium presents environmental concerns when the exterior of the airplane is refurbished. In recent years, polyaniline has been proposed as an alternative coating material. Discuss the advantages, in addition to the environmental impact, of using polyaniline rather than chromium to protect the hull. What characteristics of polyaniline are needed for application as a protective coating?

Plan
- Compare oxidized polyaniline to oxidized chromium for cycling.
- Consider the weight of polyaniline.
- Consider conduction.

Implementation
- Once oxidized, polyaniline has the potential for easy reduction, which is accomplished by applying a potential or by washing the material with an acid. Chromium oxides are nonconductive and difficult to reduce.
- Polyaniline is composed of light elements: hydrogen, carbon, and nitrogen. It will form a lightweight coating.
- Conduction is an important characteristic for polyaniline to serve as a protective coating. Whether oxygen or other oxidizers attack polyaniline or attack the metal directly, conduction supplies electrons to the metal, provided that the reduction potential of polyaniline is lower than that of titanium or aluminum.

See Exercises 23 – 26.

11.2   Formation Mechanisms

Polymer formation mechanisms fall into two broad classes: condensation and addition. These two methods produce distinctly different polymer chains, collectively called the polymer backbone. Condensation involves elements other than carbon, usually oxygen or nitrogen, at the site of condensation; these elements become incorporated in the backbone. Addition reactions almost always involve multiple bonds or strained bonds between carbon atoms, so the backbone consists of carbon atoms. An examination of the polymer's repeated unit reveals which mechanism was operative in its formation. Bonds at the end of the repeated unit that contain oxygen or nitrogen indicate that the polymer was made by condensation, whereas the presence of only carbon at the ends indicates an addition mechanism (Worked Example 11.8).

WORKED EXAMPLE 11.8 *Mechanism Identification*

Many synthetic polymers are made to imitate natural polymers. Successful imitation requires unraveling the mechanism for formation of the natural polymer. One of the earliest polymers to be imitated was silk. Another early synthetic polymer made to imitate natural rubber was polychloroprene.

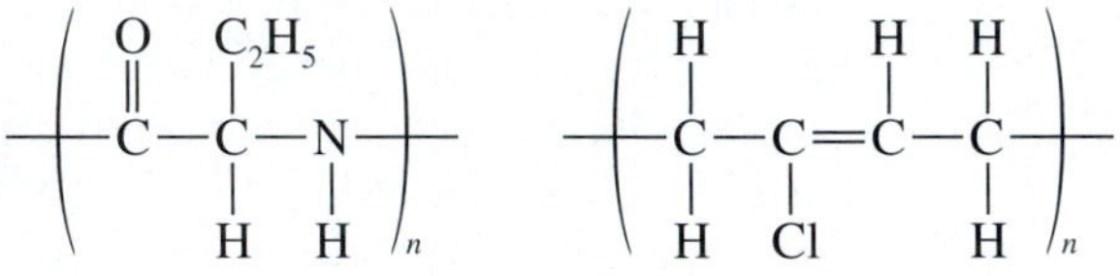

Natural silk polymer   Polychloroprene

Which mechanisms should be used to generate a polymer imitating silk? Which formation mechanism was used in generating polychloroprene?

Plan

■ Locate the elements at the ends of the repeated units.

Implementation

■ Natural silk contains a nitrogen atom at one end of the repeated unit. A silk imitation should employ a condensation mechanism. In contrast, polychloroprene contains only carbon on the ends of the repeated units. Polychloroprene was generated with an addition mechanism.

Both silk and polychloroprene are resilient polymers. The choice of formation mechanism determines the elements on the ends of the repeated units. Physical properties, such as elasticity and strength, are determined by the formation procedure and by the motifs contained in the link between the end elements.

See Exercises 28, 30–32.

Condensation Polymers

The condensation mechanism and its inverse, hydration, are extremely common particularly in organic reactions. A **condensation** reaction occurs when two molecules combine, forming a larger molecule and eliminating a small molecule. Water is the most common small molecule eliminated, although other molecules such as HCl, methanol, and ethanol are potential elimination products.

The unique feature of polymerization reactions is that the monomers have two functionalities for condensation. These functionalities determine the type of polymer formed. Several condensation polymers are listed in Table 11.2 along with their common names.

Condensation Results when two molecules come together, bond and usually eliminate a small molecule.

Table 11.2 Condensation Polymers

Name	Monomer	Monomer	Polymer
Nylon (nylon-68)	Diamine	Diacid	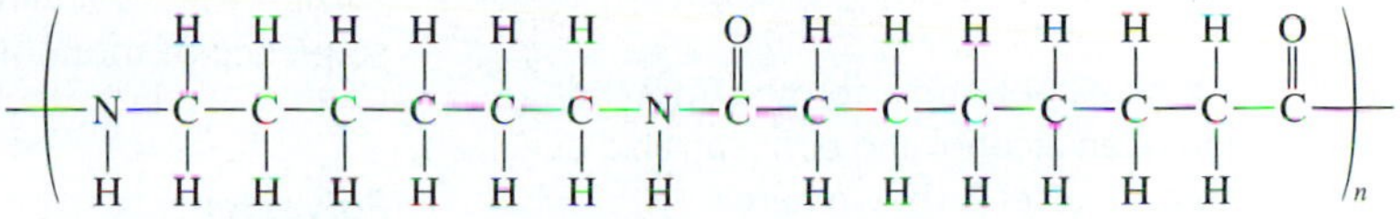 (structure shown)
Nylon (nylon-6)	Amino acid		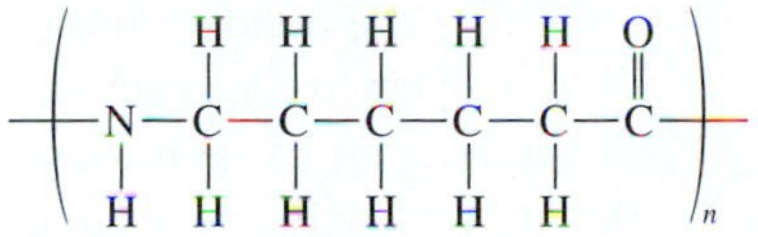 (structure shown)
Polyester (PET)	Diester	Diol	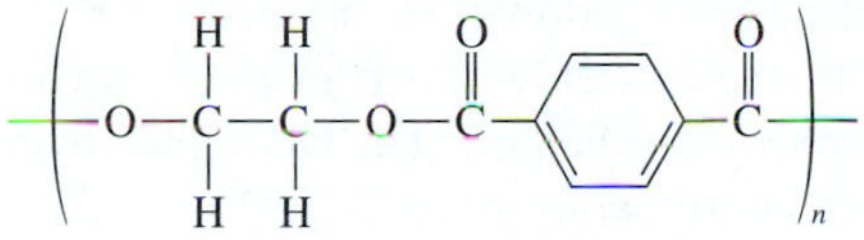 (structure shown)
Polycarbonate (Lexan)	Carbonate or carbonate ester	Diol	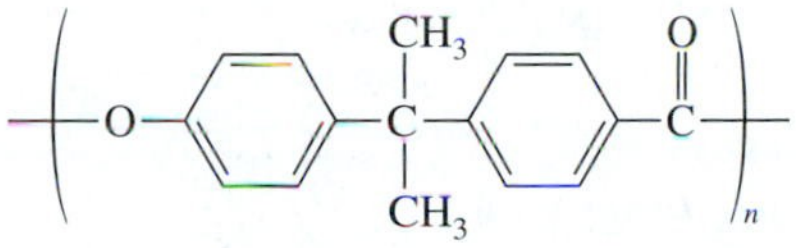 (structure shown)
Polyurethane	Diisocyanate	Diol	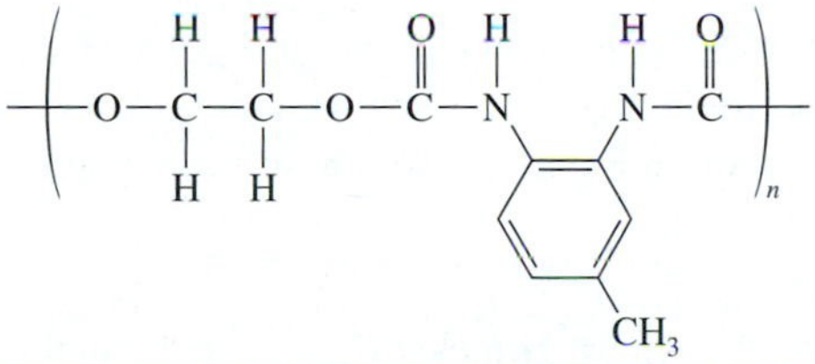 (structure shown)

Polyamides

Silk and nylon are examples of polyamides. The amide bond is formed from condensation of an amine and either an acid or acid derivative (Figure 11.39). For example, a room-temperature synthesis for nylon, substitutes —Cl for the —OH group of the acid. Condensation of an acid and an amine eliminates water; the chloride combines with the amine to eliminate HCl. The choice of acid derivative is determined by the synthesis conditions. For example, the room-temperature synthesis of nylon uses an acid chloride to eliminate the potential for deprotonation of the acid and to increase the solubility of the monomer in an organic solvent.

CONCEPT QUESTION   In the room-temperature, bench-top synthesis of nylon, it is desirable to dissolve the acid derivative in an organic solvent. What are the advantages of having the acid derivative be soluble in an organic solvent? ■

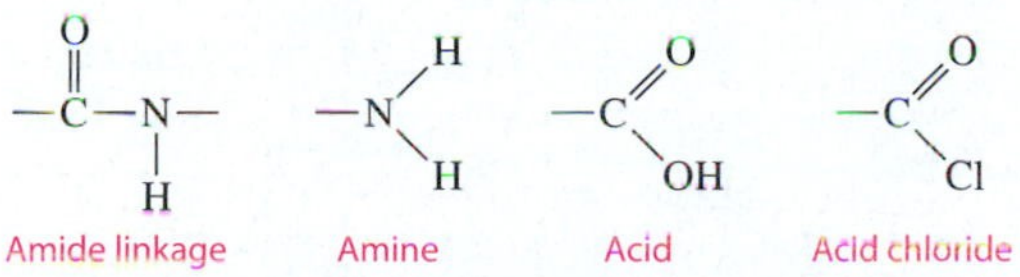

Polyamides can be formed either from a diamine plus a diacid or from an amino acid. In either case, the synthetic polymer is known as a polyamide. Biological macromolecules often involve a similar condensation reaction. For biological systems, the monomers are amino acids and the resulting linkage is known as a peptide bond. Biological macromolecules, however, have much greater variety than synthetic polymers. The variation relates to the monomers. Synthetic polymers are quite homogeneous, having only

Figure 11.39 An amide bond is formed from the condensation of an amine and either an acid or an acid derivative. An acid chloride is shown here.

Amide linkage   Amine   Acid   Acid chloride

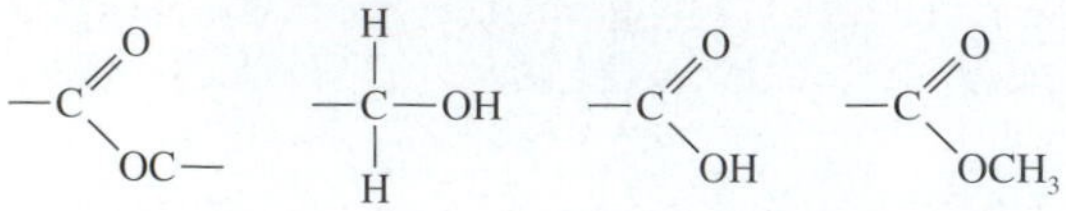

Ester linkage   Alcohol   Acid   Ester

Figure 11.40 An ester bond is formed from the condensation of an alcohol and either an acid or an acid derivative. An ester is shown here.

Ester An organic compound produced by the reaction between a carboxylic acid and an alcohol.

Alcohol An organic compound with a —COH group.

one or two monomers. Biological polypeptides are much more heterogenous, involving many different side groups on the monomers, particularly in proteins, enzymes, and hormones. The variation is not random, but rather involves a carefully regulated sequence of monomers.

Polyester

The condensation reaction that produces polyester is very similar to the reaction that produces polyamides. The **ester** linkage (Figure 11.40) contains an oxygen atom rather than the nitrogen atom of the amide. An ester is like an acid with the —OH hydrogen atom having been replaced with a hydrocarbon. Formation of an ester linkage involves the condensation of either an acid or a small ester with an alcohol. An organic **alcohol** consists of an —OH group attached to a carbon that has either hydrogen or carbon atoms filling out the other three coordination sites.

WORKED EXAMPLE 11.9   *Locating Electrons*

Oxidation numbers are helpful guides for predicting the outcome of an encounter between two molecules. Assign oxidation numbers to the acid carbon and the alcohol carbon in Figure 11.40. Draw a schematic of the interaction between the alcohol and the acid.

Plan

- Assign the two electrons of each bond to the more electronegative element in the bond.
- Count electrons: oxidation number = valence electrons − assigned electrons.
- The interaction is between negative and positive charges.

Implementation

- The *acid carbon* is less electronegative than oxygen and this carbon has the same electronegativity as the implied carbon that begins the remainder of the molecule. The *alcohol carbon* is less electronegative than the oxygen, is more electronegative than the hydrogen, and has the same electronegativity as the implied carbon.
- Acid: The only electron assigned to the acid carbon is one of the two electrons in the bond with the other carbon atom. Valence = 4, assigned = 1 ⟹ oxidation number = +3.
  Alcohol: Electrons assigned to the alcohol carbon include two electrons from each of two bonds to hydrogen atoms plus one of the two electrons in the bond with the other carbon atom. Valence = 4, assigned = 5 ⟹ oxidation number = −1.
- The acid carbon has an oxidation number of +3, indicating a center of positive charge. Oxygen has an oxidation number of −2, so the interaction is between the alcohol oxygen and the acid carbon.

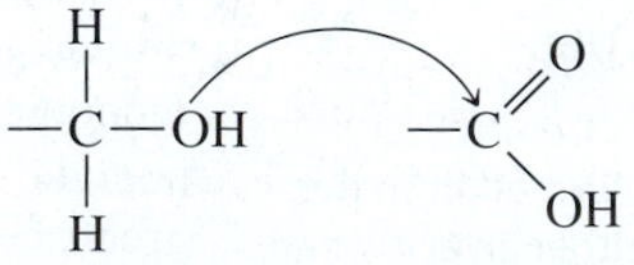

The oxidation number of the ester carbon is the same as that of the acid carbon. Thus modifying the acid to make it be a small ester results in the same interaction. Polyesters are often formed starting from small diesters.

See Exercises 37–40, 47.

Diol An organic compound having two hydroxyl groups.

Formation of a polyester requires two bifunctional monomers, a diacid or a diacid derivative and a dialcohol, known as a **diol.** The condensation of a diacid and a diol eliminates water. Modification of the diacid to make it be a small diester results in elimination of a small alcohol. Small alcohols are more volatile than water, so temperature constraints often determine the choice between a diol and a diester.

Polycarbonates

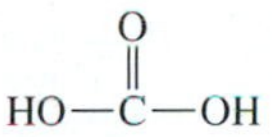

Figure 11.41 The simplest diacid is carbonic acid.

Polycarbonates represent a special case of polyester formation. The acid involved in polycarbonate formation is the simplest diacid, carbonic acid (Figure 11.41). In the synthesis of polycarbonates, carbonic acid is often stabilized by making an ester derivative. The acid carbon is in a +4 oxidation state—carbon has lost all four of its valence electrons. Carbonic acid condenses with a diol, eliminating water so that all that remains of the carbonic acid after the polymerization is the C=O group!

An example of a polycarbonate is Lexan (Figure 11.42). This strong, clear, colorless material is so strong that it is used in bulletproof windows, crash helmets, and the windows of airplanes.

Polyurethanes

Isocyanate A molecule containing the functional group —N=C=O.

Like all condensation polymers, polyurethanes are made from bifunctional molecules. One of the monomers is a diol and the other is a diisocyanate. An **isocyanate** is a molecule containing the functional group —N=C=O.

CONCEPT QUESTION Assign an oxidation number to the carbon atom in the isocyanate group. Which atom in the isocyanate interacts with the alcohol oxygen atom? ■

The carbon in the isocyanate is formally in a +4 oxidation state—very oxidized. This center of positive charge (Figure 11.43) interacts with the oxygen of the alcohol. In this case, no small molecule is eliminated. Rather, when the carbon of the isocyanate bonds to the oxygen of the alcohol, the π bond between nitrogen and carbon breaks. The alcohol releases the hydrogen atom, which shifts to attach itself to nitrogen. All coordinations are satisfied and no small molecule is released. Making the alcohol and the isocyanate bifunctional results in a polymer (Figure 11.44).

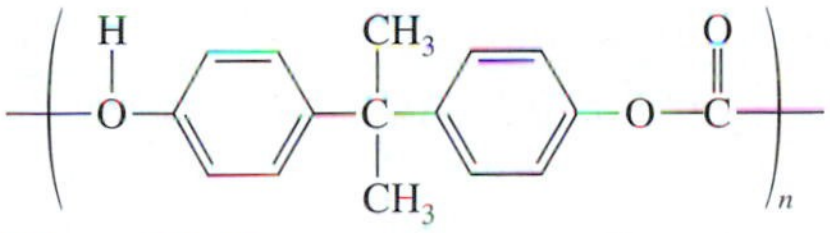

Figure 11.42 Lexan is a polycarbonate. The only remnant of the carbonic acid monomer is the C=O portion, the remainder of the repeated unit started with a diol.

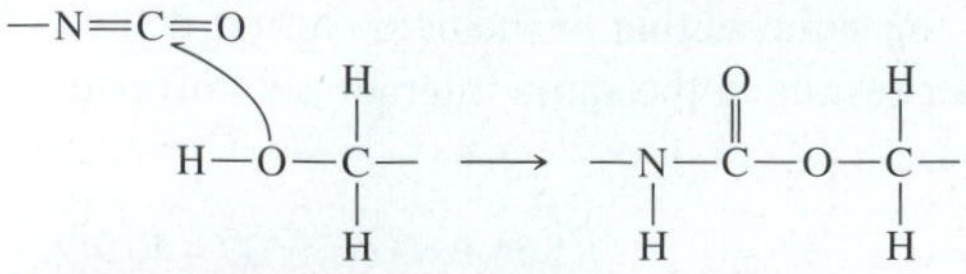

Figure 11.43 An isocyanate and an alcohol interact through attraction of the carbon of the isocyanate (formally a +4 oxidation state) and the oxygen of the alcohol. The hydrogen shifts from the oxygen of the alcohol to the carbon of the isocyanate. The product looks like an ester with a nitrogen linkage, called a carbamate ester.

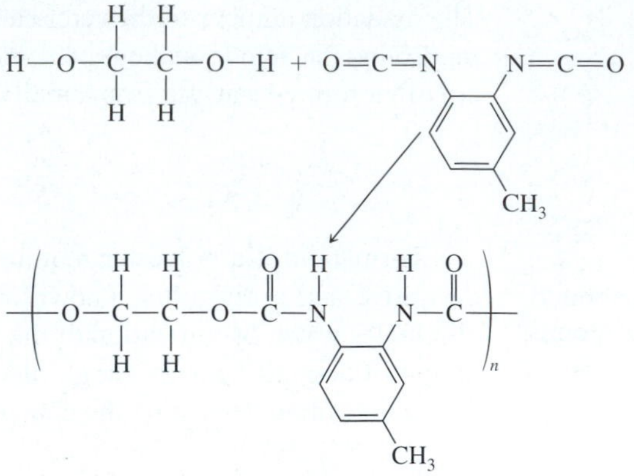

Figure 11.44 Synthesis of polyurethane from ethylene glycol, commonly used as an antifreeze in automobile radiators, and the diisocyanate. Because no small molecule is eliminated, none remains to be removed from the reaction mixture. Polyurethanes are used as tough finishes for wood floors, thereby eliminating the need for wax.

WORKED EXAMPLE 11.10   *Identifying the Monomer*

One of the first steps in planning the synthesis of a desired polymer is identification of the possible monomers. The final choice of monomer is determined by other constraints placed on the procedure, such as operating temperature, solubility in different solvents, and working conditions. Identify at least three sets of monomers that can be used to form the following polymer:

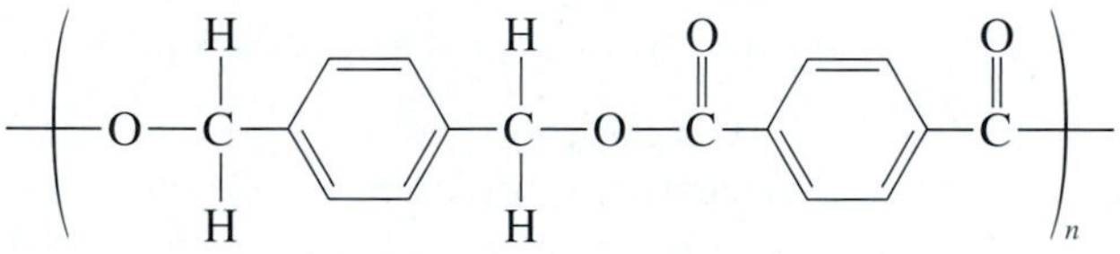

Plan

- Identify the formation mechanism.
- If a condensation mechanism is involved, insert a small molecule at the linkages.
- Identify the monomers that can be modified.

Implementation

- The ends of this polymer contain —O— and —C— linkages. With oxygen on one end, this polymer is a condensation polymer. An ester linkage appears in the middle. This polymer is made by condensation of two different monomers.
- Add water to the ends, and insert water at the ester linkage:

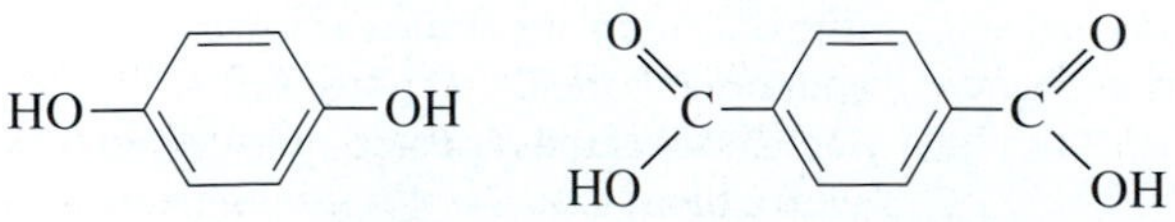

Figure 11.45 A double bond consists of two electrons in a σ bond and two electrons in a π bond. Unpairing the two electrons in the π bond forms a diradical. Each of the unpaired electrons can form a bond by pairing with an unpaired electron from another molecule.

- The diacid could be changed to a diester by replacing the —OH group with a —OCH_3, or it could be made into an acid chloride by replacing the —OH with a —Cl. Replacing it with —OCH_3 groups results in elimination of methanol during condensation. Methanol is more volatile than water, which will speed up the reaction by removal of one of the products. Replacing —OH with —Cl will generate HCl as a product. HCl is soluble in aqueous solvent.

See Exercises 30–32, 34.

Addition Polymers

Most often, addition polymers have a double-bond functional motif. This double bond opens (Figure 11.45), and each of the two electrons from the bond becomes available to pair with another electron to form a new bond. Attack at another double bond forms a bond and preserves the radical (Figure 11.46). Such a step is called a chain propagation because no loss of radical functionality occurs upon addition. Termination takes place when a pair of radicals collapses to form a bond.

$$\text{initiation} \longrightarrow \text{propagation} \longrightarrow \text{termination}$$
$$\text{(radical generation)} \quad \text{(attack at the multiple bond)} \quad \text{(pairing of two radicals)}$$

Initiating an addition reaction requires a method for generating a radical. In the synthesis of polyaniline, the radical is generated by applying a potential. This potential extracts an electron from the lone pair on aniline, leaving an odd electron. A more common method for generating the radical initiator is to add a molecule with a weak bond. Breaking the weak bond (for example, with heat) forms a radical (Figure 11.19). Once formed, the radical attacks at a double bond, opening it up and re-forming a radical that continues the reaction.

WORKED EXAMPLE 11.11   *Initiators*

Benzoyl peroxide is an excellent radical initiator. However, some applications require initiation at a lower temperature. A number of radical initiators have been devised to meet this need. One initiator is the molecule VA-044:

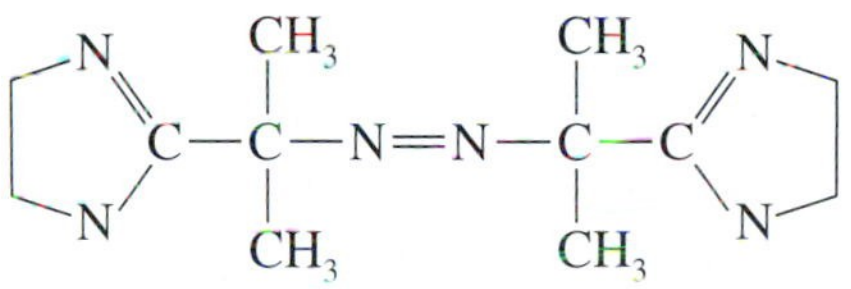

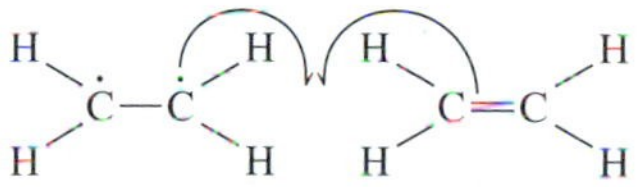

Figure 11.46 A prototypical addition reaction begins with radical attack on a double bond. The double bond opens, with one electron pairing with the radical electron to form a new σ bond and the other electron propagating the radical.

VA-044 decomposes at 44 °C, a considerably lower temperature than that for benzoyl peroxide. How does VA-044 decompose on heating?

Plan
- Identify the weakest bond.
- Examine the product after breaking the weak bond for stability of the fragment.

Implementation
- The strongest bond known is the N≡N triple bond. Molecules with this as a N=N bonded to nitrogen often extrude N_2, sometimes explosively. VA and carbon double bond. Shifting one electron from the bond between

provides one of the two electrons needed to transform the double bond into a triple bond. When such a shift occurs on both nitrogen atoms, the pair forms a triple bond.

■ The N≡N triple bond is very stable and N_2 is released. The fragment that is left

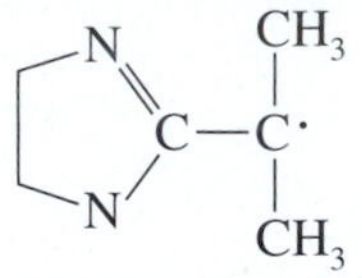

cannot easily re-form VA-004 because the N_2 molecule has been removed. The radical attacks double bonds or strained bonds to initiate polymerization.

Availability of a myriad of initiators makes it possible to generate polymers under a variety of temperature and pressure conditions.

See Exercise 59.

WORKED EXAMPLE 11.12   *Identifying Monomers from the Polymer Formed*

Polymer scientists often face the task of creating a polymer similar to a known polymer, but modified slightly so as to change its properties. For this process, it is useful to determine the monomer from which the model polymer is made. A development group targets making a more flexible polyacetylene, $(C=C)_n$, by incorporating a —CH_3 side chain into the polymer. From which monomer is polyacetylene made? Which monomer would you choose to form polyacetylene with —CH_3 side chains?

Plan

■ Identify the formation mechanism as either addition or condensation.
■ Addition polymers lose a π bond. Condensation polymers eliminate a small molecule.
■ Choose location for the —CH_3 group.

Implementation

■ Both ends of the repeated unit in polyacetylene are carbon atoms. Polyacetylene must be formed from addition.
■ Addition polymers lose a π bond, so the $C=C$ bond must have been a $C≡C$ bond. The monomer is HC≡CH, acetylene.
■ To form a polymer with —CH_3 side groups, the acetylene monomer should be modified by replacing a —H with a —CH_3. The final result should be H_3C—C≡CH.

See Exercises 30–32, 34.

Regardless of how the radical chain is initiated, propagation always preserves the radical and opens a double bond. The chain terminates when two radicals combine, as when two growing chains meet each other. Table 11.3 lists several addition polymers, the monomers from which they are formed, and some uses for each polymer.

Table 11.3 Addition Polymers

Name	Monomer	Polymer	Uses
Polyethylene	$H_2C{=}CH_2$	$[-CH_2-CH_2-]_n$	Plastic bags, plastic film
Polypropylene	$H_2C{=}CH-CH_3$	$\left(-CH_2-\overset{\displaystyle CH_3}{\underset{}{CH}}-\right)_n$	"Olefin" fibers
Polystyrene	$\underset{H}{\overset{H}{}}C{=}C\underset{H}{\overset{C_6H_5}{}}$	$\left(-CH_2-CH-\right)_n$ (phenyl)	Styrofoam cups, foam insulation
Poly(isobutylene)	$H_2C{=}C(CH_3)_2$	$\left(-CH_2-\overset{\displaystyle CH_3}{\underset{\displaystyle CH_3}{C}}-\right)_n$	Synthetic rubber
Poly(vinyl chloride)	$H_2C{=}CH-Cl$	$\left(-CH_2-\overset{\displaystyle Cl}{\underset{}{CH}}-\right)_n$	PVC pipes, hard plastics, Saran wrap
Poly(acrylonitrile)	$H_2C{=}CH-CN$	$\left(-CH_2-\overset{\displaystyle CN}{\underset{}{CH}}-\right)_n$	Orlon acrylic fibers
Poly(methyl methacrylate)	$\underset{H}{\overset{H}{}}C{=}C\underset{COOCH_3}{\overset{CH_3}{}}$	$\left(-\overset{\displaystyle H}{\underset{\displaystyle H}{C}}-\overset{\displaystyle CH_3}{\underset{\displaystyle COOCH_3}{C}}-\right)_n$	Plexiglas, Lucite
Poly(methyl cyanoacrylate)	$H_2C{=}\overset{\displaystyle }{\underset{\displaystyle CH_3}{C}}-\overset{\displaystyle O}{\overset{\|}{C}}-OCH_3$	$\left(-\overset{\displaystyle H}{\underset{\displaystyle H}{C}}-\overset{\displaystyle CN}{\underset{\displaystyle COOCH_3}{C}}-\right)_n$	Superglues
Poly(tetrafluoroethylene)	$F_2C{=}CF_2$	$[-CF_2-CF_2-]$	Teflon
Polychloroprene	$\underset{H}{\overset{H}{}}C{=}\underset{Cl}{\overset{H}{C}}-\overset{H}{C}{=}C\underset{H}{\overset{H}{}}$	$\left(-\overset{\displaystyle H}{\underset{\displaystyle H}{C}}-\overset{}{\underset{\displaystyle Cl}{C}}{=}\overset{}{\underset{\displaystyle H}{C}}-\overset{\displaystyle H}{\underset{\displaystyle H}{C}}-\right)_n$	Rubber
Polynorborene	(cyclic diene structure)	(bicyclic polymer structure)$_n$	Vibration damping rubber

11.3 Molecular Structure and Material Properties

Whether the polymer is made by condensation or addition, the material properties of a polymer are determined by a combination of the monomers, the macromolecular structure, intermolecular interactions, and any cross-linking. Macromolecular structure and intermolecular interactions were examined in the earlier discussion of polyethylene. Here the focus is on the effect of the structure of the monomers.

Flexible, resilient materials are made from monomers with linear chains or linear chains with small side groups. To make a more rigid material, phenyl rings are incorporated into the monomer. The extended molecular orbital of the phenyl ring imparts rigidity because bending the ring interferes with the sideways overlap of the p orbitals. This costs energy; hence, the ring is quite stiff. Molecules incorporating many such stiff rings are themselves relatively stiff. Examples of this property include Lexan, a polycarbonate (Figure 11.42); PET, the poly(ethylene terephthalate) used in plastic soda bottles (Figure 11.47); and Kevlar used in bulletproof vests (Figure 11.48).

To make a material that is heat resistant, the two functional ends of the molecule are connected with a portion that is resistant to oxidation. An example is the fire retardant Nomex (Figure 11.49).

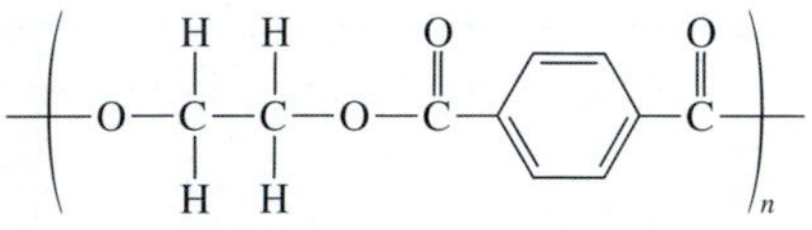

Figure 11.47 Poly(ethylene terephthalate), PET, is used in plastic soda bottles. The phenyl ring gives rigidity while the —CH_2— groups impart a measure of flexibility.

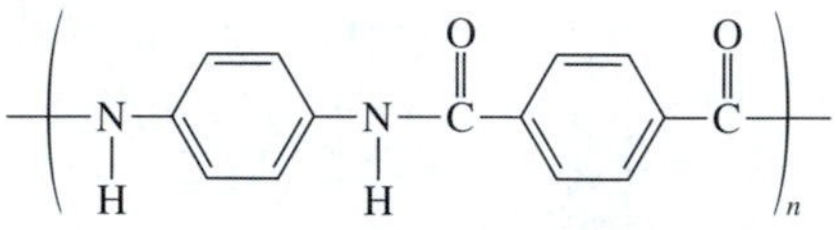

Figure 11.48 The polymer Kevlar is used in bulletproof vests. The stiffness of the vest comes from the phenyl groups, and the toughness comes from frequent intermolecular interactions due to hydrogen bonds between the C=O and N—H groups on neighboring chains.

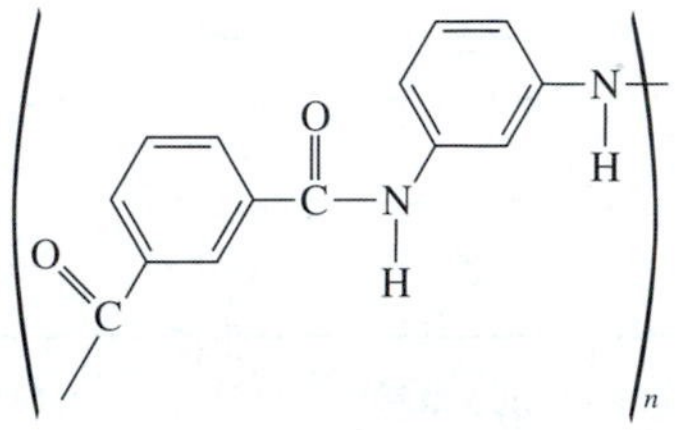

Figure 11.49 To form a good fire retardant, a polymer must be resistant to oxidation. One example is Nomex, illustrated here. Nomex incorporates phenyl rings, structures that are resistant to oxidation.

Checklist for Review

KEY TERMS

polymer (p. 378)
monomer (p. 378)
macromolecular structure (p. 378)
dalton (p. 379)
hydrocarbon (p.379)
amine group (p. 379)
  L group (p. 380)
  Lewis (p. 382)
  amide (p. 382)
  amino acid (p.

peptide bond (p. 383)
stick diagram (p. 384)
high-density polyethylene (HDPE) (p. 386)
low-density polyethylene (LDPE) (p. 386)
ethylene (p. 387)
radical (p. 388)
heterogeneous catalysis (p. 389)
coordination polymerization (p. 390)

cis (p. 392)
trans (p. 392)
vulcanization (p. 392)
cross-link (p. 392)
phenyl group (p. 395)
condensation (p. 402)
ester (p. 404)
alcohol (p. 404)
diol (p. 405)
isocyanate (p.405)

Chapter Summary

The vast majority of polymers are made by one of two mechanisms: a condensation mechanism or an addition mechanism. The condensation mechanism requires bifunctional monomers, and a small molecule is also usually produced. The addition or chain mechanism features unpairing of electrons in a multiple bond or a strained bond. One unpaired electron and a radical electron form a new bond. The addition mechanism requires an initiator, either a material that easily forms a radical or a coordination catalyst.

Material properties of polymers are determined by the constituent monomer units as well as the processing technique. Processing affects formation of side chains in the addition mechanism, as well as polymeric chain alignment. Forming a polymer into a sheet results in a film, while drawing the polymer through an orifice gives a thread. Both kinds of stress increase intermolecular interactions by aligning the molecules along their long axes.

Flexible polymers incorporate linear chains in the monomer(s), whereas more rigid materials utilize aromatic rings. Aromatic rings also increase thermal stability and resistance to oxidation.

KEY IDEA

- Extremely large molecules can be assembled by two routes. One route consists of an acid–base reaction between bifunctional monomers. The other route involves adding small molecules to a double bond or a strained bond.

CONCEPTS YOU SHOULD UNDERSTAND

- The material properties of a polymeric material are determined by the structure of the monomer in combination with the macromolecular structure and intermolecular interactions. The latter two factors are affected by the manufacture procedures (that is, pulling, drawing, or stressing).

OPERATIONAL SKILLS

- Apply principles of equilibrium and solubility to polymer problems (Worked Examples 11.1, 11.2, 11.3, and 11.5).
- Relate material properties to the structure of the monomer (Worked Example 11.4).
- Trace electron flow in polymer formation and in conduction (Worked Examples 11.6 and 11.9).
- Select polymers for applications based on properties of the monomers and bonding (Worked Example 11.7).
- Identify formation mechanisms. Identify the monomers from the structure of the polymer (Worked Examples 11.8, 11.10, and 11.12).
- Identify weak bonds in initiators (Worked Example 11.11).

Exercises

A blue exercise number indicates that the answer to that exercise appears at the back of the book.

■ SKILL BUILDING EXERCISES

11.1 Polymers

1. Determine the pH of a 1 M ammonia solution ($K_b = 1.8 \times 10^{-5}$). Which is present in greater concentration, NH_3 or NH_4^+?

2. Determine the pH of a 1 M aniline solution ($K_b = 4.2 \times 10^{-10}$). Calculate the aniline$H^+$ ion concentration. Is it greater than or less than the NH_4^+ ion concentration in Exercise 1?

3. Polyvinyl alcohol is the polymer shown below.

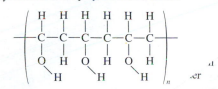

What intermolecular force describes the in~
polyvinyl alcohol molecules? Do you ~
to be soluble in water? Why or wh~

4. The polymer polyvinyl alcohol (shown in Exercise 3) can be cross-linked with borax:

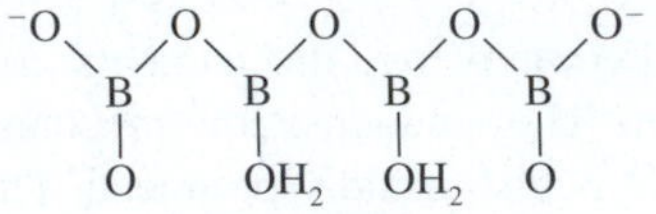

The resulting material is known as slime. Indicate the structure of slime.

5. The polymer, polychloroprene (shown below) can be cross-linked with O^{2-}, such as that released from ZnO. Indicate the structure of the cross-link.

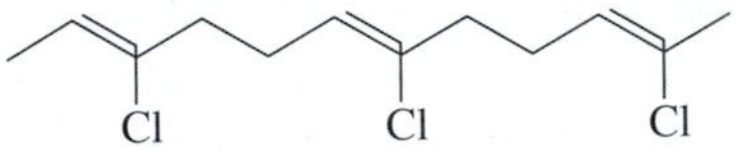

6. Which of the following liquids are held together by hydrogen bonding?
a. CCl_4 b. oil   c. water
d.

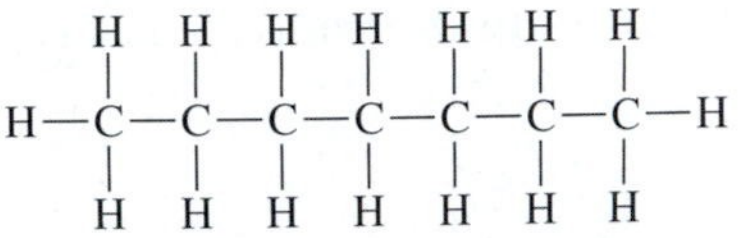

Heptane

e.

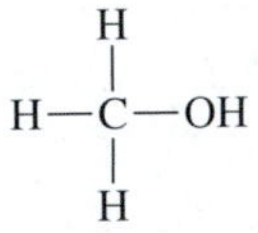

Methanol

f.

Hexane

7. Place the following forces in order of strength, from strongest to weakest.
a. dipole–dipole   b. dipole–induced dipole
c. hydrogen bonding   d. ion–dipole
e. London dispersion

8. Fill in Table 11.4 by indicating the force (H-bonding; dipole–dipole; London dispersion) that holds the polymer together.

11.2 Nylon

9. The pK_b of aniline is much less than that of ammonia. Predict the approximate value of pK_b for the diamine used in the synthesis of nylon (shown below). Justify your answer by quoting data for other amines.

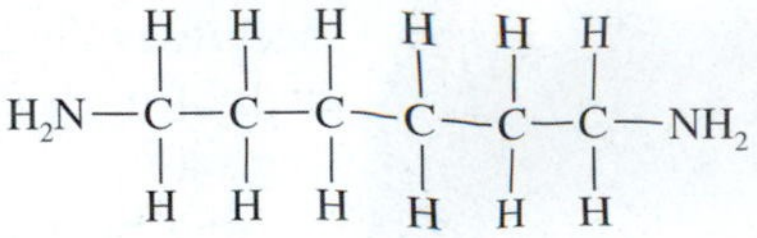

10. Nylon-6,6 is made by an acid–base condensation reaction between a diamine and a diacid.

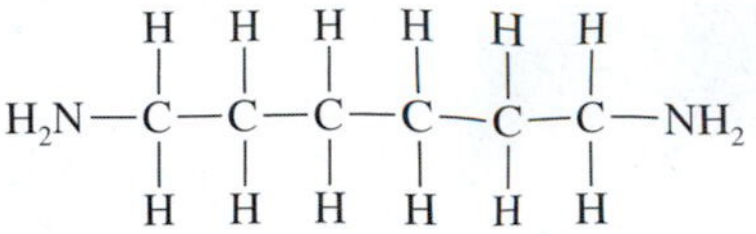

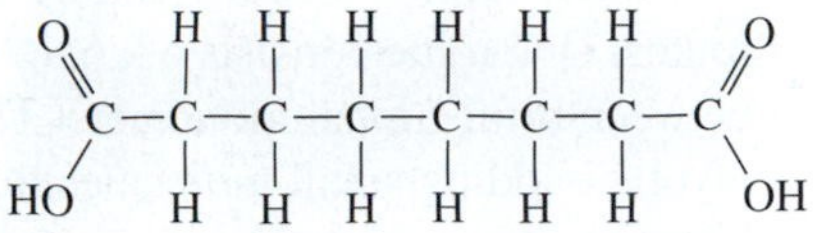

Indicate the formula for the polymer nylon-6,6. What other products are produced? Suggest whether nylon would make a good rubber. Why or why not? How might the monomer(s) or polymer be modified so that it is a better rubber?

Table 11.4

Polymer	Intermolecular Force
a. 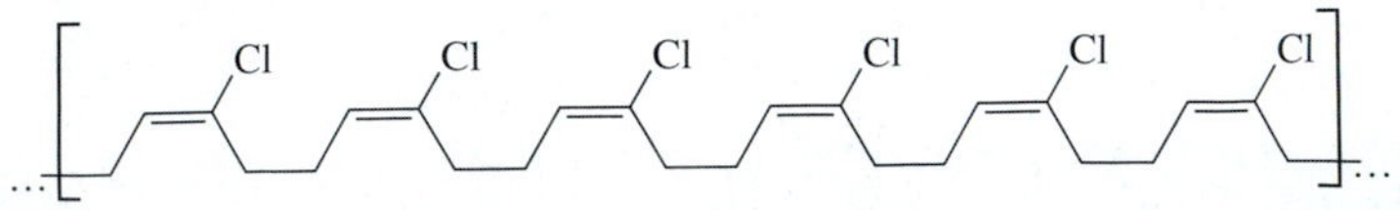	
b. 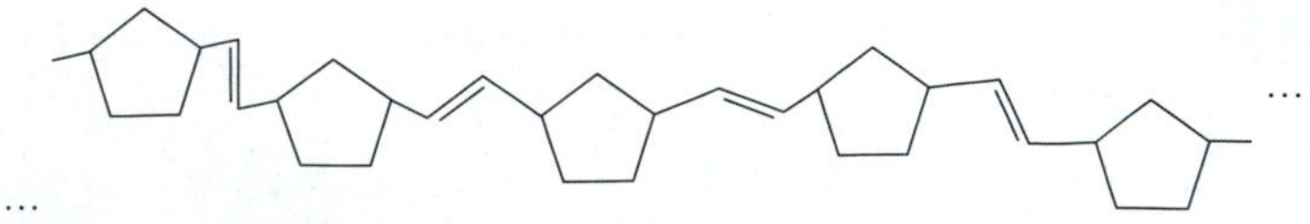	
c. 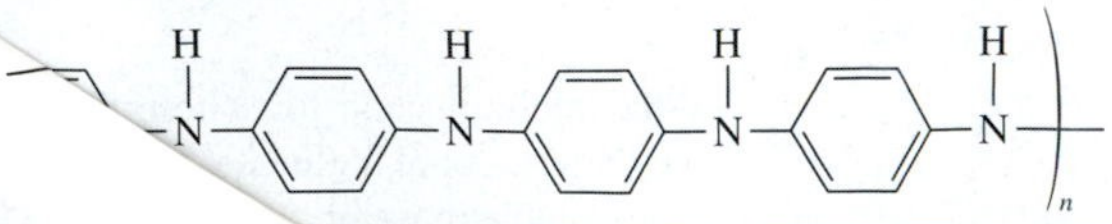	

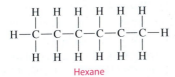

11. The polymer nylon-6 is made from the monomer

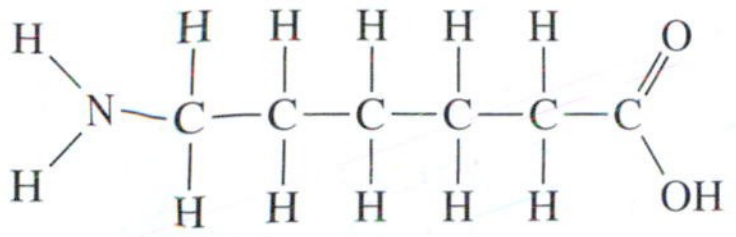

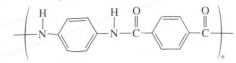

Indicate the structure of the polymer nylon-6.

12. Nylon is a polyamide. Is the polymer

also a polyamide? Which forms the more flexible fiber, nylon or the polymer shown above?

11.1 Polyethylene

13. Which polymer is high-density polyethylene?

a. · · ·

b.

14. How many ethylene molecules are required to make a polyethylene molecule of molecular weight 42,000 amu?

11.1 Rubber

15. Isoprene, the molecule in latex, has the following structure:

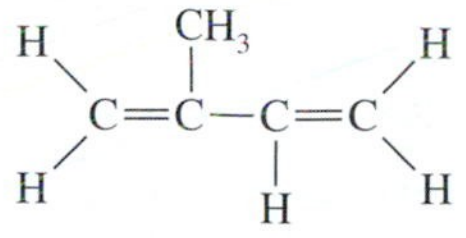

Indicate the structure of linear-chain polyisoprene.

16. Isobutylene is often used to make rubber. The structure of isobutylene is

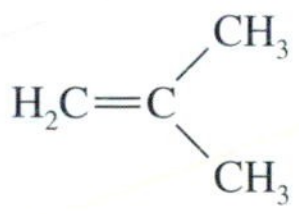

Indicate the structure of this rubber.

17. The rubber neoprene is made from the monomer

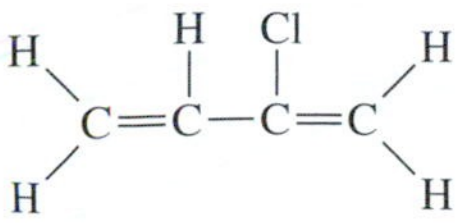

Indicate the structure of neoprene rubber.

18. Butadiene, $C=C-C=C$, is a staple of the rubber tire industry. Indicate the structure of the polymer polybutadiene. Indicate the functional feature of polybutadiene that enables it to function as a rubber.

19. In recent years, tire wear has improved greatly due to modification of butadiene rubbers with the addition of styrene. Butadiene and styrene are shown below. Indicate the average repeated unit of a 50:50 butadiene–styrene rubber. Is styrene alone capable of making a rubber? Which rubber do you expect to be stiffer, polybutadiene or the copolymer of butadiene and styrene?

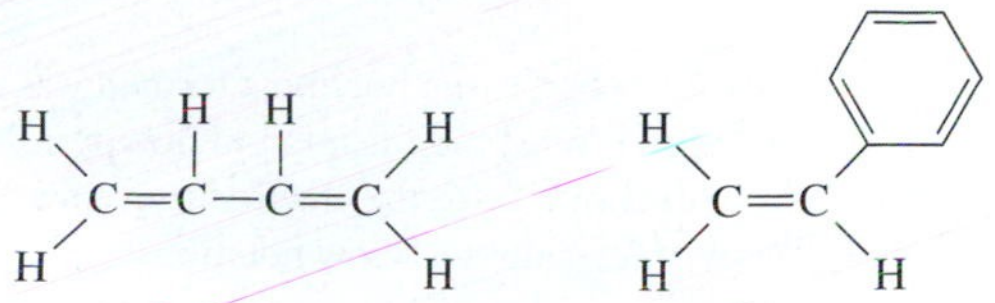

Butadiene   **Styrene**

20. Polynorborene is made from the monomer norborene.

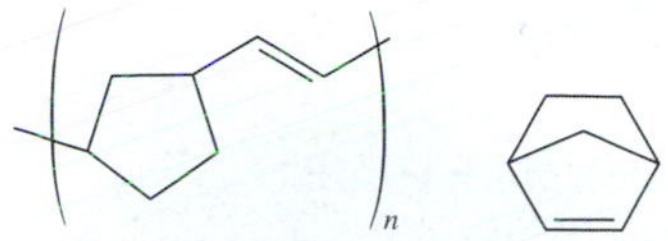

Polynorborene   **Norborene**

Indicate the formation mechanism for polynorborene. Norborene is a vibration-damping rubber. Indicate cross-linking among the polymeric chains of polynorborene.

11.1 Conducting Polymers

21. One advantage of conducting polymers is that they can be incorporated into thin, plastic films, forming flexible conducting materials. Suppose you are given the task of making a flexible product based on polyaniline. Would you use a condensation polymer or an addition polymer for the substrate? Defend your choice based on the chemistry of these alternatives.

22. Show why the emeraldine base form of polyaniline does not conduct. That is, indicate where the molecular "wire" is not connected.

23. The blue form of polyaniline can be turned green by addition of acid to the solution. Explain this observation. Is the resulting material conductive?

24. The molecule acetylene is shown below.

$$H-C\equiv C-H$$

Indicate the structure of polyacetylene.

25. It has been observed that addition of iodine ther elements results in a conductive polymer. Suggest how the mechathat could make polyacetylene condu nism for conduction.

26. The polymer shown below, pol What is the polymerizapotential to be a conducting ure of the monomer, tion mechanism? Indicate or poly(vinylphenylene) to vinylphenylene. What i be conductive?

7. The polymer polythiophene (shown below) can be made conductive. What modification is required for polythiophene to be conductive? Show the conduction using the arrow notation.

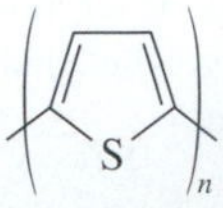

28. One of the earliest conducting polymers formed was polypyrrole (shown below). What modification of polypyrrole is needed for conduction? Trace the electron movement in conducting polypyrrole using the arrow notation.

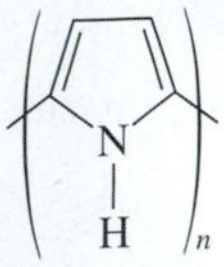

11.2 Formation Mechanisms

29. Consider the molecules ethylene chloride and ethyl chloride.

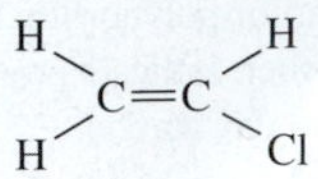

Ethylene chloride 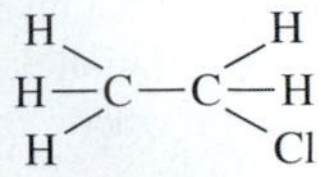   Ethyl chloride

Which of these is easily polymerized, and why?

30. The polymer shown below is a superabsorbent polymer. Is it an addition or condensation polymer? Indicate the monomer(s) from which it is made.

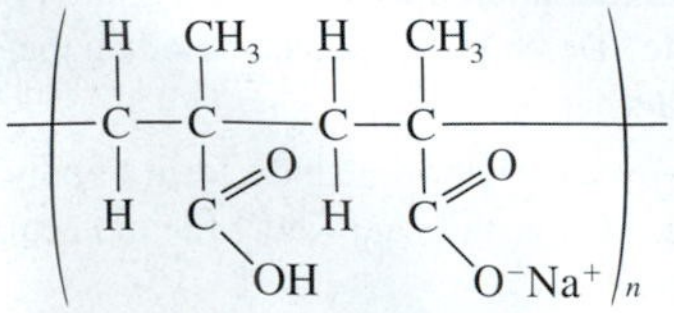

31. Indicate the interaction between water and the polymer shown in Exercise 30.

32. Cellulose (shown below) is a naturally occurring polymer. Is cellulose a condensation polymer or an addition polymer? Indicate the monomer(s) that make cellulose.

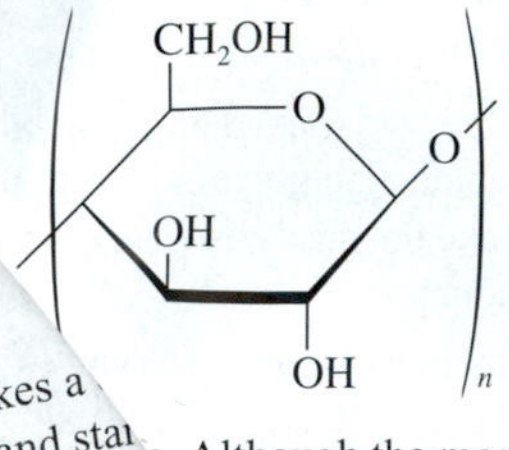

33. Shape makes a ... cellulose and sta... e. Although the monomers of including humans, ...sely related, most organisms, unfortunate implicat... cellulose is one of the ... est cellulose. This fact has biosphere, with abo... world's food supply because ...ast, starch is a sta... ...ing units of cel...ing produced annually. ...ts of many species. ...ch are shown.

Indicate the difference between the monomers for these two natural polymers.

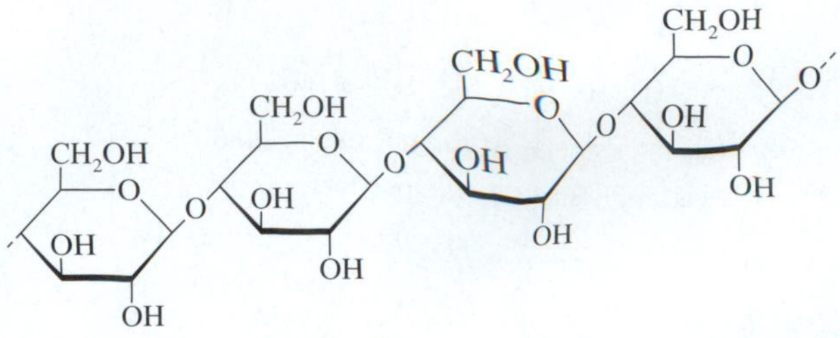

Cellulose

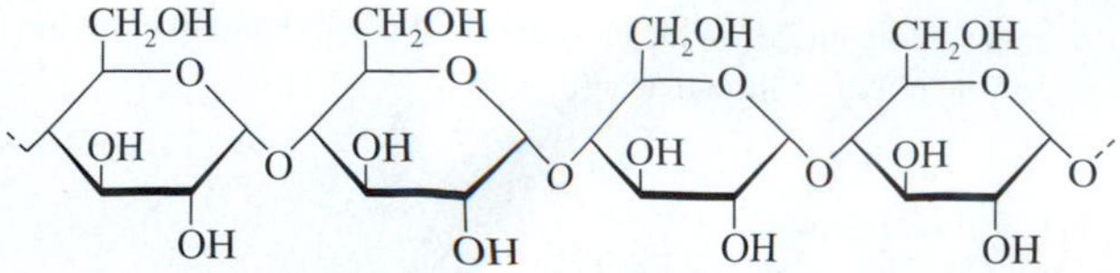

Starch

Complete the following sentence: Long strands of cellulose interact by _____. This interaction provides mechanical strength to cell walls.

Starch forms a more open macromolecular structure. Explain why starch readily absorbs water.

34. Spandex gives a great fit for sport clothing. Its structure is shown below. From what monomers is the polymer Spandex made?

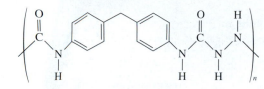

35. Recently a temperature-sensitive plastic was developed. The basic polymer is based on the copolymerization of styrene and the aniline derivative N-2-propenyl-methyl benzamine. What is the mechanism for formation of this copolymer?

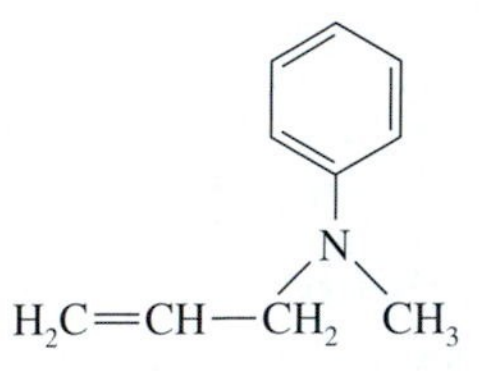

Derivative N-2-propenyl-methyl benzamine 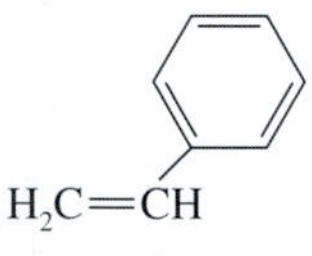   Styrene

36. Superglue is a polymer with the structure

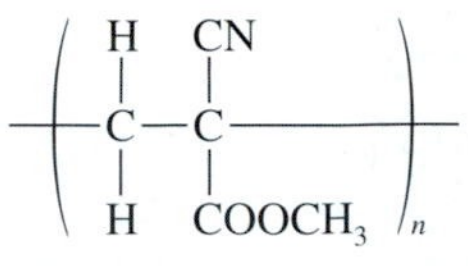

What is the formation mechanism for superglue? What is the monomer?

11.2 Condensation

37. Indicate which of the following monomers can form condensation polymers if combined with a suitable second monomer.

a.

$$\underset{HO}{\overset{O}{\|}}C-\overset{H}{\underset{H}{C}}-\overset{H}{\underset{H}{C}}-\overset{O}{\underset{OH}{\|}}C$$

b.

(benzene ring)

$H_2C=CH$

c. $H_2N-\overset{H}{\underset{H}{C}}-\overset{H}{\underset{H}{C}}-\overset{H}{\underset{H}{C}}-\overset{H}{\underset{H}{C}}-\overset{H}{\underset{H}{C}}-\overset{H}{\underset{H}{C}}-NH_2$

d. $H_3C-O-\overset{O}{\underset{}{\|}}C-$ (benzene ring) $-\overset{O}{\underset{}{\|}}C-O-CH_3$

e. $H-\overset{H}{\underset{H}{C}}-\overset{O}{\underset{H}{C}}$ with C double bonded to O and bonded to H

f. $\overset{H}{\underset{H}{}}C=C-\overset{H}{\underset{Cl}{C}}=\overset{H}{\underset{H}{C}}$

38. Which of the following are polyesters?

a. $-\!\left(\!O-\overset{H}{\underset{H}{C}}-\overset{H}{\underset{H}{C}}-O-\overset{O}{\underset{}{\|}}C-\overset{H}{\underset{H}{C}}-\overset{H}{\underset{H}{C}}-\overset{H}{\underset{H}{C}}-\overset{O}{\underset{}{\|}}C\!\right)_{n}$

b. $H-\overset{H}{\underset{H}{C}}-\overset{O}{\underset{H}{C}}$ (C double bond O, bonded H)

c. $H_3C-\overset{}{\underset{H}{N}}-\overset{O}{\underset{}{\|}}C-CH_3$

d. $H_3CO-\overset{O}{\underset{}{\|}}C-$ (benzene ring) $-\overset{O}{\underset{}{\|}}C-OCH_3$

e. $-\!\left(\!\overset{H}{\underset{H}{C}}-\overset{CH_3}{\underset{\underset{OCH_3}{C=O}}{C}}\!\right)_{n}$

f. $-\!\left(\!\overset{}{\underset{H}{N}}-\overset{H}{\underset{H}{C}}-\overset{H}{\underset{H}{C}}-\overset{H}{\underset{H}{C}}-\overset{H}{\underset{H}{C}}-\overset{H}{\underset{H}{C}}-\overset{O}{\underset{}{\|}}C\!\right)_{n}$

39. In each of the following compounds, determine the oxidation number of the indicated element(s).

a. $H-\overset{H}{\underset{H}{C}}-\overset{H}{\underset{H}{C}}-\overset{H}{\underset{H}{C}}-\overset{H}{\underset{H}{C}}-\overset{H}{\underset{H}{C}}-\overset{H}{\underset{H}{C}}-\overset{O}{\underset{OH}{C}}$

b. $H-\overset{H}{\underset{H}{C}}-\overset{H}{\underset{H}{C}}-\overset{H}{\underset{H}{C}}-\overset{H}{\underset{H}{C}}-OH$

c. $H-\overset{H}{\underset{H}{C}}-\overset{H}{\underset{H}{C}}-\overset{O}{\underset{}{\|}}C-\overset{H}{\underset{H}{C}}-H$

d. $H-\overset{H}{\underset{H}{C}}-\overset{H}{\underset{H}{C}}-\overset{H}{\underset{H}{C}}-\overset{O}{\underset{H}{C}}$

e. $H-\overset{H}{\underset{H}{C}}-\overset{H}{\underset{H}{C}}-\overset{H}{\underset{H}{C}}-NH_2$

f. $H-\overset{H}{\underset{H}{C}}-\overset{H}{\underset{H}{C}}-N=C=O$

40. Acids and alcohols condense to form a larger molecule. Indicate the interaction between the acid in Exercise 39(a) and the alcohol in Exercise 39(b). The resulting compound is called an ester. Indicate the product and the source of each of the two oxygen atoms in the resulting ester. Indicate the source of the oxygen in the water that is produced. (*Hint:* The oxidation numbers are helpful.)

41. In each of the following compounds, determine the oxidation number of the indicated element(s).

a. (benzene ring)–NH_2

b. (benzene ring)–OH

c. (benzene ring)–CH_3

d. (benzene ring)–$N=C=O$

42. Indicate the repeated unit of a polymer made from the amine and isocyanate shown below. Is the product a stiff or a flexible material? Defend your choice.

(benzene ring)–NH_2

$O=C=N-$ (benzene ring) $-N=C=O$

43. Fat is a biological material that results from the condensation of the triol glycerol with three long-chain acids. Fats are an example of a class of biological compounds called lipids. Given the structure and a long-chain acid, indicate the structure of a fat.

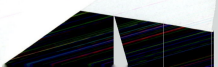

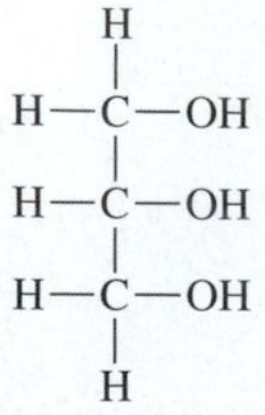

The triol, glycerol 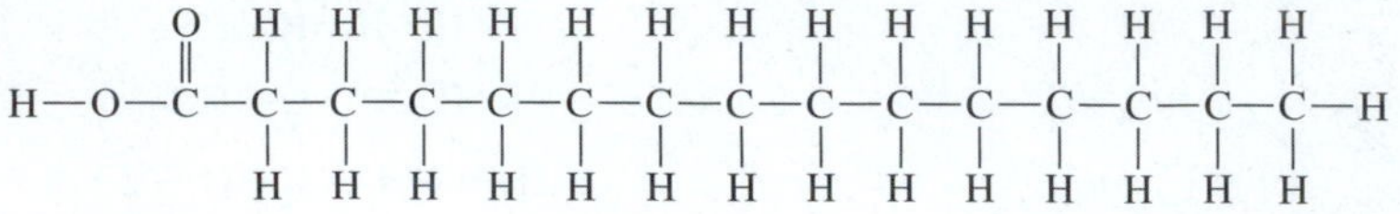   A fatty acid

44. What product is expected when the following compound is heated to 250 °C?

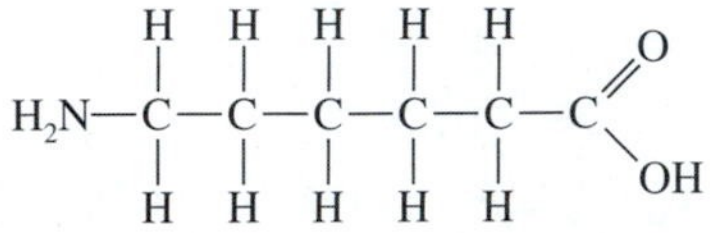

45. Draw the structure of the product formed in the following reaction. Be certain to clearly indicate the position of the labeled oxygen atoms in the products.

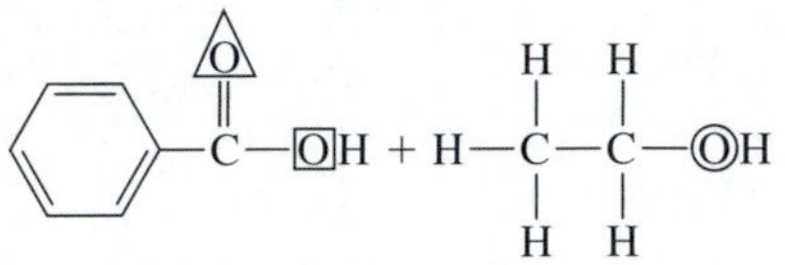

46. Which of the following are amines?

a.

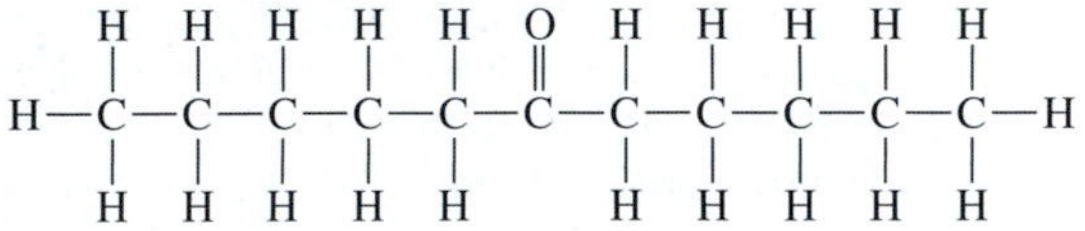

b. 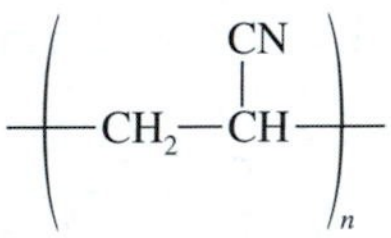   c.

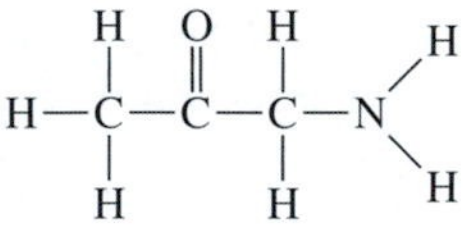

d.

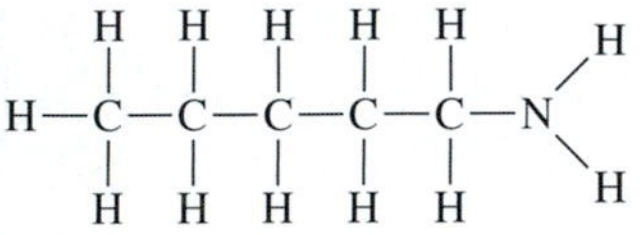

e.

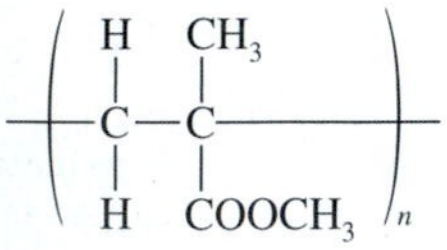

f.

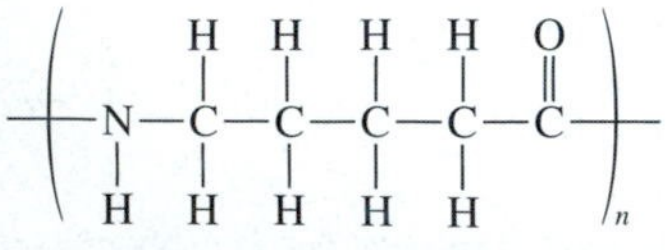

raw an acid, an amine, and an alcohol.

ester linkage. Draw an amide linkage.

a. HO following compounds, determine the oxidation
indicated.

b.

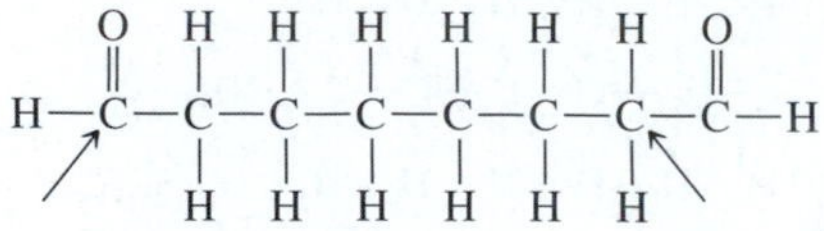

c.

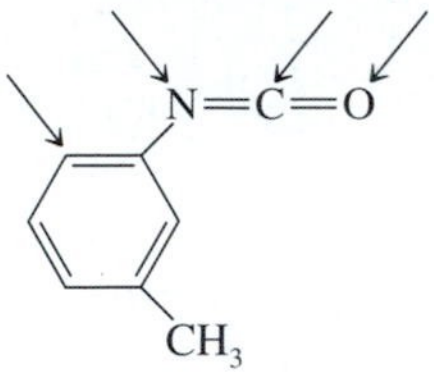

50. Polyesters can be degraded by acids, which catalyze rehydration of the ester linkage. Indicate the result of acid degradation of the following polyester.

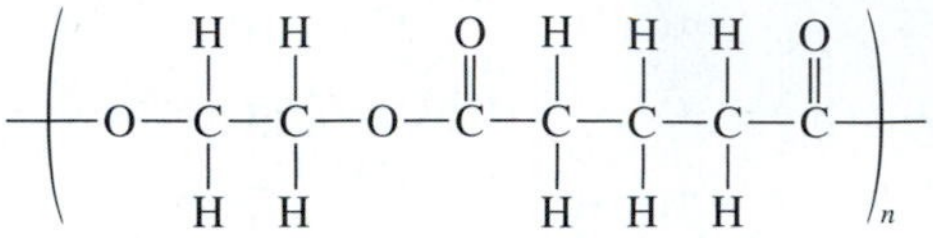

51. CDs and DVDs are based on a polycarbonate substrate. One such polycarbonate is shown below. Indicate the two monomers in this polymer. Which part of the monomer is responsible for the rigid quality of CDs and DVDs?

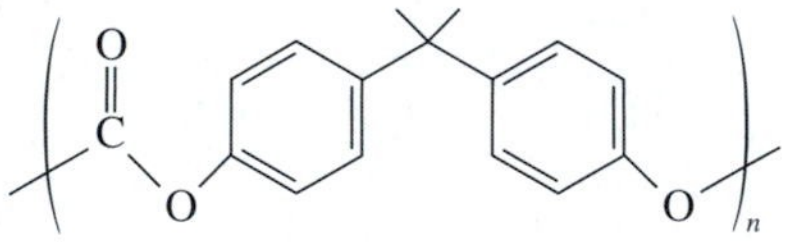

52. The polycarbonate used in CDs and DVDs (shown in Exercise 51) tends to form large, flat plates when stressed during molding. This tendency causes optical distortion and difficulties in reading. To prevent this problem, the polymer is modified as shown below. Explain how the modification prevents optical distortion.

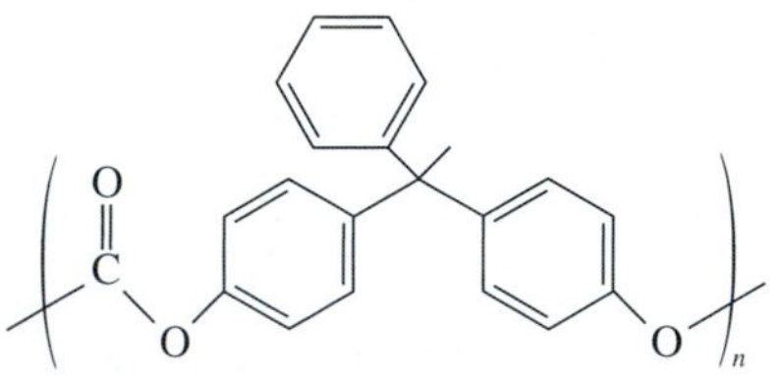

11.2 Addition Polymers

53. Teflon is made from the monomer $F_2C=CF_2$. Indicate the structure of Teflon.

54. Polypropylene is made from the monomer propylene (shown below). Indicate the structure of polypropylene.

$$
H-\underset{\underset{H}{|}}{\overset{\overset{H}{|}}{C}}-\underset{\underset{H}{|}}{\overset{\overset{H}{|}}{C}}=C\overset{H}{\underset{H}{}}
$$

55. Which of the following are candidates for formation of addition polymers?

a.
$$
\underset{HO}{\overset{O}{\|}}C-\underset{\underset{H}{|}}{\overset{\overset{H}{|}}{C}}-\underset{\underset{H}{|}}{\overset{\overset{H}{|}}{C}}-\overset{O}{\overset{\|}{C}}\underset{OH}{}
$$

b. $H_2C{=}CH$ — phenyl

c. $H_2N-\underset{\underset{H}{|}}{\overset{\overset{H}{|}}{C}}-\underset{\underset{H}{|}}{\overset{\overset{H}{|}}{C}}-\underset{\underset{H}{|}}{\overset{\overset{H}{|}}{C}}-\underset{\underset{H}{|}}{\overset{\overset{H}{|}}{C}}-\underset{\underset{H}{|}}{\overset{\overset{H}{|}}{C}}-\underset{\underset{H}{|}}{\overset{\overset{H}{|}}{C}}-NH_2$

d. $H_3C-O-\overset{O}{\overset{\|}{C}}$ — (benzene ring) — $\overset{O}{\overset{\|}{C}}-O-CH_3$

e. $H-\underset{\underset{H}{|}}{\overset{\overset{H}{|}}{C}}-C\overset{O}{\underset{H}{}}$

f. $\underset{H}{\overset{H}{}}C{=}C\underset{Cl}{\overset{H}{|}}-C{=}C\overset{H}{\underset{H}{}}$

56. Indicate all portions of the following molecule where addition can occur.

$$
\underset{H}{\overset{H}{}}C{=}C-\underset{}{\overset{\overset{H}{|}}{C}}{=}C-\underset{}{\overset{\overset{H}{|}}{C}}-\overset{O}{\overset{\|}{C}}\underset{OH}{}
$$
(with H's: $C{=}C-\overset{H}{\underset{H}{}}C{=}C-\overset{H}{\underset{H}{}}C$)

57. Which of the following polymers are made by addition reactions?

a. $\left(-N-\underset{\underset{H}{|}}{\overset{\overset{H}{|}}{C}}-\underset{\underset{H}{|}}{\overset{\overset{H}{|}}{C}}-\underset{\underset{H}{|}}{\overset{\overset{H}{|}}{C}}-\underset{\underset{H}{|}}{\overset{\overset{H}{|}}{C}}-\underset{\underset{H}{|}}{\overset{\overset{H}{|}}{C}}-\underset{\underset{H}{|}}{\overset{\overset{H}{|}}{C}}-N-\overset{O}{\overset{\|}{C}}-\underset{\underset{H}{|}}{\overset{\overset{H}{|}}{C}}-\underset{\underset{H}{|}}{\overset{\overset{H}{|}}{C}}-\underset{\underset{H}{|}}{\overset{\overset{H}{|}}{C}}-\underset{\underset{H}{|}}{\overset{\overset{H}{|}}{C}}-\underset{\underset{H}{|}}{\overset{\overset{H}{|}}{C}}-\overset{O}{\overset{\|}{C}}-\right)_n$

b. $\left(-CH_2-\underset{Cl}{\overset{|}{CH}}-\right)_n$

c. $\left(-O-\underset{\underset{H}{|}}{\overset{\overset{H}{|}}{C}}-\underset{\underset{H}{|}}{\overset{\overset{H}{|}}{C}}-O-\overset{O}{\overset{\|}{C}}-(benzene)-\overset{O}{\overset{\|}{C}}-\right)_n$

d. $\left(-\underset{\underset{H}{|}}{\overset{\overset{H}{|}}{C}}-\underset{\underset{C}{|}}{\overset{\overset{CH_3}{|}}{C}}-\right)_n$ with $C\overset{O}{\underset{OCH_3}{}}$

e. $\left(-O-(benzene)-\underset{\underset{CH^3}{|}}{\overset{\overset{CH_3}{|}}{C}}-(benzene)-O-\overset{O}{\overset{\|}{C}}-\right)_n$

f. (cyclopentane ring with vinyl group) $_n$

58. The following reactants are mixed together and heated above 100 °C. Balance and complete the reaction.

(benzene)$-\overset{O}{\overset{\|}{C}}-O-O-\overset{O}{\overset{\|}{C}}-$(benzene) $+$ $\underset{H}{\overset{H}{}}C{=}C\underset{H}{\overset{H}{|}}-C{\equiv}N$

59. Lithium ion batteries are used in applications such as cell phones and laptop computers, where high energy density and light weight are prime considerations. Most of these batteries use a polymer as a binder for the cathode and the anode. The most prevalent binder is a polymer made from vinylidene fluoride: $H_2C{=}CF_2$. Indicate the polymer obtained from vinylidene fluoride. What is the mechanism for the polymerization reaction?

60. Fluoroelastomers are enjoying a renaissance due to their robustness upon exposure to acids, bases, oils, solvents, and a wide range of temperature (-30 to 250 °C). One fluoroelastomer is a copolymer of vinylidene fluoride (see Exercise 59) and hexafluoropropylene (shown below). What is the formation mechanism for this copolymer?

$$
\underset{F}{\overset{F}{}}C{=}C\underset{F}{\overset{F}{|}}-CF_3
$$

61. The molecule referred to as V 70 can be used as a radical initiator for reactions occurring at temperatures higher than 40 °C. Indicate the decomposition of V 70. Is the decomposition easily reversed?

$$
H_3C-\underset{\underset{CH_3}{|}}{\overset{\overset{CH_3}{|}}{C}}-CH_2-\underset{\underset{CN}{|}}{\overset{\overset{CH_3}{|}}{C}}-N{=}N-\underset{\underset{CN}{|}}{\overset{\overset{CH_3}{|}}{C}}-CH_2-\underset{\underset{CH_3}{|}}{\overset{\overset{CH_3}{|}}{C}}-CH_3
$$

62. Addition can also occur at strained bonds. Identify the strained bonds in each of the following monomers and indicate the resulting polymer.

Epichlorohydrin

3,3 Bis(hydroxymethyl)oxetane

Poly[3,3-bis(hydroxymethyl)oxetane] is a ce?
Indicate the interaction between poly[3.?
methyl)oxetane] chains. What interm?
is operational in epichlorohydrin?

11.3 Molecular Structure and Material Properties

63. Polychloroprene makes a bouncy, responsive rubber. Polynorborene makes a vibration-damping rubber. Explain which rubber is appropriate for the following and why.
    a. running shoes
    b. tires
    c. gym safety mat
    d. the core of a baseball

64. Styrene has the following structure:

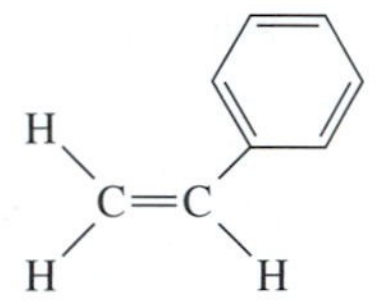

    Is styrene a candidate for forming a high-density polymer? Why or why not?

65. Polypropylene is made from the monomer propylene:

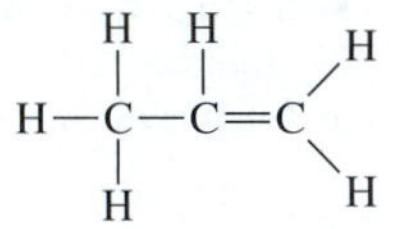

    Indicate the structure of polypropylene if the polymerization is initiated with a radical initiator. Synthesizing polypropylene with a Ziegler-Natta catalyst results in a very different structure. Indicate this structure and compare it with the polymer formed from radical initiation.

66. The molecule 2-butene has the following structure:

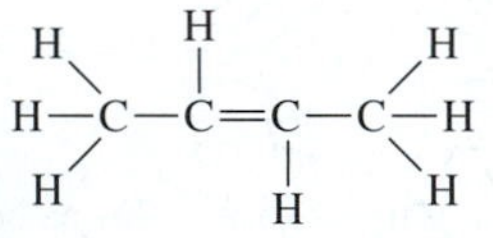

    Indicate the structure of the polymer. Is 2-butene a candidate for making rubber?

67. Based on your answer to Exercise 66, discuss the physical properties of polybutene.

68. Identify the intermolecular force between chains of polynorborene. What force operates in polychloroprene? Are these forces responsible for the rubber properties of these materials? If not, what is?

69. The polyester PET has the following structure:

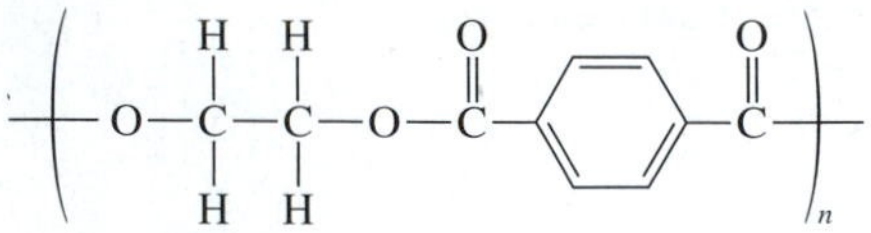

    Indicate the *two* monomers that are condensed to form PET. What functional part of the monomer is responsible for the rigidity of the polymer PET?

    flame-retardant polymer Nomex and the very ...olymer Kevlar (used in bulletproof vests) are

closely related. Comment on the flame-retardant characteristics of Kevlar.

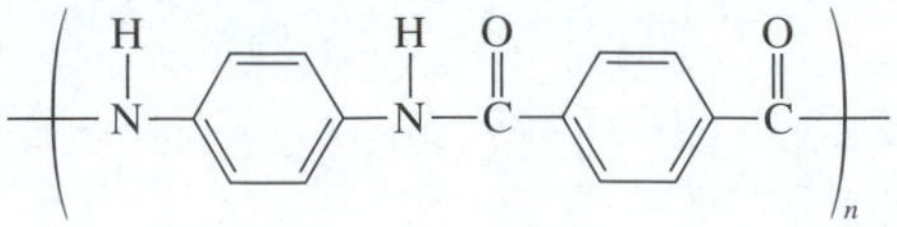

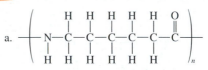

Kevlar

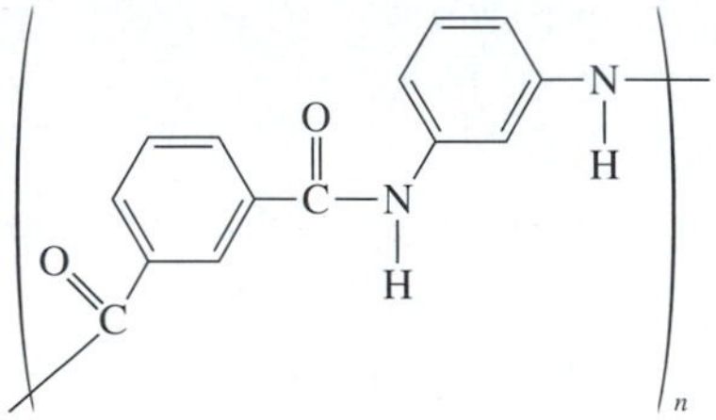

Nomex

71. Comment on the potential of Nomex for use in bulletproof vests.

72. Which of the following polymers is the most rigid? Why?

    a.

    b.

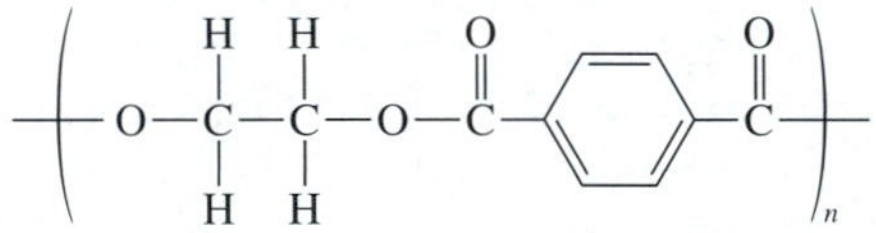

    c.

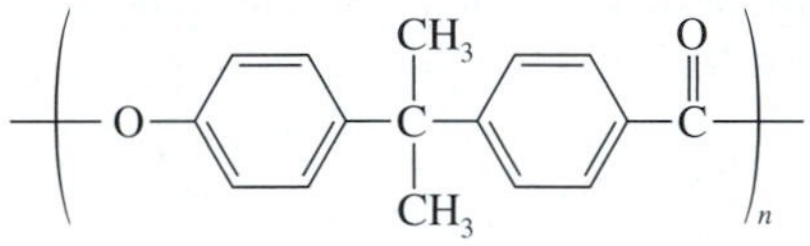

    d.

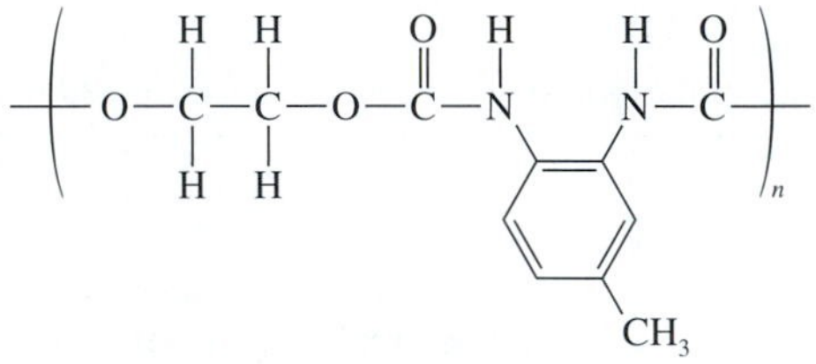

■ CONCEPTUAL EXERCISES

73. Define a polymer. Which characteristic distinguishes a polymer from other molecules?

74. Which three molecular-level properties affect the material properties of a polymer?

75. Explain why the shorthand notation $-(CH_2)_n$ does not adequately describe polyethylene.

76. The syntheses of nylon, polyesters, and polyurethane can all be classified as acid–base reactions. Explain why these are acid–base reactions. Can any of these condensation polymers be made into a rubber? If so, what is needed to make them into a rubber? If not, why not.

77. Draw a diagram indicating the relationship among the following terms: monomer, polymer, macromolecular structure, material properties, HDPE, LDPE. All terms should have at least one connection.

78. Draw a diagram indicating the relationship among the following terms: amine, amide, alcohol, ester, acid, diester, diol, diacid, diamine, condensation. All terms should have at least one connection.

✲ APPLIED EXERCISES

79. Nathaniel Wyeth invented poly(ethylene terephthalate) for use in the soft-drink bottle. PET is made from two monomers, ethylene glycol and terephthalic acid:

Indicate the repeated unit in this polymer. To make a bottle with a high burst strength, Wyeth pulled the polymer sheet, then used two sheets with pull directions perpendicular to each other. Explain how this strategy works.

80. Modern sports equipment relies heavily on polymeric components. How does the molecular-level structure impart the following properties:
   a. A soft cover for golf balls creates additional spin. Polymer used: polyisoprene.
   b. Lightweight, stiff shafts of golf clubs increase head speed. Stiffness in skis hold an edge and reduce vibrations referred to as chatter. Polymer used: Kevlar.
   c. Football helmets that are lightweight, yet strong are created from Lexan.
   d. Footballs are made waterproof with polyurethane. Polyurethane also provides the spring in tennis racket strings.

81. Absorbable sutures consist of a biodegradable polymer. One suture is based on glycolic acid, obtained from the dimer shown below (hydrogen atoms have been omitted for clarity). The dimer hydrolyzes to form the monomer. What is the structure of the monomer? What is the structure of the polymer?

82. Tough floor waxes that can be removed with ammonia cleaners have greatly simplified care of wooden flooring. One such finish is polyacrylamide, made by radical polymerization of the monomer $H_2C=CHCONH_2$. Indicate the polymer in acrylamide. The finish gets its great toughness from a cross-linking reaction that occurs with Zn^{2+} as the finish dries. Zinc is initially present as $Zn(NH_3)_4^{2+}(aq)$. Ammonia is released during the drying process. Indicate the structure of the cross-linked product. Explain how an ammonia cleaner dissolves the finish.

■ INTEGRATIVE EXERCISES

83. Is the polymer that is formed from the monomers

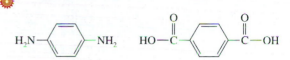

a candidate for forming a conducting polymer? Why or why not?

84. The reaction

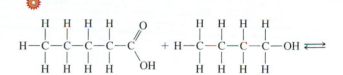

comes to equilibrium at room temperature. What functional groups are in the reactants and the products? Suggest a method that will yield more product.

85. The diamine and diacid shown below condense to form the polymer Kevlar, which is used in bulletproof vests. Indicate the repeated unit in the polymer.

How does the polymerization give the vest material its famous strength?

86. Contact lenses have gone through two major advances since they were first manufactured in 1948. The first contact lenses were hard contacts based on a polymer of methyl methacrylate (MMA). Unfortunately, polymethylmethacrylate forms a polymer that is not very permeable to gases. Because the cornea gets its oxygen supply directly from the atmosphere, a gas-impermeable lens cannot be worn for long. The first major advance consisted of incorporation of tris(trimethylsiloxysilane) (TRIS) as a copolymer. TRIS creates small pores and improves gas permeability. The second major advance was the development of a soft contact lens. The soft contact is based on a cross-linked hydrophilic polymer. A number of these lenses are now on the market, but the first was based on the monomer 2-hydroxyethyl methacrylate (HEMA).

$H_2C=CH$

MMA

$H_2C=CH$

TRIS

$H_2C=CCH_3$

HEMA

a. Indicate the structure of the MMA polymer and the MMA–TRIS copolymer.
b. Indicate the structure of the HEMA polymer.
c. Explain the interaction of water with each of the materials: MMA, MMA–TRIS, and the MMA–HEMA copolymer.

d. The MMA–HEMA copolymer absorbs 38% by weight water. How many water molecules per monomer does MMA–HEMA absorb?

87. Design a polymer that is both flexible and a flame retardant. Are these two requirements mutually exclusive? Cite properties of related polymers to support your conclusions.

88. Recently, organic conductors and dyes have received attention as potential components of flexible display devices for microelectronics. Discuss the advantages of organic materials for this application. Design a display device that solves the following issues:
a. Making an electrical connection to the conductor
b. Maintaining the connection when the display is flexed
c. Maintaining color fidelity (that is, keeping the emitter from fading, usually the result of oxidation)
Discuss the potential for combining an organic conductor with an inorganic nanoparticle for emission.

Chapter 12
Kinetics

Kinetics is about the dynamics of a reaction. Understanding the speed, twists and turns that molecules go through on the way from reactants to products often makes it possible to control the reaction rate in useful ways.

CONCEPTUAL FOCUS

■ Determine the time needed for a chemical transformation to occur and characterize dependence of the reaction time on reactant concentration and temperature.

■ Characterize the steps of a reaction and explain how a catalyst alters the reaction rate.

SKILL DEVELOPMENT OBJECTIVES

■ Determine rate laws and rate constants (Worked Examples 12.1 – 12.3, 12.5, 12.7).

■ Use rate constant to determine age of an object (Worked Examples 12.4, 12.6).

■ Identify mechanisms consistent with experimental rate laws (Worked Example 12.8).

■ Explain how a catalyst speeds up a reaction; determine activation energy; draw energy profiles (Worked Examples 12.9, 12.11).

■ Connect reaction rate and mechanism to equilibrium (Worked Example 12.10).

Chemical thermodynamics and equilibrium indicate the direction of spontaneous change and the ultimate fate of the system. Neither, however, provides any information about the time needed for a change to occur. Reactions occur over a large range of time scales (Figure 12.1), and these time scales determine applications for the reaction. If automobile air bags deployed as slowly as a car rusts, they would do little to protect the car occupants. Kinetics deals not only with the speed of a reaction but also with the steps involved in transformation of reactants into products—it gives a "look under the hood" into the inner workings of a reaction. Understanding the steps and ways that the speed of each step is affected by various factors can suggest approaches to control the speed of the process. Consider the rusting of automobiles. Several decades ago, any cars more than three or four years old carried tell-tale signs of corrosion of the body. Based on an understanding of the corrosion process, surface treatments and alloys were developed. Today, modern cars show little or no sign of corrosion even after a decade of use.

Rusting is an example of kinetic versus thermodynamic control. Thermodynamically, cars will rust; biological systems will turn into water, carbon dioxide, and a smaller pile of rust. The process, however, can take long enough that for all practical purposes it does not matter. This situation demonstrates kinetic control. Think of a raindrop falling on Pikes Peak. Controlled by potential energy, it will end up in the sea. Along the way, it can be caught in a lake basin, making a home for fish and a waterway for boats to transport goods.

12.1   Reaction Rates

Concentration

The first step in the kinetic analysis of a reaction is determining how the reaction rate depends on the various substances involved in the reaction. Suppose that a reaction is running slowly, too slowly to be useful. Can the reaction be speeded up by simply adding more of one of the reagents?

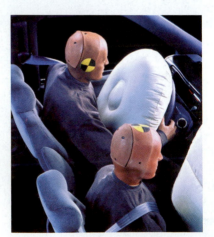
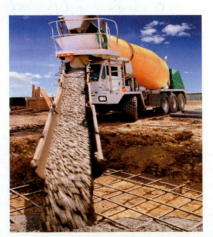
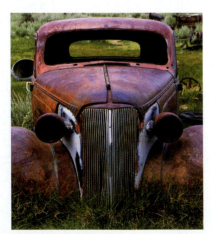

Figure 12.1 The time required for a chemical reaction often determines the application for the reaction. An automobile air bag deploys in a few milliseconds to protect the car occupants. If concrete set equally rapidly, it could not be used as the fundamental structural material it has become. The rusting of an automobile is mercifully slow, allowing the car to have a useful lifetime.

The answer to this question is not apparent from the stoichiometric equation. Indeed, there is generally *no* relationship between the stoichiometric coefficient and the dependence of the reaction speed on any given reactant or product. The rate law is *not* determined from stoichiometry, but rather is experimentally determined. The correlation between the stoichiometric equation and the reaction rate is hidden by the intricacies of the steps of the reaction. As an analogy, think of traveling from Boston to San Francisco. The speed along the path depends on the mode of travel and the trip time depends on the path. Travel on foot is much slower than travel by direct air flight, yet the origin and the destination remain the same. Travel from Boston to London by foot will not even happen; there is a barrier—the Atlantic Ocean—in the way. The details of the process are the province of kinetics.

The first issue in determining the rate of a reaction is defining precisely what is meant by *rate*. As an example, consider the following reaction:

(12.1)   cyclopropane $\longrightarrow$ propylene

Suppose the reaction starts with 1 mol cyclopropane. After 100 s, there are 0.75 mol cyclopropane and 0.25 mol propylene. The reaction rate could be defined as the rate of formation of propylene:

(12.2) $$\frac{\Delta(\text{moles propylene})}{\Delta t} = \frac{0.25 \text{ mol}}{100 \text{ s}} = 2.5 \times 10^{-3} \text{ mol/s}$$

Alternatively, the rate could be defined by the change in cyclopropane concentration:

(12.3) $$\frac{\Delta(\text{moles cyclopropane})}{\Delta t} = \frac{-0.25 \text{ mol}}{100 \text{ s}} = -2.5 \times 10^{-3} \text{ mol/s}$$

Equations (12.2) and (12.3) differ by a minus sign. By convention, the disappearance rate for reactants is defined as having a minus sign to keep the rate positive, so Equation (12.3) is modified as follows:

(12.4) $$\frac{-\Delta(\text{moles cyclopropane})}{\Delta t} = \frac{0.25 \text{ mol}}{100 \text{ s}} = 2.5 \times 10^{-3} \text{ mol/s}$$

The conversion of cyclopropane to propylene has a simple 1:1 stoichiometry.

Next, consider a reaction with different stoichiometric coefficients:

(12.5) $2ICl + H_2 \longrightarrow I_2 + 2HCl$

The disappearance rate for ICl is twice that for H_2. The definition of the reaction rate accounts for the stoichiometry of the reaction by dividing by the stoichiometric coefficient: $\frac{1}{2}$ for ICl.

In addition to stoichiometry, another more subtle issue arises when defining the reaction rate: The rate often changes with the concentration of some component, and that concentration changes with time.

CONCEPT QUESTIONS The rate of a reaction often depends on the concentrations of the substances involved in the reaction. These concentrations change as the reaction proceeds. Consider the data for Equation (12.5) given in Table 12.1. What is the average rate of the reaction from 0 to 10 seconds? What is the average rate from 4 to 6 seconds? Which average better represents the rate at 5 seconds? ∎

The answers to the Concept Questions indicate that defining a reaction rate by the average rate of disappearance of a reactant results in a reaction rate that depends on the

Table 12.1 Data for Reaction of ICl with H_2.

Time (s)	Moles ICl	Moles H_2
0	2.000	1.000
1	1.348	0.674
2	1.052	0.526
3	0.872	0.436
4	0.748	0.374
5	0.656	0.328
6	0.586	0.293
7	0.530	0.265
8	0.484	0.242
9	0.444	0.222
10	0.412	0.206

time interval chosen. In this example, the average rates at 5 seconds differ by nearly a factor of 2. However, as the time interval around 5 seconds becomes smaller, the average rates converge to a limit. Using the notation of calculus:

Reaction rate The change in concentration of a reactant or product per unit time. Also called *reaction velocity*.

(12.6)
$$\lim_{\Delta t \to 0} \frac{-^1/_2 \Delta [ICl]}{\Delta t} = \lim_{\Delta t \to 0} \frac{-\Delta [H_2]}{\Delta t} = -\frac{1}{2}\frac{d[ICl]}{dt} = -\frac{d[H_2]}{dt}$$

In general, for the reaction

(12.7)
$$aA + bB \longrightarrow pP + sS$$

the reaction rate (in biochemical systems, this is called the velocity of the reaction) is

Reaction velocity The change in concentration of a reactant or product per unit time. Also called *reaction rate*.

(12.8)
$$v = -\frac{1}{a}\frac{d[A]}{dt} = -\frac{1}{b}\frac{d[B]}{dt} = \frac{1}{p}\frac{d[P]}{dt} = \frac{1}{s}\frac{d[S]}{dt}$$

This is the basic definition of the **reaction rate** or **reaction velocity.**

Rate Law and Reaction Order

One of the goals of the kinetic analysis of a reaction is to determine the dependence of the reaction rate or velocity on the concentrations of the substances involved in the reaction. Sometimes the dependence is quite subtle, perhaps involving the concentrations of substances that do not appear explicitly among the reactants or products. For example, for the hydrolysis of sugars in dilute acid, $C_{12}H_{22}O_{11}(aq) + H_2O(l) \to 2C_6H_{12}O_6(aq)$, the rate equation is

(12.9)
$$-\frac{d[C_{12}H_{22}O_{11}]}{dt} = k[C_{12}H_{22}O_{11}][H_3O^+]$$

Here the hydrolysis rate depends on the acid concentration even though H_3O^+ is neither a reactant nor a product. In general, for the reaction

Rate law An expression that shows how the rate of reaction depends on the concentration of substances involved in the reaction.

(12.10)
$$aA + bB \longrightarrow pP + sS$$

the rate equation, or **rate law,** is

(12.11)
$$v = k_{ex}[A]^\alpha [B]^\beta [P]^\pi [S]^\sigma [I]^\tau$$

Specific rate constant An experimentally measured quantity, k_{ex}. The constant of proportionality in the rate law.

Reaction order The sum of the powers of all substances in the rate law.

where α, β, π, σ, or τ can have any positive, negative, integer, noninteger, or zero value. Specifically, the substance I does not enter the *net* equation at all; yet like H_3O^+ in sugar hydrolysis, it appears in the rate law. The value k_{ex} is an experimentally measured quantity called the **specific rate constant.** The specific rate constant is a characteristic of a reaction. For most reactions, k is a function of temperature.

The sum of the exponents in the rate law, $\alpha + \beta + \pi + \sigma + \tau$, is the **reaction order.** The order with respect to each substance is the exponent for that substance. For example, the reaction order for A in Equation (12.11) is α.

CONCEPT QUESTIONS The acidic gas HBr is formed from a reaction between gaseous hydrogen and bromine: $H_2(g) + Br_2(g) \rightarrow 2HBr(g)$. At the start of the reaction, the rate equation for this reaction is

$$\frac{d[HBr]}{dt} = k_1[H_2][Br_2]^{1/2}$$

What is the reaction order? What is the order with respect to H_2? With respect to Br_2? ■

Determining the reaction order for a substance involved in a reaction answers the question about whether changing the concentration of a substance in a reaction changes the reaction rate.

WORKED EXAMPLE 12.1   *Ester Hydrolysis*

Soap is made by hydrolyzing fats, which are esters of long-chain acids with glycerol, under strongly basic conditions. An ester, ethyl acetate, $CH_3COOC_2H_5$, is used in a kinetics experiment to determine whether increasing the pH will increase the rate of hydrolysis. It is found that doubling the hydroxide ion concentration (increasing the pH by 0.3) doubles the rate of the reaction. Doubling the ester concentration also doubles the reaction rate. What is the reaction order? Write the rate equation.

Plan

■ Determine the exponent for each substance involved.
■ Use Equation (12.11).

Implementation

■ The change in reaction rate is directly proportional to the changes in the OH^- and $CH_3COOC_2H_5$ concentrations. Thus, the exponent on each of these concentrations is 1. The reaction is first order in hydroxide ion concentration, first order in ester concentration, and second order overall.
■ $v = k_{ex}[OH^-][CH_3COC_2H_5]$

Raising the pH by adding more alkali is a relatively inexpensive way of increasing the rate of soap production.

See Exercises 6, 7.

Finding the Rate Law from Experimental Data

The rate law is not correlated with the stoichiometric coefficients, but rather must be determined experimentally. How is this done? Because rate laws can be quite complex, two methods are commonly used to simplify the concentration dependence: (1) the method of initial rates and (2) Ostwald's isolation method, also known as the flooding method.

Method of Initial Rates

At the beginning of a reaction, the concentrations of any products or substances formed along the way are very small and have little effect on the reaction rate. Thus focusing on the beginning of a reaction highlights the order with respect to the reactants. The procedure involves a divide-and-conquer method. A series of experiments are run, varying the initial concentration of one reactant at a time. Each time, the initial reaction rate is monitored. If doubling the concentration of one reactant leaves the rate unchanged, then the reaction must be independent of the concentration of that substance; if the rate doubles, the reaction must be dependent on the first power of the substance; if the rate quadruples, the reaction must second order in that substance, and so on. This technique is called the **method of initial rates.**

Method of initial rates
Procedure for determining the reaction order for substances in a reaction in which the rate of the reaction is determined while systematically varying the initial concentrations.

WORKED EXAMPLE 12.2   *Oxidation of NO*

NO is one of the nitrogen oxides generated in combustion processes such as take place in the internal combustion engine of a car. When exposed to oxygen in the air, NO is transformed into NO_2 and ultimately contributes to the generation of acid rain and smog. Understanding how this reaction depends on the concentration of NO indicates whether prevention measures based on removal of NO from automobile exhaust will improve air quality. Experiments were run on the following reaction:

$$2NO(g) + O_2(g) \longrightarrow 2NO_2(g)$$

The rate of appearance of NO_2 was measured and recorded for each experiment. Determine the order of the reaction with respect to NO and O_2. Write the rate law.

Experiment	$[NO]_0$ (mol/L)	$[O_2]_0$ (mol/L)	$\left(\dfrac{d[NO_2]}{dt}\right)_0$ mol/L·s
1	1.0×10^{-3}	1.0×10^{-3}	7×10^{-6}
2	2.0×10^{-3}	1.0×10^{-3}	28×10^{-6}
3	1.0×10^{-3}	2.0×10^{-3}	14×10^{-6}
4	3.0×10^{-3}	1.0×10^{-3}	63×10^{-6}
5	1.0×10^{-3}	3.0×10^{-3}	21×10^{-6}

The subscript 0 indicates the initial values.

Plan

■ Compare reaction rates for experiments in which only one concentration changes.
■ The reaction rate is the product of the NO and O_2 concentration dependence.

Implementation

■ Experiments 1, 2, and 4 have the same initial O_2 concentration. The initial NO concentration doubles from experiment 1 to 2 and triples from experiment 1 to 4. The initial rate quadruples from experiment 1 to 2 and increases by a factor of 9 from experiment 1 to 4. The rate depends on $[NO]^2$. Experiments 1, 3, and 5 have the same initial NO concentration. The initial O_2 concentration doubles from experiment 1 to 3 and triples from experiment 1 to 5. The initial rate doubles from experiment 1 to 3 and triples from experiment 1 to 5. The rate is proportional to the first power of the O_2 concentration.

■ The rate law is

$$\frac{d[NO_2]}{dt} = k[NO]^2[O_2]$$

Every reduction of NO emission reduces the rate of production of NO_2 by the square of that reduction. This magnification makes a focus on reduction of NO emission very worthwhile.

See Exercises 17, 18.

Ostwald's Isolation Method

The second method for determining the rate law depends on the following approximation: When a reactant is present in large excess, the concentration of that reactant hardly changes in the course of the reaction. For example, being near burning fuel or smoking a cigarette results in inhalation of NO. NO is oxidized by oxygen to NO_2, a powerful oxidant that converts Fe(II) to Fe(III) and is itself converted to HNO_2 and HNO_3, both of which cause pulmonary problems. The reaction with oxygen is

Ostwald's isolation method
A method to determine the order of a reaction with respect to a substance by introducing a large excess of another substance. Also known as *flooding* the reaction.

(12.12) $$2NO(g) + O_2(g) \longrightarrow 2NO_2(g)$$

The rate law for this reaction is

(12.13) $$\frac{d[NO_2]}{dt} = k[NO]^2[O_2]$$

Typical ambient conditions consist of atmospheric oxygen, 20% oxygen, and NO concentration in the 400 to 1000 ppm (parts per million) range. Oxygen is in much larger concentration, so the oxygen concentration can be considered a constant. With this approximation Equation (12.13) becomes

Flooding Introducing a large excess of one substance involved in a reaction for the purpose of determining the order in a remaining substance. Also called *Ostwald's isolation method*.

(12.14) $$\frac{d[NO_2]}{dt} = k[NO]^2[O_2] = k'[NO]^2$$

where $k' = k[O_2]$. The order of the reaction is reduced from third order to second order. Experiments on the oxidation rate of NO reveal the order with respect to NO. This technique is called **Ostwald's isolation method** or **flooding** the reaction.

WORKED EXAMPLE 12.3   *Recombination of Iodine*

Iodine atoms are powerful oxidants that can form when iodine vapor is exposed to light. Light causes molecular iodine, I_2, to dissociate into iodine atoms, I. Recombination of the iodine atoms occurs slowly unless another gas-phase substance is present. An experiment to determine the order of the recombination reaction for

I consists of dissociating I_2 with a strong laser pulse in the presence of a larger concentration of Ar gas. Four such experiments were done in the presence of 0.001 M Ar, with results shown in the table below. What is the order of the reaction with respect to I?

Experiment	1	2	3	4
$[I]_0$ (M)	1.0×10^{-5}	2.0×10^{-5}	4.0×10^{-5}	6.0×10^{-5}
Initial rate (mol/L · s)	8.7×10^{-4}	3.48×10^{-3}	1.39×10^{-2}	3.13×10^{-2}

Plan

- Check: Is Ar present in large excess?
- Compare the rates for the various iodine atom concentrations.

Implementation

- The Ar concentration is 10^{-3} M, one to two orders of magnitude larger than the I atom concentration. This is a large excess.
- From experiment 1 to 2, [I] doubles and the rate quadruples. From experiment 2 to 3, [I] doubles and the rate quadruples. From experiment 2 to 4, [I] triples and the rate increases by a factor of 9. The order with respect to I is 2.

$$\frac{d[I_2]}{dt} = k'[I]^2$$

This reaction is pseudo-second order in the presence of an excess of Ar.

See Exercise 21.

12.2   Rate Laws

Reaction mechanism The series of elementary steps involved in a chemical reaction.

The connection between the net equation and the rate equation is hidden by the steps of the reaction. These steps are referred to as the **reaction mechanism.** A mechanism most often consists of a series of elementary steps that are unimolecular (first order), bimolecular (second order), or, rarely, termolecular (third order). For an elementary step, called an elementary reaction, a direct connection exists between the stoichiometry and the rate law for that step. Occasionally, a reaction consists of a single elementary step. For single-step elementary reactions, and *only for single-step elementary reactions, there is a direct connection between stoichiometry and the rate law.*

First-Order Reactions

The simplest elementary reaction is a first-order reaction. An example of a reaction that consists of a single first-order step is radioactive decay. Some nuclei are unstable and spontaneously emit subatomic particles to become another element. The rate law for a first-order reaction of substance A is

(12.15)
$$\frac{d[A]}{dt} = -k[A]$$

The relationship between concentration and time is

(12.16)
$$[A]_t = [A]_0 e^{-kt}$$

where e is the base of the natural logarithms. In log form, Equation (12.16) is

(12.17) $$\log[A]_t/[A]_o = -kt/2.303$$

This equation describes a straight line. Thus a plot of $\log[A]_t/[A]_o$ against time is a straight line and the rate constant, k, is determined from the slope $= -k/2.303$.

WORKED EXAMPLE 12.4   *Dating Old Objects*

The Dead Sea Scrolls, manuscripts of books from the Old Testament, were found in 1947. Authentication of such artifacts includes dating the material to establish the age of the item. Carbon is an important building block for all living things. In particular, organisms contain the three carbon isotopes—^{12}C, ^{13}C, and ^{14}C—in their natural abundance ratio. After the organism dies, ^{12}C and ^{13}C remain because they are stable isotopes, but ^{14}C decays according to the reaction

$$^{14}_{6}C \xrightarrow{k} \; ^{14}_{7}N + \; ^{0}_{-1}e$$

where $k = 1.21 \times 10^{-4}$/yr. In living organisms, the ^{14}C decay rate is 15.3 disintegrations per minute (dpm) per gram of carbon. For the Dead Sea Scrolls, carbon-14 activity in the linen wrappings was measured and found to be 12 dpm per gram. Calculate the approximate age of the linen.

Plan
- Use Equation (12.17): $\log[A]_t/[A]_0 = -kt/2.303$.

Implementation
- $[A]_0$ is proportional to 15.3 dpm, while $[A]_t$ is proportional to 12 dpm. Rearranging Equation (12.17) to

$$t = -\frac{2.303 \times \log([A]_t/[A]_0)}{k} = -2.303 \frac{\log\left(\dfrac{12 \text{ dpm}}{15 \text{ dpm}}\right)}{1.21 \times 10^{-4} \, yr^{-1}}$$

$$= \frac{-2.303 \times -0.097}{1.21 \times 10^{-4} \, yr^{-1}} = 1840 \text{ yr}$$

Radiochemical dating is an important technique for determining many events in the history of the earth. For example, wood building materials buried in a volcanic eruption can be used to date the eruption, revealing details of the history of humankind.

See Exercises 23, 24.

Second-Order Reactions

A bimolecular *elementary* reaction has a second-order rate law. The elementary reaction has an important interpretation: Two molecules collide and form a product. The reaction of methyl bromide with hydroxide ion to produce methanol and bromide ion (Figure 12.2) is an example. This type of bimolecular collision is important in synthesis of carbon compounds because the spatial orientation around the carbon atom is well controlled. Two mirror-image arrangements are possible when four different groups are bound to a single carbon atom. Mirror-image arrangements are called enantiomers. In many natural products, only one of two possible enantiomers is found (Figure 12.3). Particularly in drug synthesis, control of the arrangement around a carbon atom is therefore critical.

$$OH^- + CH_3Br \longrightarrow CH_3OH + Br^-$$

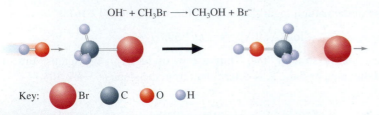

Key: Br   C   O   H

Figure 12.2 Substitution of an OH group for the bromine in methyl bromide is an example of a bimolecular reaction. OH^- in solution collides with the backside of the methyl bromide molecule, simultaneously ejecting the bromide ion and inverting the methyl group like an umbrella in the wind.

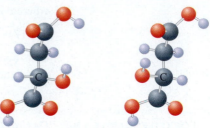

Figure 12.3 Note the spatial arrangements of the groups around the carbon atom marked with a C. The COOH, H, OH, and C_2H_2COOH groups are arranged as mirror images in the two molecules depicted here. This molecule is malic acid, an acid found in high concentration in apples and other fruits. As with most natural substances, only the form shown on the left is found in fruit.

Bimolecular elementary reactions are of two types:

$$\textbf{(12.18)} \qquad \frac{d[A]}{dt} = -k[A]^2 \quad \text{or} \quad \frac{d[A]}{dt} = -k[A][B]$$

The second can be related to the first by stoichiometry, so only the former is examined here. For a bimolecular reaction, the relationship between concentration and time is

$$\textbf{(12.19)} \qquad \frac{1}{[A]_t} - \frac{1}{[A]_o} = kt$$

The time evolution of the concentration provides a method for differentiating first- and second-order reactions. For a first-order reaction, a plot of $\log([A]_t/[A]_0)$ versus time is linear. For a second-order reaction, a plot of $1/[A]_t$ versus time is linear.

WORKED EXAMPLE 12.5   *Disappearing Smog*

The substance primarily responsible for the brown haze associated with polluted air is NO_2, nitrogen dioxide. When the air cools, the brown color diminishes. It is conjectured that the brown color disappears due to the dimerization reaction: $2NO_2(g) \rightarrow N_2O_4(g)$. Use the following data to determine the rate of this reaction, and comment on the feasibility of the conjecture.

pNO_2 (torr)	250	238	224	210
Time (s)	0	200	500	900

Plan

■ Use Equation (12.19): $\dfrac{1}{[A]_t} - \dfrac{1}{[A]_o} = kt$. Plot $1/[A]_t$ versus time, the slope $= k$.

■ Evaluate the time for the reaction: Is it on the order of a few hours or less?

Implementation

■ Plot the data.

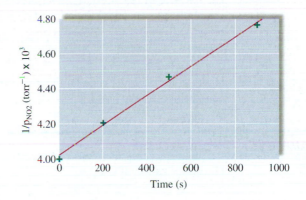

Slope $= 8.41 \times 10^{-7}$ torr$^{-1} \cdot$s^{-1}.

■ We find that 20% of the original NO_2 has dimerized in 15 minutes, which is sufficient to diminish the brown coloration of the air.

See Exercises 31, 32.

Half-life The time required for a reactant to reach half of its original concentration. In radioactive decay, the time required for the number of nuclides in a radioactive sample to reach half of the original value.

Half-Life

One way to measure the rate of a reaction is to determine how long it takes for one-half of the reactant concentration to turn into product—a time referred to as the **half-life.** For a first-order reaction, rearranging Equation (12.17) indicates

(12.20) $$t_{1/2} = -\frac{2.303 \times \log([A]_t/[A]_0)}{k} = -\frac{2.303 \times \log(0.5)}{k} = 0.693/k$$

WORKED EXAMPLE 12.6 *Evaluating Potential Applications*

Radioactive nuclei are often used in cancer therapy. To be a useful therapeutic agent, the nucleus needs to be stable enough to be administered prior to decay, yet decay rapidly enough that the patient's exposure time is limited. Three isotopes of Co are being evaluated for a certain therapy: ^{50}Co, with a decay rate of 15.75 s^{-1}; ^{55}Co, with a decay rate of 0.03953 h^{-1}; and ^{60}Co, with a decay rate of 0.1315 yr^{-1}. Determine the half-life of each isotope and suggest which is the best choice for use as a cancer treatment. Justify your answer.

Plan

■ Use Equation (12.20): $t_{1/2} = 0.693/k$.
■ Evaluate half-life: The half-life should be hours to days to be effective.

Implementation

■ The half-lives are ^{50}Co, 44 ms; ^{55}Co, 17.53 h; and ^{60}Co, 5.270 yr.
■ The best choice is ^{55}Co.

See Exercises 35, 36.

Notice that for a first-order reaction, the half-life is *independent* of the initial concentration of the reactant. This is a general characteristic of a first-order reaction. The half-life of reactions of higher orders are concentration dependent. In particular, for a second-order reaction, Equation (12.19) indicates

(12.21)
$$t_{1/2} = \frac{1}{k}\left(\frac{1}{\frac{1}{2}[A]_o} - \frac{1}{[A]_o}\right) = \frac{1}{k}\frac{1}{[A]_o}$$

Thus the half-life depends on the initial concentration: The larger the initial concentration, the shorter the half-life.

WORKED EXAMPLE 12.7   *Reaction Order*

The dimerization of C_2F_4,

$$2C_2F_4 \longrightarrow C_4F_8$$

occurs in a closed container with no C_4F_8 present initially. The C_2F_4 concentration is measured as a function of time, and the following data are obtained:

[C₂F₄] (mM)	12.00	8.52	6.60	5.40	4.44	3.84	3.55	3.36	3.31
Time (s)	0	150	300	450	600	750	900	1050	1200

Determine the order of the reaction by the method of half-lives, and calculate the rate constant.

Plan
- Plot the data, choosing a series of times as "initial" times.
- Determine the time required for the "initial" concentrations to fall to half of the initial value.
- Determine whether $t_{1/2}$ is independent of the "initial" concentration or whether $t_{1/2} \propto 1/[C_2F_4]_0$.
- Use Equation (12.20), $k = 0.693/t_{1/2}$, or Equation (12.21), $t_{1/2} = 1/k[C_2F_4]_0$.

Implementation
- Plot the data:

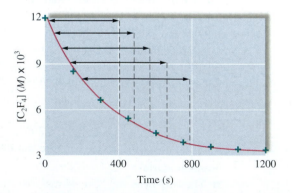

- The concentration falls from 12 mM to 6 mM in 400 s, from 11 mM to 5.5 mM in 437 s, from 10 mM to 5 mM in 480 s, from 9 mM to 4.5 mM in 534 s, and from 8 mM to 4 mM in 600 s.
- The half-life depends on the initial concentration. Check: Is $t_{1/2} \propto 1/[C_2F_4]_0$? Plot it.

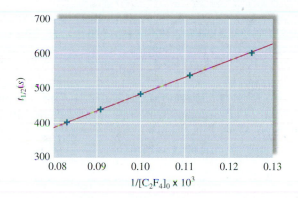

Rate-limiting reaction
Reaction in a mechanism that determines the overall rate of the reaction. Also called the *rate-limiting process* or the *rate-determining step*.

The plot of $t_{1/2}$ versus $1/[C_2F_4]_0$ is a straight line, indicating $t_{1/2} \propto [C_2F_4]_0$. Thus the reaction is second order.

■ Slope $= 1/k \Rightarrow 1/k = 4820$ s/mM $\Rightarrow k = 2.07 \times 10^{-4}$ mM/s

See Exercises 37–39.

12.3   Elementary Reactions and the Rate Law: Mechanisms

 Although a bimolecular elementary reaction has a second-order rate law, the converse is not true. A second-order reaction does not imply a bimolecular process. Similarly, a unimolecular elementary step has a first-order rate law, but a first-order rate law does not mean that the reaction consists of a unimolecular step. The reaction steps can be, and often are, more complex.

The purpose of a kinetic study is often to develop an atomic-molecular picture of the steps in a reaction—that is, the reaction mechanism. This picture is needed to determine whether one step controls the reaction rate. For example, is one step much slower than the rest? A single slow step is called the **rate-limiting reaction.** When a rate-limiting reaction exists, altering the reaction conditions can change the speed of the slow step and thus the reaction rate.

A mechanism consists of a series of elementary reactions. Thus, for a mechanistic step, there *is* a direct connection between the stoichiometry of the *elementary reaction* and the *rate law for that step.*

The requirements for a mechanism are as follows:

■ The net result of all mechanistic steps must be the equation for the net reaction; the stoichiometry must be consistent.
■ The mechanism must be consistent with the observed rate law.
■ When a single slow step is present, the rate equation involves the concentration of the reactants in that slow step only. In addition to those concentrations, the rate depends only on temperature.

An example of a reaction mechanism is the model for a first-order reaction. Consider the decomposition of azomethane, CH_3NNCH_3. The N≡N triple bond is the most stable bond known, so azomethane with its N=N bond is a somewhat unstable molecule (Figure 12.4). Azomethane decomposes according to the reaction stoichiometry

(12.22) $CH_3NNCH_3(g) \longrightarrow CH_3CH_3(g) + N_2(g)$

A measurement of the azomethane pressure as a function of time yields the following data:

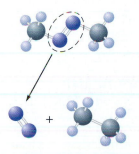

Figure 12.4 Due to the N=N bond, azomethane is an unstable molecule that decomposes to form very stable N_2 and ethane.

Time (s)	0	1000	2000	3000	4000
p (CH$_3$NNCH$_3$) (torr)	8.20	5.72	3.99	2.78	1.94
log (p/p_0)	0	−0.156	−0.313	−0.470	−0.626
Difference		0.156	0.157	0.157	0.156

Showing that $\log(p_{CH_3NNCH_3})$ is a linear function of time means that the decomposition must be a first-order reaction.

(12.23)
$$\frac{d[CH_3CH_3]}{dt} = k[CH_3NNCH_3]$$

In an effort to determine how the decomposition takes place, it was discovered that at low pressure the rate law looks quite different.

(12.24)
$$\frac{d[CH_3CH_3]}{dt} = k'[CH_3NNCH_3]^2$$

The mechanism for this reaction must be consistent with both the high-pressure and the low-pressure data as well as with the stoichiometry of Equation (12.22). A mechanism that successfully explains the characteristics of first-order reactions consists of two steps. In the bimolecular collision step,

(12.25)
$$CH_3NNCH_3 + M \underset{k_{-2}}{\overset{k_2}{\rightleftarrows}} CH_3NNCH_3^* + M$$

where M designates any molecule, including CH_3NNCH_3, that collides with the reacting molecule and deposits energy into it. The energized molecule is denoted by the asterisk (*) notation. The resulting energized molecule subsequently decomposes

(12.26)
$$CH_3NNCH_3^* \overset{k_1}{\longrightarrow} CH_3CH_3 + N_2$$

The result of Equation (12.25) followed by Equation (12.26) is the net reaction, Equation (12.22). Thus the first criterion for a mechanism is satisfied.

Mechanistic steps or elementary reactions are directly connected to the rate law for each step. For example, Equation (12.26) indicates that the energized molecule falls apart directly into products with no other substances involved. The rate law for this step,

(12.27)
$$\frac{d[CH_3CH_3]}{dt} = k_1[CH_3NNCH_3^*]$$

involves only the concentration of the energized molecule. Equation (12.27), however, cannot be the rate law for the net reaction because it involves the energized molecule, something that is neither a reactant nor a product. The energized molecule is involved in both elementary reactions. The energized molecule is formed in the forward step, Equation (12.25) through a bimolecular collision between CH_3NNCH_3 and M at a rate equal to $k_2[CH_3NNCH_3][M]$. The energized molecule loses energy in the reverse reaction via collision with M at a rate equal to $k_{-2}[CH_3NNCH_3^*][M]$. Finally, the energized molecule decomposes to products in Equation (12.26) with a rate equal to $k_1[CH_3NNCH_3^*]$. The net rate of change of the energized molecule is

(12.28)
$$\frac{d[CH_3NNCH_3^*]}{dt} = k_2[CH_3NNCH_3][M] - k_{-2}[CH_3NNCH_3^*][M] \\ - k_1[CH_3NNCH_3^*]$$

Reaction intermediate
Substance that is formed and consumed during the course of a reaction.

The energized molecule is present in only a small concentration, and it is consumed as fast as it forms. Substances like the energized molecule that are intermediaries in a reaction and never build up to a high concentration are called **reaction intermediates.** The concentration of a reaction intermediate is essentially constant during the course of reaction. A constant concentration implies that $d[\text{concentration}]/dt = 0$, so

(12.29)
$$k_2[CH_3NNCH_3][M] = k_{-2}[CH_3NNCH_3^*][M] + k_1[CH_3NNCH_3^*]$$

Steady state The approximation that the rate of formation of a reaction intermediate is zero.

Intermediates are said to be present in a **steady state** concentration. With the steady state approximation, it is possible to solve for the concentration of intermediates in terms of reactants and products. For the azomethane decomposition,

$$(12.30) \qquad [CH_3NNCH_3^*] = \frac{k_2[CH_3NNCH_3][M]}{k_{-2}[M] + k_1}$$

The elementary rate law, Equation (12.27), becomes

$$(12.31) \qquad \frac{d[CH_3CH_3]}{dt} = \frac{k_1k_2[CH_3NNCH_3][M]}{k_{-2}[M] + k_1}$$

For this rate law to apply to the decomposition of azomethane, it must simplify to a first-order reaction at high pressure and a second-order reaction at low pressure. Within the high-pressure limit, $k_{-2}[M] \gg k_1$ and

$$(12.32) \qquad \frac{d[CH_3CH_3]}{dt} \approx \frac{k_1k_2[CH_3NNCH_3][M]}{k_{-2}[M]} = \frac{k_1k_2[CH_3NNCH_3]}{k_{-2}}$$

This is a first-order reaction with a rate constant of k_1k_2/k_{-2}.

Conversely, at low pressure, $k_{-2}[M] \ll k_1$ and Equation (12.31) becomes

$$(12.33) \qquad \frac{d[CH_3CH_3]}{dt} \approx \frac{k_1k_2[CH_3NNCH_3][M]}{k_1} = k_2[CH_3NNCH_3][M]$$

If the only substance present is azomethane, then M is azomethane and Equation (12.33) reduces to

$$(12.34) \qquad \frac{d[CH_3CH_3]}{dt} = k_2[CH_3NNCH_3]^2$$

which is the observed second-order reaction at low pressure.

Rate-determining step The slowest step in a reaction mechanism, the one determining the overall rate. Also called *rate-limiting process or rate-limiting reaction*.

Rate-limiting process The slowest step in a reaction mechanism, the one determining the overall rate. Also called *rate determining step or rate-limiting reaction*.

The mechanism provides an atomic-molecular–level picture of the reaction. An azomethane molecule collides with another molecule, accumulating extra energy. The excited molecule subsequently either dissociates or loses its energy via collision. At high pressure, the excitation–deexcitation process is fast compared with the decomposition rate, and the reaction is first order. At low pressure, the excitation process is slow and the reaction is second order. In each case, the slow step determines the reaction rate. This slow step is therefore called the **rate-determining step** or **rate-limiting process.**

Knowledge of the mechanism enables a measure of control over the reaction speed. Because azomethane decomposition results in two gas-phase molecules from one, if the reaction is too fast the vessel will over-pressurize with undesirable side effects. However, at low azomethane pressure, the reaction can occur too slowly. The mechanism suggests that the rate of the low-pressure reaction can be increased by adding any other gas. This prediction is experimentally verified.

WORKED EXAMPLE 12.8   *Ozone Destruction*

The hole in the stratospheric ozone layer has received a lot of attention in recent years. Due to the amount of publicity about this phenomenon, it is easy to lose sight of the fact that ozone is an unstable molecule and has a number of natural destruction mechanisms. To sort out the natural and anthropogenic cycles, it is necessary to understand natural cycles beginning with the connection between oxygen and ozone: $2O_3(g) \rightarrow 3O_2(g)$. In the investigation it was found that the rate law for this reaction is

$$\frac{d[O_3]}{dt} = -\frac{k[O_3]^2}{[O_2]}$$

Two mechanisms are proposed:

Mechanism I

$$O_3 \underset{k_{-1}}{\overset{k_1}{\rightleftharpoons}} O_2 + O \qquad \text{Reversible}$$

$$O + O_3 \xrightarrow{k_2} 2O_2 \qquad \text{Slow step}$$

Mechanism II
$$2O_3 \longrightarrow 3O_2$$

Evaluate these mechanisms. Is either consistent with the rate law?

Plan

- Determine whether the mechanisms are consistent with the net stoichiometry.
- Evaluate the mechanisms for consistency with the rate law.

Implementation

- Mechanism II, the single-step mechanism, is clearly consistent with the net stoichiometry. The sum of the two steps in mechanism I is also consistent with the stoichiometry because the reaction intermediate, O, formed in the first step is consumed in the second step. O_3 is a reactant in both steps, giving a net two O_3. O_2 is a product of step 1, and two O_2 are produced in step 2, giving a net three O_2.
- The second mechanism consists of a single bimolecular step, so the rate law is

$$\frac{d[O_3]}{dt} = -k[O_3]^2$$

This is not consistent with the observed rate law. Mechanism II is not correct.

Mechanism I is more involved. The second step is the slow step and thus the rate-determining step. From the second step

$$\frac{d[O_3]}{dt} = -k_2[O_3][O]$$

Since O is an intermediate, it must be eliminated from the rate law. The first step provides the elimination. It is relatively fast and reversible, so the net forward and reverse reactions are equal; $k_1[O_3] = k_{-1}[O][O_2] \Rightarrow [O] = k_1[O_3]/k_{-1}[O_2]$. Substitution gives

$$\frac{d[O_3]}{dt} = -k_2[O_3]\frac{k_1[O_3]}{k_{-1}[O_2]} = -\frac{k_1 k_2}{k_{-1}}\frac{[O_3]^2}{[O_2]}$$

Identification of k with $k_1 k_2 / k_{-1}$ results in the observed rate law.

The mechanism for the ozone cycle is an example of the lack of connection between the net reaction stoichiometry and the rate law.

See Exercises 45, 46.

12.4 Modeling Reactions

Temperature and the Activation Barrier

Figure 12.5 The path from one destination to another can be blocked by a considerable barrier. To travel from one destination to the other, a vehicle must surmount this barrier.

Some reactions are very slow, such as conversion of diamond to graphite at atmospheric pressure. Other reactions are very fast, such as the deployment of an air bag. Nearly every reaction runs faster at higher temperature. The concept of an activation barrier provides an explanation of why rates differ so widely and how elevated temperature makes the reaction faster. As an analogy, think of taking a trip from Reno, Nevada, to San Francisco (Figure 12.5). San Francisco is downhill from Reno, yet a car cannot simply roll from one city to the other even if given a push to get started. A mountain range, the Sierra Nevadas, stands in the way!

Reactions are similar. Between the higher Gibbs free energy of the reactants and the lower Gibbs free energy of the products, a barrier obstructs the reaction. Just as the pioneers crossing the Donner Pass learned, this barrier can be formidable, slowing the rate of passage considerably.

APPLY IT At Halloween, many costume stores sell "light sticks." Light sticks can also be obtained at sporting goods stores—fishermen use them to mark their fishing lines. They may also be called cyalume sticks. Inside the stick is a small glass tube that is closed at both ends. When the tube is broken, the stick lights up.

Obtain three light sticks. Put one in a freezer or ice chest for several hours. Obtain some hot water, about 50 °C. Put one light stick in the hot water for about 10 minutes. Activate all three lights sticks by bending them to break the glass tube. Record your observations about the intensity of the light from each of the tubes. Observe how the intensity changes as the cold tube warms up and the warm tube cools down.

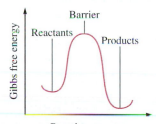

Figure 12.6 A schematic representation of the Gibbs free energy as a reaction progresses shows that the lower energy of the products relative to the reactants is separated by a high energy barrier.

The barrier between reactants and products is quite general (Figure 12.6), as there is usually at least a small repulsive barrier as two reactants approach each other. For example, as two neutral molecules approach each other, the electron clouds interact first. Electrons repel electrons, so without sufficient energy the two will not get close enough for a reaction. Think of two billiard balls being knocked together on a pool table. They collide and bounce off each other. However, if the billiard balls were shot out of a cannon at each other, . . . A pair of molecules faces a similar situation. The molecules must have a minimum energy to form a product. They may also need to arrange themselves into a favorable configuration before reaction can occur, such a rearrangement is more likely at higher temperature. The combination of configuration and repulsion forms the barrier, called the **activation barrier.** The energy required to surmount the barrier is called the **activation energy.**

The energy profile for the reaction between methyl bromide, CH_3Br, and chloride ion, Cl^- (Figure 12.7), illustrates the origin of the energy barrier for this reaction. A successful reaction features the chloride ion approaching the methyl bromide molecule

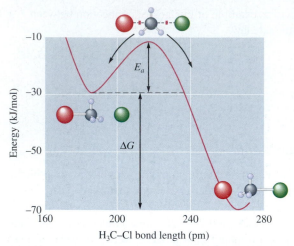

Energy (kJ/mol)

E_a

ΔG

H$_3$C–Cl bond length (pm)

Figure 12.7 The energy profile for replacement of Br in CH$_3$Br (methyl bromide) by Cl to form CH$_3$Cl (methyl chloride) shows the energy barrier between the two molecules. Methyl bromide is kinetically stable due to the barrier.

Activation barrier The threshold energy that must be overcome to produce a chemical reaction. Also called *activation energy.*

Activation energy The threshold energy that must be overcome to produce a chemical reaction. Also called *activation barrier.*

from the side opposite the bromine atom. As the chloride ion pushes the bromine out, the carbon atom becomes five coordinate. (Five-coordinate carbon atoms occur *only* in the short time that the molecule sits atop the energy barrier.) Five-coordinate carbon is unstable, but can be stabilized either by rolling downhill to the original methyl bromide configuration or by changing to the more favorable methyl chloride configuration. The barrier height and its origin differ from reaction to reaction.

In all cases, the two reacting molecules surmount the activation barrier if they have sufficient energy—most often, sufficient kinetic energy. The kinetic energy of a collection of molecules is proportional to the absolute temperature. At higher temperature, more molecules have the energy required to surmount the activation barrier and the reaction occurs more rapidly. The mathematical description of the dependence of the rate constant on temperature was formulated by the Swedish scientist Svante Arrhenius in 1889 and is known as the **Arrhenius equation:**

(12.35)
$$k = Ae^{-E_a/RT}$$

where E_a is the activation energy (Figure 12.7), R is the gas constant, T is the absolute temperature, and A is an experimentally determined preexponential factor. Both A and E_a are essentially independent of temperature. Equation (12.35) indicates that the specific rate constant *increases* fairly rapidly with temperature. Usual activation energies are in the range of 200–400 kJ/mol, so near room temperature the specific rate constant roughly doubles for every 10 °C temperature rise! To experimentally determine the activation energy, the reaction rate and the specific rate constant are determined at several temperatures (see Worked Example 12.9).

WORKED EXAMPLE 12.9 *Pure Metals*

Arrhenius equation The equation representing the rate constant as $k = Ae^{-E_a/RT}$, where A represents the product of the collision frequency and the steric factor, and $e^{-E_a/RT}$ is the fraction of collisions with sufficient energy to produce a reaction.

One method for production of pure metals is to first make the metal into a metal carbonyl (metal–CO) compound, and then to decompose the metal carbonyl in a vacuum at elevated temperature. One metal produced in this fashion is nickel:

$$Ni(CO)_4(g) \longrightarrow Ni(s) + 4CO(g)$$

The slow step in this reaction is the elimination of the first CO:

$$Ni(CO)_4 \longrightarrow Ni(CO)_3 + CO$$

This step was examined with the following results:

Temperature (K)	320.45	324.05	328.15	333.15	339.15
k (s^{-1})	0.263	0.354	0.606	1.022	1.873

What is the activation energy for the reaction?

Plan

■ Rearrange Equation (12.35) as $\ln(k) = \ln(A) - E_a/RT$. Plot $\ln(k)$ versus $1/T$.

Implementation

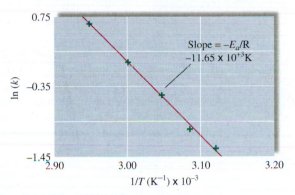

The plot of ln (k) versus $1/T$ is a straight line with slope $= -E_a/R = -11.65 \times 10^{+3}$ K. Slope $\times R$ is 96.9 kJ/mol.

Note that the slope is directly proportional to the activation energy, so a reaction with a higher activation energy has a steeper slope and a reaction with a lower activation energy has a shallower slope. Decomposition of nickel carbonyl has a relatively low activation barrier.

See Exercises 57 – 59.

Collision Theory

The first premise of collision theory states that to react, the molecules must collide with each other. An increase in the concentration of one of the molecules increases the number of collisions between the molecules. Because the number of collisions is linearly proportional to the concentration, the rate of a bimolecular reaction is proportional to the product of the concentration of each of the substances involved in the reaction.

The second premise states that to react, the two substances that collide must have the proper orientation. As an analogy, think of shaking hands (Figure 12.8). Two individuals can approach each other in many different ways. To successfully shake hands, they must be facing each other and both have the same hand extended, by custom the right hand. An approach with one left hand and one right hand extended does not end in a successful handshake.

Now consider the reaction between NO and N_2O to produce NO_2 and N_2:

(12.36) $\qquad NO(g) + N_2O(g) \longrightarrow NO_2(g) + N_2(g)$

Oxygen is transferred from N_2O to NO. Successful transfer requires that the oxygen end of the N_2O face the nitrogen end of NO in the collision (Figure 12.9). Any other

Figure 12.8 For a successful handshake, the two individuals must face each other and both extend the same hand (by custom, this is the right hand). If one person extends the left hand and the other extends the right hand, the handshake does not occur.

Figure 12.9 The reaction between NO and N_2O features a transfer of an oxygen atom from N_2O to NO. A productive collision has the nitrogen end of the NO molecule facing the oxygen end of the N_2O. Rotation of the NO so as to present the oxygen end to the N_2O molecule does not result in oxygen transfer. There are many more nonproductive orientations than productive ones.

orientation in the collision is ineffective for the transfer. Due to the restricted geometry for productive transfer, the rate of this reaction is very slow. As a consequence, N_2O is very unreactive in the atmosphere. The nonreactivity is exploited in use of N_2O as a trace gas to determine the fate of air parcels in developing atmospheric circulation models.

Geometric restrictions vary from substance to substance. Two reactants that are basically spheres, where all orientations are essentially equivalent, have less orientational restriction than two substances that have limited reaction sites. For example, the reaction between Na^+ and Cl^- in the gas phase has little orientational restriction. In contrast, reactions in biological systems often are highly restrictive, requiring the reactants to be properly oriented in narrow channels.

Activated Complex

Activated complex The arrangement of atoms found at the top of the potential energy barrier as a reaction proceeds from reactants to products. Also called *transition state.*

Transition state The arrangement of atoms found at the top of the potential energy barrier as a reaction proceeds from reactants to products. Also called *activated complex.*

Collision theory applies to molecules in the gas phase. Molecules in solution differ from those in the gas phase in that the former are surrounded by solvent molecules. Rather than flying through space, molecules in solution are continuously buffeted by the solvent molecules. When the two reacting molecules come together, they stay in the vicinity of each other for a longer period of time; the solvent molecules act like a cage holding the two colliding molecules together. If the jostling by solvent molecules results in a favorable configuration for the potential reacting molecules, the potential reactants are said to form an **activated complex** or a **transition state.**

For example, the methyl bromide to methyl chloride reaction takes place in aqueous solution. The configuration at the top of the barrier in Figure 12.7 is the transition state for this reaction. A colliding pair such as CH_3Br and Cl^- often acquires the energy needed to surmount the activation barrier to form the transition state through collisions with the surrounding solvent molecules.

12.5   Rates and Equilibrium

Equilibrium is characterized by equal *net* forward and reverse reactions, suggesting a connection between kinetics and equilibrium. Indeed, these two concepts are connected. Consider the dimerization of NO_2 to form N_2O_4:

(12.37)
$$2NO_2(g) \underset{k_r}{\overset{k_f}{\rightleftharpoons}} N_2O_4(g)$$

This elementary reaction consists of a bimolecular forward reaction and a unimolecular reverse reaction. The *net* rate of the forward reaction is

(12.38)
$$v_{forward} = k_f[NO_2]^2$$

and the *net* rate of the reverse reaction is

(12.39)
$$v_{reverse} = k_r[N_2O_4]$$

Setting the *net* rates equal to each other results in

(12.40) $$k_f[NO_2]^2 = k_r[N_2O_4] \quad \text{or} \quad \frac{k_f}{k_r} = \frac{[N_2O_4]}{[NO_2]^2} = K$$

The ratio of the forward rate constant to the reverse rate constant is equal to the equilibrium constant. Reactions with more involved mechanisms have a similar result as long as the mechanism involves a set of reactions in which every step is reversible. Then the equilibrium constant is equal to

(12.41)
$$K = \frac{k_1}{k_{-1}} \times \frac{k_2}{k_{-2}} \times \frac{k_3}{k_{-3}} \times \cdots$$

where $k_1, k_2, \ldots$ are the rate constants for the forward reactions and $k_{-1}, k_{-2}, \ldots$ are the rate constants for the reverse reactions. In kinetic terms, a large equilibrium constant results when k_f is much larger than k_r. The forward reaction builds up product faster than the products re-form the reactants. Kinetics provides a molecular-level model for equilibrium.

WORKED EXAMPLE 12.10   *Reactions in the Atmosphere*

Numerous oxides of nitrogen exist in the atmosphere and accounting for the balance among them is an important test for models of atmospheric chemistry. A proposed mechanism for the oxidation of NO to NO_2 by oxygen is given below. What is the reaction? Does this mechanism predict the equilibrium between NO and NO_2? Mechanism:

$$2NO \underset{k_{-1}}{\overset{k_1}{\rightleftarrows}} N_2O_2$$

$$O_2 + N_2O_2 \xrightarrow{k_2} 2NO_2$$

Plan
- Add steps eliminating intermediates.
- Check for reversibility.

Implementation
- N_2O_2 is the intermediate: It is produced in the first step and consumed in the second. The net reaction is

$$2NO + O_2 \longrightarrow 2NO_2$$

- The second step is not reversible, thus the net reaction is also not reversible.

In the atmosphere, there are many more reactions in addition to those shown above. Additional reactions can complete the loop between NO_2 and NO resulting in equilibrium. If observations indicate that NO and NO_2 are in equilibrium, these additional reactions need to be identified and included.

See Exercise 71.

12.6   Catalysts

Catalyst A substance that speeds up a reaction without being consumed.

Of all the methods to control the speed of a reaction, one of the most impressive involves the use of a catalyst. A **catalyst** is a substance that is neither produced nor consumed in a reaction, but that changes the rate of the reaction. For example, hydrogen peroxide, H_2O_2, is an excellent oxidant that is used to disinfect contact lenses. The typical contact lens cleaning solution is about 3% hydrogen peroxide by volume, the minimum concentration needed to kill bacteria and viruses on the lens in a reasonable time. However, such solutions are irritating to the eyes. Before a contact lens can be reinserted in the eye, the peroxide concentration must be reduced to less than 60 ppm. Contact lens cleaning systems often contain instructions for the user to soak the lens in a solution contained in a lens cup. The cup has an insert that contains platinum. Platinum (like most metals) reduces the half-life of the reaction

(12.42) $2H_2O_2(aq) \xrightarrow{Pt} 2H_2O(l) + O_2(g)$

to about three hours—much less than the many years for the uncatalyzed reaction.

Although the details of how a catalyst speeds up a particular reaction are often not well understood, it is generally believed that the catalyst reduces the activation barrier for the reaction. Think again of going from Reno to San Francisco. If a tunnel were cut through the mountain range, it would be possible to roll all the way from Reno to San Francisco. Catalysts cut such a tunnel.

CONCEPT QUESTION Suppose that a reaction has an activation barrier of 150 kJ/mol. How much lower does the activation barrier need to be to double the rate of the reaction at room temperature? ■

Catalysts are ubiquitous in today's industry. Catalysts called enzymes control the rate of many biological reactions, for example, and are responsible for more than $3.5 trillion in goods and services—a big market. Automobiles now carry catalytic converters to reduce pollution. Catalysts are important in oil refining, are used for production of many polymers and fine chemicals, and accelerate the degradation of pollutants in waterways. Increasing demand in the pharmaceutical industry has mushroomed the need for fine chemicals from a very small market a decade ago to $1.4 billion in 2004, and an annual growth of 8% is forecast for this market through at least 2007.

The pharmaceutical industry particularly needs catalysts to produce specific enantiomers, called stereoisomers, in complex molecules. Enzymes in the body respond very differently to different stereoisomers. For example, the drug thalidomide that was used in the early 1960s has two stereoisomers. One very effectively prevents morning sickness; the other, even in small dosages, causes horrible birth defects. Development of catalysts capable of producing a specific stereoisomer netted the 2001 Nobel Prize for William Knowles, Ryoji Noyori, and Barry Sharpless.

Homogeneous catalyst
A substance that speeds up the rate of a reaction without being consumed that is part of the solution. Includes such materials as enzymes, acids, and bases.

Two basic types of catalysts exist: homogeneous and heterogeneous. A **homogeneous catalyst** is part of the solution and includes such materials as enzymes, acids, and bases. An example can be found in the acid-catalyzed hydrolysis of sugar:

(12.43) $C_{12}H_{22}O_{11}(aq) + H_2O(l) \longrightarrow 2C_6H_{12}O_6(aq)$

Acid is neither produced nor consumed in this reaction, yet the reaction rate is much faster in the presence of the acid.

Heterogeneous catalyst
A substance that increases the rate of a reaction without being consumed and that is in a different phase than the reaction. For polymers, a solid surface on which additional polymerization takes place.

A **heterogeneous catalyst** is part of a separate phase. Often heterogeneous catalysts are solids, such as the platinum on a plastic support in the contact lens cleaning system.

There are two important principles to keep in mind regarding catalysts:

- A catalyst cannot make a reaction happen if the thermodynamics says that it will not. That is, if $\Delta G > 0$, the reaction will not occur.
- A catalyst cannot alter the equilibrium between reactants and products. A catalyst may shorten the time that it takes to reach equilibrium, but the *equilibrium concentrations are unaffected* by its presence.

Homogeneous Catalysts

Even small concentrations of homogeneous catalysts can have a big impact. Consider the ozone hole over Antarctica (Figure 12.10). Chlorine atoms get to the stratosphere as

Figure 12.10 The concentration of ozone over Antarctica is affected by small concentrations of chlorofluorocarbons released in the troposphere, which catalytically destroy ozone when they reach the stratosphere. This series of graphs depict the ozone concentration over Antarctica from October 1999 to October 2003 illustrating concentration in Dobson units (DU). The normal range is 300–500 Dobson units (yellow-green through red). Diminished ozone is shown in colors from blues (200–275 DU) through purple (150–200 DU) and pink (125–150 DU) to grey (100–125 DU). Purple through grey are troublesome concentrations.

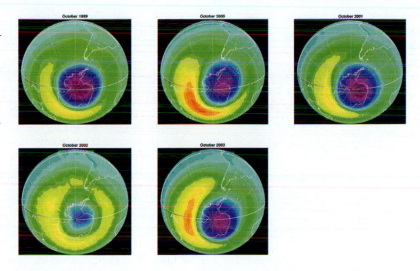

part of small carbon molecules containing chlorine and fluorine, called chlorofluorocarbons (CFCs). Because CFCs are quite stable, they do not react with substances in the troposphere, instead accumulating in the stratosphere. In the stratosphere short-wavelength radiation from the sun is quite intense, resulting in reactions like

(12.44) $$CF_2Cl_2(g) \xrightarrow{h\nu \text{ (short } \lambda)} CF_2Cl(g) + Cl(g)$$

Once released, the chlorine atom attacks ozone

(12.45) $$O_3(g) + Cl(g) \longrightarrow ClO(g) + O_2(g)$$

It has been estimated that a single chlorine atom destroys 10^5 ozone molecules before being inactivated. This phenomenon explains why a small amount of CFCs can destroy a very large amount of ozone. The remainder of the catalytic cycle for chlorine and ozone depends on OH radicals that exist throughout the atmosphere. OH reacts with ozone to form the intermediate HO_2:

(12.46) $$OH(g) + O_3(g) \longrightarrow HO_2(g) + O_2(g)$$

HO_2 converts ClO to photolabile HOCl:

(12.47) $$ClO(g) + HO_2(g) \longrightarrow HOCl(g) + O_2(g)$$

Sunlight then regenerates the chlorine atom:

(12.48) $$HOCl \xrightarrow{h\nu} Cl(g) + OH(g)$$

The net reaction is $2O_3 \longrightarrow O_2$.

Atmospheric models indicate that the chlorine cycle plus the analogous bromine cycle account for more than 70% of the ozone loss in the stratosphere. Identification of these cycles is very important, and their work in this area resulted in the 1995 Nobel Prize being awarded to Paul Crutzen, Mario Molina, and Sherwood Roland. As a consequence of the pioneering work of these scientists and others, use of CFCs has been sharply curtailed and the decline of stratospheric ozone is being reversed. This case offers a dramatic example of how kinetic analysis can affect the health and well-being of the entire planet.

Figure 12.11 The oxidation of tartaric acid to carbon dioxide, water, and oxalic acid by hydrogen peroxide is catalyzed by Co^{2+}.

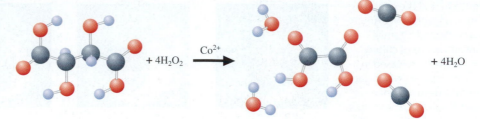

In the chlorine cycle, chlorine, ozone, and hydroxyl radicals are in the gas phase. The chlorine cycle is an example of homogeneous catalysis. The mechanism by which chlorine atoms lower the activation barrier for conversion of ozone to oxygen is not well understood. In general, homogeneous catalysts are believed to bring reactants together by forming a complex and altering the bonding in the reactants or by orienting reactants so that they achieve a more favorable configuration for reaction. An example of a catalyst that forms a complex is seen in the oxidation of tartaric acid (Figure 12.11) by hydrogen peroxide with a cobalt catalyst. The structure of the complex has not been characterized, but formation of a complex is evident from the solution color (Figure 12.12). Co^{2+} ions in aqueous solution have a characteristic pink color that changes to olive green, and then back to pink, as the reaction progresses.

Enzymes

An important subclass of homogeneous catalysts includes enzymes. As noted earlier, enzymes are responsible for making many reactions in biological systems run at rates that are practical for keeping the organism alive. The mechanism of enzyme action is complex, involving weakening of bonds by binding reactants, bringing reactants together, and lowering activation barriers. Oxidation of glucose is an important example. The net reaction for oxidation of glucose is

(12.49) $C_6H_6O_{12}(aq) + 6O_2(aq) \longrightarrow 6CO_2(aq) + 6H_2O(l)$ $\Delta H = -2802$ kJ/mol

If all of the heat were released at once, the organism would become overheated and die. Instead, a set of 10 steps releases a small amount of energy at each step, and the rate of

Figure 12.12 When solutions of tartaric acid, hydrogen peroxide, and Co^{2+} are initially mixed, the solution has the characteristic pink color of hydrated Co^{2+} ions. As the reaction progresses, the solution turns an olive green color and fizzing occurs due to release of CO_2 and O_2 gases. When the fizzing ceases, the solution returns to the original pink color.

each of these 10 steps is controlled by a different enzyme. Enzymes and the energy-shuttling molecule ATP manage the energy release.

 APPLY IT Both yeast and hydrogen peroxide (3% by volume) can be obtained in a grocery store. Obtain $^1/_8$ teaspoon of dry yeast and soften it in one tablespoon of tepid water in a glass for 7 minutes. Pour 50 mL of hydrogen peroxide solution in a glass container, add the softened yeast solution, and swirl the container. What do you observe?

The mechanism for many enzyme-catalyzed reactions was identified by Leonor Michaelis and Maud Menten in 1913. The resulting mechanism carries their names; it is called the Michaelis-Menten mechanism. If we represent the enzyme by E and the reactant, called the substrate, by S, the first step consists of formation of a complex, ES, between the enzyme and the substrate:

(12.50)
$$E + S \underset{k_{-1}}{\overset{k_1}{\rightleftharpoons}} ES$$

The complex decays in a unimolecular step to release the products and regenerate the enzyme:

(12.51)
$$ES \xrightarrow{k_2} E + \text{Products}$$

The rate of product production is

(12.52)
$$\frac{d[\text{Product}]}{dt} = k_1[ES]$$

The intermediate, the enzyme–substrate complex, depends on time as

(12.53)
$$\frac{d[ES]}{dt} = k_1[E][S] - k_{-1}[ES] - k_2[ES]$$

Complex formation is typically slow, so applying $d[ES]/dt = 0$,

(12.54)
$$(k_2 + k_{-1})[ES] = k_1[E][S]$$

At all times, the enzyme is present either as free enzyme or as the enzyme–substrate complex. Thus $[E] = [E]_0 - [ES]$, where $[E]_0$ is the initial enzyme concentration. Substituting into Equation (12.54) and rearranging,

(12.55)
$$(k_2 + k_{-1} + k_1[S])[ES] = k_1[E]_0[S]$$

The product generation rate is then

(12.56)
$$\frac{d[\text{Product}]}{dt} = \frac{k_1 k_2[E]_0[S]}{k_2 + k_{-1} + k_1[S]} = \frac{k_2[E]_0[S]}{(k_2 + k_{-1})/k_1 + [S]} = \frac{k_2[E]_0[S]}{K_M + [S]}$$

where K_M is called the Michaelis constant.

The significance of Equation (12.56) is that at low substrate concentration, the reaction rate is proportional to the substrate concentration: It is first order in the substrate. At a high substrate concentration, the enzyme is saturated with substrate and the reaction rate is independent of the substrate concentration. Disrupting the enzyme-substrate interaction can have serious consequences. For example, many poisons, including nerve gas

Table 12.2 Important Metal-Catalyzed Industrial Reactions

Metal	Process
Ni	H_2 + unsaturated oils to saturated oils
Fe	N_2 + H_2 to NH_3
Ag	ethylene + oxygen to ethylene oxide
Pt-Rh	NH_3 + O_2 to HNO_3
Ir-Rh	CO + O_2 to CO_2

and arsenic, act by binding strongly to the enzymes needed to transmit nerve impulses. The substrates cannot bind, and as a result cells and organisms die.

Heterogeneous Catalysts

Many industrial catalyst-driven reactions involve heterogeneous metal catalysts (Table 12.2). The mechanisms for the participation of these catalysts vary and are not completely understood in many cases. Two cases that are understood to some degree are the hydrogenation of ethylene and the oxidation of CO. The net reaction for hydrogenation of ethylene is

(12.57)
$$C_2H_4(g) + H_2(g) \xrightarrow{Pt} C_2H_6(g)$$

This is the prototype reaction for turning unsaturated oils, which are hydrocarbons with double bonds, into saturated oils, or hydrocarbons with single bonds. Unsaturated oils are believed to play a role in many diseases, including arteriosclerosis. Through the use of surface-sensitive techniques, scientists have developed an atomic-molecular−level picture of the first steps of this heterogeneous reaction (Figure 12.13). Ethylene "belly flops" onto the metal surface. Electrons in the metal surface combine with the electrons in the π cloud of the ethylene molecule and form bonds between each of the carbon atoms and the surface. This reduces the C=C double bond to a single bond. The double-bond activation is the first step in the hydrogenation.

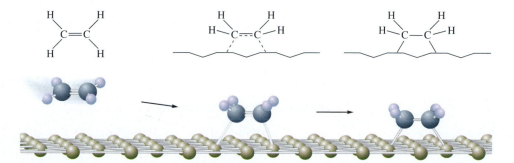

Figure 12.13 The first step in the hydrogenation of ethylene on a platinum catalyst involves an interaction between ethylene and the metal surface. The flat ethylene molecule "belly flops" onto the surface, bending the four hydrogen atoms away from the surface as the double bond weakens. Then the two carbon atoms form a bond to surface metal atoms, resulting in a single bond between the carbon atoms. Hydrogen subsequently adds to the bond, resulting in a single bond called a saturated bond.

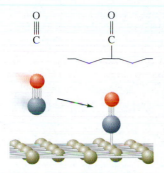

Figure 12.14 Oxidation of CO to CO_2 occurs in the catalytic converter of an automobile. In the first step in the process, CO is captured on the catalytic surface. Surface science techniques have shown that CO is captured on platinum with the carbon end down.

In an automobile catalytic converter, CO is oxidized to CO_2. One step in this process involves the capture of CO by the metal catalyst, usually platinum (Figure 12.14). Oxygen is also captured and dissociates. The dissociated oxygen and CO combine and desorb as CO_2.

The role of the metal catalysts in both the hydrogenation of ethylene and the oxidation of CO is to activate the multiple bond through interactions between electrons on the surface of the metal and electrons in the multiple bonds. Upon completion of these reactions, the products depart from the surface and the catalyst is regenerated. Many steps in this process remain poorly understood. For example, the catalyst eventually gets "poisoned." This may occur because the catalyst surface changes shape and therefore electronic states after many cycles, or the surface may not completely rid itself of adsorbed species and therefore has no room for incoming substances. Poisoning of catalysts remains a topic of intense research efforts.

CONCEPT QUESTIONS The decomposition of hydrogen peroxide can be catalyzed by several substances: the enzymes in yeast, platinum, bromide ions, and the enzymes found in raw potatoes. Which of these catalysts is a heterogeneous catalyst? Which ones are homogeneous catalysts? ■

WORKED EXAMPLE 12.11   *Getting Rid of Peroxides*

Peroxide (H_2O_2) occurs as a product of several biological processes. If not decomposed, peroxide damages tissue. To minimize this damage, the enzyme catalase catalyzes the decomposition of peroxide to water and oxygen. Write the reaction for decomposition of peroxide. ΔG for this reaction in physiological conditions is approximately -100 kJ/mol. Without catalase, the activation energy for decomposition of peroxide is 72 kJ/mol; with catalase, the activation energy is 28 kJ/mol. Sketch an energy (ΔG) profile for the uncatalyzed and the catalyzed decomposition.

Plan
■ The reactant is peroxide, H_2O_2, and the products are water and molecular oxygen.
■ Overall, ΔG decreases after going over a barrier.

Implementation
■ $2H_2O_2(aq) \rightarrow 2H_2O(l) + O_2(g)$

■

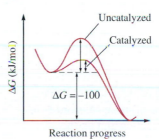

Under uncatalyzed conditions, peroxide would build up to fatal levels in biological systems.

See Exercises 75, 76.

Checklist for Review

KEY TERMS

reaction rate (p. 424)
reaction velocity (p. 424)
rate law (p. 424)
specific rate constant
   (p. 425)
reaction order (p. 425)
method of initial rates
   (p. 426)
Ostwald's isolation
   method (p. 427)
flooding (p. 427)
reaction mechanism
   (p. 428)
half-life (p. 431)
rate-limiting reaction
   (p. 433)
reaction intermediate
   (p. 434)

steady state (p. 435)
rate-determining step
   (p. 435)
rate-limiting process
   (p. 435)
activation barrier (p. 438)
activation energy (p. 438)
Arrhenius equation
   (p. 438)
activated complex (p. 440)
transition state (p. 440)
catalyst (p. 441)
homogeneous catalyst
   (p. 442)
heterogeneous catalyst
   (p. 442)

KEY EQUATIONS

First-order reaction:

$$\frac{d[A]}{dt} = -k[A] \Longrightarrow [A]_t = [A]_o e^{-kt} \qquad t_{1/2} = 0.693/k$$

Second-order reaction:

$$\frac{d[A]}{dt} = -k[A]^2 \Longrightarrow \frac{1}{[A]_t} - \frac{1}{[A]_o} = kt$$

$$t_{1/2} = 1/k[A]_o$$

$$k = Ae^{-E_a/RT}$$

$$K = \frac{k_1}{k_{-1}} \times \frac{k_2}{k_{-2}} \times \frac{k_3}{k_{-3}} \times \cdots$$

$$\frac{d[\text{Products}]}{dt} = \frac{k_1[E]_o[S]}{K_M + [S]}$$

Chapter Summary

The rate and the mechanism of a reaction both fall within the province of kinetics. The reaction rate often determines the practical applications of a reaction. For example, air bags must deploy quickly, and it is preferable for cars to rust slowly. A picture of the mechanism of a reaction can provide insight into potential ways to alter the reaction conditions so as to control the reaction rate.

The specific rate constant depends only on temperature, whereas the reaction rate often depends on the concentrations of reactants or products. Determining the dependence of the reaction rate on concentrations indicates whether addition of a reactant or a product can alter the reaction rate. The rate of a reaction can also depend on the concentrations of substances that are neither reactants nor products. These substances are called catalysts. A catalyst can be either homogeneous or heterogeneous.

KEY IDEA

The time required for a chemical reaction to occur depends on the concentrations of those substances in the rate-limiting step of the reaction and on the temperature.

CONCEPTS YOU SHOULD UNDERSTAND

■ Catalysts alter the rate of a reaction by lowering the activation barrier for the reaction. Catalysts do not alter the ultimate equilibrium fate of the reaction.

OPERATIONAL SKILLS

■ Write a rate law from reaction orders, and determine reaction orders from the rate law (Worked Example 12.1).
■ Determine reaction order from initial rates or by the isolation method. (Worked Examples 12.2 and 12.3).
■ Determine age from half-life and half-life from reaction rate (Worked Examples 12.4 and 12.6).
■ Determine rate constant from experimental data (Worked Examples 12.5 and 12.7).

- Evaluate potential mechanisms (Worked Example 12.8).
- Determine activation energy and temperature dependence of the rate constant (Worked Example 12.9).
- Connect the mechanism with equilibrium (Worked Example 12.10).
- Draw a reaction energy profile; relate the activation energy for a reaction to the activation energy of the catalyzed reaction (Worked Example 12.11).

Exercises

A blue exercise number indicates that the answer to that exercise appears at the back of the book.

■ SKILL BUILDING EXERCISES

12.1 Concentration

1. Cyanide is a toxic ion due to its strong affinity for iron ion in the respiratory enzyme cytochrome oxidase. When complexed with cyanide ion, the functioning of cytochrome oxidase ceases and the cell dies. An antidote for cyanide poisoning is the product Kelocyanor, which contains Co_2(EDTA). Co^{2+} complexes to CN^-,

$$Co^{2+}(aq) + 6CN^-(aq) \rightleftharpoons Co(CN)_6^{4-}(aq)$$

more strongly than Fe^{2+}, thereby scrubbing CN^- from the system. Many kinetic processes control the efficacy of this treatment. Write expressions for the rate of the reaction for Co^{2+} complex formation in terms of $[Co^2]$, in terms of $[CN^-]$, and in terms of $[[Co(CN)_6]^{4-}]$.

2. A magnesium fire is extremely dangerous because the reaction of magnesium with oxygen, nitrogen, or carbon dioxide is extremely exothermic. Sand is a good material to use on a magnesium fire. The chemical reaction of magnesium with CO_2 is

$$2Mg(s) + CO_2(g) \longrightarrow 2MgO(s) + C(s)$$

Write expressions for the rate of this reaction in terms of Mg, CO_2, MgO, and C.

12.1 Rate Law and Reaction Order

3. What is the overall order and the order for each of the reactants in the following reactions?

Reaction	Observed Rate Law
a. $H_2(g) + I_2(g) \rightarrow 2HI(g)$	$\frac{1}{2}\frac{d[HI]}{dt} = k[H_2][I_2]$
b. $2HI(g) \rightarrow H_2(g) + I_2(g)$	$-\frac{1}{2}\frac{d[HI]}{dt} = k[HI]^2$
c. $2N_2O_5(g) \rightarrow 4NO_2(g) + O_2(g)$	$-\frac{1}{2}\frac{d[N_2O_5]}{dt} = k[N_2O_5]$
d. $2N_2O(g) \rightarrow 2N_2(g) + O_2(g)$	$-\frac{1}{2}\frac{d[N_2O]}{dt} = k[N_2O]$
e. $CH_3Br(aq) + OH^-(aq) \rightarrow CH_3OH(aq)$	$-\frac{d[CH_3Br]}{dt} = k[CH_3Br][OH^-]$
f. $C_{12}H_{22}O_{11}(aq) + H_2O(l) \rightarrow 2C_6H_{12}O_6(aq)$	$-\frac{d[C_{12}H_{22}O_{11}]}{dt} = k[C_{12}H_{22}O_{11}][H^+]$

4. Combustion processes release NO into the atmosphere where NO is oxidized by a number of substances, including O_2. The reaction with O_2 is $2NO + O_2 \rightarrow 2NO_2$. Observation of the rate of the reaction reveals that it is first order in oxygen pressure and second order in NO pressure. Write the rate law for the oxidation of NO by O_2.

5. The reaction

$$2HCrO_4^-(aq) + 3HSO_3^-(aq) + 5H_3O^+(aq) \longrightarrow 2Cr^{3+}(aq) + 3SO_4^-(aq) + 10H_2O(l)$$

follows the rate law

$$\nu = k[HCrO_4^-][HSO_3^-]^2[H_3O^+]$$

Why do the rate law exponents differ from the stoichiometric coefficients?

6. One of the first anesthetics identified was nitrous oxide, N_2O, which was discovered in the 1840s. N_2O can be produced from NO by reduction with H_2:

$$2NO(g) + H_2(g) \longrightarrow N_2O(g) + H_2O(l)$$

The initial reaction rate was measured for several experimental runs to determine the rate law, and the following data were collected. What is the order of the reaction for NO? For H_2? What is the rate law?

Initial [NO] (M)	0.60	1.20	1.20
Initial [H_2] (M)	0.37	0.37	0.74
Initial rate (mol/L · min)	0.18	0.72	1.44

7. Several oxides of sulfur exist in aqueous sulfate solutions. One of these is $S_2O_8^{2-}$. $S_2O_8^{2-}$ can be reduced by I^-:

$$S_2O_8^{2-}(aq) + 2I^-(aq) \longrightarrow 2SO_4^{2-}(aq) + I_2(s)$$

The kinetics of this reaction were monitored in aqueous solution, and the following data were collected. What is the rate expression for the reaction? What is the overall order of the reaction?

Initial [$S_2O_8^{2-}$] (mM)	0.1	0.2	0.2
Initial [I^-] (mM)	10	10	5
Initial rate (mol/L · min)	6.5×10^{-7}	1.30×10^{-6}	6.5×10^{-7}

8. For the reaction $BrO_3^-(aq) + 5Br^-(aq) + 6H_3O^+(aq) \rightarrow 3Br_2(aq) + 9H_2O(l)$, the rate law is

$$-\frac{d[BrO_3^-]}{dt} = k[BrO_3^-][Br^-][H_3O^+]^2$$

a. What is the order of the reaction with respect to Br^-?

b. What is the relationship of the reaction order with respect to H_3O^+ and the stoichiometry with respect to H_3O^+?

c. How does the reaction rate change if the concentration of BrO_3^- is doubled?

d. What change in concentration of H_3O^+ is required to double the rate of the reaction?

9. The reaction

is first order in both reactants. Write the rate law for this reaction. Fill in the blank: Doubling the concentration of ⬡N-H _____ the reaction rate.

10. The reaction

$$CH_3Br(aq) + I^-(aq) \longrightarrow CH_3I(aq) + Br^-(aq)$$

is found to be first order in $[I^-]$ and independent of $[CH_3Br]$. Write the rate law for the reaction.

12.1 Finding the Rate Law from Experimental Data

11. The chemotherapy agent cisplatin reacts in water as follows:

The reaction rate data for cisplatin are

[cisplatin] (mM)	8	6	4	3	2
Reaction rate × 10^{-6} (M/min)	11.8	8.85	5.90	4.42	2.95

What is the order of the reaction?

12. In basic solution, the ester CH_3COOCH_3 hydrolyzes to form methanol and acetate ion:

$$CH_3COOCH_3(aq) + OH^-(aq) \longrightarrow CH_3COO^-(aq) + CH_3OH(aq)$$

The following data were obtained to determine the rate law.

Experiment	$[CH_3COOCH_3]$ (M)	$[OH^-]$ (M)	Initial Rate (mol/L·s)
1	0.050	0.050	3.45×10^{-4}
2	0.050	0.100	6.90×10^{-4}
3	0.100	0.100	1.38×10^{-3}

Write the rate law for the hydrolysis of CH_3COOCH_3.

13. Ammonia dissociates at high temperature: $NH_3(g) \rightarrow NH_2(g) + H(g)$. An experiment is run to determine the order and rate constant for this reaction. The data gathered are given in the accompanying table. What is the reaction order and rate constant?

Time (h)	0	25	50	75
$[NH_3] \times 10^{-7}$ (M)	8.00	6.75	5.84	5.15

14. Carbon monoxide poisons by binding to hemoglobin (Hb) more strongly than does dioxygen, thereby suffocating the victim. The rate of reaction of carbon monoxide with hemoglobin was studied at 20 °C, and the following data were collected.

[CO](μM)	1.00	1.00	3.00
[Hb](μM)	3.36	6.72	6.72
reaction rate ($\mu M/L \cdot s$)	0.941	1.88	5.64

a. Write the rate law for the reaction.

b. Calculate the rate constant.

15. Carbon compounds can have a rich variety of configurations of the atoms, even in relatively simple compounds. One example is the compound with empirical formula $C_2H_2Cl_2$. Two of these isomers are called *cis-* and *trans-*1, 2-dichloroethylene:

At high temperature, these two forms can interconvert. Data for the interconversion:

Time (min)	0	10	20	30
Moles *trans*	1.00	0.90	0.81	0.73

a. Determine the order of the reaction.

b. Determine the half-life of the reaction.

16. The rate of the reaction $I^-(aq) + OCl^-(aq) \rightarrow Cl^-(aq) + OI^-(aq)$ depends on the pH of the solution:

$[OH]$ (M)	1.00	0.50	0.25
Rate (M/s)	60	120	240

Determine the order with respect to hydroxide ion.

12.1 Method of Initial Rates

17. The atmospheric pollutant NO can be oxidized by chlorine gas:

$$2NO(g) + Cl_2(g) \longrightarrow 2NOCl(g)$$

The following data were obtained for the initial rate of this reaction.

Experiment	Initial p_{NO} (atm)	Initial p_{Cl_2} (atm)	Initial Rate (atm/s)
1	0.50	0.50	5.1×10^{-3}
2	1.0	1.0	4.0×10^{-2}
3	0.50	1.0	1.0×10^{-2}

Write the rate equation for this reaction.

18. The reaction $S_2O_8^{2-}(aq) + 2I^-(aq) \longrightarrow SO_4^{2-}(aq) + I_2(s)$ occurs in aqueous solution. The following data were obtained from three experiments. What is the rate law for this reaction?

Experiment	$[S_2O_8^{2-}]_0$ (M)	$[I^-]_0$ (M)	Initial Rate (mol/L · s)
1	1.0×10^{-4}	1.0×10^{-2}	6.5×10^{-7}
2	2.0×10^{-4}	1.0×10^{-2}	1.30×10^{-6}
3	2.0×10^{-4}	5.0×10^{-3}	6.5×10^{-7}

19. For the reaction $2NO(g) + O_2(g) \longrightarrow 2NO_2(g)$, the following data were obtained. Write the rate law for the reaction.

Experiment	[NO] (M)	$[O_2]$ (M)	Rate of Formation of NO_2 (M/s)
1	0.015	0.05	0.048
2	0.030	0.015	0.192
3	0.015	0.030	0.096
4	0.030	0.030	0.384

20. The gas ClO_2 dissolves in water and reacts in basic solution according to the following reaction:

$$2ClO_2(aq) + 2OH^-(aq) \longrightarrow ClO_3^-(aq) + ClO_2^-(aq) + H_2O(l)$$

Several experiments were conducted to determine the order of this reaction, and the following results were obtained.

Experiment	1	2	3
$[ClO_2]_0$ (M)	0.060	0.020	0.020
$[OH^-]_0$ (M)	0.030	0.030	0.090
k (M/s)	2.48×10^{-2}	2.76×10^{-3}	8.28×10^{-3}

Write the rate law for the reaction.

12.1 Ostwald's Isolation Method

21. Iodine atoms can exist in monatomic form, atomic I, in the gas phase. In the presence of an inert gas, I dimerizes to form I_2. The recombination rate is investigated in the presence of Ar to determine the rate law. Data on the rate of I_2 formation for various concentrations of I and Ar are listed below. Find the order of the reaction with respect to I and Ar. (See table below)

22. Hydrogen peroxide is reduced with I^- as follows:

$$H_2O_2(aq) + 3I^-(aq) + 2H^+(aq) \longrightarrow I_3^-(aq) + 2H_2O(l)$$

Doubling the I^- concentration cuts the reaction time in half. Doubling the peroxide concentration doubles the reaction rate. Write the rate law for the reaction.

12.2 Rate Laws: First-Order Reactions

23. Uranium-238 decays to lead-206 with a rate constant of 1.54×10^{-10} yr^{-1}. This decay provides a method for dating rock samples. If no other lead isotopes are present, all the lead in the rock originated as uranium. Moon rocks were dated using this method. Estimate the age of moon rocks found to have a ^{206}Pb to ^{238}U ratio of $1.00 : 1.09$.

24. Radioactive isotopes are used in the treatment of cancer. One of those isotopes is ^{192}Ir. Iridium-192 has a decay constant of 9.3×10^{-3} d^{-1}. What fraction of a ^{192}Ir sample remains after 100 days?

25. The acid-catalyzed hydrolysis of sucrose,

$$C_{12}H_{22}O_{11} + H_2O \longrightarrow 2C_6H_{12}O_6$$

is first order in sucrose concentration. After 25 h, an initial 10 g of sucrose is reduced to 3.5 g. Determine the rate constant for hydrolysis of sucrose.

26. The four-carbon molecule cyclobutane, $\boxed{}$, is unstable and decomposes to two ethylene molecules, C_2H_4. At 1000 °C, the first-order rate constant for the decomposition is 87 s^{-1}.
   a. How long does it take to reduce 0.5 g of cyclobutane to 0.05 g?
   b. How much of a 1.00-g sample remains after 10 s?
   c. What is the half-life of the reaction?

27. N_2O_5 dissociates in the atmosphere, forming N_2O_4 and O_2, with a first-order rate constant of 1.20×10^{-2} s^{-1} at 45 °C.
   a. Write the net reaction equation for the dissociation of N_2O_5.
   b. How long does it take to reduce an initial 5 mM concentration to 1 mM?
   c. How much longer is required to reduce the initial 5 mM concentration to 0.5 mM?

28. An isotope of plutonium, ^{239}Pu, decays by alpha particle emission. One μg of ^{239}Pu emits 1.4×10^5 alpha particles per minute. Determine the half-life of ^{239}Pu.

29. Automobile tires degrade through oxidation and other reactions, leading to a recommended safe storage period. Estimate how much longer the safe storage period would be if the storage temperature were lowered by 10 °C?

30. The lifetime of gas-phase substances in the environment vary significantly. The pseudo-first-order decay of two sulfur species illustrates the vast range. The decay constant for CS_2 is 8.3×10^{-2} d^{-1}, while that of OCS is $2.3 10^{-2}$ yr^{-1}. Determine the half-life (the lifetime) of each of these substances.

Experiment	1	2	3	4	5	6	7	8
$[I]_0 \times 10^{-5}$ (M)	1.0	2.0	4.0	6.0	1.0	2.0	4.0	6.0
$[Ar]_0 \times 10^{-3}$ (M)	1.0	1.0	1.0	1.0	5.0	5.0	5.0	5.0
Initial rate (mol/L · s)	8.7×10^{-4}	3.48×10^{-3}	1.39×10^{-2}	3.13×10^{-2}	4.35×10^{-3}	1.74×10^{2}	6.96×10^{-2}	1.57×10^{-1}

12.2 Rate Laws: Second-Order Reactions

31. The dimerization of C_4H_6 is second order. Analyze the following data to determine the rate constant for this reaction.

$[C_4H_6]$ (M)	0.120	0.111	0.104	0.092
Time (min)	0	200	400	800

32. The major component of compressed natural gas, a widely used fuel, is methane. In transport and usage, some methane escapes into the atmosphere. The only known significant chemical loss mechanism for atmospheric methane is reaction with OH in the elementary reaction

$$CH_4(g) + OH(g) \longrightarrow CH_3(g) + H_2O(g)$$
$$k = 6.3 \times 10^{-15} \text{ mL/molecule} \cdot s$$

a. Write the rate law for the removal of methane by OH.
b. The typical daytime concentration of OH radical is 1×10^6 radicals/mL. Determine the removal rate for methane.
c. The lifetime of methane in the atmosphere is the inverse of the removal rate. What is the typical lifetime (in days or years) of methane in the atmosphere?

33. Propane is an important fuel. In the atmosphere, the major removal mechanism for propane is reaction with OH. The first step in this process is

$$C_3H_8(g) + OH(g) \longrightarrow C_3H_7(g) + H_2O(g)$$
$$k = 1.1 \times 10^{-12} \text{ mL/molecule} \cdot s$$

a. Write the rate law for the removal of propane by OH.
b. The typical daytime concentration of OH radical is 1×10^6 radicals/mL. Determine the removal rate for propane.
c. The lifetime of propane in the atmosphere is the inverse of the removal rate. What is the typical lifetime (in days or years) of propane in the atmosphere?
d. Compare the removal rate for methane (Exercise 32) with that for propane. Comment on the wisdom of regulatory agencies focusing on control of non-methane hydrocarbons.

34. CFCs have been largely replaced by hydrogenated chlorofluorocarbons (HCFCs) because they are degraded more rapidly by OH radical, the scrubber of the troposphere.
a. Use the following information to determine the reaction rate and lifetime of HCFC-22 (CHF_2Cl).

$$[OH] = 5 \times 10^5 \text{ molecules/cm}^3$$
$$[CHF_2Cl] = 0.1 \text{ ppbv (part per billion by volume)}$$
$$k = 1.44 \times 10^{-11} \text{ cm}^3\text{/molecule} \cdot h$$

b. In comparison, the lifetime of CFC-14 (CF_3Cl) is 680 years. Determine the removal rate for CFC-14.

12.2 Half-Life

35. The radioactive isotope ^{14}C is used for dating ancient artifacts. The first-order decay constant for ^{14}C is 1.21×10^{-4} yr^{-1}. What is the half-life of this reaction?

36. At 600 K azomethane decomposes according to the reaction

$$CH_3NNCH_3 \longrightarrow C_2H_6 + N_2$$

with a first-order rate constant of 3.6×10^{-4} s^{-1}. What time is required for half of the azomethane to decompose?

37. Kinetic experiments on the hydrolysis of ethyl nitrobenzoate (ENB) in high pH solution produce the data given below. Determine the order of the reaction by the method of half-lives and determine the rate coefficient.

[ENB] (mM)	50.0	35.5	27.5	22.5	18.5	16.0	14.8	14.0	13.8
Time (s)	0	100	200	300	400	500	600	700	800

38. The discovery of radioactivity is attributed to Marie Curie, who won the Nobel Prize in physics in 1903 for her work on radiation and the Nobel Prize in chemistry in 1911 for the discovery of radium and polonium. Madam Curie is among a rare group of people who have won two Nobel Prizes in a science field. In 1921, the women of America honored Curie by giving her a gift of 1.00 g of pure radium (currently housed in Paris at the Curie Institute of France). The main isotope of radium, ^{226}Ra, has a half-life of 1.60×10^3 yr. How many grams of radium remain?

39. The skeleton of an early human referred to as *Zinjanthropus* (East African man) was found in a canyon in association with volcanic ash containing potassium minerals. An isotope of K, ^{40}K, decays to ^{40}Ar with a half-life of 1.5×10^9 yr. Any ^{40}Ar present at the time of the eruption would have boiled out of the molten lava, so all ^{40}Ar found with the skeleton must have originated as ^{40}K. The amount of ^{40}Ar in the ash was measured and found to be 0.078 of the amount of ^{40}K. How old is the skeleton?

40. NO is both an industrial pollutant and ubiquitous biomolecule. NO was named "molecule of the year" by *Science* magazine in 1992, and the 1998 Nobel Prize for medicine or physiology was awarded to R. L. Furchgott, L. Ignarro, and F. Murad for discoveries connected with the biological importance of NO. NO is oxidized by O_2:

$$2NO(aq) + O_2(aq) \longrightarrow 2NO_2(aq)$$

The rate law is

$$-\frac{d[NO]}{dt} = k[NO]^2[O_2]$$

In the gas phase, the specific rate constant is 1.42×10^4 $M^{-2} \cdot s^{-1}$. In aqueous solution, the rate constant is 8×10^6 $M^{-2} \cdot s^{-1}$.
a. Under normal conditions, $[O_2] \gg [NO]$. Write the rate law for normal conditions.
b. Calculate the half-life for NO in the gas phase for concentrations of 1% to 10 ppm at a temperature of 25 °C. (Gas-phase concentration of $O_2 = 20\%$.)
c. In aqueous solution, the solubility of oxygen is 4.6×10^{-4} M. Calculate the half-life of NO in the blood at 25 °C for concentrations of 100 μM and 1 μM.
d. A 50% blood vessel relaxation occurs with a 10 nM NO concentration. Determine the time required for an initial 100 μM concentration to reach 10 nM.

41. In the lungs, NO binds to oxyhemoglobin with a rate constant about 20 times larger than the oxidation rate (see Exercise 40). The relative reaction rates are important biologically because NO_2 oxidizes Fe(II) to Fe(III). The rate law for binding to Fe(II) is

$$-\frac{d[NO]}{dt} = k[NO]^2[Fe(II)]$$

The human body contains about 6 million red blood cells per milliliter of blood, and each red cell contains 250 million hemoglobin molecules. Each hemoglobin contains four Fe(II) atoms. If a smoker inhales NO, raising the NO content of the blood to 10 μM, what fraction of the hemoglobin is destroyed in the first minute in the absence of oxidation of NO by O_2?

42. The reaction $N_2O_5 \rightarrow NO_2 + NO_3$ takes place in the gas phase and is important in the nitrogen budget of the atmosphere. The decomposition is first order with a rate constant of $4.8 \times 10^{-4}\ s^{-1}$.
   a. Determine the half-life.
   b. A pressure of 0.5 atm of N_2O_5 is introduced in a reaction vessel. Determine the pressure after 1 min.
   c. How long does it take for the pressure to reach 0.75 atm?

43. The chemistry of cement is complex, and most of the reactions remain ill characterized. The major calcium substances in concrete are referred to as C_2S ($2CaO \cdot SiO_2$) and C_3S ($3CaO \cdot SiO_2$). Cement gets its strength from the interlocking crystallites of hydrates of these calcium species, which are admixed with iron and aluminum oxides. The pseudo-first-order hydration rates for C_2S and C_3S species are quite different, affecting the strength as well as the cure time. The half-life of the C_3S hydration reaction is measured in hours, while that for the C_2S hydration reactions is measured in days. Compare the rate of hydration of C_2S with that of C_3S.

44. The half-life of radium (^{226}Ra) is 1590 yr. Determine the value of the decay constant. What fraction of a sample of radium decays in 1 yr?

12.3 Elementary Reactions and the Rate Law: Mechanisms

45. A proposed mechanism for the oxidation of NO to NO_2,

$$2NO(g) + O_2(g) \longrightarrow 2NO_2(g)$$

is given below. Determine the predicted rate law for the oxidation.

Mechanism:

$$2NO \underset{k_{-1}}{\overset{k_1}{\rightleftarrows}} N_2O_2$$

$$O_2 + N_2O_2 \xrightarrow{k_2} 2NO_2 \qquad \text{Slow step}$$

46. The reaction of NO_2 with fluorine, $2NO_2(g) + F_2(g) \rightarrow 2FNO_2(g)$, has the following rate law:

$$-\frac{d[NO_2]}{dt} = 2[NO_2][F_2]$$

The following mechanism is proposed:

$$NO_2 + F_2 \longrightarrow FNO_2 + F \qquad \text{Slow}$$
$$NO_2 + F \longrightarrow FNO_2 \qquad \text{Fast}$$

Is the mechanism consistent with the reaction and rate law?

47. Two mechanisms are proposed for the reaction
$$OH^-(aq) + CH_3Br(aq) \rightarrow CH_3OH(aq) + Br^-(aq).$$

Mechanism I:

$$CH_3Br \longrightarrow CH_3^+ + Br^- \qquad \text{Slow}$$
$$CH_3^+ + OH^- \longrightarrow CH_3OH \qquad \text{Fast}$$

Mechanism II:

$$CH_3Br + OH^- \longrightarrow CH_3OH + Br^-$$

   a. Write the rate law for each mechanism.
   b. Show that mechanism I gives the correct net reaction.
   c. Given the following data, select the correct mechanism.

Experiment	[CH$_3$Br] (M)	[OH$^-$] (M)	Production Rate for CH$_3$OH (M/min)
1	0.200	0.200	0.015
2	0.400	0.200	0.030
3	0.400	0.400	0.030

48. Consider the mechanism for production of C from A:

$$A \xrightarrow{k_1} B + C$$
$$A + B \xrightarrow{k_2} C$$

   a. What is B called?
   b. What is the net stoichiometry?
   c. If Step 2 is the rate-determining step, what is the rate equation?

49. Consider the mechanism for production of C from A:

$$A \xrightarrow{k_1} 2B$$
$$A + B \xrightarrow{k_2} C$$

   a. What is the net stoichiometry?
   b. If Step 1 is the rate-determining step, what is the rate equation?

50. The utilization of methanol as a fuel has been delayed due to production of formaldehyde from incomplete combustion of methanol. Formaldehyde can be oxidized by Co(III). The following mechanistic steps have been identified:

$$HCHO + Co^{3+} \xrightarrow{k_1} Co^{2+} + H^+ + HCO$$
$$HCO + Co^{3+} \xrightarrow{k_2} H^+ + CO + Co^{2+}$$
$$HCO + H_2O \xrightarrow{k_3} HCHO + HO$$
$$HCO + H_2O \xrightarrow{k_4} HCOOH + H$$
$$H + HCHO \xrightarrow{k_5} H_2 + HCO$$
$$HO + HCHO \xrightarrow{k_6} H_2O + HCO$$

   a. What are the intermediates?
   b. What is the net reaction?
   c. The rate law is determined to be first order in Co^{3+} and first order in HCO. What is the rate-determining step?

51. Destruction of peroxide by iodide proceeds according to the net reaction

$$2I^-(aq) + H_2O_2(aq) + 2H_3O^+(aq) \longrightarrow I_2(aq) + 4H_2O(l)$$

The rate law for this reaction is

$$-\frac{d[H_2O_2]}{dt} = k[I^-][H_2O_2]$$

A proposed mechanism for this reaction is

Step 1: $H_2O_2 + I^- \rightarrow HOI + OH^-$
Step 2: $HOI + I^- \rightarrow I_2 + OH^-$
Step 3: $2OH^- + 2H_3O \rightarrow 4H_2O$

a. Show that the mechanism is consistent with the net reaction.
b. Which is the slow step?

52. Ammonia and hypochlorite ion, OCl^-, react in basic solution to yield the important reducing agent hydrazine, N_2H_4. The following mechanism is proposed for this reaction:

$$NH_3 + OCl^- \longrightarrow NH_2Cl + OH^- \quad \text{Slow}$$
$$NH_2Cl + NH_3 \longrightarrow N_2H_5^+ + Cl^- \quad \text{Fast}$$
$$N_2H_5^+ + OH^- \longrightarrow N_2H_4 + H_2O \quad \text{Fast}$$

a. What is the net reaction?
b. Which is the rate-limiting step?
c. What is the rate law for the reaction?
d. Which substances are reaction intermediates?

53. Gaseous iodine catalyzes the isomerization of *cis*-2-butene to *trans*-2-butene.

The following mechanism is proposed for this reaction:

The rate law for the reaction is $-\dfrac{d[\text{cis-2-butene}]}{dt} =$

$k[\text{cis-2-butene}][I_2]^{1/2}$

a. Identify the slow step(s) in the mechanism.
b. The activation barrier for the uncatalyzed reaction is 262 kJ/mol and that for the catalyzed reaction is 115 kJ/mol. How much faster is the catalyzed reaction?

54. Aircraft that fly through the stratosphere produce NO, which decomposes O_3 in the protective ozone layer according to the net reaction $O_3(g) + NO(g) \rightarrow NO_2(g) + O_2(g)$. Several mechanisms have been proposed for this reaction. Which mechanism is consistent with the rate law $d[O_3]/dt = k[O_3][NO]$?

Mechanism I:

$$NO + O_3 \longrightarrow NO_3 + O \quad \text{Slow}$$
$$NO_3 + O \longrightarrow NO_2 + O_2 \quad \text{Fast}$$

Mechanism II:

$$NO + O_3 \longrightarrow NO_2 + O_2 \quad \text{One step}$$

Mechanism III:

$$O_3 \longrightarrow O_2 + O \quad \text{Slow}$$
$$O + NO \longrightarrow NO_2 \quad \text{Fast}$$

55. In the troposphere, reaction between NO_2 and ozone produces N_2O_5 and O_2. The rate law for the reaction is $v = k[NO_2][O_3]$. Is either of the following mechanisms consistent with the rate law?

Mechanism I:

$$2NO_2 \Longleftrightarrow N_2O_4 \quad \text{Fast equilibrium}$$
$$N_2O_4 + O_3 \longrightarrow N_2O_5 + O_2 \quad \text{Slow}$$

Mechanism II:

$$NO_2 + O_3 \longrightarrow NO_3 + O_2 \quad \text{Slow}$$
$$NO_3 + NO_2 \longrightarrow N_2O_5 \quad \text{Fast}$$

56. The Chapman mechanism for generation of ozone in the stratosphere consists of the following reactions:

$$O_2 + h\nu \longrightarrow 2O$$
$$O + O_2 + M \longrightarrow O_3 + M$$
$$O_3 + h\nu \longrightarrow O_2 + O$$
$$O + O_3 \longrightarrow 2O_2$$

a. Ozone is found in a well-defined layer of the stratosphere. Explain why, citing evidence from the mechanism.
b. O and O_3 are often referred to as "odd oxygen" and treated together. Explain why.
c. Determine the steady state concentrations of O and O_3.

12.4 Modeling Reactions: Temperature and the Activation Barrier

57. Milk sours in about 20 h at 28 °C. Stored in a refrigerator at 5 °C, milk sours in about 10 days. What is the activation energy for souring of milk?

58. The reaction

$$NO(g) + O_3(g) \longrightarrow NO_2(g) + O_2(g)$$

is important in stratospheric ozone depletion. The preexponential factor for this reaction is 6.31×10^8 L/mol · s and the activation energy is 10 kJ/mol.

a. Calculate the rate constant for this reaction at 220 K (the temperature of the lower stratosphere).
b. Calculate the rate of this reaction in the troposphere (where we live) at a temperature of 300 K.
c. This bimolecular reaction is first order in NO and first order in O_3. Determine the reaction rate at 220 K if $[NO] = 1.0 \times 10^{-6}$ M and $[O_3] = 0.01 \times 10^{-6}$ M.

59. The rate of radioactive decay is independent of the temperature of the solid for temperatures easily attained in a laboratory. What does this observation imply about the activation energy of the process?

60. The hydrolysis of sucrose proceeds according to the following reaction:

$$C_{12}H_{22}O_{11}(aq) + H_2O(l) \longrightarrow 2C_6H_{12}O_6(aq)$$

The activation barrier for this reaction is 108 kJ/mol and the preexponential A factor is 1.5×10^{15} L/mol · s. Determine the rate constant for sugar hydrolysis at room temperature (25 °C) and at the temperature of the body (37 °C).

61. Diffusion is an activated process that can be described by a rate of diffusion, D:

$$D = D_o e^{(-E_d/RT)}$$

where D_o is a preexponential factor, E_d is the activation energy for diffusion, R is the gas law constant, and T is the absolute temperature. The diffusion rate for carbon into γ-Fe (FCC iron) is 9.2×10^{-12} m^2/s at 900 °C and the activation energy is 136 kJ/mol. Determine the diffusion rate for carbon into γ-Fe at 700 °C and 1100 °C. Steel objects like gears are case hardened by exposing them to a rich carbon atmosphere at high temperature. What temperature would you choose to produce a case-hardened gear with a 1-mm-high carbon outer layer: 700 °C, 900 °C, or 1100 °C? Justify your choice.

62. The anesthetic nitrous oxide, N_2O, decomposes to N_2 and O_2 with a rate constant of 2.6×10^{-11} s^{-1} at 300 °C and 2.1×10^{-10} s^{-1} at 330 °C. Calculate the activation energy for the reaction. Make a plot of ΔG versus reaction coordinate for this reaction.

63. Cyclopropane, △, is an unstable molecule that rearranges to propylene, ⋏. The rate constant has been determined at a series of temperatures:

T(K)	600	650	700	750	800	850	900
k(s^{-1})	3.30×10^{-9}	2.19×10^{-7}	7.96×10^{-6}	1.51×10^{-4}	2.74×10^{-3}	3.04×10^{-2}	2.58×10^{-1}

a. Calculate the activation energy for this reaction.
b. Estimate the value of k at 500 K.

64. The substance N_2O_5 plays an important role in the nitrogen oxide cycles in the atmosphere. Data for the temperature dependence of the gas-phase reaction,

$$N_2O_5(g) \longrightarrow 2NO_2(g) + {}^1/_2O_2(g)$$

are given in Table 12.3. Determine the activation energy of this reaction.

65. Iodine-containing molecules are often unstable at elevated temperatures. Methyl iodide, CH_3I, follows this pattern. The activation energy for the first-order, gas-phase decomposition,

$$CH_3I(g) \longrightarrow CH_3(g) + I(g)$$

is 180 kJ/mol. Determine the percent increase in dissociation from room temperature (25 °C) to 35 °C.

66. The activation energy for the decomposition of N_2O_5 in carbon tetrachloride solvent is 103 kJ/mol and has a rate constant of 6.2×10^{-4} s^{-1} at 45 °C. Determine the rate constant at 100 °C.

67. At elevated temperature, ethyl chloride, C_2H_5Cl, decomposes according to the reaction

$$C_2H_5Cl(g) \longrightarrow C_2H_4(g) + HCl(g)$$

The reaction is first order, $A = 1.6 \times 10^{14}$ s^{-1}, and $E_a = 249$ kJ/mol.
a. Determine the value of k at 450 °C.
b. Determine the fraction of ethyl chloride that decomposes in 10 min at 450 °C.
c. At what temperature is the decomposition of ethyl chloride twice as fast as it is at 450 °C?

68. Methane is a significant greenhouse gas that is found in trace amounts in the troposphere. The main removal mechanism for methane is oxidation by the OH radical.

$$OH(g) + CH_4(g) \longrightarrow H_2O(g) + CH_3(g)$$

The activation energy for this reaction is 19.5 kJ/mol. Determine the relative change in reaction rate for methane with OH at the earth's surface ($T = 295$ K) and at the top of the troposphere ($T = 220$ K).

12.4 Modeling Reactions: Collision Theory

69. The dimerization of NO_2 is a second-order reaction.
a. Determine the shape of NO_2 and the bonding in N_2O_4.
b. Draw a schematic of the collision complex for formation of N_2O_4.

70. For an elementary reaction that comes to equilibrium, the activation energy of the reverse reaction is related to that of the forward reaction. Draw a schematic reaction profile and discuss the relationship between the activation energy of the forward and reverse reactions.

12.5 Rates and Equilibrium

71. The gas-phase reactions

i. $NO_2 + NO_3 \longrightarrow N_2O_5$ $k = 2.2 \times 10^{-30}$ cm^6/molecules^2 · s

ii. $N_2O_5 \longrightarrow NO_2 + NO_3$

come to equilibrium at 300 K with an equilibrium constant, K_c, of 1.63×10^{-6}. Determine the rate constant for reaction (ii).

Table 12.3 Data for Exercise 64

T(K)	273	298	308	318	328	338
k(s^{-1})	7.87×10^{-7}	3.46×10^{-5}	1.35×10^{-4}	4.98×10^{-4}	1.50×10^{-3}	4.87×10^{-3}

72. The equilibrium constant is a constant for any reaction and varies only with the temperature. Using an elementary reaction

$$A + B \rightleftharpoons C$$

with a forward rate constant k_f and a reverse rate constant k_r, explain the temperature dependence of the equilibrium constant.

12.6 Catalysts

73. The decomposition of ammonia on W occurs as follows: $2NH_3(g) \rightarrow N_2(g) + 3H_2(g)$. The rate is independent of the NH_3 concentration. What is the reaction order with respect to ammonia? What is the catalyst?

74. A mechanism for the decomposition of H_2O_2 in a solution containing HBr consists of two steps:

Step 1: $2Br^- + H_2O_2 + 2H^+ \rightleftharpoons Br_2 + 2H_2O$
Step 2: $Br_2 + H_2O_2 \rightleftharpoons 2Br^- + 2H^+ + O_2$

a. Write the net reaction.
b. Is either Br^- or H^+ a catalyst?
c. Classify the catalyst as homogeneous or heterogeneous.

75. Approximately 20 billion lb of NO is produced annually in the United States for the manufacture of nitric acid for fertilizers and explosives. Industrially, NO is produced by oxidation of ammonia:

$$4NH_3(g) + 5O_2(g) \xrightarrow[900°\ C]{\text{Pt-Rh Catalyst}} 4NO(g) + 6H_2O(g)$$

This reaction is followed by air oxidation to NO_2 and dissolving in water to produce HNO_3 and NO. The unreacted NO is recycled for further nitric acid production. The 1909 Nobel Prize recognized Ostwald for his work on the development of this important process.

a. Is the Pt-Rh catalyst homogeneous or heterogeneous?
b. The reaction rate is independent of the ammonia and oxygen partial pressures. Suggest a mechanism that accounts for this observation.
c. $\Delta G < 0$ for this reaction. Sketch a reaction energy (ΔG) profile for both the catalyzed and uncatalyzed reactions.

76. The enzyme catalase catalyzes the decomposition of peroxide. The uncatalyzed decomposition of peroxide has an activation energy of 72 kJ/mol, and the catalyzed reaction has an activation energy of 28 kJ/mol.

a. How much faster is the catalyzed reaction compared with the uncatalyzed reaction at room temperature (298 K)?
b. How much faster is the catalyzed reaction at 37 °C, normal body temperature?

77. CO_2 is a by-product of digestion of sugars, primarily glucose, that fuel living organisms. In the body, CO_2 is produced in muscle tissue when work is done. The CO_2 is transferred to the blood using an enzyme called carbonic anhydrase. It is estimated that one enzyme molecule hydrates 10^6 CO_2 molecules per second. How many kilograms of CO_2 per liter are hydrated in 1 h by 1 μM enzyme? How many kilograms of sugar fuel are required to produce this mass of CO_2?

78. When a coil of red-hot platinum is exposed to the air, it cools quickly. When exposed to ammonia gas, the coil continues to glow red-hot. Explain this observation.

■ CONCEPTUAL EXERCISES

79. Draw a diagram indicating the relationships among the following terms: reaction rate, rate law, specific rate constant, reaction mechanism, reaction intermediate, and reaction order. All terms should have at least one connection.

80. Draw a diagram indicating the relationships among the following terms: transition state, activation energy, catalyst, activated complex, reaction rate, and Arrhenius equation. All terms should have at least one connection.

☀ APPLIED EXERCISES

81. Antacid tablets produce a familiar fizz when dropped in water. These pills contain a variety of ingredients depending on the manufacturer, but all contain sodium bicarbonate and a mild acid. The sequence of reactions is

$$NaHCO_3(aq) \longrightarrow Na^+(aq) + HCO_3^-(aq)$$
$$HCO_3^-(aq) + H_3O^+(aq) \rightleftharpoons H_2CO_3(aq) + H_2O(l)$$
$$H_2CO_3(aq) \longrightarrow H_2O(l) + CO_2(g)$$

An antacid tablet was dissolved in 10 mL water at 20 °C and the following data were obtained:

Time (s)	0	20	30	50	70	90	160
Volume CO_2 (mL)	0	20	30.2	40.9	48.9	54.2	62.1

a. What is the order of the reaction?
b. What is the rate constant?
c. Another experiment was performed at 14 °C and the rate constant, $k_{14\ °C}$, was determined to be one-half the rate constant at 20 °C. What is the activation energy for the decarboxylation of antacid?
d. Predict the rate constant at 26 °C.
e. Based on these results, should antacids be dissolved in cold, tepid, or warm water? Discuss the reasons for your choice.

82. Commercial fishing fleets have found that deep-sea fish such as tuna and swordfish are attracted to the light emitted by cyalume sticks in night water. Long-line fishermen attach light sticks to their monofilament long lines, which can extend for 80 miles and hold up to 3000 hooks. The light also aids in locating the line should the filament be cut. The energy to produce the light results from the reaction between hydrogen peroxide and phenyl oxalate ester:

The energetic molecule, C_2O_4, collides with the fluorescent substance, decomposing to CO_2:

Step 2:

$+ \text{ fluorescer } \xrightarrow{k_2} \text{ fluorescer* } + 2CO_2$

The excited fluorescer emits light as it returns to the ground state:

Step 3: $\text{fluorescer*} \xrightarrow{k_3} \text{fluorescer} + hv$

a. Is peroxide a catalyst? What kind of reaction is Step 1?
b. Identify the intermediates in the reaction mechanism.
c. Under certain conditions, the reaction is observed to be first order. Identify the slow step.
d. Assuming a steady state concentration for the intermediates, write the rate law for the fluorescence.

■ INTEGRATIVE EXERCISES

83. Why does it take longer to boil an egg in the Rocky Mountains (elevation 11,000 ft) than in Boston (elevation: sea level)?

84. The insoluble salt HgI_2 crystallizes in two different forms: a rhombic crystal that is lemon yellow and a tetragonal crystal that is orange. The color of the product from addition of a solution of KI to aqueous $Hg(NO_3)_2$ changes over time. The relevant equilibria are

$$Hg^{2+}(aq) + 2I^-(aq)$$

It is found that $k_1 > k_2$ and $\dfrac{k_2}{k_{-2}} > \dfrac{k_1}{k_{-1}}$.

a. What is the color of the solid precipitate immediately after the reaction begins?
b. Which color product is favored at equilibrium? (*Hint:* Which equilibrium constant is larger, that for formation of the yellow or the orange product?)
c. Make a schematic sketch of the Gibbs free energy, starting with Hg^{2+} and I^- in solution, and ending with the yellow solid compared with ending with the orange solid.

85. In this exercise, you are required to obtain some supplies and make observations to answer the questions. Obtain methylene blue from an aquatic supply store (used for a disease called ich), some Cu(II) sulfate sold as "Had-a-Snail"

(also available at an aquatic supply store), some vitamin C tablets, some table salt, and a 2-L soda bottle. Fill the bottle two-thirds full with water, add 5 mL vitamin C powder, 4 drops methylene blue, and 2.5 mL salt. Cap and shake the bottle.

i. Add 1.25 mL Cu(II) sulfate. Cap the bottle and shake it for 20 seconds. Record your observations.
ii. Let the solution stand for 5 minutes. Observe the color particularly at the solution–air interface. Repeat five or six times.
iii. Repeat, leaving out one reactant at a time. Record your observations.
iv. Repeat the experiment, increasing the vitamin C amount to 10 mL.
v. Repeat the original experiment, increasing the methylene blue amount to 8 drops.
vi. Repeat the original experiment, decreasing the methylene blue amount to 2 drops.
vii. Repeat the original experiment, increasing the Cu(II) sulfate amount to 4 mL.
viii. Repeat the original experiment but add 5 mL baking soda ($NaHCO_3$).

The blue color is due to an oxidized form of methylene blue; the clear color is due to the reduced form. Answer the following questions.

a. Which substance is responsible for reduction of methylene blue?
b. Which substance is responsible for oxidation of methylene blue?
c. Does Cu(II) affect the rate of the reaction? What is the role of Cu(II) in this reaction?
d. Does the reaction rate depend on the concentration of methylene blue? On the concentration of Cu(II)?
e. What observation provides information on the role of oxygen in this reaction? Suggest an experiment to determine the importance of oxygen.

86. A catalytic converter in an automobile converts NO to N_2 and O_2 and completes the oxidation of fuels.

a. Write the reaction equation for oxidation of octane, C_8H_{18}, a component of gasoline.
b. It is observed that the catalytic conversion of incompletely burned fuel is independent of oxygen pressure. Suggest a slow step for the reaction.
c. Draw a reaction profile for conversion of NO to N_2 and O_2.
d. Suggest how the platinum catalyst acts in the conversion of NO.

Appendix A
Data Tables

Table A.1 Acid Ionization Constants, Weak Acids at 25 °C

Acid	Reaction	K_a
Acetic	$CH_3COOH(aq) + H_2O \rightarrow H_3O^+(aq) + CH_3COO^-(aq)$	1.76×10^{-5}
Arsenic	$H_3AsO_4(aq) + H_2O \rightarrow H_3O^+(aq) + H_2AsO_4^-(aq)$	$5.62 \times 10^{-3} = K_1$
	$H_2AsO_4^-(aq) + H_2O \rightarrow H_3O^+(aq) + HAsO_4^{2-}(aq)$	$1.7 \times 10^{-7} = K_2$
	$H_2AsO_4^{2-}(aq) + H_2O \rightarrow H_3O^+(aq) + AsO_4^{3-}(aq)$	$3.95 \times 10^{-12} = K_3$
Arsenous	$H_3AsO_3(aq) + H_2O \rightarrow H_3O^+(aq) + H_2AsO_3^-(aq)$	$6.0 \times 10^{-10} = K_1$
	$H_2AsO_3^-(aq) + H_2O \rightarrow H_3O^+(aq) + HAsO_3^{2-}(aq)$	$3.0 \times 10^{-14} = K_2$
Benzoic	$C_6H_5COOH(aq) + H_2O \rightarrow H_3O^+(aq) + C_6H_5COO^-(aq)$	6.46×10^{-5}
Boric	$B(OH)_3(aq) + H_2O \rightarrow H_3O^+(aq) + BO(OH)_2^-(aq)$	$7.3 \times 10^{-10} = K_1$
	$BO(OH)_2^-(aq) + H_2O \rightarrow H_3O^+(aq) + BO_2(OH)^{2-}(aq)$	$1.8 \times 10^{-13} = K_2$
	$BO_2(OH)^{2-}(aq) + H_2O \rightarrow H_3O^+(aq) + BO_3^{3-}(aq)$	$1.6 \times 10^{-14} = K_3$
Carbonic	$H_2CO_3(aq) + H_2O \rightarrow H_3O^+(aq) + HCO_3^-(aq)$	$4.3 \times 10^{-7} = K_1$
	$HCO_3^-(aq) + H_2O \rightarrow H_3O^+(aq) + CO_3^{2-}(aq)$	$5.6 \times 10^{-11} = K_2$
Citric	$C_3H_5O(COOH)_3(aq) + H_2O \rightarrow H_3O^+(aq) + C_4H_5O_3(COOH)_2^-(aq)$	$7.1 \times 10^{-4} = K_1$
	$C_4H_5O_3(COOH)_2^-(aq) + H_2O \rightarrow H_3O^+(aq) + C_5H_5O_5COOH^{2-}(aq)$	$1.68 \times 10^{-5} = K_2$
	$C_5H_5O_5(COOH)^{2-}(aq) + H_2O \rightarrow H_3O^+(aq) + C_6H_5O_7^{3-}(aq)$	$4.0 \times 10^{-7} = K_3$
Chromic	$H_2CrO_4(aq) + H_2O \rightarrow H_3O^+(aq) + HCrO_4^-(aq)$	$1.8 \times 10^{-1} = K_1$
	$HCrO_4^-(aq) + H_2O \rightarrow H_3O^+(aq) + CrO_4^{2-}(aq)$	$3.2 \times 10^{-7} = K_2$
Cyanic	$HOCN(aq) + H_2O \rightarrow H_3O^+(aq) + OCN^-(aq)$	3.5×10^{-4}
Formic	$HCOOH(aq) + H_2O \rightarrow H_3O^+(aq) + HCOO^-(aq)$	1.77×10^{-4}
Hydrazoic	$HN_3(aq) + H_2O \rightarrow H_3O^+(aq) + N_3^-(aq)$	1.9×10^{-5}
Hydrocyanic	$HCN(aq) + H_2O \rightarrow H_3O^+(aq) + CN^-(aq)$	4.93×10^{-10}
Hydrofluoric	$HF(aq) + H_2O \rightarrow H_3O^+(aq) + F^-(aq)$	3.53×10^{-4}
Hydrogen peroxide	$H_2O_2(aq) + H_2O \rightarrow H_3O^+(aq) + HO_2^-(aq)$	2.4×10^{-12}
Hydrosulfuric	$H_2S(aq) + H_2O \rightarrow H_3O^+(aq) + HS^-(aq)$	$9.1 \times 10^{-8} = K_1$
	$HS^-(aq) + H_2O \rightarrow H_3O^+(aq) + S^{2-}(aq)$	$1.1 \times 10^{-12} = K_2$
Hypobromous	$HOBr(aq) + H_2O \rightarrow H_3O^+(aq) + OBr^-(aq)$	2.06×10^{-9}
Hypochlorous	$HOCl(aq) + H_2O \rightarrow H_3O^+(aq) + OCl^-(aq)$	2.95×10^{-8}
Hypoiodous	$HOI(aq) + H_2O \rightarrow H_3O^+(aq) + OI^-(aq)$	2.3×10^{-11}
Nitrous	$HNO_2(aq) + H_2O \rightarrow H_3O^+(aq) + NO_2^-(aq)$	4.5×10^{-4}
Oxalic	$(COOH)_2(aq) + H_2O \rightarrow H_3O^+(aq) + COOCOOH^-(aq)$	$5.9 \times 10^{-2} = K_1$
	$COOCOOH^-(aq) + H_2O \rightarrow H_3O^+(aq) + (COO)_2^{2-}(aq)$	$6.4 \times 10^{-5} = K_2$
Phenol	$C_6H_5OH(aq) + H_2O \rightarrow H_3O^+(aq) + C_6H_5O^-(aq)$	1.28×10^{-10}
Phosphoric	$H_3PO_4(aq) + H_2O \rightarrow H_3O^+(aq) + H_2PO_4^-(aq)$	$7.52 \times 10^{-3} = K_1$
	$H_2PO_4^-(aq) + H_2O \rightarrow H_3O^+(aq) + HPO_4^{2-}(aq)$	$6.23 \times 10^{-8} = K_2$
	$HPO_4^{2-}(aq) + H_2O \rightarrow H_3O^+(aq) + PO_4^{3-}(aq)$	$2.2 \times 10^{-13} = K_3$
Phosphorous	$H_3PO_3(aq) + H_2O \rightarrow H_3O^+(aq) + H_2PO_3^-(aq)$	$1.0 \times 10^{-2} = K_1$
	$H_2PO_3^-(aq) + H_2O \rightarrow H_3O^+(aq) + HPO_3^{2-}(aq)$	$2.6 \times 10^{-7} = K_2$
Propanoic	$C_2H_5COOH(aq) + H_2O \rightarrow H_3O^+(aq) + C_2H_5COO^-(aq)$	1.34×10^{-5}
Selenic	$H_2SeO_4(aq) + H_2O \rightarrow H_3O^+(aq) + HSeO_4^-(aq)$	Very large $= K_1$
	$HSeO_4^-(aq) + H_2O \rightarrow H_3O^+(aq) + SeO_4^{2-}(aq)$	$1.2 \times 10^{-2} = K_2$

(Continued)

Table A.1 Acid Ionization Constants, Weak Acids at 25 °C *(Continued)*

Acid	Reaction	K_a
Selenous	$H_2SeO_3(aq) + H_2O \rightarrow H_3O^+(aq) + HSeO_3^-(aq)$	$3.5 \times 10^{-2} = K_1$
	$HSeO_3^-(aq) + H_2O \rightarrow H_3O^+(aq) + SeO_3^{2-}(aq)$	$4.9 \times 10^{-8} = K_2$
Sulfuric	$H_2SO_4(aq) + H_2O \rightarrow H_3O^+(aq) + HSO_4^-(aq)$	Very large $= K_1$
	$HSO_4^-(aq) + H_2O \rightarrow H_3O^+(aq) + SO_4^{2-}(aq)$	$1.2 \times 10^{-2} = K_2$
Sulfurous	$H_2SO_3(aq) + H_2O \rightarrow H_3O^+(aq) + HSO_3^-(aq)$	$1.54 \times 10^{-2} = K_1$
	$HSO_3^-(aq) + H_2O \rightarrow H_3O^+(aq) + SO_3^{2-}(aq)$	$1.02 \times 10^{-7} = K_2$
Tellourous	$H_2TeO_3(aq) + H_2O \rightarrow H_3O^+(aq) + HTeO_3^-(aq)$	$3.3 \times 10^{-3} = K_1$
	$HTeO_3^-(aq) + H_2O \rightarrow H_3O^+(aq) + TeO_3^{2-}(aq)$	$2 \times 10^{-8} = K_2$

Table A.2 K_a of Metal Ions

Ion	K_a	Ion	K_a
$Fe^{3+}(aq)$	6×10^{-3}	$Cu^{2+}(aq)$	3×10^{-8}
$Sn^{2+}(aq)$	4×10^{-4}	$Pb^{2+}(aq)$	3×10^{-8}
$Cr^{3+}(aq)$	1×10^{-4}	$Zn^{2+}(aq)$	1×10^{-9}
$Al^{3+}(aq)$	1×10^{-5}	$Co^{2+}(aq)$	2×10^{-10}
$Be^{2+}(aq)$	4×10^{-6}	$Ni^{2+}(aq)$	1×10^{-10}

Table A.3 Base Ionization Constants at 25 °C

Base	Reaction	K_b
Ammonia	$NH_3(aq) + H_2O \rightarrow NH_4^+(aq) + OH^-(aq)$	1.79×10^{-5}
Aniline	$C_6H_5NH(aq)_2 + H_2O \rightarrow C_6H_5NH_4^+(aq) + OH^-(aq)$	4.27×10^{-10}
Diethylamine	$(C_2H_5)_2NH(aq) + H_2O \rightarrow (C_2H_5)_2NH_2^+(aq) + OH^-(aq)$	3.1×10^{-4}
Dimethylamine	$(CH_3)_2NH(aq) + H_2O \rightarrow (CH_3)_2NH_2^+(aq) + OH^-(aq)$	5.4×10^{-4}
Ethylenediamine	$H_2N(CH_2)NH_2(aq) + H_2O \rightarrow H_2N(CH_2)_2NH_3^+(aq) + OH^-(aq)$	$5.15 \times 10^{-4} = K_1$
	$H_2N(CH_2)_2NH_3^+(aq) + H_2O \rightarrow H_3N(CH_2)_2NH_3^{2+}(aq) + OH^-(aq)$	$3.66 \times 10^{-7} = K_2$
Hydrazine	$N_2H_4(aq) + H_2O \rightarrow N_2H_5^+(aq) + OH^-(aq)$	$1.7 \times 10^{-6} = K_1$
	$N_2H_5^+(aq) + H_2O \rightarrow N_2H_6^{2+}(aq) + OH^-(aq)$	$8.9 \times 10^{-16} = K_2$
Hydroxylamine	$NH_2OH(aq) + H_2O \rightarrow NH_3OH^+(aq) + OH^-(aq)$	1.07×10^{-8}
Methylamine	$CH_3NH_2(aq) + H_2O \rightarrow CH_3NH_3^+(aq) + OH^-(aq)$	4.5×10^{-4}
Pyridine	$C_5H_5N(aq) + H_2O \rightarrow C_5H_5NH^+(aq) + OH^-(aq)$	1.77×10^{-9}
Trimethylamine	$(CH_3)_3N(aq) + H_2O \rightarrow (CH_3)_3NH^+(aq) + OH^-(aq)$	6.45×10^{-5}
Triethylamine	$(C_2H_5)_3N(aq) + H_2O \rightarrow (C_2H_5)_3NH^+(aq) + OH^-(aq)$	1.02×10^{-3}

Table A.4 Water Ionization at Various Temperatures

Water Reaction	$2H_2O \rightarrow H_3O^+(aq) + OH^-(aq)$
Temperature (°C)	K_w
0	1.13×10^{-15}
10	2.92×10^{-15}
25	1.00×10^{-14}
37 (body temperature)	2.38×10^{-14}
45	4.02×10^{-14}
60	9.6×10^{-14}

Table A.5 Complex Ion Formation Constants at 25 °C

Complex	Formation Constant	Reaction
$[AgBr_2]^-$	1.3×10^7	$Ag^+ + 2Br^- \rightarrow [AgBr_2]^-$
$[AgCl_2]^-$	2.5×10^5	$Ag^+ + 2Cl^- \rightarrow [AgCl_2]^-$
$[Ag(CN)_2]^-$	5.6×10^{18}	$Ag^+ + 2CN^- \rightarrow [Ag(CN)_2]^-$
$[AgEDTA]^{3-}$	2.1×10^7	$Ag^+ + EDTA^{4-} \rightarrow [AgEDTA]^{3-}$
$[Ag(en)]^+$	1.0×10^4	$Ag^+ + en \rightarrow [Ag(en)]^+$
$[Ag(NH_3)_2]^+$	1.6×10^7	$Ag^+ + 2NH_3 \rightarrow [Ag(NH_3)_2]^+$
$[Ag(SO_3)_2]^{3-}$	3.3×10^7	$Ag^+ + 2SO_3^{2-} \rightarrow [Ag(SO_3)_2]^{3-}$
$[Ag(S_2O_3)_2]^{3-}$	1.6×10^{13}	$Ag^+ + 2S_2O_3^{2-} \rightarrow [Ag(S_2O_3)_2]^{3-}$
$[AlEDTA]^-$	1.3×10^{16}	$Al^{3+} + EDTA^{4-} \rightarrow [AlEDTA]^-$
$[AlF_6]^{3-}$	5.0×10^{23}	$Al^{+3} + 6F^- \rightarrow [AlF_6]^{3-}$
$[Al(OH)_4]^-$	7.7×10^{33}	$Al^{+3} + 4OH^- \rightarrow [Al(OH)_4]^-$
$[Au(CN)_2]^-$	2.0×10^{38}	$Au^{+3} + 2CN^- \rightarrow [Au(CN)_2]^-$
$[CaEDTA]^{2-}$	5.0×10^{10}	$Ca^{2+} + EDTA^{4-} \rightarrow [CaEDTA]^{2-}$
$[Cd(CN)_4]^{2-}$	1.3×10^{17}	$Cd^{+2} + 4CN^- \rightarrow [Cd(CN)_4]^{2-}$
$[CdCl_4]^{2-}$	1.0×10^4	$Cd^{+2} + 4Cl^- \rightarrow [Cd(Cl)_4]^{2-}$
$[Cd(NH_3)_4]^{2+}$	1.0×10^7	$Cd^{+2} + 4NH_3 \rightarrow [Cd(NH_3)_4]^{2+}$
$[Co(CN)_6]^{4-}$	1.3×10^{19}	$Co^{2+} + 6CN^- \rightarrow [Co(CN)_6]^{4-}$
$[Co(en)_3]^{2+}$	6.7×10^{13}	$Co^{+2} + 3en \rightarrow [Co(en)_3]^{+2}$
$[Co(en)_3]^{3+}$	5×10^{48}	$Co^{+3} + 3en \rightarrow [Co(en)_3]^{+3}$
$[Co(NH_3)_6]^{2+}$	7.7×10^4	$Co^{+2} + 6NH_3 \rightarrow [Co(NH_3)_6]^{2+}$
$[Co(NH_3)_6]^{3+}$	5×10^{33}	$Co^{+3} + 6NH_3 \rightarrow [Co(NH_3)_6]^{+3}$
Co_2EDTA	2×10^{16}	$2Co^{2+} + EDTA^{4-} \rightarrow Co_2EDTA$
$[Cu(CN)_2]^-$	1.0×10^{17}	$Cu^{+1} + 2CN^- \rightarrow [Cu(CN)_2]^-$
$[Cu(C_2O_4)_2]^{2-}$	2.1×10^{10}	$Cu^{2+} + 2C_2O_4^{2-} \rightarrow [Cu(C_2O_4)_2]^{2-}$
$[CuCl_2]^-$	1.0×10^4	$Cu^{+1} + 2Cl^- \rightarrow [Cu(Cl)_2]^-$
$[CuEDTA]^{2-}$	6.3×10^{18}	$Cu^{2+} + EDTA^{4-} \rightarrow [CuEDTA]^{2-}$
$[Cu(NH_3)_2]^+$	7.1×10^{10}	$Cu^{+1} + 2NH_3 \rightarrow [Cu(NH_3)_2]^+$
$[Cu(en)_2]^{2+}$	4.4×10^{18}	$Cu^{+2} + 2en \rightarrow [Cu(en)_2]^{2+}$
$[Cu(NH_3)_4]^{2+}$	1.1×10^{12}	$Cu^{+2} + 4NH_3 \rightarrow [Cu(NH_3)_4]^{2+}$
$[Fe(CN)_6]^{4-}$	7.7×10^{36}	$Fe^{+2} + 6CN^- \rightarrow [Fe(CN)_6]^{4-}$
$[Fe(CN)_6]^{3-}$	7.7×10^{43}	$Fe^{+3} + 6CN^- \rightarrow [Fe(CN)_6]^{3-}$
$[Fe(C_2O_4)_3]^{3-}$	1.7×10^{20}	$Fe^{3+} + 3C_2O_4^{2-} \rightarrow [Fe(C_2O_4)_3]^{3-}$
$[FeEDTA]^{2-}$	2.1×10^{14}	$Fe^{2+} + EDTA^{4-} \rightarrow [FeEDTA]^{2-}$
$[FeEDTA]^-$	1.3×10^{25}	$Fe^{3+} + EDTA^{4-} \rightarrow [FeEDTA]^-$
$[FeF_6]^{3-}$	2×10^{15}	$Fe^{3+} + 6F^- \rightarrow [FeF_6]^{3-}$
$[FeSCN]^{2+}$	1×10^3	$Fe^{3+} + SCN^- \rightarrow [FeSCN]^{2+}$
$[HgBr_4]^{2-}$	4.3×10^{21}	$Hg^{2+} + 4Br^- \rightarrow [HgBr_4]^{2-}$
$[Hg(CN)_4]^{2-}$	2.5×10^{41}	$Hg^{2+} + 4CN^- \rightarrow [Hg(CN)_4]^{2-}$
$[HgCl_4]^{2-}$	1.2×10^{16}	$Hg^{2+} + 4Cl^- \rightarrow [Hg(Cl)_4]^{2-}$
$[HgEDTA]^{2-}$	6.3×10^{21}	$Hg^{2+} + EDTA^{4-} \rightarrow [HgEDTA]^2$
$[HgI_4]^{2-}$	1.9×10^{20}	$Hg^{2+} + 4I^- \rightarrow [HgI_4]^{2-}$
$[MgEDTA]^{2-}$	4.9×10^8	$Mg^{2+} + EDTA^{4-} \rightarrow [MgEDTA]^{2-}$
$[MnEDTA]^{2-}$	6.2×10^{13}	$Mn^{2+} + EDTA^{4-} \rightarrow [MnEDTA]^{2-}$
$[Ni(CN)_4]^{2-}$	1.0×10^{31}	$Ni^{+2} + 4CN^- \rightarrow [Ni(CN)_4]^{2-}$
$[NiEDTA]^{2-}$	4.2×10^{18}	$Ni^{2+} + EDTA^{4-} \rightarrow [NiEDTA]^{2-}$
$[Ni(en)_3]^{2+}$	2×10^{18}	$Ni^{+2} + 3en \rightarrow [Ni(en)_3]^{2+}$

(Continued)

Table A.5 Complex Ion Formation Constants at 25 °C *(Continued)*

Complex	Formation Constant	Reaction
$[Ni(NH_3)_6]^{2+}$	4.07×10^8	$Ni^{+2} + 6NH_3 \rightarrow [Ni(NH_3)_6]^{2+}$
$[Pb(CH_3COOH)_4]^{2-}$	1.2×10^2	$Pb^{2+} + 4CH_3COO^- \rightarrow [Pb(CH_3COOH)_4]^{2-}$
$[Pb(CN)_4]^{2-}$	2×10^{10}	$Pb^{2+} + 4CN^- \rightarrow [Pb(CN)_4]^{2-}$
$[PbCl_3]^-$	2.4×10^1	$Pb^{2+} + 3Cl^- \rightarrow [PbCl_3]^-$
$[PbEDTA]^{2-}$	1.1×10^{18}	$Pb^{2+} + EDTA^{4-} \rightarrow [PbEDTA]^{2-}$
$[PbI_3]^-$	2.8×10^5	$Pb^{2+} + 3I^- \rightarrow [PbI_3]^-$
$[SnCl_4]^{2-}$	3.1×10^1	$Sn^{2+} + 4Cl^- \rightarrow [SnCl_4]^{2-}$
$[SnF_6]^{2-}$	1×10^{18}	$Sn^{4+} + 6F^- \rightarrow [SnF_6]^{2-}$
$[Zn(CN)_4]^{2-}$	7.7×10^{16}	$Zn^{2+} + 4CN^- \rightarrow [Zn(CN)_4]^{2-}$
$[Zn(OH)_4]^{2-}$	2.8×10^{16}	$Zn^{2+} + 4OH^- \rightarrow [Zn(OH)_4]^{2-}$
$[Zn(NH_3)_4]^{2+}$	2.9×10^{10}	$Zn^{2+} + 4NH_3 \rightarrow [Zn(NH_3)_4]^{2+}$

Note: Metal ions in aqueous solution are always coordinated to water. The water is not shown in the reactions given here.

Table A.6 Gas-Phase Equilibrium Constants

	T	K_c	K_p
$N_2 + 3H_2 \rightarrow 2NH_3$	300 °C	9.60	4.34×10^{-3}
$2SO_3 \rightarrow 2SO_2 + O_2$	1000 K	4.08×10^{-3}	0.355
$CO + Cl_2 \rightarrow COCl_2$	100 °C	4.57×10^9	
$N_2 + O_2 \rightarrow 2NO$	25 °C	1×10^{-30}	1×10^{-30}
$H_2 + I_2(g) \rightarrow 2HI$	298 K	794	
$H_2 + I_2(g) \rightarrow 2HI$	700 K	54	
$N_2O_4 \rightarrow 2NO_2$	100 °C	0.212	
$2NO_2 \rightarrow N_2O_4$	100 °C	4.72	
$2NO_2 \rightarrow N_2O_4$	25 °C	164.8	

Table A.7 Solubility Products of Inorganic Compounds at 25 °C

Compound	K_{sp}	Compound	K_{sp}	Compound	K_{sp}
Ag_3AsO_4	1.1×10^{-20}	$AlPO_4$	1.3×10^{-20}	$CdCO_3$	2.5×10^{-14}
$AgBr$	3.3×10^{-13}	$AuBr$	5.0×10^{-17}	$Cd(OH)_2$	1.2×10^{-14}
$AgCl$	1.8×10^{-10}	$AuBr_3$	4.0×10^{-36}	CdS	3.6×10^{-29}
$AgCN$	1.2×10^{-16}	$AuCl$	2.0×10^{-13}	Co_2S_3	2.6×10^{-124}
Ag_2CO_3	8.1×10^{-12}	$AuCl_3$	3.2×10^{-25}	$Cu_3(AsO_4)_2$	7.6×10^{-36}
Ag_2CrO_4	9.0×10^{-12}	AuI	1.6×10^{-23}	$CuBr$	5.3×10^{-9}
$Ag_4[Fe(CN)_6]$	1.6×10^{-41}	AuI_3	1.0×10^{-46}	$CuCl$	1.9×10^{-7}
Ag_2O	2.0×10^{-8}	$Au(OH)_3$	1.0×10^{-53}	$CuCN$	3.2×10^{-20}
AgI	1.5×10^{-16}	$BaCO_3$	8.1×10^{-9}	$CuCO_3$	2.5×10^{-10}
$AgIO_3$	3×10^{-8}	$BaSO_4$	1.08×10^{-10}	$Cu_2[Fe(CN)_6]$	1.3×10^{-16}
Ag_3PO_4	1.3×10^{-20}	$CaCO_3$	1.0×10^{-8}	CuI	5.1×10^{-12}
Ag_2S	1.0×10^{-49}	$Ca(OH)_2$	5.02×10^{-6}	Cu_2O	1.0×10^{-14}
$AgSCN$	1.0×10^{-12}	$Ca_3(PO_4)_2$	1.0×10^{-25}	$Cu(OH)_2$	1.6×10^{-19}
Ag_2SO_3	1.5×10^{-14}	$CaSO_4$	2.45×10^{-5}	CuS	8.7×10^{-36}
Ag_2SO_4	1.7×10^{-5}	$Cd_3(AsO_4)_2$	2.2×10^{-32}	Cu_2S	1.6×10^{-48}
$Al(OH)_3$	3.7×10^{-32}	$Cd(CN)_2$	1.0×10^{-8}	$CuSCN$	1.6×10^{-11}

(Continued)

Table A.7 Solubility Products of Inorganic Compounds at 25 °C   *(Continued)*

Compound	K_{sp}	Compound	K_{sp}	Compound	K_{sp}
$FeCO_3$	3.5×10^{-11}	$MgNH_4PO_4$	2.5×10^{-12}	$PbSeO_4$	1.5×10^{-7}
$Fe_4[Fe(CN)_6]_3$	3.0×10^{-41}	$Mn_3(AsO_4)_2$	1.9×10^{-11}	$PbSO_4$	1.8×10^{-8}
$Fe(OH)_2$	7.9×10^{-15}	$MnCO_3$	1.8×10^{-11}	SnI_2	1.0×10^{-4}
$Fe(OH)_3$	6.3×10^{-38}	$Mn(OH)_2$	4.6×10^{-14}	$Sn(OH)_2$	2.0×10^{-26}
FeS	4.9×10^{-18}	$Mn(OH)_3$	$\sim 1.0 \times 10^{-36}$	$Sn(OH)_4$	1.0×10^{-57}
Fe_2S_3	1.4×10^{-88}	MnS	5.1×10^{-15}	SnS	1.0×10^{-28}
Hg_2Br_2	1.3×10^{-22}	$Ni_3(AsO_4)_2$	1.9×10^{-26}	SnS_2	1.0×10^{-70}
Hg_2Cl_2	1.1×10^{-18}	$NiCO_3$	6.6×10^{-9}	$Sr_3(AsO_4)_2$	1.3×10^{-18}
$Hg(CN)_2$	3.0×10^{-23}	$Ni(CN)_2$	3.0×10^{-23}	$SrCO_3$	9.4×10^{-10}
Hg_2CO_3	8.9×10^{-17}	$Ni(OH)_2$	2.8×10^{-16}	$SrC_2O_4 \cdot 2H_2O$	5.6×10^{-8}
Hg_2CrO_4	5.0×10^{-9}	$NiS(\alpha)$	3.0×10^{-21}	$SrCrO_4$	3.6×10^{-5}
HgI_2	4.0×10^{-29}	$NiS(\beta)$	1.0×10^{-26}	$Sr(OH)_2 \cdot 8H_2O$	3.2×10^{-4}
Hg_2I_2	4.5×10^{-29}	$NiS(\gamma)$	2.0×10^{-28}	$Sr_3(PO_4)_2$	1.0×10^{-31}
$Hg(OH)_2$	2.5×10^{-26}	$Pb_3(AsO_4)_2$	4.1×10^{-36}	$SrSO_3$	4.0×10^{-8}
$Hg_2O \cdot H_2O$	1.6×10^{-23}	$PbBr_2$	6.3×10^{-6}	$SrSO_4$	2.8×10^{-7}
HgS	3.0×10^{-53}	$PbCl_2$	1.7×10^{-5}	$Zn_3(AsO_4)_2$	1.1×10^{-27}
Hg_2S	5.8×10^{-44}	$PbCO_3$	1.5×10^{-13}	$ZnCO_3$	1.5×10^{-11}
Hg_2SO_4	6.8×10^{-7}	$PbCrO_4$	1.8×10^{-14}	$Zn(CN)_2$	8.0×10^{-12}
$Mg_3(AsO_4)_2$	2.1×10^{-20}	PbF_2	3.7×10^{-8}	$Zn_2[Fe(CN)_6]$	4.1×10^{-16}
$MgCO_3 \cdot 3H_2O$	4.0×10^{-5}	PbI_2	8.7×10^{-9}	$Zn(OH)_2$	4.5×10^{-17}
MgC_2O_4	8.6×10^{-5}	$Pb(OH)_2$	2.8×10^{-16}	$Zn_3(PO_4)_2$	9.1×10^{-33}
MgF_2	6.4×10^{-9}	$Pb_3(PO_4)_2$	3.0×10^{-44}	ZnS	1.1×10^{-21}
$Mg(OH)_2$	1.5×10^{-11}	PbS	8.4×10^{-28}		

Table A.8 Ionization Energies of the Elements

Atomic Number		Valence Configuration	I	II	III	IV	V	VI	VII	VIII
1	H	$1s^1$	13.595							
2	He	$1s^2$	24.580	54.40						
3	Li	$2s^1$	5.390	75.6193	122.420					
4	Be	$2s^2$	9.320	18.206	153.850	217.657				
5	B	$2s^2 2p^1$	8.296	25.149	37.920	259.298	340.127			
6	C	$2s^2 2p^2$	11.264	24.376	47.864	64.476	391.99	489.84		
7	N	$2s^2 2p^3$	14.54	29.605	47.426	77.450	97.863	551.925	666.83	
8	O	$2s^2 2p^4$	13.614	35.146	54.934	77.394	113.873	138.080	739.114	871.12
9	F	$2s^2 2p^5$	17.42	34.98	62.646	87.23	114.214	157.117	185.139	953.6
10	Ne	$2s^2 2p^6$	21.559	41.07	64	97.16	126.4	157.91		
11	Na	$3s^1$	5.138	47.29	71.65	98.88	138.6	172.36	208.44	264.155
12	Mg	$3s^2$	7.644	15.03	80.12	109.29	141.23	186.86	225.31	265.97
13	Al	$3s^2 3p^1$	5.984	18.823	28.44	119.96	153.77	190.42	241.93	285.13
14	Si	$3s^2 3p^2$	8.149	16.34	33.46	45.13	166.73	205.11	246.41	303.87
15	P	$3s^2 3p^3$	11.0	19.65	30.156	51.354	65.007	220.414	263.31	309.26
16	S	$3s^2 3p^4$	10.357	23.4	35.0	47.29	72.5	88.029	380.99	328.80
17	Cl	$3s^2 3p^5$	13.01	23.80	39.90	53.5	67.80	96.7	114.27	348.3
18	Ar	$3s^2 3p^6$	15.755	27.62	40.90	59.79	75.0	91.3	124.0	143.46

(Continued)

Table A.8 Ionization Energies of the Elements *(Continued)*

Atomic Number		Valence Configuration	I	II	III	IV	V	VI	VII	VIII
19	K	$4s$	4.339	31.81	46	60.90		99.7	118	155
20	Ca	$4s^2$	6.111	11.87	51.21	67	84.39		128	147
21	Sc	$3d^14s^2$	6.56	12.98	24.75	73.9	92	111.1		159
22	Ti	$3d^24s^2$	6.83	13.57	28.14	43.24	99.8	120	140.8	
23	V	$3d^34s^2$	6.74	14.65	29.7	48	65.2	128.9	151	173.7
24	Cr	$3d^54s^1$	6.76	16.49	ca31	ca50	ca73			
25	Mn	$3d^54s^2$	7.43	15.64	ca32	ca52	ca76			
26	Fe	$3d^64s^2$	7.90	16.18						
27	Co	$3d^74s^2$	7.86	17.05						
28	Ni	$3d^84s^2$	7.63	18.15						
29	Cu	$3d^{10}4s^1$	7.72	20.29	29.5					
30	Zn	$3d^{10}4s^2$	9.39	17.96	40.0					
31	Ga	$3d^{10}4s^23p^1$	6.00	20.51	30.6	63.8				
32	Ge	$3d^{10}4s^23p^2$	7.88	15.93	34.07	45.5	93.0			
33	As	$3d^{10}4s^23p^3$	9.81	20.2	28.0	49.9	62.5			
34	Se	$3d^{10}4s^23p^4$	9.75	21.5	33.9	52.7	72.8	81.4		
35	Br	$3d^{10}4s^23p^5$	11.84	21.6	25.7	ca50				
36	Kr	$3d^{10}4s^23p^6$	14.00	24.56	36.8	ca68				
37	Rb	$5s^1$	4.18	27.5	ca47	ca80				
38	Sr	$5s^2$	5.69	11.03						
39	Y	$4d^15s^2$	6.38	12.23	20.4					
40	Zr	$4d^25s^2$	6.84	12.92	24.00	33.8				
41	Nb	$4d^45s^1$	6.88	13.90	24.2					
42	Mo	$4d^55s^1$	7.13	15.72						
43	Tc	$4d^55s^2$	7.23	14.87						
44	Ru	$4d^75s^1$	7.36	16.60						
45	Rh	$4d^85s^1$	7.46	15.92						
46	Pd	$4d^{10}5s^0$	8.33	19.42						
47	Ag	$4d^{10}5s^1$	7.57	21.48	35.9					
48	Cd	$4d^{10}5s^2$	8.99	16.90	38					
49	In	$4d^{10}5s^24p^1$	5.78	18.83	27.9	57.8				
50	Sn	$4d^{10}5s^24p^2$	7.33	14.63	30.5	39.4	80.7			
51	Sb	$4d^{10}5s^24p^3$	8.64	19	24.7	44.0	55.5			
52	Te	$4d^{10}5s^24p^4$	9.01	21.5	30.5	37.7	60.0	ca72		
53	I	$4d^{10}5s^24p^5$	10.44	19.0						
54	Xe	$4d^{10}5s^24p^6$	12.13	21.21	32.0	ca46	ca76			
55	Cs	$6s^1$	3.89	25.1	ca35	ca51	ca58			
56	Ba	$6s^2$	5.21	10.00						
57	La	$5d^16s^2$	5.61	11.43	20.4					
72	Hf	$5d^24f^{14}6s^2$	5.5	14.9						
73	Ta	$5d^34f^{14}6s^2$	7.7							
74	W	$5d^44f^{14}6s^2$	7.98							
75	Re	$5d^54f^{14}6s^2$	7.87							
76	Os	$5d^64f^{14}6s^2$	8.7							
77	Ir	$5d^74f^{14}6s^2$	9.2							
78	Pt	$5d^94f^{14}6s^1$	9.0	18.56	34.5	ca72	ca82			
79	Au	$5d^{10}4f^{14}6s^1$	9.22	20.5	29.7	50.5				
80	Hg	$5d^{10}4f^{14}6s^2$	10.43	18.75	31.9	42.11	69.4			

(Continued)

Table A.8 Ionization Energies of the Elements *(Continued)*

Atomic Number		Valence Configuration	I	II	III	IV	V	VI	VII	VIII
81	Tl	$5d^{10}4f^{14}6s^25p^1$	6.11	20.42	25.42	45.1	55.7			
82	Pb	$5d^{10}4f^{14}6s^25p^2$	7.42	15.03						
83	Bi	$5d^{10}4f^{14}6s^25p^3$	7.29	19.3						
84	Po	$5d^{10}4f^{14}6s^25p^4$	8.43							
85	At	$5d^{10}4f^{14}6s^25p^5$								
86	Rn	$5d^{10}4f^{14}6s^25p^6$	10.74							
87	Fr	$7s^1$								
88	Ra	$7s^2$	5.28	10.14						

Table A.9 Standard Reduction Potentials

Reduction Half-Reaction	$E°$ (V)	Reduction Half-Reaction	$E°$ (V)
$F_2 + 2e^- \rightarrow 2F^-$	+2.87	$Fe^{3+} + 3e^- \rightarrow Fe$	−0.036
$NiO_2 + 4H^+ + 2e^- \rightarrow Ni^{2+} + 2H_2O$	+1.93	$Pb^{2+} + 2e^- \rightarrow Pb$	−0.1263
$PbO_2 + SO_4^{2-} + 4H^+ + 2e^- \rightarrow PbSO_4 + 2H_2O$	+1.685	$Sn^{2+} + 2e^- \rightarrow Sn$	−0.1364
$2HOCl + 2H^+ + 2e^- \rightarrow Cl_2 + 2H_2O$	+1.63	$AgI + e^- \rightarrow Ag + I^-$	−0.1519
$2HOBr + 2H^+ + 2e^- \rightarrow Br_2 + 2H_2O$	+1.59	$Cu(OH)_2 + 2e^- \rightarrow Cu + 2OH^-$	−0.224
$Mn_2O_3 + 6H^+ + 2e^- \rightarrow 2Mn^{2+} + 3H_2O$	+1.485	$Ni^{2+} + 2e^- \rightarrow Ni$	−0.23
$2HOI + 2H^+ + 2e^- \rightarrow I_2 + 2H_2O$	+1.45	$V_2O_5 + 10H^+ + 10e^- \rightarrow 2V + 5H_2O$	−0.242
$Au^{3+} + 3e^- \rightarrow Au$	+1.42	$PbSO_4 + 2e^- \rightarrow Pb + SO_4^{2-}$	−0.356
$Cl_2(g) + 2e^- \rightarrow 2Cl^-$	+1.3583	$Cd^{2+} + 2e^- \rightarrow Cd$	−0.4026
$O_2 + 4H^+ + 4e^- \rightarrow 2H_2O$	+1.229	$Fe^{2+} + 2e^- \rightarrow Fe$	−0.409
$Pt^{2+} + 2e^- \rightarrow Pt$	+1.2	$PbO + H_2O + 2e^- \rightarrow Pb + 2OH^-$	−0.576
$Br_2(aq) + 2e^- \rightarrow 2Br^-$	+1.087	$Ni(OH)_2 + 2e^- \rightarrow Ni + 2OH^-$	−0.66
$Br_2(l) + 2e^- \rightarrow 2Br^-$	+1.065	$Ag_2S + 2e^- \rightarrow 2Ag + S^{2-}$	−0.7051
$Hg^{2+} + 2e^- \rightarrow Hg$	+0.851	$2MnO_2(s) + 2e^- + H_2O \rightarrow Mn_2O_3(s) + 2OH^-$	−0.73
$Ag^+ + e^- \rightarrow Ag$	+0.7996	$Cr^{3+} + 3e^- \rightarrow Cr$	−0.74
$Fe^{3+} + e^- \rightarrow Fe^{2+}$	+0.770	$Cd(OH)_2 + 2e^- \rightarrow Cd + 2OH^-$	−0.761
$I_2 + 2e^- \rightarrow 2I^-$	+0.535	$Zn^{2+} + 2e^- \rightarrow Zn$	−0.7628
$NiO_2 + 2H_2O + 2e^- \rightarrow Ni(OH)_2 + 2OH^-$	+0.49	$2H_2O + 2e^- \rightarrow H_2 + 2OH^-$	−0.8277
$O_2 + 2H_2O + 4e^- \rightarrow 4OH^-$	+0.401	$Mn^{2+} + 2e^- \rightarrow Mn$	−1.185
$Ag_2O + H_2O + 2e^- \rightarrow 2Ag + 2OH^-$	+0.342	$Cr(OH)_3 + 3e^- \rightarrow Cr + 3OH^-$	−1.3
$Cu^{2+} + 2e^- \rightarrow Cu$	+0.3402	$Mn(OH)_2 + 2e^- \rightarrow Mn + 2OH^-$	−1.47
$AgCl + e^- \rightarrow Ag + Cl^-$	+0.2223	$Cr_2O_3 + 3H_2O + 6e^- \rightarrow 2Cr + 6OH^-$	−1.48
$HgO + H_2O + 2e^- \rightarrow Hg + 2OH^-$	+0.0984	$Be^{2+} + 2e^- \rightarrow Be$	−1.70
$AgBr + e^- \rightarrow Ag + Br^-$	+0.0713	$Al^{3+} + 3e^- \rightarrow Al$	−1.706
$2H^+ + 2e^- \rightarrow H_2$	0.000	$Mg^{2+} + 2e^- \rightarrow Mg$	−2.375
		$Mg(OH)_2 + 2e^- \rightarrow Mg + 2OH^-$	−2.67
		$Na^+ + e^- \rightarrow Na$	−2.709
		$Ca^{2+} + 2e^- \rightarrow Ca$	−2.76
		$Sr^{2+} + 2e^- \rightarrow Sr$	−2.89
		$Ba^{2+} + 2e^- \rightarrow Ba$	−2.90
		$Cs^+ + e^- \rightarrow Cs$	−2.923
		$K^+ + e^- \rightarrow K$	−2.924
		$Rb^+ + e^- \rightarrow Rb$	−2.925
		$Ca(OH)_2 + 2e^- \rightarrow Ca + 2OH^-$	−3.02
		$Li^+ + e^- \rightarrow Li$	−3.045

Table A.10 Electron Affinity of the Elements

Symbol	Z	Electron Affinity (eV)	(kJ/mol)	Symbol	Z	Electron Affinity (eV)	(kJ/mol)	Symbol	Z	Electron Affinity (eV)	(kJ/mol)
H	1	0.754188	72.7678	Mn	25	~0	~0	In	49	0.30	29
He	2	~0	~0	Fe	26	0.151	14.6	Sn	50	1.112	107.3
Li	3	0.9180	88.57	Co	27	0.662	63.9	Sb	51	1.046	100.9
Be	4	~0	~0	Ni	28	1.156	111.5	Te	52	1.9708	190.15
B	5	0.277	26.7	Cu	29	1.235	119.2	I	53	3.05900	295.148
C	6	1.2629	121.85	Zn	30	~0	~0	Xe	54	~0	~0
N	7	~0	~0	Ga	31	0.30	29	Cs	55	0.47162	45.504
O	8	1.46110	140.974	Ge	32	1.23	119	Ba	56	0.15	14
F	9	3.4012	328.160	As	33	0.81	78	La	57	0.52	50
Ne	10	NA	NA	Se	34	2.02065	194.962	Hf	72	~0	~0
Na	11	0.548262	52.8991	Br	35	3.363	324.5	Ta	73	0.322	31.1
Mg	12	~0	~0	Kr	36	~0	~0	W	74	0.815	78.6
Al	13	0.441	42.5	Rb	37	0.48592	46.884	Re	75	0.150	14.5
Si	14	1.385	133.6	Sr	38	0.148	14.3	Os	76	1.10	106
P	15	0.7464	72.02	Y	39	0.30	29	Ir	77	1.564	150.9
S	16	2.077082	200.4073	Zr	40	0.426	41.1	Pt	78	2.128	205.3
Cl	17	3.61266	348.567	Nb	41	0.893	86.2	Au	79	2.3088	222.76
Ar	18	~0	~0	Mo	42	0.748	72.2	Hg	80	~0	~0
K	19	0.50147	48.384	Tc	43	0.55	53	Tl	81	0.21	20
Ca	20	0.02455	2.369	Ru	44	1.05	101	Pb	82	0.364	35.1
Sc	21	0.1880	18.14	Rh	45	1.137	109.7	Bi	83	0.946	91.3
Ti	22	0.079	7.6	Pd	46	0.562	54.2	Po	84	1.86	180
V	23	0.524	50.6	Ag	47	1.302	125.6	At	85	2.80	270
Cr	24	0.6660	64.26	Cd	48	~0	~0	Rn	86	~0	~0

Table A.11 Enthalpy of Formation for Selected Oxides

Oxide	$\Delta H_{formation}$ (kJ/mol)	Oxide	$\Delta H_{formation}$ (kJ/mol)	Oxide	$\Delta H_{formation}$ (kJ/mol)
Ag_2O	−30.58	Ga_2O_3	−1080	$NO(g)$	91.3
$Al_2O_3(\alpha)$	−1675.7	GeO_2	−536.8	$NO_2(g)$	33.2
Au_2O_3	80.7	$H_2O(l)$	−285.8	$N_2O_4(g)$	9.66
B_2O_3	−1260	HgO (red)	−90.8	$N_2O(g)$	81.55
BaO	−558.1	Hg_2O	−88.62	N_2O_5	15.1
BeO	−610	In_2O_3	−930.9	Na_2O	−415
CaO	−634.9	IrO_2	−167.8	NiO	−244
CdO	−254.6	K_2O	−361	PbO (red)	−219
$Cl_2O(g)$	76.1	Li_2O	−595.8	PbO_2	−276.6
$CO_2(g)$	−393.5	MgO	−601.6	PdO	−85.3
CoO	−239	MnO	−385	RaO	−523
Cr_2O_3	−1128	MnO_2	−520.9	Rb_2O	−330
Cs_2O	−317	Mn_2O_3	−959	RhO	−90.8
CuO	−157.3	Mn_3O_4	−1386.6	RuO_2	−220
$F_2O(g)$	23	MoO_2	−544	$SO_2(g)$	−296.9
Fe_2O_3	−824.2	MoO_3	−754.50	$SO_3(g)$	−395.18

(Continued)

Table A.11 Enthalpy of Formation for Selected Oxides   (Continued)

Oxide	$\Delta H_{formation}$ (kJ/mol)	Oxide	$\Delta H_{formation}$ (kJ/mol)	Oxide	$\Delta H_{formation}$ (kJ/mol)
SeO_2	−230.1	Ta_2O_5	−2091.6	V_2O_4	−1439
SiO_2 (quartz)	−910.94	TeO_2	−325.1	V_2O_5	−1561
SnO	−286	TiO_2 (rutile)	−912	ZnO	−350.5
SnO_2	−580.7	V_2O_2	−836.8	ZrO	−1080
SrO	−590.4	V_2O_3	−1213		

Table A.12 Band Gap in Selected Semiconductors

Material	Band Gap (eV)	Material	Band Gap (eV)	Material	Band Gap (eV)
Elemental		**II – VI**		**III – V**	
C	5.4	ZnS	3.7	AlN	5.9
SiC	2.3	ZnSe	2.7	AlP	2.5
Si	1.12	ZnTe	2.3	AlAs	2.2
Ge	0.67	CdS	2.42	AlSb	1.6
α-Sn	0.08	CdSe	1.8	GaN	3.3
		CdTe	1.5	GaP	2.8
		PbS	0.37	GaAs	1.35
		PbSe	0.26	GaSb	0.7
		PbTe	0.2	InN	2.4
				InP	1.3
				InAs	0.4
				InSb	0.2

Note: Oxide semiconductors are quite ionic, so the band gap transition is from an oxygen band to the metal band. Therefore these oxides do not follow the same trend and are not included here.

Table A.13 Band Gap in Oxide Semiconductors (eV)

Periodic table with band gap values (eV) for oxide semiconductors:

1	2	3	4	5	6	7	8	9	10	11	12	13	14	15	16	17	18
1 H																	2 He
3 Li	4 Be 10.4											5 B 7	6 C	7 N	8 O	9 F	10 Ne
11 Na	12 Mg 7.7											13 Al 9.5	14 Si 11	15 P	16 S	17 Cl	18 Ar
19 K	20 Ca 7.7	21 Sc (+3) 6.0	22 Ti (+4) 3.3	23 V (+5) 2.34	24 Cr (+3) 1.68	25 Mn (+2) 3.7	26 Fe (+3) 2.34	27 Co (+2) 0.47	28 Ni (+2) 3.7	29 Cu (+1) 2.02	30 Zn (+2) 3.35	31 Ga 4.54	32 Ge 5.56	33 As 4.	34 Se	35 Br	36 Kr
37 Rb	38 Sr 5.77	39 Y (+3) 5.6	40 Zr (+4) 4.99	41 Nb (+5) 3.48	42 Mo (+6) 2.8	43 Tc	44 Ru	45 Rh	46 Pd (+2) 1.5	47 Ag (+1) 1.2	48 Cd (+2) 2.3	49 In 2.6	50 Sn 2.7	51 Sb 3.31	52 Te 3.	53 I	54 Xe
55 Cs	56 Ba 5.13	57 La (+3) 1.05	72 Hf (+4) 5.55	73 Ta (+5) 4.6	74 W (+6) 2.8	75 Re (+6) 2.3	76 Os	77 Ir	78 Pt (+2) 0.2	79 Au	80 Hg (+2) 2.48	81 Tl 2	82 Pb 1.9	83 Bi 2.6	84 Po	85 At	86 Rn
87 Fr	88 Ra	89 Ac															

Table A.14 $\Delta H_{\text{hydration}}$ (kJ/mol), Radius (pm), and $1/r$ ($\times$ 100 pm^{-1}) for Selected Ions

Ion	ΔH	r	$1/r$	Ion	ΔH	r	$1/r$	Ion	ΔH	r	$1/r$
Li^+	-515	76	1.32	Be^{2+}	-2487	109	0.92	Cr^{2+}	-1904	77	1.30
Na^+	-405	102	0.98	Mg^{2+}	-1922	72	1.39	Mn^{2+}	-1841	67	1.49
K^+	-321	138	0.72	Ca^{2+}	-1592	100	1.00	Fe^{2+}	-1950	69	1.45
Rb^+	-296	152	0.66	Sr^{2+}	-1445	118	0.85	Co^{2+}	-1996	70	1.43
Cs^+	-263	167	0.60	Ba^{2+}	-1304	135	0.74	Ni^{2+}	-2105	69	1.45
								Cu^{2+}	-2100	73	1.37
								Zn^{2+}	-2050	74	1.35
				F^-	-506	133	0.75				
Ag^+	-375	115	0.87	Cl^-	-364	184	0.54	Al^{3+}	-4660	53	2.22
NH_4^+	-336	204	0.49	Br^-	-337	196	0.51	Fe^{3+}	-4430	60	1.67
H_3O^+	-335	203	0.49	I^-	-296	220	0.45	Cr^{3+}	-4368	61	1.64
				OH^-	-335	205	0.49	Mn^{3+}	-3374	61	1.64

Table A.15 Heat of Formation

Substance	ΔH_f° (kJ/mol)	ΔG_f° (kJ/mol)	S° (J/K·mol)	Substance	ΔH_f° (kJ/mol)	ΔG_f° (kJ/mol)	S° (J/K·mol)
Acetic acid(l)	-484.3	-389.9	159.8	$C_4H_{10}O(l)$, 2-butanol	-342.6		214.9
Al	0		28.3	$C_6H_{12}O_6(s)$, glucose	-1273.3		
Al_2O_3($crystal$)	-1675.7	-1582.3	50.9	$C_6H_{12}O_6(s)$, fructose	-1265.6		
$Br_2(l)$	0		152.2	$C_8H_{18}(l)$, octane	-250.1		
C($graphite$)	0		5.74	$C_{12}H_{22}O_{11}$, sucrose	-2226.1		
$CH_3(g)$, methyl radical	145.6			Ca	0		25.9
$CH_2O(g)$, formaldehyde	-108.6	-102.5	218.8	$Ca^{2+}(aq)$	-542.8	-553.6	-53.1
$CH_2O_2(l)$, formic acid	-425.0	-361.4	129	$CaCl_2(aq)$	-877.1	-816	59.8
$CH_3OH(g)$, methanol	-201	-162.3	239.9	$CaCl_2$($crystal$)	-795.4	-748.8	108.4
$CH_3OH(l)$, methanol	-239.2	-166.6	126.8	CaO	-634.9	603.3	38.1
$CH_4(g)$, methane	-74.6	-50.5	186.3	$Cl^-(aq)$	-167.2	-131.2	56.5
$CO(g)$	-110.5	-137.2	197.7	$Cl_2(g)$	0		223.1
$CO_2(g)$	-393.5	-394.4	213.8	Cr_2O_3($crystal$)	-1139.7	-1058.1	81.2
$C_2H_2(g)$, acetylene	227.4	209.9	200.9	Cu	0		33.2
$C_2H_5OH(l)$, ethanol	-277.6	-174.8	160.7	CuO	-157.3	-129.7	42.6
$C_2H_5COOH(l)$, acetic acid	-510.7		191	$F_2(g)$	0		202.78
$C_2H_6(g)$, ethane	-84.0	-32.0	229.2	Fe_2O_3($crystal$)	-824.2	-742.2	60.3
C_3H_6O, propaldehyde	-185.6		304.5	FeS(s)	-100.0	100.4	
$C_3H_8(g)$, propane	-103.8	-23.4	270.3	$FeS_2(s)$	-178.2	-166.9	52.9
$C_3H_8(l)$, propane	-120.9			$H_2(g)$	0		130.7
$C_3H_7OH(l)$	-302.6		193.6	$H_2O(g)$	-241.8	-228.6	188.8
n-$C_3H_7OH(g)$	-255.1		322.6	$H_2O(l)$	-285.8	-237.1	70
n-$C_3H_7OH(l)$	-302.6		193.6	$H_2S(g)$	-20.6	-33.4	205.8
$C_4H_{10}(g)$, butane	-125.7			HBr(g)	-36.3	-53.4	198.7
$C_4H_{10}(l)$, butane	-147.3			HCl(aq)	-167.159	-131.228	56.5
$C_4H_{10}O(g)$, 2-butanol	-292.8		359.5	HCl(g)	-92.307	-95.299	186.91
$C_4H_{10}O(l)$, diethyl ether	-279.5		172.4	HF(aq)	-332.63	-278.9	-13.8
$C_4H_{10}O(g)$, diethyl ether	-252.1		342.7	HF(g)	-271.1	-273.2	173.8

(Continued)

Table A.15 Heat of Formation   (Continued)

Substance	ΔH_f° (kJ/mol)	ΔG_f° (kJ/mol)	S° (J/K · mol)	Substance	ΔH_f° (kJ/mol)	ΔG_f° (kJ/mol)	S° (J/K · mol)
$HI(g)$	26.5	1.7	206.6	$NO_3^-(aq)$	−207.4	−111.3	146.4
Hg	0		75.9	$NaCl(s)$	−441.2	−384.1	50.5
HgO	−90.8	−58.5	70.3	$O_2(g)$	0		205.2
$I_2(s)$	0		116.135	$O_3(g)$	142.7	163.2	236.9
$Mg(s)$	0		32.7	$OH^-(aq)$	−230	−157.2	−10.8
$MgO(s)$	−610.6	−569.3	37.2	$S(rhombic, crystal)$	0		32.1
$Mn_2O_3(s)$	−959.0	−881.1	110.5	$SiCl_4(g)$	−657.01	−616.98	330.73
$MnO_2(s)$	−520	−465.1	53.1	$SiF_4(g)$	−1614.94	−1572.65	282.49
$N_2(g)$	0		191.6	$SiO_2(s, quartz)$	−910.94	−856.64	41.84
$N_2O_4(g)$	11.1	99.8	304.4	$Sn(\beta)(s)$	−2.1	0.1	44.1
$NH_3(g)$	−45.9	−16.4	192.8	$SnO(s)$	−280.7	−251.9	−8.4
NH_4^+	−132.5	−79.3	113.4	$Ti(s)$	0		30.7
$NH_4NO_3(aq)$	−339.9	−190.6	259.8	Zn	0		41.6
$NH_4NO_3(crystal)$	−365.6	−183.9	151.1	$Zn^{2+}(aq)$	−153.9	−147.1	−112.1
$NO(g)$	91.3	87.6	210.8	ZnO	−350.5	−320.5	43.7
$NO_2(g)$	33.2	51.3	240.1				

Table A.16 Heat of Formation of Aqueous Solutions

Substance	ΔH_f° (kJ/mol)	ΔG_f° (kJ/mol)	S° (J/K · mol)	Substance	ΔH_f° (kJ/mol)	ΔG_f° (kJ/mol)	S° (J/K · mol)
Cations				Na^+	−240.1	−261.9	59.0
				Ni^{2+}	−54.0	−45.6	−128.9
Ag^+	105.6	77.1	72.7	Pb^{2+}	−1.7	−24.4	10.5
Al^{3+}	−531.0	−485.0	−321.7	Rb^+	−251.2	−284.0	121.5
Ba^{2+}	−537.6	−560.8	9.6	Sc^{3+}	−614.2	−586.6	−255.0
Be^{2+}	−382.8	−379.7	−129.7	Sn^{2+}	−8.8	−27.2	−17.0
Ca^{2+}	−542.8	−553.6	−53.1	Zn^{2+}	−153.9	−147.1	−112.1
Cd^{2+}	−75.9	−77.6	−73.2				
Co^{2+}	−58.2	−54.4	113.0	**Anions**			
Co^{3+}	92.0	134.0	−305.0	Br^-	−121.6	−104.0	85.4
Cr^{2+}	−143.5			CH_3COO^-	−486.0	−369.3	86.6
Cs^+	−258.3	−292.0	133.1	Cl^-	−167.2	−131.2	56.5
Cu^+	71.1	50.0	40.6	CN^-	153.6	17.4	94.1
Cu^{2+}	64.8	65.5	−99.6	CO_3^{2-}	−677.1	−527.8	−56.9
Fe^{2+}	−89.1	−78.9	−137.7	CrO_4^{2-}	−881.2	−727.8	50.2
Fe^{3+}	−48.5	−4.7	−315.9	$Cr_2O_7^{2-}$	−1490.3	−1301.1	261.9
H^+	0	0	0	F^-	−332.6	−278.8	−13.8
Hg^{2+}	171.1	164.4	−32.2	HCO_3^-	−692.0	−568.8	91.2
Hg_2^{2+}	172.4	153.5	84.5	HSO_3^-	−626.2	−527.7	139.7
K^+	−252.4	−283.3	102.5	HSO_4^-	−887.3	−755.9	131.8
Li^+	−278.5	−293.3	13.4	$H_2PO_4^-$	−1296.3	−1130.2	90.4
Mg^{2+}	−466.9	−454.8	−138.1	I^-	−55.2	−51.6	111.3
Mn^{2+}	−220.8	−228.1	−73.6	MnO_4^-	−541.4	−447.2	191.2
NH_4^+	−132.5	−79.3	113.4				

(Continued)

Table A.16 Heat of Formation of Aqueous Solutions   *(Continued)*

Substance	ΔH_f° (kJ/mol)	ΔG_f° (kJ/mol)	S° (J/K · mol)	Substance	ΔH_f° (kJ/mol)	ΔG_f° (kJ/mol)	S° (J/K · mol)
MnO_4^{2-}	−653.0	−500.7	59.0	$CaCO_3$	−1220.0	−1081.4	−110.0
NO_2^-	−104.6	−32.2	123.0	$CaCl_2$	−877.1	−816.0	59.8
NO_3^-	−207.4	−111.3	146.4	CaF	−1208.1	−1111.2	−80.8
OH^-	−230.0	−157.2	−10.8	$CaSO_4$	−1452.1	−1298.1	−33.1
PO_4^{3-}	−1277.4	−1018.7	−220.5	$Cd(NO_3)_2$	−490.6	−300.1	219.7
S^{2-}	33.1	85.8	−14.6	$Cu(NO_3)_2$	−350.0	−157.0	193.3
SCN^-	76.4	92.7	144.3	$CuSO_4$	−844.5	−679.0	−79.5
SO_3^{2-}	−635.5	−486.5	−29.0	HCl	−167.2	−131.2	56.5
SO_4^{2-}	−909.3	−744.5	20.1	HF	−332.6	−278.8	−13.8
Neutral				HNO_3	−207.4	−111.3	146.4
$AgCl$	−61.6	−54.1	129.3	H_2SO_4	−909.3	−744.5	20.1
$AlCl_3$	−1033.0	−879.0	−152.3	NH_4NO_3	−339.9	−190.6	259.8
$Al_2(SO_4)_3$	−3791.0	−3205.0	−583.2	NH_4OH	−362.5	−236.5	102.5
$BaSO_4$	−1446.9	−1305.3	29.7	$NaCl$	−407.3	−393.1	115.5
CH_3COOH	−486.0	−369.3	86.6	$MgCl_2$	−801.2	−717.1	−25.1
CH_3COONH_4	−618.5	−448.6	200.0	$Pb(NO_3)_2$	−416.3	−246.9	303.3
CH_3COONa	−726.1	−631.2	145.6				

Table A.17 Heat of Atomization

Substance	H_a°	Substance	H_a°	Substance	ΔH (kJ/mol)	ΔG (kJ/mol)	S° (J/K · mol)
Ag	284.9	H	218.0	$Br^-(aq)$	232.9	186	93
Al	329.7	Hf	619.0	$C_2H_4(g)$	2253	2083	556
As	302.5	Hg	61.4	$C_2H_6(g)$	2826	2591	775
Au	368.2	I	106.8	$CaCO_3(s)$	2850		703
B	565.0	In	243.0	$CaO(s)$	1062		276
Ba	177.8	Ir	669.0	$CH_3OH(g)$	2039	1877	538
Be	324.0	K	89.0	$CH_3OH(l)$	2077	1880	651
Bi	209.6	Li	159.3	$CH_4(g)$	1664	1533	431
Br	111.9	Mg	147.1	$Cl^-(aq)$	288.3	236	108
C	716.7	Mn	283.3	$Cl_2(g)$	242.6		107
Ca	177.8	Mo	658.1	$CO(g)$	1076	1038	121
Cd	111.8	N	472.7	$CO_2(g)$	1608	1527	266
Ce	423.0	Na	107.5	$F^-(aq)$	412.1		173
Cl	121.3	Nb	721.3	$H^+(aq)$		203	115
Co	428.4	Ni	430.1	$H_2(g)$	436		98
Cr	397.0	O	249.2	$H_2O(g)$	927.2	867	201
Cs	76.5	Os	787.0	$H_2O(l)$	971.2	875	320
Cu	337.4	P	316.5	$H_2S(g)$	734	679	191
Er	317.7	Pb	195.2	$H_2SO_4(aq)$	2619		1021
F	79.4	Pd	376.6	$HBr(g)$	365.7	339	91
Fe	415.5	Pt	565.7	$HCl(g)$	431.8	403	93
Ge	372.0	Pu	364.4	$HF(g)$	568.6	538	99

(Continued)

Table A.17 Heat of Atomization *(Continued)*

Substance	H_a°	Substance	H_a°	Substance	ΔH (kJ/mol)	ΔG (kJ/mol)	S° (J/K · mol)
Rb	80.9	Te	196.6	$HNO_3(aq)$	1645	1465	605
Rc	774.0	Th	602.0	$MgCO_3(s)$	2725		724
Rh	556.0	Ti	473.0	$MgO(s)$	998.5		283
Ru	650.6	Tl	182.2	$Mg(OH)_2(s)$	2007		636
S	277.2	U	533.0	$N_2(g)$	945.4		115
Sb	264.4	V	514.2	$N_2O_4(g)$	1932	1740	647
Sc	377.8	W	849.8	$NH_3(aq)$	1207	1092	
Se	227.2	Y	424.7	$NH_3(g)$	1173	1082	304
Si	450.0	Yb	152.1	$NH_4^+(aq)$	1477	1348	
Sn	301.2	Zn	130.4	$NH_4Cl(s)$	1780		681
Sr	163.6	Zr	608.8	$NO(g)$	631.9	600	103
Ta	782.0			$NO_2(g)$	937.1	867	235
				$O_2(g)$	498.4		117
				$O_3(g)$	604.6		244
				$OH^-(aq)$	697.2	592	287
				$SO_2(g)$	1072	1002	242
				$SO_3(g)$	1421		394
				$SO_4^{2-}(aq)$	2183		792

Table A.18 Crystal Lattice Energy

Substance	$\Delta H_{crystal\ lattice}$ (kJ/mol)	Substance	$\Delta H_{crystal\ lattice}$ (kJ/mol)
AgBr	897	LiCl	834
AgCl	610	LiF	1030
$AgNO_3$	820	LiI	756
AgOH	918	$LiNO_3$	848
$AlCl_3$	5376	LiOH	1021
$Al(OH)_3$	5627	$Mg(OH)_2$	2870
$BaSO_4$	2469	$MgCl_2$	2477
$CaCl_2$	2268	MgF_2	2922
CaF_2	2597	$MgNO_3$	2481
$Ca(NO_3)_2$	2268	MgO	3356
CaO	3414	NH_4NO_3	661
$Ca(OH)_2$	2506	$(NH_4)_2SO_4$	1766
$CaSO_4$	2489	NaBr	732
$Cd(NO_3)_2$	2238	NaCl	769
CsCl	652	NaF	910
CsI	611	NaI	701
Cs_2SO_4	1596	$NaNO_3$	755
KBr	671	NaOH	887
KCl	701	SnO	3652
KF	808	SnO_2	11807
KI	646	$Sn(OH)_4$	9188
KNO_3	685	$Zn(NO_3)_2$	2376
KOH	789	$Zn(OH)_2$	2795
LiBr	788	ZnO	4142

Table A.19 Heat Capacity

Substance	Heat Capacity (J/mol · K)
Al	21.33
Al_2O_3	51.12
Cu	19.86
$H_2O(l)$	4.184

Table A.20 Heat of Combustion

Substance	$\Delta H_{combustion}$ (kJ/mol)
C	-393.5
CO	-283
H_2	-285.8
CH_4	-890.8
C_2H_2	-1301.1
C_2H_6	-1560.7
C_3H_8	-2219.2
H_3COH	-726.1
C_2H_5OH	-1366.8
$C_6H_{12}O_6$	-2802

Table A.21 Percent Ionic Character in a Single Chemical Bond

Difference in electronegativity	0.1	0.2	0.3	0.4	0.5	0.6	0.7	0.8	0.9	1.0	1.1	1.2	1.3	1.4	1.5	1.6	1.7	1.8	1.9	2.0	2.1	2.2	2.3	2.4	2.5	2.6	2.7	2.8	2.9	3.0
Percent ionic character	0.5	1	2	4	6	9	12	15	19	22	26	30	34	39	43	47	51	55	59	63	67	70	74	76	79	82	84	86	88	89

Table A.22 Madelung Constants

Substance	Ion Type	Based on Crystal Form	Madelung Constant	Coordination
NaCl	M^+, X^-	FCC	1.7475	6
CsCl	M^+, X^-	BCC	1.76267	8
CaF_2	$M^{2+}, 2X^-$	Cubic	2.51939	8/4
Zinc blende	M^{2+}, X^{2-}	FCC	1.63806	4
Wurtzite	M^{2+}, X^{2-}	HCP	1.641	4

Table A.23 Oxide Density

Oxide	Density (g/cm³)
Cr_2O_3	5.22
SnO	6.45
VO	5.758
V_2O_3	4.87
VO_2	4.339
V_2O_5	3.335
ZnO	5.6

Appendix B

Fundamental Constants

Acceleration of gravity	$g = 9.80665 \text{ m/s}^2$
Avogadro's number	$N = 6.022137 \times 10^{23}$ particles/mol
Atomic mass unit	$\text{amu} = 1.660540 \times 10^{-24} \text{ g}$
Boltzmann's constant	$k = 1.380662 \times 10^{-23} \text{ J/K}$
Charge of an electron	$e = 1.60218 \times 10^{-19} \text{ C}$
Coulomb	$1 \text{ C} = 1 \text{ amp} \cdot \text{s}$
Faraday's constant	$F = 96,485.31 \text{ C/mol}$
Gas constant	$R = 0.08206 \text{ L} \cdot \text{atm/mol} \cdot \text{K}$
	$= 1.987 \text{ cal/mol} \cdot \text{K}$
	$= 8.3145 \text{ J/mol} \cdot \text{K}$
Ion product for water at 25 °C	$K_{\text{w}} = 1.0 \times 10^{-14}$
Mass of a neutron	$1.6749 \times 10^{-24} \text{ g}$
Mass of a proton	$1.6726 \times 10^{-24} \text{ g}$
Planck's constant	$h = 6.6262 \times 10^{-34} \text{ J} \cdot \text{s}$
	$= 6.6262 \times 10^{-27} \text{ erg} \cdot \text{s}$
Rest mass of the electron	$m_{\text{e, rest}} = 0.00054858 \text{ amu}$
	$= 9.1094 \times 10^{-28} \text{ g}$
Rydberg constant	$R = 1.097 \times 10^7 \text{ m}^{-1}$
Speed of light	$c = 2.9979 \times 10^8 \text{ m/s}$
Vacuum permittivity	$\varepsilon_{\text{o}} = 8.854 \times 10^{-12} \text{ C}^2/\text{J} \cdot \text{m}$

Appendix C

Relationships

Mass and Weight
$1 \text{ kg} = 1000 \text{ g} = 2.205 \text{ lb}$
$1 \text{ g} = 1000 \text{ mg}$
$1 \text{ lb} = 453.59 \text{ g}$
$1 \text{ g} = 6.022 \times 10^{23} \text{ amu}$
$1 \text{ pt} = 16 \text{ oz} = 1 \text{ lb}$

Volume
$1 \text{ L} = 1000 \text{ mL}$
$kilo = 10^3$
$mega = 10^6$
$giga = 10^9$
$nano = 10^{-9}$
$pico = 10^{-12}$
$micro = 10^{-6}$

Pressure
1 Atmosphere
$= 760 \text{ torr}$
$= 760 \text{ mm Hg}$
$= 1.01325 \times 10^5 \text{ Pa}$
$= 1.01325 \times 10^5 \text{ N/m}^2$
$= 14.70 \text{ lb/in}^2$
$1 \text{ lb/in}^2 = 1 \text{ psi} = 6894 \text{ N/m}^2$
$1 \text{ bar} = 10^5 \text{ Pa}$
$1000 \text{ psi} = 6.894 \text{ MPa}$

Force
$1 \text{ lb} = 4.448 \text{ N}$

Dipole
$1 \text{ Debye} = 3.335641 \times 10^{-30} \text{ C} \cdot \text{m}$

Length
$1 \text{ m} = 100 \text{ cm}$
$1 \text{ km} = 1000 \text{ m}$
$1 \text{ Å} = 1.0 \times 10^{-10} \text{ m}$
$1 \text{ Å} = 1.0 \times 10^{-8} \text{ cm}$
$1 \text{ ft} = 12 \text{ in}$
$1 \text{ in} = 2.54 \text{ cm}$

Energy
$1 \text{ cal} = 4.184 \text{ J} = 4.129 \text{ L} \cdot \text{atm}$
$1 \text{ J} = 1 \times 10^7 \text{ erg}$
$1 \text{ eV} = 1.6022 \times 10^{-19} \text{ J}$
$1 \text{ eV} = 96.487 \text{ kJ/mol}$
$1 \text{ J} = 1 \text{ kg} \cdot \text{m}^2/\text{s}^2$
$1 \text{ J} = 1 \text{ C} \cdot \text{V}$
$1 \text{ N} \cdot \text{m} = 1 \text{ J}$

Temperature
$°F = [°C \times (9/5)] + 32°$
$°C = (°F - 32°) \times (5/9)$
$K = °C + 273.15$
$°C = K - 273.15$

Electrical
$1 \text{ amp} = 1 \text{ C/s} = 1 \text{ V}/\Omega$
$1 \text{ W} = 1 \text{ J/s} = \text{amp}^2 \cdot \Omega$

Gases at Room Temperature and 1 atm
24.45388 L/mol
0.04089331 mol/L
$2.4626509 \times 10^{22} \text{ molecules/L}$

Appendix D

List of Elements

Sorted by Number

#	Sym	Name	#	Sym	Name
1	H	Hydrogen	43	Tc	Technetium
2	He	Helium	44	Ru	Ruthenium
3	Li	Lithium	45	Rh	Rhodium
4	Be	Beryllium	46	Pd	Palladium
5	B	Boron	47	Ag	Silver
6	C	Carbon	48	Cd	Cadmium
7	N	Nitrogen	49	In	Indium
8	O	Oxygen	50	Sn	Tin
9	F	Fluorine	51	Sb	Antimony
10	Ne	Neon	52	Te	Tellurium
11	Na	Sodium	53	I	Iodine
12	Mg	Magnesium	54	Xe	Xenon
13	Al	Aluminum	55	Cs	Cesium
14	Si	Silicon	56	Ba	Barium
15	P	Phosphorus	57	La	Lanthanum
16	S	Sulfur	58	Ce	Cerium
17	Cl	Chlorine	59	Pr	Praseodymium
18	Ar	Argon	60	Nd	Neodymium
19	K	Potassium	61	Pm	Promethium
20	Ca	Calcium	62	Sm	Samarium
21	Sc	Scandium	63	Eu	Europium
22	Ti	Titanium	64	Gd	Gadolinium
23	V	Vanadium	65	Tb	Terbium
24	Cr	Chromium	66	Dy	Dysprosium
25	Mn	Manganese	67	Ho	Holmium
26	Fe	Iron	68	Er	Erbium
27	Co	Cobalt	69	Tm	Thulium
28	Ni	Nickel	70	Yb	Ytterbium
29	Cu	Copper	71	Lu	Lutetium
30	Zn	Zinc	72	Hf	Hafnium
31	Ga	Gallium	73	Ta	Tantalum
32	Ge	Germanium	74	W	Tungsten
33	As	Arsenic	75	Re	Rhenium
34	Se	Selenium	76	Os	Osmium
35	Br	Bromine	77	Ir	Iridium
36	Kr	Krypton	78	Pt	Platinum
37	Rb	Rubidium	79	Au	Gold
38	Sr	Strontium	80	Hg	Mercury
39	Y	Yttrium	81	Tl	Thallium
40	Zr	Zirconium	82	Pb	Lead
41	Nb	Niobium	83	Bi	Bismuth
42	Mo	Molybdenum	84	Po	Polonium

Alphabetical by Symbol

#	Sym	Name	#	Sym	Name
89	Ac	Actinium	80	Hg	Mercury
47	Ag	Silver	67	Ho	Holmium
13	Al	Aluminum	108	Hs	Hassium
95	Am	Americium	53	I	Iodine
18	Ar	Argon	49	In	Indium
33	As	Arsenic	77	Ir	Iridium
85	At	Astatine	19	K	Potassium
5	B	Boron	36	Kr	Krypton
79	Au	Gold	57	La	Lanthanum
56	Ba	Barium	3	Li	Lithium
4	Be	Beryllium	103	Lr	Lawrencium
107	Bh	Bohrium	71	Lu	Lutetium
83	Bi	Bismuth	101	Md	Mendelevium
97	Bk	Berkelium	12	Mg	Magnesium
35	Br	Bromine	25	Mn	Manganese
6	C	Carbon	42	Mo	Molybdenum
20	Ca	Calcium	109	Mt	Meitnerium
48	Cd	Cadmium	7	N	Nitrogen
58	Ce	Cerium	11	Na	Sodium
98	Cf	Californium	41	Nb	Niobium
17	Cl	Chlorine	60	Nd	Neodymium
96	Cm	Curium	10	Ne	Neon
27	Co	Cobalt	28	Ni	Nickel
24	Cr	Chromium	102	No	Nobelium
55	Cs	Cesium	93	Np	Neptunium
29	Cu	Copper	8	O	Oxygen
105	Db	Dubnium	76	Os	Osmium
110	Ds	Darmstadium	15	P	Phosphorus
66	Dy	Dysprosium	91	Pa	Protactinium
68	Er	Erbium	82	Pb	Lead
99	Es	Einsteinium	46	Pd	Palladium
63	Eu	Europium	61	Pm	Promethium
9	F	Fluorine	84	Po	Polonium
26	Fe	Iron	59	Pr	Praseodymium
100	Fm	Fermium	78	Pt	Platinum
87	Fr	Francium	94	Pu	Plutonium
31	Ga	Gallium	88	Ra	Radium
64	Gd	Gadolinium	37	Rb	Rubidium
32	Ge	Germanium	75	Re	Rhenium
1	H	Hydrogen	104	Rf	Rutherfordium
2	He	Helium	45	Rh	Rhodium
72	Hf	Hafnium	86	Rn	Radon

Alphabetical by Name

#	Sym	Name	#	Sym	Name
89	Ac	Actinium	1	H	Hydrogen
13	Al	Aluminum	49	In	Indium
95	Am	Americium	53	I	Iodine
51	Sb	Antimony	77	Ir	Iridium
18	Ar	Argon	26	Fe	Iron
33	As	Arsenic	36	Kr	Krypton
85	At	Astatine	57	La	Lanthanum
56	Ba	Barium	103	Lr	Lawrencium
97	Bk	Berkelium	82	Pb	Lead
4	Be	Beryllium	3	Li	Lithium
83	Bi	Bismuth	71	Lu	Lutetium
107	Bh	Bohrium	12	Mg	Magnesium
5	B	Boron	25	Mn	Manganese
35	Br	Bromine	109	Mt	Meitnerium
48	Cd	Cadmium	101	Md	Mendelevium
20	Ca	Calcium	80	Hg	Mercury
98	Cf	Californium	42	Mo	Molybdenum
6	C	Carbon	60	Nd	Neodymium
58	Ce	Cerium	10	Ne	Neon
55	Cs	Cesium	93	Np	Neptunium
17	Cl	Chlorine	28	Ni	Nickel
24	Cr	Chromium	41	Nb	Niobium
27	Co	Cobalt	7	N	Nitrogen
29	Cu	Copper	102	No	Nobelium
96	Cm	Curium	76	Os	Osmium
110	Ds	Darmstadium	8	O	Oxygen
105	Db	Dubnium	46	Pd	Palladium
66	Dy	Dysprosium	15	P	Phosphorus
99	Es	Einsteinium	78	Pt	Platinum
68	Er	Erbium	94	Pu	Plutonium
63	Eu	Europium	84	Po	Polonium
100	Fm	Fermium	19	K	Potassium
9	F	Fluorine	59	Pr	Praseodymium
87	Fr	Francium	61	Pm	Promethium
64	Gd	Gadolinium	91	Pa	Protactinium
31	Ga	Gallium	88	Ra	Radium
32	Ge	Germanium	86	Rn	Radon
79	Au	Gold	75	Re	Rhenium
72	Hf	Hafnium	45	Rh	Rhodium
108	Hs	Hassium	37	Rb	Rubidium
2	He	Helium	44	Ru	Ruthenium
67	Ho	Holmium	104	Rf	Rutherfordium

Sorted by Number		Alphabetical by Symbol		Alphabetical by Name	
85 At Astatine	99 Es Einsteinium	44 Ru Ruthenium	22 Ti Titanium	62 Sm Samarium	90 Th Thorium
86 Rn Radon	100 Fm Fermium	51 Sb Antimony	81 Tl Thallium	21 Sc Scandium	69 Tm Thulium
87 Fr Francium	101 Md Mendelevium	21 Sc Scandium	69 Tm Thulium	106 Sg Seaborgium	50 Sn Tin
88 Ra Radium	102 No Nobelium	34 Se Selenium	73 Ta Tantalum	34 Se Selenium	22 Ti Titanium
89 Ac Actinium	103 Lr Lawrencium	106 Sg Seaborgium	92 U Uranium	14 Si Silicon	74 W Tungsten
90 Th Thorium	104 Rf Rutherfordium	16 S Sulfur	112 Uub Unununbium	47 Ag Silver	112 Uub Unununbium
91 Pa Protactinium	105 Db Dubnium	14 Si Silicon	111 Uuu Unununium	11 Na Sodium	111 Uuu Unununium
92 U Uranium	106 Sg Seaborgium	62 Sm Samarium	23 V Vanadium	38 Sr Strontium	92 U Uranium
93 Np Neptunium	107 Bh Bohrium	50 Sn Tin	74 W Tungsten	16 S Sulfur	23 V Vanadium
94 Pu Plutonium	108 Hs Hassium	38 Sr Strontium	54 Xe Xenon	73 Ta Tantalum	54 Xe Xenon
95 Am Americium	109 Mt Meitnerium	65 Tb Terbium	39 Y Yttrium	43 Tc Technetium	70 Yb Ytterbium
96 Cm Curium	110 Ds Darmstadium	43 Tc Technetium	70 Yb Ytterbium	52 Te Tellurium	39 Y Yttrium
97 Bk Berkelium	111 Uuu Unununium	52 Te Tellurium	30 Zn Zinc	65 Tb Terbium	30 Zn Zinc
98 Cf Californium	112 Uub Unununbium	90 Th Thorium	40 Zr Zirconium	81 Tl Thallium	40 Zr Zirconium

Legend (key):

- Atomic number — 4
- Ionization energy (kJ/mol) — 899.5
- Electron affinity (kJ/mol) — ~0
- Electronegativity (Paulings) — 1.57
- Symbol — **Be**
- Covalent radius (pm) — 90
- Atomic radius (pm) — 140
- Resistivity (μΩ cm) — 3.76

Each cell lists: Atomic number; Ionization energy / Electron affinity / Electronegativity; Symbol; Covalent radius / Atomic radius / Resistivity.

Period 1

IA	VIIIA
1 **H** — 1312.032 / 72.7678 / 2.20; 32 / 79 / —	2 **He** — 2372.3 / — / —; 93 / 49 / —

Period 2

IA	IIA	IIIA	IVA	VA	VIA	VIIA	VIIIA
3 **Li** 520.2147 / 88.57 / 0.98; 123 / 205 / 9.55	4 **Be** 899.5 / ~0 / 1.57; 90 / 140 / 3.76	5 **B** 800.63 / 26.7 / 2.04; 82 / 117 / 10^12	6 **C** 1086.4 / 121.85 / 2.55; 77 / 91 / 27×10^8	7 **N** 1402.3 / ~0 / 3.04; 75 / 75 / —	8 **O** 1313.9 / 140.974 / 3.44; 73 / 65 / —	9 **F** 1681.0 / 328.160 / 3.98; 72 / 57 / —	10 **Ne** 2080.6 / — / —; 71 / 51 / —

Period 3

IA	IIA	IIIA	IVA	VA	VIA	VIIA	VIIIA
11 **Na** 495.839 / 52.8991 / 0.93; 154 / 223 / 4.93	12 **Mg** 737.7 / ~0 / 1.31; 136 / 172 / 4.51	13 **Al** 577.53 / 42.5 / 1.61; 118 / 182 / 2.733	14 **Si** 786.5 / 133.6 / 1.90; 111 / 146 / 3×10^6	15 **P** 1011.8 / 72.02 / 2.19; 106 / 123 / 10^17	16 **S** 999.6 / 200.4073 / 2.58; 102 / 109 / 10^23	17 **Cl** 1251.2 / 348.567 / 3.16; 99 / 97 / —	18 **Ar** 1520.6 / — / —; 98 / 88 / —

Period 4

IIIB	IVB	VB	VIB	VIIB	VIIIB	VIIIB	VIIIB	IB	IIB
21 **Sc** 633.1 / 18.14 / 1.36; 144 / 209 / 56.2	22 **Ti** 658.8 / 7.6 / 1.54; 132 / 200 / 39	23 **V** 650.9 / 50.6 / 1.63; 122 / 192 / 20.2	24 **Cr** 652.8 / 64.26 / 1.66; 118 / 185 / 12.7	25 **Mn** 717.3 / ~0 / 1.55; 117 / 179 / 144	26 **Fe** 762.5 / 14.6 / 1.83; 117 / 172 / 9.98	27 **Co** 760.4 / 63.9 / 1.88; 116 / 167 / 5.6	28 **Ni** 737.1 / 111.5 / 1.91; 115 / 162 / 7.20	29 **Cu** 745.5 / 119.2 / 1.90; 117 / 157 / 1.725	30 **Zn** 906.4 / ~0 / 1.65; 125 / 153 / 6.06

Group IA–IIA and IIIA–VIIIA of Period 4:
19 **K** 418.8042 / 48.384 / 0.82; 203 / 277 / 7.47 · 20 **Ca** 589.8 / 2.369 / 1.00; 174 / 223 / 3.45 · 31 **Ga** 578.8 / 29 / 1.81; 126 / 181 / 13.6 · 32 **Ge** 762.2 / 119 / 2.01; 122 / 152 / 10^7 · 33 **As** 944.4 / 78 / 2.18; 120 / 133 / 33.3 · 34 **Se** 640.9 / 194.962 / 2.55; 116 / 122 / 10^6 · 35 **Br** 1139.8 / 324.5 / 2.96; 114 / 112 / 10^15 · 36 **Kr** 1350.7 / — / —; 189 / 103 / —

Period 5

IIIB	IVB	VB	VIB	VIIB	VIIIB	VIIIB	VIIIB	IB	IIB
39 **Y** 599.8 / 29 / 1.22; 162 / 227 / 59.6	40 **Zr** 640.1 / 41.1 / 1.33; 145 / 216 / 43.3	41 **Nb** 652.1 / 86.2 / 1.6; 134 / 208 / 15.2	42 **Mo** 684.3 / 72.2 / 2.16; 130 / 201 / 5.52	43 **Tc** 702.4 / 53 / 2.10; 127 / 195 / 23	44 **Ru** 710.2 / 101 / 2.2; 125 / 189 / 7.1	45 **Rh** 719.7 / 109.7 / 2.28; 125 / 183* / 4.3	46 **Pd** 804.4 / 54.2 / 2.20; 128 / 179 / 10.8	47 **Ag** 731.0 / 125.6 / 1.93; 134 / 175 / 1.629	48 **Cd** 867.7 / ~0 / 1.69; 141 / 171 / 6.8

Group IA–IIA and IIIA–VIIIA of Period 5:
37 **Rb** 403.0262 / 46.884 / 0.82; 216 / 298 / 13.3 · 38 **Sr** 549.5 / 14.3 / 0.95; 191 / 245 / 13.5 · 49 **In** 558.3 / 29 / 1.78; 144 / 200 / 8.0 · 50 **Sn** 708.6 / 107.3 / 1.96; 141 / 172 / 11.5 · 51 **Sb** 830.6 / 100.9 / 2.05; 140 / 153 / 39.0 · 52 **Te** 869.6 / 190.15 / 2.1; 136 / 142 / 10^5 · 53 **I** 1008.4 / 295.148 / 2.66; 133 / 132 / 10^15 · 54 **Xe** 1170.3 / — / 2.60; 131 / 124 / —

Period 6

IIIB	IVB	VB	VIB	VIIB	VIIIB	VIIIB	VIIIB	IB	IIB
57 **La**† 538.1 / 50 / 1.10; 125 / 274 / 61.5	72 **Hf** 658.5 / ~0 / 1.3; 144 / 216 / 34.0	73 **Ta** 728.4 / 31.1 / 1.5; 134 / 209 / 13.5	74 **W** 758.7 / 78.6 / 1.7; 130 / 202 / 5.44	75 **Re** 755.8 / 14.5 / 1.9; 128 / 197 / 17.2	76 **Os** 814.2 / 106 / 2.2; 126 / 192 / 8.1	77 **Ir** 865.2 / 150.9 / 2.20; 127 / 187 / 4.7	78 **Pt** 864.4 / 205.3 / 2.2; 130 / 183 / 10.8	79 **Au** 890.1 / 222.76 / 2.4; 134 / 179 / 2.271	80 **Hg** 1007 / ~0 / 1.9; 149 / 176 / 94.1

Group IA–IIA and IIIA–VIIIA of Period 6:
55 **Cs** 375.699 / 45.504 / 0.79; 235 / 334 / 21.0 · 56 **Ba** 502.8 / 14 / 0.89; 198 / 278 / 34.3 · 81 **Tl** 589.3 / 20 / 1.8; 148 / 208 / 15 · 82 **Pb** 715.6 / 35.1 / 1.8; 147 / 181 / 21.3 · 83 **Bi** 702.9 / 91.3 / 1.9; 146 / 163 / 107 · 84 **Po** 812.1 / 180 / 2.0; 153 / 153 / 40 · 85 **At** — / 270 / 2.2; 147 / 143 / — · 86 **Rn** 1037.0 / — / —; — / 134 / —

Period 7

87 **Fr** 392.95 / 44 / 0.7; — / — / — · 88 **Ra** 509.3 / 0.9 / ; — / — / — · 89 **Ac**‡ 498 / / 1.1; — / — / — · 104 **Rf** · 105 **Db** · 106 **Sg** · 107 **Bh** · 108 **Hs** · 109 **Mt** · 110 **Ds** · 111 **Uuu** · 112 **Uub** · 114

† Lanthanides

Element	IE / EA / EN	cov / at / res
58 **Ce**	534.4 / / 1.12	165 / 270 / 78
59 **Pr**	527.2 / / 1.13	165 / 267 / 70.0
60 **Nd**	506.5 / / 1.14	164 / 267 / 64.3
61 **Pm**	538.4 / / 1.13	163 / 262 / 75est
62 **Sm**	544.5 / / 1.17	162 / 259 / 94.0
63 **Eu**	547.1 / / 1.2	185 / 256 / 90.0
64 **Gd**	593.4 / / 1.20	161 / 254 / 131
65 **Tb**	565.8 / / 1.2	159 / 251 / 115
66 **Dy**	573.0 / / 1.22	159 / 249 / 92.6
67 **Ho**	581.0 / / 1.23	158 / 247 / 81.4
68 **Er**	589.0 / / 1.24	157 / 245 / 86
69 **Tm**	596.7 / / 1.25	156 / 242 / 67.6
70 **Yb**	603.4 / / 1.1	170 / 240 / 25.0
71 **Lu**	523.5 / / 1.0	156 / 225 / 58.2

‡ Actinides

Element	IE / EA / EN	cov / at / res
90 **Th**	608.5 / / 1.3	165 / — / 14.7
91 **Pa**	568.3 / / 1.5	— / — / 17.7
92 **U**	597.6 / / 1.7	142 / — / 28
93 **Np**	604.5 / / 1.3	— / 108 / 141
94 **Pu**	581.4 / / 1.3	— / — /
95 **Am**	576.4 / / 1.3	— / — /
96 **Cm**	580.8 / /	— / — /
97 **Bk**	601.1 / /	— / — /
98 **Cf**	607.8 / /	— / — /
99 **Es**	619.4 / /	— / — /
100 **Fm**	627.1 / /	— / — /
101 **Md**	634.9 / /	— / — /
102 **No**	641.6 / /	— / — /
103 **Lr**	— / /	— / — /

Periodic Table of the Elements

Legend (key box — Beryllium):

Symbol	Be
Atomic mass(g)	9.01218
Valence	2
Density (g/cm³)*	1.848
Crystal structure	(hexagonal)
Boiling point (K)	2744
Melting point (K)	1560
Electron configuration	$1s^2 2s^2$
Name	Beryllium

*Gases at 273 K, 1 atm (g/L)

Green symbol = Gas
Blue symbol = Liquid

Crystal structure symbols:
- Cubic
- BCC (body-centered cubic)
- FCC (face-centered cubic)
- HCP (hexagonal close pack)
- Diamond
- Tetragonal
- Rhombohedral

(..) = Most stable isotope
Italic = Not found naturally on earth
† = Gas
‡ = Liquid

Group IA / IIA and transition metals (main groups)

Element	Symbol	Atomic mass	Valence	Density	Boiling pt (K)	Melting pt (K)	Configuration
Hydrogen	H	1.00794	1	0.08988	20.28	13.81	$1s^1$
Lithium	Li	6.941	1	0.534	1615	453.6	$1s^2 2s^1$
Sodium	Na	22.98977	1	0.971	1156	370.87	$[Ne]3s^1$
Potassium	K	39.0983	1	0.862	1032.3	336.43	$[Ar]4s^1$
Rubidium	Rb	85.4678	1	1.532	961	312.46	$[Kr]5s^1$
Cesium	Cs	132.9054	1	1.873	944	301.59	$[Xe]6s^1$
Francium	Fr	(223)	1	—	950	300	$[Rn]7s^1$
Beryllium	Be	9.01218	2	1.848	2744	1560	$1s^2 2s^2$
Magnesium	Mg	24.305	2	1.738	1363	923	$[Ne]3s^2$
Calcium	Ca	40.078	2	1.55	1757	1115	$[Ar]4s^2$
Strontium	Sr	87.62	2	2.54	1655	1050	$[Kr]5s^2$
Barium	Ba	137.33	2	3.5	2170	1000	$[Xe]6s^2$
Radium	Ra	226	2	5	1809	973	$[Rn]7s^2$

Group IIIB – IB

Element	Symbol	Atomic mass	Valence	Density	Boiling pt (K)	Melting pt (K)	Configuration
Scandium	Sc	44.9559	3	2.989	3109	1814	$[Ar]3d^1 4s^2$
Yttrium	Y	88.9059	3	4.469	3618	1795	$[Kr]4d^1 5s^2$
Lanthanum	La†	138.9055	3	6.145	3737	1191	$[Xe]5d^1 6s^2$
Actinium	Ac‡	(227)	3	10.07	3473	1323	$[Rn]6d^1 7s^2$
Titanium	Ti	47.867	4,3	4.54	3560	1941	$[Ar]3d^2 4s^2$
Zirconium	Zr	91.224	4	6.506	4682	2128	$[Kr]4d^2 5s^2$
Hafnium	Hf	178.49	4	13.31	4876	2506	$[Xe]4f^{14}5d^2 6s^2$
Rutherfordium	Rf	(261)	—	—	—	—	$[Rn]5f^{14}6d^2 7s^2$
Vanadium	V	50.9415	5,2,3,4	6.11	3680	2183	$[Ar]3d^3 4s^2$
Niobium	Nb	92.9064	5,3	8.57	5017	2750	$[Kr]4d^4 5s^1$
Tantalum	Ta	180.9479	5	16.654	5731	3290	$[Xe]4f^{14}5d^3 6s^2$
Dubnium	Db	(262)	—	—	—	—	$[Rn]5f^{14}6d^3 7s^2$
Chromium	Cr	51.996	3,2,6	7.19	2944	2180	$[Ar]3d^5 4s^1$
Molybdenum	Mo	95.94	6,2,3,4,5	10.22	4912	2896	$[Kr]4d^5 5s^1$
Tungsten	W	183.84	6,2,3,4,5	19.3	5828	3695	$[Xe]4f^{14}5d^4 6s^2$
Seaborgium	Sg	(263)	—	—	—	—	$[Rn]5f^{14}6d^4 7s^2$
Manganese	Mn	54.9380	2,3,4,5,6,7	7.33	2334	1519	$[Ar]3d^5 4s^2$
Technetium	Tc	(98)	7	11.5	4538	2430	$[Kr]4d^5 5s^2$
Rhenium	Re	186.207	7,2,4,6	21.02	5869	3459	$[Xe]4f^{14}5d^5 6s^2$
Bohrium	Bh	(262)	—	—	—	—	$[Rn]5f^{14}6d^5 7s^2$
Iron	Fe	55.845	2,3	7.874	3200	1811	$[Ar]3d^6 4s^2$
Ruthenium	Ru	101.07	3,4,6,8	12.41	4423	2607	$[Kr]4d^7 5s^1$
Osmium	Os	190.3	2,3,4,6,8	22.57	5285	3306	$[Xe]4f^{14}5d^6 6s^2$
Hassium	Hs	(265)	—	—	—	—	$[Rn]5f^{14}6d^6 7s^2$
Cobalt	Co	58.9332	2,3	8.90	3134	1768	$[Ar]3d^7 4s^2$
Rhodium	Rh	102.9055	3,4	12.41	3968	2237	$[Kr]4d^8 5s^1$
Iridium	Ir	192.217	2,3,4,6	22.42	4701	2719	$[Xe]4f^{14}5d^7 6s^2$
Meitnerium	Mt	(266)	—	—	—	—	$[Rn]5f^{14}6d^7 7s^2$
Nickel	Ni	58.6934	2,3	8.902	3186	1728	$[Ar]3d^8 4s^2$
Palladium	Pd	106.42	2,4	12.0	3236	1828.1	$[Kr]4d^{10}$
Platinum	Pt	195.08	2,4	21.45	4098	2041.6	$[Xe]4f^{14}5d^9 6s^1$
Darmstadtium	Ds	(271)	—	—	—	—	$[Rn]5f^{14}6d^8 7s^2$
Copper	Cu	63.546	2,1	8.96	2835	1357.8	$[Ar]3d^{10}4s^1$
Silver	Ag	107.8682	1	10.50	2436	1234.93	$[Kr]4d^{10}5s^1$
Gold	Au	196.9665	3,1	19.3	3129	1337.3	$[Xe]4f^{14}5d^{10}6s^1$
Unununium	Uuu	(272)	—	—	—	—	(Unununium)
Zinc	Zn	65.39	2	7.133	1180	692.68	$[Ar]3d^{10}4s^2$
Cadmium	Cd	112.41	2	8.65	1040	594.22	$[Kr]4d^{10}5s^2$
Mercury	Hg	200.59	2,1	13.546	629.88	234.16	$[Xe]4f^{14}5d^{10}6s^2$
Ununbium	Uub	(285)	—	—	—	—	(Ununbium)

Groups IIIA – VIIIA

Element	Symbol	Atomic mass	Valence	Density	Boiling pt (K)	Melting pt (K)	Configuration
Boron	B	10.81	3	2.34	4273	2348	$1s^2 2s^2 2p^1$
Aluminum	Al	26.98154	3	2.6989	2792	933.47	$[Ne]3s^2 3p^1$
Gallium	Ga	69.723	3	6.095	2477	302.915	$[Ar]3d^{10}4s^2 4p^1$
Indium	In	114.818	3	7.31	2345	429.77	$[Kr]4d^{10}5s^2 5p^1$
Thallium	Tl	204.383	1,3	11.85	1746	577	$[Xe]4f^{14}5d^{10}6s^2 6p^1$
Carbon	C	12.011	4,2	1.9	4470*	3823	$1s^2 2s^2 2p^2$
Silicon	Si	28.0855	4	2.33	3538	1687	$[Ne]3s^2 3p^2$
Germanium	Ge	72.61	4	5.323	3106	1211.4	$[Ar]3d^{10}4s^2 4p^2$
Tin	Sn	118.710	4,2	7.31	2875	505.08	$[Kr]4d^{10}5s^2 5p^2$
Lead	Pb	207.2	2,4	11.35	2022	600.61	$[Xe]4f^{14}5d^{10}6s^2 6p^2$
(Ununquadium)		(285)					
Nitrogen	N	14.0067	±3,2,4,5	1.2506	77.4	63	$1s^2 2s^2 2p^3$
Phosphorus	P	30.97376	±3,4,5	1.82	553.6	317.3	$[Ne]3s^2 3p^3$
Arsenic	As	74.9216	±3,5	5.73	876 subl	1090	$[Ar]3d^{10}4s^2 4p^3$
Antimony	Sb	121.760	±3,5	6.691	1860	903.78	$[Kr]4d^{10}5s^2 5p^3$
Bismuth	Bi	208.9804	3,5	9.747	1837	544.6	$[Xe]4f^{14}5d^{10}6s^2 6p^3$
Oxygen	O	15.9994	-2	1.429*	90.2	50.36	$1s^2 2s^2 2p^4$
Sulfur	S	32.066	±2,4,6	2.07	717.7	388.36	$[Ne]3s^2 3p^4$
Selenium	Se	78.96	±2,4,6	4.79	958	494	$[Ar]3d^{10}4s^2 4p^4$
Tellurium	Te	127.60	±2,4,6	6.24	1261	722.65	$[Kr]4d^{10}5s^2 5p^4$
Polonium	Po	(209)	4,2,6	9.32	1235	527	$[Xe]4f^{14}5d^{10}6s^2 6p^4$
Fluorine	F	18.99840	-1	1.696	85.03	53.53	$1s^2 2s^2 2p^5$
Chlorine	Cl	35.4527	±1,3,5,7	3.12	239.1	171.6	$[Ne]3s^2 3p^5$
Bromine	Br	79.904	±1,5	3.12	330.93	266.0	$[Ar]3d^{10}4s^2 4p^5$
Iodine	I	126.9045	±1,5,7	4.93	457.5	386	$[Kr]4d^{10}5s^2 5p^5$
Astatine	At	(210)	±1,3,5,7	—	610	575	$[Xe]4f^{14}5d^{10}6s^2 6p^5$
Helium	He	4.00260	0	0.1785	4.22	0.95	$1s^2$
Neon	Ne	20.1797	0	0.8990	27.07	24.56	$1s^2 2s^2 2p^6$
Argon	Ar	39.948	0	1.784*	87.30	83.81	$[Ne]3s^2 3p^6$
Krypton	Kr	83.80	0	3.733*	119.93	115.79	$[Ar]3d^{10}4s^2 4p^6$
Xenon	Xe	131.29	0	5.87	165	161.40	$[Kr]4d^{10}5s^2 5p^6$
Radon	Rn	(222)	0	9.73	211	202	$[Xe]4f^{14}5d^{10}6s^2 6p^6$

Lanthanide series

Element	Symbol	Atomic mass	Valence	Density	Boiling pt (K)	Melting pt (K)	Configuration
Cerium	Ce	140.115	3,4	6.770	3697	1071	$[Xe]4f^1 5d^1 6s^2$
Praseodymium	Pr	140.9077	3,4	6.77	3793	1204	$[Xe]4f^3 6s^2$
Neodymium	Nd	144.24	3	7.008	3347	1294	$[Xe]4f^4 6s^2$
Promethium	Pm	(145)	3	7.264	3273	1315	$[Xe]4f^5 6s^2$
Samarium	Sm	150.36	3,2	7.52	2067	1347	$[Xe]4f^6 6s^2$
Europium	Eu	151.965	3,2	5.244	1802	1095	$[Xe]4f^7 6s^2$
Gadolinium	Gd	157.25	3	7.90	3546	1586	$[Xe]4f^7 5d^1 6s^2$
Terbium	Tb	158.9253	3,4	8.230	3503	1629	$[Xe]4f^9 6s^2$
Dysprosium	Dy	162.50	3	8.551	2840.5	1685	$[Xe]4f^{10}6s^2$
Holmium	Ho	164.9303	3	8.795	2973	1747	$[Xe]4f^{11}6s^2$
Erbium	Er	167.26	3	9.066	3141	1802	$[Xe]4f^{12}6s^2$
Thulium	Tm	168.9342	3,2	9.321	2223	1818	$[Xe]4f^{13}6s^2$
Ytterbium	Yb	173.04	3,2	6.66	1469	1092	$[Xe]4f^{14}6s^2$
Lutetium	Lu	174.967	3	9.841	3675	1936	$[Xe]4f^{14}5d^1 6s^2$

Actinide series

Element	Symbol	Atomic mass	Valence	Density	Boiling pt (K)	Melting pt (K)	Configuration
Thorium	Th	232.0381	4	11.72	5061	2023	$[Rn]6d^2 7s^2$
Protactinium	Pa	231.0359	5,4	15.37	—	1845	$[Rn]5f^2 6d^1 7s^2$
Uranium	U	238.029	6,3,4,5	18.95	4404	1408	$[Rn]5f^3 6d^1 7s^2$
Neptunium	Np	237	5,3,4,6	20.25	4175	917	$[Rn]5f^4 6d^1 7s^2$
Plutonium	Pu	(244)	4,3,5,6	19.84	3501	913	$[Rn]5f^6 7s^2$
Americium	Am	(243)	3,4,5,6	13.67	2284	1449	$[Rn]5f^7 7s^2$
Curium	Cm	(247)	3	13.51	—	1618	$[Rn]5f^7 6d^1 7s^2$
Berkelium	Bk	(247)	3,4	14 (est.)	—	1323	$[Rn]5f^9 7s^2$
Californium	Cf	(251)	3	—	—	1173	$[Rn]5f^{10}7s^2$
Einsteinium	Es	(252)	3	—	—	1133	$[Rn]5f^{11}7s^2$
Fermium	Fm	(257)	3	—	—	1800	$[Rn]5f^{12}7s^2$
Mendelevium	Md	(258)	3	—	—	1100	$[Rn]5f^{13}7s^2$
Nobelium	No	(259)	2,3	—	—	—	$[Rn]5f^{14}7s^2$
Lawrencium	Lr	(262)	3	—	—	—	$[Rn]5f^{14}6d^1 7s^2$

(..) = Most stable isotope   Italic = Not found naturally on earth   † = Gas   ‡ = Liquid

Answers to Selected Exercises

Chapter 1

1. Avogadro's number is the number of particles in a mole: 6.022×10^{23} particles/mole. A mole contains Avogadro's number of particles.

3. (a) $250 \text{ mL} \times \dfrac{1 \text{ g}}{\text{mL}} \times \dfrac{1 \text{ mol}}{18 \text{ g}} \times \dfrac{6.022 \times 10^{23} \text{ molecules}}{\text{mol}}$

$= 8.36 \times 10^{24}$ molecules

(b) Surface area of earth: $4\pi r^2 = 4\pi(4000 \text{ mi})^2 \times \left(\dfrac{8 \text{ km}}{5 \text{ mi}}\right)^2$

$= 5.1 \times 10^8 \text{ km}^2$

Volume of oceans $= 5 \text{ km} \times 5.1 \times 10^8 \text{ km}^2 \times {}^3/_4 = 1.9 \times 10^9 \text{ km}^3$. $1 \text{ km} = 10^3 \text{ m} = 10^5 \text{ cm}$, so the volume of the oceans is $1.9 \times 10^{24} \text{ cm}^3 = 1.9 \times 10^{24} \text{ mL}$. A cup is 250 mL, so the oceans contain 7.6×10^{21} cups of water.

(c) There are many more molecules in a cup of water, about three orders of magnitude more (roughly 1000 times as many), than cups of water in the oceans.

5. 0.1246 nm or 124.6 pm

7. 6.022×10^{23} diameters $\times \dfrac{2 \text{ radii}}{\text{diameter}} \times \dfrac{134 \text{ pm}}{\text{radii}} \times \dfrac{1 \text{ m}}{10^{12} \text{ pm}}$

$= 1.6 \times 10^{14} \text{ m} = 1.6 \times 10^{11} \text{ km}$

9. (a) g/cm^3 or g/L; (b) molecules/cm^3 or molecules/L; (c) atoms/cm^3 or atoms/L

11. $0.41 \text{ g/tablet} \times 80 \text{ tablets} = 33 \text{ g}$
$0.41 \text{ g/tablet} \times 6.022 \times 10^{23} \text{ tablets} = 2.5 \times 10^{23} \text{ g}$

This is about 1/25 the mass of the earth!

13. $159 \text{ mg} \div 39.1 \text{ g/mol} = 4.07 \text{ mmol } (4.07 \times 10^{-3} \text{ mol})$

15. (a) The ratio of H to Cl in hydrogen chloride must be 1:1 since Avogadro's hypothesis is that equal volumes of all gases contain the same number of particles. (b) The number of particles of hydrogen chloride produced is twice the number of particles of hydrogen or chlorine. Each volume of hydrogen must contain twice as many hydrogen atoms as an equal volume of hydrogen chloride; the same is true for chlorine. Simple whole-number ratios mean that the formula for hydrogen must be H_2, chlorine must be Cl_2, and hydrogen chloride must be HCl.

17. There are equal numbers of molecules of hydrogen and fluorine. The number of molecules of hydrogen fluoride produced is twice the number of molecules of hydrogen or fluorine reacted. Consistent with these data, the formula of hydrogen is H_2, fluorine is F_2, and hydrogen fluoride is HF.

19. The information given is pressure and volume. The problem involves determining the initial volume from the dimensions. The final volume is determined by using the pressure drop and the initial volume. The number of balloons is determined by using this final volume and the volume per balloon.

Determine initial volume: A cylinder 450 cm^2 by $1 \text{ m} = 100 \text{ cm}$ has a volume of

$450 \text{ cm}^2 \times 100 \text{ cm} = 4.5 \times 10^4 \text{ cm}^3$

Determine the final volume:

$V_2 = \dfrac{P_1 V_1}{P_2} = \dfrac{(40 \text{ atm})(4.5 \times 10^4 \text{ cm}^3)}{1 \text{ atm}} = 1.8 \times 10^6 \text{ cm}^3$

Determine the volume per balloon:

30-cm diameter = 15-cm radius

$\text{Volume} = \dfrac{4}{3}\pi r^3 = \dfrac{4}{3}\pi(15 \text{ cm})^3 = 1.4 \times 10^4 \text{ cm}^3$

Number of balloons:

$\dfrac{\text{Total Volume}}{\text{Volume per Balloon}} = \dfrac{1.8 \times 10^6 \text{ cm}^3}{1.4 \times 10^4 \text{ cm}^3/\text{balloon}} = 128 \text{ balloons}$

The 129th balloon is only partly filled.

21. Information given is the initial pressure and the volume ratio. The information needed is the final pressure.

$P_2 = P_1 \dfrac{V_1}{V_2} = P_1 \times \dfrac{1}{0.2} = 0.85 \text{ atm} \times \dfrac{1}{0.2} = 4.3 \text{ atm}$

23. Tire pressure is measured relative to 1 atm; $22 \text{ lb/in}^2 + 15 \text{ lb/in}^2 = 37 \text{ lb/in}^2$. The initial pressure and both initial and final temperatures are given. However, the temperature is given in °C, so it must be converted to K by addition of 273. The final pressure is needed.

$P_2 = \dfrac{P_1 T_2}{T_1} = 37 \text{ lb/in}^2 \times \dfrac{320 \text{ K}}{273 \text{ K}} = 43 \text{ lb/in}^2$

$43 \text{ lb/in}^2 - 15 \text{ lb/in}^2 = 28 \text{ psig}$ (pounds per square inch gauge— that is, over atmospheric).

25. The information given is the initial volume and the initial and final temperatures. The information needed is the final volume. The initial volume needs to be calculated from the given radius.

Initial volume:

$V_1 = \dfrac{4}{3}\pi r^3 = \dfrac{4}{3}\pi(5 \text{ cm})^3 = 524 \text{ cm}^3$

$V_2 = V_1 \times \dfrac{T_2}{T_1} = 524 \text{ cm}^3 \dfrac{270 \text{ K}}{293 \text{ K}} = 483 \text{ cm}^3$

27. One mole of gas occupies 22.4 L at STP, so 2 L holds 2/22.4 mol or 0.089 mol.

29. (a) Pressure increases by a factor of 1.5, which decreases the volume. Temperature increases, which increases the volume, but the temperature increase is from 473 K to 573 K, a smaller factor than the pressure increase. Hence, the volume decreases.

(b) Pressure decreases by about half, which roughly doubles the volume. Temperature decreases, which decreases the volume. The temperature decreases by a factor of (247 K)/(523 K), or about $^1/_2$. The volume stays approximately the same. Calculating it exactly:

$\dfrac{V_2}{V_1} = \dfrac{P_1 T_2}{P_2 T_1} = \dfrac{760 \text{ Torr}}{350 \text{ Torr}} \times \dfrac{247 \text{ K}}{523 \text{ K}} = 1.03$

(c) Pressure decreases from 2 atm to 1 atm, doubling the volume. Temperature increases from 300 K to 600 K, again doubling the volume. The volume will increase greatly.

31. (a) Molecular density = $\dfrac{\text{density}}{\text{molecular weight}} \times N_A$

$$= \frac{2.66 \text{ g/L}}{64.06 \text{ g/mol}} \times \frac{6.022 \times 10^{23} \text{ molecules}}{\text{mol}}$$

$$= 2.50 \times 10^{22} \text{ molecule/L}$$

(b) There are 2 atoms of O per molecule of SO_2, thus there are 5.00×10^{22} atoms of oxygen per liter of SO_2 at 20 °C and 1 atm. Molecular density is 2.69×10^{22} molecules/L. The atomic density is $3 \times 2.69 \times 10^{22}$ atoms/L = 8.07×10^{22} atoms/L.

33. 12 °C is 285 K; 40 °C is 313 K.

$$V_f = \frac{T_f P_i}{T_i P_f} = \frac{(313 \text{ K})(750 \text{ torr})}{(285 \text{ K})(720 \text{ torr})} V_i \times 200 \text{ L} = 229 \text{ L}$$

35. 30 °C is 303 K; −20 °C is 253 K. Is 303/1 greater than or less than 253/0.6? 303/1 = 303; 253/0.6 = 422. Thus the volume increases.

37. Molecular density is 2.690×10^{22} molecules/L. Atomic density is $3 \times 2.69 \times 10^{22}$ atoms/L = 8.070×10^{22} atoms/L.

39. $V = \dfrac{nRT}{P}$

$$= \frac{(1 \text{ mol}) \times (0.08206 \text{ L} \cdot \text{atm/K} \cdot \text{mol}) \times (273.15 \text{ K})}{1 \text{ atm}}$$

$$= 22.4 \text{ L}$$

41. Unit analysis indicates that there are several conversions. Because the required answer is in feet, the distance in meters must be converted to feet: 1 in. is exactly 2.54 cm.

Conversion of meters to feet:

Know m		**Need** ft

$$m \longrightarrow cm \longrightarrow in. \longrightarrow ft$$

$$1 \text{ m} = 1 \text{ m} \times \frac{100 \text{ cm}}{1 \text{ m}} \times \frac{1 \text{ in.}}{2.54 \text{ cm}} \times \frac{1 \text{ ft}}{12 \text{ in.}} = 3.2808 \text{ ft}$$

$$h = \frac{1 \text{ atm}}{(1 \text{ g/cm}^3)(9.80665 \text{ m/s}^2)} \times \frac{1.01 \times 10^5 \text{ N/m}^2}{1 \text{ atm}}$$

$$\times \frac{1 \text{ kg} \cdot \text{m/s}^2}{1 \text{ N}} \times \frac{1000 \text{ g}}{\text{kg}} \times \left(\frac{\text{m}}{100 \text{ cm}}\right)^3$$

$$\times \frac{3.2808 \text{ ft}}{\text{m}} = 33.789 \text{ ft}$$

43. First convert pints to liters; then 22.4 L at STP is 1 mol. 1 qt = 2 pt = 0.9463 L. ³/₄ pt is about 0.35 L. 0.35 L/(22.4 L/mol) = 0.016 mol, about 1×10^{22} molecules.

45. The average molecular weight of air is $(0.8 \times 28.0 \text{ g/mol}) + (0.2 \times 32.0 \text{ g/mol}) = 28.8$ g/mol. At 30 KM, 45,000 m³ of air weighs

$$(45000 \text{ m}^3)\left(\frac{273 \text{ K}}{227 \text{ K}}\right)\left(\frac{8.36 \text{ torr}}{760 \text{ torr}}\right)\left(\frac{1 \text{ mol}}{22.4 \text{ L}}\right)\left(\frac{1 \text{ L}}{1000 \text{ cm}^3}\right)$$

$$\left(\frac{100 \text{ cm}}{\text{m}}\right)^3\left(\frac{28.8 \text{ g}}{\text{mol}}\right) = 765 \text{ kg}$$

The payload is 50 kg, so the air inside the balloon must be heated sufficiently to reduce the total mass to 765 kg − 50 kg = 715 kg. So the balloon must be heated to

$$T = \frac{(765 \text{ kg})(227 \text{ K})}{715 \text{ kg}} = 243 \text{ K, or } 16 \text{ °C above ambient.}$$

47. Water is more dense than ice at 0° C. This is proven by the fact that ice floats in water.

49. There are six carbon atoms and six hydrogen atoms in benzene.

51. There are fourteen carbon atoms and ten hydrogen atoms in benzoyl peroxide. The other element found in benzoyl peroxide is oxygen (four atoms).

53. Molar mass of CCl_4 is $12 + (4 \times 35.5)$ g/mol = 154 g/mol. Of this, 142 g is due to Cl. Thus 142/154 × 1 kg is required, or 922 g of Cl_2.

55. The atomic mass of hydrogen is 1.0 g/mol and that of bromine is 79.9 g/mol. Thus hydrogen accounts for $\dfrac{1 \text{ g}}{80.9 \text{ g}}$ of HBr. To produce 2.5 mg of HBr requires $\dfrac{1 \text{ g}}{80.9 \text{ g}} \times 2.5 \text{ mg} = 0.031$ mg of hydrogen.

57. $4Na + O_2 \rightarrow 2Na_2O$

59. $3H_2 + N_2 \rightarrow 2NH_3$

61. 1.5 volumes of O_2 are produced from a volume of O_3.

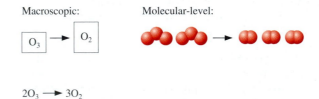

Macroscopic: Molecular-level:

$2O_3 \longrightarrow 3O_2$

63. Nitrous oxide, N_2O; nitric oxide, NO; nitrogen dioxide, NO_2; dinitrogen tetroxide, N_2O_4; dinitrogen pentoxide, N_2O_5

65. The mass of $Al_2O_3 = (2 \times 26.98) + (3 \times 16) = 101.96$ g/mol. Of this, 48 g is due to O_2, so to oxidize 1 kg of Al requires $(48/101.96) = 471$ g of O_2. If only 4.7 g of O_2 is incorporated into the object, only 1% of the aluminum has oxidized.

67. Molecular mass of octane is $(8 \times 12.01) + (18 \times 1.01) = 114.26$ g/mol.

Reaction: $C_8H_{18} + 12.5O_2 \rightarrow 8CO_2 + 9H_2O$

or $2C_8H_{18} + 25O_2 \rightarrow 16CO_2 + 18H_2O$

Mass of O_2 required: $(12.5 \times 32)/114.26 \times 1$ lb $= 3.50$ lb

Mass of CO_2 produced: $(8 \times 44)/114.26 \times 1$ lb $= 3.08$ lb $= 1400$ g CO_2

69. Molecular Reactions: $2C + O_2 \rightarrow 2CO$

$$Fe_2O_3 + 3CO \longrightarrow 2Fe + 3CO_2.$$

Three carbon atoms are required for every two iron atoms produced. This ratio requires 3×12 g $= 36$ g of C for 2×55.85 g $= 111.7$ g of Fe. A ton of Fe requires $36/111.7 \times 1$ ton $= 0.322$ ton $= 644.6$ lb of C.

The ratio of CO_2 to Fe produced is $(3 \times 44)/(2 \times 55.85) = 1.18$ ton of CO_2.

71. (a) is both a net ionic and an ionic equation, (b) is a total ionic equation, (c) is both a net ionic and an ionic equation, and (d) is a total ionic equation.

(b) net ionic: $H^+ + OH^- \rightarrow H_2O$

molecular: $2HBr + Ba(OH)_2 \rightarrow 2H_2O + BaBr_2$

(d) net ionic: $H_2SO_4 + Ni^{2+} \rightarrow NiSO_4 + 2H^+$

molecular: $H_2SO_4 + NiCl_2 \rightarrow NiSO_4 + 2HCl$

73. First write the balanced reaction to determine the proportions:

$$NH_3 + HNO_3 \longrightarrow NH_4NO_3$$

Determine the number of moles of ammonium nitrate in 50 lb, which is also the number of moles of nitric acid required. Multiplying by the molecular mass of nitric acid yields the number of pounds of nitric acid needed.

Formula mass of ammonium nitrate: 2×14.0067 g $= 28.0134$ g for N

$$4 \times 1.0079 \text{ g} = 4.0316 \text{ g for H}$$

$$3 \times 15.999 \text{ g} = 47.997 \text{ for O}$$

$$\text{Total} = 80.042 \text{ g/mol}$$

Number of moles in 50 lb is

$$50 \text{ lb} \times \frac{453.59 \text{ g}}{\text{lb}} \times \frac{1 \text{ mol}}{80.04 \text{ g}} = 283.3 \text{ mol}$$

Molecular mass of HNO_3:

1.0079 g from H

14.0067 g from N

3×15.999 g $= 47.997$ g from O

Total $= 63.012$ g/mol

$$63.012 \text{ g/mol} \times 283.34 \text{ mol} \times \frac{1 \text{ lb}}{453.59 \text{ g}}$$

$$= 39.36 \text{ lb of nitric acid}$$

An alternative approach in this case is to determine the fraction of the molar mass of NH_4NO_3 that is due to HNO_3 $(63.012/80.042 = 0.7872)$. Multiply this times 50 lb $= 39.36$ lb. This strategy works because all of the nitric acid is incorporated into the ammonium nitrate.

75. One volume of nitrogen contains the same number of molecules as one volume of oxygen. Two volumes of nitric oxide con-

tain twice as many molecules. Thus the ratio of nitrogen atoms in nitrogen gas to nitrogen atoms in nitric oxide is $2:1$. A similar ratio holds for oxygen. The simplest whole-number ratio consistent with this is for nitrogen gas to be N_2, oxygen gas to be O_2, and nitric oxide to be NO.

Macroscopic: Molecular-level:

$$O_2 + N_2 \longrightarrow 2NO$$

77. K.E. $= \frac{1}{2} mv^2 = \frac{1}{2}$ 1 mg $\times \dfrac{1 \text{ kg}}{10^6 \text{ mg}}$ (3 mi/h)$^2 \times \left(\dfrac{1 \text{ h}}{3600 \text{ s}}\right)$

$$\times \left(\frac{1609 \text{ m}}{\text{mi}}\right)^2 = 8.99 \times 10^{-7} \text{ J}$$

The wing tip traces a semicircle with a radius of 3 mm; the length of the semicircle is $(3 \text{ mm})\pi = 9.425$ mm. Beating at 1000 times per second, the velocity is 9425 mm/s.

K.E. $= \frac{1}{2} mv^2$

$$= \frac{1}{2} 1 \mu g \times \frac{1 \text{ kg}}{10^9 \mu g} (8425 \text{ mm/s})^2 \times \left(\frac{1 \text{ m}}{1000 \text{ mm}}\right)^2$$

$$\times \left(\frac{1 \text{ h}}{3600 \text{ s}}\right)^2 \times \left(\frac{1609 \text{ m}}{\text{mi}}\right)^2 = 7.09 \times 10^{-9} \text{ J}$$

The energy of the wing tip is two orders of magnitude smaller than the kinetic energy of the mosquito in the air. Only 1% of the energy is expended to keep the wing tip aloft.

79. P.E. $= mgh = 140$ lb $\times \dfrac{0.4536 \text{ kg}}{\text{lb}} \times (9.81 \text{ m/s}^2)(3 \text{ m})$

$$= 1870 \text{ J} = 447 \text{ cal}$$

The energy comes from the food that the person eats. Expending 0.447 Cal per flight, it takes 224 flights to expend 100 Cal.

81. P.E. $= -k\dfrac{q_1 q_2}{r}$

$$= -\frac{1}{4\pi \times 8.854 \times 10^{-12} \text{ C}^2/\text{J}\cdot\text{m}}$$

$$\times \frac{(1.60218 \times 10^{-19} \text{ C})(-1.60218 \times 10^{-19} \text{ C})}{52.9 \times 10^{-12} \text{ m}}$$

$$= 4.36 \times 10^{-18} \text{ J}$$

Energies only become of measurable size when a large number of atoms is considered.

83. The gravitational potential for a 10-g mass 1 m above the earth: $= mgh = (0.010 \text{ kg}) \times (9.81 \text{ m/s}^2)(1 \text{ m}) = 0.098$ J

$$q_2 = -\frac{r \times \text{P.E.}}{kq_1}$$

$$= -\frac{4\pi \cdot 8.854 \times 10^{-12} \text{ J}^{-1}\text{C}^2\text{m}^{-1}(1 \text{ m})(0.098 \text{ J})}{(-4.3 \times 10^5 \text{ C})}$$

$$= 2.54 \times 10^{-17} \text{ C}$$

85. 1 W = 1 J/s, so 6 mJ falls on the screen in one minute. The electric field contains 3 mJ of energy.

87. Many diagrams are possible. One example is shown here.

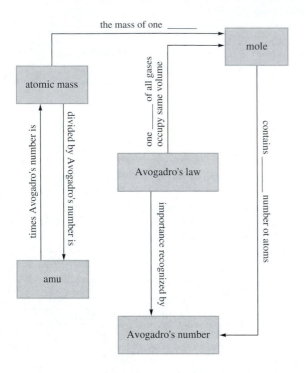

89. (a) The density of water at 25° C (room temperature) is 1 g/cm³. The molecular mass of water is 18 g/mol, so the molar volume is 18 cm³/mol.
(b) At 0° C and 1 atm pressure, one mole of gas occupies 22.4 L.
(c) 400° F is (400 − 32) × 5/9° C = 204° C. At 204° C the volume is

$$V_{400° \text{ F}} = \frac{477}{273} \times V_{0° \text{C}} = \frac{477}{273} \times 22.4 \text{ L} = 39.1 \text{ L.}$$

At 212° F = 100° C = 373 K, the volume of steam is 30.6 L. There is approximately 25% more loft with baking at 400° F. In addition, at 400° F water turns to steam faster and a bit more completely than at 212° C.

91. The mass per liter for H_2 is 2 g/22.4 L = 0.089 g/L. The mass per liter for He is 4 g/22.4 L = 0.179 g/L. Air is 78% N_2, which contributes 0.78 × 28 g = 21.84 g per 22.4 L. Oxygen contributes 0.22 × 32 g = 7.04 g per 22.4 L, so the mass of air is

$$\frac{(21.84 \text{ g} + 7.04 \text{ g})}{22.4 \text{ L}} = \frac{28.88 \text{ g}}{22.4 \text{ L}} = 1.29 \text{ g/L}$$

The buoyant force for H_2 is 0.089 g/L − 1.29 g/L = −1.20 g/L. (The negative sign indicates that the force is upward, or a lift.)
The buoyant force for He is 0.179 g/L − 1.29 g/L = −1.11 g/L. The conclusion is that He provides 92.5% of the lift of hydrogen. The builders of the *Hindenburg* took a terrible risk for very little gain.

93. $\left(\dfrac{2.6989 \text{ g}}{\text{cm}^3}\right) \times \left((66.67 \text{ yd} \times \dfrac{36 \text{ in.}}{\text{yd}} \times \dfrac{2.54 \text{ cm}}{\text{in.}})\right.$

$\left. \times (12 \text{ in.} \times \dfrac{2.54 \text{ cm}}{\text{in.}}) \times (6.5 \times 10^{-4} \text{ in.} \times \dfrac{2.54 \text{ cm}}{\text{in.}})\right)$

$= 830 \text{ g}$

Area of foil = $(66.67 \text{ yd} \times \dfrac{36 \text{ in.}}{\text{yd}}) \times 12 \text{ in.} = 2.88 \times 10^4 \text{ in}^2$

Mass per square inch = 830 g/2.88 × 10⁴ in² = 0.0288 g/in²

$0.0288 \text{ g/in}^2 \times \dfrac{1 \text{ mol}}{26.98 \text{ g}} \times \dfrac{6.022 \times 10^{23} \text{ atoms}}{\text{mol}}$

$= 6.43 \times 10^{20} \text{ atoms/in}^2 \text{ of foil.}$

The radius of Al is 118 pm, so there are

$\dfrac{1 \text{ in.}}{(118 \text{ pm} \times 2)/\text{atom}} \times \dfrac{1 \text{ pm}}{10^{-10} \text{ cm}} \times \dfrac{1 \text{ cm}}{2.54 \text{ in.}}$

$= 1.67 \times 10^7 \text{ atoms in 1 in.}$

$\Rightarrow 2.79 \times 10^{14} \text{ atoms per in}^2 \text{ layer of atoms}$

$\Rightarrow \dfrac{6.43 \times 10^{20} \text{ atoms/in}^2 \text{ foil}}{2.79 \times 10^{14} \text{ atom/layer}} = 2.30 \times 10^6 \text{ layers}$

95. (a) The oil rises when heated, so its density must decrease to be less than that of water. (b) The density of water decreases as it is heated. (c) The thermal coefficient of expansion of oil must be greater than that of water given that both the oil and the water are heated by the bulb, yet the oil rises in the water.

Chapter 2

1. The third period consists of 8 elements: Na−Ar (11−18). The fifth row consists of 18 elements: Rb−Xe (37−54). The number of elements in each period of the periodic table follows the pattern: 2, 8, 8, 18, 18, 36, . . .

3. The evidence for bonding being electrical in nature is the result that completing a circuit through some molten material (either a molten salt or a solution) causes the material to break down into its constituent elements.

5. The primary experiment demonstrating that matter consists of a nonuniform distribution of density is Rutherford's scattering experiment. A uniform distribution cannot produce such an anisotropic distribution of scattered particles. There must be lumps or areas where matter is very dense to give the extreme back scatter of a few α particles.

7. The mass of the electron is approximately 10^{-3} that of the nucleus. If the nucleus is represented by 2000 lb, then the electron mass is 2 lb. If the nucleus is 2 m in diameter, then the electron fills a volume of about 10^4 times larger, or 20 km. The mass of the proton is about 2000 times larger than the mass of the electron. The radius of the proton is about 10^{-4} that of the electron cloud, so the volume is about 10^{-12} the volume of the electron. Thus the density of the proton is about 2×10^{15} larger than the density of the electron.

9. About 1/2000 of the mass is due to electrons, or about 0.1 g. The radius of the nucleus is about 10^{-4} the radius of the atom, so the volume of the nucleus is about 10^{-12} that of the atom. The total volume is 200 cm³, so the volume of the nuclei is about 2×10^{-10} cm³, or about 200 μm³.

11.
$$\frac{26.98 \text{ g}}{\text{mol}} \times \frac{1 \text{ cm}^3}{2.702 \text{ g}} = 9.985 \text{ cm}^3/\text{mol}$$

In a cube 1 cm on a side, the volume is 1 cm³. In this volume there is 0.10 mol Al atoms. The volume occupied by the nuclei is $(10^{-4})^3 \times 1 \text{ cm}^3 = 10^{-12} \text{ cm}^3$.

13. The notations are $_{26}^{54}\text{Fe}$, $_{26}^{56}\text{Fe}$, $_{26}^{57}\text{Fe}$, and $_{26}^{58}\text{Fe}$, respectively.

15. The highest-energy wave is the one with the shortest wavelength: wave d. The highest-intensity wave is the one with the greatest amplitude, wave b. The wavelengths are (a) 0.6 cm, (b) 0.35 cm, (c) 0.2 cm, and (d) 0.1 cm. There are two nodes per wavelength, so wave (a) has a node every 0.3 cm or 333 nodes/m; wave (b) has a node every 0.17 cm or 588 nodes/m; wave (c) has a node every 0.1 cm or 1000 nodes/m; and wave (d) has a node every 0.05 cm or 2000 nodes/m.

17. The wavelength would be 4 m. The frequency is c/λ or $7.5 \times 10^7 \text{ s}^{-1}$. This wave would be found in the FM radio wave region.

19. (a) Frequency $= \nu = c/\lambda = 4.469 \times 10^{14} \text{ s}^{-1}$. (b) Energy $= E = hc/\lambda = 2.963 \times 10^{-19} \text{ J} = 1.850 \text{ eV}$. (c) 670.8 nm is red.

21. $\lambda = c/\nu = 590 \text{ nm}$; $3.37 \times 10^{-19} \text{ J} = 2.0 \text{ eV}$

23. The Balmer equation is an empirical equation because it mathematically fits the observational data but is not based on a physical model.

25. These lines are in the visible spectrum and are part of the Balmer series. The first, 656.3 nm, is the $3 \rightarrow 2$ transition. The next lines are the $4 \rightarrow 2$, $5 \rightarrow 2$, and $6 \rightarrow 2$ transitions. The next line in the series, the $7 \rightarrow 2$ transition, occurs at

$$\frac{1}{\lambda} = R\left(\frac{1}{n_1^2} - \frac{1}{n_2^2}\right) = 1.097 \times 10^7 \text{ m}^{-1}\left(\frac{1}{2^2} - \frac{1}{7^2}\right)$$
$$= 1.097 \times 10^7 \text{ m}^{-1}(0.23) \text{ m}^{-1} = 2.52 \times 10^6 \text{ m}^{-1} \Longrightarrow$$
$$\lambda = 397.0 \text{ nm}$$

27.
$$\frac{1}{\lambda} = R\left(\frac{1}{n_1^2} - \frac{1}{n_2^2}\right) \Longrightarrow \frac{1}{R\lambda} = \left(\frac{1}{n_1^2} - \frac{1}{n_2^2}\right)$$
$$= \frac{1}{(1.097 \times 10^7 \text{ m}^{-1}) \times 121.6 \times 10^{-9} \text{ m}}$$
$$= (0.7497) \Longrightarrow n_1 = 1 \text{ and } n_2 = 2.$$

The hydrogen atom is in the ground state before the absorption and in the first excited state after absorbing the photon.

29. (a) From $n = 4$, the electron can relax to $n = 3$, 2, or 1. Once on $n = 3$, it can relax to $n = 2$ or 1. From $n = 2$, it can relax to $n = 1$. This is a total of six lines.

(b) The frequency of the lines is given by $\nu = Rc\left(\frac{1}{n_1^2} - \frac{1}{n_2^2}\right)$; the frequencies are $(1.097 \times 10^7 \text{ m}^{-1})(2.9979 \times 10^8 \text{ m} \cdot \text{s}^{-1}) = 3.289 \times 10^{15} \text{ s}^{-1}$ times the factor from the n's or (in units of s^{-1}):

$n = 4 \rightarrow 3$	$4 \rightarrow 2$	$4 \rightarrow 1$
1.60×10^{14}	6.15×10^{14}	3.09×10^{15}

$3 \rightarrow 2$	$3 \rightarrow 1$	$2 \rightarrow 1$
4.57×10^{14}	2.92×10^{15}	2.47×10^{15}

(c) The energy-level diagram is shown here. Of these transitions, only $n = 4 \rightarrow 2$ and $n = 3 \rightarrow 2$ are in the visible region of the spectrum.

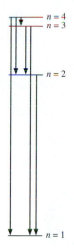

31. $m\nu r = n\hbar \Rightarrow \nu = n\hbar/mr$.

33. From Exercise 31: $\Rightarrow \nu = n\hbar/mr$.

$$\text{K.E.} = \frac{1}{2}m\nu^2 = \frac{1}{2}\frac{mn^2\hbar^2}{m^2r^2} = \frac{1}{2}\frac{mn^2\hbar^2 e^4 m^2}{m^2 16\pi^2 \varepsilon_o^2 \hbar^4 n^4}$$
$$= \frac{1}{2}\frac{me^4}{16\pi^2 \varepsilon_o^2 \hbar^2 n^2}$$

35. Energy $= \text{K.E.} + \text{P.E.} = -\frac{1}{2}\frac{e^4 m}{16\pi^2 \varepsilon_o^2 \hbar^2 n^2} \Longrightarrow R$
$$= -\frac{1}{2}\frac{e^4 m}{16\pi^2 \varepsilon_o^2 \hbar^2}$$

37. (a) The electron energy gets closer to zero with increasing n; it varies as $1/n^2$.
(b) The distance from the nucleus increases with increasing n; it varies as n^2.

39. Most of the lines in the spectrum of hydrogen correspond to the electron absorbing less than 13.6 eV. At 13.6 eV, hydrogen ionizes. Energy from ground state to ionization = 13.6 eV. All transitions shown below are lower in energy than ionization.

41. (a) 97.25 nm is in the short UV range of the electromagnetic spectrum. (b) Find n such that

$$\frac{1}{\lambda} = R\left(1 - \frac{1}{n^2}\right) \implies n = \left(\sqrt{1 - \frac{1}{R\lambda}}\right)^{-1}$$

$$= \left(\sqrt{1 - \frac{1}{(1.097 \times 10^7 m^{-1})(97.25 \times 10^{-9} m)}}\right)^{-1} = 4$$

(c) The ionization energy of the electron in the $n = 4$ state is 13.6 eV/16 = 0.85 eV.

43. (a) An atom absorbs a photon by the electron being lifted to a higher-energy state. (b) Emission corresponds to the high-energy electron relaxing to a lower-energy state.

(c)

Photon in

(d) The schematic for photon emission is the opposite of that for absorption: The atom starts out large and shrinks as the energy needed to maintain the electron in a high-energy orbit is emitted.

45. The energies of the remaining three visible transitions are 2.55 eV, 2.86 eV, and 3.02 eV, respectively. The wavelengths are 486.2 nm, 434.1 nm, and 410.2 nm, respectively.

47. The frequency of this transition is 3.16×10^{15} Hz, which is deep in the ultraviolet part of the spectrum.

49. This line is in the visible region of the spectrum, so it must be part of the Balmer series. Determine $1/\lambda$ in meters:

$$\frac{1}{\lambda} = \frac{1}{656.28\ nm} \times \frac{1\ nm}{10^{-9}\ m} = 1.52 \times 10^6\ m^{-1}$$

The Rydberg constant is $1.097 \times 10^7\ m^{-1}$.

$$1.52 \times 10^6\ m^{-1}/1.097 \times 10^7\ m^{-1} = 0.138 = \left(\frac{1}{2^2} - \frac{1}{3^2}\right).$$

The transition is from $n = 3$ to $n = 2$, which is in the Balmer series. The $n = 2$ state is not the ground state. To return to the ground state, the atom must emit a photon of energy (13.6 − 3.4) eV or 10.2 eV.

51. The energy of the $n = 1$ level is −13.59 eV. The energy of $n = 3$ is −13.59/(3²) eV = −1.51 eV. The difference is 12.08 eV.

53. The Bohr model does not explain how the electron fills most of the volume of the atom. It applies only to hydrogen and hydrogen-like atoms (so it applies to no other neutral element). It does not provide a model for the number of elements in each row of the periodic table. The most important results of the wave model are a model for how the electron fills the volume of an atom and an explanation of why there is a limit to the number of electrons that can go into any shell.

55. Wave amplitude, ψ, is an indication of the displacement of the wave and its phase. Density, $|\psi|^2$, is a measure of the probability of locating an electron with a given wave at a given location.

57. The mode has two circular nodes and one linear node. With three nodes, this mode is more energetic than the one with only two nodes. It is $n = 4$, $\ell = 1$, or a $4p$ mode.

59. The 2s orbital has a change of phase from the center to the outer edge, the 3p orbital has both a change of phase from the cen-

ter to the outer edge and a node through the center, and the $3d$ orbital has two nodes through the center. The relative sizes are $1:2.25:2.25$ for the $2s$, $3p$, and $3d$ orbitals respectively.

2s   3p   3d

61. A $4s$ orbital has three nodes, all spherical. A $5s$ orbital has four nodes, all spherical. The $4s$ orbital has three nodes starting from the center outward: crest, node, trough, node, crest, node, and trough. The $5s$ starting from the center outward: crest, node, trough, node, crest, node, trough, node, and crest.

63. A $4s$ orbital has three nodal surfaces, and a $3d$ orbital has two nodal surfaces. A $4s$ orbital has three spherical nodes and no planar nodes. A $3d$ orbital has two planar nodes, and no spherical nodes.

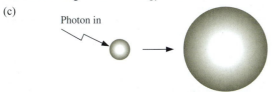

4s   3d

65. Three quantum numbers are required to specify an orbital: n, the principal quantum number; ℓ, the angular momentum or azimuthal quantum number; and m_ℓ, the magnetic quantum number.

67. (a) n indicates the number of nodes ($= n - 1$); (b) $\ell =$ number of planar nodes; (c) m_ℓ orientation of planar nodes

69. (a) A $2d$ orbital cannot occur because d implies two planar nodes, but $n = 2$ can have only one node. (b) A g orbital has four planar nodes, and $n = 6$ can have five nodes, so $6g$ can occur. (c) A p orbital has one planar node, and $n = 6$ can have five nodes, so $6p$ can occur.

71. The first three cannot occur: (a) and (c) because ℓ must be $n - 1$ or less, and (b) because m_ℓ is restricted to values between $+\ell$ and $-\ell$. The set in (a) would work if n were increased to five, (b) would work if m_ℓ were changed to zero, and (c) would work if n were changed to five and, since m_s cannot equal zero, change m_s to $\pm\frac{1}{2}$.

73.

Shell Number	Maximum Number of Nodes	Number of Angular Nodes	n	ℓ	m_ℓ	m_s	Electrons in Shell
2	1	0	2	0	0	$\pm\frac{1}{2}$	8
	1	1	2	1	1	$\pm\frac{1}{2}$	
	1	1	2	1	0	$\pm\frac{1}{2}$	
	1	1	2	1	−1	$\pm\frac{1}{2}$	

75.

	n	ℓ	m_ℓ	m_s
(a) $2s$	2	0	0	$\pm\frac{1}{2}$
(b) $2p$	2	1	1	$\pm\frac{1}{2}$
		1	0	$\pm\frac{1}{2}$
		1	-1	$\pm\frac{1}{2}$
(c) $3d$	3	2	2	$\pm\frac{1}{2}$
		2	1	$\pm\frac{1}{2}$
		2	0	$\pm\frac{1}{2}$
		2	-1	$\pm\frac{1}{2}$
		2	-2	$\pm\frac{1}{2}$

The maximum number of electrons is two electrons in a $2s$ orbital, six electrons in a $2p$ orbital, and ten electrons in a $3d$ orbital.

77. (a) The d orbitals hold a maximum of ten electrons.
(b) The f orbitals hold a maximum of 14 electrons.
(c) The h orbitals hold a maximum of 22 electrons.
Explanation: A d subshell can have five values of $\ell (\pm 2, \pm 1, 0)$, and each can have $m_s = \pm\frac{1}{2}$. $5 \times 2 = 10$. An f subshell can have seven values of $\ell (\pm 3, \pm 2, \pm 1, 0)$, and each can have $m_s = \pm\frac{1}{2}$. $7 \times 2 = 14$. An h subshell can have 11 values of $\ell (\pm 5, \pm 4, \pm 3, \pm 2, \pm 1, 0)$, and each can have $m_s = \pm\frac{1}{2}$. $11 \times 2 = 22$.

79. In a $3d$ orbital, $n = 3$, $\ell = 2$. m_ℓ can be 2, 1, 0, -1, -2, and $m_s = \pm\frac{1}{2}$. The possible quantum numbers are $(3, 2, 2, \frac{1}{2})$, $(3, 2, 2, -\frac{1}{2})$, $(3, 2, 1, \frac{1}{2})$, $(3, 2, 1, -\frac{1}{2})$, $(3, 2, 0, \frac{1}{2})$, $(3, 2, 0, -\frac{1}{2})$, $(3, 2, -1, \frac{1}{2})$, $(3, 2, -1, -\frac{1}{2})$, $(3, 2, -2, \frac{1}{2})$, and $(3, 2, -2, -\frac{1}{2})$.

81.

Orbital	Total Nodes	Planar Nodes	Spherical Nodes
$4s$	3	0	3
$3p$	2	1	1
$6f$	5	3	2
$10g$	9	4	5

83. The radial amplitude is plotted below. The smallest n for which a p orbital can exist is $n = 2$.

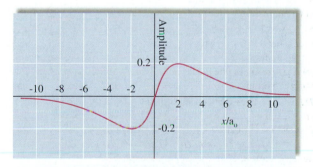

85. To be a sphere, the angular dependence of the density must be a constant. Thus, the problem is to show that the sum of the square of the amplitude of the p_x, p_y, and p_z orbitals does not depend on θ or ϕ.

$$|p_x|^2 = \frac{3}{4\pi} \sin^2\theta \cos^2\phi; \quad |p_y|^2 = \frac{3}{4\pi} \sin^2\theta \sin^2\phi;$$

$$|p_z|^2 = \frac{3}{4\pi} \cos^2\theta$$

Adding $|p_x|^2$ and $|p_y|^2$ gives $\dfrac{3}{4\pi} \sin^2\theta$. Adding this to $|p_z|^2$ yields $\dfrac{3}{4\pi}$. This is a constant, so the sum is a sphere.

87. The factors include the shell number, the number of valence electrons, the number of d-orbital electrons in the $n - 1$ shell, and a finer variation from a half-filled subshell.

89. The energy of an orbital depends only on the shell number for hydrogen because there are no other electrons. However, in carbon, there are other electrons. The s-orbital electron penetrates to the nucleus more effectively than a p-orbital electron, so it is lower in energy.

91. $P > K > Al > Na$

Element	P	K	Al	Na
Ionization energy	10.486	4.34	5.986	5.139
Shell	3	4	3	3
Effective charge	$3\sqrt{\dfrac{10.486}{13.6}}$ $= 2.63$	$4\sqrt{\dfrac{4.34}{13.6}}$ $= 2.26$	$3\sqrt{\dfrac{5.986}{13.6}}$ $= 1.99$	$3\sqrt{\dfrac{5.139}{13.6}}$ $= 1.84$

93. The $3p$ to $3s$ transition in Na occurs at about 589 nm, which is in the yellow region of the spectrum. In contrast, 383 nm is well into the UV region. The transition is higher energy in Mg because the core electrons plus the one other valence electron shield the nuclear charge of Mg less well than the core electrons shield the nuclear charge of Na.

   The ionization energy of Mg is 7.646 eV. The energy of the $3p \rightarrow 3s$ transition (383 nm) is 3.24 eV. Thus the ionization energy of the $3p$ electron is 7.646 eV $-$ 3.24 eV $=$ 4.41 eV. Z_{eff} is 1.71 for an electron in a $3p$ orbital in Mg compared with $Z_{eff} = 2.25$ for an electron in a $3s$ orbital.

95. The Pauli exclusion principle states that no two electrons can have the same set of quantum numbers. Along with the quantum mechanical model, it helps determine the configuration of the elements. The quantum mechanical model indicates the number of orbitals in each subshell and the number of subshells in each shell. Together with the Pauli exclusion principle, it indicates the number of electrons that occupy any given set of orbitals and any given shell.

97. Core electrons are not considered when discussing the chemistry of the elements because they are held too tightly to be involved in ordinary chemical reactions and therefore are not given up to participate in bonding. Core electrons consist of a filled spherical shell, so they shield the nuclear charge rather effectively. This makes it unlikely that the element will participate in chemistry by taking on another electron, which would have to be outside this stable configuration.

99.

Element	Fe	Sn	W	Ni	Ba
Valence electrons	8	4	6	10	2

The $4d$-orbital-electrons in Sn are not considered valence electrons. The $4f$ orbital electrons in W are not considered valence electrons.

101. F has only two core electrons. Ge has 28 core electrons, which include the $3d$ orbital electrons since they do not participate in the chemistry of Ge. The valence configuration of F is $2s^22p^5$. The valence configuration of Ge is $4s^24p^2$.

103. Elements in a group in the periodic table all have similar valence configurations. Elements in succeeding periods have more core electrons. Similar valence configuration is the basis of the periodic law: Elements with a common valence configuration have similar chemical and physical properties.

105. The effective charge increases down the column. This pattern occurs because there are increasingly more charges to be shielded. As the shell number increases, the electrons are spread over a larger volume and, for shells four and higher, the lower-numbered shells are incomplete.

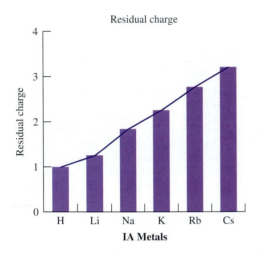

Residual charge

IA Metals

107. If 10% of the energy comes out in the visible region, then there are 7.5 watts of visible energy. Since 7.5 watts is 7.5 J/s, in one second 7.5 J is emitted. Use $E = hc/\lambda$ to determine the energy of one 500-nm photon, then divide to determine the number of 500-nm photons in 7.5 J.

$$E = \frac{hc}{\lambda} = \frac{(6.6262 \times 10^{-34} \text{ J·s})(2.9979 \times 10^8 \text{ m/s})}{(500 \text{ nm})(1.0 \times 10^{-9} \text{ m/nm})}$$
$$= 3.97 \times 10^{-19} \text{ J}$$

Thus 1.89×10^{19} photons are emitted per second.

109. $0.800 \text{ g}/107.8682 \text{ g/mol} = 7.42 \times 10^{-3} \text{ mol}$
To deposit one mole, the current must flow for $1/7.42 \times 10^{-3}$ times as long: $(1/7.42 \times 10^{-3}) \times \frac{1}{2} \text{ h} = 67.4 \text{ h}$. An amp is one C/s. In 30 min there are $30 \times 60 = 1800$ s. The current is

$$\frac{(7.42 \times 10^{-3} \text{ mol})(1.602 \times 10^{-19} \text{ C/electron})(6.022 \times 10^{23} \text{ electron/mol})}{1800 \text{ s}}$$
$$= 0.400 \text{ amp or 400 milliamp.}$$

111. $\lambda = h/p = h/mv$

$$\frac{6.6262 \times 10^{-34} \text{ J·s}}{(9.1094 \times 10^{-28} \text{ g})(400000 \text{ m·s}^{-1})} \times \frac{1000 \text{ g}}{\text{kg}} = 1.82 \text{ nm}$$

Nitrogen: 391 nm is beyond the visible; 470.0 nm is green.
Oxygen: 557.7 nm is yellow; 630 nm is red.

113. The yellow color is due to sodium, which is often added in salt to foods or is present in foods naturally. The yellow color

indicates that two electronic states in the sodium atom are separated by an energy equal to a yellow photon, approximately 2.25 eV.

115. Hydrogen's placement in Group IA is supported by having only one electron and by forming 1:1 compounds with the halogens. Hydrogen is also like the halogens (Group VIIA) in that it forms a 2:1 compound with oxygen, gain of one electron gives it the same number of electrons as a noble gas, and its first ionization energy is similar to that of the halogens. Like the halogens, hydrogen is a diatomic gas in its elemental form. Because hydrogen has properties of both Groups IA and VIIA, it is sometimes placed halfway between them. Alternatively, because the properties of hydrogen are in many ways unique, this element gets a unique placement in the middle of the top of the table.

117. The difference in the path length to the pit and that to the base plane is twice the depth of the pit. If the pit is $\frac{1}{4}\lambda$ deep, then the difference in the two path lengths is $\frac{1}{2}\lambda$. At $\frac{1}{2}\lambda$ path difference, the wave reflecting off the pit will be exactly out of phase with that reflecting off the base plane. The two waves will destructively interfere. The pit length of 0.83 μm is 830 nm, or 1.064 times the wavelength of the light used to read the information. If the same proportion applies, then a 635-nm read laser should be able to read pits that are 680 nm or 0.68 μm long. The pits in a DVD are 0.82 times smaller than those on a CD, so the DVD information is stored in a track at 1.22 times the density of that on a CD. Information storage is in a plane, so the DVD can store $(1.22)^2 = 1.49$, or slightly less than one and a half times as much information as a CD.

119.

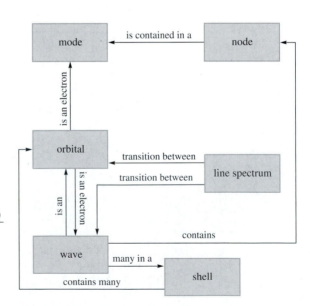

121. Many diagrams are possible. Two examples are shown here

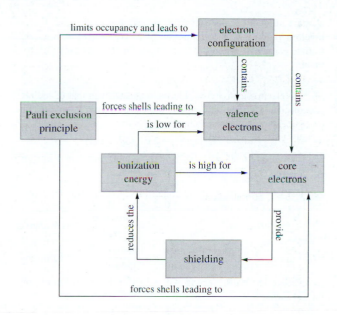

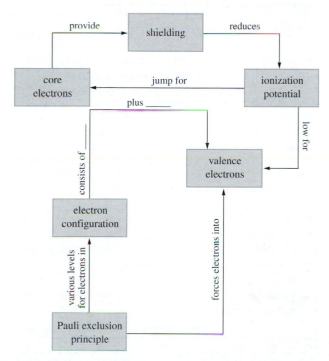

Chapter 3

1. All of the noble gases were missing from the table because they were not found in any compounds. Several years after Mendeléeff published his table, the noble gases were isolated from air.

3. The term "coinage metal" refers to the three elements Cu, Ag, and Au, which have been used for coins since very early times. These three elements form Group IB in the transition elements.

5. The increase of about 4 amu from a halogen to the next alkali metal when the general trend is about two units indicates a group is missing.

7. There are about 93 mass units between K and Cs, and the atomic number increases by 36 between K and Cs: 18 elements in Period 4 and 18 in Period 5. The mass increase should be approximately evenly split among the intervening elements. Thus, Rb should have an atomic mass of about $39 + 46 = 85$ amu. The actual atomic mass is 85.5 amu.

9. (a) 3; (b) 6; (c) 2; (d) 3

11. There are 44 possibilities. Here are some examples.

Name	Symbol	Period	Group
nitrogen	N	2	VA
oxygen	O	2	VIA
hydrogen	H	1	IA
chlorine	Cl	3	VIIA
iodine	I	5	VIIA

13. There are about 88 possibilities. Here are some examples.

Name	Symbol	Period	Group
sodium	Na	3	IA
potassium	K	4	IA
aluminum	Al	3	IIIA
iron	Fe	4	VIIIB
lead	Pb	6	IVA

15. Any five of the following:

Name	Symbol	Period	Group
boron	B	2	IIIA
silicon	Si	3	IVA
germanium	Ge	4	IVA
arsenic	As	4	VA
antimony	Sb	5	VA
tellurium	Te	5	VIA

17. For the atom, shielding refers to the effective charge holding the outermost electron. Since the atom is neutral, this outer electron must necessarily provide the last charge to balance that of the nucleus. Thus there are n charges in the nucleus and $n - 1$ negatively charged electrons without the last one. The last electron must necessarily be attracted by at least one charge. The logic is similar for the ion; without the last electron, the ion would be neutral. If the electrons of the atom perfectly shielded the last electron, the effective charge would be zero.

19. Many ions are isoelectronic with Ar. The ions closest to argon are Cl^-, K^+, and Ca^{2+}.

Ion	Cl^-	K^+	Ca^{2+}
Parent atom electron configuration	$[Ne]3s^23p^5$	$[Ar]4s^1$	$[Ar]4s^2$

21. Cr: $[Ar]4s^13d^5$; Cu: $[Ar]4s^13d^{10}$; Mo: $[Kr]5s^14d^5$; Co^{2+}: $[Ar]3d^7$

23. S: $[Ne]3s^23p^4$; S^{2-}: $[Ne]3s^23p^6$. S^{2-} is isoelectronic with Ar. Al^{3+}: $[Ne]$. Al^{3+} is isoelectronic with Ne.

25. Cl: $[Ne]3s^23p^5$; Cl^-: $[Ne]3s^23p^6 = [Ar]$. Argon is isoelectronic with Cl^-.

27. (a) AlF_3; (b) Ca_3N_2; (c) GeH_4; (d) PH_3

29. The ionization energy of S is expected to be less than that of P because S has a single electron outside a half-filled set of p orbitals. The ionization energy of S is 10.360 eV and that of P is 10.486 eV, so the prediction is correct.

31. The highest first-ionization energy should be found on the upper-right portion of the table: He. The lowest first-ionization energy should be found on the lower-left portion: Fr.

33. Cs

35. Ionization energy generally increases across the period due to the increased nuclear charge and increasingly ineffective shielding by same-shell electrons. Dips occur at Al and at S because each of these elements has a single electron outside a spherical configuration: a filled s orbital for Al and a half-filled set of p orbitals for S.

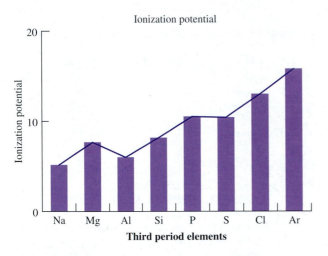

Ionization potential

37. The ionization energy of the anion is the electron affinity of the atom. For Br this is 2.96 eV, for I this is 2.66 eV. For Br^-, $n = 4$. For I^-, $n = 5$.

For Br^- $Z_{eff} = 4\sqrt{\dfrac{2.96\ eV}{13.6\ eV}} = 1.89$;

for I^- $Z_{eff} = 5\sqrt{\dfrac{2.66\ eV}{13.6\ eV}} = 2.21$

Compared to 1.00 for F^- and 1.546 for Cl^-, each of these is slightly larger than that of Cl^-.

39. Electron affinity: $atom(g) + e^-(g) \rightarrow ion^-(g) + energy$. The element with the highest electron affinity is fluorine.

41. The halogens have very high electron affinities because they are one electron short of complete s and p valence subshells.

43. The electron affinity of P is less than that of S because P has a half-filled p subshell.

45. The candidates are the Group IIA elements, the Group IIB elements, and the Group VIIB elements. Any four of these are acceptable.

47. Electronegativity is the tendency to attract electrons in a compound or solid. Electron affinity is the energy with which an

element attracts an electron to it *in isolation in the gas phase.* Electronegativity is a combination of electron affinity and ionization energy. It smoothes the bumps in the electron affinity trend.

49. $H < C < O < F$

51. The electronegativity of chlorine is greater than that of iodine, so molecular chlorine will capture the electron from I^- to become Cl^-, generating molecular iodine.

$$Cl_2(g) + 2I^-(aq) \longrightarrow 2\ Cl^-(aq) + I_2(s)$$

53. There are $(2.6989\ g/cm^3)/(26.98154\ g/mol)$ moles in $1\ cm^3$ of Al, or 0.10003 moles in a cube 1 cm on a side.

55. $5.264 \times 10^{22}\ atom/cm^3 \times \dfrac{mol}{6.022 \times 10^{23}\ atom}$
$$\times \dfrac{69.723\ g}{mol} = 6.09\ g/cm^3.$$

57. $1\ cm \times \dfrac{10^{10}\ pm}{1\ cm} \times \dfrac{1\ radius}{117\ pm} \times \dfrac{1\ atom}{2\ radii} = 4.3 \times 10^8\ atoms$

on an edge. The top surface is $(4.3 \times 10^8)^2 = 1.8 \times 10^{15}$ atoms on the top surface.

Atoms in the cube $= 1\ cm^3 \times \dfrac{8.96\ g}{1\ cm^3} \times \dfrac{mol}{63.5\ g}$

$\times \dfrac{6.022 \times 10^{23}\ atoms}{mol} = 8.5 \times 10^{22}$ atoms or about $1/10$ a mole.

59. (a) Rb is larger due to filling of electrons into a higher shell number. (b) Ga is larger due to beginning to fill the p subshell. The elements K and Rb should have very similar properties because both have a single valence s electron.

61. Ionization energy and size are correlated because an atom is small if the electrons are drawn very close to the nucleus. The nucleus is positively charged, and the electrons are negatively charged, so the closer they are, the stronger is the attraction between them. This attraction must be overcome to ionize an atom, so the smaller an element is, the more energy it takes to ionize it.

63. Ba; it is a member of the highest-numbered period and is a Group IIA element.

65. S^{2-} is expected to be larger than Ar because Ar has two additional nuclear charges to attract the electrons.

67. K, Br^-, Al^{3+}, Ge

69. For the elements around Tc, the following data are known: Atomic mass (amu)

			Density (g/cm³)			Melting Point (K)		
Cr	**Mn**	**Fe**	**Cr**	**Mn**	**Fe**	**Cr**	**Mn**	**Fe**
52.0	55.0	55.8	7.19	7.33	7.87	2180	1519	1811

Mo		**Ru**	**Mo**		**Ru**	**Mo**		**Ru**
95.9		101.0	10.2		11.4	2896		2237

W	**Re**	**Os**	**W**	**Re**	**Os**	**W**	**Re**	**Os**
183.8	186.2	190.3	19.3	21.0	22.6	3695	3459	3306

From these trends, the atomic mass is predicted to be about 98.5 amu, the density to be $10.8\ g/cm^3$, and the melting point to be 2350 K. The actual values are: 98 amu, $11.5\ g/cm^3$, and 2430 K.

71.

Element	F	Ge	Ag	Hg	Xe	B
Metal			X	X		
Nonmetal	X				X	
Semimetal		X				X

73. The essential points are that the atoms in the solid are in a regular array while those in the liquid have more random spacing and arrangement. The distances between atoms in both phases are nearly the same. In the liquid, atoms randomly jostle and slide past each other. In the solid, the atoms vibrate back and forth, but maintain the same neighbors.

75. (a) IA; (b) IA; (c) Zn; (d) Mg; (e) Y; (f) Sr.

77. The important points here are that in the straight rod, the atoms are in a regular array with some vacancy defects. As the rod is bent, atoms at the inside of the bend get closer while those on the outside move farther away, and more vacancies occur on the outside of the bend. As the bend increases, vacancies converge, creating a crack.

79. At 2125 K, the melting point of Zr is even higher than that of Ti. Therefore, it is expected to be less flexible than is Ti.

81. Since Na is much softer than Zn, the melting point is expected to be lower. The actual melting point is 371 K compared with that of Zn, 693 K, just as expected.

83. The melting points of Zn and Ni are 693 K and 1726 K, respectively. The melting point of Mg falls in between these values. Therefore Mg is expected to be harder than Zn but softer than Ni.

85. The atomic density of Al is $(2.6989 \text{ g/cm}^3)/(26.98154 \text{ g/mol}) \times (6.022 \times 10^{23} \text{ atoms/mol}) = 6.022 \times 10^{22}$ atoms/cm^3

87. The missing element is Ge. Mendeléeff predicted 18 elements from Si to Ge and a mass gain of about 45 mass units for successive elements in a group. This predicts a mass of $28 + 45$ amu $= 73$ amu from Si, or $119 - 45$ amu $= 74$ amu from Sn, compared with the actual mass of 72.61 amu. For the density, Mendeléeff probably relied on the density trend in the next two periods:

Fifth period

	Cd	In	Sn	Sb
$\rho(\text{g/cm}^3)$	8.65	7.31	7.31	6.691

Sixth period

	Hg	Tl	Pb	Bi
$\rho(\text{g/cm}^3)$	13.546	11.85	11.35	9.747

Prediction for the fourth period

	Zn	?	?	As
$\rho(\text{g/cm}^3)$	7.133	~6.5	~6.5	5.73

There is a substantial drop in density at the end of the transition elements. The next two elements have nearly the same density, followed by an additional drop in Group VA. The two missing elements should have a density of about 6.5 g/cm^3. The actual density of Ge is 5.323 g/cm^3.

89. There is a gain in atomic mass of approximately 45 amu for elements in the same group from one period to the next. The mass of Y is 88.90 amu. Subtracting 45 amu results in a predicted

mass of 44 amu, which is very near the 44.96 amu actual mass. Mendeléeff probably used the trend from the elements before and after Sc for the density and compared it with the period below it. In the period below, the densities are as follows:

Fifth period

	Rb	Sr	Y	Zr
$\rho(\text{g/cm}^3)$	1.5	2.5	4.5	6.5
ρ relative to IA (approx)		double	triple	quadruple

Sixth period

	Cs	Ba	La	unknown
$\rho(\text{g/cm}^3)$	1.9	3.5	6.1	
ρ relative to IA (approx)		double	triple	

Prediction for the fourth period

	K	Ca	?	Ti
$\rho(\text{g/cm}^3)$	0.9	1.5	~3	4.5
ρ relative to IA		double	triple	quadruple

The actual density of Sc is 2.989 (g/cm^3), very close to 3.0.

91. Unit analysis:

Atomic density $= (\text{g/cm}^3) \times (\text{moles/g})$
$\times (\text{atoms/mol}) = \text{atoms/cm}^3$

$\Rightarrow$ atomic density $=$ density $\times (1/\text{atomic mass})$
$\times$ (Avogadro's number)

Na: $0.971 \text{ g/cm}^3 \times (1 \text{ mol}/22.98977 \text{ g}) \times$
$(6.022 \times 10^{23} \text{ atoms/mol}) = 2.54 \times 10^{22}$ atoms/cm^3
Mg: $1.738 \text{ g/cm}^3 \times (1 \text{ mol}/24.305 \text{ g}) \times$
$(6.022 \times 10^{23} \text{ atoms/mol}) = 4.36 \times 10^{22}$ atoms/cm^3
Al: $2.6989 \text{ g/cm}^3 \times (1 \text{ mol}/26.98 \text{ g}) \times$
$(6.022 \times 10^{23} \text{ atoms/mol}) = 6.024 \times 10^{22}$ atoms/cm^3

The density is about 1.5 times larger for Mg than for Na and is about 2.5 times greater for Al as for Na.

93. (a) IIA; (b) IB; (c) Cs; (d) Hf; (e) Fe.

95. The plot is shown below. The general trend is increasing density with atomic number until the end of the transition series, where the

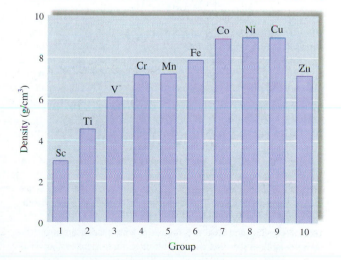

density drops significantly. There is also an arrest in the steady increase of density at the fifth element, Mn.

97. To be a sphere, the angular dependence of the density must be a constant. Thus the problem is to show that the sum of the square of the amplitudes of the p_x, p_y, and p_z orbitals does not depend on θ or ϕ.

$$|p_x|^2 = \frac{3}{4\pi} \sin^2\theta \cos^2\phi; \quad |p_y|^2 = \frac{3}{4\pi} \sin^2\theta \cos^2\phi;$$

$$|p_z|^2 = \frac{3}{4\pi} \cos^2\theta$$

Adding $|p_x|^2$ and $|p_y|^2$ gives $\frac{3}{4\pi} \sin^2\theta$. Adding this to $|p_z|^2$ yields $\frac{3}{4\pi}$. This is a constant, so the result is a sphere.

99. Many diagrams are possible. This is one example.

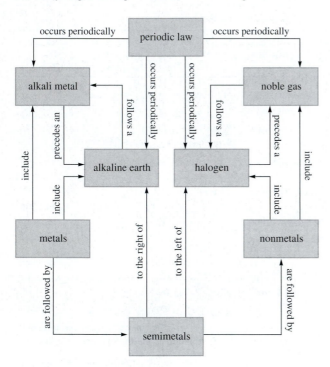

101. The density of Cu is nearly three times that of Al, so copper wires are heavier. The melting point of Al is nearly 50 °C lower than that of Cu, so Al melts easier. Most importantly, the resistivity of Al is 1.6 times that of Cu. Resistance translates into heating of the wire. So the higher resistance of an Al wire along with its lower melting point means that Al wires melt and heat the surrounding insulation: both bad for household wiring.

103. (a) Water adds $0.9970 \text{ g} \cdot \text{cm}^{-3} \times 250 \text{ mL} = 249.3$ g; total mass 439.3 g. (b) Alcohol adds $0.7893 \text{ g} \cdot \text{cm}^{-3} \times 250 \text{ mL} = 197.3$ g; total mass 387.3 g. (c) Mercury adds $13.546 \text{ g} \cdot \text{cm}^{-3} \times 250 \text{ mL} = 3386.5$ g; total mass 3576.5 g.

Mercury in the beaker is nearly an order of magnitude more dense than is water or alcohol. The denser material thus sinks to the bottom of the lake or stream.

105. Silver, tin, and lead are indicated with shaded areas on the periodic table in the next column. Of these three elements, silver is the most chemically inert, being one of the three coinage metals. It is a good choice for soldering an electric circuit because it will remain as elemental silver, retaining the excellent conduction properties of a metal. Lead is somewhat unreactive, melts at a low temperature (about 300 °C), and is much less expensive than silver. However, lead has deleterious health effects, so its use on water pipes is now banned.

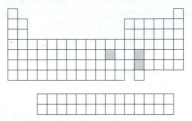

107. The interatomic interactions in diamond must be stronger and involve directed bonds. In contrast, bonds in graphite must be rather diffuse and weak.

Chapter 4

1. Na: $\frac{0.971 \text{ g}}{\text{cm}^3} \times \frac{1 \text{ mol}}{22.99 \text{ g}} = 4.22 \times 10^{-2} \text{ mol/cm}^3$

Convert to atomic density by multiplying by Avogadro's number:
$4.22 \times 10^{-2} \text{ mol/cm}^3 \times 6.022 \times 10^{23} \text{ atoms/mol} = 2.54 \times 10^{22} \text{ atoms/cm}^3$

Volume occupied by atom plus associated space is the inverse:
$1/(2.54 \times 10^{22} \text{ atoms/cm}^3) = 3.94 \times 10^{-23} \text{ cm}^3/\text{atom}$

Na is a BCC lattice, so 68% of this volume is occupied by the atom:
$(3.94 \times 10^{-23} \text{ cm}^3/\text{atom}) \times 0.68 = 2.68 \times 10^{-23} \text{ cm}^3$ occupied by the atom

$$\text{Atomic radius} = \sqrt[3]{\frac{\text{atomic volume}}{4\pi/3}}$$

$$\sqrt[3]{\frac{2.68 \times 10^{-23}}{\frac{4}{3}\pi}} = 1.86 \times 10^{-8} \text{ cm} = 186 \text{ pm}$$

The quantum mechanical radius is 223 pm.

For Mg:

$$\left[\frac{1.738 \text{ g}}{\text{cm}^3} \times \frac{1 \text{ mol}}{24.305 \text{ g}} \times \frac{6.022 \times 10^{23} \text{ atoms}}{\text{mol}} \right]^{-1} \times 0.74$$

$$= 1.72 \times 10^{-23} \text{ cm}^3/\text{atom}$$

Atomic radius = 160 pm. The quantum mechanical radius is 172 pm.

In each case, the quantum mechanical radius is larger, as it must be. The quantum mechanical radius is for the atom with an intact electron cloud, while the volume taken up in the solid reflects the fact that part of the electron density is contributed to the sea in forming the metallic solid. Stripping away part of the electron density decreases the size of the atom.

3. Molar density is mol/cm³. Divide mass density by molar mass:

$$\frac{8.96 \text{ g/cm}^3}{63.546 \text{ g/mol}} = 0.141 \text{ mol/cm}^3 \Longrightarrow 8.49 \times 10^{22} \text{ atoms/cm}^3$$

$$\Longrightarrow 1.18 \times 10^{-23} \text{ cm}^3/\text{atom}$$

Volume of an atom $= \frac{4}{3}\pi r^3 \Longrightarrow r = 1.41 \times 10^{-8}$ cm $= 141$ pm
If the 74% packing is considered, then the 1.18×10^{-23} cm^3 consists of 74% atom and 26% space, so the atom volume is $0.74 \times 1.18 \times 10^{-23}$ cm$^3 = 8.73 \times 10^{-24}$ cm^3, and $r = 1.28 \times 10^{-8}$ cm $= 128$ pm.

5. Molar density = mass density/molecular mass.

Molar density Na $= 4.22 \times 10^{-2}$ mol/cm^3
Molar density Mg $= 7.15 \times 10^{-2}$ mol/cm^3
Molar density Al $= 0.100$ mol/cm^3

The molar density of Al is 2.5 times that of Na and nearly 1.5 times that of Mg. Ratio molar density Al:Na = 2.37; ratio molar density of Mg:Na = 1.69. The density increases significantly from group to group across the Group IA, IIA, and IIIA elements.

7. Zn, Cd, and Hg all have filled $(n - 1)$ d subshells [and Hg has a filled $(n - 2)$ f subshell], while Cr and Mo each have only a half-filled $(n - 1)$ d subshell. The half-filled subshell shields the core charge less effectively, yielding a stronger attraction for the electron sea, and hence a higher melting point than Zn, Cd, and Hg with the filled subshell. W has a s^2 d^4 configuration and an even less effective shield for the core charge; thus it has an even greater attraction for electrons of the sea. W has nearly the highest melting point of any element known.

9. Cu, Ag, and Au are transition elements and have 10 added nuclear charges compared with K, Rb, and Cs. Along with the 10 extra nuclear charges come 10 additional electrons. However, the electrons do not completely shield the added nuclear charge so, within the electron sea model, the ions of Cu, Ag, and Au have a greater attraction for the electrons of the sea.

11. For simple cubic packing, the cube edge dimension is $2 \times r$, where r is the atom radius. With one atom at each of the eight corners, there is one complete atom within the unit cell. The atom volume $= \frac{4}{3}\pi r^3$; the cube volume $= 8r^3$. Thus the packing efficiency is

$$\frac{\frac{4}{3}\pi r^3}{8r^3} \times 100\% = 52.4\%$$

The coordination number in simple cubic packing is 6.

13. In a BCC lattice, there are two atoms per unit cell. These atoms just touch along the body diagonal. Letting the side of the unit cell be a, the body diagonal is $\sqrt{3}\,a$, so $\sqrt{3}\,a = 4r$, where r is the radius of the atom. Within the unit cell, the atoms occupy a volume equal to $2 \times \frac{4}{3}\pi r^3$. The unit cell volume is $a^3 = \left(\frac{1}{\sqrt{3}}\right)^3 r^3$. Thus the packing percentage is

$$\frac{2 \times \frac{4}{3}\pi r^3}{\frac{64}{3\sqrt{3}} r^3} \times 100\% = 68\%$$

15. Molar density is mol/cm^3. Divide mass density by molar mass:

$$\frac{19.3 \text{ g/cm}^3}{183.84 \text{ g/mol}} = 0.105 \text{ mol/cm}^3 \Longrightarrow 6.32 \times 10^{22} \text{ atoms/cm}^3$$
$$\Longrightarrow \text{net } 1.58 \times 10^{-23} \text{ cm}^3/\text{atom}$$

Considering the 68% packing efficiency, the 1.58×10^{-23} cm^3 consists of 68% atom and 32% space, so the atom volume is $0.68 \times 1.58 \times 10^{-23}$ cm$^3 = 1.074 \times 10^{-23}$ cm^3, and $r = 1.37 \times 10^{-8}$ cm $= 137$ pm. The unit cell body diagonal is $4r = 548$ pm. Body diagonal = unit cell edge $\times \sqrt{3}$, $\Longrightarrow$ edge = 316 pm.

17. The diamond lattice is not a close pack lattice. There are many more-efficient ways of packing spheres into three-dimensional space than the packing in diamond. There are large channels in a hard-sphere model of diamond.

19. The coordination number in body-centered cubic packing is 8. The mass is determined as follows: From the unit cube dimension, determine the volume of the unit cube. From the volume and the density, determine the mass of the unit cube. Divide the unit cube mass by the number of atoms in a unit cube (2).

Unit cube volume $= (533.3 \text{ pm})^3 = (533.3 \times 10^{-10} \text{ cm})^3$
$$= 1.517 \times 10^{-22} \text{ cm}^3$$

Unit cube mass $= (0.862 \text{ g/cm}^3) \times (1.517 \times 10^{-22} \text{ cm}^3)$
$$= 1.31 \times 10^{-22} \text{ g}$$

Atomic mass of K: $(0.655 \times 10^{-22} \text{ g}) \times (6.022 \times 10^{23} \text{ atoms/mol})$
$$= 39.4 \text{ g/mol}$$

21. Elastic distortion corresponds to a bending of the angles between three adjacent atoms and either stretching or compressing of bonds. On the macroscopic level, the object bends or stretches under a stress and returns to its original shape upon removal of the stress. Plastic distortion corresponds to sliding along slip systems, yielding a permanent change of shape that remains after the stress is removed. Typical metal objects plastically stretch by only about 10% prior to catastrophic failure. Damascus steel is highly unusual in that it can stretch by 11 times (1100%) without failure.

23. V: $4s^2 3d^3$; Cr: $4s^1 3d^5$; Mn: $4s^2 3d^5$.
The elastic modulus is the slope of the stress-strain curve: the amount of stretch produced with a given applied strain. With a relatively high slope, a lot of stress is required to produce a little strain. This result is due to bond stretching and angle bending. The half-filled d subshell provides maximum directionality to the bonding with a nuclear charge that is insufficient to pull the d-orbital electrons into the core.

25. The density of Au is 19.3 g/cm$^3 \Longrightarrow 1$ g is 0.0518 cm^3. If this covers an area of 1 cm^2, then the thickness is 0.0518 cm.
The radius of a Au atom in gold is

$$\frac{19.3 \text{ g/cm}^3}{196.97 \text{ g/mol}} = 0.09798 \text{ mol/cm}^3 \Longrightarrow 5.90 \times 10^{22} \text{ atoms/cm}^3$$
$$\Longrightarrow 1.695 \times 10^{-23} \text{ cm}^3/\text{atom}$$
$$74\% \text{ packing} \Longrightarrow 1.25 \times 10^{-23} \text{ cm}^3 \text{ per atom}$$
$$\Longrightarrow r = 144 \text{ pm, diameter is } 288 \text{ pm}$$
$$\Longrightarrow \frac{0.0518 \text{ cm}}{288 \text{ pm}} \times \frac{10^{-2} \text{ m}}{\text{cm}} \times \frac{1 \text{ pm}}{10^{-12} \text{ m}}$$
$$= 1.80 \times 10^4 \text{ atoms thick}$$

27. The elastic modulus is the slope of the stress-strain curve.

$$\text{Elastic modulus} = \frac{\Delta \text{ stress}}{\Delta \text{ strain}} = \frac{79.79 \text{ g/mm}^2}{0.9 \text{ pm}/246 \text{ pm}} = 21.8 \text{ kg/mm}^2$$

29. The atomic density of the sixth-row elements is nearly the same as that of the fifth-row transition elements, implying that the sizes of the fifth- and sixth-row elements are nearly the same. Sixth-row elements are expected to be larger because the valence shell is one shell higher. However, the sixth period has an added 14 nuclear charges that accompany filling of the 14 f-orbital electrons. Since the f-orbital electrons have three intersecting nodal

planes at the nucleus, these electrons do not shield the added nuclear charges effectively. Thus, the sixth-row transition elements feel more of the nuclear charge and the valence electrons are drawn toward the nucleus, making the atoms relatively small. The result is an atom that is about the same size as the fifth-period elements.

31. Cu, Ag, and Au all have electron configurations that feature one s-orbital valence electron and a filled d subshell. The s valence orbital can accommodate two electrons, so the electron affinities of all of these elements are significant, providing an attraction for an extra electron. With space to accommodate that electron, it moves easily from atom to atom and the resistivity is very low.

33. Resistance of the Cu wire is

$$R/\ell = \rho/A = [(1.725 \ \mu\Omega\cdot cm)(0.035 \ cm^2)] \times (1 \ cm/10^{-2} m)$$
$$= 4930 \ \mu\Omega/m$$
$$\text{Power} = (20 \ amp)^2(0.004930 \ \Omega/m) = 1.972 \ W/m$$

The resistivity of Al is three times that of Cu. Resistivity is proportional to resistance, so the Al wire of the same gauge has three times the resistance of the Cu wire. Power is directly proportional to resistance, so the Al wire dissipates three times the power of the Cu wire. The resistance is directly proportional to the cross-sectional area, and the area is the square of the radius. To cut the power dissipation by a factor of 3, the radius must increase by a factor $\sqrt{3} = 1.7$, or 0.36 cm in diameter.

35.
$$\text{Moles Ag} \propto (92.5/107.9) = 0.857$$
$$\text{Moles Cu} \propto (7.5/63.55) = 0.118$$
$$\text{Atomic \% Ag} = [0.857/(0.857 + 0.118)] \times 100\% = 87.9\%$$
$$\text{Atomic \% Cu} = 12.1\%$$

37.
$$\text{Atoms of Au} \propto 14/196.97 = 0.0711$$
$$\text{Atoms Ag} \propto 10/107.9 = 0.0927$$
$$\text{Atom ratio (Au:Ag)} = 0.0711/0.0927 = 0.767$$

The radius of a gold atom in elemental gold is

$$\frac{19.3 \ g/cm^3}{196.97 \ g/mol} = 0.09798 \ mol/cm^3 \Longrightarrow 5.90 \times 10^{22} \ atoms/cm^3$$
$$\Longrightarrow 1.695 \times 10^{-23} \ cm^3/atom$$
$$74\% \ packing \Longrightarrow 1.25 \times 10^{-23} \ cm^3 \ per \ atom \Longrightarrow r = 144 \ pm$$

The gold unit cell is an FCC lattice. The face diagonal is

$$4r = \sqrt{2} \ times \ the \ edge \Longrightarrow edge = 4 \times 144 \ pm/\sqrt{2} = 407 \ pm$$

The radius of a silver atom is

$$\frac{10.50 \ g/cm^3}{107.9 \ g/mol} = 0.0973 \ mol/cm^3 \Longrightarrow 5.86 \times 10^{22} \ atoms/cm^3$$
$$\Longrightarrow 1.71 \times 10^{-23} \ cm^3/atom$$
$$74\% \ packing \Longrightarrow 1.27 \times 10^{-23} \ cm^3 \ per \ atom \Longrightarrow r = 145 \ pm$$

The silver unit cell is an FCC lattice. The face diagonal is

$$4r = \sqrt{2} \ times \ the \ edge \Longrightarrow edge = 4 \times 145 \ pm/\sqrt{2} = 410 \ pm$$

The majority element in a 14-carat gold ring is gold. Gold and silver atoms have nearly the same radius, and both crystallize in the same crystal structure. The silver atoms do not strain the gold lattice.

39. *Solid solution:* Only Cu satisfies all three Hume-Rothery rules. Al, Cu, Co, Cr, and Fe make it on size. Of these, Al and Cu (and Co is close) satisfy the crystal structure rule. Only Cu also makes it on electronegativity.

Substitutional impurities: Those elements that are close in size can make substitutional impurities: Al, Cu, Co, Cr, and Fe.

Interstitial impurities: We need to determine the size of the octahedral hole—the largest hole in an FCC lattice. The size of the largest atom that fits into an octahedral hole is determined as

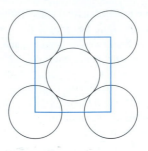
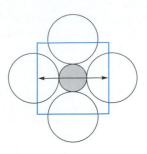

follows. The face center atoms delimit the octahedral hole in the unit cube, as shown in the figure. Atoms in the faces just touch. Let the cube edge dimension be a. The relationship to the atom radius is $\sqrt{2}a = 4r$, where r is the atom radius. Slicing the unit cube through the center shows where the octahedral hole sits. The distance across the center of the square is the same as the cube edge: $a = \frac{4}{\sqrt{2}}r$. The space for the hole has a diameter of $a - 2r$: $\frac{4}{\sqrt{2}}r - 2r = 0.83r$. The radius of the hole is $0.42r$. Hence the radius of the interstitial atom must be less than $0.42r$. If $r = 115$ pm, then the atom must be 48.3 pm or smaller. Only hydrogen fits in this space. Note that Ni is used as a hydrogenation catalyst. Storage of hydrogen in the interstitial spaces is part of the hydrogenation mechanism.

41. To form a substitutional alloy, the size, crystal structure, and electronegativity of the combining elements have to be close. Fe has a BCC structure. Some of the Group IA and IIA metals form BCC structures but they are all much less electronegative than Fe. Thus V, Cr, Nb, Mo, Ta, and W are candidates from the crystal structure point of view. The fifth- and sixth-period elements are all larger than Fe, so the choice comes down to V or Cr. Both V and Cr are less electronegative than Fe, so Fe is unlikely to form a substitutional alloy with any other element. Pd forms an FCC lattice, so it has many candidates from the crystal structure point of view. The electronegativity of Pd is 2.2, similar to that of Rh, Ir, and Pt. The radius of Pd is 128 pm; Rh, Ir, and Pt are close in terms of that parameter. Thus the likely alloy formers are Rh, Ir, and Pt.

43.

	Size	Structure	Electronegativity
Cu	117	FCC	1.90
Ag	134	FCC	1.93

Cu and Ag are acceptable in all categories except size: Ag is much larger. The melting point of Cu is 1358 K, while that of Ag is 1235 K. 780 °C = 1053 K; 1053 K is less than the melting point of either Cu or Ag. This situation arises only when the two elements form a low-melting eutectic; they do not form an alloy.

45. An alloy that is 5% Ni and 95% Cu has a density that is just slightly less than that of Cu. In fact, within the significant digits given, it has the same density as Cu. Ni is responsible for 5% of the mass of 1 cm³, or 0.448 g/cm³; 0.448 g Ni = 7.63×10^{-3} mol, Ni = 4.60×10^{21} atoms/cm³.

47. (a) The radius of a lead atom is

$$\frac{11.35 \ g/cm^3}{207.2 \ g/mol} = 0.0548 \ mol/cm^3 \Longrightarrow 3.30 \times 10^{22} \ atoms/cm^3$$
$$\Longrightarrow 3.03 \times 10^{-23} \ cm^3/atom$$
$$74\% \ packing \Longrightarrow 2.24 \times 10^{-23} \ cm^3/atom \Longrightarrow r = 175 \ pm$$

(b) A lead atom is slightly larger than a tin atom, so lead will be found where the space is larger—that is, the body center.

49. 1400 °C: all liquid;
1300 °C: two phases—some liquid, some solid;
1200 °C: all solid

51. (a) Cu;
(b) Ca;
(c) Ag;
(d) Ba

53. (a) The first solid appears at 960 °C.
(b) The solid contains more copper.
(c) The last solid forms at about 775 °C.

55. In this composition range, cooling the solid slowly to 790 °C results in a homogeneous solid alloy. If the solid is then cooled rapidly, separation into the silver-rich and copper-rich solid phases does not occur. Instead, the alloy is frozen in. Because the solution is uniform, it will have a uniform look and be rather hard due to the size differential between copper and silver. The hardness enables it to be polished strenuously, and thus to take on a uniform sheen.

57. 2 g C per 100 g Fe. 0.1667 atoms C/1.79 atoms Fe = 9% C, 91% Fe. (There are two Fe atoms per unit cell, so about one unit cell in five has an interstitial C atom at the maximum solubility.)

59. The radius of W is 130 pm, while that of Fe is 117 pm; thus the octahedral holes in W are larger. The diagonal of the W lattice unit cube is 4 × 130 pm = 520 pm. The cube edge is 520 pm/$\sqrt{3}$ = 300 pm. The face diagonal is 368 pm. The radius of the hole is 184 − 130 = 54 pm. In the other dimension, the distance to the body center atom is 150 pm; the radius of the space is 150 − 130 = 20 pm. Thus the space is 54 × 20 pm in size. The radius of carbon is 77 pm, so it is also squeezed in a W lattice. Incorporation of carbon into tungsten will alter the properties of W in about the same way as inclusion of C alters the properties of Fe to form steel.

61. The spaces in the diamond lattice are much larger than those in the FCC and BCC lattices. Both the FCC and BCC lattices host atoms that are much smaller than the host element. In the diamond lattice, the spaces are nearly the same size as the host element.

63. Atoms along the body diagonal just touch. The body diagonal is $\sqrt{3}$ × 291 = 504 pm. Two atoms are in a BCC cell along the diagonal; therefore, the size is 4r. The radius of Cr in the lattice is 126 pm. The diameter in the short direction is (291 − 252) = 39 pm. The long direction is determined from the face diagonal. The face diagonal is $\sqrt{2}$ times the cube edge = 412 pm. The hole is (412 − 2r) pm = (411 − 252) pm = 159 pm. The radius is 159 pm/2 = 80 pm. Thus the dimensions are 80 pm × 20 pm. No atom fits in this hole without producing distortion. The largest atom causing no distortion in the long direction is nitrogen at 75 pm.

65. Ni and Ti will not form an alloy because their electronegativities are greatly different, they have different crystal structures, and their atomic radii differ by more than 15%.

67. The filter is shaped into its unfolded form (pictured in Figure 4.46) at high temperature. It is subsequently cooled to below the austenite to martensite transition to become flexible and soft martensite. It is then compacted to fit into a catheter to insert into the target vein. Body temperature is above the transition to austenite and the device unfolds to the open shape shown. The feet and the basket anchor the device to the vein walls and the bird's nest basket catches clots preventing them from traveling and damaging the patient.

69. β-Sn is packed more tightly, so it is more stable at high pressure.

71. In the 49 × 18 pm space, no atom is small enough to fit without stress. As discussed in Worked Example 6.7, a larger atom will lead to distortion of the BCC iron lattice. This distortion will change the properties of the solid and inhibit the plastic flow, making the lattice stronger but possibly brittle.

73. Many diagrams are acceptable. One example is shown here.

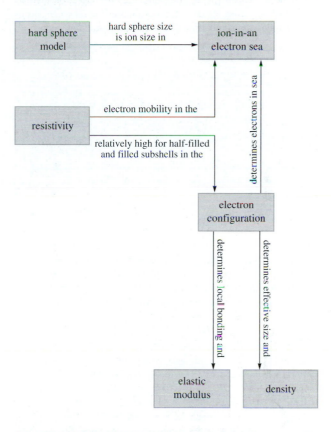

75. The resistivity of tungsten is low enough to allow current to pass, but high enough to dissipate energy as the electrons pass. Energy dissipation results in heating of the wire high enough to emit visible light. The high melting point is essential to keep the filament from melting when heated. Melting would break the circuit, so that the bulb would go out.

77.

Element	Sc	Ti	V	Cr	Mn	Fe	Co	Ni	Cu	Zn
MP (K)	1814	1941	2183	2180	1519	1811	1768	1728	1358	693

The data are shown graphically on page 496. The steady increase that begins with K is due to increasingly more electrons being contributed to the electron sea. The steady upward trend is interrupted at Cr with the electron configuration $4s^1 3d^5$. Drawing in one of the s electrons to complete a half-filled d subshell reduces the electron density in the electron sea, thereby decreasing cohesion and lowering the melting point. The melting point has a major dip at Mn with its $4s^2 3d^5$ configuration. This spherical distribution of electrons is all that each Mn atom needs, so each Mn atom has little attraction for additional electrons (electron affinity ≈ 0). The low attraction for electrons makes attraction for other Mn atoms low; the melting point is also low. After Mn, the Fe-Co-Ni-Cu series increasingly draws the d electrons into the core, making fewer electrons available for the electron sea, and leading to less cohesion and a lower melting point. The melting point of Zn is particularly low, even lower than that of Ca. Zn has nearly no attraction for an additional electron (electron affinity ≈ 0), little electron sharing, little attraction for other Zn atoms and their electrons, and, therefore, a low melting point.

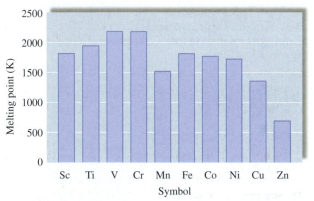

79. The response of Sn is that of a typical metal: beginning with an elastic response that reflects the bonding between atoms and continuing to a plastic response as atoms slide along slip systems. In contrast, martensitic nitinol responds to a stress by shifting from one of 24 equivalent configurations to another that aligns with the stress.

81. The radius of the carbon atom is 70 pm, while the interstitial space is 18×49 pm. This creates a $\Delta\ell$ of $2 \times (70 - 18)$ pm = 104 pm in one direction and $2 \times (70 - 49)$ pm = 42 pm in the other direction. The length over which this strain is absorbed is four cells by three cells. The length of the unit cell is 270 pm, so four cells is 4×270 pm = 1080 pm and that for three cells is 810 pm. Assume that the larger $\Delta\ell$ is accommodated in the longer direction. The strain is then $\Delta\ell/\ell = (104/1080) \times 100\% = 9.6\%$ and $(42/810) \times 100\% = 5.2\%$ in the two directions. Both of these levels exceed the normal elastic limit: The BCC lattice becomes distorted due to this large stain.

Chapter 5

1. (a) O; (b) N; (c) Sb; (d) Se. The larger electron density is around the atom with the larger electronegativity.

3. (a) H and O; (b) Ga and P; (c) Zn and Se; (d) Al and N

5. Covalent molecules are held together by covalent bonds in which each element attains an octet. Interaction between such molecules is weak due to each element having an octet and the molecule as a whole being uncharged. In contrast, ionic materials consist of a collection of ions that are attracted to each other by coulombic forces, which are very strong.

7. (a) ionic; (b) covalent; (c) covalent; (d) covalent, but just barely.

9. The coordination number of Na in NaCl is 6: Each Na^+ ion is surrounded by six Cl^- ions. The NaCl unit cell is an FCC lattice of Na^+ ions interpenetrated with an FCC structure of Cl^- ions. There are $\frac{1}{8} \times 8$ corner ions plus $\frac{1}{2} \times 6$ face-center ions, or four Na^+ ions per unit cell. For electrical neutrality there must also be four Cl^- ions ($\frac{1}{4} \times 12$ edge Cl^- ions plus one in the center). MgO simply replaces Na^+ with Mg^{2+} and Cl^- with O^{2-}, so there are four Mg^{2+} and four O^{2-} ions per

unit cell. The electronegativity difference in NaCl is 2.23, so it is about 72% ionic. For MgO, the electronegativity difference is 2.13, so it is 68% ionic, slightly less than NaCl. For NaCl, there are

$$\frac{2.17 \text{ g/cm}^3}{58.5 \text{ g/mol}} \times \frac{6.022 \times 10^{23} \text{ NaCl}}{\text{mol}} = 2.24 \times 10^{22} \text{ NaCl/cm}^3$$

For MgO, there are

$$\frac{3.6 \text{ g/cm}^3}{40.3 \text{ g/mol}} \times \frac{6.022 \times 10^{23} \text{ MgO}}{\text{mol}} = 5.4 \times 10^{22} \text{ MgO/cm}^3.$$

The unit density of MgO is much greater than that of NaCl. In MgO, the ion charges are $+2$ and -2, whereas in NaCl the charges are $+1$ and -1. The larger charge in MgO results in a stronger attraction between the ions. As a consequence the distance between Mg and O is smaller than the distance between Na and Cl. A shorter bond distance means a higher unit density.

11. The Lewis dot structure is shown here. CF_2Cl_2 is very stable because the carbon atom is surrounded by highly electronegative and small F and Cl atoms.

$$\ddot{\text{Cl}} : \overset{\displaystyle :\ddot{\text{F}}:}{\underset{\displaystyle :\ddot{\text{F}}:}{\text{C}}} : \ddot{\text{Cl}} :$$

13. Nitrogen needs three electrons to fill its outer shell—to attain an octet. Sharing an electron with each of three other atoms fills the octet; therefore, nitrogen forms three bonds. The remaining two electrons form a lone pair.

$$\overset{\displaystyle ::}{\underset{\displaystyle :\ddot{\text{F}}:}{:\ddot{\text{F}}:\ddot{\text{N}}:\ddot{\text{F}}:}}$$

15. A σ_s molecular orbital has no nodes. A σ_s^* molecular orbital has one node. A σ_{p_z} orbital has two nodes through the two nuclei. A $\sigma_{p_z}^*$ orbital has three nodes: two through the two nuclei and one between the two nuclei. A π orbital has one node along the line connecting the nuclei. A π^* orbital has two nodes: one through the two nuclei and one on the axis between the two nuclei. In energy order from lowest to highest, these are σ_s, σ_s^*, σ_{p_z}, π, π^*, and $\sigma_{p_z}^*$. Note that the π orbital, even though it has only one node, is higher in energy than the σ_{p_z} orbital because the σ_{p_z} orbital has its density between the two nuclei.

17. The molecular-orbital energy-level diagram is shown below. The bond order = $(5 - 2)/2 = 1\frac{1}{2}$.

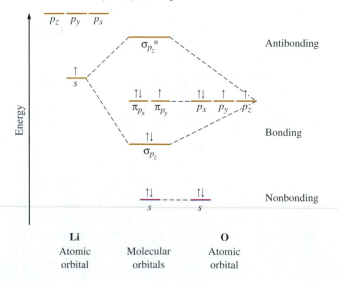

19. The bond order in CO is $(8 - 2)/2 = 3$. This a very strong triple bond.

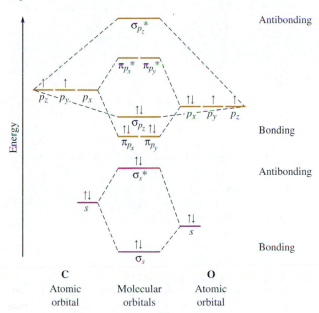

C
Atomic orbital

Molecular orbitals

O
Atomic orbital

21. The energy-level diagrams are shown on page 498. The bond orders are 2 for O_2, $1\frac{1}{2}$ for O_2^-, and 1 for O_2^{2-}. O_2 is the most stable of the three species. O_2 and O_2^- are paramagnetic; O_2^{2-} is diamagnetic.

23. (a) The energy-level diagram for NO is shown below. The bond order is $2\frac{1}{2}$. (b) The energy-level diagram for O_2 adds one antibonding electron to that; O_2^{2-} adds two more antibonding electrons. The bond order is 1. (c) CN^- is isoelectronic with N_2; the bond order is 3. (d) CO has one fewer electron than NO, and that electron comes from an antibonding orbital. CO has a triple bond.

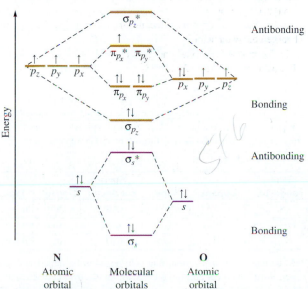

N
Atomic orbital

Molecular orbitals

O
Atomic orbital

25. The bonding combination will be higher in energy than the similar $s-s$ bonding combination due to the node at one of the nuclei in the $s-p$ combination.

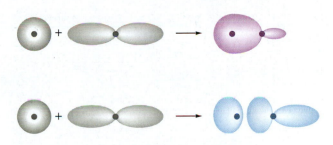

27. The major evidence for the band picture is that the electronic states must be extended for metals to conduct. There must be an energy gap for semiconducting materials to be semiconductors and for insulators to insulate.

29.

s bonding	sp bonding	sp^2 bonding	sp^3 bonding	s anti- bonding	sp^3 anti- bonding
1	2	3	4	1	4

31.

Number of equivalent bonds	3	4	2
Hybridization	sp^2	sp^3	sp

33. Size order (smallest to largest): P < Al < As < Ga < In. Se is more electronegative than Te. Al is more electronegative than In. Cd is more electronegative than Sn. O is more electronegative than N.

35. Diamond is colorless because it transmits all wavelengths of visible light. It transmits all colors because the band gap is greater in energy than the shortest wavelength of visible light. Diamond with a B impurity has holes in the valence band. Electrons can be excited into these holes when energy in the visible region is absorbed. Similarly, diamond with N impurities has electrons in the conduction band. These can also be excited with visible light. As a consequence, both N and B impurities impart color to the diamond lattice.

37. P is more electronegative than Al. Se is more electronegative than Zn. P is more electronegative than Ga. P is more electronegative than Sb.

39. Since ZnS has a band gap beyond the visible region, an element that lowers the band gap is required. Such an element should be larger or less electronegative than S. To make a semiconductor with Zn, the element should be chosen from sulfur's group (VIA). Se and Te are good choices.

41. Beginning with the GaP diode, the wavelength emitted becomes longer as more As is substituted for P. As is a larger atom than P, so the interaction with Ga is weaker for As than it is for P. A weaker interaction means that the band gap is smaller, the energy lost by the electron as it drops from the conduction band to the valence band is smaller, and thus the wavelength is longer.

43. The solid appears black because it absorbs all visible wavelengths. This means that the band-gap energy in this semiconductor is less than the energy of visible light—that is, less than $1.8-1.9$ eV.

45. An AlN diode would be beyond the visible range. Mixing in some P will lower the energy of the band gap, moving the light into the visible region. AlP has a band gap of 2.5 eV, almost in the green region of the spectrum, so some N would have to be retained to hit the blue region.

47. 1 g of Si contains 1/28 mol of Si or $(1/28) \times 6.022 \times 10^{23}$ atoms $= 0.215 \times 10^{23}$ atoms. 1 part in 10^9 of that is 2.15×10^{13}

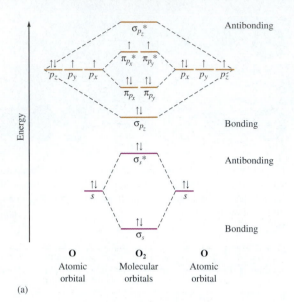

(a)

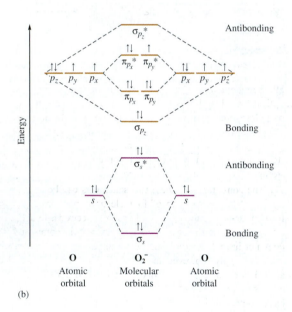

(b)

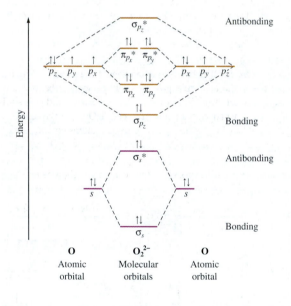

atoms; 2.15×10^{13} impurity atoms are in each gram of Si. The density of Si is 2.33 g/cm^3.

$$\frac{2.33 \text{ g/cm}^3}{28 \text{ g/mol}} \times \frac{6.022 \times 10^{23} \text{ atoms}}{\text{mol}} \times \frac{1 \text{ impurity}}{10^9 \text{ Si}}$$
$$= 5.01 \times 10^{13} \text{ impurity atoms/cm}^3$$

49.

$$E = \frac{(6.6262 \times 10^{-34} \text{ J·s})(2.9979 \times 10^8 \text{ m/s})}{650 \text{ nm}}$$
$$\times \frac{1 \text{ nm}}{10^{-9} \text{ m}} \times \frac{1 \text{ eV}}{1.6022 \times 10^{-19} \text{ J}} = 1.91 \text{ eV}$$

The feature size is $4 \times 650 \text{ nm} = 2600 \text{ nm}$ along the track, and the dimension between tracks is 6500 nm. A feature is 2.6 μm by 6.5 μm. This puts 3846 features in 1 cm along the direction of the tracks and 1538 tracks in 1 cm, or 5.9×10^6 features/cm^2. There are four bits in a byte, and eight bytes in a word, so this is 185 KB/cm^2.

51. To make a p-type semiconductor, the valence band needs to be not quite full. To be not quite full, an element with fewer valence electrons than Si needs to be used. Of those listed, Ga, In, and Zn qualify.

53.

Semi-conductor	Dopant	Type	Semi-conductor	Dopant	Type
Si	Al	p-type	AlP	S	n-type
Ge	P	n-type	ZnSe	Sb	n-type
GaAs	Cd	p-type	CdTe	In	p-type

55. To make a $p–n$ junction, one side needs to be doped with an electron-rich element and the other with an electron-poor element. The n side needs the electron-rich element, any one to the right of Group IV. The p side needs an element to the left of Group IV. To forward-bias this diode, negative potential is applied to the electron-rich side and positive potential to the electron-poor side. The negative terminal should be attached to the electron-rich side. In the transformer, the diode conducts only when forward biased. AC current consists of alternating forward and negative biasing, so in a transformer the diode passes only the forward-biased part of the AC current. The result no longer alternates polarity and is a DC current.

57. Ag has the highest conductivity. Hg has nearly the highest resistivity of the metals, so Ag's resistivity is lower. Both semiconductors, Si and Ge, have lower conductivity. Of these two, Ge has slightly greater resistivity, hence it has the lowest conductivity of the four.

59. For a metal, the increase in resistivity is due primarily to the mobility of the carriers—that is, the scattering of the electrons as they pass through the lattice. In a pure semiconductor, the electrons need to acquire an energy equal to the band gap to conduct. Doping makes spaces (either electrons in the conduction band or holes in the valence band) for charge movement. This changes the resistivity by orders of magnitude.

61. The band diagrams for Zn and Cd show a bond due to the valence p orbitals that is near in energy to the s- and d-orbital bands. Electrons easily jump from one band to the other, so resistivity is low. In contrast, the additional nuclear charges of the Period 6 elements due to filling of the 14 f subshell electrons stabilize the valence s and d bands, opening a bit bigger gap to the valence p band. Jumping over this slight gap takes energy, which is manifested as a resistance.

63. TiO_2 is expected to be a semiconductor. The general principle is that satisfying the octet rule produces a filled valence band and an empty conduction band. Ti has four valence electrons and O is sufficiently electronegative to remove all four. TiO_2 is nearly an ionic material. Adding four electrons to two oxygen atoms completes each of these octets. Ti^{4+} has an Ar core, and hence a complete octet as well.

65. Many diagrams are possible. One example is shown here.

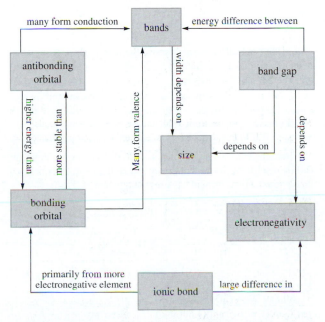

67. Many diagrams are possible. One example is shown at the top of the next column.

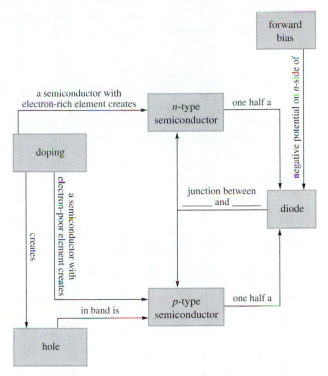

69. Semiconductors can store and retrieve data on a scale of optical wavelengths: 450 nm × 450 nm for a feature. Magnetic particles store data on a size scale given by the characteristic particle size: (100 nm × 20 nm)/0.4 = 100 nm × 50 nm. If the magnetic read head can read on a similar scale, magnetic media can store about 40 times more data per unit area. (All of today's storage media operate in two dimensions.)

71. The consistency of these two statements can be understood as follows. With an average of four electrons per atom, each atom forms four bonds and each bond has two electrons. In a 1:1 semiconductor, the atom acquires eight electrons in its compound. Al_2O_3 does not have 1:1 stoichiometry, so the average of four electrons does not apply; however, the octet rule still applies. When each atom has an octet, the valence band is filled and the conduction band is empty; the material is a semiconductor. In Al_2O_3 each atom has an octet. Hence, Al_2O_3 is expected to be a semiconductor.

Chapter 6

1.

	ONF	NF₃	GeCl₄	SCl₂
Electrons needed	24	32	40	24
Valence electrons	$6 + 5 + 7 = 18$	$5 + (3 \times 7) = 26$	$4 + (4 \times 7) = 32$	$6 + (2 \times 7) = 20$
Bonding pairs	3	3	4	2
	:O::N:F:	:F:N:F: :F:	:Cl: :Cl:Ge:Cl: :Cl:	:Cl:S:Cl:

3. 40 electrons needed; 30 valence electrons plus 2 from negative charges available $\Rightarrow$ 4 bonding pairs.

$$\left[\begin{array}{c} :\ddot{O}: \\ :\ddot{O}:S:\ddot{O}: \\ :\ddot{O}: \end{array}\right]^{2-}$$

5. Ozone: number of electrons required = 24; number of valence electrons available = 18; number of bonds required = 3. In molecular oxygen, the two oxygen atoms are connected by a double bond. In ozone, the oxygen atoms are linked by $1\frac{1}{2}$ bonds. The bond order in ozone is less than in molecular oxygen, the bond is weaker, hence ozone is more reactive.

$$:\ddot{O}:\ddot{O}::\ddot{O}: \qquad :\ddot{O}::\ddot{O}:\ddot{O}:$$

7. Number of electrons required = 24; number of valence electrons available = 16; number of bonds required = 4.

(a) $:O:::C:\ddot{O}:$ (b) $:\ddot{O}::C::\ddot{O}:$ (c) $:\ddot{O}:C:::O:$

Formal charges:

	O	C	O
(a)	+1	0	−1
(b)	0	0	0
(c)	−1	0	+1

Structure (b) has a formal charge of zero on all atoms, it is the best structure.

9. Number of electrons required = 32; number of valence electrons available = 24; number of bonds required = 4.

(a) $:\ddot{O}:C:\ddot{C}l:$ (b) $:O::C:\ddot{C}l:$
  $\quad:\ddot{C}l:$ $\qquad\quad:\ddot{C}l:$

Formal charges:

	O	C	Cl	Cl (lower)
(a)	−1	0	0	+1
(b)	0	0	0	0

Structure (b) has a formal charge of zero on all atoms and is therefore the best structure.

11. Number of electrons required = 24; number of valence electrons available = 16; number of bonds required = 4.

$:\ddot{S}::C::\ddot{N}: \qquad :S:::C:\ddot{N}: \qquad :\ddot{S}:C:::N:$
$\quad\quad$ A $\quad\quad\quad\quad\quad$ B $\quad\quad\quad\quad\quad$ C

Formal charges:

	S	C	N
A	0	0	−1
B	+1	0	−2
C	−1	0	0

Of these, A and C have the lowest formal charges. A is the more favorable of the two because it has the negative charge on the more electronegative element.

13.

$$:\ddot{O}::N:\ddot{O}: \qquad :\ddot{O}::N:\ddot{O}: \qquad \begin{array}{c} :\ddot{O}::N:\ddot{O}: \\ | \\ :\ddot{O}::N:\ddot{O}: \end{array}$$
$\quad\quad :\ddot{O}:$
$\quad\quad NO_3{}^- \qquad\qquad NO_2 \qquad\qquad N_2O_4$

NO_2 is an exception to the octet rule: It is an odd-electron molecule. The odd electron in NO_2 pairs up with another electron from a second NO_2 when NO_2 forms N_2O_4. This is the force driving the dimerization.

15.

$$H:\ddot{N}:H \qquad :\ddot{F}:B:\ddot{F}:$$
$\quad\;\; H \qquad\qquad\quad :\ddot{F}:$

NH_3 features a lone pair on the nitrogen atom. BF_3 features a boron atom that is an electron pair short of an octet. When these two get together, the lone pair on nitrogen fills in the missing pair on boron, enabling all atoms to satisfy the octet rule.

17.

$$\begin{array}{c} :\ddot{C}l:P:\ddot{C}l: \\ :\ddot{C}l: \end{array} \quad \begin{array}{c} :\ddot{C}l: \;\; :\ddot{C}l: \\ \backslash \;/ \\ P \\ :\ddot{C}l: \;|\; :\ddot{C}l: \\ :\ddot{C}l: \end{array} \quad \begin{array}{c} :\ddot{F}:B:\ddot{F}: \\ :\ddot{F}: \end{array} \quad \begin{array}{c} :\ddot{F}:\ddot{A}s:\ddot{F}: \\ :\ddot{F}: \end{array}$$
$\quad\;\; PCl_3 \qquad\qquad PCl_5 \qquad\qquad BF_3 \qquad\quad AsF_3$

Both PCl_3 and AsF_3 have lone pairs on the central atom.

19.

$$H:\ddot{O}:\ddot{O}:H$$

Peroxide is a bent molecule. The strength of the O—O bond in peroxide is weaker than the O=O bond in O_2.

21.

$$H:\dot{C}:H$$
$\quad\;\; H$

The methyl radical is planar because the odd electron is unpaired in an unhybridized p orbital.

23. The left nitrogen atom has a lone pair, so it has tetrahedral electronic geometry and is pyramidal. The right nitrogen atom, which carries the positive charge, is trigonal planar. The pentagonal ring is planar at each of the vertices, so the molecule is planar overall.

25. Acetylene is a symmetrical molecule, so it has no dipole moment.

27. Benzene has no dipole moment; acetone has a dipole moment; carbon disulfide is linear and has no dipole moment; ethanol has a dipole moment; acetic acid has a dipole moment.

29.

31. The charge separation is

$$\frac{(0.110 \text{ D}) \times (3.335641 \times 10^{-30} \text{ C·m/D})}{(113 \text{ pm})(1 \text{ m}/10^{12} \text{ pm})(1.60 \times 10^{-19} \text{ C/electron})}$$
$$= 0.0203 \text{ electron}$$

33.

$$\text{OH: } \frac{(1.668 \text{ D}) \times (3.335641 \times 10^{-30} \text{ C·m/D})}{(97 \text{ pm})(1 \text{ m}/10^{12} \text{ pm})(1.60 \times 10^{-19} \text{ C/electron})}$$
$$= 0.36 \text{ electron}$$

For water, use geometry to determine the center of the positive charge: 58.8 pm from the oxygen atom.

$$\text{H}_2\text{O: } \frac{(1.854 \text{ D}) \times (3.335641 \times 10^{-30} \text{ C·m/D})}{(58.8 \text{ pm})(1 \text{ m}/10^{12} \text{ pm})(1.60 \times 10^{-19} \text{ C/electron})}$$
$$= 0.657 \text{ electron}$$

In water, the electron transfer is divided between the two hydrogen atoms, so 0.33 electron is transferred from each hydrogen atom to oxygen. The transfer is less complete in water than in OH⁻. The electronegativity of oxygen is divided between the two hydrogen atoms in water.

35. The vector sum of the CH bonds is a null vector. The vector sums of the NH bonds in ammonia and the OH bonds in water are nonzero. The vector sum of the NH bonds in ammonia and OH bonds in water locates the center of the positive charge. The center of the negative charge is the vector sum of the lone-pair orbitals. The result is a dipole.

37. The carbon in formic acid is sp^2 hybridized.

39. In nitric acid, nitrogen is sp^2 hybridized.

41. There are four bonds between Ti and the four Cl atoms. This requires four orbitals. The valence orbitals in Ti are $4s$ and $3d$, so the hybridization is sd^3.

43. The central oxygen atom in ozone and the sulfur atom in SO_2 both have the same hybridization. Both bond the two attached oxygen atoms and must have an orbital for the lone pair. Thus, they are sp^2 hybridized.

45. The sp^2 carbon atoms are indicated with circles with a solid line. The sp^3 atoms are circled with a dotted line. The silicon atoms are all sp^3 hybridized.

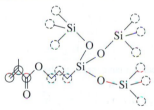

47. An inverse correlation exists between the size of the ion and the melting and boiling points. The size of the halide ions are shown in the table. The melting point of NaAs is estimated to be 550–600 °C and the boiling point to be 1220–1250 °C.

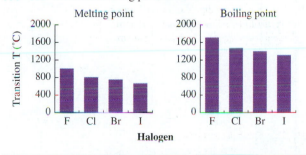

Ion	F⁻	Cl⁻	Br⁻	I⁻	As⁻
Radius (pm)	133	184	196	220	240

49. NaCl is the most ionic substance and therefore conducts electricity best when melted. The poorest conductor is CCl_4 which is not even polar.

51.

$$\text{KCl: } E =$$
$$\frac{(9.0 \times 10^9 \text{ N·m}^2/\text{C}^2)(1.60 \times 10^{-19} \text{ C})(-1.60 \times 10^{-19} \text{ C})}{(2.67 \times 10^{-10} \text{ m})}$$
$$= -8.6 \times 10^{-19} \text{ J or } -520 \text{ kJ/mol}$$

$$\text{MgO: } E =$$
$$\frac{(9.0 \times 10^9 \text{ N·m}^2/\text{C}^2)(3.20 \times 10^{-19} \text{ C})(-3.20 \times 10^{-19} \text{ C})}{(1.75 \times 10^{-10} \text{ m})}$$
$$= -53 \times 10^{-19} \text{ J or } -3200 \text{ kJ/mol}$$

$$\text{CaO: } E =$$
$$\frac{(9.0 \times 10^9 \text{ N·m}^2/\text{C}^2)(3.20 \times 10^{-19} \text{ C})(-3.20 \times 10^{-19} \text{ C})}{(1.82 \times 10^{-10} \text{ m})}$$
$$= -51 \times 10^{-19} \text{ J or } -3100 \text{ kJ/mol}$$

$$\text{NH}_4\text{Cl: } E =$$
$$\frac{(9.0 \times 10^9 \text{ N·m}^2/\text{C}^2)(1.60 \times 10^{-19} \text{ C})(-1.60 \times 10^{-19} \text{ C})}{(3.03 \times 10^{-10} \text{ m})}$$
$$= -7.6 \times 10^{-19} \text{ J or } -460 \text{ kJ/mol}$$

KCl and NH_4Cl are comparable to NaCl; MgO and CaO are approximately four times larger due to a doubling of both the positive and negative charges.

53. The next term has a denominator of $\sqrt{3}r$, the distance across the cube diagonal. The numerator is $8 \times q_{Cl^-} q_{Na^+}$.

$$\text{Lattice energy} = -1.747 \, E_{NaCl} N_A$$

where E_{NaCl} is the energy for one NaCl molecule in the gas phase.

55.

$$\text{Lattice energy} = \frac{1.747 q^+ q^-}{r} \times \frac{\text{Lattice energy}_{NaCl}}{\text{Lattice energy}_{NaBr}}$$
$$= \frac{790 \text{ kJ/mol}}{754 \text{ kJ/mol}} = \frac{1/270 \text{ pm}}{1/r}$$

$$\implies r_{NaBr} = \frac{790 \times 270}{745} \text{ pm} = 290 \text{ pm}$$

57. As the anion size becomes larger, the distance between the anion and the cation increases, and the interaction gets weaker, falling off as the inverse of the distance. The melting point of NaCl is higher than that of DMIM-Cl due to the weaker interaction in the DMIM-Cl salt due to the larger size of the DMIM cation.

59. The charge separation (the amount of electron transferred) is

$$\text{CO} = \frac{(0.12 \text{ D}) \times (3.335641 \times 10^{-30} \text{ C·m/D})}{(182.2 \text{ pm})(1 \text{ m}/10^{12} \text{ pm})(1.60 \times 10^{-19} \text{ C/electron})}$$
$$= 0.014 \text{ electron}$$

ClF: 0.11 electron; NaCl: 0.80 electron; CsCl: 0.75 electron; OH: 0.36 electron

61. The cotton towel absorbs water better due to the large number of —OH groups on the cellulose chain. The —OH groups can form hydrogen bonds with water. Polyester does not have a group that can hydrogen-bond with water.

63. Molecule A is unable to form a hydrogen bond because no hydrogen atoms are attached to the nitrogen. Molecule B cannot form hydrogen bonds for the same reason. Only molecule C can form a hydrogen bond via the —OH group.

65. Only NO_2 and N_2O_5 have dipoles. The geometry around C in CCl_4 is tetrahedral with no resultant dipole. The geometry around S in SO_3 is trigonal planar with no resultant dipole. NO_2 is an -odd-electron molecule with only seven electrons around the nitrogen. It is bent planar and has a dipole. The structure of N_2O_5 is shown below. The geometry around each of the nitrogen atoms is trigonal planar, the electronic geometry around the central oxygen is tetrahedral, and the molecular geometry is bent.

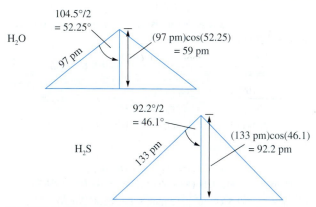

67.

H:Ö:H

H H
| |
HO—C—C—OH
| |
H H

H:Ö:H

69. For water, use geometry (see drawing) to determine the center of the positive charge: 59 pm from the oxygen atom.

$$H_2O: \frac{(1.854 \text{ D}) \times (3.335641 \times 10^{-30} \text{ C} \cdot \text{m/D})}{(59 \text{ pm})(1 \text{ m}/10^{12} \text{ pm})(1.60 \times 10^{-19} \text{ C/electron})}$$

$$= 0.65 \text{ electron}$$

H_2O

104.5°/2
= 52.25°

97 pm

(97 pm)cos(52.25)
= 59 pm

H_2S

92.2°/2
= 46.1°

133 pm

(133 pm)cos(46.1)
= 92.2 pm

For H_2S, the distance from the center of the positive charge to the center of the negative charge (see drawing) is 92.2 pm.

$$H_2S: \frac{(0.97 \text{ D}) \times (3.335641 \times 10^{-30} \text{ C} \cdot \text{m/D})}{(92.2 \text{ pm})(1 \text{ m}/10^{12} \text{ pm})(1.60 \times 10^{-19} \text{ C/electron})}$$

$$= 0.22 \text{ electron}$$

Sulfur is much less electronegative than oxygen, so the extent of electron transfer is less in H_2S than in H_2O.

71. The mole fractions are all quite small, so to a good approximation the mole fraction can be multiplied by the number of water molecules in 1 cm^3 to determine the number of dissolved gas molecules per cubic centimeter. The density of water is 1 g/cm^3, so the number of water molecules per cubic centimeter is

$$\frac{1 \text{ g/cm}^3}{18 \text{ g/mol}} \times \frac{6.022 \times 10^{23} \text{ molecules}}{\text{mol}}$$

$$= 3.34 \times 10^{22} \text{ } H_2O \text{ molecules/cm}^3$$

There are 2.35×10^{17} He atoms/cm^3; 2.81×10^{17} Ne atoms/cm^3; 8.18×10^{16} Ar atoms/cm^3; 1.68×10^{17} Kr atoms/cm^3; and 3.02×10^{17} Xe atoms/cm^3. The expected trend is that the gas becomes more soluble as the size, and therefore polarizability, increases. The small atoms, He and Ne, run counter to this trend.

73. Among substances with a similar structure, the amount of energy needed to break one molecule free from its neighbors is largely dependent on the contact area. This area increases with mass. For example, with hydrocarbon molecules, the molecular mass goes up 14 mass units per CH_2 group. Among noble gases, the polarizability increases as the atom size increases, and so does the atomic mass.

75. Sites for hydrogen bonding are shown with solid-line circles (TRIS has no hydrogen bonding sites), dipoles are indicated with dotted circles, and the remainder of each molecule interacts through London dispersion forces.

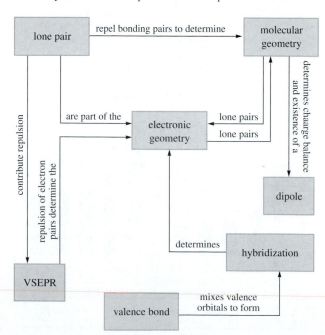

77. CFC-113 has a small dipole moment due to the two fluorine atoms on one end and only one flourine atom on the other end. The chlorine atoms are not as electronegative as fluorine, so they do not balance it. The intermolecular force is a dipole–dipole force. Hence CFC-113 and water interact via dipole–dipole forces.

79. Many answers are acceptable. One example is shown here.

lone pair	repel bonding pairs to determine	molecular geometry

contribute repulsion

repulsion of electron pairs determine the

are part of the

electronic geometry

lone pairs

lone pairs

determines chaarge balance and existence of a

dipole

determines

hybridization

VSEPR

valence bond

mixes valence orbitals to form

81. Each tetrahedral Si center has a dipole moment. However, the random twisting of the flexible backbone results in, at most, a weak dipole for the molecule. The oxygen atoms in the backbone are attached to relatively large Si atoms and are surrounded by —CH_3 groups that hang off the backbone and prevent water from closely interacting with the oxygen atoms. Water therefore does not stick to the windshield, improving visibility and preventing frost from sticking to the windows.

83. The main constituent of soda, beer, or champagne is water. The forces involved in forming a bubble are the outward force of the gas and the inward force of the hydrogen bonding of water. CO_2, the gas that constitutes the bubble, is a nonpolar molecule, so it disrupts the hydrogen bonding of water molecules. To minimize unfavorable interactions between CO_2 and water, the bubble adopts a shape that minimizes the surface-to-volume ratio—that is, a sphere.

Chapter 7

1. Condensation of water is exothermic. Twice as much heat is released on condensation of 20 g of water as is released on condensation of 10 g.

3. (a) Bond formation releases energy; this reaction is exothermic. (b) Formation of water is also combustion of H_2. Combustion is exothermic, so formation of water is also exothermic. (c) Sublimation takes a solid and makes it a gas. This reaction requires energy, so it is endothermic. (d) Evaporation takes a liquid and turns it into a gas. This reaction requires energy, so it is endothermic.

5. When liquid water freezes, it releases energy equal to the heat of fusion. This energy warms the surroundings, including the fruit and the trees. If the temperature drops to $-6\,°C$, the local tree needs to warm by only 2 K to keep the sap from freezing. This requires enough heat to warm 5 L of water by 2 K. 5 L = 5000 g. Heating by 2 K requires

$$5000\text{ g} \times \frac{4.184\text{ J}}{\text{g}\cdot\text{K}} \times 2\text{ K} = 40\text{ kJ}$$

Each liter of added water releases

$$1000\text{ g} \times \frac{4.184\text{ J}}{\text{g}\cdot\text{K}} \times 13\text{ K} = 50\text{ kJ}$$

of heat in cooling from 13 °C to 0 °C, plus an additional

$$1000\text{ g} \times \frac{1\text{ mol}}{18\text{ g}} \times \frac{6\text{ kJ}}{\text{mol}} = 300\text{ kJ}$$

upon freezing. If the added water does not freeze, 0.72 L is required for each tree. If the added water does freeze, only 104 mL is required. The latent heat of fusion really helps.

7. (a)
$$\Delta H = 2 \times \Delta H_f(\text{MgO}) - \Delta H_f(\text{CO}_2)$$
$$= 2\text{ mol} \times (-610.6\text{ kJ/mol}) - 1\text{ mol}$$
$$\times (-393.5\text{ kJ/mol}) = -827.7\text{ kJ}$$

(b) CO_2 will not put out a magnesium fire because CO_2 combines with Mg in an exothermic reaction.

9. (a) This is false. The temperature does not decrease upon freezing. (b) This is false. Heat is released to the environment. (c) This is false. Heat is released to the environment.

11. (a) The enthalpy for condensation of 20 g of water is twice that for 10 g. (b) Same magnitude, opposite sign.

13. The enthalpy of the reaction $\text{Fe}_2\text{O}_3(s) + \text{Al}(s) \rightarrow \text{Al}_2\text{O}_3(s) + \text{Fe}(s)$ is $-1675.7\text{ kJ} + 824.2\text{ kJ} = -851.5\text{ kJ}$. The heat capacity of water is 4.184 J/g, so 203.5 kg of water is heated by this reaction.

15. Supercooled steam is steam at temperatures less than 100 °C. Condensation of 100 °C steam releases the enthalpy of vaporization and the heat needed to cool the 100 °C water to 90 °C. Condensation of supercooled steam can be thought of as a three-step process: (1) warming the steam to 100 °C (endothermic) followed by (2) condensation of steam at 100 °C and (3) cooling the resulting water to 90 °C. The last two steps are the same as for condensation of 100 °C steam. Because the first step is endothermic, less heat is released in condensing supercooled steam.

17. Digestion is combustion.

$$\text{C}_{12}\text{H}_{22}\text{O}_{11}(s) + 12\text{O}_2(g) \longrightarrow 12\text{CO}_2(g) + 11\text{H}_2\text{O}(l)$$
$$\Delta H = 11\text{ mol} \times (-285.8\text{ kJ/mol}) + 12\text{ mol} \times (-393.5\text{ kJ/mol})$$
$$- 1\text{ mol} \times (-2226.1\text{ kJ/mol}) = -5639.7\text{ kJ/mol}$$

$$\text{Calories from sugar} = -5639.7\text{ kJ/mol} \times \frac{1\text{ mol}}{342\text{ g}}$$
$$\times \frac{1\text{ kCal}}{4.184\text{ kJ}} \times \frac{1\text{ food calorie}}{\text{kCal}} \times 22\text{ g}$$
$$= 86.6\text{ food calories}$$

There are 23.3 food calories from fat and protein.

19.
$$\text{C}_8\text{H}_{18}(l) + 12\tfrac{1}{2}\text{O}_2(g) \longrightarrow 8\text{CO}_2(g) + 9\text{H}_2\text{O}(l)$$
$$\Delta H = 8 \times \Delta H_f(\text{CO}_2) + 9 \times \Delta H_f(\text{H}_2\text{O})$$
$$- \Delta H_f(\text{C}_8\text{H}_{18}) - \Delta H_f(\text{O}_2)$$
$$= 8\text{ mol} \times (-393.5\text{ kJ/mol}) + 9\text{ mol} \times (-285.8\text{ kJ/mol})$$
$$- 1\text{ mol} \times (250.1\text{ kJ/mol}) - 1\text{ mol} \times (0)$$
$$= -3148\text{ kJ} - 2572\text{ kJ} - 250.1\text{ kJ}$$
$$= -5970\text{ kJ per mol octane}$$

The molecular weight of octane is 114.232 g/mol and the density is 0.6989 g/mL. Thus octane produces

$$\frac{5970\text{ kJ}}{\text{mol}} \times \frac{1\text{ mol}}{114.232\text{ g}} \times \frac{0.6989\text{ g}}{\text{mL}} = 36.52\text{ kJ/mL}$$

$$\frac{1 \times 10^6\text{ kJ}}{\text{day}} \times \frac{1\text{ mL}}{36.52\text{ kJ}} \times \frac{1\text{ L}}{1000\text{ mL}} = 27.38\text{ L/day}$$

30% efficiency $\Rightarrow$ multiply by 1/0.3: 91.27 L/day

21. Use the equation:

$$\Delta H = \Sigma H_{f,\text{ products}} - \Sigma H_{f,\text{ reactants}}$$
Methane is CH_4: $\text{CH}_4(g) + 2\text{O}_2(g) \longrightarrow \text{CO}_2(g) + 2\text{H}_2\text{O}(l)$
1 mol $\times (-393.5\text{ kJ/mol}) + 2\text{ mol} \times (-285.8\text{ kJ/mol}) - 1\text{ mol}$
$\times (-74.6\text{ kJ/mol}) = -890.5\text{ kJ/mol or} -55.66\text{ kJ/g}$

Octane: $-5470.1\text{ kJ/mol and} -47.87\text{ kJ/g}$

Methanol: $-726.1\text{ kJ/mol and} -22.66\text{ kJ/g}$

Ethanol: $-1366.8\text{ kJ/mol and} -29.66\text{ kJ/g}$

The most efficient fuel is the one that produces the greatest enthalpy per gram, which is methane. The alcohols are all partially oxidized already and yield less energy per gram.

23. Combustion of methanol:

$$\text{CH}_3\text{OH}(l) + 1\tfrac{1}{2}\text{O}_2(g) \longrightarrow \text{CO}_2(g) + 2\text{H}_2\text{O}(l)$$
$$\Delta H = 2\text{ mol} \times (-285.8\text{ kJ/mol}) + 1\text{ mol} \times (-393.5\text{ kJ/mol})$$
$$- 1\text{ mol} \times (-239.2\text{ kJ/mol}) = -725.9\text{ kJ}$$

Fuel cell reaction:

$$CH_3OH(l) + \tfrac{1}{2}O_2(g) \longrightarrow H_2CO(g) + H_2O(l)$$

$$\Delta H = 1 \text{ mol} \times (-108.6 \text{ kJ/mol}) + 1 \text{ mol} \times (-285.8 \text{ kJ/mol})$$
$$- 1 \text{ mol} \times (-239.2 \text{ kJ/mol}) = -155.2 \text{ kJ}$$

A fuel cell produces far less heat than does combustion.

25. (a) $3O_2(g) + h\nu \rightarrow 2O_3(g)$. The heat of formation of ozone is not listed; however, the heats of atomization of both O_2 and O_3 are listed. ΔH for the reaction is 3 mol $\times$ (498.4 kJ/mol) − 2 mol $\times$ (604.6 kJ/mol) = 286 kJ.

(b) The reaction forming ozone is endothermic, with the needed energy being supplied by the sun. When ozone is converted back to oxygen, this heat is released, heating the surrounding air to produce a higher temperature for the lower stratosphere than for the upper troposphere just below it. Essentially, the sun's photons are converted to heat in the $O_2 \rightarrow O_3 \rightarrow O_2$ cycle.

27. Volume of the balloon is $\dfrac{4}{3}\pi r^3 = 524 \text{ cm}^3 = 0.524 \text{ L}$.

$$w = -P\Delta V = -1 \text{ atm} \times 0.524 \text{ L}$$
$$= 0.524 \text{ L} \cdot \text{atm} \times \frac{8.3145 \text{ J/mol} \cdot \text{K}}{0.08206 \text{ L} \cdot \text{atm/mol} \cdot \text{K}} = 53.1 \text{ J}$$

29. Combustion of methane:

$$CH_4(g) + 2O_2(g) \longrightarrow CO_2(g) + 2H_2O(l)$$

Combustion of methanol:

$$2CH_3OH(l) + 3O_2(g) \longrightarrow 2CO_2(g) + 4H_2O(l)$$

Work for methane $= -1 \text{ atm} \times 22.4 \text{ L/mol} \times -2 \text{ mol}$
$$= 44.8 \text{ L} \cdot \text{atm}$$

Work for methanol $= -1 \text{ atm} \times 22.4 \text{ L/mol} \times -1 \text{ mol}$
$$= 22.4 \text{ L} \cdot \text{atm}$$

The work produced is positive, which means that the system energy increases in the combustion process. Less work is needed to combust methanol than to combust methane. The difference is partly due to the oxygen already incorporated in the methanol and partly due to methanol being a condensed fuel while methane is a gas. If the water produced is also a gas, the combustion of methane neither produces nor consumes any work. The combustion of methanol liquid would produce 3×22.4 L · atm of work.

31.

$$\text{Expansion per day} = \frac{0.75 \text{ pint}}{\text{breath}} \times \frac{20 \text{ breaths}}{\text{min}} \times \frac{60 \text{ min}}{\text{h}}$$
$$\times \frac{24 \text{ h}}{\text{day}} \times \frac{0.454 \text{ L}}{\text{pt}} = 9800 \text{ L}$$

$$w = P\Delta V = -1 \text{ atm} \times 9800 \text{ L} \times \frac{8.3145 \text{ J}}{0.08206 \text{ L} \cdot \text{atm}} = 990 \text{ kJ/day}$$

33. Volume of the pastry prior to puffing:

$$(8 \text{ in}) \times (11 \text{ in}) \times (0.75 \text{ in}) \times (2.54 \text{ cm/in})^3 = 1081 \text{ cm}^3$$

After puffing:

$$V = 4324 \text{ cm}^3$$

$$w = -(1 \text{ atm}) \times (4.324 \text{ L}) = -4.32 \text{ L} \cdot \text{atm} = -438 \text{ J}$$

The puff pastry is more successful if the oven is preheated because the moisture in the pastry quickly turns to steam prior to the dough hardening as it bakes. If the dough hardens first, some of the steam escapes and does not puff the pastry.

35. $\Delta H = (4.184 \text{ J/g} \cdot \text{K})(250 \text{ g})(37 \text{ K}) = 38.7 \text{ kJ}$

Lifting the groceries for three flights of stairs raises the energy by

$$mgh = (20 \text{ lb}) \times (0.45359 \text{ kg/lb})(9.8 \text{ m/s}^2)(10 \text{ m}) = 890 \text{ J}$$

More energy is expended in drinking the ice water.

37. 990 kJ is required for breathing each day. If the sucrose-to-work conversion is 50% efficient, 1980 kJ must be consumed. Combustion of sucrose:

$$C_{12}H_{22}O_{11}(s) + 17\tfrac{1}{2}O_2(g) \longrightarrow 12CO_2(g) + 11H_2O(l)$$

$$\Delta H = -5640 \text{ kJ/mol of sucrose} = 16.5 \text{ kJ/g}$$

120 g of sucrose is required.

39. Digestion of sugar:

$$C_{12}H_{22}O_{11}(s) + 17\tfrac{1}{2}O_2(g) \longrightarrow 12CO_2(g) + 11H_2O(l)$$

$$\Delta H = -5640 \text{ kJ/mol of sugar}$$

Digestion of alcohol, C_2H_5OH:

$$C_2H_5OH(l) + 3O_2(g) \longrightarrow 2CO_2(g) + 3H_2O(l)$$

$$\Delta H = -1366.8 \text{ kJ/mol}$$

On a per-gram basis, sugar produces 16.5 kJ/g; alcohol produces 29.7 kJ/g. Alcohol produces far more energy per gram than does sugar. The mass of sugar consumed is three times the mass of alcohol consumed, so the person loses more weight by giving up the candy bar snack.

41. $\Delta H = 1 \text{ mol} \times (-824.2 \text{ kJ/mol}) - 2 \text{ mol} \times (-266.5 \text{ kJ/mol})$
$$= -291.2 \text{ kJ}$$

43.

	Methane	Ethane	Propane	Butane
ΔH per mole (kJ)	−890.5	−1560.4	−2202.8	−2877.3
ΔH per carbon (kJ)	−890.5	−780.2	−724.3	−719.3
ΔH per gram (kJ)	−55.66	−53.01	−50.06	−49.61

The heat content per mole increases as the number of carbon atoms increases. However, on a per-carbon-atom or per-mole basis, the heat of combustion decreases with increasing size. The most efficient storage of energy therefore occurs for those molecules with the larger H : C ratio.

45. 2 mol $\times$ (−1657.7 kJ/mol) − 3 mol $\times$ (−520 kJ/mol)
$$= -1755 \text{ kJ}$$

The process is exothermic.

47. It is predicted that formation of CdS is more exothermic than is formation of CdTe. The wider band gap in CdS indicates a stronger interaction between Cd and S than between Cd and Te. A stronger interaction means that more heat is released as the atoms come together.

49.

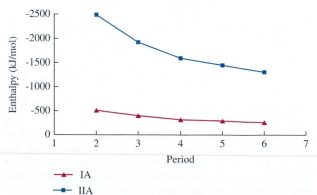

As the ion size increases (going down a series), the hydration enthalpy decreases.

51. The heat of formation of a solution of $BaSO_4$ is -1446.9 kJ/mol. Dissolving $BaSO_4$ is exothermic. The enthalpy of solution contributes to the solubility as $\Delta G = \Delta H - T\Delta S$, $\Delta G < 0$ is spontaneous, and ΔH is negative.

53. Calculation of the atomization enthalpy of $CaCl_2$ follows the process:

$$CaCl_2(s) \longrightarrow Ca(s) + Cl_2(g) \longrightarrow Ca(atom) + 2Cl(atom)$$

ΔH for this process $= \Delta H(CaCl_2) + \Delta H_{atom}(Ca) + 2\Delta H_{atom}(Cl)$
$= -1$ mol $\times (-795.4$ kJ/mol$) + 1$ mol $\times (177.8$ kJ/mol$)$
$+ 2$ mol $\times (121.3$ kJ/mol$) = 1215.8$ kJ/mol $CaCl_2$

For NaCl:

$$NaCl(s) \longrightarrow Na(s) + \tfrac{1}{2}Cl_2(g) \longrightarrow Na(atom) + Cl(atom)$$

ΔH for this process $= \Delta H(NaCl) + \Delta H_{atom}(Na) + \Delta H_{atom}(Cl)$
$= -1$ mol $\times (-411.2$ kJ/mol$) + 1$ mol $\times (107.5$ kJ/mol$)$
$+ 1$ mol $\times (121.3$ kJ/mol$) = 640$ kJ/mol NaCl

Much more energy is required to separate $CaCl_2$ into its atoms than to separate NaCl into its atoms, reflecting the double charge on the Ca ion.

55.

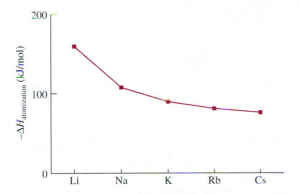

The atomization enthalpy decreases as the atom size increases: As atoms become larger, they interact less strongly because they do not approach one another as closely. The atomization enthalpy for Fr is predicted to be 65–70 kJ/mol.

57. When a solid melts, the entropy increases. When a solid sublimes, the entropy increases much more than for melting the solid. The entropy change for condensing a gas is equal in magnitude but opposite in sign to that for vaporizing a liquid.

59. (a) Iotalamic acid is $380.7/613.7 \times 100\% = 62\%$ iodine by mass.

(b) The solution with the higher ionic strength has greater entropy. This greater entropy drives the process of pulling water out of the surrounding tissue in an attempt to balance the ionic strength of the tissue and the blood.

(c) The energy required to pump the blood is provided by the muscles of the heart as well as the muscles of the artery walls. When muscles work, they burn fuel (mainly glucose). Combustion is exothermic, which leads to the warm sensation.

61. (a) $\Delta H = 1$ mol $\times (-1675.7$ kJ/mol$) - 1$ mol $\times (-1139.7$ kJ/mol$) = -536$ kJ produces 2 mol Cr or 104 g, so the heat per gram is 5.15 kJ/g.

(b) The process is exothermic. Unless there is a large loss of entropy, the process will be spontaneous.

(c) The products are Al_2O_3 and 2Cr with a combined heat capacity of $(1/2 \times 51.12 + 19.86)$ J/K per mole of Cr or 45.42 J/K per mol Cr = 0.8735 J/K per g Cr. To melt Cr, the temperature must rise to 2180 K from a room temperature of 300 K; that is, the temperature must increase by 1880 K. With a heat capacity of 0.8735 J/K, 0.8735 J·$K^{-1} \times 1880$ K = 1.6422 kJ per g Cr. The heat produced is more than three times that amount, so the Cr will be molten.

63. Consider the reaction $2NO_2 \rightarrow N_2O_4$. ΔG for this reaction is 1 mol $\times (99.8$ kJ/mol$) - 2$ mol $\times 51.3$ kJ/mol $= -2.8$ kJ, so the process is spontaneous at room temperature. At room temperature, N_2O_4 is slightly more stable than is NO_2.

65. (a) Evaluate ΔG from $\Delta H - T\Delta S$.

$$\Delta H = [3 \times (-393.5) - 2 \times (-824.2)] \text{ kJ} = 467.9 \text{ kJ}$$
$$\Delta S = [3 \times (213.8) + 4 \times (27.3) - 2 \times (87.4) - 3 \times (5.74)] \text{ J/K}$$
$$= 558 \text{ J/K}$$

Since both ΔH and ΔS are positive, the temperature determines the spontaneity of the process.

$$\Delta G = 467.9 \text{ kJ} - (298)(0.558) \text{ kJ} = 301.6 \text{ kJ}$$

Since ΔG is positive, the process is not spontaneous at room temperature.

(b) Heating the reaction makes the process spontaneous as long as

$$T > \frac{467.9}{0.558} \text{ K} = 838.5 \text{ K}$$

67. (a) ΔH_f(gas) $- \Delta H_f$(liquid)
2-butanol: $(-292.8 + 342.6)$ kJ $= 49.8$ kJ/mol
diethyl ether: $(-252.1 + 279.5)$ kJ/mol $= 27.4$ kJ/mol

(b) 2-butanol: $\Delta S = (214.9 - 359.5)$ J/K·mol $= -144.6$ J/K·mol
diethyl ether: $\Delta S = (172.4 - 342.7)$ J/K·mol $= -170.3$ J/K·mol

(c) The boiling point is that temperature at which $\Delta H = T\Delta S$, or $T = \Delta H/\Delta S$. For 2-butanol, $T_b = \dfrac{49.8}{0.1446}$ K $= 344.4$ K. For diethyl ether, $T_b = \dfrac{27.4}{0.1703}$ K $= 160.9$ K.

69. Synthesis of ATP requires input of 34.5 kJ; synthesizing 38 ATP molecules requires an input of 1311 kJ. Oxidation of glucose releases ΔH of 2802.5 kJ/mol; ΔS is 471.6 J/K·mol. $\Delta G = [-2802.5 - (298)(0.4716)]$ kJ $= -2943$ kJ/mol. Coupling these two reactions results in a negative $\Delta G = -1632$ kJ. Coupling makes the process spontaneous.

71. Coupling the two reactions yields the reaction

$$Cu_2O + C \longrightarrow 2Cu + CO$$

ΔG for this reaction is $\Delta G_f(CO) - \Delta G_f(Cu_2O) = 1$ mol $\times (-143.8$ kJ/mol$) - 1$ mol $\times (-140$ kJ/mol$) = -3.8$ kJ/mol. Coupling the reactions makes production of metallic copper just spontaneous at 375 K.

73. Many diagrams are possible. One example is shown on page 506.

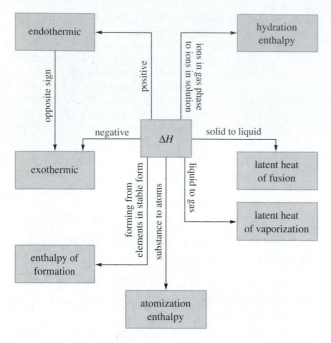

75.

$$C(graphite) + O_2(g) \xrightarrow{\Delta H_{atomization}} C(atoms)$$

$$+ 2O(atoms) -1 \xrightarrow{-\Delta H_{atomization}} CO_2(g)$$

$$\Delta H_f = \Delta H_{atomization}(graphite) + \Delta H_{atomization}(O_2)$$
$$- \Delta H_{atomization}(CO_2)$$
$$= 716.7\ kJ + 498.4\ kJ - 1608\ kJ$$
$$= -393\ kJ/mol4\ of\ CO_2\ formed.$$

The enthalpy of atomization is always positive because the attraction energy between the bonded atoms needs to be overcome to free the atoms.

77. If the container starts at room temperature, 25 °C, then the temperature rise is 15 °C = 15 K. The heat capacity of water 1 cal/K·g, so the heat required is

$$\Delta H = \frac{1\ cal}{K \cdot g} \times 250\ g \times 15\ K \times \frac{4.184\ J}{cal} = 16\ kJ$$

Generation of 16 kJ requires 0.027 mol CaO or 1.5 g CaO. The 1.5 g will easily fit into a small packet to heat the can contents, so the scheme is feasible. The limitations of the scheme relate to any leak of the CaO solution into the contents of the can. CaO is a strongly basic substance, which will probably seriously affect the taste of the can's contents.

79. Determine the heat needed from the temperature change and the heat capacity. Convert °F to °C; multiply the temperature increase by the heat capacity of water = 1 cal/g·K. Convert body mass from pounds to grams.
Determine the heat produced: Write the glucose oxidation reaction, and get ΔH data.
Divide the heat needed by heat per mole to determine the number of moles needed, and convert to grams.

Heat needed: 65 °F is 18.3 °C; 98.6 °F is 37 °C; ΔT is 19 K. 140-lb body is 64 kg (1 lb = 453.59 g)
Heat needed = (64 kg) × (1000 g/kg) × (1 cal/g·K) × 19 K
$$\times (4.184\ J/cal) \times (1\ kJ/1000\ J) = 4970\ kJ$$
Glucose oxidation reaction:
$$C_6H_{12}O_6(s) + 6O_2(g) \longrightarrow 6CO_2(g) + 6H_2O(l)$$
$$\Delta H = 6\ mol \times (-393.5\ kJ/mol) + 6\ mol \times (-285.8\ kJ/mol)$$
$$- 1\ mol \times (1273.3\ kJ/mol) = -5349.1\ kJ$$
Moles needed = $(-5349.1\ kJ)/(1273.3\ kJ/mol)$ = 4.201 mol.
Half is used to heat $\Rightarrow$ moles needed = 8.402 mol
Grams needed = (168 g/mol) × 8.402 mol = 1412 g

Chapter 8

1. (a) $K = \dfrac{[SO_3]^2}{[SO_2]^2[O_2]}$

(b) $K_p = K_c(RT)^{\Delta n}$; $\Delta n = -1$, so $K_p = 1.67 \times 10^{-4}$ at $T = 298\ K$

(c) The ratio is 3.34×10^{-5}; direct oxidation by O_2 accounts for little of the oxidation of SO_2.

3. (a) $K = \dfrac{[H_3O^+][HCO_3^-]}{[H_2CO_3]}$

(b) $K = [Cd^{2+}][S^{2-}]$

(c) $K = \dfrac{[H_3O^+]^2\,[NO_3^-]^2 p(NO)}{(p(NO_2))^3}$

5.
$$CuCO_3(s) \Longleftrightarrow Cu^{2+}(aq) + CO_3^{2-}(aq)$$
$$Cu(OH)_2(s) \Longleftrightarrow Cu^{2+}(aq) + 2OH^-(aq)$$
$$K_{sp}(CuCO_3) = [Cu^{2+}][CO_3^{2-}]$$
$$K_{sp}(Cu(OH)_2) = [Cu^{2+}][OH^-]^2$$

7. Using the density of water (1 g/mL), 1 mL of water contains 5.556×10^{-2} moles of water.

$$\text{mol fr } O_2 \text{ in water} = \frac{\text{mol } O_2}{\text{mol } O_2 + \text{mol } H_2O}$$
$$= \frac{\text{mol } O_2}{\text{mol } O_2 + 5.556 \times 10^{-2}} \approx \frac{\text{mol } O_2}{5.556 \times 10^{-2}}$$

$\Longrightarrow$ mol O_2 in water =
$(5.556 \times 10^{-2}\ mol/mL) \times (2.301 \times 10^{-5}) = 1.278 \times 10^{-6}\ mol/mL$
The concentration of oxygen molecules in the atmosphere, is

$$\frac{1\ mol}{22.4\ L} \times 0.22 \times \frac{1\ L}{1000\ mL} = 9.821 \times 10^{-6}\ mol/mL$$

As you can see, there is only about eight times more oxygen in air than in water.

9. Sodium salts are soluble, so the concentration of $HOOCCOO^-$ is 0.01 M.

11. Chloride salts are mostly soluble, so NH_4Cl is soluble. It dissociates to yield NH_4^+ and Cl^- ions. NH_4^+ is the conjugate acid of the base NH_4OH. The equilibrium is $NH_4^+ + 2H_2O \rightleftharpoons NH_4OH + H_3O^+$.

$$K = \frac{[H_3O^+][NH_4OH]}{[NH_4^+]} = \frac{[H_3O^+][NH_4OH]}{[NH_4^+]} \times \frac{[OH^-]}{[OH^-]} = \frac{K_w}{K_B}$$
$$= \frac{1 \times 10^{-14}}{1.79 \times 10^{-5}} = 5.59 \times 10^{-10}$$

$$NH_4^+ + 2H_2O \Longleftrightarrow NH_4OH + H_3O^+$$

Initial:	0.1		
Change:	$-x$	x	x
End:	$0.1 - x$	x	x

$x^2 = 5.59 \times 10^{-11} \Longrightarrow x = 7.48 \times 10^{-6} \Longrightarrow [OH^-]$
$$= 1.34 \times 10^{-9} \, M$$

13. (a) The molecular weight of citric acid is 192, so 1 g (1000 mg) is 5.2×10^{-3} mol; the initial concentration is 0.10 M. The ionization reaction is

$$\text{Citric acid}(aq) + H_2O(l) \rightleftharpoons \text{citrate}^-(aq) + H_3O^+(aq)$$

Initial:	0.10		
Change:	$-x$	x	x
End:	$0.10 - x$	x	x

$$K_a = 7.1 \times 10^{-4} = \frac{[H_3O^+][\text{citrate}^-]}{[\text{citric acid}]} = \frac{[x][x]}{[0.10 - x]} \approx \frac{x^2}{0.10}$$

$$\Longrightarrow x = 8.4 \times 10^{-3} \Longrightarrow pH = 2.1$$

(b) Sodium salts are soluble, so the $NaHCO_3$ dissociates to yield Na^+ and HCO_3^-. The HCO_3^- ion is protonated by the H_3O^+:

$$HCO_3^-(aq) + H_3O^+(aq) \rightleftharpoons H_2CO_3(aq)$$

(c) $HCO_3^-(aq) + H_3O^+(aq) \rightleftharpoons H_2CO_3(aq) \rightleftharpoons H_2O(l) + CO_2(g)$

15. HCl dissociates completely, which gives $[H_3O^+] = 1$ and $pH = 0$. For HF, $K_a = 3.53 \times 10^{-4}$. Use the equilibrium constant to find the concentration of $[H_3O^+]$:

$$HF(aq) + H_2O(l) \rightleftharpoons H_3O^+(aq) + F^-(aq)$$

$$K = \frac{[H_3O^+][F^-]}{[HF]} = \frac{x^2}{1 - x} \approx \frac{x^2}{1} = 3.53 \times 10^{-4}$$

$$\Longrightarrow x = 0.0188 \Longrightarrow pH = 1.73$$

Therefore, HCl is a stronger acid than HF.

17. $K_{sp}(Zn(OH)_2) = 4.5 \times 10^{-17}$. With $[Zn^{2+}]$ at 0.20 M, $[OH]^2 = 4.5 \times 10^{-17} \times 0.20 = 9.0 \times 10^{-18} \Longrightarrow [OH]$
$$= 3.0 \times 10^{-9} \Longrightarrow pOH = 8.5$$

19. (a) The discoloring reaction involves formation of PbS: $Pb^{2+} + S^{2-} \rightleftharpoons PbS$. Note that these are not aqueous solutions, but rather this is a solid-state reaction.
(b) The discoloration is due to formation of PbS.

21. $Cd^{2+}(aq) + S^{2-}(aq) \rightleftharpoons CdS(s)$; $K_{sp} = 3.6 \times 10^{-29}$. To limit $[Cd^{2+}]$ to $10^{-9} \, M$, $[S^{2-}] = 3.6 \times 10^{-29}/10^{-9} = 3.6 \times 10^{-20} \, M$.

23. $K_{sp}(PbCl_2) = 1.6 \times 10^{-5}$
$K_{sp}(PbBr_2) = 6.3 \times 10^{-6}$
$K_{sp}(PbI_2) = 8.7 \times 10^{-9}$

$$PbX_2(s) \text{ (where X can be Cl, Br, or I)} \rightleftharpoons Pb^{2+}(aq) + 2X^-(aq)$$
$$K_{sp} = [Pb^{2+}][X^-]^2 = (x)(2x)^2 = 4x^3$$
$$x = \text{molar solubility} = \sqrt[3]{\frac{K_{sp}}{4}}$$

x for $PbCl_2 = 0.0159$ mol/L
x for $PbBr_2 = 0.0116$ mol/L
x for $PbI_2 = 0.00130$ mol/L

As you go down Group VII, the molar solubility decreases.

25. The NO_3^- ion forms a soluble combination with both Sn^{2+} and Na^+; Na^+ forms a soluble combination with S^{2-}. However, Sn^{2+} and S^{2-} form an insoluble combination, SnS. The solubility product for SnS is 1×10^{-28}. $Q = 10^{-6}$, which is much larger than K_{sp}, so there will be many more moles of solid than ions left in solution.

27. $$Cu^{2+}(aq) + 2OH^-(aq) \rightleftharpoons Cu(OH)_2(s)$$
$$K_{sp} = [Cu^{2+}][OH^-]^2$$

29. Increasing the Cl^- ion concentration by 100-fold decreases the Ag^+ concentration to 1/100, or 0.01 of the original concentration since $[Ag^+] \times [Cl^-]$ is a constant.

31. Buffering the solution at pH 5 means that $[H^+] = 10^{-5} \, M \Rightarrow [OH^-] = 10^{-9} \, M$; this value does not change. $K_{sp} (Cu(OH)_2) = 1.6 \times 10^{-19} = [Cu^{2+}][OH]^2 \Rightarrow [Cu^{2+}] = 1.6 \times 10^{-19}/(1 \times 10^{-9})^2 = 0.16 \, M$.

33. $Pb(NO_3)_2$ is a soluble salt, as is Na_2S. Similarly, $NaNO_3$ is a soluble salt. The insoluble material here is PbS. With equal molar Pb^{2+} and S^{2-} starting concentrations, the ending concentrations are also equal.

$$PbS(s) \rightleftharpoons Pb^{2+}(aq) + S^{2-}(aq)$$
$$K_{sp}(PbS) = 8.4 \times 10^{-28}$$
$$[Pb^{2+}] = [S^{2-}] = \sqrt{8.4 \times 10^{-28}} \, M = 2.9 \times 10^{-14} \, M$$

35. (a) The molecular weight of potassium bitartrate is 188 g/mol. 1 g is 5.32×10^{-3} moles in 162 mL $\Rightarrow$ solubility of 0.0328 mol/L. $K_{sp} = (0.0328)^2 = 1.076 \times 10^{-3}$.
(b) In ethanol, the solubility of potassium bitartrate is 6.03×10^{-4} mol/L. $K_{sp} = (6.03 \times 10^{-4})^2 = 3.64 \times 10^{-7}$.
(c) Potassium bitartrate is nearly 3000 times more soluble in water than it is in alcohol. Wine is about 10% ethanol by volume, so the potassium bitartrate is about 300 times less soluble in wine than it is in the aqueous solution from which wine is made. If the potassium tartrate is near the solubility limit in the original grape juice, then it will precipitate out in the wine.

37. (a) The molecular weight of Ag_2CrO_4 is 332 g/mol. 0.0030 g is 9.04×10^{-6} mol $\Rightarrow [Ag^+] = 2 \times \frac{9.04 \times 10^{-6} \text{ mol}}{100 \text{ mL}} \times \frac{1000 \text{ mL}}{L} = 1.81 \times 10^{-4} \, M$; $[CrO_4^{2-}] = \frac{9.04 \times 10^{-6} \text{ mol}}{100 \text{ mL}} \times \frac{1000 \text{ mL}}{L} = 9.04 \times 10^{-5} \, M$. $K_{sp} = [Ag^+]^2[CrO_4^{2-}] = 2.96 \times 10^{-12}$.
(b) With $[Ag^+] = 0.01 \, M$, the solubility is the CrO_4^{2-} concentration.
$$[CrO_4^{2-}] = (2.96 \times 10^{-12})(0.01 \, M) = 2.96 \times 10^{-14} \text{ mol/L}.$$
(c) In 0.01 M K_2CrO_4, $[CrO_4^{2-}] = 0.01 \, M \Rightarrow [Ag^+] = \sqrt{2.96 \times 10^{-12}/0.01} = 1.72 \times 10^{-5} \, M$.

39. The spectator ions are NO_3^- and K^+.

41. There is only one spectator ion, Na^+. The ion H_3CCOO^- and the acid H_3CCOOH are involved in the acid equilibrium.

43. (a) Spectator ions: NO_3^- and Na^+; $CdS(s) \rightleftharpoons Cd^{2+}(aq) + S^{2-}(aq)$
(b) Spectator ions: NO_3, K^+, and Na^+; $AgBr(s) \rightleftharpoons Ag^+(aq) + Br^-(aq)$; $AgI(s) \rightleftharpoons Ag^+(aq) + I^-(aq)$
(c) Spectator ion: Cl^-; $Cu(OH)_2(s) + 2H_3O^+(aq) \rightleftharpoons Cu^{2+}(aq) + 4H_2O(l)$
(d) Spectator ion: Cl^-; $Fe(OH)_2(s) + 2H_3O^+(aq) \rightleftharpoons Fe^{2+}(aq) + 4H_2O(l)$
(e) Spectator ions: NO_3^- and Na^+; $Cu(OH)_2(s) \rightleftharpoons Cu^{2+}(aq) + 2OH^-(aq)$

45. (a) No spectator ions; $CH_3COOH(aq) + H_2O(l) \rightleftharpoons CH_3COO^-(aq) + H_3O^+(aq)$
(b) No spectator ions; $HCOOH + H_2O(l) \rightleftharpoons HCOO^-(aq) + H_3O^+(aq)$

47. HF should be combined with a soluble salt containing F^- ions: NaF and KF are good candidates. If approximately equal molar amounts of NaF and HF are combined, the resulting pH is about $pK_a = -\log(3.53 \times 10^{-4}) = 3.5$.

49. Initial concentration of $C_6H_5NH_3^+(aq) = 0.01\ M$.

$$C_6H_5NH_3^+(aq) + 2H_2O(l) \rightleftharpoons C_6H_5NH_3OH(aq) + H_3O^+(aq)$$

Initial:	0.01		
Change:	$-x$	x	x
End:	$0.01 - x$	x	x

$$K = \frac{[H_3O^+][C_6H_5NH_3OH]}{[C_6H_5NH_3^+]} = \frac{[H_3O^+][C_6H_5NH_3OH]}{[C_6H_5NH_3^+]}$$

$$\times \frac{[OH^-]}{[OH^-]} = \frac{K_w}{K_b\,(C_6H_5NH_3OH)}$$

$$= \frac{1 \times 10^{-14}}{4.27 \times 10^{-10}} = 2.34 \times 10^{-5} \Longrightarrow x^2 = 2.34 \times 10^{-7}$$

$$\Longrightarrow x = 4.84 \times 10^{-4} \Longrightarrow [OH^-] = 2.06 \times 10^{-11}\ M$$

51. This is a buffer problem. Addition of the completely ionized base, NaOH, adds 0.1 mol of OH^- to the solution, converting 0.1 mol of acetic acid to acetate ion. The initial concentrations are acetic acid, 0.025 M, and acetate ion, 0.1 M. Use the Henderson-Hasselbalch acid equation:

$$pH = pK_a + \log\left(\frac{[salt]}{[acid]}\right)$$

pK_a for acetic acid is $-\log(1.76 \times 10^{-5}) = 4.75$; $\log(0.1/0.025) = 0.60 \Rightarrow pH = 5.35$.

53. To buffer in the basic range, choose a base as the buffer. To buffer in the 7.8–8.3 range pOH is in the range 6.2–5.7; the pK_b of the base should be about 6 ($K_b \sim 10^{-6}$). There are several candidates; the environmentally reasonable ones include hydrazine and ammonia. With hydrazine, $K_b = 1.7 \times 10^{-6}$, and $pK_b = 5.77$. To adjust this to 6.2 requires $\log([salt]/[base]) = 0.43$ or 2.7 times as much salt as base. To adjust to 5.7 requires $\log([salt]/[base]) = -0.07$ or 0.85 times as much salt as base. For an ammonia buffer, $K_b = 1.79 \times 10^{-5}$, and $pK_b = 4.75$. To adjust this to 6.2 requires $\log([salt]/[base]) = 1.45$ or 28 times as much salt as base, which is on the edge of the buffering capacity. To adjust to 5.7 requires $\log([salt]/[base]) = 0.95$ or 8.9 times as much salt as base.

Alternatively, the buffer could be based on the ionization constant of a weak acid with $K \sim 10^{-8}$. Acid candidates are carbonic acid, $K_a = 4.3 \times 10^{-7}$; H_2S, $K_a = 9.1 \times 10^{-8}$; the second ionization of phosphoric acid, $K_2 = 6.23 \times 10^{-8}$; selenous acid, $K_2 = 4.9 \times 10^{-8}$; and tellourous acid, $K_2 = 2 \times 10^{-8}$.

The buffer system most often used in tropical fish tanks is the second ionization of phosphoric acid, $K_2 = 6.23 \times 10^{-8}$. $pK_2 = 7.21$. To adjust the pH to 7.8 requires $\log([salt]/[acid]) = 0.59$ or 3.9 times as much salt (e.g., Na_2HPO_4) as acid (e.g., NaH_2PO_4).

55. Sodium salts are soluble, so the initial concentrations are

$$HPO_4^{2-}: \frac{100}{250} \times 0.1\ M = 0.04\ M$$

$$H_2PO_4^-: \frac{150}{250} \times 0.1\ M = 0.06\ M$$

$$K_2 = 6.23 \times 10^{-8}$$

$$pH = pK_a + \log\left(\frac{[salt]}{[acid]}\right) = 7.21 + \log\left(\frac{0.04}{0.06}\right) = 7.03$$

$$\Longrightarrow pOH = 6.97 \Longrightarrow [OH^-] = 1.07 \times 10^{-7}\ M$$

57. The $HCOO^-$ ion concentration is the same as the HCOONa concentration since sodium salts are soluble. $[HCOO^-] = 0.01\ M$. This is a buffer problem. K_a of HCOOH is 1.77×10^{-4}.

$$pH = pK_a + \log\left(\frac{[salt]}{[acid]}\right) = 3.75 + \log\left(\frac{0.01}{0.1}\right) = 2.75$$

59. Both vinegar and lemon juice contain acid, H_3O^+, which is the active ingredient. Both acids are weak acids, so

$$HA(aq) + H_2O(l) \rightleftharpoons A^-(aq) + H_3O^+(aq)$$

For $CuCO_3$:

$$CuCO_3(s) + 2H_3O^+(aq) \rightleftharpoons Cu^{2+}(aq)$$
$$+ H_2CO_3(aq) + 2H_2O(l)$$
$$H_2CO_3(aq) \rightleftharpoons CO_2(g) + H_2O(l)$$

For $Cu(OH)_2$:

$$Cu(OH)_2(s) + 2H_3O^+(aq) \rightleftharpoons Cu^{2+}(aq) + 4H_2O(l)$$

For $CuCO_3$, $K_{sp} = 2.5 \times 10^{-10}$; for $Cu(OH)_2$, $K_{sp} = 1.6 \times 10^{-19}$. The $Cu(OH)_2$ equilibria involve soluble species, while the $CuCO_3$ equilibria involve generation of CO_2 gas. Evolution of the gas means that the concentrations shift to the right, so 1 M acetic acid dissolves 1 mol/L $CuCO_3$. The equilibria involved with $Cu(OH)_2$ dissolve less than the full 1 mol/L. Thus $CuCO_3$ is more soluble.

61. With a H_2CO_3/HCO_3^- buffer system, if the capacity of the buffer is exceeded and the blood becomes more acidic, the reaction

$$H_2CO_3(aq) + H_2O(l) \rightleftharpoons HCO_3^-(aq) + H_3O^+(aq)$$

tips the concentrations in favor of the reactants. The reaction

$$CO_2(g) + H_2O(l) \rightleftharpoons H_2CO_3(aq)$$

also tips the concentrations in favor of the reactants. In this case, the amount of $CO_2(g)$ in the blood exceeds the solubility and gas bubbles are formed. The result can range from painful to fatal.

Conversely, if the blood becomes too basic, the reaction concentrations tip to the right, with the reaction

$$HCO_3^-(aq) + H_2O(l) \rightleftharpoons CO_3^{2-}(aq) + H_3O^+(aq)$$

producing a greater CO_3^{2-} ion concentration. Since the blood also contains Ca^{2+} ions (needed to regulate heart and other muscles, among other functions), the insoluble material $CaCO_3$ forms. While this development is good for bone, it is a disaster for the shorter-term vital function of pumping blood around the body. The pH therefore needs to remain in a narrow range to ensure the proper functioning of the body.

63. (a) pH $= 7 =$ pOH; $[OH^-] = 1 \times 10^{-7}$ and $K_{sp} = [Al^{3+}][OH^-]^3 = 3.7 \times 10^{-33} = [Al^{3+}](10^{-7})^3 \Rightarrow [Al^{3+}] = 3.7 \times 10^{-12}\ M$. This is a very low concentration.
(b) pH $= 5 \Rightarrow$ pOH $= 9$; $[OH^-] = 1 \times 10^{-9}$ and $[Al^{3+}] = 3.7 \times 10^{-6}\ M$, or six orders of magnitude more than at pH $= 7$.

65. $CaCl_2$ and Na_2SO_4 are both soluble salts, so the initial concentrations are

$$[Ca^{2+}] = \frac{50}{100} \times 0.001\ M = 0.0005\ M$$

$$[SO_4^{2-}] = \frac{50}{100} \times 0.001\ M = 0.0005$$

$$Q = (0.0005) \times (0.0005) = 2.5 \times 10^{-7}$$
$$K_{sp} (CaSO_4) = 2.45 \times 10^{-5}$$

Since $Q < K_{sp}$, $CaSO_4$ will not precipitate.

67. (a) Calculate ΔG of the reaction. Since ΔG for $H_2(g)$ is unknown, ΔG must be determined from $\Delta G = \Delta H - T\Delta S$:

$\Delta H = 1$ mol $\times$ (−201 kJ/mol) − 1 mol $\times$ (−110.5 kJ/mol)
$\quad$ − 2 mol $\times$ (0) = −90.5 kJ

$\Delta S = 1$ mol $\times$ (239.9 J/K·mol) − 1 mol $\times$ (197.7 J/K·mol)
$\quad$ − 2 mol $\times$ (130.7 J/K·mol) = −219.2 J/K

$\Delta G = \Delta H - T\Delta S = -90.5$ kJ − (298 K) $\times$ (−0.2192 kJ/K)
$\quad$ = −25.2 kJ

$K = \exp(-\Delta G/RT) = 2.59 \times 10^4$

(b) ΔH for the reaction is negative $\Rightarrow$ reaction is exothermic. Raising the temperature will favor reactants, thus K is smaller at higher temperatures.

69. (a) No, the concentrations change, but the relative concentrations are constant. (b) No. (c) No. (d) Yes, only the temperature changes the value of the equilibrium constant.

71. The solubility of any gas is proportional to the pressure. Under the sea, the pressure is higher, so the solubility of N_2 in blood is greater there. As the diver surfaces, the pressure decreases, so N_2 becomes less soluble. As N_2 comes out of the solution, it forms bubbles of gas in the blood. This is painful. He is less soluble in the blood than is N_2, so less He comes out of solution as the diver surfaces, reducing the pain associated with bubble formation.

73. (a) Increasing pressure favors the side of the reaction with fewer gas-phase molecules. In this case that is the product, so more products form. K is constant.
(b) Addition of product causes more reactant to form. K is constant.
(c) Increasing the volume is the same as decreasing the pressure. It will favor the side of the reaction with more gas-phase molecules, so more reactant will form. K is constant.
(d) Increasing the temperature is equivalent to adding a reactant since the reaction is endothermic; hence more products will form. K increases.

75. (a) Increasing the vessel size is equivalent to decreasing the pressure. Decreasing pressure favors the side of the reaction with more gas-phase molecules, in this case the product. The yield increases.
(b) Increasing the partial pressure of water adds a product, so the concentrations shift to favor reactants. The yield decreases.
(c) Increasing the partial pressure of oxygen adds a reactant, so the yield increases.

77. True. Raising the temperature adds a reactant if the reaction is endothermic and adds a product if the reaction is exothermic. Adding a reactant produces more product, and K increases. Adding a product produces more reactant, and K decreases.

79. At a higher temperature, the amount of products increases relative to the amount of reactants (the equilibrium constant increases). Increasing the temperature adds heat, which drives the reaction toward products. Heat must be a reactant, which means that the reaction is endothermic.

81. When rapidly expanded, the balloon feels warm. Based on this observation, if the stretched balloon were heated, it would contract. The weight will lift.

83. $K = \dfrac{[NH_3]^2}{[N_2][H_2]^3}$

High pressure favors the side of the reaction with fewer moles of gas—here the products. It is desirable to run the reaction at high pressure to increase the ammonia yield.

85. (a) No effect; there are the same number of gas-phase molecules on both sides of the reaction.
(b) Increasing the pressure decreases the yield; there are more gas-phase molecules on the product side than on the reactant side.
(c) Increasing the pressure increases the yield; there are more gas-phase molecules on the reactant side than on the product side.

87. (a) The equilibrium concentrations are 2.4 molecules of A and 47.6 molecules of B. It takes 65 cycles to reach equilibrium.
(b) The equilibrium concentrations are the same and the number of cycles that it takes to reach equilibrium is also the same.

89. (a) The consistent interpretation from all observations is as follows:
(i) $AW(aq) + BX(aq) \rightleftharpoons AX(s) + B^+(aq) + W^-(aq)$
(ii) $CW(aq) + BX(aq) \rightleftharpoons CX(s) + B^+(aq) + W^-(aq)$
(iii) $AW(aq) + BY(aq) \rightleftharpoons A^+(aq) + W^-(aq) +$
$\quad B^+(aq) + Y^-(aq)$
(iv) $AZ(aq) + BX(aq) \rightleftharpoons AX(s) + Z^-(aq) + B^+(aq)$
(b) always, B; sometimes, A; probably insoluble, C.
(c) soluble, W, Y, Z; insoluble, X

91. (a) $CO_2(g) + H_2O(l) \rightleftharpoons H_2CO_3(aq)$; $H_2CO_3(aq) + 2H_2O(l) \rightleftharpoons 2H_3O^+(aq) + CO_3^{2-}(aq)$
(b) $CH_3COOH(aq) + H_2O(l) \rightleftharpoons H_3O^+(aq) + CH_3COO^-(aq)$
(c) $CaCO_3(s) + 2H_3O^+(aq) \rightleftharpoons H_2CO_3(aq) + H_2O(l) + Ca^{2+}(aq)$

Chapter 9

1. Neither water nor CO_2 stop oxidation of magnesium, so neither can put out a magnesium fire. Sand should be used to smother the fire.

3.

	N₂O	NO	N₂O₃	NO₂	N₂O₄	N₂O₅	NH₃	N₂H₂	HNO₃	HNO₂
Oxidation number of nitrogen	+1	+2	+3	+4	+4	+5	−3	−1	+5	+3

5. Sn: $5s^2 4d^{10} 5p^2$. Sn^{2+} gives away two electrons to form $5s^2 4d^{10}$. Sn^{4+} gives up four electrons, forming $4d^{10}$. This reaction is more common because giving away these four electrons leaves only a tight, stable valence shell consisting of a filled d subshell.

7. Fluorine is the most electronegative element in the second row (indeed, in the entire periodic table) and lithium is the least electronegative element.

9. francium (Fr)

11. (a) Compounds of Group IIA metals are ionic since the ionization energies are low and the core has a rare gas configuration.
(b) Group IIA metals exist primarily in a +2 ionization state since the first and second ionization energies are low and the electron affinity is the lowest in the table (with the exception of the noble gases).

13. These elements have the strongest potential for reduction and the highest electronegativities. They are located in the upper-right portion of the periodic table. The strongest reducing agents are the elements with the lowest potential for oxidation and the lowest electronegativities. They are located in the lower-left portion of the periodic table. The least active metals are found on the right side of the transition metals (Sb, Cu, Ag, Hg, Pt, Au).

15. The four compounds of fluorine with iodine are IF, IF_3, IF_5, and IF_7.

17. (a) is a decomposition reaction, not a redox reaction. (b) is a redox reaction: Hg is reduced from $+2$ to 0 and oxygen is oxidized from -2 to 0.

19. When heated in excess oxygen, each of these elements forms the oxide with the highest oxidation state.

	Zn	Ti	In	Cd	Sr	Cs	Sn	Pb
Highest oxidation state	2	4	3	2	2	1	4	4
Formula for the oxide	ZnO	TiO_2	In_2O_3	CdO	SrO	Cs_2O	SnO_2	PbO_2

21. The halogens are all quite electronegative, with a great propensity to gain one electron. Many metallic elements in the environment can supply this electron.

23. Gold is quite unreactive; it is found at the bottom of the reactivity list. If gold ions encounter any other metal in elemental form, gold will be reduced and the other metal will be oxidized. The opposite is true of sodium. Most other elemental metal ions or hydrogen in water oxidize sodium to sodium ions, leaving the other metal or hydrogen in elemental form.

25. The reduction potential of Ni is less than that of Ag, so Ni will replace Ag^+ ions in solution to form Ag metal and Ni^{2+} ions.

$$2Ag^+(aq) + Ni(s) \longrightarrow Ni^{2+}(aq) + 2Ag(s)$$

27. The reduction potential of Fe is greater than that of Al, so Al will replace Fe in Fe_2O_3.

$$Fe_2O_3(s) + 2Al(s) \longrightarrow Al_2O_3(s) + 2Fe(s)$$

29. Lead has a lower reduction potential than does Ag but a higher reduction potential than Sn. Pb will therefore oxidize Sn but not Ag.

$$Pb^{2+}(aq) + Sn(s) \longrightarrow Sn^{2+}(aq) + Pb(s)$$

31. Ca reacts with cold water, Al with steam, and Pb only with acid. In cold, aqueous solution, the following reaction occurs:

$$Ca(s) + 2H_2O(l) \longrightarrow H_2(g) + 2OH^-(aq) + Ca^{2+}(aq)$$

33. Cr is the more active metal, so it is a better reducing agent. The Cr^{3+} ion is not as good an oxidizing agent. The better oxidizing agent is Co^{3+}.

35. Al is the more active metal, so the Al wire corrodes if either the Cu or the Al wire comes in contact with an oxidant such as O_2.

37. Cr is a more active metal and has a lower reduction potential than does Fe. When in contact with an oxidant, Cr oxidizes in preference to Fe. The CrO formed makes a protective coating on the object.

39. Of the three metals, Fe has the lowest reduction potential, so Fe will corrode.

41. This arrangement saves the steel pipe because both Al and Mg are more active metals, so they corrode in preference to the steel pipe. The copper conductor is insulated because otherwise any oxygen or other oxidant that attacks the Cu will also cause corrosion and deterioration of the stake. The stake would have to be replaced more often.

43. Iron is a more active metal than is copper. Hence, when any oxidant attacked any part of the copper cladding of the Statue of Liberty, it caused corrosion of the iron struts inside the statue.

45. An oxidation reaction features loss of one or more electrons, while a reduction reaction features gain of one or more electrons. Oxidation and reduction are opposites. Oxidation occurs at the anode, and reduction at the cathode.

47. (a) $(0.770 + 2.375)$ V $= 3.145$ V; (b) $(0.7996 + 0.409)$ V $= 1.209$ V; (c) $(0.401 + 2.67)$ V $= 3.07$ V; (d) $(-1.47 + 0.7051)$ V $= -0.765$ V

49. Anode: $Zn \rightarrow Zn^{2+} + 2e^-$; potential $= 0.7628$ V

Cathode: $HgO + H_2O + 2e^- \rightarrow Hg + 2OH^-$; potential $= 0.0984$ V; Cell potential $= (0.7628 + 0.0984)$ V $= 0.8612$ V

51. The Zn reaction involves Zn and K_2ZnO_2. It is balanced in steps:

Skeleton	$Zn \longrightarrow K_2ZnO_2$
Balance material oxidized or reduced	$Zn + 2KOH \longrightarrow K_2ZnO_2$
Balance electrons (Zn: $0 \longrightarrow +2$)	$Zn + 2KOH \longrightarrow K_2ZnO_2 + 2e^-$
Balance charge (with OH^-)	$Zn + 2OH^- + 2KOH \longrightarrow K_2ZnO_2 + 2e^-$
Balance H and O with water	$Zn + 2OH^- + 2KOH \longrightarrow K_2ZnO_2 + 2e^- + 2H_2O$

The other reaction has only Hg specified. In the basic environment, this will be HgO or $Hg(OH)_2$ depending on the water content. The battery is a miniature version and KOH is a paste, so water is not plentiful.

Skeleton	$HgO \longrightarrow Hg$
Balance material oxidized or reduced	$HgO \longrightarrow Hg$
Balance electrons (Hg: $+2 \longrightarrow 0$)	$HgO + 2e^- \longrightarrow Hg$
Balance charge (with OH^-)	$HgO + 2e^- \longrightarrow Hg + 2OH^-$
Balance H and O with water	$HgO + 2e^- + H_2O \longrightarrow Hg + 2OH^-$

The overall reaction is obtained by adding these two reactions:
$$Zn(s) + 2KOH(aq) + HgO(s) \longrightarrow K_2ZnO_2(aq) + H_2O(l) + Hg(l)$$

53. The oxidizing power of the halogens goes in order: I < Br < Cl < F. Hence iodide can be oxidized with bromine, leaving the bromide, chloride, and fluoride. Similarly, bromide can be oxidized with chlorine, leaving the chloride and fluoride. Next, the chloride could be oxidized with either fluorine or oxygen. Elemental fluorine is obtained with electrolysis.

55. (a) $Cu(s) + 2Ag^+(aq) \rightarrow Cu^{2+}(aq) + 2Ag(s)$
(b) $2Na(s) + 2H_2O(l) \rightarrow 2Na^+(aq) + 2OH^-(aq) + H_2(g)$
(c) $2Ag(s) + Zn^{2+}(aq) \rightarrow 2Ag^+(aq) + Zn(s)$

57. C to CO:

Skeleton	$C \longrightarrow CO$
Balance material oxidized or reduced	$C \longrightarrow CO$
Balance electrons	$C \longrightarrow CO + 2e^-$
Balance charge	$C + 2OH^- \longrightarrow CO + 2e^-$
Balance H and O with water	$C + 2OH^- + \longrightarrow CO + 2e^- + H_2O$

C to CO_2:

Skeleton	$C \longrightarrow CO_2$
Balance material oxidized or reduced	$C \longrightarrow CO_2$
Balance electrons	$C \longrightarrow CO_2 + 4e^-$
Balance charge	$C + 4OH^- \longrightarrow CO_2 + 4e^-$
Balance H and O with water	$C + 4OH^- + \longrightarrow CO_2 + 4e^- + 2H_2O$

O_2 reaction:

Skeleton	$O_2 \longrightarrow OH^-$
Balance material oxidized or reduced	$O_2 \longrightarrow 2OH^-$
Balance electrons	$O_2 + 4e^- \longrightarrow 2OH^-$
Balance charge	$O_2 + 4e^- \longrightarrow 2OH^- + 2OH^-$
Balance H and O with water	$O_2 + 4e^- + 2H_2O \longrightarrow 4OH^-$

Reaction for oxidation to CO: $2C(coke) + O_2(g) \rightarrow 2CO(g)$

Reaction for oxidation to CO_2: $C(coke) + O_2(g) \rightarrow CO_2(g)$

59.

Oxidation	Reduction
$C \longrightarrow CO_2$	$PbO \longrightarrow Pb$

Balance species oxidized/reduced:

Already balanced	Already balanced

Balance electrons:

$0 \qquad\quad +4$	$+2 \qquad\quad 0$
$C \longrightarrow CO_2 + 4e^-$	$PbO + 2e^- \longrightarrow Pb$

Balance charge with H^+:

$C \longrightarrow CO_2 + 4e^- + 4H^+$	$PbO + 2e^- + 2H^+ \longrightarrow Pb$

Balance H and O with water:

$2H_2O + C \longrightarrow CO_2 + 4e^- + 4H^+$	$PbO + 2e^- + 2H^+ \longrightarrow Pb + H_2O$

Multiply the lead reaction by 2 and add:

$$2PbO + 4e^- + 4H^+ \longrightarrow 2Pb + 2H_2O$$
$$2H_2O + C \longrightarrow CO_2 + 4e^- + 4H^+$$

Resultant: $2PbO(s) + C(coke) \longrightarrow 2Pb(s) + CO_2(g)$

61. Oxidation reactions: $Al \rightarrow Al^{3+} + 3e^-$; $Cr \rightarrow Cr^{3+} + 3e^-$; $Fe \rightarrow Fe^{3+} + 3e^-$; reduction reaction $2H^+ + 2e^- \rightarrow H_2$
Balanced reactions:
$2Al(s) + 6H_3O^+(aq) \rightarrow 2Al^{3+}(aq) + 3H_2(g) + 6H_2O(l)$
$2Cr(s) + 6H_3O^+(aq) \rightarrow 2Cr^{3+}(aq) + 3H_2(g) + 6H_2O(l)$;
$2Fe(s) + 6H_3O^+(aq) \rightarrow 2Fe^{3+}(aq) + 3H_2(g) + 6H_2O(l)$

63. $2Rb(s) + 2H_2O(l) \rightarrow 2Rb^+(aq) + 2OH^-(aq) + H_2(g)$
$Ba(s) + 2H_2O(l) \rightarrow Ba^{2+}(aq) + 2OH^-(aq) + H_2(g)$
$Cu(s) + Fe^{2+}(aq) \rightarrow$ no reaction; Cu is a less active metal than is Fe

65. $2K(s) + 2H_2O(l) \longrightarrow 2K^+(aq) + H_2(g) + 2OH^-(aq)$
Fire results when the exothermicity of the reaction ignites the hydrogen. Hydrogen combining with the oxygen of air gives the flame:
$$2H_2(g) + O_2(g) \longrightarrow 2H_2O(l)$$

67. Reduction: $2e^- + H_2O + Cu_2O \longrightarrow 2Cu + 2OH^-$
Oxidation: $4OH^- + Cu_2S \longrightarrow SO_2 + 4e^- + 2H_2O + 2Cu$

69. Reduction: $Fe_2O_3 + 6e^- + 3H_2O \longrightarrow 2Fe + 6OH^-$
or $Fe_2O_3 + 6e^- + 6H^+ \longrightarrow 2Fe + 3H_2O$
Oxidation: $C + 2OH^- \longrightarrow CO + 2e^- + H_2O$
or $C + H_2O \longrightarrow CO + 2e^- + 2H^+$

71. Oxidation reaction, anode: $Zn \rightarrow Zn^{2+} + 2e^-$
Reduction, cathode: $Cl_2 + 2e^- \rightarrow 2Cl^-$
The potential is $(+0.7628 + 1.3583)$ V = 2.1211 V

73. The cathode reaction is the reduction of oxygen: $O_2 + 2H_2O + 4e^- \rightarrow 4OH^-$. A good ion for completing the circuit is OH^-. The system could be designed with some OH^- ion in solid form on the bottom of an aluminum cup. The solid form could be NaOH, KOH, or another Group IA metal hydroxide.

75. Anode: $2Cl^- \longrightarrow Cl_2 + 2e^-$
Cathode: $Mg^{2+} + 2e^- \longrightarrow Mg$

77. 5.0 g Ca is 5.0/40.1 = 0.12 mol Ca. This requires 0.24 mol of electrons. Electrons are 96,485.31 C/mol, so 2.3×10^4 C are required. A current of 10 A is 10 C/s, so 2300 s or 38.3 min is required.

79. A piece of cold worked metal has multiple grain boundaries, some of which penetrate to the surface. All metals are more corrosive at grain or defect boundaries, so a piece of cold worked metal will be more prone to corrosion than an annealed one.

81. Al would provide Fe with galvanic protection in the same way as Zn. The advantage of Zn is that it forms a more compatible lattice with the Fe, so it adheres better. In addition, Al is more expensive to produce than is Zn.

83. (a) Tin forms a protective oxide coat, so the joints of the woodsman should not have seized. (b) A tin can consists of a tin coat over a steel core. The tin protects only as long as the coating remains intact. Any crack or scratch exposes the steel, which quickly rusts. In this case, an oil coat would help. (c) Density of Sn: 7.31 g/cm³; density of SnO: 6.45 g/cm³. The Pilling-Bedworth ratio is 1.28, so the SnO coat is protective.

85. Half-reactions: $Ag^+ + e^- \longrightarrow Ag$; potential = 0.7996 V
$Pb \longrightarrow Pb^{2+} + 2e^-$; potential = -0.1263 V
Reaction: $Pb + 2Ag^+ \rightleftharpoons Pb^{2+} + 2Ag$
Cell potential = $(0.7996 - 0.1263)$ V = 0.6733 V.
Use $\Delta G° = -nFE = -(2 \text{ mol}) \times (96485 \text{ C/mol}) \times (0.6733 \text{ V}) \times (1 \text{ J/C} \cdot \text{V}) = -129.9$ kJ

87. The potential difference determines the ion concentration difference across the membrane as $E° = (RT/nF) \ln K$. With no potential difference, $K = 1$: The concentrations on both sides of the membrane are the same. With a potential difference of 70 mV, the concentration ratio becomes

$$K = \exp\left(\frac{nFE°}{RT}\right) = \exp\left(\frac{(1 \text{ mol})(96{,}485 \text{ C/mol})(0.070 \text{ V})}{(8.3145 \text{ J/K} \cdot \text{mol})(311 \text{ K})}\right)$$
$$= 13.62$$

With a potential difference of 50 mV, the concentration ratio is 6.46. The sodium ion concentration gradient has dropped by nearly a factor of 2.

89. Al is more active than both Zn and Cr, so a similar process will work.

$$3ZnO(s) + 2Al(l) \longrightarrow 3Zn(l) + Al_2O_3(s)$$

91. Oxidation half-reaction: $2Cl^- \rightarrow Cl_2(g) + 2e^-$
Reduction half-reaction: $K^+ + e^- \rightarrow K(s)$

93. Many diagrams are possible.
One example is shown here.

95. CO_2 is heavier than air: The molecular weight of CO_2 is 44, while that of air is about 28 (the mass of N_2). CO_2 thus blankets the fire and eliminates oxygen, one of the ingredients needed for combustion. The other mechanism is a cooling of the burning material. The liberation of CO_2 from baking soda is endothermic—it absorbs heat. High temperatures are required to support combustion.

97. The oxidizing agent is HOOH.
Oxidations:

$$2H_7C_4SH \longrightarrow H_7C_4SSC_4H_7 + 2e^- + 2H^+$$
$$2H_{11}C_5SH \longrightarrow H_{11}C_5SSC_5H_{11} + 2e^- + 2H^+$$

Reduction:

$$HOOH + 2e^- + 2H^+ \longrightarrow 2H_2O$$

Overall reactions:

$$2H_7C_4SH + HOOH \longrightarrow H_7C_4SSC_4H_7 + 2H_2O$$
$$2H_{11}C_5SH + HOOH \longrightarrow H_{11}C_5SSC_5H_{11} + 2H_2O$$

99. Plan: Determine the equilibrium constant from $\Delta G°$; determine $E°$ from $\Delta G°$.

$$CH_3OH(l) + 1\tfrac{1}{2}O_2(g) \rightleftharpoons CO_2(g) + 2H_2O(l)$$
$$\Delta G° = \Delta H° - T\Delta S$$

$\Delta H° = 2$ mol $\times$ (-285.8 kJ/mol) + 1 mol $\times$ (-393.5 kJ/mol)
$\quad - 1$ mol $\times$ (-239.2 kJ/mol) $= -725.9$ kJ

$\Delta S = 2$ mol $\times$ (70 J/K·mol) + 1 mol (213.8 J/K·mol) $-$ 1 mol
$\quad \times$ (126.8 J/K·mol) $-$ 1.5 mol $\times$ (205.2 J/K·mol)
$$= -80.8 \text{ J/K}$$

$\Delta G° = -701.8$ kJ for a mol of reaction as written.
$$\Delta G° = -RT \ln K$$

$$\Longrightarrow K = \exp\left(\frac{-\Delta G°}{RT}\right)$$

$$= \exp\left(-\frac{-701.8 \text{ kJ/mol}}{(8.3145 \times 10^{-3} \text{ kJ/mol·K})(298 \text{ K})}\right)$$

$$= 1.04 \times 10^{123}$$

$$E° = -\Delta G°/nF = -\frac{-702 \text{ kJ/mol} \times 1000 \text{ J/kJ}}{6 \times (96,485 \text{ C/mol}) \times 1 \text{ J/C·V}} = 1.21 \text{ V}$$

Note: In methanol, the oxidation state of carbon is -2, and in CO_2 it is $+4$. Thus six electrons are exchanged.
For partial oxidation to formaldehyde, the reaction is

$$CH_3OH(l) + \tfrac{1}{2}O_2(g) \rightleftharpoons CH_2O(g) + H_2O(l)$$

$\Delta H° = 1$ mol $\times$ (-285.8 kJ/mol) + 1 mol $\times$ (-108.6 kJ/mol)
$\quad - 1$ mol $\times$ (-239.2 kJ/mol) $= -155.2$ kJ

$\Delta S = 1$ mol $\times$ (70 J/K·mol) + 1 mol (218.8 J/K·mol) $-$ 1 mol
$\quad \times$ (126.8 J/K·mol) $-$ 0.5 mol $\times$ (205.2 J/K·mol)
$$= 59.4 \text{ J/K}$$

$\Delta G° = -172.9$ kJ for a mol of reaction as written.
$$\Delta G° = -RT \ln K$$

$$\Longrightarrow K = \exp\left(\frac{-\Delta G°}{RT}\right)$$

$$= \exp\left(-\frac{-172.9 \text{ kJ/mol}}{(8.3145 \times 10^{-3} \text{ kJ/mol·K})(298 \text{ K})}\right)$$

$$= 2.02 \times 10^{30}$$

$$E° = -\Delta G°/nF = -\frac{-172.9 \text{ kJ/mol} \times 1000 \text{ J/kJ}}{2 \times (96,485 \text{ C/mol}) \times 1 \text{ J/C·V}} = 0.896 \text{ V}$$

Note: In methanol, the oxidation state of carbon is -2, and in formaldehyde it is zero. Thus, two electrons are exchanged. The potential for oxidation to formaldehyde is a substantial fraction of that for complete oxidation.

Chapter 10

1. The strength of the interaction is deduced from the colors. A gem appears red if it absorbs the complement of red light, or green light plus blue light (cyan). A gem appears green if it absorbs the complement of green light, or red light plus blue light. The lower-energy color is red. Thus emeralds absorb red light, and rubies absorb cyan light. Cyan light is higher in energy than is red light. The color indicates that the interaction in ruby is stronger than it is in emerald. A weaker interaction in emerald is consistent with emerald being a less dense and softer material.

3.

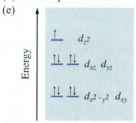

5. (a) $NaCN(s) \rightarrow Na^+(aq) + CN^-(aq)$;
$Ag^+(aq) + 2CN^-(aq) \rightleftharpoons [Ag(CN)_2]^-(aq)$;
$2Ag^+(aq) + Zn(s) \rightarrow 2Ag(s) + Zn^{2+}(aq)$
(b) The oxidation state of silver in $[Ag(CN)_2]^-$ is $+1$.
(c) The coordination number of silver is 2.
(d) The complex is linear.
(e)

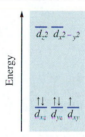

7. In CrO_4^{2-}, chromium is in a $+6$ oxidation state; in MnO_4^-, manganese is in a $+7$ oxidation state. CrO_4^{2-} is tetrahedral, as is MnO_4^-. The coordination number of both metals is 4.

9. The color of the glaze consists of those colors that are reflected back at the observer; the color absorbed is its complement. To appear red-yellow, green and higher-energy colors are absorbed. To appear green, red and blue light is absorbed. The lower-energy light is the red light. In a reducing atmosphere, glaze is produced that absorbs low-energy red light. The reducing atmosphere produces the lower oxidation state of iron: Fe^{2+}. In an oxidizing atmosphere, higher-energy green light is absorbed. In the oxidizing atmosphere, the higher oxidation state is produced: Fe^{3+}. The higher oxidation state interacts more strongly with its ligands.

11. A larger lattice spacing implies a weaker interaction and absorption of light closer to the red end of the spectrum. The glaze with Zn^{2+} appears green, so it absorbs red light plus blue light of which red light is the lower in energy. Glazes with Pb^{2+} appear orange or red, so they absorb green light and blue light. Of these, green and blue light are higher in energy than is red light. Thus the Pb^{2+} ion results in a lattice with smaller spacing; Zn^{2+} produces a lattice with larger spacing.

13. The electron configuration of Ti^{4+} is $[Ar]$. There are no d-orbital electrons to make a transition upon absorption of visible light.

15. The Cu^{2+} ion acts as a sensor of the interactions between the oxygen lattice and the Cu^{2+} ion. The $MgAl_2O_4$ lattice results in a longer wavelength absorption, so the interaction is weaker. The lattice spacing in $MgAl_2O_4$ is larger than that in MgO alone.

17. Upon exiting the lungs, blood is saturated with O_2. The red color of this complex indicates that blue light plus green light is absorbed. When coming from the muscles, blood is laden with CO_2. The blue color of this complex indicates that red light plus green light is absorbed. Blue light plus green light is higher in energy than is red light plus green light, indicating that the interaction between Fe^{2+} and O_2 is stronger than that between Fe^{2+} and CO_2.

19. (a) Since $[Fe(NCS)_6]^{3-}$ appears red, it absorbs blue and green light.
(b) The oxidation state of iron in $[Fe(NCS)_6]^{3-}$ is $+3$. The electron configuration of Fe^{3+} is $[Ar]d^5$. Since the absorption is of a high energy, the interaction is strong and the splitting is large.

(c) It is likely that $[Fe(NCS)_6]^{3-}$ is a low spin complex since the absorption is high in energy.

21. (a) $NaOH(s) \rightarrow Na^+(aq) + OH^-(aq)$;
$Zn(NO_3)_2(s) \rightarrow Zn^{2+}(aq) + NO_3^-(aq)$
$Zn^{2+}(aq) + OH^-(aq) \rightleftharpoons [Zn(OH)]^+(aq)$
(b) $[Zn(OH)]^+(aq) + OH^-(aq) \rightleftharpoons Zn(OH)_2(s)$
(c) $Zn(OH)_2(s) + 2OH^-(aq) \rightleftharpoons [Zn(OH)_4]^{2-}(aq)$

23. The first processes is $Cu^{2+}(aq) + 2OH^-(aq) \rightleftharpoons Cu(OH)_2(s)$. This is followed by $Cu(OH)_2(s) + 4NH_3(aq) \rightleftharpoons [Cu(NH_3)_4]^{2+}(aq)$. $Cu(OH)_2$ is a blue solid; $[Cu(NH_3)_4]^{2+}$ is a deep blue complex.

25. NH_4OH is hydrated NH_3. NH_3 forms a complex with Ag^+. The equilibria are

$$NH_3(aq) + H_2O(l) \rightleftharpoons NH_4^+(aq) + OH^-(aq)$$
$$AgCl(s) \rightleftharpoons Ag^+(aq) + Cl^-(aq)$$
$$[Ag(NH_3)_2]^+(aq) \rightleftharpoons Ag^+(aq) + 2NH_3(aq)$$

Addition of nitric acid to the solution shifts the first equilibrium concentrations toward products, as the H_3O^+ generated from nitric acid combines with the OH^- from ammonia. This decreases the NH_3 concentration, which in turn decreases the $[Ag(NH_3)_2]^+$ concentration and increases the Ag^+ ion concentration. Increasing the Ag^+ ion concentration results in $[Ag^+][Cl^-]$ exceeding the solubility product, so more AgCl forms.

27. The final concentrations are $0.05\,M$ Ag^+ and $0.05\,M$ $[Ag(NH_3)_2]^+$.

$$K = \frac{[[Ag(NH_3)_2]^+]}{[Ag^+][NH_3]^2} = \frac{1}{[NH_3]^2} = 1.6 \times 10^7 \implies [NH_3]$$

$$= 2.5 \times 10^{-4}\,M$$

29. $AgBr(s) + 2NH_3(aq) \rightleftharpoons [Ag(NH_3)_2]^+(aq) + Br^-(aq)$ Note: No calculation is required, only the reaction.

31. The formation constant for $[Cd(CN)_4]^{2-}$ is 1.3×10^{17}, about 10 orders of magnitude larger than the formation constant for $[Cd(NH_3)_4]^{2+}$. Since the initial CN^- concentration is only one order of magnitude smaller than the initial NH_3 concentration, most of the Cd^{2+} will be tied up as $[Cd(CN)_4]^{2-}$.

33. One mole of Fe^{3+} combines with six moles of CN^- to form the complex $[Fe(CN)_6]^{3-}$. Since Fe^{3+} is in large excess, most will remain as $Fe^{3+}(aq)$.

35. Potassium salts are soluble.

$$Al^{3+}(aq) + 6F^-(aq) \rightleftharpoons [AlF_6]^{3-}(aq)$$

			1
Start:			1
Change:	x	$6x$	$-x$
End:	x	$6x$	$1-x$

37.
$$Cd^{2+}(aq) + 4CN^-(aq) \rightleftharpoons [Cd(CN)_4]^{2-}(aq)$$
$$Cd^{2+}(aq) + 4NH_3(aq) \rightleftharpoons [Cd(NH_3)_4]^{2+}(aq)$$

The formation constant for $[Cd(CN)_4]^{2-}$ is 10 orders of magnitude larger than that for $[Cd(NH_3)_4]^{2+}$, so the $[Cd(NH_3)_4]^{2+}$ solution has a larger Cd^{2+} concentration.

39. $NiCO_3(s) + 4CN^-(aq) \rightleftharpoons [Ni(CN)_4]^{2-}(aq) + CO_3^{2-}(aq)$ This requires essentially 0.004 mol CN^-, but no more, since the formation constant for $[Ni(CN)_4]^{2-}$ is 1.0×10^{31}, much larger than K_{sp} for $NiCO_3$, 6.6×10^{-9}. (K for the above reaction is 6.6×10^{22}.)

41.

$$[Cu(H_2O)_2]^+(aq) + 2NH_3(aq) \rightleftharpoons [Cu(NH_3)_2]^+(aq) + 2H_2O(l)$$

$$[Cu(H_2O)_4]^{2+}(aq) + 4NH_3(aq) \rightleftharpoons [Cu(NH_3)_4]^{2+}(aq) + 4H_2O(l)$$

For $[Cu(NH_3)_2]^+$:

$$[Cu(H_2O)_2]^+(aq) + 2NH_3(aq) \rightleftharpoons [Cu(NH_3)_2]^+(aq) + 2H_2O(l)$$

Start:	0.001	0.5	
Completion:		0.498	0.001
Change:	x	$2x$	$-x$
End:	x	$0.498 + 2x$	$0.001 - x$

$$K = \frac{0.001 - x}{x(0.498 + 2x)^2} \approx \frac{0.001}{x(0.498)^2} = 7.1 \times 10^{10} \implies$$

$$x = 5.7 \times 10^{-14} = [Cu(H_2O)_2]^+$$

For $[Cu(NH_3)_4]^{2+}$:

$$K = \frac{0.001 - x}{x(0.496 + 4x)^4} \approx \frac{(0.001)}{x(0.496)^4} = 2.0 \times 10^{-10}$$

$$1.5 \times 10^{-14} = [Cu(H_2O)_4]^{2+}$$

43. The reaction is $Co^{2+}(aq) + 6NH_3(aq) \rightleftharpoons [Co(NH_3)_6]^{2+}(aq)$, and similarly for the Co^{3+} complex.

$$Co^{2+}(aq) + 6NH_3(aq) \rightleftharpoons [Co(NH_3)_6]^{2+}(aq)$$

Start:	0.001	0.5	
Completion:		0.494	0.001
Change:	x	$6x$	$-x$
End:	x	$0.494 + 6x$	$0.001 - x$

$$K_f([Co(NH_3)_6]^{2+}) = 7.7 \times 10^4$$

$$K = \frac{0.001 - x}{x(0.494 + 6x)^6} \approx \frac{(0.001)}{x(0.494)^6} \implies x = 8.9 \times 10^{-7}$$

$$= [Co(H_2O)_6]^{2+}$$

$K_f([Co(NH_3)_6]^{3+}) = 5 \times 10^{33}$, so $[Co(H_2O)_6]^{3+} = 1.4 \times 10^{-35}$

Co^{3+} forms a much stronger complex with ammonia than does Co^{2+}.

45. (a) $Mn^{2+}(g) + 6H_2O(l) \rightarrow [Mn(H_2O)_6]^{2+}(aq)$; $Ni^{2+}(g) + 6H_2O(l) \rightarrow [Ni(H_2O)_6]^{2+}(aq)$

(b) The reactions are exothermic because the negative end of the water molecule is strongly attracted to the positive charge on the metal ion.

47. Na^+: $\Delta T = \dfrac{4.05\,kJ}{4.184\,J/K \cdot g} \times \dfrac{1000\,J}{kJ} \times \dfrac{1}{1000\,g} = 0.97\,K$

Mg^{2+}: $4.59\,K$; Al^{3+}: $11.1\,K$

49. $Fe_2(SO_4)_3$ is an ionic salt containing Fe^{3+} ions. Fe^{3+} has five d-orbital electrons. All of these have the same energy, so all are unpaired. $Fe_2(SO_4)_3$ is strongly paramagnetic, so it is strongly attracted to a magnet. $FeSO_4$ is also an ionic solid, but the Fe^{3+} ion has six d-orbital electrons. Since the d orbitals all have the same energy, four of the electrons are unpaired, and $FeSO_4$ is also strongly paramagnetic. $K_3Fe(CN)_6$ consists of the ions K^+ and $[Fe(CN)_6]^{3-}$. The Fe^{3+} ion is in a strong octahedral field created by the six CN^- ions. The five d-orbital electrons are therefore in the lower-energy set of orbitals, one is unpaired, and $K_3Fe(CN)_6$ is weakly paramagnetic. $K_4Fe(CN)_6$ consists of K^+ and $Fe(CN)_6^{4-}$ ions. Iron is in a $+2$ oxidation state and has six d-orbital electrons. All are in the lower-energy set of orbitals and are paired. $K_4Fe(CN)_6$ is therefore diamagnetic.

51. $[Co(NH_3)_6]^{2+}$ is an octahedral complex and ammonia is a fairly strong ligand. In the $+2$ oxidation state, cobalt has seven electrons in the d orbitals: six paired in the lower-energy set and one unpaired electron in the higher-energy set. As such $[Co(NH_3)_6]^{2+}$ is a paramagnetic compound and thus gains weight in a magnetic field (is attracted to it). However, $[Co(NH_3)_6]^{2+}$ can oxidize in air to $[Co(NH_3)_6]^{3+}$. The Co^{3+} ion has only six electrons in the d orbitals, and all are paired in the lower-energy set. With all electrons paired, $[Co(NH_3)_6]^{3+}$ is diamagnetic and weakly repelled from a magnetic field; thus it loses weight in the field.

53. $[FeCl_6]^{3-}$ has iron in a $+3$ oxidation state; hence iron has a d^5 configuration. As a high spin complex, $[FeCl_6]^{3-}$ has all five electrons unpaired. $[Fe(CN)_6]^{3-}$ also has iron in a $+3$ oxidation state. Since the complex is low spin, the five electrons are all in the lower-energy set of orbitals, four electrons are in two spin-paired sets, and the fifth electron is unpaired. Thus $[Fe(CN)_6]^{3-}$ has one unpaired electron.

55.

Ion	Cr^{3+}	Cr^{2+}	Mn^{2+}	Fe^{2+}	Co^{3+}	Co^{2+}
Number of d electrons	3	4	5	6	6	7
Unpaired in a weak field	3	4	5	4	4	3
Unpaired in a strong field	3	2	1	0	0	1

57.

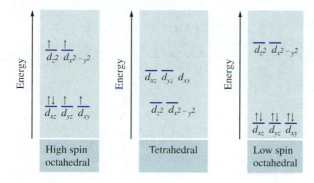

59. $[Fe(H_2O)_6]^{2+}$ has four unpaired electrons and is paramagnetic. For the paramagnetism to disappear, all electrons must be paired. This is not possible for six electrons in a tetrahedral complex. Hence the 1,10-phenanthroline complex must be octahedral and 1,10-phenanthroline is a strong ligand.

| High spin octahedral | Tetrahedral | Low spin octahedral |

61. The three magnetic elements are iron, cobalt, and nickel. As pure nickel, the Canadian coin is magnetic. The Cu–Ni alloy in the U.S. coin lacks the large-scale domains of aligned, unpaired electrons, so it is nonmagnetic. The coins can be distinguished with a magnet.

63. (a) The cobalt ion is in a +2 oxidation state and EDTA carries a −4 charge. Co_2EDTA has a $(+2) + (+2) + (−4) = 0$ charge.

(b) The formation constant for $[Co(CN)_6]^{4−}$ is 1.3×10^{19}. The reaction for replacement of EDTA by $CN^−$ is

$$Co_2EDTA(aq) + 12CN^−(aq) \rightleftharpoons 2[Co(CN)_6]^{4−}(aq) + EDTA^{4−}(aq)$$

$$K = \frac{(K_f[Co(CN)_6]^{4−})^2}{K_f(Co_2EDTA)} = \frac{(1.3 \times 10^{19})^2}{2.0 \times 10^{16}} = 8.5 \times 10^{21}$$

favoring the formation of the cyanide complex.

65. Binding of oxygen to hemoglobin is an equilibrium process described by the following reactions:

$$Hb(aq) + 4O_2(aq) \rightleftharpoons Hb \cdot 4O_2(aq)$$
$$O_2(g) \rightleftharpoons O_2(aq)$$

where Hb stands for hemoglobin. If the oxygen partial pressure falls, the amount of oxygen in the blood (aqueous solution) drops, which induces hemoglobin to release more of its oxygen.

(a) In the lungs, $p(O_2) = 110$ torr, $Hb \cdot O_2 + Hb = 100\%$, $Hb \cdot O_2 = 97\%$

$$K = \frac{[Hb \cdot 4O_2]}{[Hb][O_2]^4} = \frac{0.97}{0.03 \times \left(\frac{110}{760}\right)^4} = 7.4 \times 10^4$$

(b) Mild exercise:

$$K = \frac{0.75}{0.25 \times \left(\frac{40}{760}\right)^4} = 3.9 \times 10^5$$

(c) Vigorous exercise:

$$K = \frac{0.50}{0.50 \times \left(\frac{20}{760}\right)^4} = 2.1 \times 10^6$$

67.
$$Hb(aq) + 4O_2(aq) \rightleftharpoons Hb \cdot 4O_2(aq)$$
$$O_2(g) \rightleftharpoons O_2(aq)$$
$$My(aq) + O_2(aq) \rightleftharpoons My \cdot O_2(aq)$$

where Hb stands for hemoglobin and My stands for myoglobin. When exercising mildly, the oxygen partial pressure is 40 torr and hemoglobin releases a portion of its oxygen load. When exercising vigorously, the oxygen partial pressure drops to 20 torr. Hemoglobin exhibits a synergistic unloading of oxygen: The lower the oxygen partial pressure, the more easily hemoglobin unloads its oxygen. Under vigorous exercise, a larger fraction of hemoglobin's oxygen is available for myoglobin to bind. In relative terms, myoglobin has a greater level of oxygenation due to hemoglobin losing more of its oxygen.

69. CO binds 250 times more strongly than does O_2, implying that at 0.44 torr CO would completely replace oxygen. Since the lethal replacement pressure is half of that, CO is fatal at only 0.22 torr.

71. Many diagrams are possible. One example is shown here.

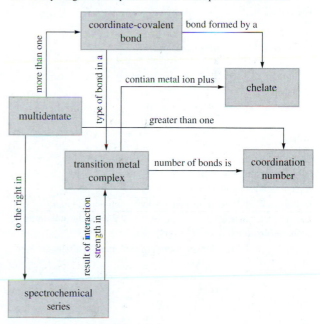

73. EDTA is an effective hexadentate complex that could sequester metal ions, thereby preventing them from catalyzing the oxidation of foodstuffs.

Metal ion	Cu^{2+}	Fe^{2+}/Fe^{3+}	Ni^{2+}	Ca^{2+}
K_f	6.3×10^{18}	$2.1 \times 10^{14}/1.3 \times 10^{25}$	4.2×10^{18}	5.0×10^{10}

The largest formation constant is for Fe^{3+}, which will be very effectively sequestered by EDTA. Indeed, all of these metals form stronger complexes with $EDTA^{4-}$ than does Ca^{2+}.

75. The splitting in tetrahedral complexes is shown

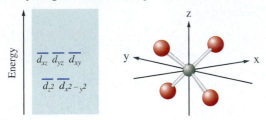

For the complex to be low spin, the splitting between the levels needs to be greater than the repulsive pairing energy. The observation that tetrahedral complexes are high spin indicates that the splitting is relatively small. Splitting is due to repulsion between an electron in the d orbitals and the negative charge on lone pairs of the ligands. A small splitting indicates that the interaction is weak. This is the expected result since the ligands in a tetrahedral complex are not directly in line with an electron in any of the d orbitals.

77. The electron configuration of Pd^{2+} is $[Kr]4d^8$. To form a square planar complex, the ligands must interact strongly with the metal ion. A strong interaction leads to a large energy splitting between levels and pairing of electrons in the lower-energy levels. A square planar complex features a single empty high-energy orbital. With large splitting, this orbital is empty for a metal ion with a d^8 configuration; all other orbitals contain paired electrons. Thus $[Pd(CN)_4]^{2-}$ is diamagnetic.

Chapter 11

1.
$$NH_3(aq) + H_2O(l) \rightleftharpoons NH_4^+(aq) + OH^-(aq)$$

Initial:	1		
Change:	$-x$	x	x
Equilibrium:	$1 - x$	x	x

$$K = \frac{x^2}{1-x} \approx \frac{x^2}{1} = 1.8 \times 10^{-5} \implies x = 4.2 \times 10^{-3}, pOH$$
$$= 2.4, pH = 11.6$$

$[NH_3]$ is essentially $1\,M$, while $[NH_4^+]$ is $4.2 \times 10^{-3}\,M$. NH_3 is present in larger concentration.

3. Polyvinyl alcohol molecules are hydrogen bonded to each other. This polymer should be soluble in water due to the multitude of hydrogen-bonding sites on the polymer.

5. The π bond forms a coordinate-covalent bond with the Zn^{2+} ion.

7. (d) ion–dipole > (c) hydrogen bonding > (a) dipole–dipole > (b) dipole–induced dipole > (e) London dispersion

9. There are no π bonds in the diamine to interact with the lone pair on nitrogen, so it will interact with water in much the same way and to the same extent as does ammonia. The pK_b for the diamine used in nylon is approximately the same as that of ammonia.

11.

13. Polymer (a) is high-density polyethylene.

15.

17.

19.

Styrene alone cannot make a rubber because it lacks the functionality to enable cross-linking. The copolymer is stiffer due to the sliding resistance provided by the pendant phenyl ring.

21. A condensation polymer is the better choice due to hydrogen-bonding interactions between the condensation polymer substrate and the N—H group of polyaniline. With an addition polymer, the interaction between polyaniline and the polymer is a London dispersion force. Because it is weaker, it could lead to flaking of the polyaniline coating.

23. The difference between the blue form of polyaniline and the green form is one H^+ per two aniline units. Addition of H^+ provides the missing piece. The resulting material is the green emeraldine salt, which is conductive.

25. Fluorine or oxygen can also make polyacetylene conductive. The electronegative element removes electrons from the alternating double bonds so that there is a hole for electron movement; the conduction band is not filled. Conduction occurs by electrons moving in one direction and the "hole" in the other direction.

27. For polythiophene to conduct, an electron needs to be removed. That is, it needs to be partially oxidized. Removing half an electron per thiophene unit results in a conductive polymer.

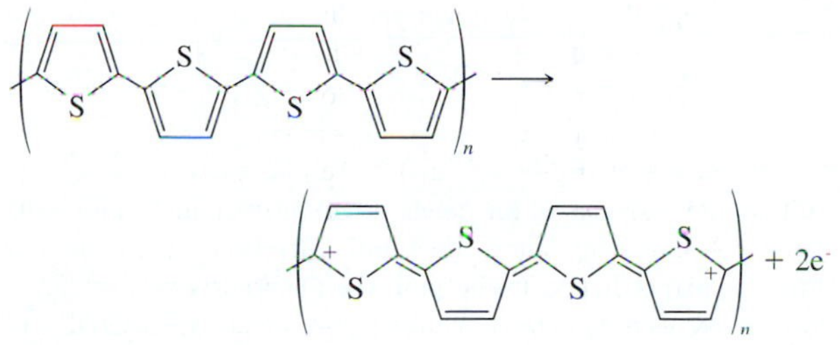

Conduction:

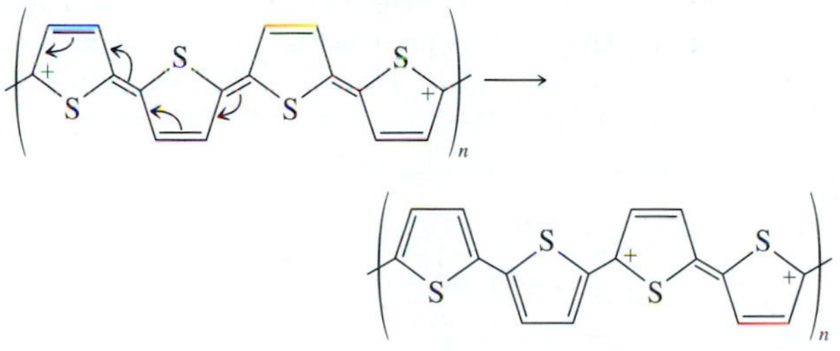

29. Of these molecules, only ethylene chloride can be polymerized due to the double bond. Ethyl chloride lacks a functional group that can be polymerized.

31.

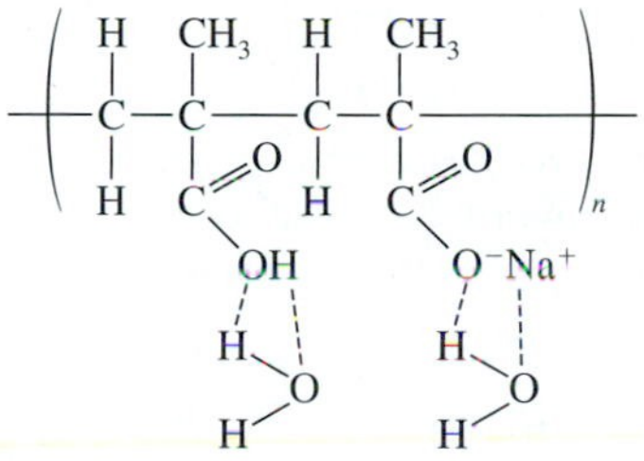

33. The difference between starch and cellulose lies in the geometry of the carbon at the linking end. In cellulose, the oxygen is above the ring and hydrogen is below it. In starch, these two atoms are switched. As a result, in starch sugar units all lie on the same side of the oxygen atoms, whereas in cellulose they are diagonal.
Long strands of cellulose interact by hydrogen bonding.
Starch readily absorbs water due to the numerous —OH groups that interact with water via hydrogen bonding.

35. Both monomers have a double bond, which is the functional group that enables them to be polymerized.

37. (a), (c), (d)

39. (a) $+3$; (b) -2; (c) $+2$; (d) $+1$; (e) -1; (f) (left to right) $+1$, -3, $+4$

41. (a) C is $+1$, N is -3; (b) C is $+1$, O is -2; (c) C is 0, top C is -3; (d) N is -3, C is $+4$

43.

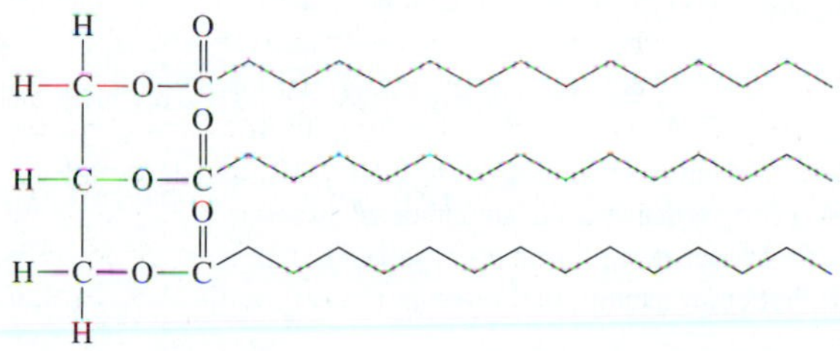

45.

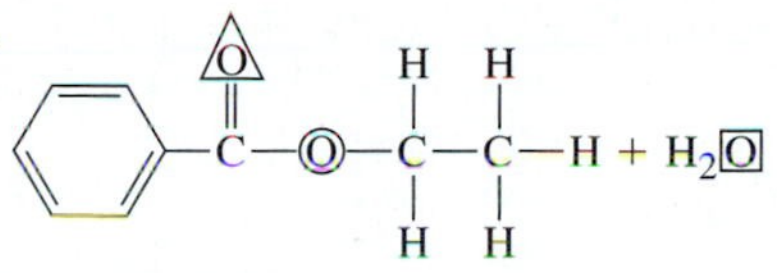

47.

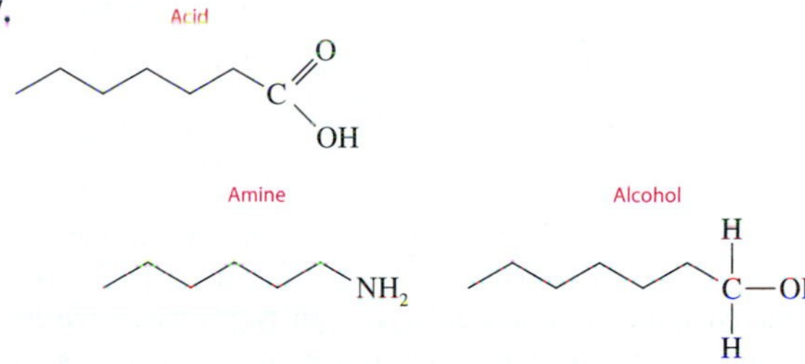

49. (a) C is $+4$; (b) C is $+1$, C is -2; (c) C is -1, N is -3, C is $+4$, O is -2

51. One monomer is H_2CO_3. The other is

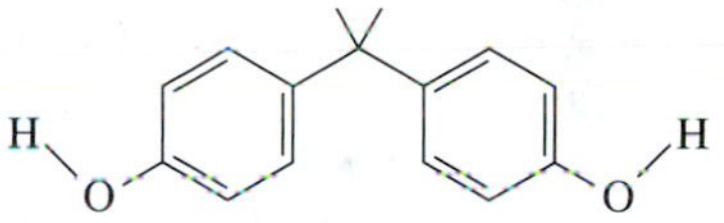

The latter monomer imparts the rigid quality due to the two phenyl rings in the backbone.

53. $-(CF_2-CF_2)_n-$

55. (b), (f)

57. (b), (d), (f)

59. The polymer is shown below. The mechanism is an addition mechanism.

61.

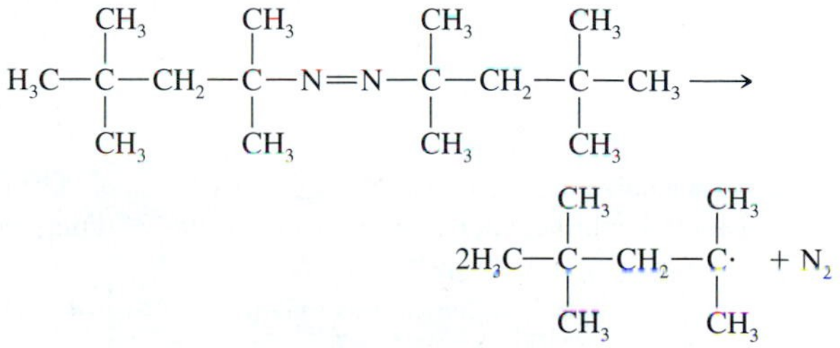

The decomposition reaction is not reversible because the very stable, diatomic, gas-phase molecule $N\equiv N$ is generated.

63. (a) Running shoes should bounce a bit to keep the knees from absorbing the shock of impact with the road.
(b) Tires should damp vibrations to maximize contact with the pavement.
(c) A gym safety mat should be energy- or vibration-absorbing so that the energy of the impact is absorbed by the mat rather than the athlete.
(d) The batter wants the baseball to spring off the bat and fly far. The pitcher wants the baseball's core as energy-damping as possible.

65.

Radical initiated

Zeigler-Natta catalyst

67. Polybutene is likely to be a soft solid consisting of intertwined long-chain and branch-chain molecules. Intermolecular interaction occurs via London dispersion forces, so polybutene is likely to have a putty-like consistency.

69.

HO—C—C—OH HO—C⟨benzene ring⟩C—OH

Diol   Diacid

The rigidity of PET drives from the phenyl ring in the diacid.

71. Nomex is not as good a polymer for bulletproof vests because part of the strength and resiliency of Kevlar derives from intermolecular hydrogen bonds. Nomex has a bent geometry that makes for very inefficient hydrogen bonding. Nomex is not expected to form a very good fiber.

73. A polymer is a molecule consisting of repeating units: "many mers." The distinguishing characteristics of a polymer are that it is a very large molecule and it consists of repeating small units.

75. The shorthand notation indicates the repeating unit in the polymer, but gives no information about either the branching of the long chain or the macromolecular configuration. Both strongly affect the properties of the polymer.

77. Many diagrams are possible. One example is shown here.

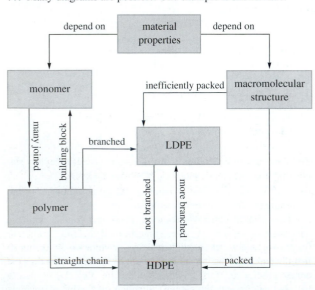

79. The repeated unit is

Pulling the polymer sheet tends to align the molecules with the long direction along the pull direction. Thus, in the direction perpendicular to the pull, the material is very strong—covalent bonds need to be broken to tear or break the material. Parallel to the chains, the material is much less strong—only weaker dispersive forces need to be overcome to tear the material in this direction. Combining two sheets with the pull direction in a perpendicular orientation aligns strong covalent bonds in both directions. The material is thus very strong in both directions.

81. The monomer is the acid–alcohol

HO—C—C⟨=O, OH⟩

The polymer of this material is a polyester:

⟨O—C—C=O⟩_n

83. This polymer is not a candidate for a conducting polymer because the conjugation is broken at the amide bond.

85. The polymer Kevlar is

⟨N—⟨benzene⟩—N—C—⟨benzene⟩—C⟩_n

To impart strength, the Kevlar fiber is drawn during formation to align the hydrogen-bonding functional groups, N—H and C=O, and thereby maximize hydrogen bonding. To give the vest strength in both directions, alternate sheets of Kevlar are overlaid with perpendicular drawing directions.

87. To be flexible, the polymer backbone needs to have a linear hydrocarbon chain portion. To be a good flame retardant, the polymer backbone needs to incorporate phenyl rings, as they resist oxidation. Phenyl rings also impart rigidity, so flame-retardant properties and flexibility are somewhat mutually exclusive. The flame retardant Nomex has phenyl rings and an amide bond. The amide enables the polymer to have a measure of flexibility.

Chapter 12

1. Let k_f be the specific forward rate constant and k_r be the reverse rate constant.

$$\frac{d[[Co(CN)_6]^{4-}]}{dt} = k_f[Co^{2+}][CN^-]^6 - k_r[[Co(CN)_6]^{4-}]$$

3. (a) second order overall, first order in H_2, first order in I_2
(b) second order overall, second order in HI
(c) first order overall, first order in N_2O_5
(d) first order overall, first order in N_2O

(e) second order overall, first order in CH_3Br, first order in OH (f) second order overall, first order in $C_{12}H_{22}O_{11}$, first order in H^+

5. The stoichiometric coefficients are determined by conservation of mass; the rate law coefficients are determined by the mechanism. There is no reason to expect them to be related.

7. Doubling the initial $[S_2O_8^{2-}]$ doubles the reaction rate; the reaction is first order in $S_2O_8^{2-}$. Doubling the initial I^- concentration doubles the reaction rate; the reaction is first order in I^-. The reaction is second order overall. Rate law:

$$-\frac{d[S_2O_8^{2-}]}{dt} = k[S_2O_8^{2-}][I^-]$$

9.   DNC   NCH

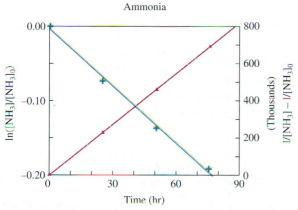

Rate law: $-\dfrac{d[DNC]}{dt} = k[DNC][NCH]$

Doubling the concentration of ⬡N-H <u>doubles</u> the reaction rate.

11. Doubling the cisplatin concentration doubles the reaction rate. The reaction is first order in cisplatin.

13. A graph of $\ln([NH_3]/[NH_3]_0)$ is a straight line if the reaction is first order; $1/[NH_3]$ is a straight line if the reaction is second order.

Ammonia

(plot with axes: left axis $\ln([NH_3]/[NH_3]_0)$ from 0.00 to −0.20; right axis $1/[NH_3] - 1/[NH_3]_0$ (Thousands) from 0 to 800; horizontal axis Time (hr) from 0 to 90)

▲ Second order
✛ First order

A plot of $[NH_3]^{-1} - ([NH_3]_0)^{-1}$ is a straight line, so the reaction is second order. $[NH_3]^{-1} - (NH_3]_0)^{-1} = kt \Rightarrow k = 9.22 \times 10^3$ M/h.

15. (a) In every 10-min interval, 90% of the initial concentration remains; thus the reaction is first order. (The 90% life is a constant, so the half-life is as well.)
(b) The half-life is 6.6 times the 90% life, or 66 min $= t_{1/2}$.

17. Comparing experiments 1 and 3, $[Cl_2]$ doubles and the initial rate doubles; the reaction is first order in $[Cl_2]$. Comparing experiments

2 and 3, [NO] doubles and the rate quadruples; the reaction is second order in [NO].

$$-\frac{d[Cl_2]}{dt} = k[NO]^2[Cl_2]$$

19. Experiments 3 and 4: Doubling [NO] quadruples the rate; the reaction is second order in [NO]. Experiments 2 and 4: Doubling O_2 doubles the rate; the reaction is first order in $[O_2]$.

$$-\frac{1}{2}\frac{d[NO]}{dt} = k[NO]^2[O_2]$$

21. Increasing the [I] concentration by a factor of 2, 4, or 6 raises the rate by 4, 16, or 36, respectively; the reaction is second order in [I]. Increasing [Ar] by a factor of 5 increases the rate by a factor of 5; the reaction is first order in [Ar].

23. The initial concentration of ^{238}U is the sum of the remaining ^{238}U and the ^{206}Pb. This is proportional to $2.09 \Rightarrow -\ln(1.09/2.09)/(1.54 \times 10^{-10} \text{ yr}^{-1}) = t = 4.23 \times 10^9$ yr.

25. $k = -\dfrac{\ln\left(\dfrac{3.5 \text{ g}}{10 \text{ g}}\right)}{25 \text{ h}} = 4.2 \times 10^{-2} \text{ h}^{-1}$

27. (a) $2N_2O_5 \rightarrow 2N_2O_4 + O_2$

(b) $t = -\dfrac{\ln\left(\dfrac{1 \text{ m}M}{5 \text{ m}M}\right)}{1.20 \times 10^{-2} \text{ s}^{-1}} = 134$ s

(c) $t = -\dfrac{\ln\left(\dfrac{0.5 \text{ m}M}{5 \text{ m}M}\right)}{1.20 \times 10^{-2} \text{ s}^{-1}} = 192$ s, or 58 s longer

29. Near room temperature, typical reaction rates double for every 10 °C temperature rise. Lowering the temperature by 10 °C reduces the reaction rate by about half. The half-life is inversely proportional to the rate constant, so reducing the rate by one-half increases the half-life by a factor of 2. The tires can be safely stored twice as long.

31. Plot $\dfrac{1}{[C_4H_6]_t} - \dfrac{1}{[C_4H_6]_0}$ versus time. Slope $= k = 3.2 \times 10^{-3}$ $M^{-1} \cdot \text{m}^{-1}$.

33. (a) $-\dfrac{d[C_3H_8]}{dt} = [OH][C_3H_8]$

(b) Removal rate $= (1.1 \times 10^{-12} \text{ mL/molecule} \cdot \text{s}) \times (1 \times 10^6 \text{ molecules/mL}) = 1.1 \times 10^{-6} \text{ s}^{-1}$
(c) Lifetime is 9.0×10^5 s. The typical lifetime is 250 h, or just over 10 d.
(d) The lifetime of propane in the atmosphere is far shorter than that of methane. The major regulatory efforts should be directed at methane.

35. $t_{1/2} = 0.693/k = 5727$ yr

37. The method of half-lives consists of determining whether the half-life depends on the initial concentration. If the half-life does not depend on concentration, then the reaction is first order; otherwise, it is a higher order. The plot shows clearly that $t_{1/2}$ is not independent of initial concentration. Check: Is $t_{1/2} \propto 1/[ENB]$?

The half-life is proportional to 1/concentration$_0$. The slope is 1.67×10^4 s·mM = k.

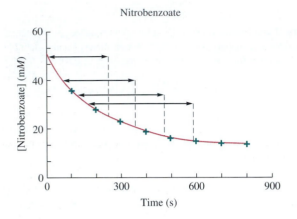

Nitrobenzoate

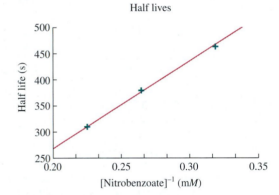

Half lives

39. [K]$_o$ is proportional to the total Ar + K = 1.078.

$$[K]_t = 1.000 - 0.078 = 0.922$$

$$k = 0.693/1.5 \times 10^9 \text{ yr} = 4.62 \times 10^{-10} \text{ yr}^{-1}$$

$$t = -\frac{\ln\left([A]\big/[A]_0\right)}{k} = -\frac{\ln\left([0.922]\big/1.078\right)}{4.62 \times 10^{-10} \text{ yr}^{-1}} = 1.653 \times 10^8 \text{ yr}$$

41. The concentration of Fe(II) in blood is

$$\frac{6 \times 10^6 \text{ cells}}{\text{mL}} \times \frac{250 \times 10^6 \text{ Hb}}{\text{cell}} \times \frac{4 \text{ Fe(II) atoms}}{\text{Hb}}$$

$$\times \frac{1 \text{ mol}}{6.022 \times 10^{23} \text{ atoms}} = 1 \times 10^{-8} \text{ M}$$

At 10 μM, NO is present in much larger concentration than is Fe(II). Oxidation of Fe(II) is thus a pseudo-first-order process with a rate constant of $(8 \times 10^6 \text{ M}^{-2} \text{ s}^{-1}) \times 20 \times (1 \times 10^{-5} \text{ M})^2 = 1.6 \times 10^{-2} \text{ s}^{-1}$.

Use $\frac{[A]_t}{[A]_0} = \exp(-kt)$. In the first minute, $kt = 0.96 \Rightarrow$ [Fe(II)]$_t$ = 0.38[Fe(II)]$_0 \Rightarrow 38\%$ destroyed.

With oxidation by O_2, the half-life of NO with an initial 10 μM concentration is about one-half minute. Since the rate of oxidation

of hemoglobin decreases as the square of the NO concentration, oxidation of NO by O_2 is very important.

43. $k = 0.693/t_{1/2}$. For C3S, $k \sim$ h^{-1}. For C2S, $k \sim$ d^{-1} or about 30 times smaller.

45. The mechanism is determined by the slow step, so the rate law is

$$-\tfrac{1}{2}\frac{d[\text{NO}]}{dt} = k[\text{O}_2][\text{N}_2\text{O}_4]$$

47. (a) Mechanism I:

$$-\frac{d[\text{CH}_3\text{Br}]}{dt} = [\text{CH}_3\text{Br}]$$

Mechanism II:

$$-\frac{d[\text{CH}_3\text{Br}]}{dt} = [\text{CH}_3\text{Br}][\text{OH}^-]$$

(b) Net reaction for mechanism I: CH$_3$Br + OH$^-$ → Br$^-$ + CH$_3$OH because the CH$_3^+$ produced in Step 1 is consumed in Step 2 by OH$^-$.

(c) Doubling [CH$_3$Br] doubles the rate; doubling OH$^-$ does not change the rate. Mechanism I is consistent with the data.

49. (a) 3A → C.

(b) $-\dfrac{1}{3}\dfrac{d[\text{A}]}{dt} = k_1\,[\text{A}]$

51.

$$\text{H}_2\text{O}_2 + \text{I}^- \longrightarrow \text{HOI} + \text{OH}^-$$
$$\text{HOI} + \text{I}^- \longrightarrow \text{I}_2 + \text{OH}^-$$
$$2\text{OH}^- + 2\text{H}_3\text{O}^+ \longrightarrow 4\text{H}_2\text{O}$$

Net reaction:

$$2\text{I}^-(aq) + \text{H}_2\text{O}_2(aq) + 2\text{H}_3\text{O}^+(aq) \longrightarrow \text{I}_2(aq) + 4\text{H}_2\text{O}(l)$$

(b) Step 1 is the slow step.

53. (a) The first two steps are the slow steps.

(b) Use $k_{\text{cat}}/k = \exp(E_a - E_{\text{cat}})/RT = 5.8 \times 10^{25}$ times as fast at room temperature.

55. Mechanism II is consistent with the rate law.

57. Use $k = \exp(-E_a/RT)$.

$$E_a = \frac{\ln(1/10)R}{\left(\dfrac{1}{278} - \dfrac{1}{301}\right)} = 69.7 \text{ kJ/mol}$$

59. The activation energy must be very high so that $\exp(-E_a/RT)$ is always near 1 and is affected little by temperature.

61. At 700 °C, $k = 5.25 \times 10^{-13}$ m^2/s. At 1100 °C, $k = 7.04 \times 10^{-11}$ m^2/s. For every 200 °C temperature rise, the diffusion rate goes up by an order of magnitude. Because control of the diffusion depth is not an issue (the rate is sufficiently slow that the depth can be controlled at all these temperatures), the processing time will decrease by an order of magnitude for every 200 °C temperature rise. Thus the process should be carried out at 1100 °C.

63. (a) The plot of ln (k) versus $-1/RT$ should yield a straight line with slope equal to E_a.

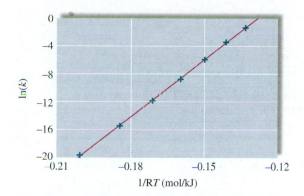

From the slope $E_a = 271.8$ kJ/mol.

(b) Use $k = \exp(-E_a/RT)$. At 500 K, $k = 6.11 \times 10^{-14}$ s^{-1}.

65. The production is proportional to the rate constant.

$$\frac{k(T_1)}{k(T_2)} = \exp\left(-\frac{E_a}{R}\left(\frac{1}{T_1} - \frac{1}{T_2}\right)\right)$$

Ratio = 10.6

67. (a) $k = A \exp(-E_a/RT)$

$k = 1.64 \times 10^{-4}$ s^{-1}

(b) Use $[A]_t = [A]_0 e^{-kt}$. Percent of ethyl chloride left after 10 min = 91.9% $\Rightarrow$ 8.1% decomposed.

(c)
$$\frac{k(T_1)}{k(T_2)} = \exp\left(-\frac{E_a}{R}\left(\frac{1}{T_1} - \frac{1}{T_2}\right)\right)$$

$T_2 = 735$ K = 462 °C

69. (a) NO_2 contains 17 valence electrons, and 24 are needed for full octets. NO_2 is bonded by three bonds and contains a single unpaired electron. NO_2 is bent planar. The formation of N_2O_4 entails joining the two NO_2 molecules at the lone electron on the nitrogen.

(b)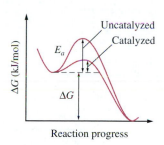

71. $K_c = \dfrac{[N_2O_5]}{[NO_2][NO_3]} = \dfrac{k_i}{k_{ii}}$. Convert from molecules/cm^3 to M = mol/L:

$$1\,\frac{\text{molecule}}{\text{cm}^3} \times \frac{1000\ \text{cm}^3}{\text{L}} \times \frac{1\ \text{mol}}{6.022 \times 10^{23}\ \text{molecules}}$$
$$= 1.66 \times 10^{-21}\ M$$

Convert K from concentration in M to molecules/cm^3. Multiply K by 1.66×10^{-21}.

$$K = 2.71 \times 10^{-27}$$

$$k_{ii} = k_i/K = \frac{(2.22 \times 10^{-30}\ \text{cm}^6/\text{molecules}^2 \cdot \text{s})}{2.71 \times 10^{-27}}$$
$$= 8.20 \times 10^{-4}\ \text{cm}^3/\text{molecule} \cdot \text{s}$$

73. The order with respect to ammonia is zero. The catalyst is W.

75. (a) Heterogeneous catalyst

(b) Many mechanisms can be suggested. All must have rapid ammonia and oxygen uptake, followed by a rate-determining step consisting of bond breaking.

Step 1: $NH_3(g) \longrightarrow NH_3(adsorbed)$

Step 2: $NH_3(adsorbed) \longrightarrow NH_2(adsorbed) + H(adsorbed)$

Step 3: $O_2(g) \longrightarrow 2O(adsorbed)$

Step 4: $O(adsorbed) + NH_2(adsorbed) \longrightarrow NOH(adsorbed) + OH(adsorbed)$

Step 5: $OH(adsorbed) + H(adsorbed) \longrightarrow H_2O(g)$

Step 6: $NOH(adsorbed) + O(adsorbed) \longrightarrow NO(g) + OH(adsorbed)$

Of these steps, neither the first nor the third can be rate-determining.

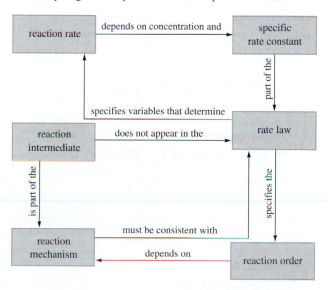

77. The enzyme concentration is 6.022×10^{17} molecules/L. The enzymes oxidize 6.022×10^{23} CO_2 molecules/s, or 1 mol/s. In 1 h, this is 3600 mol. At 44 g/mol, this is 158 kg CO_2. Glucose is $C_6H_{12}O_6$, so it produces 6 CO_2 molecules per sugar molecule. Thus 600 mol of sugar is consumed per hour. The molar mass of glucose is 180 g/mol, so 108 kg glucose is consumed per hour.

79. Many diagrams are possible. One example is shown here.

81. (a) The volume of CO_2 generated is proportional to the amount of HCO_3^- reacted. The first step is to translate the CO_2 volume data into $[HCO_3^-]$. The total CO_2 volume is proportional

Time (s)	0	20	30	50	70	90	160
Volume CO_2 (mL)	0	20	30.2	40.9	48.9	54.2	62.1
Millimoles CO_2	0	0.83	1.26	1.70	2.03	2.25	2.58
$[HCO_3]^-$	0.27	0.18	0.14	0.098	0.064	0.042	0.009

to $[HCO_3^-]_0$ (graph i). Then $[HCO_3^-]_t$ is proportional to the difference between the total CO_2 volume and the CO_2 volume at time t. Convert volume of CO_2 into moles using the ideal gas law.

If a plot of $\ln([HCO_3^-]/[HCO_3^-]_0)$ versus time is a straight line, the reaction is first order. If a plot of $1/([HCO_3^-])$ versus time is a straight line, then the reaction is second order.

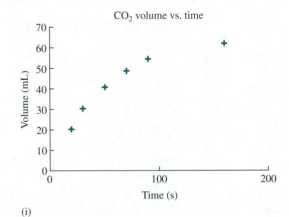

(i)

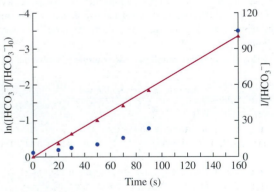

▲ $\ln([HCO_3^-]/[HCO_3^-]_0)$
● $1/[HCO_3^-]$

(ii)

Plot (ii) shows that the reaction is first order.

(b) $\ln[HCO_3^-]/[HCO_3^-]_0] = -kt$. The slope is $-k \Rightarrow k = 2.13 \times 10^{-2} \text{ s}^{-1}$.

(c) Use $k = A \exp(-E_a/RT)$;

$$\frac{k_{14\,°C}}{k_{20\,°C}} = \frac{\exp(-E_a/R \times 287)}{\exp(-E_a/R \times 293)} = 0.5$$

$$\Longrightarrow E_a = -R \times \left(\frac{1}{287} - \frac{1}{293}\right) \ln(0.5) = 80.8 \text{ kJ/mol}$$

(d) $k_{26\,°C} = k_{20\,°C} \times \exp\left(-\frac{E_a}{R}\left(\frac{1}{299} - \frac{1}{293}\right)\right) = 0.038 \text{ s}^{-1}$

(e) Part of the beneficial effect of an antacid tablet comes from the neutralization of stomach acid. The tablet accomplishes this partly by the reaction of $H_3CO_3^-$ with acid, producing $CO_2(g)$. Hence, it is desirable to prevent HCO_3^- from evolving CO_2 prior to entering the stomach. Antacids should be dissolved in cold water.

83. The temperature at which water boils depends on the pressure. In the Rockies, water boils at a lower temperature and the reactions that are involved in boiling an egg occur more slowly at this lower temperature.

85. (a) vitamin C
(b) air or O_2
(c) Yes, Cu^{2+} is a catalyst.
(d) Yes and yes; the reaction rate depends on both methylene blue and Cu(II).
(e) Observation of the blue color at the interface for an unstirred solution. The solution could be purged of all O_2, capped tightly, and shaken. If oxidation is due to O_2, the blue color will not appear. If it is due to another cause, the blue color will appear.

Glossary

Following the definitions of each term in the Glossary is the number of the section in which the term is discussed.

Acid (Arrhenius definition) An acid is a substance that contains H and produces H^+ in aqueous solutions. (8.4)

Acid group The —COOH portion of an organic molecule. (11.1)

Acid ionization constant, K_a The equilibrium constant for ionization of an acid. (8.4)

Actinides The group of 14 elements following actinium in the periodic table. (2.1)

Activated complex The arrangement of atoms found at the top of the potential energy barrier as a reaction proceeds from reactants to products. Also called *transition state*. (12.4)

Activation barrier The threshold energy that must be overcome to produce a chemical reaction. Also called *activation energy*. (12.4)

Activation energy The threshold energy that must be overcome to produce a chemical reaction. Also called *activation barrier*. (12.4)

Activity The effective concentration or pressure of a substance: For the solvent, activity equals one; for a solid, activity equals one. (8.4)

Activity series A list of metallic elements in order of the tendency to oxidize. (9.3)

Alcohol An organic compound with a —COH group. (11.2)

Alkali metals The IA metals. Members of the group: Li, Na, K, Rb, Cs, Fr. (3.1)

Alkaline earth An IIA metal. Member of the group: Be, Mg, Ca, Sr, Ba, Ra. (3.1)

Allotropes Two or more forms of an element that differ in the ways the atoms are linked, e.g. carbon as graphite and diamond. (3.4)

Alloy A homogeneous solid solution of two or more elements having metallic properties. (4.3)

Amide A bond which results from condensation of an amine and an acid. (11.1)

Amine Ammonia that has one or more hydrogen atoms replaced by a hydrocarbon. (10.3)

Amine group An organic compound with an —NH_2 group. (11.1)

Amino acid A molecule containing both an amine and an acid functional group. (11.1)

Amplitude Magnitude of the wave function of the electron wave at a point in space. (2.4)

Anion An atom or group of atoms with a negative charge. (1.6)

Anode The electrode in a galvanic cell at which oxidation occurs. (9.3)

Antibonding molecular orbital A molecular orbital that, when occupied, raises the energy of the molecule and weakens the bonding. Higher-energy orbital with a node between the nuclei. (5.1)

Aqueous solution A homogeneous solution with water as the major component. (1.5)

Arrhenius equation The equation representing the rate constant as $k = Ae^{-E_a/RT}$, where A represents the product of the collision frequency and the steric factor, and $e^{-E_a/RT}$ is the fraction of collisions with sufficient energy to produce a reaction. (12.4)

Atom The smallest particle that an element can be subdivided into and retain the chemical properties of that element. (1.2)

Atomic mass The weighted average mass, in atomic mass units, of the atoms in a naturally occurring substance. (1.2)

Atomic mass unit (amu) 1.660540×10^{-24} g, 1/12 the mass of carbon-12, approximately the mass of a hydrogen atom. (1.2)

Atomic number, Z The number of protons in the nucleus of the atom; determines the identity of the element. (2.2)

Atomic radius The radius of the sphere encompassing 90% of the electron density for the free atom. (3.3)

Austenite A cubic form of carbon steel. (4.4)

Avogadro's law Equal volumes of gases at the same temperature and pressure contain the same number of particles. (1.3)

Avogadro's number, (N_a) The number of atoms in exactly 12 grams of pure ^{12}C, equal to 6.022×10^{23}. (1.2)

Azimuthal A nodal plane cutting an arc through the bounding sphere. Associated orbital has a nodal plane through the nucleus. (2.4)

Azimuthal quantum number, ℓ Second quantum number that is equal to the number of planar (azimuthal) nodes. $\ell \leq n - 1$. (2.4)

Ball-and-stick model A molecular model that represents atoms as balls (ball size not significant) and bonds as sticks showing bond relationships clearly. (1.4)

Band A closely spaced set of energy levels, usually in a solid material. Small-energy difference between successive orbitals within a bonding set. (5.1)

Band gap A range of energies for which there are no allowed electronic states in a solid. Energy difference between the valence band and the conduction band in a semiconductor. (5.1)

Body-centered cubic, BCC An arrangement of atoms with atoms on each corner of a cube and on the body center of the unit cube. (4.1)

Bond order An indication of bonding strength. It is $\frac{1}{2}$ the number of bonding electrons minus $\frac{1}{2}$ the number of antibonding electrons. (5.1)

Bonding molecular orbital A molecular orbital that, when occupied, lowers the energy of the molecule, characterized by a large electron density between two nuclei. (5.1)

Box diagram Graphic representation of the valence electrons in which each orbital is represented as a box and electrons in the orbital are shown as arrows. (2.5)

Boyle's law The volume of a given sample of gas at constant temperature varies inversely with the pressure. (1.3)

Buffer A solution containing a weak acid and its salt (or a weak base and its salt) in approximately equal concentrations. (8.4)

Buffered solution A solution that resists a change in its pH when either hydroxide ions or protons are added. (8.4)

Catalyst A substance that speeds up a reaction without being consumed. (12.6)

Cathode The electrode in a galvanic cell at which reduction occurs. (9.3)

Cation An atom or group of atoms with a positive charge. (1.6)

Cell potential (E) Potential that results in a driving force in a galvanic cell that pulls electrons from the reducing agent in one compartment to the oxidizing agent in the other. Also called *electromotive force* or *cell voltage*. (9.3)

Cell voltage (V) Potential that pulls electrons from the anode to the cathode. Also called *cell potential* (E) or the *electromotive force*. (9.3)

Charles's law The volume of a given sample of gas at constant pressure is directly proportional to the absolute temperature (K). (1.3)

Chelate A polydentate ligand bound to a metal ion. (10.3)

Cis Arrangement of two functional groups on each end of a double bond in which both groups are on the same side of the double bond with hydrogen atoms on the other side. (11.1)

Close pack Crystal structure in which the atoms are densely packed, occupying the least total volume in a *crystal lattice*. (4.1)

Common ion effect An ion appearing in more than one simultaneous equilibrium. (8.4)

Complex ion A charged species consisting of a metal ion surrounded by ligands. (10.2)

Complex The combination of a metal ion and the ligands joined to the metal by coordinate covalent bonds. (10.2)

Compound A substance with constant composition that can be broken down into elements by chemical processes. (1.2)

Concept map A visual representation of the relationships or connections between concepts. (1.8)

Condensation Results when two molecules come together, bond and usually eliminate a small molecule. (11.2)

Condensed structural formula A compact version of a structural formula showing how atoms are grouped together. (1.4)

Conduction band An incompletely filled band in a solid. The empty or nearly empty band of states next highest in energy to the filled band in a semiconductor. (5.1)

Conjugate acid The species formed when a proton is added to a base. (8.4)

Conjugate base What remains of an acid molecule after a proton ionizes. (8.4)

Conservation of energy Energy can be converted from one form to another but can be neither created nor destroyed so the total energy remains constant. (1.7)

Continuous spectrum Continuous bands of color produced, e.g. by a prism being inserted into a shaft of white light. (2.3)

Coordinate-covalent bond A bond consisting of two electrons from one bonding molecule/ion/atom and none from the second. (10.1)

Coordination number For complexes, the number of electron pairs donated to a transition metal ion. (4.1)

Coordination number Number of nearest neighbors or the number of atoms bonded to the central atom. (4.1)

Coordination polymerization Polymerization mechanism with addition to a double bond occurring via coordination to a solid catalyst. (11.1)

Core electrons Inner electrons in an atom; electrons of the previous rare gas, plus any filled d or f subshells. (2.5)

Coulombic energy $E = (9.00 \times 10^{18}\ \text{J} \cdot \text{m/C}^2)(q_1 q_2)/r$ where E is the energy of interaction between a pair of ions, expressed in joules; r is the distance between the ion centers in nm; and q_1 and q_2 are the numerical ion charges. (1.7)

Covalent bond A shared pair of electrons. (5.1)

Cross-link Covalent bonds holding two polymer chains together. (11.1)

Crystal lattice energy, $\Delta H_{\text{x'tal lattice}}$ Energy released when ions in the gas phase combine forming a crystalline lattice. (7.2)

Crystal lattice The repeating pattern of many solid materials, such as metals, when viewed at the atomic level. (4.1)

Dalton Expression for mass. One dalton is one amu. (11.1)

Defect Site with atoms out of place. (4.2)

Density isosurface Represents the molecular structure with a surface of constant electron density. (1.4)

Density potential Charge distribution showing charge with colors from red for the most negatively charged regions to green for neutral regions to blue for the most positively charged regions. (1.4)

Dentate Literally, "having teeth." Lone pairs in a ligand that coordinate to a metal ion. (10.3)

Diamond structure Tetrahedral arrangement of atoms with every atom identically bonded to four other atoms. (4.3)

Diol An organic compound having two hydroxyl groups. (11.2)

Dipole Result of separation of the centers of positive and negative charges. (6.1)

Dipole-dipole interaction Interaction between molecules having a permanent dipole when hydrogen bonding is not involved. (6.2)

Disperse To separate. (2.3)

Doping Addition of a low concentration of a second substance into an otherwise pure solid material. (5.1)

Ductile The ability to be drawn into a wire. (3.4)

Effective nuclear charge, Z_{eff} The effective charge holding an electron in the atom under a Bohr model of the atom, i.e. all charge treated as a point charge at the center of the atom. (2.5)

Elastic deformation Distortion of a solid such that it returns to the original shape when the distorting force is removed. (4.1)

Elastic modulus Constant of proportionality between stress and strain, linear portion of the stress-strain plot. Elastic modulus, $E = \sigma/\varepsilon$. (4.2)

Electrode A solid electrical conductor through which an electric current enters or leaves an electrolytic cell or other medium. Location for oxidation or reduction in an electrochemical cell. (2.2, 9.3)

Electrolysis The splitting of a compound into its components using electricity. (2.2)

Electromagnetic energy A form of energy carried by virtually perpendicular oscillating electric and magnetic fields. (1.7)

Electromagnetic radiation Radiant energy consisting of oscillating and magnetic fields; includes visible light, x rays, and radio waves. (2.3)

Electromotive force (emf) (misnomer, a potential) Potential for taking on electrons. (9.3)

Electron affinity Measure of the energy of an atom's attraction for an additional electron. Energy released in the reaction $\text{Atom}(g) + e^-(g) \longrightarrow \text{Ion}^-(g) + \text{energy}$. (3.2)

Electron configuration Arrangement of electrons in an atom or ion. (2.5)

Electron density surface Encompasses 90% of the electron density – the sum of the square of the wave amplitude. (5.1)

Electron sea Pool of negative charges in an extended wave surrounding positive ions. (4.1)

Electron volt, eV Energy required to move an electron from a region of a given potential to a region one volt higher in potential. (2.4)

Electronegativity The tendency of an atom in a molecule to attract electrons in a molecule or a solid, i.e. in competition with other atoms. (3.2)

Electrostatic charge density Represents the charge on the molecular surface with colors: red for negative, green for neutral, and blue for positive. (5.1)

Empirical formula Simplest whole number ratio that represents the composition of the substance. (1.4)

Endothermic A process in which heat is absorbed. (7.2)

Energy level Allowed energy for an electron in an atom. (2.3)

Energy-level diagram Illustrates energy levels. (2.3)

Enthalpy change, ΔH Heat transferred for constant-pressure processes. Also known as *heat of reaction*. (7.1)

Enthalpy of atomization Energy required to transform a sample into atoms in the gas phase. (7.2)

Enthalpy of formation, ΔH_f° Enthalpy change in the reaction consisting of forming a substance from its elements in their most stable form at 25 °C and one atmosphere pressure. (7.2)

Entropy, S Thermodynamic function that measures the energy dispersal in a system. (7.3)

Equilibrium constant, K_c Ratio of concentrations. Specifically, the product of the product concentrations divided by the product of the reactant concentrations. More exactly, it is the product of the product *activities* divided by the product of the reactant *activities*. (8.1)

Equilibrium Lowest Gibbs free energy for a chemical reaction in a closed container. Also called *equilibrium state*. (8.1)

Ester An organic compound produced by the reaction between a carboxylic acid and an alcohol. (11.2)

Ethylene The molecule $H_2C{=}CH_2$.

Exothermic A process in which heat is released. (7.2)

Excited states Higher energy orbits. (2.4)

Extensive Any property that depends on the amount of material. (7.2)

Face-centered cubic, FCC Arrangement of atoms in a solid in which the unit cube consists of atoms on every corner plus one in each face of the cube. Atoms in one layer are packed in equilateral triangles such that any one atom is surrounded by six other atoms in that layer. Layered in an ABCABC repeating pattern. (4.1)

Family A vertical column of elements in the periodic table showing similar properties. Also called a *group* of elements. (2.1)

Ferromagnetic The ability to be permanently magnetized; having large domains of aligned spins. (10.5)

Flooding Introducing a large excess of one substance involved in a reaction for the purpose of determining the order in a remaining substance. Also called *Ostwald's isolation method*. (12.1)

Formal charge Difference between the number of valence electrons for an element and the number assigned the element in a compound: Number of valence electrons−number of lone pair electrons −$\frac{1}{2}$ number of shared electrons. (6.1)

Formation constant, K_f The equilibrium constant for forming a complex ion from its constituent metal ion and ligands. (10.4)

Forward bias Application of a potential to a *p-n* junction with the negative pole on the *n* side and the positive pole on the *p* side. The conducting configuration. (5.1)

Free energy, G The energy that is free to do work at constant temperature and pressure. Determines the direction of spontaneous change, $\Delta G < 0$ is spontaneous for reactions at constant temperature and pressure. Also known as *Gibbs free energy*. (7.3)

Frequency, ν The number of waves (cycles) per second that pass a given point in space. (2.3)

Fuel cell A primary electrochemical cell in which the reactants are continuously fed while the cell is in use. (9.5)

Fundamental wave The longest wavelength for a standing wave that can occur between two fixed ends. (2.3)

Galvanic cell A device in which chemical energy from a spontaneous redox reaction is changed to electrical energy that can be used to do work (a battery). (9.3)

Gas constant, R The proportionality constant in the ideal gas law; $0.08206 \; L \cdot atm/K \cdot mol$ or $8.3145 \; J/K \cdot mol$. (1.3)

Gibbs free energy, G A thermodynamic function equal to the enthalpy (H) minus the product of the entropy (S) and the Kelvin temperature (T); $G = H - TS$. Also known as *free energy*. (7.3)

Grain boundary Region where two adjacent crystallites; each of which is a regular array, are out of register with each other. (4.2)

Ground state The lowest possible energy state of an atom or molecule. (2.4)

Group A vertical column of elements in the periodic table showing similar properties. Also called a *family* of elements. (2.1)

Half reaction One part of an oxidation–reduction reaction. One half of the reaction represents oxidation; the other represents reduction. (9.3)

Half-life The time required for a reactant to reach half of its original concentration. In radioactive decay, the time required for the number of nuclides in a radioactive sample to reach half of the original value. (12.2)

Halogen Members of the family: F, Cl, Br, I, At. (3.1)

Hard-sphere model Picture of atoms as rigid and having definite edges. (4.1)

Heat capacity, C_p, The ratio of the heat supplied to the temperature rise per unit mass or mole (7.2)

Heat of reaction, ΔH, Heat produced during constant-pressure processes. Also known as *enthalpy change*. (7.1)

Heat, q Energy transferred in a constant volume process. (7.2)

Henderson-Hasselbach equation Equation relating the pH of a buffer to the pK_a of a weak acid (or pK_b of a weak base) and the acid and salt concentrations. (8.4)

Hess's law The energy change in any process is the same irrespective of the path followed. (7.2)

Heterogeneous catalysis Process in which a substance in a different phase (often solid) increases the rate of a reaction without being consumed in the reaction. (11.1)

Heterogeneous catalyst A substance that increases the rate of a reaction without being consumed and that is in a different phase than the reaction. For polymers, a solid surface on which addition polymerization takes place. (11.1, 12.4)

Heterogeneous mixture A mixture in which the individual components lie in distinct, macroscopic regions. (4.4)

Heterogenous equilibria Equilibria involving reactants and/or products in more than one phase. (8.2)

Hexagonal close pack, HCP Arrangement of a solid in which the atoms within a layer are packed together in equilateral triangles such that each atom is surrounded by six other atoms in that layer. Layer above and below are directly over each other for an ABAB repeat pattern. (4.1)

High spin complex A complex in which unpaired electrons occupy the higher-energy set of orbitals rather than pairing in the lower-energy set. (10.5)

High-density polyethylene (HDPE) The form of polyethylene that consists of unbranched chains resulting in efficient packing and a high density. (11.1)

Hole, h^+ Space in the valence band of a semiconductor. (5.1)

Holes Spaces between atoms in a hard-sphere model of a solid. (10.1)

Homogenous catalyst A substance that speeds up the rate of a reaction without being consumed that is part of the solution. Includes such materials as enzymes, acids, and bases. (12.6)

Homogenous equilibria An equilibrium system where all reactants and products are in the same phase. (8.2)

Homogenous solution Solution with uniform physical and chemical characteristics. (4.4)

Hume-Rothery rules Rules stating that to form a substitutional alloy over the entire composition range, two elements must differ by less than 15% in their atomic radius, have the same crystal structure, and have similar electronegativity. (4.4)

Hybrid orbital An orbital consisting of a combination of atomic orbitals. (5.1)

Hydration enthalpy, $\Delta H_{hydration}$ The energy evolved when an ion in the gas phase is surrounded by water molecules to form an aqueous solution. (7.2, 10.4)

Hydrocarbon A compound composed of carbon and hydrogen. (11.1)

Hydrogen bond A strong intermolecular interaction that occurs when hydrogen is between two very electronegative atoms (nitrogen, oxygen, or flourine). (6.2)

Hydronium Ion The H_3O^+ ion; a hydrated proton. (8.4)

Hydrophilic Water loving: Readily absorbing water. (6.3)

Hydrophobic Water hating: Repelling, tending not to combine with water. (6.3)

Hydroxide ion The OH^- ion. (8.4)

Ideal gas A gas with no interactions between the particles. (1.3)

Ideal gas law An equation of state for a gas, where the state of the gas is its condition at a given time; expressed by $PV = nRT$, where P = pressure, V = volume, n = moles of the gas, R = the universal gas constant, and T = absolute temperature. This equation expresses behavior approached by real gases at moderate T and P. (1.3)

Induced dipole An instantaneous dipole created when the motion of an electron cloud causes the electron density to shift slightly with respect to the atomic cores of the molecule. (6.2)

Induced dipole-induced dipole An intermolecular interaction resulting from an induced dipole inducing a dipole in a neighboring molecule. The resulting interaction force is called *London force.* (6.2)

Intensive Any property that does not depend on the sample size. (7.2)

Internal energy A property of a system that can be changed by a flow of work, heat or both. (7.2)

Interstitial atom Guest atom that fits into interstitial space. (4.4)

Interstitial spaces Spaces between atoms. (4.4)

Ion An atom or a group of atoms with a charge. (1.4)

Ion-dipole interaction Interaction whenever an ion interacts with a polar molecule. (6.2)

Ionic bond Bond in which one or more electrons transfer from an element with low electronegativity to one with a high electronegativity; the resulting ions hold together due to the attraction between an anion and a cation. (5.1)

Ion-ion interaction Interaction between two ions. (6.2)

Ionization energy The energy required to remove an electron from an atom, ion, or molecule. Also known as *binding energy* or *ionization potential*. (2.4)

Ion-product constant for water, K_w The equilibrium constant for the auto-ionization of water; $K_w = [H^+][OH^-]$. At 25 °C, K_w equals 1.0×10^{-14}. (8.4)

Isocyanate A molecule containing the functional group: —N=C=O. (11.2)

Kinetic energy ($\frac{1}{2}mv^2$) Energy due to the motion of an object; dependent on the mass of the object and the square of its velocity. (1.7)

Lanthanides (actinides) A group of 14 elements following lanthanum (actinium) in the periodic table. (2.1)

Latent heat of fusion The enthalpy change that occurs upon melting a solid at its melting point at a pressure of one atmosphere. (7.2)

Latent heat of vaporization The enthalpy change that accompanies vaporizing one mole of a liquid at a pressure of one atmosphere. (7.2)

Law of definite proportions A law stating that when two elements form a series of compounds, the ratios of the masses of the second element that combine with 1 g of the first elements can always be reduced to small whole numbers. (1.5)

Law of multiple proportions A law stating that when two elements form a series of compounds, the ratios of the masses of the second element that combine with 1 gram of the first element can always be reduced to small whole numbers. (1.5)

Le Châtelier's Principle If a stress is applied to a system at equilibrium, the equilibrium concentrations change in a direction that tends to reduce the stress. (8.7)

Lewis acid An electron pair acceptor. (11.1)

Lewis base An electron pair donor. (11.1)

Lewis dot structure A diagram of atoms or molecules depicting valence electrons as dots. (3.2)

Ligand A molecule or ion that bonds to a transition metal ion via a coordinate covalent bond. (10.2)

Limiting reactant The reactant that is present in sufficiently low concentration so that it limits the extent of the reaction. (1.5)

Line defect Row of adjacent atoms shifted slightly from the regular array position. (4.2)

Line formula Representation of an organic compound where bonds are shown as lines, carbon atoms (not shown explicitly) occur at the junction of lines, hydrogen atoms fill required valence, and other atoms are represented by their chemical symbol. (1.4)

Line spectrum Emitted light consisting of discrete wavelengths. (2.3)

London force An intermolecular force resulting from induced dipoles on neighboring molecules. Also called an *induced dipole-induced dipole* force. (6.2)

Lone pair A pair of electrons not involved in bonding. (6.1)

Low spin complex A complex in which electrons pair in the lower-energy set of orbitals rather than unpair by occupying the higher-energy set of orbitals. (10.5)

Low-density polyethylene (LDPE) The form of polyethylene that consists of branched chains resulting in inefficient packing and a low density. (11.1)

Macromolecular structure Spatial relationship of one part of a large molecule to other parts of the same molecule. (11.1)

Magnetic quantum number, m_ℓ The quantum number relating to the orientation of an orbital in space relative to the other orbitals with the same ℓ quantum number. It can have integral values between ℓ and $-\ell$, including zero. (2.4)

Magneto resistance effect A change in resistance that occurs when the data sensor passes near a magnetized region of a magnetic disk. (10.5)

Main group Elements in the groups labeled IA, IIA, IIIA, IVA, VA, VIA, VIIA, and VIIIA in the periodic table. (2.1)

Malleable Capable of being shaped or formed or pounded into thin sheets, as by hammering or pressure. (3.4)

Martensite A tetragonal form of carbon steel formed when austenite is cooled rapidly. (4.4)

Metal An element that is malleable and ductile. Metals are good conductors of heat and electricity. (3.4)

Metallic bonding Nondirectional bonding consisting of a collection of cations floating in an electron sea held together by the coulombic attraction between the cations and the electron sea. (4.1)

Mass density Mass Per Unit Volume. Typical Units: g/cm³ (Solids and Liquids) or (gases) g/L. (1.3)

Metal A material that is malleable and ductile. Most metals are good conductors of heat and electricity. (3.4)

Metallurgy The process of separating a metal from its ore and preparing it for use. (9.7)

Method of initial rates Procedure for determining the reaction order for substances in a reaction in which the rate of the reaction determined while systematically varying the initial concentrations. (12.1)

Miscible Describes two liquids forming a uniform mixture. (6.1)

Mode Refers to the entire wave. (2.4)

Molar mass Mass per mole of an atom or molecule. (1.2)

Molar density Moles per unit volume. (1.3)

Mole (mol) The number equal to the number of carbon atoms in exactly 12 grams of pure ^{12}C: Avogadro's number. One mole represents 6.022×10^{23} units. (1.2)

Molecular equation An equation representing a reaction in aqueous solution showing the reactants and products in undissociated form, whether they are strong or weak electrolytes. (1.5)

Molecular orbital An electron orbital spread over a molecule; the square gives the probability of finding the electron at a particular location. (5.1)

Molecular stoichiometry Quantitative relationship among the atoms that constitute a molecule. (1.5)

Molecule A bonded collection of two or more atoms of the same or different elements. The smallest particle of a compound having the chemical properties of the compound. (1.2)

Monomer A small molecule that is the building block for a polymer. (11.1)

Multidentate Multi-toothed; a ligand having more than one lone pair that can form a coordinate covalent bond. (10.3)

Negative phase The trough of a wave. (2.4)

Nernst equation An equation expressing the cell potential in terms of the concentrations of the substances involved in the cell reaction. $E = E^{\circ} - (RT/nF) \ln Q$. (9.5)

Net ionic equation An equation for a reaction in aqueous solution, where strong electrolytes are written as ions, showing only those components that are directly involved in the chemical change. (1.5)

Neutron An uncharged particle of mass slightly greater than the proton. (2.2)

Noble gas A Group VIIIA element (3.1)

Node Point of an orbital having zero electron probability. (2.4)

Nomenclature The system of naming compounds. (1.6)

Nonmetal An element that is neither malleable nor ductile. (3.4)

_n_-type semiconductor A semiconductor with conduction characterized by the motion of electrons. (5.1)

Nucleus The small, dense center of positive charge in an atom that is responsible for most of the mass. (2.2)

Octahedral holes A cavity in a crystalline lattice that is surrounded by six atoms. Six atoms define the vertices of an octahedral solid. (10.1)

Octet Eight valence electrons in an s^2p^6 configuration. (3.2)

Octet rule In a compound, atoms tend to achieve a noble gas configuration by acquiring or sharing electrons. (3.2)

Orbital A specific wave function for an electron in an atom. The square of this function gives the probability distribution for the electron. (2.4)

Ore Natural mineral source of a metal : rock. (9.7)

Ostwald's isolation method A method to determine the order of a reaction with respect to a substance by introducing a large excess of another substance. Also known as _flooding_ the reaction. (12.1)

Overlap rule The energy separation between two electronic states that are the same or close in energy in separate atoms depends on the spatial overlap of the two orbitals. (5.1)

Oxidation number Charge on an atom due to the difference between the number of valence electrons and the number of electrons resulting from assigning all electrons in a bond to the more electronegative element in the bond. (9.2)

Oxidation Removal of one or more electrons. (7.4, 9.1)

Oxidation state The charge on an element resulting from electrons being assigned to the more electronegative element in a bond. (9.2)

Pauli exclusion principle No more than two electrons with opposite spins can occupy one orbital: no two electrons can have the same set of quantum numbers. (2.5)

Peptide bond The bond resulting from the condensation reaction between amino acids. (11.1)

Period Row in the periodic table. (2.1)

Periodic law The properties of the elements are periodic functions of their atomic numbers. (2.2)

pH scale A convenient way to represent solution acidity; pH = $-\log[H^+]$ (log base 10). (8.4)

Phase A physical state of matter described by a characteristic relationship between the atoms or molecules; examples include solid, liquid, gas, semiconducting or metallic ion. (4.3)

Phase boundary Line separating two phases in a phase diagram. (4.3)

Phase change An alteration in the spatial relationship among atoms or molecules in a substance, e.g. liquid to solid. (3.4)

Phase diagram A graphical representation of the conditions for stability of phases of a system. (4.3)

Phase transition Change from one phase of a substance to another phase. (3.4, 4.3)

Phenyl group The benzene molecule minus one hydrogen atom: —C_6H_5. (11.1)

Pilling-Bedworth ratio, P-B ratio Ratio of the density of metal atoms in the elemental metal to the density of metal atoms in the metal oxide. (9.5)

Plastic deformation A distortion of a solid past the elastic deformation stage. The solid remains in an altered shape after the distorting stress is removed. (4.1)

p-n junction, diode Region of transition from a semiconductor with vacancies in the valence band to one with electrons in the conduction band. (5.1)

Point defect Missing atom or a single impurity atom in an otherwise regular array. (4.2)

Polarizability Ease with which the centers of positive and negative charge can be displaced from each other. (6.1)

Polymer A molecule consisting of many repeat units that are all the same. (11.1)

Positive phase The crest of a wave. (2.4)

Potential energy Energy due to position or composition. (1.7)

Precipitation reaction A reaction in which ions in a solution come together to form a solid. (1.5)

Principal quantum number, n The quantum number relating to the size and energy of an orbital; it can have any positive integer value. (2.4)

Probability density, $|\psi|^2$ Measure of the probability of finding the electron in a small volume. (2.4)

Product Final substance in a chemical reaction. (1.5)

Proton A positively charged particle in an atomic nucleus. (2.2)

p-type semiconductor A semiconductor with conductivity characterized by the movement of positively charged holes. (5.1)

Quantum mechanical model Mathematical formulation of the electron as a wave. (2.4)

Quantum numbers Compact notation for specifying orbitals. (2.4)

Radical An atom, molecule, or ion with at least one unpaired electron. (11.1)

Rate law An expression that shows how the rate of reaction depends on the concentration of substances involved in the reaction. (12.1)

Rate-determining step The slowest step in a reaction mechanism, the one determining the overall rate. Also called *rate-limiting process or rate-limiting reaction*. (12.3)

Rate-limiting process The slowest step in a reaction mechanism, the one determining the overall rate. Also called *rate determining step or rate-limiting reaction*. (12.3)

Rate-limiting reaction Reaction in a mechanism that determines the overall rate of the reaction. Also called the *rate-limiting process* or the *rate-determining step*. (12.3)

Reactant Starting substance in a chemical reaction. (1.5)

Reaction Transformation of one or more molecules into a different set of molecules. A physical reaction involves a change in physical state with no change in molecular identity. (1.5)

Reaction intermediate Substance that is formed and consumed during the course of a reaction. (12.3)

Reaction mechanism The series of elementary steps involved in a chemical reaction. (12.2)

Reaction order The sum of the powers of all substances in the rate law. (12.1)

Reaction quotient, Q The ratio of the product of product concentrations raised to their stoichiometric coefficients to the product of reactant concentrations raised to their stoichiometric coefficients. (8.6)

Reaction rate The change in concentration of a reactant or product per unit time. Also called *reaction velocity*. (12.1)

Reaction stoichiometry Quantitative relationship among the molecules involved in a reaction. See *stoichiometry*. (1.5)

Reaction velocity The change in concentration of a reactant or product per unit time. Also called *reaction rate*. (12.1)

Redox Contraction of oxidation and reduction. (9.1)

Reduction Addition of electrons. (9.1)

Resistivity Tendency not to conduct; Inverse of conductivity. Resistance $\times$ (cross sectional area)/length. Typical units: $\mu\Omega$cm. (4.2)

Resonance structures Blending of two or more Lewis structures that differ only in the arrangement of the electrons. (6.1)

Reverse bias Application of a potential to a *p*-n junction with the negative pole on the *p* side and the positive pole on the *n* side. The conducting configuration. (5.1)

Rydberg constant Fundamental constant that relates to the energy states of hydrogen, $R = 1.097 \times 10^7 \, \text{m}^{-1}$. (2.4)

Sacrificial anode Metal of lower reduction potential which protects an object made of a metal with a higher reduction potential. (9.5)

Salt bridge A concentrated salt solution in a gel that provides a conducting path between two compartments of an electrochemical cell. (9.3)

Saturated solution A solution containing ions in contact with the solid salt of the ions. (8.4)

Semimetal, metalloid The smallest set of elements falling between the metals and nonmetals in the periodic table. (3.4)

Shell All electron orbitals with a common value of *n*. (2.4)

Shielding Diminishment of the nuclear charge felt by an electron due to the presence of other electrons. (2.5)

Slip system Combination of a favored slide direction and a close pack plane. (4.2)

Solubility limit The amount of s substance that can be dissolved in a given amount of solvent. (4.4)

Solubility product The equilibrium product of the concentrations of the ions in a salt raised to their stoichiometric coefficients. (8.4)

Solubility The amount of a substance that dissolves in a given volume of solvent at a given temperature. (8.4)

Solute Substance dissolved in a liquid to form a solution. (8.2)

Solvent The dissolving medium in a solution. (8.2)

Space-filling model A model of a molecule showing the relative sizes of the atoms and their relative orientations. (1.4)

Specific rate constant An experimentally measured quantity k_{ex}. The constant of proportionality in the rate law. (12.1)

Spectator ion An ion that does not participate in a reaction. It is present to maintain overall electrical neutrality. (1.5, 8.4)

Spectrochemical series A series of ligands with increasing *d* orbital splitting; increasingly shorter wavelength absorption. (10.4)

Spectrum The collection of frequencies or energies emitted by an atom, ion, molecule, or solid. (2.3)

Spin magnetic quantum number, m_s Fourth quantum number that contributes a small energy correction: values $\pm \frac{1}{2}$. (2.4)

Spontaneous process A process that occurs naturally without application of an external force. (7.3)

Standard reduction potential ($E°$) The potential for a reduction reaction to occur at 25 °C with all ions present in 1 *M* concentration. Measured with respect to the reduction of H^+ to H_2. (9.3)

Standard temperature and pressure (STP) The condition 0 °C and 1 atm of pressure; at STP one mole of an ideal gas occupies 22.4 L. (1.3)

Standing wave A stationary sustained wave such as on a string of a musical instrument. (2.3)

State function A property that depends only on the state of the system. It is independent of the pathway to arrive at that state. (7.2)

State Describes the location of the electron relative to the nucleus. (2.4)

Steady state The approximation that the rate of formation of a reaction intermediate is zero. (12.3)

Stick diagram Representation of an organic molecule consisting of lines representing bonds, carbon atoms occur at the junction between lines, hetero atoms are shown explicitly and there are a sufficient number of hydrogen atoms to fill out the four coordination of carbon. (11.1)

Stoichiometry Quantitative relationship between the quantities of reactants consumed and products formed in a chemical reaction. (1.5)

Strain, ε Elongation per unit length, $\Delta \ell / \ell$. (4.2)

Stress hardening Making a metallic solid less flexible by introducing many grain boundaries, often by drawing the metal through a dye, compressing it between rollers, or pounding it. (4.2)

Stress, σ Applied load per unit area, F/A_0. (4.2)

Strong field ligand A ligand that produces a large splitting between sets of *d* orbitals in a transition metal ion; a splitting sufficiently large that electrons pair in the lower energy set of orbitals rather than occupying the higher energy set. (10.5)

Structural formula Indicates how atoms are linked together, but does not contain the three-dimensional information of the ball-and-stick model. (1.4)

Subshell All the orbitals of a given shell, *n*, with a common value of ℓ. (2.4)

Substitutional alloy Formation of a solid solution by atoms of one element substituting for the other. (4.4)

Surface tension A force that resists an increase in the area of the surface. (6.3)

Tetrahedral holes A cavity in a crystalline lattice that is surrounded by four atoms. (10.1)

Tetragonal solid Crystalline solid characterized by a unit cell with 90° angles and length in two directions being equal. The third length is unique. (4.3)

Third law of thermodynamics The entropy of a perfect crystal at 0 K is zero. (7.3)

Torr Another name for millimeter of mercury (mm Hg). (1.3)

Total ionic equation An equation representing a reaction in aqueous solution showing all ions present in the solution. (1.5)

Trans Arrangement of two functional groups on each end of a double bond in which the groups are on the opposite side of the double bond. (11.1)

Transition elements Denotes those elements from Sc to Zn, Y to Cd, La, Hf to Hg, and Ac, Rf to the not-yet-discovered element #112. (2.1).

Transition energy Energy difference between two electronic states. (2.3)

Transition metal complex A substance consisting of a transition metal ion surrounded by ligands bound with coordinate covalent bonds. (10.2)

Transition state The arrangement of atoms found at the top of the potential energy barrier as a reaction proceeds from reactants to products. Also called *activated complex*. (12.4)

Unit cell The smallest unit that when stacked together infinitely in three dimensions reproduces the solid. (4.1)

Valence band The band of states filled with valence electrons in a semiconductor. (5.1)

Valence electrons Those electrons beyond the previous rare gas core minus any filled d or f orbitals. (2.5)

Valence shell electron pair-repulsion (VSEPR) Method for determining the shape of a molecule or molecular fragment. All valence pairs repel to establish the electronic geometry. Bonded electrons determine the molecular shape. (6.1)

Vulcanization A process in which sulfur is added to rubber and the mixture is heated, causing crosslinking of the polymer chains and thus adding strength and resiliency to the rubber. (11.1)

Wave function, ψ A function of the coordinates of an electron's position in three-dimensional space that describes the properties of the electron. (2.4)

Wavelength, λ The distance between two consecutive peaks or troughs in a wave. (2.3)

Weak field ligand A ligand that produces only a small splitting between sets of d orbitals in a metal ion; splitting sufficiently small that electrons occupy the higher energy set of orbitals rather than pairing with electrons in the lower energy set of orbitals. (10.5)

Work, w Force acting over a distance. For processes with gases: $dw = -P dV$ (7.2)

Zinc blende Variation on the diamond structure composed of two elements; every atom of one type is surrounded by four atoms of the second type. (5.1)

Index

Photo Credits

Chapter 1: p. xxii: Dale Chihuly; p. 1 (left): Neal Preston/CORBIS; p. 1 (middle): Mehau Kulyk/SPL/Photo Researchers, Inc.; p. 3 (jet): Jim Ross/NASA; p. 3 (bridge): John Wang/PhotoDisc Red/Getty Images; p. 3 (coins): Layne Kennedy/CORBIS; p. 3 (copper wires): Index Stock/Alamy; p. 3 (pails): Anne Domdey/CORBIS. **Chapter 2:** p. 39: IBM Corporation, Research Division/Almaden Research Center; p. 41 (left): Ilya Repin/Scala/Art Resource; p. 41 (right): Courtesy of Smithsonian Institution Libraries, Washington, D.C.; p. 42 (bike gears): Tom Pantages; p. 42 (silicon wafers): Will & Deni McIntyre/Photo Researchers, Inc.; p. 42 (cutting sodium metal): Charles D. Winters/Photo Researchers, Inc.; p. 42 (magnesium ribbon): Science Photo Library/Photo Researchers, Inc.; p. 42 (copper pipes): Think-Stock/Getty Images; p. 42 (silver coins): CORBIS; p. 42 (gold coins): Brand X Pictures/Getty Images; p. 42 (iodine): Science Photo Library/Photo Researchers, Inc.; p. 44: Science Photo Library/Photo Researchers, Inc.; p. 45: The Cavendish Laboratory/Cambridge University; p. 46: Bettmann/CORBIS; p. 49: PhotoLink/PhotoDisc/PictureQuest; p. 52 (left): Joel Gordon; p. 52 (right): Richard Treptow/Photo Researchers, Inc.; p. 55: Emilio Segre Visuals Archives/AIP; p. 59: Emilio Segre Visuals Archives/AIP; p. 62: Science Museum/Science & Society Picture Library. **Chapter 3:** p. 86: Botanica/Getty Images; p. 90: The Bancroft Library; p. 104 (top, left): Charles D. Winters/Photo Researchers, Inc.; p. 104 (top, middle): Jim Ross/NASA; p. 104 (top, right): Elizabeth Whiting & Associates/Alamy; p. 104 (bottom, left): Charles D. Winters/Photo Researchers, Inc.; p. 104 (bottom, right): Tom Pantages; p. 105 (top, left): M. Kulyk/Photo Researchers, Inc.; p. 105 (bottom, left): John Connell/Index Stock Imagery; p. 105 (bottom, middle): Sally Brown/Index Stock Imagery; p. 105 (bottom, right): Mike Powell/Getty Images. **Chapter 4:** p. 118: Peter Menzel/Stock Boston; p. 122: The M.C. Escher Company, Baarn, Holland, © 2004. All rights reserved; p. 123: The M. C. Escher Company, Baarn, Holland, © 2004. All

rights reserved; p. 126: Vecco Instruments; p. 136 (left): From P.G. Shewman, *Transformations in Metals,* © 1969, McGraw-Hill, N.Y. Photo by C. S. Smith; p. 136 (right): From Marc Andre Meyers and Krishan Kumar Chawla, *Mechanical Metallurgy: Principles & Applications,* © 1984, Prentice Hall; p. 149: Derek Morgan/Ancient-east.com; p. 150: Photo courtesy of Bard Peripheral Vascular/C. R. Bard, Inc. **Chapter 5:** p. 158: Pascal Goetgheluck/Photo Researchers, Inc.; p. 161: Arnold Fisher/Photo Researchers, Inc.; p. 170: Richard Megna/Fundamental Photographs; p. 176: Ava Helen and Linus Pauling Papers/OSU Foundation–Pauling Library Account. **Chapter 6:** p. 197: © Felice Frankel. **Chapter 7:** p. 233: Chuck Doswell/Visuals Unlimited; p. 252: CORBIS; p. 260 (left): Ahn Young-joon/AP Photo; p. 260 (right): Elan Sun Star/Index Stock Imagery. **Chapter 8:** p. 270: Paul Barton/CORBIS; p. 283 (top): Dr. David S. Goodsell; p. 283 (bottom): CVRI/SPL/Photo Researchers, Inc.; p. 290: David Muench/CORBIS; p. 297: Keith Dannemiller/CORBIS. **Chapter 9:** p. 305: Jeff J. Daly/Visuals Unlimited; p. 306 (left): Jeff Greenberg/Index Stock Imagery; p. 306 (middle): Jim McGuire/Index Stock Imagery/PictureQuest; p. 306 (right): Brand X Pictures/Fotosearch Stock Photography; p. 308 (middle): CORBIS; p. 337 (top): Jan Butchofsky-Houser/CORBIS; p. 337 (bottom): E. R. Degginger/Color-Pic. **Chapter 10:** p. 346: James L. Stanfield/National Geographic/Getty Images; p. 347 (left): Luis Veiga/The Image Bank/Getty Images; p. 347 (right): Comstock Images. **Chapter 11:** p. 377: Gary Braasch/The Image Bank/Getty Images; p. 378 (top, left): Paul S. Souders/CORBIS; p. 378 (top, right): Courtesy of The Dow Chemical Company; p. 378 (bottom, left): Aaron Haupt/Photo Researchers, Inc.; p. 378 (bottom, right): Photo 24/Getty Images; p. 384: Charles D. Winters/Photo Researchers, Inc. **Chapter 12:** p. 421: Robert Landau/CORBIS; p. 422 (left): Don Johnston/Stone/Getty Images; p. 422 (middle): Lester Lefkowitz/The Image Bank/Getty Images; p. 422 (right): CORBIS; p. 443 (all photos): NASA.

Note: Portions of the Chapter opener photos for Chapters 1, 2, 4, 6 and 11 also appear on the Contents pages v–ix.